ENVIRONMENTAL WASTE MANAGEMENT

ENVIRONMENTAL WASTE MANAGEMENT

Edited by

Ram Chandra

CRC Press
Taylor & Francis Group
Boca Raton London New York

CRC Press is an imprint of the
Taylor & Francis Group, an **informa** business

CRC Press
Taylor & Francis Group
6000 Broken Sound Parkway NW, Suite 300
Boca Raton, FL 33487-2742

© 2015 by Taylor & Francis Group, LLC
CRC Press is an imprint of Taylor & Francis Group, an Informa business

No claim to original U.S. Government works

International Standard Book Number-13: 978-0-367-57548-9 (Paperback)

Library of Congress Cataloging-in-Publication Data

Environmental waste management / Ram Chandra, editor.
        pages cm
    "A CRC title."
    Includes bibliographical references and index.
    ISBN 978-1-4987-2474-6 (alk. paper)
    1. Hazardous wastes. 2. Hazardous wastes--Biodegradation. I. Chandra, Ram (Biotechnology professor)

TD1030.E58 2016
628.4--dc23   2015009165

Visit the Taylor & Francis Web site at
http://www.taylorandfrancis.com

and the CRC Press Web site at
http://www.crcpress.com

Contents

Preface

Rapid industrialization has resulted in the generation of huge quantities of hazardous waste, both solid as well as liquid, from industrial sectors such as sugar, pulp and paper, tanneries, distilleries and textiles, petroleum hydrocarbon and agrochemicals, etc. However, the safe disposal and proper management and utilization of hazardous waste present not only a challenge for the country but a threat to the scientific society as well. The management and recycling of industrial waste are essential for the sustainable development of society. Despite regulatory guidelines for pollution control measures, these wastes are being dumped on land or discharged into water bodies without adequate treatment, which causes environmental pollution and health hazards. In-depth knowledge of the physicochemical properties of various industrial waste and their chemical composition and environmental health hazards are still riddles to research. Therefore, it is essential to update knowledge and information regarding all hazardous industrial waste. Keeping the facts above in mind, a number of experts from various universities, national research laboratories, and industries have shared their specialized knowledge in environmental microbiology and biotechnology for monitoring various industrial waste and environmental pollutants in order to update the information available to students, scientists, and researchers. The chapters in this book cover numerous topics including biocomposting of press mud, treatment of pulp and paper mill wastewater, biodegradation of agrochemicals, and bioenergy production from industrial waste for safe recycling. The current knowledge regarding the persistent organic pollutants (POPs) discharged from various industrial wastes is also described in detail. This book emphasizes the relationship of metagenomics with POPs present in the sugarcane molasses-based distillery waste and pulp paper mill wastewater after secondary treatment. However, the fate of the metabolic products of various hazardous pollutants is unknown. The role of bioreactors for industrial wastewater treatment is presented, which currently is needed for the treatment of complex industrial wastewater and optimization of treatment parameters. The microbes for plastic degradation are pertinent to environmental pollution management nationally, and it is a global concern. Furthermore, the health hazard of hospital waste is also a challenge due to the outbreak of diseases by pathogenic bacteria. The biodegradation of several organophosphates and organohalides is still unknown. Moreover, the environmental fate of metabolic products of organophosphates and organohalides is still a subject of research. This book describes in detail the biotransformation and biodegradation of organophosphates and organohalides in the environment by different bacterial populations. The role of metallothioneins for metal homeostasis and tolerance are discussed. This book will benefit a wide range of readers including students, researchers, and consulting professionals in biotechnology, microbiology, biochemistry, and molecular biology. It will also be an important tool in describing waste management techniques.

Editor

Ram Chandra is currently head and senior principal scientist at Environmental Microbiology Division, Indian Institute of Toxicology Research (IITR), Lucknow, Uttar Pradesh, India. He earned his BSc (Hons) in 1984, and MSc in 1987 from Banaras Hindu University, Uttar Pradesh, India. Subsequently, he earned his PhD in 1994. He started his career as Scientist B at the Industrial Toxicology Research Centre, Lucknow in the field of biotechnology in 1989. He became a senior principal scientist (Scientist F) in 2009 in environmental microbiology at the Indian Institute of Toxicology Research (IITR), Lucknow.

Dr. Chandra became a professor and head of the Department of Environmental Microbiology (2011) at Babasaheb Bhimrao Ambedkar Central University, Lucknow. His leading work includes bacterial degradation of lignin from pulp paper mill waste and molasses melanoidin from distillery waste. He has published more than 90 original research papers in national and international peer-reviewed journals from Elsevier-USA, Springer-USA, Taylor & Francis-USA, and John Wiley & Sons-USA. In addition, he has published 19 book chapters and 2 books. He has vast experience in strategic R&D management preparation of scientific reports. Professor Chandra has earned awards for writing 25 popular scientific articles in Hindi. He has presented more than 65 national and international conference papers in microbiology, biotechnology, and environmental biology. He is a life member of various scientific societies. Dr. Chandra has also trained scientists from Germany and Nigeria under the TWAS-CSIR Fellowship program. He has chaired various scientific sessions in different scientific conferences and has been a guest reviewer for various national and international journals in his discipline. He was a member of the delegation team that visited Japan for a study on environmental protection from industrial waste. Dr. Chandra is a member of the American Society for Microbiology (ASM), USA and life member of the National Academy of Sciences, Allahabad, India (NASI). Based on his outstanding contribution to the field of environmental microbiology and environmental biotechnology, Professor Chandra has been named Fellow of the Academy of Environmental Biology (FAEB), Fellow of the Association of Microbiologists of India (FAMI), and Fellow of the Biotech Research Society of India (FBRSI).

Contributors

Arokiaswamy Robert Antony
Department of Microbiology
Bharathidasan University
Tamil Nadu, India

Ram Naresh Bharagava
Department of Environmental
    Microbiology
School for Environmental Sciences
Babasaheb Bhimrao Ambedkar University
    (A Central University)
Uttar Pradesh, India

Nishi Kant Bhardwaj
Avantha Centre for Industrial Research &
    Development
Paper Mill Campus
Haryana, India

Rima Biswas
Wastewater Technology Division
CSIR-National Environmental Engineering
    Research Institute
Maharashtra, India

Debasis Chakrabarty
Ecotoxicology and Bioremediation Division
CSIR-National Botanical Research Institute
Uttar Pradesh, India

Ram Chandra
Department of Environmental
    Microbiology
School for Environmental Sciences
Babasaheb Bhimrao Ambedkar University
    (A Central University)
Uttar Pradesh, India

Togarcheti Sarat Chandra
Environmental Biotechnology Division
CSIR-National Environmental Engineering
    Research Institute
Maharashtra, India

Sekar Chandru
Department of Microbiology
Bharathidasan University
Tamil Nadu, India

Amar Jyoti Das
Department of Environmental
    Microbiology
School for Environmental Sciences
Babasaheb Bhimrao Ambedkar University
    (A Central University)
Uttar Pradesh, India

Pallavi Datta
Department of Biological Science
Rani Durgavati University
Madhya Pradesh, India

Rajendran Sangeetha Devi
Department of Microbiology
Bharathidasan University
Tamil Nadu, India

Sanjay Dwivedi
Ecotoxicology and Bioremediation Division
CSIR-National Botanical Research Institute
Uttar Pradesh, India

Ahmed ElMekawy
Genetic Engineering and Biotechnology
    Research Institute
University of Sadat City
Sadat City, Egypt

Shazia Faridi
Department of Microbiology
University of Delhi South Campus
New Delhi, India

Neelam Gautam
Ecotoxicology and Bioremediation Division
CSIR-National Botanical Research Institute
Uttar Pradesh, India

Arumugam Gnanamani
Microbiology Division
CSIR-CLRI (Central Leather Research
 Institute)
Tamil Nadu, India

Sanjay P. Govindwar
Department of Biochemistry
Shivaji University
Maharashtra, India

Suvidha Gupta
Avantha Centre for Industrial Research
 & Development
Paper Mill Campus
Haryana, India

Niti B. Jadeja
Environmental Genomics Division
CSIR-National Environmental Engineering
 Research Institute
Maharashtra, India

Kanthaiah Kannan
Department of Microbiology
Bharathidasan University
Tamil Nadu, India

Velu Rajesh Kannan
Department of Microbiology
Bharathidasan University
Tamil Nadu, India

Atya Kapley
Environmental Genomics Division
CSIR-National Environmental Engineering
 Research Institute
Maharashtra, India

Varadharajan Kavitha
Microbiology Division
CSIR-CLRI (Central Leather Research
 Institute)
Tamil Nadu, India

Rahul V. Khandare
Department of Biochemistry
Shivaji University
Maharashtra, India

Rajesh Kumar
Department of Environmental Microbiology
School for Environmental Sciences
Babasaheb Bhimrao Ambedkar University
 (A Central University)
Uttar Pradesh, India

Vineet Kumar
Department of Environmental Microbiology
School for Environmental Sciences
Babasaheb Bhimrao Ambedkar University
 (A Central University)
Uttar Pradesh, India

Shatrohan Lal
Department of Environmental Microbiology
School for Environmental Sciences
Babasaheb Bhimrao Ambedkar University
 (A Central University)
Uttar Pradesh, India

Sameena Malik
Environmental Biotechnology Division
CSIR-National Environmental Engineering
 Research Institute
Maharashtra, India

Seema Mishra
Ecotoxicology and Bioremediation Division
National Botanical Research Institute
 (NBRI)
Uttar Pradesh, India

Gunda Mohanakrishna
Separation & Conversion Technologies
VITO - Flemish Institute for Technological
 Research
Mol, Belgium

Sandeep Mudliar
Plant Cell Biotechnology Division
CSIR-CFTRI
Karnataka, India

Tapas Nandy
Wastewater Technology Division
CSIR-National Environmental Engineering
 Research Institute
Maharashtra, India

Krishnan Natarajan
Department of Microbiology
Bharathidasan University
Tamil Nadu, India

Duraisamy Nivas
Department of Microbiology
Bharathidasan University
Tamil Nadu, India

Bamidele T. Odumosu
Department of Biosciences and
    Biotechnology
Babcock University
Ilishan-Remo Ogun State, Nigeria

Kiran Padoley
Environmental Biotechnology Division
CSIR-National Environmental Engineering
    Research Institute
Maharashtra, India

Vasundhara Paliwal
Environmental Genomics Division
National Environmental Engineering
    Research Institute
CSIR-National Environmental Engineering
    Research Institute
Maharastra, India

Ram Awatar Pandey
Environmental Biotechnology Division
CSIR-National Environmental Engineering
    Research Institute
Maharashtra, India

Deepak Pant
Separation & Conversion Technologies
VITO - Flemish Institute for Technological
    Research
Mol, Belgium

Hemant J. Purohit
Environmental Genomics Division
CSIR-National Environmental Engineering
    Research Institute
Maharastra, India

Tulasi Satyanarayana
Department of Microbiology
University of Delhi South Campus
New Delhi, India

Gaurav Saxena
Department of Environmental
    Microbiology
School for Environmental Sciences
Babasaheb Bhimrao Ambedkar University
    (A Central University)
Uttar Pradesh, India

Abhinav Sharma
Environmental Biotechnology Division
CSIR-National Environmental Engineering
    Research Institute
Maharashtra, India

Pradyumna Kumar Singh
Ecotoxicology and Bioremediation
    Division
CSIR-National Botanical Research Institute
Uttar Pradesh, India

Surendra Singh
Department of Biological Science
Rani Durgavati University
Madhya Pradesh, India

Sandipam Srikanth
Separation & Conversion Technologies
VITO - Flemish Institute for Technological
    Research
Mol, Belgium

Prachi Tembhekar
Environmental Biotechnology Division
CSIR-National Environmental Engineering
    Research Institute
Maharastra, India

Preeti Tripathi
Ecotoxicology and Bioremediation
    Division
National Botanical Research Institute
    (NBRI)
Uttar Pradesh, India

Rudra Deo Tripathi
Ecotoxicology and Bioremediation Division
National Botanical Research Institute
(NBRI)
Uttar Pradesh, India

Sangeeta Yadav
Department of Environmental Microbiology
School for Environmental Sciences
Babasaheb Bhimrao Ambedkar University
(A Central University)
Uttar Pradesh, India

Sheelu Yadav
Department of Environmental Microbiology
School for Environmental Sciences
Babasaheb Bhimrao Ambedkar University
(A Central University)
Uttar Pradesh, India

Trilok C. Yadav
Environmental Genomics Division
CSIR-National Environmental Engineering
Research Institute
Maharastra, India

1

The Use of PMDE with Sugar Industries Pressmud for Composting: A Green Technology for Safe Disposal in the Environment

Ram Chandra and Sangeeta Yadav

CONTENTS

1.1 Introduction

Composting is a sustainable waste management technique in developing countries. It is an eco-friendly approach for bioconversion into value added products which may be utilized as plant nutrients. It also reduces disposal and pollution problems arising from distillery effluent. Composting is an aerobic, thermophilic, and controlled microbial bio-oxidation process resulting in a product rich in humus which is used as a fertilizer. The oxidation produces a transient thermophilic stage which is followed by a period of cooling of degrading organic matter. The resulting material is held at ambient temperatures for maturation purposes which results in a stable, volume-reduced, hygienic, humus-like material which is beneficial to soil and plants. Composting is a popular option adopted by several Indian distilleries attached to sugar mills with adequate land availability. The wastewater of distilleries, that is, spent wash, either directly or after biomethanation is sprayed in a controlled manner on sugarcane pressmud. The effluent discharged after the biomethanation process is known as postmethanated distillery effluent (PMDE).

The rapid growth of distilleries in India has resulted in a substantial increase in the industrial pollutant load. There are more than 315 distilleries in India producing 3.25×10^9 L of alcohol and 4.04×10^{11} L of effluent each year. The industrial waste generated by various distillery units is posing a serious threat to adjoining aquatic and terrestrial habitats due to the practice of discharging waste into nearby wastewater courses and land. Distillery effluents have a high-biological oxygen demand (BOD), chemical oxygen demand (COD), phenols, and heavy metals. The color of the effluent persists even after anaerobic treatment and causes a serious threat to environment. The water bodies receiving color waste get colored and the penetration of light in aquatic ecosystems is reduced, which in turn affects aquatic life. PMDE is an effective organic liquid fertilizer after appropriate dilution (Bharagava et al., 2008; Chandra et al., 2009). PMDE could be recycled in agriculture both as irrigation water and as a source of plant nutrients. It contains a large amount of organic carbon, K, Ca, Mg, Cl, and SO_4 and moderate amount of N and P and traces of Zn, Cu, Fe, and Mn (Mahimairaja and Bolan, 2004).

There are a number of large-scale distilleries integrated with sugar mills in India. The waste products from sugar mills comprise bagasse (the residue from sugarcane crushing), pressmud (the mud and dirt residue from juice clarification), and molasses (the final residue from the sugar crystallization section) as shown in Figure 1.1. Bagasse is used in paper manufacturing and as fuel in boilers; molasses as raw material in distilleries for alcohol production. These days pressmud has been used in the composting of wastewater generated by distilleries in India because it is a source of nutrients. Pressmud is a solid residue, obtained from sugarcane juice before the crystallization of sugar. It is a soft, spongy, lightweight, amorphous, dark brown to black colored material (Figure 1.1a). About 3% pressmud is produced from a total quantity of crushed cane. Pressmud is a rich source of organic carbon, nitrogen, phosphate, potassium (NPK), and other micronutrients. It generally contains 60%–85% moisture (w/w) and the chemical composition depends on the

FIGURE 1.1
(**See color insert.**) View of byproducts of the sugar industries, pressmud (a) molasses (b), and bagasses (c and d).

cane variety, soil condition, nutrients applied in the field, process of clarification adopted and other environmental factors. Comparative physicochemical properties of pressmud and distillery wastewater are shown in Table 1.1.

Pressmud from sugar factories typically contains 71% moisture, 9% ash, and 20% volatile solids (VSs), with 74%–75% organic matter found on solids. Hence, it may be the best source material for microbial growth and for use in the treatment of distillery waste. The distillery effluent based compost can be prepared by using pressmud and the compost can be enriched with the use of rock phosphate, gypsum, yeast sludge, bagasse, sugarcane trash, boiler ash, coir pith, and water hyacinth. First, the pressmud is spread in the compost yard to form a heap of 1.5 m height, 3.5 m width, and 300 m length. Ten liters of bacterial culture, diluted with water, in the ratio of 1:10 is sufficient for a tone of pressmud. A consortium of efficient microbial decomposers, namely, *Phanerocheate chrysosporium, Trichurus spiralis, Pacelomyces fusisporus, Trichoderma* spp., and so on, are sprayed on the pressmud and mixed thoroughly using an aerotiller which makes the pressmud aerable and enhances the process of decomposition.

After three days, distillery effluent is sprayed on the heaps to a moisture level of 60% and the pressmud heaps are allowed to absorb the effluent for 4–5 h. The heaps are then thoroughly mixed by an aerotiller as shown in Figure 1.2. When the moisture level drops below 30%–40%, the effluent is sprayed again, mixed with pressmud, and the heaps are once again formed.

Effluent can be sprayed once or twice in a week depending on the moisture content of the pressmud heaps. Mixing of effluent and heap formation will be repeated for 8 weeks so

TABLE 1.1

Characteristics of Pressmud and Wastewater Generated from a Distillery

Parameters	Pressmud (Joshi and Sharma, 2010)	Distillery Spent Wash	Post-Methanated Distillery Effluent
Color (Co–pt)	—	Dark brown (150,000)	Dark brown (70,000)
Odor	—	Jaggery smell	Mild sulfur smell
pH	7.66 ± 0.047	3.9 ± 0.25	8.3 ± 0.310
Temperature (°C)	44.5 ± 1.098	90 ± 2.0	35 ± 1.2
T.S.	—	103,084 ± 5.50	34,317 ± 455
T.D.S.	—	77,776 ± 3768	20,022 ± 438
T.S.S.	—	25,308 ± 1201	14,276 ± 16
COD	—	90,000 ± 231	58,018 ± 185
BOD	—	42,000 ± 123	29,120 ± 265
Chloride	—	2200 ± 105	1300 ± 60.5
Phenol	—	4.20 ± 1.8	1.65 ± 0.76
Sulfate	2.297 ± 0.123	5,760 ± 260	13656 ± 21.23
Phosphate	1.80 ± 0.157	5.36 ± 0.168	1.16 ± 0.15
Total nitrogen	1.13 ± 0.095	2,800 ± 130	568 ± 23
Total organic carbon	—	25,368 ± 1.060	10,904 ± 0.34
Water holding capacity	78.2 ± 1.323	—	—
Moisture content	54.9 ± 8.993	—	—
Carbon content	29.67 ± 1.058	—	—
Organic matter	51.2 ± 1.827	—	—
Electrical conductivity (S/cm)	1.77 ± 0.092	—	—
Metals	—	—	—
Cadmium	—	0.02 ± 0.00	2.281 ± 0.067
Chromium	—	0.192 ± 0.008	0.440 ± 0.013
Iron	—	6.312 ± 0.210	84.01 ± 1.980
Nickel	—	0.1706 ± 0.006	1.241 ± 0.037
Copper	—	0.961 ± 0.001	0.955 ± 0.022
Lead	—	0.945 ± 0.002	4.446 ± 0.064
Zinc	—	2.012 ± 0.001	4.631 ± 0.108
Manganese	—	0.214 ± 0.001	2.112 ± 0.045

Note: Mean ± SD, $n = 3$. All values of distillery wastewater are reported in mg/L and pressmud in % except pH, temperature, odor, and color. TS: total solid, TDS: total dissolve solid, TSS: total suspended solid, COD: chemical oxygen demand, BOD: biological oxygen demand.

that the pressmud and effluent proportion reaches an optimum ratio of 1:3. The heaps are then allowed to cure for a month. The compost obtained from this process is neutral in pH with an electric conductivity (EC) of 3.12–6.40 dS/m. It contains 1.53% N, 1.50% P, 3.10% K, 300 ppm Fe, 130 ppm Cu, 180 ppm Mn, and 220 ppm Zn. The organic carbon and C:N ratio reduces from 36% to 18% and from 28.12% to 16.3%, respectively. The technology of using distillery effluent for the composting of pressmud, pressmud along with sugarcane trash and coir waste, pressmud plus bagasse ash, and city garbage has been successfully used in several places.

Composting involves the conversion of organic residues of distillery effluent into manure. It is largely a microbiological process based on the activities of several bacteria,

Effluent   Pressmud

FIGURE 1.2
(See color insert.) Mixing of effluent and pressmud with the help of an aerotiller.

actinomycetes, and fungi. During the composting of effluent with pressmud, microorganisms that firstly colonize in the heap are mesophilic bacteria, actinomycetes, fungi, and protozoa. They grow between 10°C and 40°C and break down easily degradable components such as sugars and amino acids. The degradation of fresh matter starts when the heap is formed. Then due to the oxidative action of microorganisms the temperature increases. Although there is a drop in pH at the very beginning of composting, caused by the formation of volatile fatty acids (VFAs), the subsequent degradation of acids brings about an increase in pH. When the temperature of a waste heap reaches 40–60°C, thermophilic microorganisms replace mesophilic ones. The second phase is called the thermophilic phase and can last several weeks. It is the active phase of composting: most of the organic matter is degraded and consequently most oxygen is consumed in this phase. The degradation of lignin and more persistent compounds also start during this phase. Indeed, the optimum temperature for thermophilic micro-fungi and actinomycetes which mainly degrade lignin is 40–50°C. Above 80°C, these microorganisms cannot grow and lignin degradation is slowed down. After the thermophilic phase, the peak of degradation of fresh organic matter, the microbial activity decreases, as does the temperature. This is termed the cooling phase. The compost maturation phase then begins when the compost temperature falls to that of the ambient air. During this phase, mesophilic microorganisms colonize the compost heap and slowly degrade complex organic compounds such as lignin. This last phase is important because a humus-like substance is produced in this phase to form mature compost. The microbial succession in different steps of composting is shown in Figure 1.3.

Compostable substrates (feedstocks) in effluents contain metabolizable carbon which will enhance microbial diversity and activity during composting and will promote the degradation of xenobiotic or persistent organic compounds, such as melanoidin (the brown colored carbonyl-amino acid Maillard product of sugar industries), phenolics, and organochlorines. Metallic pollutants are not degraded during composting but may be converted into organic species that are less bioavailable. Selvamurugan et al. (2013) conducted a field experiment to study the impact of biomethanated distillery spent wash and pressmud biocompost with inorganic fertilizers in various proportions on the yield attributes, yield, oil, and protein content of groundnut.

Results of the field experiment revealed that the yield of groundnut was improved by the application of biomethanated distillery spent wash and pressmud biocompost compared to recommended NPK as chemical fertilizers. Davamani et al. (2006) also reported that biocompost significantly enhanced the yield and yield components and the juice quality

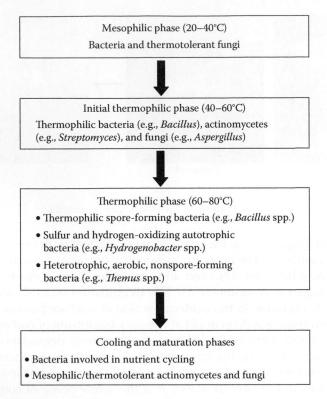

FIGURE 1.3
Microbial succession during composting.

of sugarcane. Bhalerao et al. (2006) observed that increased nutrient uptake by sugarcane is due to the use of spent wash and pressmud compost. The application of biomethanated distillery spent wash and pressmud biocompost substantially increased the microflora and enzyme activities of the soil throughout the crop growth period of sugarcane. The increased microbial biomass and enzymatic activities in sugarcane grown soil expedited mineralization of biomethanated distillery spent wash and biocompost, nutrient cycling and formation of organic matter and soil structure. Thus, biocomposting not only solves disposal problems but also helps in saving the cost of chemical fertilizers. Many workers have studied the effect of spent wash as a source of plant nutrients; however, very little information is available on the use of PMDE and sugarcane pressmud in composting.

1.2 The Principles of the Composting Process

Composting is the process by which complex organic materials are transformed into a material with environmentally beneficial applications. As said earlier, the various microorganisms such as bacteria, actinomyces, and fungi are used to break down the organic compounds into simpler substances. The composting process can transform large quantities of organic material into compost in a relatively short time by properly managing moisture, air and nutrients. During composting, the microorganisms consume oxygen and feed on organic matter. Active composting generates a considerable amount of heat and large quantities of carbon dioxide (CO_2) and water vapor are released into the air. The CO_2 and water losses can amount to half the weight of the initial organic materials, so composting reduces both the volume and mass of the raw materials while transforming them into a beneficial humus-like material.

$$\text{Organic compounds} + O_2 \xrightarrow{\text{Microbial metabolism}} \text{Stabilized organic waste}$$
$$\text{material} + CO_2 + 2H_2O + \text{Energy}$$

This conversion is not achieved through a single reaction, but through a series of reactions. These reactions serve not only to liberate significant quantities of energy, but also to form a large number of organic intermediates that serve as starting points for other synthetic reactions. The two possible modes of energy yielding metabolism for heterotrophic microorganisms are respiration and fermentation. Respiration can be either aerobic or anaerobic. Aerobic respiration is preferred over anaerobic respiration and fermentation for composting because it is more efficient, generates more energy, operates at higher temperatures, and does not produce odorous compounds. Aerobes can also use a greater variety of organic compounds as a source of energy that results in more complete degradation and stabilization of the compost material. In anaerobic respiration, the microorganisms use electron acceptors other than O_2, such as nitrates (NO_3^-), sulfates (SO_4^{2-}), and carbonates (CO_3^{2-}) to obtain energy. Their use of these alternate electron acceptors in the energy-yielding metabolism produces odorous or undesirable compounds, such as hydrogen sulfide (H_2S) and methane (CH_4). Anaerobic respiration also leads to the formation of organic acid intermediates that tend to accumulate and are detrimental to aerobic microorganisms. Aerobic respiration also forms organic acid intermediates, but these intermediates are readily consumed by subsequent reactions so that they do not pose as significant a potential for odors as anaerobic respiration. Fermentation is the simplest means of energy generation. It does not require oxygen and is quite inefficient. Most of the carbon decomposed through fermentation is converted to end-products, not cell substituent, while liberating only a small amount of energy. In addition, nitrogenous organic residue is broken down to obtain the nitrogen necessary for the synthesis of cellular material in heterotrophic microorganisms. Nitrogenous organic residues or proteins undergo enzymatic oxidation (digestion) to form complex amino compounds through a process called aminization. Carbon dioxide (CO_2), energy, and other by-products are also produced.

$$\text{Proteins} + O_2 \rightarrow \text{Complex amino compounds} + CO_2 + \text{Energy} + \text{Other products}$$

The complex amino compounds formed can then be synthesized into microorganisms or undergo additional decomposition into simpler products. The products of the digestion of proteins and complex amino acids can only be used in the synthesis of new cellular material if sufficient carbon is available. If not enough carbon or energy is available to incorporate these amino compounds into the cells, unstable nitrogen forms accumulate through the process of ammonification. As the ammonia group is characteristic of amino acids, ammonia (NH_3) or ammonium ions (NH_4^+) accumulates. The ammonium compound that is formed interconverts between the two forms depending on the pH and temperature of the heap. This interconversion between NH_3 and NH_4^+ is described by the reaction shown below. Acidic conditions (pH < 7) promote the formation of NH_4^+, while basic conditions promote the formation of NH_3. Elevated temperature also favors the formation of NH_3 and because of the low vapor pressure of NH_3, it generally results in gaseous NH_3 emissions from the heap. Another key chemical transformation of the composting process is nitrification, the process by which ammonia or ammonium ions are oxidized to nitrates. Nitrification is a two-step process. In the first step, NH_4^+–N is oxidized to form nitrites (NO_2^-) through the action of autotrophic bacteria that use the energy produced by this conversion. The nitrites are then rapidly converted to nitrates (NO_3^-) by a different group of microorganisms called nitrifying bacteria. The reactions are as follows:

$$NH_4^+ + 11/2O_2 \rightarrow NO_2^- + H_2O + 2H^+ + \text{Energy}$$
$$NO_2 + 11/2O_2 \rightarrow NO_3^- + \text{Energy}$$

Nitrification occurs during the curing period. Since nitrites (NO_2^-) are toxic to plants and nitrates (NO_3^-) are the form of nitrogen most usable in plant metabolism, enough time must be allowed for the curing period so nitrates are the final nitrogen product in the compost. In addition, because nitrification requires oxygen, proper aeration of the compost pile must be maintained during curing. Another important nitrogen transformation is denitrification. Denitrification occurs in oxygen-depleted environments. It can be carried out by either aerobic or anaerobic bacteria. If denitrification is carried out by aerobic bacteria, the reaction is as follows:

$$NO_3^- \rightarrow NO_2^- \rightarrow N_2O \rightarrow N_2(\text{gas})$$

If denitrification is carried out by anaerobic bacteria, the general reaction is

$$HNO_3^- + H_2 \rightarrow NH_2 + N_2O$$

As nitrous oxide (N_2O) is an odorous compound and results in the loss of beneficial nitrate–nitrogen, denitrification is not desired and can be avoided by maintaining aerobic heap conditions. This is accomplished with proper aeration.

1.3 Categories of Composting

Composting and vermicomposting are two of the best-known processes for the biological stabilization of organic wastes.

1.3.1 Microbial Composting/Thermophilic Composting

Composting involves the accelerated degradation of organic matter by microorganisms under controlled conditions, in which the organic material undergoes a characteristic thermophilic stage that allows sanitization of the waste by the elimination of pathogenic microorganisms. It may call hot composting or thermophilic composting. Two phases can

FIGURE 1.4
View of thermophilic composting of pressmud.

be distinguished in composting: (i) the thermophilic stage, where decomposition takes place more intensively and which therefore constitutes the active phase of composting and (ii) the maturing stage which is marked by a decrease in temperature to the mesophilic range and where the remaining organic compounds are degraded at a slower rate. A view of thermophilic composting is shown in Figure 1.4.

The duration of the active phase depends on the characteristics of the waste (amount of easily decomposable substances) and on the management of the controlling parameters (aeration and watering). The extent of the maturation phase is also variable and it is normally marked by the disappearance of the phytotoxical compounds. Loss of nitrogen through volatilization of NH_3 during the thermophilic stage of the process is one of the major drawbacks of the process. Thermophilic composting involves an important heating stage. This heat is caused by microbial metabolism and is dependent on the size of the heap, C:N ratio of the materials, moisture content, and aeration. During this heating stage, temperatures will ideally be in the 140°F (60°C) range, but will often be higher or lower. This type of composting typically follows the "batch" model—that is to say all the materials for the heap are piled up at one time and no more is added. In order to establish a sustained heating phase a "critical mass" of materials is required. Composting can be achieved when materials in the heap have a C:N of between 20:1 and 40:1, but ideally it should be between 25:1 and 30:1. A list of the important abiotic parameters associated with the success of the composting process and the range in which they should preferably remain, is presented in Table 1.2.

1.3.1.1 Advantages of Composting

1. Proceeds relatively quickly under ideal conditions.
2. Weed seeds and pathogens are killed during the process.
3. Proceeds easily in cold weather on a large scale.

TABLE 1.2

Key Parameters That Influence the Composting Process and Their Optimum Values

Parameter	Optimum Value for Composting
C:N ratio of the feed	25:1 to 35:1
Particle size	10 mm for agitated systems and forced aeration, 50 mm for long heaps and natural aeration
Moisture content	50%–60% (higher values when bulking agents are used)
Air flow	0.6–1.8 m³ air/day/kg volatile solids during thermophilic stage, or maintain oxygen level at 10% or higher
Temperature	55–60°C held for 3 days
Agitation	No agitation to periodic turning in simple systems and short bursts of vigorous agitation in mechanized systems
pH control	Normally not necessary
Heap size	Any length, 1.5 and 2.5 m wide for heaps using natural aeration. With forced aeration, heap size depends on need to avoid overheating
Activators	Use of efficient cellulolytic fungi and biofertilizers

1.3.1.2 Disadvantages of Composting

1. Requires attention and is labor intensive.
2. May require some stock-piling until sufficient materials are available for a "batch."
3. Heating can lead to considerable nitrogen loss.
4. Heat can kill off many beneficial microbes.

1.3.2 Vermicomposting

Vermicomposting is a simple biotechnological process of composting, in which certain species of earthworms is used to enhance the process of waste conversion and produce a better end product. Vermicomposting is somewhat similar to thermophilic composting, but one of the major differences is that it involves the joint action of earthworms and microorganisms (whereas the other process relies solely on microbes). Although it is the microorganisms that biochemically degrade the organic matter, earthworms are the crucial drivers of the process, as they aerate, condition, and fragment the substrate, thereby drastically altering the microbial activity. Figure 1.5 shows the vermicomposting.

Vermicomposting is also a much cooler process working best at 59–86°F (15–30°C). The relationship between temperature and time in thermophilic and vermicomposting is shown in Figure 1.6. Vermicomposting typically follows the "continuous" composting model that is to say materials are added continuously (usually in smaller amounts). Earthworms act as mechanical blenders and by comminuting the organic matter they modify its physical and chemical status by gradually reducing the ratio of C:N and increasing the surface area exposed to microorganisms—thus making conditions much more favorable for microbial activity and further decomposition.

Therefore, two phases can also be distinguished: (i) an active phase where the earthworms process the waste modifying its physical state and microbial composition and (ii) a maturation-like phase marked by the displacement of the earthworms toward fresher layers of undigested waste, where the microbes take over in the decomposition of the waste. As in composting, the duration of the active phase is not fixed and will depend on the

FIGURE 1.5
(See color insert.) View of vermicomposting.

species and density of earthworms, the main drivers of the process, and their ability to ingest the waste (ingestion rate). Vermicomposting is not fully adapted to the industrial scale and since the temperature is always in the mesophilic range, pathogen removal is not ensured, although some studies have provided evidence of suppression of pathogens. In some cases, organic residues require pretreatment before vermicomposting as they may

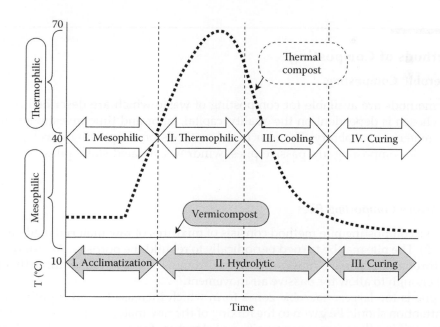

FIGURE 1.6
Time versus temperature curves for thermophilic composting and vermicomposting. Phases for thermal compost are adapted from Chefetz et al. (1996) and those for vermicompost are adapted from Benitez et al. (2000).

contain substances that are toxic for earthworms, such as acidic compounds. The combination of composting and vermicomposting has recently been considered as a more appropriate technique to treat complex wastewater.

1.3.2.1 Advantages of Vermicomposting

1. Is less labor-intensive—no turning/aerating necessary (worm activity helps to mix, fragment, and aerate materials).
2. Higher moisture content not an issue.
3. Cooler temperatures help to conserve nitrogen.
4. Under ideal conditions, waste can be processed very quickly.
5. Materials can be constantly added (no need to stock pile in preparation for next "batch").
6. Size of system unimportant—ideally suited for both indoors and outdoors.

1.3.2.2 Disadvantages of Vermicomposting

1. More space required to process similar amounts as thermophilic composting—need to be careful with amount added (since excess heat will kill worms).
2. Outdoor systems much more limited by cold weather.
3. Worms need to be separated from compost.
4. Worms (although quite resilient) do require some attention and proper care.

1.4 Methods of Composting

1.4.1 Aerobic Composting

Various methods are available for composting of waste which are described below. The method chosen is dependent on the quality, capital, labor and time investment, and land and raw material availability. The four broad methods of composting developed for use in large-scale composting are passive piles, windrows, aerated static piles, and in-vessel systems.

1.4.1.1 Passive Composting Piles

The passive composting pile method consists of mixtures of raw material made into a pile (Figure 1.7). The pile may be turned periodically to rebuild the porosity. Aeration is accomplished through the passive movement of air through the pile. This requires that the pile be small enough to allow for passive air movement.

If the pile is too large, anaerobic zones form which increase the odor problem. Hence, special attention should be given to the mixing of the raw material. The mix must be capable of maintaining the necessary porosity and structure for adequate aeration throughout the entire composting period. The passive composting method requires minimal labor and equipment. This method is often used to compost leaves.

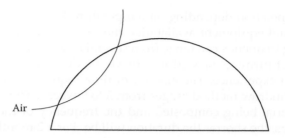

FIGURE 1.7
Passive composting piles.

1.4.1.2 Windrow Composting

In windrow composting piles are elongated and arranged in parallel fashion (Figure 1.8). These piles are turned regularly. Raw material is either mixed before pile formation or mixed as a part of pile formation. Windrow shapes and sizes vary depending on the climate and equipment and on the material used. Typically, windrows are 6–10 feet high, 15–20 feet wide and are up to several hundred feet long. A wet climate requires a windrow shape that allows moisture runoff. Moisture may be maintained by the effluent. A concave top may be required in drier climates to collect water and maintain pile moisture. Smaller windrows experience greater heat loss, while larger piles run the risk of anaerobic zones and odors. Dense material, such as manure, should be piled at a lower height than fluffy material, such as leaves. Bucket loaders and backhoes can produce higher windrows than turning machines. Windrows are aerated by passive aeration as in the passive composting method. The porosity necessary for adequate passive aeration is maintained by regularly turning the windrows. Turning windrows also serves to mix the material, and releases heat, water vapor and gases, and composts material more evenly.

As significant amounts of heat are released on turning the windrow, turning prevents excessive temperature accumulation within the windrow. Turnings are more frequent during the initial stages of composting. The schedule of turnings during composting varies

FIGURE 1.8
View of windrow composting (a) and mixing with the help of an aerotiller (b).

from operation to operation depending on temperature levels in the pile, consistency of the manure, labor and equipment availability, the season and how soon the compost is needed. The turning frequency can range from several times weekly to monthly. The number and frequency of turnings needed to achieve the desired quality of compost is best determined through experience. The amount of time required to finish the composting process using the windrow method ranges from 3 to 9 weeks. The duration is dependent on the type of material being composted and the frequency of the turnings. The more frequent the turnings, the shorter the duration will be. For a 2-month composting period, five to seven turnings are typical. Curing generally lasts at least 1 month. The windrow method is the most widely used for industrial wastewater such as distilleries and pulp paper mill industries.

This method can be applicable for diverse waste, including industrial waste, yard trimmings, grease, liquids, and animal by-products (such as fish and poultry waste). In a warm, arid climate, windrows are sometimes covered or placed under a shelter to prevent water from evaporating. In the rainy season, the shapes of the pile can be adjusted so that water runs off the top of the pile rather than being absorbed into the pile. Windrow composting can also work in cold climates.

1.4.1.2.1 Passively Aerated Windrows

Passively aerated windrows are not turned. Aeration is accomplished solely through the passive movement of air through perforated pipes embedded in the base layer of the pile as shown in Figure 1.9. Passively aerated windrows are different from turned windrows due to the presence of the perforated basal layer and the top layer in windrow construction. The base layer is typically composed of straw, peat moss, or finished compost. The main characteristic of the base layer is that it is porous so that the air that comes through the pipes is evenly distributed. The base layer also helps to insulate the pile and absorb moisture. The top layer is composed of peat moss or finished compost and serves several functions. The first function is to retain odors through the affinity of peat moss and finished compost for the molecules that cause odors. The top layer also deters flies and retains moisture and ammonia. Initial construction of this type of windrow requires more labor than other windrow methods.

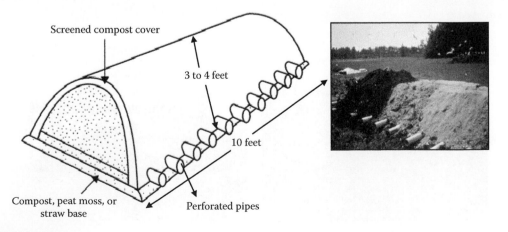

Screened compost cover

3 to 4 feet

10 feet

Compost, peat moss, or straw base

Perforated pipes

FIGURE 1.9
(See color insert.) Passively aerated windrows.

Once the windrow is formed, however, the labor requirement necessary is to primarily monitor the temperature and porosity of the pile. As in the passive composting system, the key element is to formulate a mix with good porosity and structure to allow for adequate aeration. Passive aeration also requires that the piles not be as high as those for the windrow method. The typical height is 3–4 feet with a width of about 10 feet. The bottom and top layers should each be about 6 inches thick.

1.4.1.2.2 *Aerated Static Pile*

Aerated static piles are similar to passively aerated windrows. The main difference between passively aerated windrows and an aerated static piles is that the aerated static pile uses blowers that either suction air from the pile or blow air into the pile using positive pressure (Figure 1.10). The suction method of aeration allows better odor control than positive pressure aeration, particularly if the air is directed through an odor filter. An odor filter is essentially a pile of finished compost that has an affinity for odor causing molecules. The blowers used for aeration serve not only to provide oxygen, but also to provide cooling. Blowers can be run continuously or at intervals. When operated at intervals, the blowers are activated either at set time intervals or based on compost temperature. Temperature-set blowers are turned off when the compost cools below a particular temperature. Blower aeration with temperature control allows for greater process control than windrow turning.

A forced aeration static pile has a base layer and top layer much like the passively aerated windrow. The purpose of the base layer for the aerated static pile is to distribute air evenly either as it enters or leaves the aeration pipes. This requires porous material, such as wood chips or straw. The top layer is generally composed of finished compost or sawdust to absorb odors, deter flies, and retain moisture, ammonia, and heat. As with all static piles, the initial mix and pile formation must have proper porosity and structure for adequate air distribution and even composting. A decay-resistant bulking agent is required to provide the necessary porosity. Wood chips are a good example of a bulking agent. They undergo minimal degradation during the composting process and can be screened from the finished compost and reused. The use of forced aeration also requires additional calculations. The size of the blower as well as the number, length, diameter, and types of pipes to use for adequate aeration must be determined. Pipes and blowers interfere with pile formation and cleanup operations. Aerated static piles are not commonly used for farm-scale composting operations.

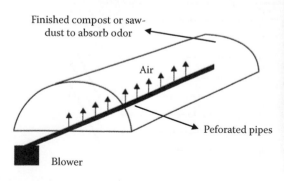

FIGURE 1.10
(**See color insert.**) Aerated static pile.

1.4.1.3 In-Vessel Systems

1.4.1.3.1 Bin System

Bin composting uses either constructed wooden bins, unused storage bins or some other appropriate vessel either with or without a roof. Some bins have aeration systems similar to those of forced aeration static piles. The same principles as forced aeration piles apply to these. The material in nonaerated bins must be turned regularly to maintain aerobic composting. Bin composting is perhaps the simplest in-vessel method. The materials are contained by walls and usually a roof. The bin may simply be wooden slatted walls (with or without a roof), a grain bin, or a bulk storage building. Bins can also eliminate weather problems, contain odors, and provide better temperature control. Bin composting methods operate in a similar way to the aerated static pile method. They include some means of forced aeration in the floor of the bin and little or no turning of the materials.

Occasional remixing of the material in the bins can invigorate the process. Where several bins are used, the composting materials can be moved periodically from one bin to the next in succession (Figure 1.11).

1.4.1.3.2 Rectangular Agitated Bed

The rectangular agitated bed method uses long, narrow beds to compost in and an automated turner for periodic turning (Figure 1.12). The turner is supported on rails that are mounted on either side of the bed for its whole length. As the turner moves along the bed, the compost is turned and moved a set distance until it is ejected at the end of the bed.

In some systems, blowers are also used to force air into the beds. The duration of the composting process is determined by the length of the bed and the turning frequency. An extended curing period is generally required.

1.4.1.3.3 Silo

The silo method is a rapid composting method that requires a prolonged curing stage (Figure 1.13). Compost material is loaded into the silo at the top and removed from the bottom using a borer. Aeration is provided through the base of the silo so that air is

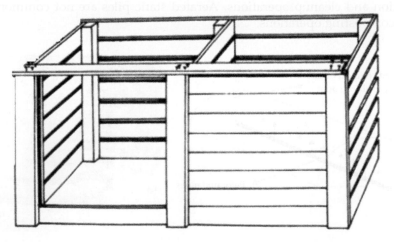

FIGURE 1.11
Bin system of composting.

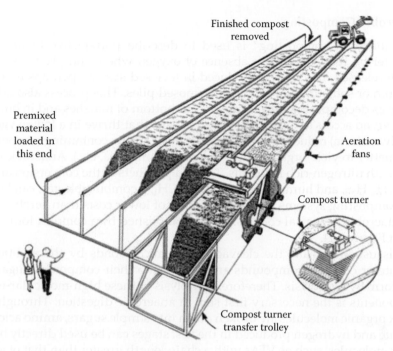

FIGURE 1.12
Rectangular agitated bed methods of composting.

forced upward through the compost material. Outlet air can be collected from the top and directed to an odor treatment system, such as a biofilter.

1.4.1.3.4 Rotating Tube

The rotating tube is a method that can be used where small amounts of waste require composting. The compost mix is loaded into the upper part of the tube. The mix rests on the first baffle plate. When the tube has filled from the first baffle plate to the top of the tube, it is rotated to aerate the compost mix and to empty the tube above the first baffle plate. This allows additional compost mix to be loaded in the tube. Ideally, the tube is operated so the composting process is complete by the time the material exits the tube. Tube size will be limited to what can be rotated when it is filled to capacity.

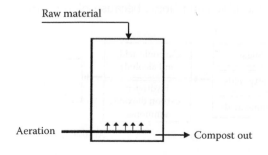

FIGURE 1.13
Silo composting method.

1.4.2 Anaerobic Composting

The term "anaerobic composting" is used to describe putrefactive breakdown of the organic matter by reduction in the absence of oxygen where end products such as CH_4 and H_2S are released. It is mostly produced in a closed system, perhaps in a plastic bag or a sealed bin or sometimes even in open exposed piles. This process also takes place in nature, such as decomposing organic mud at the bottom of marshes and in buried organic materials with no access to oxygen. Microorganisms that thrive in a low-oxygen environment (mostly bacteria) reduce nitrogen-containing or sulfur-containing compounds found in organic matter to yield organic acids and gases (ammonia, etc.). Anaerobic composting works best with nitrogen-rich materials. The end products of the conversion are new cells, CO_2, CH_4, NH_3, H_2S, and humus or compost. As CH_4, a combustible gas, can be recovered, anaerobic composting systems offer the potential of lower costs than aerobic composting systems. Anaerobic biological conversion is accomplished by a complex four step process as shown in Figure 1.14.

Hydrolysis usually means the cleavage of chemical bonds by the addition of water. Carbohydrate or complex compounds are broken into their component sugar molecules or simpler form by hydrolysis. Therefore, hydrolysis of these high-molecular-weight polymeric components is the necessary first step in anaerobic digestion. Through hydrolysis the complex organic molecules are broken down into simple sugars, amino acids, and fatty acids. Acetate and hydrogen produced in the first stages can be used directly by methanogens. Other molecules, such as VFAs with a chain length greater than that of acetate must first be catabolized into compounds that can be directly used by methanogens. The biological process of acidogenesis results in further breakdown of the remaining components by acidogenic (fermentative) bacteria. Here, VFAs are created, along with ammonia, carbon dioxide and hydrogen sulfide as well as other by-products. The third stage of anaerobic digestion is acetogenesis. Here, simple molecules created through the acidogenesis phase are further digested by acetogens to produce largely acetic acid, as well as carbon dioxide and hydrogen. The terminal stage of anaerobic digestion is the biological process of methanogenesis. Here, methanogens use the intermediate products of the preceding stages and convert them into methane, carbon dioxide, and water.

The environmental requirements for anaerobic composting with respect to C:N ratio and particle size are similar to aerobic composting. Moisture content requirements range from 90%–96% (low solids) to 70%–80% (high solids). When materials are composted anaerobically and not covered with water, the odor nuisance may be quite severe. However, if material is kept submerged, gases dissolve in the water and are usually released slowly into the atmosphere. If the water is replaced from time to time when removing some of the material, no serious nuisance is created. The pathogenic organisms gradually die in the organic mass in anaerobic conditions, because of the unfavorable environment and

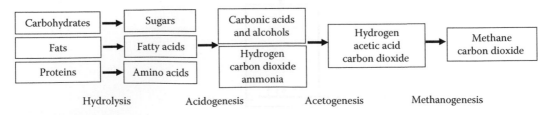

FIGURE 1.14
Different process stages of anaerobic digestion.

due to biological antagonisms. Their disappearance is slow and the material must be held for periods of 6 months to a year to ensure relatively complete destruction of *Ascaris* eggs.

1.5 Microorganism in Compost

Within a compost pile a variety of microbial populations develops in response to the different levels of temperature, moisture, oxygen, and pH. This microbial diversity enables the composting process to continue despite the constantly changing environmental and nutritional conditions within the pile. The microorganisms responsible for composting degrade a broad range of compounds from the simplest form of sugars to complex persistent compounds. Temperature levels and available food supply generally have the greatest influence in determining what class and species of organisms make up the microbial population at a particular time.

Decomposition proceeds rapidly in the initial stages of composting because of the abundant supply of readily degradable material. The material is characterized by a low molecular weight and a simple chemical structure. It is water soluble and can pass easily through the cell wall of the organisms, which allows it to be metabolized by a broad range of nonspecialized organisms. As the readily degradable material is consumed and the supply diminishes, more complex, less degradable material begins to be decomposed. This material is characterized by a high molecular weight, and a polymeric (long chain) chemical structure that cannot pass directly into the cells. The material must be broken down into smaller components through the action of extracellular enzymes. Not all of the microorganisms present in the compost pile can produce these enzymes, particularly simple organisms, such as bacteria. Such decomposition requires more specialized organisms, such as fungi. After the polymeric material is hydrolyzed into smaller components by these specialized organisms, the resulting fragments can then be used by the nonspecialized organisms. Microorganisms that are present in a compost pile are in three major classes: bacteria, fungi, and actinomycetes. The microorganisms within a compost pile can be psychrophilic, mesophilic, or thermophilic depending on the temperature range within which they experience optimal growth rates. Scanning of compost by electron microscope showing the diversity of microorganisms is shown in Figure 1.15. The psychrophilic temperature range is defined as being below 50°F, mesophilic between 50°F and 105°F and thermophilic between 105°F and 160°F.

1.5.1 Bacteria

Bacteria are small, simple organisms present primarily during the early stages of the composting period. They are responsible for much of the initial decomposition and include a wide range of organisms that can survive in many different environmental conditions. Although they are small relative to fungi and actinomycetes, they are present in significantly greater numbers. Bacteria are fast decomposers. Bacteria function optimally within a pH range of 6–7.5 and are less tolerant of low moisture conditions than other types of microorganisms. Some bacteria form endospores that enable them to withstand unfavorable environmental conditions, such as high temperature or low moisture. When the environment becomes more favorable for survival, the endospores germinate and the bacteria become active again. This feature of certain bacteria helps to continue the composting process during the cooling

FIGURE 1.15
Compost under an electron microscope showing a diverse ecosystem of flourishing bacteria and fungi.

phase that follows peak thermophilic temperatures. They may also utilize more complex materials, or may exploit substances released from the less degradable materials due to the extracellular enzyme activities of other organisms. Among bacteria that occur commonly in aerobically decomposing substrate are species of *Bacillus, Cellulomonas, Pseudomonas,* and *Klebsiella,* while *Clostridium* occurs substantially in anaerobic conditions. Typical bacteria of the thermophilic phase are species of *Bacillus,* for example, *B. subtilis, B. licheniformis,* and *B. circulans.* Many thermophilic species of *Thermus* have been isolated from compost at temperatures as high as 65°C and even 82°C. *Nitrosomonas* spp. and *Nitrobacter* spp. are the ammonium oxidizing and nitrite oxidizing bacteria, respectively, present in the compost heap. Establishment of a large population of denitrifying bacteria suggests that some anaerobic microhabitat exists within the compost piles. These microhabitats could have been developed within the piles partially due to the initial high water content (65%) of the piles and partially because of the rich contents of organic matter and nitrogen present in the substrate, which promote microbial activity to the extent of causing depletion in the O_2 content in isolated pockets within the piles. Some microbial genera capable of denitrification are *Bacillus, Flavobacterium,* and *Pseudomonas.* Mesophilic microorganisms are partially killed or poorly active during the thermogenic stage (40–60°C). The diversity decreased as temperature increased, with a shift from *Pseudomonas, Achromobacter, Flavobacterium, Micrococcus,* and *Bacillus* to being dominated by *Bacillus.* Bacteria related to *B. schlegelii, Hydrogenobacter* spp. and particularly to the genus Thermus (*Th. thermophilus, Th. aquaticus*) appear to be the main active microbes in hot compost (65–80°C). Bacterial survival in high temperature composting material is possible through the formation of microcolonies. Mesophiles are likely to contribute little to compost degradation at these temperatures. Microbial fermentation of carbohydrates generally results in an increase in acidity. *Clostridium* species commonly ferment glucose to yield butyl and ethyl alcohols and certain acids. *Lactobacillus lactis* yields almost entirely lactic acid, while *Lactobacillus bevis* yields lactic and acetic acids, ethyl alcohol and carbon dioxide.

1.5.2 Fungi

Fungi are members of a large group of eukaryotic organisms and are larger organisms than bacteria. They form networks of individual cells in strands called filaments. Fungi tend to be present in the later stages of composting because of the nature of the material they decompose.

Fungi are less sensitive than bacteria to environments with low moisture and pH, but because most fungi are obligate aerobes (requiring oxygen to grow), they have a lower tolerance for low-oxygen environments than bacteria. Most fungi are eliminated by high temperatures, but they commonly recover when temperatures are moderate and the remaining substrates are predominantly cellulose or lignin. Considering that fungi cannot survive in temperatures greater than 140°F and that they are responsible for much of the decay of resistant material indicates that excessive temperatures are detrimental to the composting process in terms of complete degradation. While, high temperature levels are desirable for pathogen destruction, they must be controlled to reduce the destruction of beneficial organisms and their subsequent effect on the completion of decomposition. The most commonly observed species of celluloytic fungi in composting materials are *Aspergillus*, *Penicillium*, *Rhizopus*, *Fusarium*, *Chaetomonium*, *Trichoderma*, *Alternaria*, and *Cladiosporium* (Figure 1.16). Some of the species of *Paecilomyces* and *Sporotrichum* have also been identified as efficient degraders of lignocellulosic wastes. White rot fungi are known to be the most efficient lignolytic microorganisms. *Phanerochaete chrysosporium* is probably the best suited microorganism with this activity and it is often used as a reference. Among other well-known white rot fungi, *Coriolus versicolor* show even higher efficiency and a wider range of lignolytic activities together with an important celluloytic activity. *Phanerochaete flavidoalba* causes preferential loss of lignin rather than of cellulose and it is more efficient than *Ph. chrysosporium* on paper mill effluents. The most important among these are white-rot fungi belonging to Basidiomycetes. Species of *Polyporus*, *Pleurotus*, *Collybia*, *Poria*, *Fomes*, *Trametes*, *Sporotrichum*, *Cyathus*, and *Coriolus* have also been found to degrade lignin. Other important

FIGURE 1.16
(See color insert.) Some fungi-like *Trichoderma* (yellow color) and *Aspergillus* (white color) growing on compost of distillery effluent and pressmud.

factors are sources of C and N and the pH. During composting, temperatures above 55°C discourage fungal growth. Fungi are excluded during the earlier high temperature stage of the composting process. A moderately high level of nitrogen is needed for fungal growth, although some fungi, mainly wood-rotting fungi, grow at low nitrogen levels. Indeed, a low nutrient nitrogen level is often a prerequisite for lignin degradation. However, low nutrient nitrogen is a rate limiting factor for the degradation of cellulose. Most fungi prefer an acidic environment but tolerate a wide range of pH, with the exception of the basidiomycotina, which do not grow well above pH 7.5. The majority of the fungi are mesophiles, which grow between 5°C and 37°C, with an optimum temperature of 25–35°C. However, in the compost environment the elevated temperature means that the small group of thermophilic fungi is an important biodegradation agent. Thermophilic fungi that have been found growing in lignocellulose substrate or compost are *Taloromyces emersonii*, *Ta. thermophilus*, *Thermoascus auranticus*, and *Thermomyces lanuginosus*. However, a few basidiomycotina grow well at elevated temperatures. *Phanerochaete chyrsoporium* (*Sporotrichum pulverulentum*) is a white-rot fungus with an optimum temperature of 36–40°C and maximum temperature of 46–49°C. *Ganoderma colosum* is another white-rot fungus that is still capable of growing at 45°C and has an optimum temperature of 40°C. In the genus *Coprinus* there are some species that have an optimum temperature of above 40°C. A thermophilic *Ascomycotina*, *Thermoascus aurantiacus*, has a high-lignolytic capacity and has been isolated from compost.

1.5.3 Actinomycetes

Actinomycetes are the third major class of microorganisms that play a major role in composting. Actinomycetes are technically bacteria because of their structure and size, but are similar to fungi in that they form filaments and are able to use a variety of substrates. Actinomycetes can degrade organic acids, sugars, starches, hemicelluloses, celluloses, proteins, polypeptides, amino acids, and even complex compounds such as lignins. They also produce extracellular proteases and can lyse other bacteria. Actinomycetes are more prevalent in the later stages of composting when most of the easily degradable compounds have been degraded, the moisture levels have decreased and the pH has become less acidic.

They are primarily strict aerobic saprophytes and are common in many environments. Their ubiquity is a result of their ability to utilize a wide range of carbon sources and to sporulate prolifically. Actinomycetes colonize more slowly than bacteria and fungi. Colonization is minimal in areas that are poorly aerated. They also appear during the thermophilic phase as well as during the cooling and maturation phase of composting and can occasionally become so numerous that they are visible as a white film on the surface of the compost. The genera of the thermophilic actinomycetes isolated from compost include *Nocardia*, *Streptomyces*, *Thermoactinomyces*, and *Micromonospora*. Actinomycetes are able to degrade some cellulose and solubilize lignin and they tolerate higher temperatures and pH than do fungi. The actinomycetes are thus well placed to exploit the compost environment as the piles cool in the immediate post peak heat phase. During the cooling stage of composting, actinomycetes actively degrade hemicellulose in the compost. With an optimum growth between 25–30°C and pH of 5–9, these microorganisms are the most significant group of microbes in the degradation of relatively complex, recalcitrant polymers. As actinomycetes develop more slowly than most bacteria or fungi, they are ineffective competitors when nutrient levels are high, but become more competitive as nutrient levels decrease. *Actinomycetes thermophilus*, *Streptomyces*, and *Micromonospora* spp. are common in compost. Although optimum growth temperatures fall in the mesophilic range, obligate thermophiles such as *Thermoactinomycetes* and *Saccharomonospora* spp. have

TABLE 1.3

Diversity of Microbes Present in Composting

Bacteria	Fungi	Actinomycetes
Alcaligenes faecalis	*Aspergillus fumigates*	*Actinobifida chromogena*
Bacillus brevis	*Humicola grisea*	*Microbispora ispora*
B. circulans complex	*Hu. insolens*	*Micropolyspora faeni*
B. Licheniformis	*Hu. lanuginose*	*Nocardia* sp.
B. megaterium	*Malbranchea pulchella*	*Pseudocardia thermophilia*
B. pumilus	*Myriococcum thermophilum*	*Steptomyces rectus*
B. megaterium	*Paecilomyces variotti*	*St. thermofuscus*
B. pumilux	*Papulaspora thermophilim*	*St. thermoviolaceus*
B. sphaericus	*Sporotrichum thermophile*	*St. thermovulgaris*
B. stearothermophilus	—	*St. violaceus-ruber*
B. subtilis	—	*Thermoactinomyces sacchari*
Clostridium thermocellum	—	*Thermoa. Vulgaris*
Escherichia coli	—	*Thermomonospora curvata*
Flavobacterium sp.	—	*Thermom. Viridis*
Pseudomonas sp.	—	—
Serratia sp.	—	—
Thermos sp.	—	—

Source: Adapted from Palmisano AC and Barlaz MA. 1996. *Microbiology of Solid Waste*, pp. 125–127. CRC Press Inc., Boca Raton, FL.

been isolated. Certain species of actinomycetes are more tolerant of high temperatures, becoming increasingly active as temperatures approach to 60°C. The diversity of microbes present in composting is shown in Table 1.3.

1.5.4 Higher Organisms

Higher organisms begin to invade the compost pile once the pile temperatures cool to suitable levels. These organisms include protozoa, rotifers, and nematodes. They consume the bacterial and fungal biomass and aid in the degradation of complex compounds. These higher organisms contribute to the disease suppressive qualities of compost.

1.5.5 Pathogens in Composting

One of the aspects of composting that makes it an attractive alternative to the direct application of untreated manure is the high degree of pathogen destruction that is possible with a well-managed composting operation. The pathogen content of the compost is important because improperly treated compost can be a source of pathogens to the environment and, as such, a threat to humans and animals. This depends on the type of pathogen involved. The type and quantity of pathogens in the initial compost mix are dependent on the waste that is being composted. Animal pathogens are in manure and on plant residue that has come into contact with any manure. Plant pathogens are in plant residue. Pathogenic microorganisms that may be in compost include bacteria, viruses, fungi, and parasites. Although parasites and viruses cannot reproduce apart from their host, they can often survive for extended periods. If they are not killed during the composting process, they can survive until the compost is land applied. At that time, they may infect a new host. Even if their numbers are reduced in the composting process, their population may recover and increase given enough time and if conditions permit. Therefore, it is not enough to reduce their numbers. Pathogenic bacteria and fungi must be killed in mature compost. Conditions unfavorable to pathogenic growth include a lack of assimilable organic matter and a pile with moisture content of less than 30%. As such conditions are difficult to achieve in mature compost, as many pathogens as possible should be destroyed during the composting process. Pathogens can be destroyed by heat, competition, destruction of nutrients, antibiosis, and time. Antibiosis is the process by which a microorganism releases a substance that, in low concentrations, either interferes with the growth of another microbe or kills it. Most pathogens do not grow at the optimum temperatures for composting. As such, exposure to high (thermophilic) temperature kills them. The few exceptions to this are among fungal plant pathogens. Some of these pathogens can withstand temperatures over 180°F. Most pathogens originating from animals cannot survive above the 130–160°F temperature range. Pathogens can also be destroyed as a result of competition with the indigenous microbial population for nutrients and space. Pathogens are at a disadvantage because they are not as well adapted to the environment as the indigenous population and their numbers are insignificant relative to the indigenous population. Pathogens must compete with the indigenous microorganisms for sites of attachment on the waste particles. Nutrient requirements of pathogenic microorganisms are specific. If their key nutrients are used by the competing indigenous microbial population, then the pathogens are deprived of nutrients, and they will die. Good pathogen destruction is possible with the various composting methods if the windrows or piles are managed correctly. The two essential elements in achieving good pathogen destruction are

- All of the material must be exposed to lethal conditions either simultaneously or successively.
- The exposure must last for a sufficient amount of time to maximize its effectiveness.

Pathogens are killed through the process of turning. During turning the innermost layers that have the highest temperature levels and greatest degree of pathogen destruction are exchanged. The outermost layers that have not been exposed to these lethal conditions are then allowed to reheat so that all material within the pile is exposed to the lethal temperature conditions. In reality, however, the outermost and innermost layers are not simply exchanged, but are instead thoroughly mixed so that the innermost layer is recontaminated with pathogens from the outermost layer. To counteract the effects of recontamination and ensure complete pathogen destruction, either the frequency of turning or the duration of active composting must be increased.

1.6 Factors Influencing Composting

A given nutrient in waste can be utilized only if it is available to active microbes. Availability may be chemical and physical. A nutrient is chemically available to microbes if it is a part of a molecule that is vulnerable to attack by the microbes. Usually, breakdown of compounds or nutrients is accomplished enzymatically by microbes that either possess the necessary enzyme or can synthesize it. Physical availability is interpreted in terms of accessibility to microbes. Accessibility is a function of the ratio of mass or volume to surface area of a waste particle, which in turn is determined by particle size. Factors which influencing composting are divided into two major group that is, nutrient factors and environmental factors.

1.6.1 Nutrient Factors Influencing Composting

1.6.1.1 Macronutrients and Micronutrients

Nutrients can be grouped into the categories "macronutrients" and "micronutrients." The macronutrients include carbon (C), nitrogen (N), phosphorus (P), calcium (Ca), and potassium (K). However, the required amounts of Ca and K are much less than those of C, N, and P. As they are required only in trace amounts, hence they are frequently referred to as the "essential trace elements." If their concentrations are above a trace they become toxic for organisms. Among the essential trace elements are magnesium (Mg), manganese (Mn), cobalt (Co), iron (Fe), and sulfur (S). Most trace elements have a role in cellular metabolism. The substrate is the source of the essential macronutrients and micronutrients. Although an element of uncertainty exists, economic reality dictates that waste constitutes most or all of the substrate in compost practice. Any uncertainty is due to variation in the availability of some nutrients to the microbes. Variation in availability, in turn, arises from differences in the resistance of certain organic molecules to microbial attack. Variations in resistance lead to variations in the rate at which the process advances. Examples of resistant materials are lignin (wood), chitin, several forms of cellulose, cholorinated hydrocharbones, and many other persistent compounds.

1.6.1.2 Carbon-to-Nitrogen Ratio (C:N)

The carbon-to-nitrogen ratio (C:N) is a major nutrient factor. Based on the relative demands for carbon and nitrogen in cellular processes, the theoretical ratio is 25:1. The ratio is weighted in favor of carbon, because the uses for carbon outnumber those for nitrogen in microbial metabolism and the synthesis of cellular materials. Thus, not only is carbon utilized in cell wall or membrane formation, protoplasm and storage products synthesis, an appreciable amount is oxidized to CO_2 in metabolic activities. On the other hand, nitrogen has only one major use as a nutrient, an essential constituent of protoplasm. In compost practice, it is in the order of 20:1 to 25:1. The general experience is that the rate of decomposition declines when the C:N exceeds that range. On the other hand, nitrogen probably will be lost at ratios lower than 20:1. The loss could be due to the conversion of the surplus nitrogen into ammonia–N. The high temperatures and pH levels characteristic of composting during the active stage could induce the volatilization of the ammonia. In a developing country, an unfavorably high C:N can be lowered by adding nitrogenous waste to the compost feedstock. If economics permit, it also can be lowered by adding a chemical nitrogen fertilizer, such as urea or ammonium sulfate. Conversely, a carbonaceous waste can be used to elevate a low C:N.

The relative proportion of carbon and nitrogen is a major controlling factor in the composting process. Carbon serves primarily as an energy source for the microorganisms, while a small fraction of the carbon is incorporated to the microbial cells. Nitrogen is critical for microbial population growth, as it is a constituent of protein that forms over 50% of dry bacterial cell mass. If nitrogen is limited, microbial populations will remain small and it will take longer to decompose the available carbon. Excess nitrogen, beyond the microbial requirements, is often lost from the system as ammonia gas. In the composting process, the substrate should achieve a C:N ratio of 30:1 for stimulating degradation and immobilization of nitrogen. A balanced carbon to nitrogen (C:N) ratio of 25:1–30:1 is ideal for an active compost pile. C:N ratios of as low as 20:1 or as high as 40:1 also produce good quality finished compost.

If C:N < 20:1

Excess nitrogen will let off gas into the atmosphere as NH_3 or N_2O, resulting in an undesirable odor

If C:N > 40:1

Nitrogen mineralization generally occurs in two phases, a rapid exponential immobilization or mineralization phase, followed by a slow linear mineralization phase. Nitrogen mineralization is the process by which organic nitrogen is converted to the plant available inorganic form such as ammonium and nitrate. The C:N ratio of the substrate determines whether immobilization or mineralization will dominate in the early stages of composting. The rate of inorganic N release to the soil from composted manure depends on the rate of decomposition of the organic matter and on subsequent turnover of the decomposed C and N in the soil. Release of plant available N from manure in the soil is controlled by the balance of N immobilization and mineralization, which in turn is controlled, to a large extent, by the C:N ratio of the decomposing organic material. The decomposition rate (i.e., composting process) slows down.

1.6.2 Environmental Factors Influencing Composting

The principal environmental factors that affect the compost process are temperature, pH, moisture, and aeration. The rate and the extent of decomposition are proportional to the degree that each nutritional and environmental factor approaches optimum. A deficiency in any one factor would limit the rate and extent of composting, in other words, the deficient factor is a limiting factor.

1.6.2.1 Moisture

Moisture is one of the composting variables that affect microbial activities, as it provides a medium for the transport of dissolved nutrients required for the metabolic and physiological activities of microorganisms. It is essential for the decomposition process, as most of the decomposition occurs in the thin liquid films on the surfaces of particles. Moisture content of 60%–70% is generally considered ideal to start with. At later stages of decomposition, the ideal moisture content may be 50%–60%. Moisture management requires a balance between microbial activity and oxygen supply. Very low (<30%) or high moisture content (>75%) inhibits microbial activities due to early dehydration or anaerobiosis. Excess moisture will fill many of the pores between particles with water, thereby limiting oxygen transport. This in turn creates anaerobic conditions and brings about putrefaction, resulting in a disagreeable odor and undesirable products. On the other hand, if the composting substrate is supplied with insufficient water, the growth and proliferation of microorganisms as well as the rate of decomposition of the organic material will slow down or even stop. It is important, therefore, to ensure adequate moisture in each layer of the compost heap.

1.6.2.2 Oxygen and Temperature

The decomposition process of pollutants present in industrial waste is also affected by oxygen and temperature. The temperature within a composting mass determines the rate at which many of the biological processes take place and plays a selective role in the development and the succession of the microbiological communities. Temperature and oxygen fluctuate in response to microbial activity, which consumes oxygen and generates heat. Both are linked by a common mechanism of control:aeration. Inadequate oxygen may lead to the growth of anaerobic microorganisms, which can produce odoros compounds. Usually, in an aerobic system, the temperature rises to 50–60°C in just a few days and can even go up to 80°C in some cases. If done correctly, a compost pile will heat to high temperatures within 24–48 h. If it does not, the pile is too wet or too dry or there is not enough green material (or nitrogen) present. The high temperature rise in the compost pile destroys weed seeds, pathogenic microorganisms and worms and prevents fly breeding. This and the generation of antibiotics during composting drastically reduce pathogens in the final compost. A temperature in the range of 55–65°C ensures destruction of pathogenic organisms. A temperature of 65°C for at least 30 min is considered a critical threshold for plant pathogens. Human pathogens are also inactivated at high temperatures. The heat resistance of human pathogens increases markedly under dry conditions. Therefore, wet conditions must prevail in the compost pile. The maximum temperature of the composting process reaches 60–80°C, the temperature level where many microorganisms become less active. At the top of the pile, the temperature is slightly lower due to conductive heat loss from the top to the surroundings. Over time, the temperature gradually drops as the degradation rate of organic matter becomes less. This course in composting will result in adequate stabilization of organic matter, drying of the compost, and killing of pathogens and weeds. Low temperature typically indicates low aerobic activity in the composting pile. Temperature alone is not a fool proof indicator of aerobic activity, as it is a result of heat production and heat removal. Lack of aerobic activity can only be confirmed by measuring the oxygen content within the compost bed. To attain temperatures high enough for heat activation throughout in compost, the vessel has to be insulated to retain the heat produced. High temperature combined with high exchange rates of the air will increase the ammonia losses. In a composting pile, however, the rate of degradation is a result of metabolic activity of a mixed microbial population that may originally include microorganisms with different temperature optima. These microorganisms adapt to the environmental temperature during composting and have a collective temperature optimum at which respiration from the microbial community is highest. Not only is microbial metabolism highly temperature dependent, but it also dramatically influences the population dynamics (e.g., composition and density) of microbes. Temperature increase within composting materials is a function of initial temperature, metabolic heat evolution, and heat conservation. Indeed, temperatures of composting material below 20°C have been demonstrated to significantly slow or even stop the composting process. Temperature in excess of 60°C has also been shown to reduce the activity of the microbial community and above this temperature, microbial activity declines as the thermophilic optimum of microorganisms is surpassed. If the temperatures reach 82°C, the microbial community is severely impeded.

1.6.2.3 Aeration

Aerobic organisms survive only in presence of air or O_2. Aeration is necessary in high temperature aerobic composting for rapid odor-free decomposition. Aeration is also useful in reducing high initial moisture content in composting materials. Several different aeration techniques can be used. Turning material is the most common method of aeration when composting is done in piles. Hand turning of the compost piles is most commonly used for small garden operations. Mechanical turning or static piles with a forced air system are most economical in large municipal or commercial operations. The most important consideration in turning compost, apart from aeration is to ensure that material on the outside of the pile is turned into the center where it will be subject to high temperatures. In hand turning with forks, this can be easily accomplished. For piles or windrows on top of the ground, material from the outer layers can be placed on the inside of the new pile. For static piles with a forced air system finished compost or a physical "cover" can be placed on the composting material, ensuring it reaches high temperatures uniformly. Volume reduces during the compost process.

1.6.2.4 pH

pH is another parameter that greatly affects the composting process. The range of pH values suitable for bacterial development is 6.0–7.5, while fungi prefer an environment in the range of pH 5.5–8.0. An initial phase characterized by a low pH is often observed during the composting of organic wastes and perhaps especially in the case of easily degradable energy-rich materials. This is due to the formation of carbon dioxide and VFAs. With the subsequent evolution of CO_2 and utilization of VFAs, the pH begins to rise and may even reach values exceeding 8.0. Organic acids are produced during the decomposition of organic matter, but their existence is only transitory. Problems may arise if the material obtained undergoes putrefaction, as appreciable amounts of troublesome organic acids are produced during anaerobic decomposition and may produce malodor. However, a rise in pH beyond 7.5 could make the environment alkaline which may cause loss of nitrogen as ammonia. The growth of active microorganisms is inhibited by a temperature of above about 40°C if short chain fatty acids and low pH are present. Microbial tolerance to thermophilic temperature is reduced by the combination of low pH and increasing concentrations of fatty acids. The optimum pH range for decomposition is between 6.5 and 8.5. The pH affects the potential for beneficial bacteria to colonize composts; below pH 5.0, bacterial biocontrol agents are inhibited. To curtail excessive ammonia loss, Hoitink and Kuter (1986) suggest that the pH should be below 7.4 in aerated composting systems. pH is an indicator of aeration levels within a composting pile. Well-aerated compost piles generally have a high pH, whereas piles with anaerobic conditions have decreased pH values. The decrease in pH during the initial period of composting is expected because of the acids formed during the metabolism of readily available carbohydrates.

1.6.2.5 Electrical Conductivity

Generally, it is found that electrical conductivity (EC) increases during composting as VSs are degraded and the amount of water-soluble salts increases on a total solids (TS) basis. At lower pH values, negatively charged surface sites of organic matter are occupied by protons, which thus lowers cation exchange capacity (CEC). A decrease in CEC results in a lower adsorption of cations to organic matter and thus an increase in EC.

1.7 Use of Composed Distillery Waste and Pressmud for Enhancing Vermicomposting

The combination of composting and vermicomposting is a more effective organic waste management strategy than either process by itself. Precomposting of distillery waste and pressmud through windrow composting followed by vermicomposting by addition of other urban waste products will produce a highly stabilized end product in half the time as composting or vermicomposting by itself. pH and moisture maintenance is easy through this combination of technologies. Most of the organic pollutants may be degraded or transformed into simpler compounds during thermophilic composting which is growth supportive for the vermicomposting worms. During vermicomposting most persistent compounds and pathogenic organisms may be reduced. It will be more economically efficient if the composting and vermicomposting process are managed and operated by the local people themselves. This process will provide more job opportunities and better organic manure from agricultural waste. This process is also helpful for treatment of broad range complex pollutants.

1.8 Advantages and Prospects

Composting of industrial as well as municipal solid waste is an economical, environmentally friendly wealth creating and sustainable technique. Managed composting leading to the production of compost acts as a fertilizer which substitutes chemical fertilizer. The use of chemical fertilizer could lead to groundwater pollution. But the use of compost discourages this water pollution. However, composting requires proper handling and appropriate technology for its sustainability. Hence, a managed composting operation promotes clean and readily finished products, minimizes any nuisance potential and is simple to run. There is a reduction in landfill space where composting is operated as a waste management technique. There is also reduced surface and groundwater contamination, which is a phenomenon in the case of landfill. According to WHO, 900 million people experience diarrhea or contact diseases such as typhoid and cholera through contaminated water. Waste management and composting enhances the recycling of materials, and has low transportation cost. In composting there is a minimal emission of greenhouse gases with a subsequent effect on climate change and global warming. Moreover, addition of compost to soil reduces soil erosion as well as improves soil structure, aeration and water retention. Composting carried out by the appropriate waste management authorities is very useful for the world's developing countries. This action will lead to waste reductions at landfills, job creation, and production of organically produced food crops. Organic agriculture has continued to gain more ground all over the world for the sustainability and safety of its farm produce. The crops produced from organic agriculture are expensive. But with the increased supply of compost fertilizer in large quantity and low price to farmers, the price of organic food could drop drastically.

1.9 Challenges of Composting/Research Needed/Work to Be Done

There are several problems linked to the composting of industrial waste which either affect the process or the environment. In the case of industrial waste there are many recalcitrant pollutants present which may not decompose. In addition, compost might disperse a bad odor when it has not yet matured. Larger piles are not advisable because they are harder to

TABLE 1.4

Challenges of Composting

Human Challenges

1.	Policy/institution	Political will, government support, stakeholder support
2.	Human acceptance (source separation and site situation)	Accept the idea, accept separation of waste, accept site situation
3.	Finances	Costs need to be manageable, government's will to pay
4.	Markets and final end use	While end uses are vast in principle, tangibly what are the markets, willingness to pay
5.	Capacity: technical and human	The local ability to implement composting, labor, infrastructure for collection, equipment required (temperature and grinder)

Process Challenges

1.	Choosing an appropriate method	Pile, windrow, anaerobic digestion, household or centralized, sources separation or mixed waste
2.	Choosing a location	Flat land, consider storm water runoff and compost runoff
3.	Creating and managing the composting pile	• The chemical makeup of the feedstock: C:N ratio • Oxygen availability: facilitate oxygen availability • The physical properties of the feedstock and pile: size of feedstock and porosity is important to consider. May need to grind large pieces • The moisture content of the pile: the pile can neither be too moist or too dry 45%–65% • Height of the pile: if it is too small it may not maintain heat, if it is too big, oxygen availability may become problematic 1–1.5 m • Temperature: ideal temperature 40–65°C must be maintained to promote the desired bacteria. If the temperature is too low, the process is slowed and if the temperature is too high, important microorganisms will die
4.	Heavy metal content and persistent organic compounds	They are not easily degradable or bio-transformable
5.	Stability: temperature testing	• Temperature can be used as an indicator to assess if the composting process has been complete and is stable; compost should be near ambient temperature and not reheat • The compost should have achieved sufficient heat to kill bacteria: turning while the temperature is high to ensure that all bacteria were killed
6.	Other testing	• pH can also be an indicator for stability. Acceptable pH for stable compost is 6–8 • Electroconductivity tests • Water solubility • Oxygen consumption • Plant growth • Pathogen analysis

Place-Specific Challenges

1.	Management	Local management
2.	Local weather patterns	Precipitation and typhoons are potential local factors which effect implementing composting, e.g., in the Hainan project
3.	Waste characteristics	• Amount of waste • Type of waste • Changes in waste: seasonal and structural

manipulate, therefore composting needs a bigger area. The bacteria that break down organic compounds into compost do not operate well in freezing temperatures so composting is not very effective in winter. However, the activity of the bacteria helps increase the temperature within the compost pile even if the environment is cold, yet it slows down the process if the environment is not hot enough for decomposition, especially if the organic matter has a lot of pathogens present . Developing countries face three types of challenges with respect to composting: human challenges, technological/process challenges, and place-specific challenges. Human challenges include policy and institutional aspects, human acceptance, finances, markets and final end use, and capacity in terms of lack of technical expertize. Process issues include choosing an appropriate method, choosing a location, creating and managing the compost pile, heavy metal and POPs content and the maturity of the compost. Finally, place-specific characteristics may pertain to both human and process issues; these include management, community response, capacity, local weather patterns, and waste characteristics. Details of the challenges of composting are shown in Table 1.4.

Industrial waste has so many recalcitrant pollutants which may not decompose. Hence, two step composting that is, thermophilic composting and vermicomposting as described earlier may be an effective process for composting. Regular monitoring of composting in terms of persistent organic pollutants, pathogenic bacterial monitoring, and nutrient content of compost is required.

References

Benitez E, Nogales R, Masciandaro G, and Ceccanti B. 2000. Isolation by isoelectric focusing of humic-urease complexes from earthworm (Eisenia fetida)-processed sewage sludges. *Biology and Fertility of Soils* 31: 489–493.

Bhalerao VP, Jadhav MB, and Bhoi PG. 2006. Effect of spent wash, press mud and compost on soil properties, yield and quality of sugarcane. *Indian Sugar* 40(6): 57–65.

Bharagava RN, Chandra R, and Rai V. 2008. Phytoextraction of trace elements and physiological changes in Indian mustard plants (*Brassica nigra* L.) grown in post methanated distillery effluent (PMDE) irrigated soil. *Bioresource Technology* 99: 8316–8324.

Chandra R, Bharagava RN, Yadav S, and Mohan D. 2009. Accumulation and distribution of toxic metals in wheat (*Triticum aestivum* L.) and Indian mustard (*Brassica campestris* L.) irrigated with distillery and tannery effluents. *Journal of Hazardous Materials* 162: 1514–1521.

Chefetz B, Hatcher PG, Hadar Y, and Chen Y. 1996. Chemical and biological characterization of organic matter during composting of municipal solid waste. *Journal of Environmental Quality* 25: 776–785.

Davamani V, Lourduraj AC, and Singaram P. 2006. Effect of sugar and distillery wastes on nutrient status, yield and quality of turmeric. *Crop Research Hisar* 32(3): 563–567.

Hoitink HAJ and Kuter GA. 1986. Effects of composts in growth media on soilborne pathogens. In: Y. Chen and Y. Avnimelech (eds.), *The Role of Organic Matter in Modern Agriculture*. Martinus Nijhoff Publishers, Dordrecht, the Netherlands, pp. 289–306.

Mahimairaja S and Bolan NS. 2004. Problems and prospects of agricultural use of distillery spent wash in India. Super Soil 2004, *3rd Australian New Zealand Soils Conference*, December 5–9, 2004. University of Sydney, Australia.

Palmisano AC and Barlaz MA. 1996. *Microbiology of Solid Waste*, pp. 125–127. CRC Press Inc., Carporate BLVD, N.W. Boca Raton, FL.

Selvamurugan M, Doraisamy P, and Maheswari M. 2013. Biomethanated distillery spent wash and press-mud biocompost as sources of plant nutrients for groundnut (*Arachis hypogaea* L.). *Journal of Applied and Natural Science* 5(2): 328–334.

2

Advances in the Treatment of Pulp and Paper Mill Wastewater

Suvidha Gupta and Nishi Kant Bhardwaj

CONTENTS

2.1 Introduction

The pulp, paper, and paper products industry is one of the largest and growing industry in the world. The production of pulp and paper has increased worldwide and will continue to increase in the near future. There have been gradual changes in wood-based pulp and paper manufacturing processes. The use of wood for paper making began with the development of mechanical pulping in Europe in 1840s and first commercial sulfite pulp mill was set-up in Sweden in late nineteenth century. The production of paper and paperboard worldwide is dominated by the United States of America, Canada, Japan, Sweden, Finland, and China. Due to closing down of a large number of paper manufacturing capacities in North America and increasing capacity building in Asia (particularly China), the latter accounts for over a third of world's paper manufacture, followed by Europe (30%) and North America (25%). The demand for paper is unequally dispersed as 22% of the world's population in the United States of America, Europe, and Japan consume 72% of the global paper. The industry is very capital intensive with low profit margin, which limits upgradation of technology and innovation in the industry (Dugal, 2009; Lamberg et al., 2013). The pulp and paper industry plays a vital role in the progress of socioeconomic aspects, while it is associated with significant environmental concerns due to its large footprints on environmental resources.

The metals and chemical industries are extremely water intensive followed by pulp- and paper-making process, which is ranked third in terms of freshwater consumption. The fresh water consumption in pulp and paper production varies greatly from one mill to another based on the raw material used and technology adopted. It can be as high as 60 m^3/ton of product in most modern and efficient operational mill (Thompson et al., 2001) in comparison with the global best specific water consumption of 28.7 m^3/ton for wood-based pulp and paper mill. The water consumption in wood-based pulp and paper mills in India is 40–150 m^3/ton of product (NPC, 2006; MoEF, 2010). High consumption of water is largely attributed to the use of old technology/equipment and poor water management practices. Each pulp- and paper-making process utilizes large amounts of water, which reappears in the form of wastewater.

2.2 Overview of the Pulp and Paper Manufacturing Processes

Manufacturing of paper consists of two major processes: pulping and bleaching. Mechanical and chemical pulping are the major pulping processes in pulp and paper industry. Mechanical pulping is highly energy intensive and generates high yield of pulp with poor brightness stability and strength properties. Whereas, high bright paper with good strength properties is obtained by chemical pulping but the process is not environment friendly and gives low yield. The chemical pulping of wood results in degradation of lignin network and removal of soluble fractions of lignin to produce unbleached pulp (cellulose 80%–90%, hemicelluloses 10%–15%, and residual lignin 2.5%–4%). Residual lignin is accountable for the unwanted dark color and photo yellowing of the pulp. The residual lignin in the unbleached pulp is oxidized during bleaching to achieve the desirable brightness of pulp while protecting the mechanical properties of pulp (Gaspar et al., 2003). Processes of the manufacture of paper in an integrated kraft pulp and paper mill can be divided into four stages, namely, wood preparation, pulping and bleaching, paper making, and chemical recovery (Figure 2.1). The wood logs are first debarked and chipped into small, uniform size chips for pulping.

2.2.1 Pulping Process

At the pulping stage, wood or other fiber material is treated with chemicals to separate the fiber fraction from other constituents of wood. Chemical pulping using kraft process is the most prevalent throughout the world. It uses sodium sulfide (Na_2S) and sodium hydroxide (NaOH) as pulping chemicals. The liquor (white liquor) comprising these two chemicals is mixed with the wood chips in a reaction vessel (digester). The products consist of wood fibers (pulp) and liquor that contains the dissolved lignin; typically termed as black liquor. The chemical process results in approximately 50% yield of pulp. The residual lignin content in pulp is measured in terms of "Kappa number," which is directly proportional to the lignin content of the pulp. The liquor generated in the pulping stage is processed at recovery boiler for recovery of chemicals and heat. Spillages of liquor from pulping stage contribute to color, high chemical oxygen demand (COD), and reducing sulfur compounds to wastewater. Recovery and recycling of the spillages in pulping process can reduce the pollution load to effluent treatment plant (ETP) to a significant level as this liquor contains degraded lignin compounds that are hard to biodegrade.

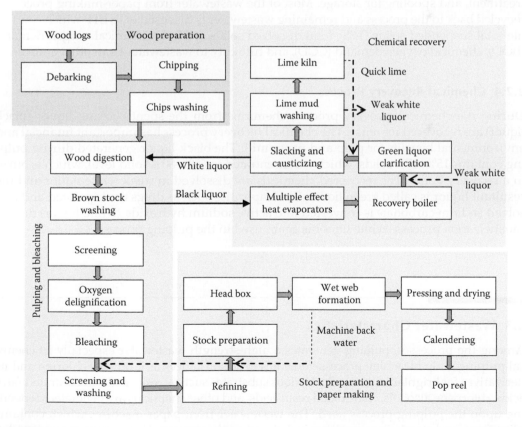

FIGURE 2.1
(**See color insert.**) Process flow diagram for integrated pulp and paper manufacturing.

2.2.2 Bleaching Process

The bleaching of unbleached chemical pulp is carried out with elemental chlorine or its derivatives such as chlorine dioxide (ClO_2) or hypochlorite ($NaOCl$ and $Ca(OCl)_2$). Until the end of twentieth century nearly every chemical pulp mill was using elemental chlorine as the first bleaching stage. Because of environmental concerns arising from the formation of organochlorine compounds, pulp mills now use partially ClO_2-substituted chlorination or elemental chlorine-free bleaching technologies. During bleaching, pulp is processed through three to five stages of chemical reaction and water washing. Bleaching stages generally alternate between acid and alkaline conditions. Chemical reactions take place with lignin during the acid stage; the alkaline stages extract the reaction products of lignin through dissolution. At the washing stages, reaction products are removed. Bleaching stage effluent is the main source of color and toxicity to wastewater in pulp and paper industry.

2.2.3 Paper Making

At the final stage, the stock is prepared by refining the pulp, and the paper is manufactured. Stock preparation includes dispersion of pulp in water, refining to increase surface area and addition of wet-end chemicals. Wet-end operation includes the formation of paper sheet from wet pulp stock, whereas dry-end operation includes drying of paper, surface treatment, and spooling for storage. Most of the wastewater from paper-making process is recycled back to the process and remaining wastewater is disposed of to ETP which contribute total suspended solids (TSS), total dissolved solids (TDS), biochemical oxygen demand (BOD), chemical oxygen demand (COD), and turbidity to wastewater from other sources.

2.2.4 Chemical Recovery Process

During the chemical recovery process, chemicals from the spent cooking liquor (black liquor) are recovered for reuse. The chemical recovery process has important financial and environmental benefits for pulp and paper mill. The black liquor generated during pulping contains 15%–17% solids, which is concentrated to 65%–80% solid level, which is burnt in a recovery boiler. The recovered chemicals are dissolved in weak white liquor and the resultant liquor is called green liquor. The impurities such as dregs are separated and dissolved sodium carbonate is converted into active sodium hydroxide (white liquor) during causticization process. White liquor is again used in the pulping process.

2.3 Wastewater Characteristics

Among the processes, pulping generates a high-strength wastewater especially in chemical pulping. Pulp bleaching process generates toxic substances as it utilizes chlorine and its derivatives for brightening the pulp. Various substrates such as resin acids, unsaturated fatty acids, diterpene alcohols, chlorinated resin acids, and other chemicals are generated depending upon the pulping process used. The wastewater from paper machine, which contains filler, fines, and starch, is generally recycled back to the process. The pollutants generated at various stages of the pulping and paper-making processes are given in Table 2.1.

TABLE 2.1

Potential Water Pollutants from Pulp- and Paper-Making Processes

Source	Pollutants
Wood handling/debarking and chip washing	Solids, BOD, COD, color
Chip digester and liquor evaporator condensate	BOD, COD, color, reduced sulfur compounds
Wastewater from pulp screening, thickening, and cleaning	Suspended solids, BOD, COD
Bleach plant filtrates	BOD, COD, color, TDS, organochlorine compounds
Paper machine water	Suspended solids, TDS, BOD, COD
Fiber and liquor spills	Solids, BOD, COD, color

Source: Adapted from USEPA, 2002. *EPA Office of Compliance Sector Notebook Project: Profile of Pulp and Paper Industry.* 2nd ed., Washington, DC, USA, EPA/310-R-02-002.

Wastewater from the pulp and paper mill contains a broad spectrum of organic and inorganic substances. It typically consists of fibrous suspended solids and dissolved organic compounds in high concentration; both low- and high-molecular weight compounds are present in dissolved form. Generally, small organic molecules exert BOD, whereas lignin and its derivatives cause a chemical oxygen demand and attribute color. Wastewater generated during bleaching of pulp is responsible for most of the color, organic matter, and toxicity in the wastewater to ETP. The paper industry effluent is characterized by high BOD, COD, suspended solids, dissolved solids, AOX, color, toxicity, and high concentration of nutrients (phosphorus and nitrogen), which cause eutrophication in receiving water bodies (Pokhrel and Viraraghavan, 2004). During the production of one ton of paper, 100 kg of color-imparting substances and 2–4 kg of organochlorines are generated in the bleach plant effluent (Kansal et al., 2008). The first two stages of bleaching of pulp typically contribute over 90% of the total color to bleach plant effluent. A characteristic of wastewater for the manufacture of different types of forest product is given in Table 2.2 (EC, 2001; Pokhrel and Viraraghavan, 2004).

2.3.1 Origin of Organochlorine Compounds

Formation of organochlorine compounds in natural ecosystem is well documented. More than 1500 organohalogen compounds have been identified (Biester et al., 2004). The pulp and paper mill is one of the artificial or man made sources of organochlorine compounds in recipient waterways (Zheng and Allen, 1996; Ali and Sreekrishnan, 2000).

TABLE 2.2

Characteristics of Wastewater for the Manufacture of Forest Products

Pulp Mill	Flow (m³)	TSS (kg)	COD (kg)	BOD$_5$ (kg)	AOX (kg)
Bleach hardwood ECF	30–52	2–5	17–35	4–5	0.1–0.2
Bleach softwood ECF	46–73	n/a	38–41	13–17	0.2–0.5
Integrated ECF	61–77	12–13	37–42	11–19	0.2–0.3
Kraft unbleach	25–60	10–25	20–50	6–15	n/a
Kraft bleach	60–90	10–40	100–140	n/a	n/a
Agro based	200–250	50–100	1000–1100	300–350	n/a
RCF based	10–30	5–10	40–60	20–30	n/a

Note: n/a, no data available. Load given is for 1 ton air-dried product.

Organochlorine compounds are formed during the bleaching of wood pulp for the removal of lignin with chlorine (Cl_2) and chlorine derivatives such as hypochlorite and chlorine dioxide (ClO_2) (Roy et al., 2004). Native lignin is a polymer comprising p-coumaryl and coniferyl alcohols. The hardwood lignin is guaiacyl–syringyl lignin formed by copolymerization of coniferyl and sinapyl alcohols. First-stage acidic bleaching using chlorine derivatives followed by alkaline extraction stage remove 75%–90% of lignin present in the unbleached pulp. Oxidation, substitution, and addition reaction of chlorine derivative with various reactive species in pulp leads to breakdown of most of the lignin and formation of organochlorine compounds. During bleaching of Eucalyptus pulp (kappa no. 18.7) with chlorine-dioxide-substituted chlorination (CD) as first acidic stage, followed by oxidative alkaline extraction (EO) stage, different organochlorine compounds, namely, chloroactone, chlorophenol, chlorocatecol, chloroguiacol, and chlorosyringol substituted with one to five chlorine atoms per molecule were generated in significant quantity (Table 2.3; Figure 2.2). Pentachlorophenol, which is the most toxic compound among all the chlorophenolic compounds, was washed out in alkaline condition in extraction stage wastewater (Roy et al., 2004). Organochlorine compounds in water and wastewater can be monitored by several techniques; among them, the one based on adsorbable organic halide (AOX) is the most commonly used (Zheng and Allen, 1996; Barroca et al., 2001). A tiny fraction (1%–3%) of the total organochlorines is lipophilic and is the cause of environmental concern. Collectively, these are termed as extractable organic halide (EOX) compounds. Chlorophenolic compounds, dioxins and dibenzofurans fall in this category (Bajpai and Bajpai, 1997). Typically, bleaching processes that result in the formation of 2,3,7,8-tetrachlorodibenzodioxin (TCDD) and 2,3,7,8-tetrachlorodibenzofuran (TCDF) also generate the higher substituted tri-, tetra-, and penta-chlorinated compounds (Freire et al., 2003; Roy et al., 2004).

2.3.2 Toxicity of Wastewater from Bleaching Plants

Toxicity of wastewater depends upon the total organically bound chlorine, extractive content of the raw material, and extent of removal of extractives during different processes. The high toxicity and low biodegradability of resins, long chain fatty acids, terpenes, tannins, lignins, and chlorinated compounds are mainly responsible for toxic effects of paper mill wastewater (Buzzini et al., 2007). Among the organochlorine compounds derived from lignin, 80% are of high-molecular weight and the remaining 20% compounds are of low-molecular weight (LMW, <1000). Out of 20% of LMW compounds, about 19% are hydrophilic and can be metabolized, while 1% is extractable only by nonpolar solvents and are called as extractable organic halide. The compounds from this fraction bioaccumulate through food chain and are potentially toxic (Berry et al., 1991). The biodegradation rate of organochlorine compounds decreases with the increase in the degree of chlorine substitution. Dichlorophenol, trichlorophenol, pentachlorophenol, chlorinated furans, and chlorocatechols are carcinogenic and strongly mutagenic.

The pulp and paper industry effluents affect reproductive physiology of fish through endocrine disruption. They cause androgenic activity and result in masculinization and sex reversal of female fish. The pulp mill effluents have been reported to induce mixed function oxygenase enzymes in liver, reduction in gonad size, depression of circulating sex steroids, increase in plasma vitellogenin, and reduced ovarian development in fish under both laboratory and field conditions (Orrego et al., 2011).

TABLE 2.3

Potential Organochlorine Compounds in Bleach Plant Effluents

Organochlorine Compound	Concentration (μg/L)	
	C_D Stage	E_O Stage
Chloroacetone		
1,1,1-Trichloroacetone	565.0	36.0
1,1,3-Trichloroacetone	1950.0	6.4
Hexachloroacetone	48.8	—
Chlorophenolics		
2-Chlorophenol	—	504
2,4-Dichlorophenol	—	4.0
2,6-Dichlorophenol	3.0	2
2,4,6-Trichlorophenol	12.0	22
Tetrachlorophenol	—	1.0
Pentachlorophenol	—	5.0
3-Chlorocatecol	8.8	—
4-Chlorocatecol	14.0	15.0
3,4-Dichlorocatecol/3,6-dichlorocatecol	85.5	2.2
3,5-Dichlorocatecol	3.5	—
4,5-Dichlorocatecol	18.0	1.5
3,4,6-Trichlorocatecol	33.0	14.0
3,4,5-Trichlorocatecol	68.0	0.75
Tetrachlorocatecol	166.0	1.9
4-Chloroguiacol/5-chloroguiacol	0.7	—
3,4-Dichloroguiacol	0.5	4.5
4,5-Dichloroguiacol/5,6-dichloroguiacol	—	12.5
3,4,6-Trichloroguiacol/3,5,6-trichloroguiacol	0.4	11.0
3,4,5-Trichloroguiacol	12.0	182.0
4,5,6-Trichloroguiacol	8.0	—
Tetrachloroguiacol	2.3	51.6
3-Chlorosyringol	17.2	—
Trichlorosyringol	13.3	164.0
Monochlorophenol	40.7	688.0
Dichlorophenol	120.7	197.1
Trichlorophenol	146.7	393.8
Tetrachlorophenol	163.8	54.5
Pentachlorophenol	—	5.0

The treated wastewater from pulp and paper industry causes slime growth, scum formation, color problems, and thermal impacts to the receiving water bodies. The discharge of wastewater is also responsible for increase of toxic substances in the water resources, causing death of the zooplankton and fish, and affecting the terrestrial ecosystem.

The growing public awareness regarding impact of these pollutants and stringent regulations by government authorities are forcing the industry to treat wastewater to the required compliance level before discharging into the environment. On the one hand, the quality of raw water is deteriorating as well as its availability is continuously diminishing,

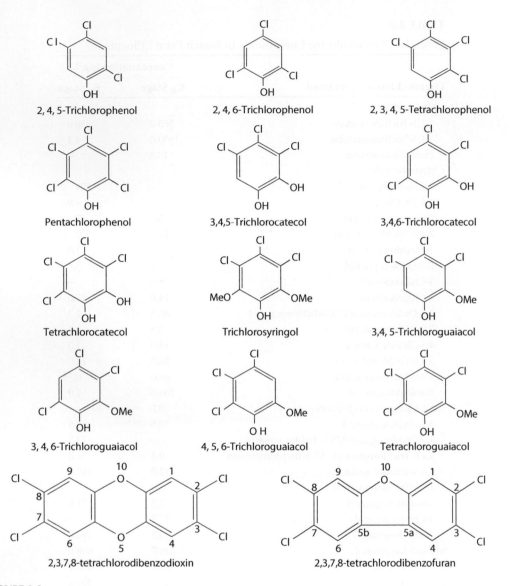

FIGURE 2.2
Organochlorine compounds in bleach plant wastewater.

while on the other hand, demand for domestic use is increasing. The socioeconomic factors are forcing the industry to recycle the maximum of its wastewater. The pretreatment of organic rich wastewater to a considerable degree is prerequisite for recycling of wastewater as such or after other treatments. In both scientific and industrial communities, there has been a growing interest in the best available technologies for the bleaching of chemical pulps in order to reduce the discharge of color causing and toxic compounds in the liquid effluents. The advanced approaches being used or identified for reducing the organic pollutants and treatment of wastewater in pulp and paper industry are shown in Figure 2.3.

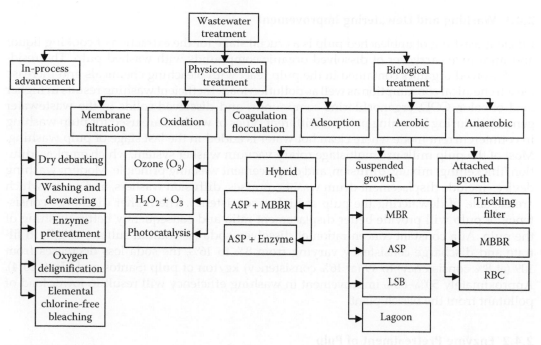

FIGURE 2.3
Advance processes for better quality of wastewater in pulp and paper mills.

2.4 In-Process Advancement for Reduction in Pollution Load

For the past two to three decades, there has been increasing pressure on the pulp and paper industry throughout the world to minimize water consumption and reduce environmental load due to growing concerns over environment issues, reduced availability of fresh water, and stringent environmental legislations.

The advancement in pulp and paper manufacturing processes and technologies has resulted in reduction in pollution load. Few major development and modification include change of wet wood debarking to dry debarking, which has resulted in reduction in water consumption and low organic load in chemical and mechanical pulping. Generation of COD and wastewater during dry debarking was reduced to 0.2–2 kg/m³ and 0.1–0.5 m³/ton of wood, respectively, from 4 to 6 kg/m³ and 0.6 to 2 m³/ton of wood, respectively, in wet debarking (EC, 2001). Advancement in brown stock washing and dewatering equipment has resulted in less carryover of black liquor to bleach plant; enzyme pretreatment of pulp has potential for oxidation of lignin and its derivative compounds at reduced consumption of chemicals in subsequent bleaching stages; implementation of one or two stage oxygen delignification stages prior to bleaching has resulted in reduction in bleach chemical consumption and subsequent low organochlorine load to effluent treatment plant; use of chlorine dioxide or ozone instead of elemental chlorine has resulted in drastic reduction in organochlorine and color-causing compounds in wastewater; recycling of alkali extraction stage wastewater generated during bleaching of pulp to chemical recovery plant has resulted in lower lignin derivatives in wastewater to ETP.

2.4.1 Washing and Dewatering Improvement

Efficient washing of unbleached pulp is a crucial stage for the extraction of cooking liquor and minimize carryover of dissolved organic compounds with washed pulp. The residual dissolved organics contained in the pulp react with bleaching chemicals and increase bleach chemical consumption as well as pollution load. Inefficient washing results in higher load of color, COD, organochlorine compounds, and dissolved solids in the wastewater generated during bleaching process. Different types of washers are used for pulp washing in countercurrent mode where clean hot water is added in the last stage of pulp washing. Most of the pulp mills use multistage rotary vacuum washer system, which employs dilution/thickening, mixing, diffusion, and displacement washing principle. Modern washing devices include displacement drum washer, presses, diffusion washers, and so on, which are capable of dewatering the pulp at higher consistency. The higher discharge consistency of pulp will provide better displacement ratio and consequently better washing of the pulp. At a constant concentration of dissolved soda (as sodium sulfate) 10 g/L in filtrate and discharge consistency varying from 8% to 16%, the soda loss decreases from 109 (at 8% consistency) to 49 (at 16% consistency) kg/ton of pulp (Santos and Hart, 2014). Approximately 50%–60% improvement in washing efficiency will result in lower load of pollutant from the bleach plant.

2.4.2 Enzyme Pretreatment of Pulp

The application of enzymes as alternative to chlorine-based chemicals for pulp bleaching is simple, cost-effective, and eco-friendly. Xylanase and lacasse enzymes have been investigated as potential agents for bio-bleaching of kraft pulp. Xylanases hydrolyze xylan layer present in the plant cell wall and influence the structural integrity that facilitate lignin removal during succeeding bleaching stages. Xylanase application can decrease pulp hexenuronic acid content, which adversely affects pulp bleachability and increases chemical consumption. Lacasses directly act on lignin, oxidize and make it water soluble. Lacasses attack primarily the phenolic groups by simultaneous reduction of O_2 to H_2O_2 through phenoxy radicals formation as intermediate (Niku-Paavola et al., 1994). The main barricade for application of enzymes is their high sensitivity toward pH and temperature. Continual efforts for the production of thermo stable and alkaline tolerant enzymes may make bleaching process more environment friendly.

2.4.3 Oxygen Delignification

The extended delignification of pulp by extended cooking at high concentration of alkali or oxygen delignification can achieve lower kappa level of pulp. In oxygen delignification, the washed unbleached pulp is allowed to react with molecular oxygen at higher pressure and temperature under alkaline condition; phenolic groups in the lignin are ionized at high pH and attacked by molecular oxygen, which depolymerize lignin to LMW compounds. These compounds are solubilized in alkaline liquor and extracted out from the pulp. Oxygen delignification extract about 45%–60% of the lignin from unbleached pulp that remains after the cooking process. The extracted liquor containing dissolved lignin after countercurrent washing of the pulp is returned to liquor cycle for recovery of chemicals and heat. Implementation of oxygen delignification stage can decrease the pollution load considerably from the bleaching stage. The combination of extended delignification along with oxygen delignification can reduce color and COD load in bleach plant wastewater by 60%–65% (Pulliam, 1991). Table 2.4 summarizes the reduction in COD load from the bleach plant for hardwood and softwood using different delignification technologies. Most of the new pulp mills come with oxygen delignification process and old mills are upgrading the process for one- or two-stage oxygen delignification.

2.4.4 Elemental Chlorine-Free Bleaching

Chlorine and its derivatives, namely, chlorine dioxide and sodium hypochlorite are used as oxidizing agent for bleaching of the pulp. Nonchlorine chemicals, namely, oxygen, hydrogen peroxide, and ozone are also used separately or in combination with chlorine compounds for multistage bleaching of pulp. Being cheaper and easily available, elemental chlorine (Cl_2) has been the major bleaching agent in first-stage acidic bleaching of pulp throughout the world. Due to substitution mode of action and subsequently generation of a large amount of chlorinated organic compounds in the bleaching wastewater, strenuous efforts have been made to replace elemental chlorine with other bleaching agents.

Use of chlorine dioxide in elemental chlorine-free bleaching sequences is environmentally much more benign than Cl_2. The quantity of chlorinated toxic compounds in the wastewater from elemental chlorine-free (ECF) bleaching sequence is many folds lower than the earlier one. The generation of pollutant during bleaching of pulp (with and without oxygen delignification) with chlorine and ECF bleaching sequences is given in Table 2.5. Wastewater from oxygen-chlorine dioxide-extraction-chlorine dioxide (ODED) sequence had lowest pollution load when compared with other sequences due to the reduction of initial kappa number during oxygen delignification stage. During kraft pulp bleaching, a significant amount of organic matter is dissolved in the first two bleaching stages and had highest pollutant load for each sequence. The concentration of AOX compounds in conventional elemental chlorine bleaching of kraft pulp varies in the range 2–5 kg Cl/ton pulp, whereas the same in ECF bleaching wastewater varies in the range 0.2–1.0 kg Cl/ton pulp.

TABLE 2.4

COD Load for Hardwood and Softwood with Different Delignification Processes

Delignification Process	Kappa Number		COD Load (kg/ton)	
	Hardwood	Softwood	Hardwood	Softwood
Conventional cooking	14–22	30–35	28–44	60–70
Conventional cooking + oxygen delignification	13–15	18–20	26–30	36–40
Extended/modified cooking	14–16	18–22	28–32	36–44
Extended cooking + oxygen delignification	8–10	8–12	16–20	16–24

Source: Adapted from EC, 2001. Integrated prevention pollution and control (IPPC), Reference document on best-available techniques in the pulp and paper industry. European Commission, Brussels.

TABLE 2.5

Generation of Pollutants during Different Bleaching
Sequences

Bleaching Sequence	Stage	Quantity (kg/ton Oven Dried Pulp)			
		BOD	COD	Color	AOX
	C_D	3.0	14.1	13.4	2.20
C_DED	E	2.2	11.7	28.3	0.30
	D	1.1	1.8	0.6	0.10
	D_0	3.8	15.8	15.6	0.32
DED	E	3.1	12.8	18.4	0.28
	D_1	0.9	3.0	1.3	0.07
	D_0	3.2	12.2	14.1	0.27
DE_PD	E_P	2.5	10.1	16.7	0.23
	D_1	0.8	2.3	1.1	0.05
	D_0	1.8	6.5	1.8	0.16
ODED	E	1.5	5.1	2.0	0.14
	D_1	0.8	1.6	0.9	0.04

2.5 End-of-Pipe Treatment of Wastewater

Improved fiber retention, better in-plant utilization of raw materials, and use of efficient and environment friendly processes/technologies are effective means of reducing pollution at site. End-of-pipe treatment is preferred choice as there is no interference with the production process and updated process/technologies are available. In the last decade, the amount of water consumption in pulp and paper mills has been drastically reduced. External treatment of wastewater is generally performed in four stages, namely, pretreatment, primary, secondary, and tertiary treatment. Pretreatment of pulp and paper mill wastewater may include equalization, neutralization, and chemical treatment. The prechemical treatment may include coagulation–flocculation or oxidation of high organic load stream (bleach plant wastewater) or massive lime treatment of influent to effluent treatment plant. The treatment of wastewater is usually carried out by means of screening and sedimentation to remove suspended solids (i.e., primary treatment) followed by biological oxidation to remove suspended and dissolved organic material (i.e., secondary treatment). Any treatment beyond primary and secondary treatments is usually termed as tertiary treatment. The sludge generated during different stages is thickened and dewatered prior to disposal. The schematic diagram of effluent treatment by activated sludge process (ASP) in pulp and paper industry is illustrated in Figure 2.4.

Primary treatment generally refers to the methodology for removing suspended solids from effluents. In the pulp and paper mill, solid removal is always accompanied by some reduction in COD, BOD, and toxicity. Screening is often used as a preliminary step to remove relatively large floating or suspended particles from an effluent stream and is performed using bar screens.

Sedimentation or gravity settling is the most common process used to separate the fibrous material from raw wastewater. Sedimentation is carried out in a holding pond or basin with sufficient retention time for settling the solid particles. Ideal sedimentation unit provides quiescent flow to allow the settlable solids to move to the bottom. Solids

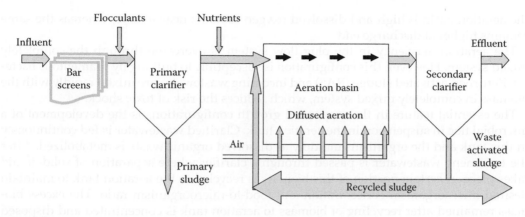

FIGURE 2.4
Schematic diagram of typical effluent treatment plant in pulp and paper mills.

are pushed to the center sump with a sludge scraper. Depending on the characteristics of the solids, the concentration of underflow may vary between 1.5% and 6% (Smook, 1982). In addition to the basic function of separation of solids, primary clarifier also acts as an equalization basin-cum-shock absorber.

Secondary or biological treatment is the heart of the wastewater purification process in pulp and paper industry. Under aerobic conditions, microorganisms (mostly bacteria) consume oxygen to convert organic waste into the ultimate end products—carbon dioxide and water. An important aspect of biological oxidation process is to provide adequate aeration and mixing for intimate contact between large concentration of microorganisms and the substrate.

2.5.1 Aerobic Biological Processes

Aerobic biological oxidation can be accomplished by various means depending on the characteristics of the wastewater, the area available for external treatment, and the required degree of BOD removal. The aerobic process is divided into three main categories, namely, suspended growth process, attached growth process, and hybrid process. ASP, which is a suspended growth process, is widely adopted in pulp and paper industry throughout the world. With increase in production, reduction in water consumption, and day by day stringent norms for discharge of wastewater, capacity of wastewater treatment is being increased by incorporating packed bed reactor, which is an attached growth process.

2.5.1.1 Activated-Sludge Process

The ASP is the most popular, well-adapted, high-rate suspended growth biological process for treatment of industrial and municipal wastewaters. The process was named activated sludge by Ardern and Lockett in 1913, as it involved the production of an activated mass of microorganisms capable of aerobic stabilization of organic material in wastewater. ASP is differentiated based on mixing arrangement as "plug flow" and "completely mixed" process (Tchobanoglous et al., 2003). The plug flow system generates less filamentous bacterial mass and produces settling sludge. This configuration often runs inefficiently from uneven load distribution along the reactors. Thus, demand of oxygen at the inlet of the aeration basin is high and dissolved oxygen remains near to zero, whereas the same remains higher at discharge end.

The aeration deficiency in the plug flow system is overcome through the completely mixed system. However, this configuration is susceptible to bulking by filamentous bacteria. Returned activated sludge (RAS) and incoming wastewater are mixed rapidly with the biomass in completely mixed system, which reduces the risk of toxic shock.

The essential feature in the suspended growth configurations is the development of a microbial floc in suspension in the aeration tank. Clarified wastewater is fed continuously in this tank and the organisms multiply as dissolved organic waste is metabolized. After the treatment, wastewater is passed through a clarifier where separation of solid–liquid takes place. A certain portion of this biomass is recycled to the aeration tank to maintain desired microorganism concentration and food-to-microorganism ratio. The excess biomass remained after recycling of biomass to aeration tank is concentrated and disposed off. Compared with the aeration lagoon, the ASP has some disadvantages; it is sensitive to changes in the characteristics of the wastewater, the requirement of nutrients is relatively higher, and settling aids are sometimes required for proper clarification.

Activated sludge process is widely used to treat pulp and paper mill effluent. The treatment can generally achieve relatively higher reduction of BOD and toxicity (Diez et al., 2002). Partial removal of AOX compounds in the biological system has also been recognized (Reeve, 1991; Taghipour and Evans, 1996).

The removal of organochlorine compounds during biological treatment is achieved through biological dechlorination, biosorption, and volatilization (Gupta et al., 2013). The wastewater from integrated pulp and paper mill following $C_DE_{OP}D_1D_2$ bleaching sequence had 0.26 ± 0.04 mg/L EOX compounds, which were only 1.6% of the AOX compounds in the wastewater. There was $36.5 \pm 4.8\%$ and $48.6 \pm 5.7\%$ removal of AOX and EOX compounds during biological treatment. EOX compounds were mostly sorbed on the biosludge (43.7%) and only 4.9% of the compounds were mineralized during microbial degradation, whereas 34.1% AOX compounds were mineralized and only 2.4% AOX compounds were sorbed on biosludge (Table 2.6). It clearly revealed that the lipophilic EOX compounds were difficult to remove by conventional biological treatment. Microorganisms hydrolyze the biodegradable dissolved and particulate organic matter into simple end products and subsequently oxidize those with generation of additional biomass. Excess cells are separated from the purified water in a concentrated form called waste activated sludge.

TABLE 2.6

Removal Mechanism for AOX and EOX Compounds during Biological Treatment

A	Reactor Operating Conditions		
	HRT (h)	8.0 ± 1.3	
	COD removal (g/day)	5.61 ± 1.6	
	Sludge yield (g/day)	1.74	
B	**AOX and EOX Removal**	**Removal of AOX**	**Removal of EOX**
	Outlet AOX and EOX (mg/L)	10.45 ± 1.29	0.13 ± 0.01
	Sludge AOX and EOX (mg/kg)	4146 ± 953	1177 ± 264
	AOX and EOX removal (%)	36.5 ± 4.8	48.6 ± 5.7
	Sorption on sludge (%)	2.4	43.7
	AOX and EOX mineralization (%)	34.1	4.9

Note: Influent AOX and EOX concentration: 16.52 ± 2.12 and 0.26 ± 0.04, respectively.

2.5.1.2 Low-Sludge Bioprocess

In the low-sludge bioprocess (LSB), no biomass is recycled to bioreactor as in ASP to maintain the concentration of biomass in the aeration basin. The hydraulic retention time (HRT) is maintained equivalent to sludge retention time (SRT). The LSB process has been used for minimizing the biosludge generation during biological treatment as well as treatment of wastewater from pulp and paper mills (Lee and Welander, 1996; Chakrabarti et al., 2012). The retention time in the reactor is optimized based on the growth cycle of microorganisms. The growth of bacteria is coupled with binary fission where division of one bacterium generates two daughter cells. The bacterial growth can be divided into four main phases, that is, lag, log, stationary, and decline phase. The initial lag phase is a period of slow growth during which the bacteria are adapting to the conditions in the fresh medium. This is followed by a log phase, where growth is exponential, doubling every replication cycle. Stationary phase occurs when the rate of multiplication equals the rate of death and generally nutrients become limiting at this stage. Logarithmic decline or death phase occurs when cells die faster than they are replaced. The HRT in LSB process should be more than the log phase to ensure the reproduction of organisms in the system. As there is no recycling of biosludge in the process, the mean SRT varies from 6 to 18 h. The biomass concentration in the bioreactor remains only 200–300 mg/L, whereas the same is maintained 2000–4000 mg/L in ASP. The biological treatment efficiency of LSB has been reported inferior than ASP but LSB may be coupled with chemical treatment using coagulant and flocculant to meet the discharge norms for wastewater (Figure 2.5). The chemical sludge can be settled along with biosludge as there is no need of recycling of biosludge to bioreactor (Chakrabarti et al., 2012).

2.5.1.3 Membrane Bioreactor

The technology of membrane bioreactor (MBR) originated from modification of conventional activated sludge treatment process, using a combination of two basic processes—biological degradation and membrane filtration (instead of clarification). The separation of biosludge by membranes is obviously more selective than settling, since separation is independent of the sludge flocculation and settling ability, which overcomes the problem of solid–liquid separation due to bulking of sludge in ASP. In the case of pulp and paper industry, most of the reported activity of MBR is at laboratory scale, with few pilot-scale

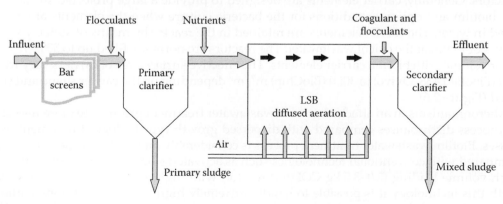

FIGURE 2.5
Schematic diagram of low-sludge bioprocess (LSB) coupled with chemical treatment.

trials and full-scale installations (Lerner et al., 2007; Zhang et al., 2009; Simstich et al., 2012). Several types and configurations of membranes have been used for MBR applications. The configurations include plate and frame, rotary disk, tubular, and hollow fiber, while the membrane material could be organic (polysulfone, polyethylene, polyolefin, etc.), metallic, or inorganic (ceramic). The membranes are usually in the microfiltration and ultrafiltration range with the pore size between 0.01 and 0.4 µm. These membranes are either used in submerged (internal) or in side stream (external) configurations. In comparison with ASP, advantages associated with the MBRs are production of effluent almost free of TSS and bacteria, much higher MLSS concentration in the bioreactor, higher SRT, small bioreactor size, and low excess biomass yields (Cicek et al., 1998). The MBR process produces an effluent of much better quality in terms of organic matter and suspended solids. The disadvantages of the technology include high cost, declining flux due to membrane fouling, and limited membrane life span.

2.5.1.4 The Moving Bed Biofilm Reactor System

When communities of microorganisms grow on surfaces, they are called biofilms; the core of the process is the biofilm elements on which attached growth of microorganisms takes place. The microorganisms secrete exo-polymeric gelatinous material, which helps them to attach themselves on solid surface and allow a little to move inside the material. The attachment of microorganisms on solid surface begins in patches and spread over to entire surface of support. Further growth is carried on by the production of new cells covering the first layer. The thickness of the biofilm layer varies according to the type of support and treatment process. There are different types of attached growth systems such as the moving bed biofilm reactor (MBBR), trickling filters, rotating biological contactors, fluidized bed reactors, fixed media submerged biofilters, and so on and all have their advantages and disadvantages. MBBR technology is one of the advanced biological treatment processes and has been implemented in several existing pulp and paper mills. MBBR technology was developed in Norway in the early 1990s and was a huge success worldwide for treatment of industrial and municipal wastewaters (Rusten et al., 2006).

The solid carriers in MBBR process are made of polyethylene having a density of 0.90–0.96 g/cm^3, which allow easy movement of the carrier in the completely mixed reactors. The biofilm carrier elements are kept suspended in the water using diffused air in aerobic reactors. Generally, carrier elements are designed to provide a large protected surface for the biofilm and optimal conditions for the bacteria culture when the elements are circulated in water. The carrier elements are retained in the reactor by means of suitably sized sieves or plates at the exit of wastewater. The reactors are normally filled up to 50%–70% of their volume with biofilm carrier elements. The effective surface area of carriers vary from 380 (FlooBed 438, Ovivo) to 3000 (BioChip) m^2/m^3 depending on the carrier shape and size used (Figure 2.6).

Microorganisms in an attached growth wastewater treatment process are more resistant to process disturbances compared with dispersed growth-type biological treatment processes. Biofilm wastewater treatment process is considerably more robust especially when compared with conventional technologies such as activated sludge. The process combines high volume loading (2.0–3.5 kg COD/m^3 reactor) with high removal efficiency of COD. With this technology it is possible to handle extremely high loading conditions without any problems of clogging, and treat industrial wastewater on a relatively small footprint. MBBR works in the up-flow stream direction at temperature ranging from psycrophilic to

| Biochip media | Biofilm chip P | K3 media | FlooBed 438 |
| 22 mm, 3000 m^2/m^3 | 45 mm, 1200 m^2/m^3 | 25 mm, 500 m^2/m^3 | 44 mm, 380 m^2/m^3 |

FIGURE 2.6
(See color insert.) Commonly used carriers (with dia and effective surface areas) in MBBR.

thermophilic. Generally two reactors in parallel or in continuous mode are used for treatment of wastewater from pulp and paper mill by MBBR process.

The main advantage associated with the MBBR is the low set-up cost of the process with low foot print. MBBR can be made highly specific for a particular toxicant/pollutant by adding specific strain/biomass or setting specific operating conditions. The requirement of nutrients (viz., nitrogen and phosphorus) is very less and there is no need of recycling of sludge to bioreactor.

2.5.1.5 Hybrid Bioprocess

Expansion of production capacity to cope-up the increasing demand of paper and paper board with the same or marginally higher water permit has forced pulp and paper industry to treat the wastewater with increased organic load in the existing or partially modified plant. As biological treatment are the cheapest and MBBR has low foot print, MBBR has been implemented in several existing pulp and paper mills as an augmentation to the existing ASP-based plants. The purpose of this add-on of biological process is either to handle additional pollutants load due to expansion or to meet more stringent discharge standards or both in some cases. Adding an MBBR prior to the existing ASP also works as shock absorber and neutralize the organic and hydraulic shocks, which gives a stable performance of ASP with stable sludge volume index below 100 mL/g. Performance of ASP, MBBRMBBR, and LSB as individual and coupled with ASP was evaluated for the biodegradation of wastewater from pulp and paper mills. The performance of the two stage process was comparable with the conventional ASP at 754 ± 28 mg/L soluble chemical oxygen demand of influent (Table 2.7). The pH and color of the feed wastewater were 7.0 and 1676 ± 134 Pt–Co unit, respectively. There was an increase in alkalinity and pH of the treated wastewater after biological treatment. The study revealed that ASP is the most robust process for treatment of wastewater from pulp and paper mills. MBBR may be coupled to ASP as shock absorber and to augment the capacity of biological treatment process. LSB is a novel approach and can be used along with tertiary chemical treatment. The analysis of morphology of biosludge revealed that floc size in ASP was large when compared with LSB–ASP and MBBR–ASP (Figure 2.7). The presence of higher microbial population such as rotifers, protozoa, and absence of filamentous organisms was responsible for excellent settling nature of biomass (low values of sludge volume index) in ASP.

It is well established that lignins can be degraded by certain enzymes produced by white rot fungi, for example, lignin peroxidase (LiP), manganese peroxidase (MnP), and laccase. The direct use of white rot fungus for treatment of wastewater from pulp and

TABLE 2.7

Operating Conditions and Performance of Different Bioprocesses

Parameter	ASP	LSB + ASP		MBBR + ASP	
pH (outlet)	7.8 ± 0.3	8.0 ± 0.2	7.9 ± 0.2	8.3 ± 0.2	7.8 ± 0.1
Temperature (°C)	37.1 ± 0.5	36.4 ± 0.7	36.6 ± 0.6	33.4 ± 1.2	36.8 ± 0.3
DO (mg/L)	2.3 ± 0.7	4.5 ± 0.8	2.0 ± 1.1	4.9 ± 0.7	2.5 ± 1.3
HRT (h)	8.1 ± 0.2	13.1 ± 0.2	8.3 ± 0.1	4.1 ± 0.1	8.1 ± 0.1
MLSS (g/L)	4.45 ± 0.50	0.234	4.52 ± 0.50	—	4.48 ± 0.52
Organic (%)	84.1 ± 0.8	—	79.3 ± 1.0	—	79.7 ± 0.6
SVI (mL/g)	9 ± 1	—	28 ± 3	—	33 ± 4
COD (mg/L)	257 ± 26	337 ± 39	232 ± 14	373 ± 31	240 ± 31
CODs reduction (%)	66.3 ± 3.0	53.0 ± 3.1	69.2 ± 1.5	49.4 ± 3.4	68.6 ± 3.5
Color reduction (%)	40.5 ± 1.6	27.0 ± 5.4	43.1 ± 8.0	11.8	36.3 ± 5.8
AOX reduction (%)	41.6 ± 4.6	33.8 ± 1.0	47.1	36.7 ± 3.3	37.1 ± 6.5
F/M	0.39 ± 0.04	—	0.34 ± 0.06	—	0.44 ± 0.06

Note: Influent CODs, AOX, and color: 754 ± 28, 16.1 ± 0.6 mg/L, and 1676 ± 134 PCU, respectively.

paper mills is not feasible due to process limitation, namely, slow degradation of pollutant requiring long HRT, low pH condition requirement, and necessity of easy digestible carbon source for the fungal growth, and so on. Application of lignin-degrading enzymes directly or as immobilized form prior to biological treatment has also been evaluated. The enzymes degrade the high-molecular weight lignin derivatives to LMW compounds, which are further degraded in biological treatment. This process also has low capital cost requirement with negligible foot print. Plant scale trials in pulp mill were performed using consortia of lignin-degrading enzymes along with mediator; both the chemicals were added directly at inlet of aeration basin. There was 15%–16% higher removal of color-causing compounds in the treated wastewater; higher reduction in color clearly indicated better efficacy of the enzyme system for mineralization of recalcitrant compounds.

2.5.2 Physicochemical Process

The biological treatment of wastewater is supposed to be the cheapest considering the cost of material for physicochemical treatment. Physicochemical treatments are preferred for handling the specific stream or improving the quality of wastewater after biological treatment.

2.5.2.1 Membrane Filtration

The closure of water loop in kraft mill bleach plant for reduction in color and TDS in wastewater involves reduction in water consumption and recycling of bleach plant wastewater, such as alkaline and acid filtrates. Recycling of these wastewaters leads to scale formation, build-up of nonprocess chemicals in recovery cycle, increased load on multieffect evaporator, and loss of operational flexibility. Combination of filtration process coupled with biological process may be a root to reduce the organic load from bleach plant in pulp and paper mill. Ultrafiltration, nanofiltration, and reverse osmosis (RO)

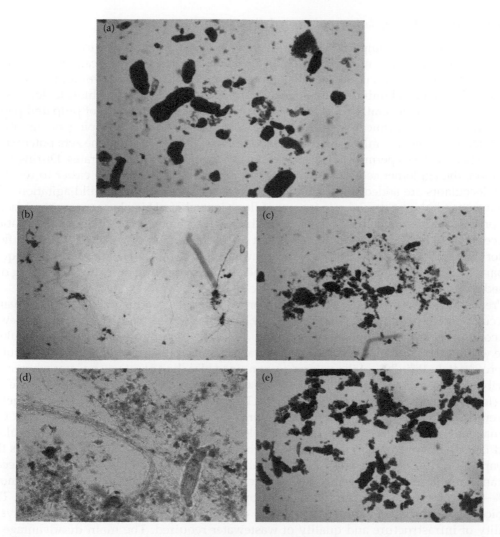

FIGURE 2.7
(See color insert.) Microphotograph of biosludge in different bioreactors. (a) Conventional activated-sludge process (ASP); (b) low-sludge bioprocess (LSB); (c) LSB followed by ASP; (d) moving bed biofilm reactor (MBBR); (e) MBBR followed by ASP.

have been applied for treatment of bleach plant wastewater (Onate et al., 2012; Shukla et al., 2013). Ultrafiltration allows removal of emulsified, suspended, colloidal, and high-molecular-mass material. The LMW compounds pass through the membrane, causing the larger material to become more concentrated. The membranes used for treatment of bleach plant wastewater usually have a cut-off value of 6000–8000 Daltons. The permeate of ultrafiltration is fed to nanofiltration where more finely suspended particles that remain in permeate are removed. RO is used for removal of dissolved material. After RO, the permeate is as good as raw water and can be recycled to process. The issue with the process is handling of reject having high organic and toxic load, which can be evaporated and used as fuel or can be treated by anaerobic biological treatment to recover energy (Sivard et al., 2012).

2.5.2.2 Chemical Coagulation and Flocculation

The wastewater from pulp and paper mills have cationic demand due to the presence of negatively charged lignin derivatives and fiber fines. Applications of cationic coagulant followed by anionic flocculant precipitate the dissolved lignin derivatives and fiber fines. Chemicals such as aluminum sulfate, polyaluminum chloride, ferric chloride, ferric sulfate, and lime are generally used as inorganic coagulant for treatment of pulp and paper mill wastewater. Chemical treatment using coagulant involves clumping of dissolved material by charge neutralization under rapid mixing. The lowering of the zeta potential at the particle surfaces permits particles to come closer and form agglomerates. During flocculation, the agglomerated small particles form larger flocs, which are easier to remove. The flocculants are added with rapid mixing and flocs are formed at mild agitation. The mixing speed plays a crucial role for size of floc. Control of pH is also an important factor for proper coagulation and floc formation. The reaction time for coagulation–flocculation process varies between 10 and 30 min based on the formation of required nature of floc. After chemical treatment, the treated wastewater is fed into a clarifier for solid–liquid separation. The separation of flocs is performed either by settling or by flotation using diffused air, based on the size and nature of floc.

Prechemical treatment of wastewater (prior to biological treatment) by coagulation and flocculation is a well-established technology, suitable for pulp and paper mill wastewater containing large amounts of colloids and dissolved compounds (Verenich et al., 2001). The wastewater generated during alkali extraction stage of bleaching of pulp contributed approximately 18%–20% of the volume to combined bleached plant wastewater whereas, it attributed 42%–44% of color and 23%–25% of COD to the combined bleach plant wastewater (Joss et al., 2007; Chakrabarti et. al., 2012). This waste stream can, therefore, be targeted to remove color and recalcitrant COD at source in mills where improved color and COD management is required. Similarly, postbiological chemical treatment using coagulant and flocculant are common to meet the discharge norms. The biological treatment is the cheapest process for removal of biodegradable organic compounds, and residual recalcitrant compounds can be removed in postchemical treatment. It is possible to remove 70%–80% color and COD from wastewater by coagulation and flocculation process. The choice between the pre- and posttreatment depends upon the cost of treatment, availability of infrastructure and quality of wastewater required. The main disadvantage of this process is handling of generated sludge as there is transfer of pollutants from liquid phase to solid phase.

2.5.2.3 Adsorption

Adsorption is an effective means of reducing the concentration of colloidal and dissolved organic material in wastewater, and has gained a wider application due to low cost of treatment and being simple, versatile, and robust in nature. The success of an adsorption process depends upon the choice of an adsorbent and its efficacy for removal of particular compounds. Several adsorbents, namely, activated charcoal, zeolites, silica gel, activated alumina, and so on have been evaluated for treatment of wastewater. Adsorption on activated carbon has been studied extensively for removal of color from pulp and paper mill wastewater due to the high specific surface area and porous structure of activated carbon. Commercial activated carbon can be prepared from lignite, bituminous coal, and hardwood; application of commercial activated carbon for wastewater treatment may not be economical considering the high cost of the same. Several waste material such as coconut shell, coconut husk, sugarcane bagasse, wood, pulp mill residue, and the like have been used for preparation of activated carbon (Amuda and Ibrahim, 2006; Tan et al., 2008; Qureshi et al., 2008). A total of 2–20 kg/m^3 dosage of activated carbon with specific surface area of 200–1000 m^2/g has been used for treatment of wastewater from pulp and paper mills (Chakradhar and Shrivastava, 2004; Tantemsapya et al., 2004; Farhan et al., 2013). The treatment of wastewater may be performed in a column using granular activated carbon or powdered carbon in a tank followed by clarification or filtration. The efficiency of treatment of wastewater depends on retention time, pH, dosage, effective surface area, and pore size of adsorbent. Also, 50%–90% removal of color and COD has been achieved at higher dosage or retention time of wastewater. Recovery of carbon is also necessary to make this process economically attractive. The exhausted carbon is removed from the process, dried, and heated for reuse. Chemical activation of regenerated adsorbent may also be required for adsorption of specific pollutant. Addition of some fresh adsorbent (5%–20%) to the system is required to compensate the loss during regeneration of carbon. Thus, the use of low-cost adsorbents derived from agricultural and industrial solid wastes for wastewater treatment and cost-effective process for recovery of exhausted adsorbent is today's attraction.

2.5.2.4 Chemical Oxidation

The organic compounds in wastewater can be oxidized partially in LMW compounds or completely in simple end products. The most common oxidizing agents used for treatment of pulp and paper mill wastewater are ozone and hydrogen peroxide in combination with catalyst. The effectiveness of ozone as oxidizing agent is well established for oxidation of refractory organic compounds (Freire et al., 2000). Compounds with a greater electron density are more susceptible to ozone attack. Ozonation of pulp and paper mill wastewater convert high-molecular weight compounds into LMW compounds having higher percentage of oxygen in the chemical structure and are more amenable to biodegradation (Assalin et al., 2004). For ozone treatment, ozone is produced at site using pure oxygen or air. The oxidation efficacy depends on the mass transfer of ozone to wastewater. Similar to chemical treatment, the oxidation of wastewater can be performed as pre-, post-, or insplit mode with the biological treatment. Pretreatment of alkaline bleach plant effluent can enhance the biodegradability of wastewater by 30%–40%, which is associated with degradation of refractory organic material into LMW compounds (Bijan and Mohseni, 2004). The posttreatment of wastewater will require biological finisher for removal of generated biodegradable organic compounds and have additional and cheap COD reduction (Kaindl, 2010). Advanced treatment with ozone makes it possible to produce high-grade papers on an environmentally sensitive location. Running costs can be minimized by exact control of ozone dosage on demand. The positive effect of decolorization makes it possible to reuse treated water in the paper production process.

2.5.2.5 Photocatalysis

The photocatalysis is an emerging technique for the remediation of both aquatic and atmospheric pollutants. The photocatalyst may lead to the complete mineralization of organic pollutants to simple end products. Various catalysts (TiO_2, ZnO, MgO, WO_3, Fe_2O_3, CdS, and ZnS) are being explored for photocatalysis. Among them, TiO_2 is the most widely used photocatalyst because it is chemically and biologically inert, photostable, nontoxic, low cost, and relatively easy to produce and use (Thiruvenkatachari et al., 2008). TiO_2 is capable of absorbing photons energy in the near UV region (<387 nm). The sun produces 0.2–0.3 M photons/m^2h^1 (300–400 nm) with a characteristic UV flux of 20–30 W/m^2 near the Earth's surface. Thus, sunlight can be utilized as an economical and ecologically sensible source of light photocatalytic degradation of wastewater using TiO_2 as catalyst. The exposure of TiO_2 to UV radiation leads to the excitement of electrons from the valence band to the conduction band leaving behind highly oxidative holes in valence band and generation of radicals (mainly $OH^•$). Pollutants are oxidized by holes on TiO_2 surface as well as by radicals in the solution (Carp et al., 2004). The hydroxylation of aromatic compounds by $OH^•$ radicals lead to consecutive oxidation/addition and ultimately ring opening. The intermediates generated, mainly aldehydes and carboxylic acids, are further carboxylated to form harmless CO_2 and H_2O. Photocatalysis for wastewater treatment can be carried out in two types of reactors depending on the state of the catalyst: (1) catalyst particles suspended in solution and (2) catalyst immobilized onto an inert support, that is, activated carbon, mesoporous clays, fibers, and membrane. The slurry reactors are more preferred owing to high surface area of catalyst per unit volume available for adsorption of pollutants. They allow more number of photons to hit the catalyst and are easier to make and maintain.

2.6 Conclusion

The pulp and paper industry is an important part of the socioeconomic development, and consumption of paper will continue to increase in developing countries. There have been gradual improvements in the processes as well as utilization of natural resources to make the industry sustainable and environment friendly. Although pulp- and paper-making process is extremely water intensive, advancements in the processes and recycling practices have reduced the same to a great extent. The water consumption in pulp and paper industry was 200–250 m^3/ton of product up to late of the last century, which has gone down to 30–150 m^3/ton of product in the recent times.

The recent advancements in the last two decades, for example, shifting from wet to dry debarking, efficient washing and dewatering equipment for brown stock, extended delignification technologies, chlorine-free bleaching sequences, and so on have reduced the pollution load in terms of color causing and toxic compounds to a significant level while producing better quality of product without harmful chemicals. Industry has also made a significant improvement for degradation of recalcitrant compounds in wastewater. Adoptions of different types of pre/postchemical treatments and advancement in biological processes have made it possible to meet the discharge norms for treated wastewater.

2.7 Perspective

Paper is the basic need of day-to-day life and its demand will continue to increase due to diversified applications. Although environmental issues related to pulp and paper industry are a burning issue, accessibility to the best-available technologies and increasing awareness can generate wastewater to meet future stringent disposal norms along with environmental protection. In-process advancement will continue to identify better environment friendly pulping and bleaching processes, and development of novel techniques for wastewater treatment will help in recycling the treated wastewater. Identification of microbial strains or consortia and development of bioreactor for application of these consortia may decompose the residual recalcitrant compounds in wastewater. Treated wastewater from pulp and paper industry is being used for irrigation and development of green belt. Better quality of wastewater from the industry may also be a useful source of water for other applications.

References

Ali, M and Sreekrishnan, TR. 2000. Anaerobic treatment of agricultural residue based pulp and paper mill effluents for AOX and COD reduction. *Process Biochemistry* 36(1): 25–29.

Amuda, OS and Ibrahim, AO. 2006. Industrial wastewater treatment using natural material as adsorbent. *African Journal of Biotechnology* 5(16): 1483–1487.

Assalin, MR, Rosa, MA, and Duran, N. 2004. Remediation of kraft effluent by ozonation: Effect of applied ozone concentration and initial pH. *Ozone: Science and Engineering* 26(3): 317–322.

Bajpai, P and Bajpai, PK. 1997. Reduction of organochlorine compounds in bleach plant effluents. In: T. Scheper (eds.), *Biotechnology in the Pulp and Paper Industry*. Springer Berlin, Heidelberg, pp. 213–259.

Barroca, MJMC, Seco, IM, Fernandes, PMM, Ferreira, LMGA, and Castro, JAAM. 2001. Reduction of AOX in the bleach plant of a pulp mill. *Environmental Science and Technology* 35(21): 4390–4393.

Berry, RM, Luthe, CE, Voss, RH, and Wrist, PE. 1991. The effects of recent changes in bleached softwood kraft mill technology on organochlorine emissions: An international perspective. *Pulp and Paper Canada* 92(6): 43–54.

Biester, H, Keppler, F, Putschew, A, Martinez-Cortizas, A, and Petri, M. 2004. Halogen retention, organohalogens, and the role of organic matter decomposition on halogen enrichment in two Chilean peat bogs. *Environmental Science and Technology* 38(7): 1984–1991.

Bijan, L and Mohseni, M. 2004. Using ozone to reduce recalcitrant compounds and to enhance biodegradability of pulp and paper effluents. *Water Science and Technology* 50(3): 173–182.

Buzzini, AP, Patrizzi, LJ, Motheo, AJ, and Pires, EC. 2007. Preliminary evaluation of the electrochemical and chemical coagulation processes in the post-treatment of effluent from an upflow anaerobic sludge blanket (UASB) reactor. *Journal of Environmental Management* 85(4): 847–857.

Carp, O, Huisman, CL, and Reller, A. 2004. Photo induced reactivity of titanium dioxide. *Progress in Solid State Chemistry* 32: 33–177.

Chakrabarti, SK, Gupta, S, Purwar, M, Bhist, SC, and Varadhan, R. 2012. Combined chemical-biological treatment of pulp and paper mill effluent. *IPPTA Journal* 242: 160–163.

Chakradhar, B and Shrivastava, S. 2004. Colour removal of pulp and paper effluents. *Indian Journal of Chemical Technology* 11(5): 617–621.

Cicek, N, Juan, PF, Suidan, MT, and Urbain, V. 1998. Using a membrane bioreactor to reclaim wastewater. *Journal of the American Water Works Association* 90(11): 105–113.

Diez, MC, Castillo, G, Aguilar, L, Vidal, G, and Mora, ML. 2002. Operational factors and nutrient effects on activated sludge treatment of *Pinus radiata* kraft mill wastewater. *Bioresource Technology* 83(2): 131–138.

Dugal, HS. 2009. The paper industry—Where is it going? Past, present, future. In: *9th International Technical Conference on Pulp, Paper and Allied Industry*, December, 4–6, New Delhi, India, pp. 1–11.

EC, 2001. Integrated prevention pollution and control (IPPC), Reference document on best available techniques in the pulp and paper industry. European Commission, Brussels.

Farhan, M, Wahid, A, Kanwal, A, and Bell, JNB. 2013. Synthesis of activated carbon from tree sawdust and its usage for diminution of color and COD of paper-mill effluents. *Pakistan Journal of Botany* 45(S1): 521–527.

Freire, CSR, Silvestre, AJD, and Neto, CP. 2003. Carbohydrate-derived chlorinated compounds in ECF bleaching of hardwood pulps: Formation, degradation, and contribution to AOX in a bleached kraft pulp mill. *Environmental Science and Technology* 37(4): 811–814.

Freire, RS, Kunz, A, and Duran, N. 2000. Some chemical and toxicological aspects about paper mill effluent treatment with ozone. *Environmental Technology* 21(6): 717–721.

Gaspar, A, Evtuguin, DV, and Neto, CP. 2003. Oxygen bleaching of kraft pulp catalysed by Mn(III)-substituted polyoxometalates. *Applied Catalysis A: General* 239(1): 157–168.

Gupta, S, Chakrabarti, SK, and Singh, S. 2013. Effect of ozonation on degradation of organochlorine compounds in biosludge of pulp and paper industry. *Ozone: Science and Engineering* 35(2): 109–115.

Joss, EN, McGrouther, KG, and Slade, AH. 2007. Comparison of the efficacy of oxidative processes and flocculation for the removal of colour from Eop effluent. *Water Science and Technology* 55(6): 57–64.

Kaindl, N. 2010. Upgrading of an activated sludge wastewater treatment plant by adding a moving bed biofilm reactor as pretreatment and ozonation followed by biofiltration for enhanced COD reduction: Design and operation experience. *Water Science and Technology* 62(11): 2710–2719.

Kansal, SK, Singh, M, and Sud, D. 2008. Effluent quality at kraft/soda agro-based paper mills and its treatment using a heterogeneous photocatalytic system. *Desalination* 228(1): 183–190.

Lamberg, JA, Ojala, J, Peltoniemi, M, and Sarkka, T. 1800. Research on evolution and global history of paper and pulp industry: An introduction. *The Evolution of Global Paper Industry* 2050: 1–18.

Lee, NM and Welander, T. 1996. Reducing sludge production in aerobic wastewater treatment through manipulation of the ecosystem. *Water Research* 30(8): 1781–1790.

Lerner, M, Stahl, N, and Galil, NI. 2007. Comparative study of MBR and activated sludge in the treatment of paper mill wastewater. *Water Science and Technology* 55(6): 23–29.

MoEF, 2010. *Technical EIA Guidance Manual for Pulp and Paper Industry.* IL&FS Ecosmart Limited, Hyderabad, India.

Niku-Paavola, ML, Ranua, M, Suurnakki, A, and Kantelinen, A. 1994. Effects of lignin modifying enzymes on pine kraft pulp. *Bioresource Technology* 50(1): 73–77.

NPC (National Productivity Council), 2006. Development of guidelines for water conservation in pulp and paper sector. Final Report, India, pp. 52–54.

Ojala, J, Lamberg JA, Peltoniemi M, Sarkka T, and Voutilainen M. 2013. The evolution of global paper industry. *O Papel* 74(10): 51–54.

Onate, E, Salazar, C, and Zaror C. 2012. Performance assessment of membrane separation technologies to recover water and chemical resources from segregated ECF bleaching effluents in *Pinus radiata* kraft pulp production. In: *Proceeding 10th IWA Symposium on Forest Industry Wastewater*, Concepcion, Chile, January 8–11, 2012, p. 64.

Orrego, R, Pandelides, Z, Guchardi, J, and Holdway, D. 2011. Effects of pulp and paper mill effluent extracts on liver anaerobic and aerobic metabolic enzymes in rainbow trout. *Ecotoxicology and Environmental Safety* 74(4): 761–768.

Pokhrel, D and Viraraghavan, T. 2004. Treatment of pulp and paper mill wastewater—A review. *Science of the Total Environment* 333(1): 37–58.

Pulliam, TL. 1991. End of pipe treatment cost can be minimized in process design stage. In: Ferguson, K. (Ed.), *Environmental Solutions*. Miller Freeman, USA, pp. 39–41.

Qureshi, K, Bhatti, I, Kazi, R, and Ansari, AK. 2008. Physical and chemical analysis of activated carbon prepared from sugarcane bagasse and use for sugar decolorisation. *International Journal of Chemical and Biomolecular Engineering* 1(3): 145–149.

Reeve, DW. 1991. Organochlorine in bleached kraft pulp. *Tappi Journal* 74(2): 123–126.

Roy, M, Chakrabarti, SK, Bharadwaj, NK, Chandra, S, Kumar, S, Singh, S, Bajpai, PK, and Jauhari, MB. 2004. Characterization of chlorinated organic material in Eucalyptus pulp bleaching effluents. *Journal of Scientific and Industrial Research* 63(6): 527–535.

Rusten, B, Eikebrokk, B, Ulgenes, Y, and Lygren, E. 2006. Design and operations of the Kaldnes moving bed biofilm reactors. *Aquacultural Engineering* 34(3): 322–331.

Santos, RB and Hart, PW. 2014. Brownstock washing—A review of the literature. *Tappi Journal* 13(01): 9–19.

Shukla, SK, Kumar, V, Kim, T, and Bansal, MC. 2013. Membrane filtration of chlorination and extraction stage bleach plant effluent in Indian paper industry. *Clean Technologies and Environmental Policy* 15(2): 235–243.

Simstich, B, Beimfohr, C, and Horn, H. 2012. Lab scale experiments using a submerged MBR under thermophilic aerobic conditions for the treatment of paper mill deinking wastewater. *Bioresource Technology* 122: 11–16.

Sivard, A, Malmaeus, M, Almemark, M, Karlsson, M, Ericsson, T, and Simon, O. 2012. Effects of further treatment stages at pulp and paper industries. In: *Proceeding 10th IWA Symposium on Forest Industry Wastewater*, Concepcion, Chile, January 8–11, 2012, p. 24.

Smook, GA. 1982. *Handbook for Pulp and Paper Technologists*. Joint Textbook Committee of the Paper Industry. TAPPI, USA, pp. 384–388.

Taghipour, F and Evans, GJ. 1996. Radiolytic elimination of organochlorine in pulp mill effluent. *Environmental Science and Technology* 30(5): 1558–1564.

Tan, IAW, Ahmad, AL, and Hameed, BH. 2008. Preparation of activated carbon from coconut husk: Optimization study on removal of 2,4,6-trichlorophenol using response surface methodology. *Journal of Hazardous Materials* 153(1): 709–717.

Tantemsapya, N, Wirojanagud, W, and Sakolchai, S. 2004. Removal of color, COD and lignin of pulp and paper wastewater using wood ash. *Songklanakarin Journal of Science and Technology* 26(1): 1–12.

Tchobanoglous, G, Burton, FL, and Stensel, HD. 2003. *Wastewater Engineering: Treatment and Reuse*, 4th ed., Metcalf & Eddy Inc., Tata McGraw-Hill Publishing, New Delhi, India, pp. 555–588, 661–747.

Thiruvenkatachari, R, Vigneswaran, S, and Moon, IS. 2008. A review on UV/TiO$_2$ photocatalytic oxidation process. *Korean Journal of Chemical Engineering* 25(1): 64–72.

Thompson, G, Swain, J, Kay, M, and Forster, CF. 2001. The treatment of pulp and paper mill effluent: A review. *Bioresource Technology* 77(3): 275–286.

USEPA, 2002. *EPA Office of Compliance Sector Notebook Project: Profile of Pulp and Paper Industry*, 2nd ed., Washington, DC, USA, EPA/310-R-02-002.

Verenich, S, Laari, A, Nissen, M, and Kallas, J. 2001. Combination of coagulation and catalytic wet oxidation for the treatment of pulp and paper mill effluents. *Water Science and Technology* 44(5): 145–152.

Zhang, Y, Ye, CMF, Kong, Y, and Li, H. 2009. The treatment of wastewater of paper mill with integrated membrane process. *Desalination* 236(1): 349–356.

Zheng, Y and Allen, DG. 1996. Biological dechlorination of model organochlorine compounds in bleached kraft mill effluents. *Environmental Science and Technology* 30(6): 1890–1895.

3

The Role of Cyanobacteria in the Biodegradation of Agrochemical Waste

Surendra Singh and Pallavi Datta

CONTENTS

3.1 Introduction

The advent of modern agriculture has to a great extent succeeded in preventing mass starvation of human population in both developed and developing countries. Agricultural practices include use of large amount of agrochemicals, which includes different classes of pesticides such as herbicides, insecticides, fungicides, and the like, and chemical fertilizers. The routine application of these agrochemicals has important ecological effect. The indiscriminate use of pesticides has resulted in serious harm and problems to humans as well as to the biodiversity (Gavrilescu, 2005; Hussain et al., 2009). They damage a wide range of beneficial microorganisms due to their persistence in the environment. Thus, more and more use of these chemicals has created environmental hazards and widespread pollution.

The use of man-made "xenobiotic" chemicals has led effort to implement new technologies to reduce these contaminants from the environment. However, commonly used pollution treatment methods (e.g., land-filling, recycling, pyrolysis, and incineration) for the remediation of contaminated sites have also had adverse effects on the environment, which can lead to the formation of toxic intermediates (Debarati et al., 2005). Furthermore, these methods are more expensive and sometimes difficult to execute, especially in extensive agricultural areas, as for instance pesticides (Jain et al., 2005). One promising treatment method is to exploit the ability of microorganisms to remove pollutants from contaminated sites, an alternative treatment strategy that is effective, minimally hazardous, economical, versatile, and environment friendly, is the process known as bioremediation (Finley et al., 2010).

Biodegradation is the breakdown of a substance catalyzed by enzymes *in vitro* or *in vivo*. The biodegradation of pesticides is often complex and involves a series of biochemical reactions. Although many enzymes efficiently catalyze the biodegradation of pesticides, the full understanding of the biodegradation pathway often requires new investigations. Several pesticide biodegradation studies have shown only the total of degraded pesticide, but have not investigated in depth the new biotransformed products and their fate in the environment.

Thereafter, scientists have been exploring the microbial diversity of areas in search of organisms that have the ability to degrade a wide range of agrochemicals. Several studies were made to understand the ecology, physiology, biochemistry, and genetic basis of microbial degradation (Mishra et al., 2001). Genes/enzymes, which provide microorganisms with the ability to degrade pesticides, have been identified and characterized, thus providing potentiality of microbes in biodegradation. The ability of these organisms to reduce the concentration of pesticides is directly linked to their long-term adaptation to environments where these compounds exist. Moreover, genetic engineering may be used to enhance the performance of such microorganisms that have the properties for biodegradation (Schroll et al., 2004).

The biological methods are advantageous to decontaminate areas that have been polluted by pesticides. These methods consider the thousands of microorganisms in the environment that in order to survive seek for alternatives to eliminate the pesticides that were sprayed. Many native microorganisms develop complex and effective metabolic pathways that permit the biodegradation of toxic substances that are released into the environment. Although the metabolic process is lengthy, it is a more viable alternative for removing the sources of xenobiotic compounds and pollution (Diaz, 2004; Schoefs et al., 2004; Finley et al., 2010).

Cyanobacteria are free-living phototrophic microorganisms. They can derive energy from sunlight and carbon from the air. Some cyanobacteria are also able to fix atmospheric nitrogen and are therefore especially inexpensive to maintain (Carr and Whitton, 1982; Castenholz and Waterbury, 1989). Filamentous cyanobacteria, including nitrogen-fixing strains that combine aerobic metabolism in their vegetative cells with anaerobic metabolism in their differentiated cells called heterocysts (Wolk et al., 1994), are widespread in many ecosystems, including polluted ones (Fogg, 1987; Sorkhoh et al., 1992). The abundance of cyanobacteria in polluted waters has stimulated research into characterizing the tolerance of these organisms to a variety of pollutants. Cyanobacteria have been reported to play an important role on bioremediation processes; for example, removal of the heavy metals from polluted water (El-Enany and Issa, 2000), biotransformation of mercury (Hg[II]) (Lefebvre et al., 2007), degradation of methyl parathion (Barton et al., 2004), crude oil (Chaillan et al., 2006), aromatic hydrocarbons (Narro et al., 1992), and xenobiotics (Megharaj et al., 1987). Recombinant filamentous cyanobacteria that degrade organic pollutants have been reported (Kuritz and Wolk, 1995).

In this chapter, we review the ability of cyanobacteria to degrade chlorinated and organophosphorus pesticide groups proposing the use of this microorganism to be considered for low-cost, low-maintenance remediation of agrochemicals.

3.2 Microorganisms Involved in Biodegradation

Different microorganisms have been used to biotransform pesticides. A fraction of the soil biota can quickly develop the ability to degrade certain pesticides, when they are continuously applied to the soil. These chemicals provide an adequate carbon source and electron donors for certain soil microorganisms (Torres, 2003), thereby generating a method for the treatment of pesticide-contaminated sites (Qiu et al., 2007). However, the transformation of such compounds depends not only on the presence of microorganisms with appropriate degrading enzymes but also on a wide range of environmental parameters (Alves et al., 2010). Additionally, some physiological, ecological, biochemical, and molecular aspects play an important role in the microbial transformation of pollutants (Iranzo et al., 2001; Becker and Seagren, 2010).

There are different sources of microorganisms with the ability to degrade pesticides. As pesticides are mainly applied to agricultural crops, soil is most affected by these chemicals. Industry's effluent sediment, sewage sludge, activated sludge, wastewater, natural waters, sediments, areas surrounding the manufacture of pesticides, and even some live organisms are also affected. In general, microorganisms that have been identified as pesticide degraders have been isolated from a wide variety of sites contaminated with some type of pesticide. At present, in different laboratories around the world there are collections of microorganisms characterized by their identification, growth, and degradation of pesticides. The isolation and characterization of microorganisms that are able to degrade pesticides makes it possible to utilize new tools to restore polluted environments or to treat wastes before their final disposition (Ortiz-Hernández et al., 2011).

3.3 Mechanisms for Pesticide Biodegradation

Pesticides belong to a category of chemicals used worldwide to prevent or control pests, diseases, weeds, and other plant pathogens in an effort to reduce or eliminate yield losses and maintain high product quality (Damalas and Eleftherohorinos, 2011). The positive aspect of the application of pesticides has resulted in enhanced crop/food productivity and a drastic reduction of vector-borne diseases (Damalas, 2009). Chemical pesticides can be classified in different ways, but they are most commonly classified according to their chemical composition. This method allows the uniform and scientific grouping of pesticides to establish a correlation between structure, activity, toxicity, and degradation mechanisms, among other characteristics. Table 3.1 shows the most important pesticides and their general characteristics.

Biodegradation is a process that involves the complete breakdown of an organic compound in its inorganic constituents. The microbial transformation may be driven by energy needs or a need to detoxify the pollutants, or it may be fortuitous in nature (cometabolism) (Becker and Seagren, 2010). The search for pollutant-degrading microorganisms, understanding their genetics and biochemistry, and developing methods for their application in the field have become an important human endeavor (Megharaj et al., 2011). The ubiquitous nature of microorganisms, their numbers and large biomass relative to other living organisms on earth, their more diverse catalytic mechanisms (Paul et al., 2005), and their ability to function even in the absence of oxygen and other

TABLE 3.1

General Characteristics of Some Pesticides

Pesticides	Characteristics	Main Composition
Organochlorines	Soluble in lipids, they accumulate in fatty tissue of animals, are transferred through the food chain; toxic to a variety of animals, long-term persistent.	Carbon atoms, chlorine, hydrogen, and occasionally oxygen. They are nonpolar and lipophilic.
Organophosphates	Soluble in organic solvents but also in water. They infiltrate reaching groundwater, less persistent than chlorinated hydrocarbons; some affect the central nervous system. They are absorbed by plants and then transferred to leaves and stems, which are the supply of leaf-eating insects or feed on wise.	Possess central phosphorus atom in the molecule. In relation with organochlorines, these compounds are more stable and less toxic in the environment. The organophosphate pesticides can be aliphatic, cyclic, and heterocyclic.
Carbamates	Carbamate acid derivatives; kill a limited spectrum of insects, but are highly toxic to vertebrates. Relatively low persistence.	Chemical structure based on a plant alkaloid *Physostigma venenosum*.
Pyrethroids	Affect the nervous system; are less persistent than other pesticides; are the safest in terms of their use, some are used as household insecticides.	Compounds similar to the synthetic pyrethrins (alkaloids obtained from petals of *Chysanthemun cinerariefolium*).
Biological	Only the *Bacillus thuringiensis* (Bt) and its subspecies are used with some frequency; are applied against forest pests and crops, particularly against butterflies. Also affect other caterpillars.	Viruses, microorganisms, or their metabolic products.

Source: Adapted from Badii M and Landeros J. 2007. *CULCYT/Toxicología de Plaguicidas* 4(19): 21–34.

extreme conditions are greatly important in the use of microorganisms for the degradation of pesticides.

In natural environments, biodegradation involves the transfer of substrates and products within a well-coordinated microbial community, a process referred to as metabolic cooperation (Abraham et al., 2002). Microorganisms have the ability to interact both chemically and physically with substances, leading to structural changes or the complete degradation of the target molecule. Pesticides interact with soil organisms and their metabolic activities and may alter the physiological and biochemical behavior of soil microbes.

Enzymes are central to the biology of many pesticides (Riya and Jagatpati, 2012). Applying enzymes to transform or degrade pesticides is an innovative treatment technique for the removal of these chemicals from polluted environments. Enzyme-catalyzed degradation of a pesticide may be more effective than existing chemical methods.

Enzymes are central to the mode of action of many pesticides: some pesticides are activated *in situ* by enzymatic action, and many pesticides function by targeting particular enzymes with essential physiological roles. Enzymes are also involved in the degradation of pesticide compounds, both in the target organism, through intrinsic detoxification mechanisms and evolved metabolic resistance, and in the wider environment, via biodegradation by soil and water microorganisms (Scott et al., 2008). Trigo et al. (2009) suggested that (i) the central metabolism of the global biodegradation networks involves transferases, isomerases, hydrolases, and ligases; (ii) linear pathways converging on particular intermediates form a funnel topology; (iii) the novel reactions exist in the exterior part of the network; and (iv) the possible pathway between compounds and the central metabolism can be arrived at by considering all the required enzymes in a given organism and the intermediate compounds (Ramakrishnan et al. 2011).

The metabolism of pesticides may involve a three-phase process. In Phase I metabolism, the initial properties of a parent compound are transformed through oxidation, reduction, or hydrolysis to generally produce a more water-soluble and usually a less-toxic product than the parent. The second phase involves the conjugation of a pesticide or pesticide metabolite to a sugar or amino acid, which increases the water solubility and reduces toxicity compared with the parent pesticide. The third phase involves conversion of Phase II metabolites into secondary conjugates, which are also nontoxic. Fungi and bacteria are involved in these processes and produce intracellular or extracellular enzymes including hydrolytic enzymes, peroxidases, oxygenases, and so on (Ortiz-Hernández et al., 2011).

Due to the diversity of chemicals used in pesticides, the biochemistry of pesticide bioremediation requires a wide range of catalytic mechanisms, and therefore a wide range of enzyme classes. Information for some pesticide-degrading enzymes is given in Table 3.2.

Among the enzymes that degrade pesticides, the hydrolases catalyze the hydrolysis of several major biochemical classes of pesticide (esters, peptide bonds, carbon–halide bonds, ureas, thioesters, etc.) and generally operate in the absence of redox cofactors, making them ideal candidates for all of the current bioremediation strategies (Scott et al., 2008). In this group, we can find the phosphotriesterases (PTEs), which are one of the most important classes (Chino-Flores et al., 2012). These enzymes have been isolated from different microorganisms that hydrolyze and detoxify organophosphate pesticides (OPs). This reaction reduces OP toxicity by decreasing the ability of OP to inactivate AchE (Porzio et al., 2007; Theriot and Grunden, 2010). The first isolated PTE belongs to *Pseudomonas diminuta* MG; this enzyme shows a highly catalytic activity toward OPs. The PTEs are encoded by a gene called *opd* (organophosphate-degrading). *Flavobacterium* ATCC 27551 contains the *opd* gene that encode a PTE (Latifi et al., 2012). These enzymes specifically hydrolyze phosphoester bonds, such as P–O, P–F, P–NC, and P–S, and the hydrolysis mechanism involves a water molecule at the phosphorus center. This enzyme has a potential use for the cleaning of organophosphorus pesticide-contaminated environments (Ortiz-Hernández et al., 2003). There are other enzymes involved in the overall degradation of apesticide. Esterases are enzymes that catalyze the hydrolysis of carboxylic esters (carboxyesterases), amides (amidases), phosphate esters (phosphatases), and so on (Bansal, 2012). In the reaction catalyzed by esterases, a wide range of ester substrates can be hydrolyzed into their alcohol and acid components as follows:

$$R\!\!=\!\!O\!-\!OCH_3 + H_2OR\!\!=\!\!O\!-\!OH + CH_3OH$$

Many insecticides (organophosphates, carbamates, and pyrethroids) have a carboxylic ester component, and the enzymes capable of hydrolyzing this type of ester bond are known as carboxylesterases.

Oxidoreductases are a broad group of enzymes that catalyzes the transfer of electrons from one molecule (the reductant or electron donor) to another (the oxidant or electron acceptor). Many of these enzymes require additional cofactors to act as electron donors, electron acceptors, or both. These enzymes have applications in bioremediation, during which they catalyze an oxidation/reduction reaction by including molecular oxygen (O_2) as the electron acceptor. In these reactions, oxygen is reduced to water (H_2O) or hydrogen peroxide (H_2O_2). The oxidases are a subclass of the oxidoreductases (Scott et al., 2008).

TABLE 3.2

Microbial Enzymes in Pesticide Biodegradation

Enzyme		Organism	Pesticide	Bioremediation Strategy
Oxidoreductases	Gox	*Pseudomonas* sp. LBr, *Agrobacterium* strain T10	Glyphosate	Plant
Monooxygenases	ESd	*Mycobacterium* sp.	Endosulfan and endosulfato	—
	Ese	*Arthrobacter* sp.	Endosulfan, aldrin, malation, DDDT, and endosulfate	—
	Cyp1A1/1A2	Rats	Atrazine, norflurazon, and isoproturon	Plant
	Cyp76B1	*Helianthus tuberosus*	Linuron, chlortoluron, and isoproturon	Plant
	P450	*Pseudomonas putida*	Hexachlorobenzene and pentachlorobenzene	—
Dioxygenases	TOD	*Pseudomonas putida*	Herbicides trifluralin	—
	E3	*Lucilia cuprina*	Synthetic pyrethroids and insecticides phosphotriester	—
Phosphodiesterases	PdeA	*Delftia acidovorans*	Organophosphorus compounds	—
Phosphotriesterases	OPH, OpdA	*Agrobacterium radiobacter*, *Pseudomonas diminuta*, *Flavobacterium* sp.	Insecticides phosphotriester: parathion, methyl parathion, malathion, coumaphos, others.	Bioremediation and free enzymes
Phosphonatase	Phn	*Escherichia coli*, *Sinorhizobium meliloti*	Organophosphorus compounds	—
Haloalkane dehalogenases	LinB	*Sphingobium* sp., *Shingomonas* sp.	Hexachlorocyclohexane (β and δ isomers)	Bioaugmentation
	AtzA	*Pseudomonas* sp. ADP	Herbicides chloro-*S*-trazine	Plants and bacteria
	TrzN	*Nocardioides* sp.	Herbicides chloro-*S*-trazine	—
	LinA	*Sphingobium* sp., *Shingomonas* sp.	Hexachlorocyclohexane (γ isomers)	Bioaugmentation
	TfdA	*Ralstonia eutropha*	2,4-Dichlorophenoxyacetic acid and pyridyl-oxyacetic	Plant
		Pseudomonas maltophilia	Dicamba	Plant
	Glp A&B	*Pseudomonas pseudomallei*		Organophosphorus compounds
	hocA	*Pseudomonas monteilli*		Organophosphorus compounds
	mpd	*Pleisomonas* sp.		Organophosphorus compounds

Source: Adapted from Singh BK and Walker A. 2006. *FEMS Microbiological Review* 30: 428–471; Scott C et al. 2008. *Indian Journal of Microbiology* 48: 65–79; Riya P and Jagatpati T. 2012. *World Journal of Science and Technology* 2: 36–41.

3.4 Ecology of Cyanobacteria

Cyanobacterial habitats consist of rice fields, where pesticides are applied directly for crop protection and surface waters that receive agricultural waste streams from various sources. Developing countries, with 75% of global population, bear the major impact of agricultural practices that use pesticides (El Sebae, 1993). More than 90% of the global amount of pesticides used for rice production is applied on Asian rice paddies (Abdullah et al., 1997). Nitrogen-fixing filamentous cyanobacteria are common in rice fields, where they exist naturally or are introduced by soil fertilization practices using cultures of *Anabaena–Azolla* or individual cyanobacteria (Singh, 1973; Subramanian et al., 1994; Quesada and Valiente, 1996; Kannaiyan et al., 1997). The diversity and ecology of cyanobacteria on rice paddies were characterized by different methods. Quesada and Valiente (1996), used traditional observational method, and Eskew et al. (1993) and Gebhardt and Nierzwicki-Bauer (1991), used molecular methods. Species of nitrogen-fixing cyanobacteria *Nostoc, Anabaena,* and *Aulosira* were observed most often and isolated from rice paddies in India (Bhashkar et al., 1992).

3.5 Degradation of Chlorinated Pesticides by Cyanobacteria

Lindane is one of the most efficient broad-specificity pesticides used for rice crop protection in rice-producing countries (Abdullah et al., 1997). The pesticide persists in the environment (Alexander, 1994) and can be detected in the air, rain, and surface water at 90% of sites, long after its application (Majewski and Capel, 1995). Ninety percent of lindane applied to rice field dissipated within 2 weeks, possibly due to atmospheric transport (Abdullah et al., 1997). Singh (1973) first reported the tolerance level of laboratory strains and rice field isolates of *Cylindrospermurn* sp., *Aulosira fertilissima,* and *Plectonema boryanum* to commercial preparations of lindane in concentrations up to 80 µg/mL. Under optimized laboratory conditions, *Anabaena* sp. and *Nostoc ellipsosporum* degraded lindane from 0.5 µg/mL to 1.0 ng/mL within 40–50 h. Lindane was also degraded like other cyanobacterial strains such as *Anabaena* sp. P30, *Calothrix* sp. ATCC29112, *Fischerella muscicola* UT1829, *Fisherella* sp. CALU926, *N. muscorum* UT387, *Nostoc pameloides* UT1627, *Nostoc* sp. strains GSV39, 40, and 236, *Phormidium uncinatum, Pl. boryanum, Plectonema* sp., and *Synechococcus* sp. PCC 7942 (Kuritz et al., 1997). The rates of lindane degradation by these strains were different; however, a band corresponding to pentachlorocyclohexene as an intermediate product was present in all cases. Since these strains belong to three different taxonomic groups (Rippka et al., 1979), it is evident that the ability to degrade lindane is common among cyanobacterial strains.

The strains of nitrogen-fixing filamentous cyanobacteria *Anabaena* sp. and *N. ellipsosporum* required the presence of nitrate in the medium for lindane degradation (Kuritz and Wolk, 1995), while ammonium inhibited the process without affecting the culture growth rate. Lindane degradation by cyanobacteria did not occur in the absence of light (Kuritz et al., 1997). Degradation of lindane by nitrogen-fixing cyanobacteria is regulated by nitrate and negatively regulated by ammonium (Kuritz et al., 1997).

3.5.1 Mechanism of Lindane Degradation by Wild-Type and Genetically Altered Cultures of *Anabaena* sp.

Kuritz and Wolk (1995) demonstrated that two filamentous cyanobacteria (*Anabaena* sp. strain PCC 7120 and *N. ellipsosporum*) have a natural ability to degrade a highly chlorinated aliphatic pesticide, lindane (γ-hexachlorocyclohexane) and presented quantitative evidence that this ability can be enhanced by genetic engineering and provided qualitative evidence that these two strains can be genetically engineered to degrade another chlorinated pollutant, 4-chlorobenzoate.

In the presence of 5 mM nitrate, the wild-type *Anabaena* sp. was able to degrade lindane from a concentration of 0.5 µg/mL to ca. 1.0 ng/mL within 1.5–3 days, depending on the initial cell density. In the absence of nitrate, lindane was degraded much more slowly (Figure 3.1). In the absence of nitrate, the *Anabaena* sp. bearing pRL634 degraded lindane more rapidly than did the wild-type *Anabaena* sp. (Figure 3.1).

Gas chromatography (GC) showed the presence of several degradation products of lindane by wild-type *Anabaena* sp. A major peak (Figure 3.2) was identified as γ-pentachlorocyclohexene by comparison of its mass spectrum with that of the authentic compound and by its gas chromatographic retention time, 4.10 ± 0.02 min. The substances produced by wild-type *Anabaena* sp. strain PCC 7120 in the presence of nitrate and by the strain bearing pRL634 in the absence of nitrate and that have retention time cochromatographed. A minor peak (Figure 3.2) was identified as 1,2,4-trichlorobenzene by comparison of its mass spectrum with that of authentic 1,2,4-trichlorobenzene and its retention time, 1.90 ± 0.02 min, in the gas chromatograph. A ca. fourfold-less-abundant product with a retention time of 2.09 ± 0.02 min was similarly identified as 1,2,3-trichlorobenzene.

The stimulation of the rate of degradation of lindane by nitrate observed may be attributable to increased availability of nitrogen to nitrate-supplemented cultures. The idea that vegetative cells grown on N_2-derived nitrogen may differ metabolically from vegetative cells grown on nitrate, in ways other than nitrate-metabolizing enzymes per se, is supported by the finding (Fleming and Haselkorn, 1974) that at least eight proteins are

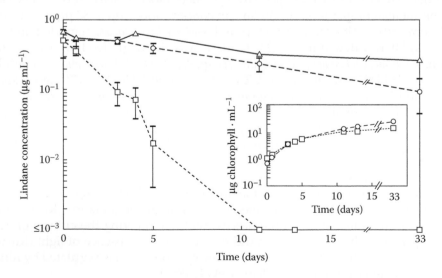

FIGURE 3.1

Degradation of lindane by wild-type *Anabaena* sp. strain PCC 7120 (o), by a derivative strain bearing pRL634 (□), by boiled cells and uninoculated medium (△; single samples).

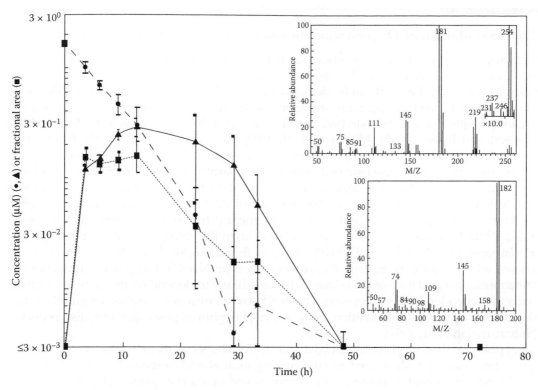

FIGURE 3.2
Time course of production of the substances identified as γ-pentachlorocyclohexene (■); the mass spectrum of the cyanobacterial product is shown in the upper inset) and 1,2,4-trichlorobenzene (▲); the mass spectrum of the cyanobacterial product is shown in the lower inset) upon degradation of lindane (●) by wild-type *Anabaena* sp. strain PCC 7120 grown in the presence of nitrate.

synthesized by nitrate grown cultures of *Anabaena* sp. strain PCC 7120 that are absent from N_2-grown cultures. Alternatively, perhaps some protein involved in the transport or reduction of nitrate is also required for the transport or metabolism of lindane.

3.5.2 Mechanism of Biodegradation of 2,4-D by Cyanobacteria, *Anabaena fertilissima, Au. fertilissima,* and *Westiellopsis prolifia*

Gas chromatography–mass spectrometry (GC–MS) study and molecular characterization by 16S rDNA amplification were carried out to evaluate differential effects of 2,4-D ethyl ester on *An. fertilissima, Au. fertilissima,* and *W. prolifia* (Kumar et al., 2013). Each organism has its own capacity to degrade pesticide into various subgroups. Different subgroups like 2,4-D methyl ester, 2,4-D isobutyl ester, isobutyric acid allyl ester, 3-bromobutyric acid, 2,4-D butyl ester, hydroxyurea, trifluroacetic acid, 2-methyl propyl ester, acetic acid 2-propenyl ester, and acetic acid (2,3-dichlorophenoxy) were transformed from 2,4-D ethyl ester. The results obtained by 16S rDNA sequencing confirmed that 16S rDNA region of *An. fertilissima* was more affected by 2,4-D ethyl ester as there was no homology in the region of 39 base pairs, in addition, several mismatches and gaps were observed, whereas less difference in 16S rDNA was observed in case of *Au. fertilissima* and *W. prolifia.*

3.6 Degradation of Organophosphorus Pesticides by Cyanobacteria

Organophosphorus compounds also have been commonly used for agricultural applications, including rice. They are less environmentally persistent than organochlorine compounds; however, they still can be detected in air and water due to heavy use (Majewski and Capel, 1995). At working concentrations and at concentrations present in wastewaters, organophosphorus compounds had no effect on growth of cyanobacteria (Singh, 1973; Doggett and Rhodes, 1991; Megharaj et al., 1994; Subramanian et al., 1994). Axenic cultures of *Nostoc, Oscillatoria* and *Phomidium* isolated from soil enriched with methyl parathion, grew in media supplemented with methyl parathion or other organophosphorus pesticides as a sole source of organic phosphorus and nitrate (Megharaj et al., 1987; Orus and Marco; 1991, Megharaj et al., 1994; Subramanian et al., 1994) and utilized phosphorus from the pesticide for growth and development (Megharaj et al., 1994; Subramanian et al., 1994). Cyanobacteria also were able to oxidize the nitrogroup of para-nitrophenol accompanied by the release of nitrite into growth media (Megharaj et al., 1994). It is not known which enzymatic system of cyanobacteria is involved in this process; however, the metabolism/assimilation of the released nitrite is likely to depend on the activity of nitrite reductase encoded by the nir operon. The link between nitrogen metabolism and the effectiveness of phosphorus utilization from organophosphorus pesticides was also reported (Subramanian et al., 1994).

Glyphosate (*N*-phosphonomethylglycine), a broad-spectrum, nonselective, postemergence herbicide is widely used in suppressing annual and perennial weeds in agricultural lands, ornamental and residential gardens and in aquatic systems (Lipok et al., 2010, Arunakumara et al., 2013).

Due to its widespread usage over the years, glyphosate is now considered to be the most studied organophosphonate (Lipok et al., 2009). The presence of chemically and thermally stable C–P bond, a characteristic feature of glyphosate (Singh and Walker, 2006) is a matter of frequent concern, because the C–P linkage is found to be heavily resistant to nonbiological degradation in the environment (Hayes et al., 2000). As per the published literature, biodegradation of glyphosate is believed to be done basically by soil microorganisms and the process can be described under two different metabolic pathways (Duke, 2011). One process involves in splitting the glyphosate C–N bond by the action of glyphosate oxidoreductase (GOX) enzyme to produce aminomethylphosphonic acid (AMPA) and glyoxylate (Schuette, 1998). In fact, glyoxylate is not only a metabolite derived of glyphosate degradation, but also a plant endogenous metabolite involved in different metabolic pathways (Rojano-Delgado et al., 2010). By the action of the enzyme C–P lyase, AMPA, the other main metabolite is degraded to methylamine, which ultimately generates formaldehyde by the action of methyl-amine dehydrogenase enzyme (Lerbs et al., 1990). Formaldehyde quickly reacts with water and/or hydroxyl radicals to form methanol. Thus, the ultimate yield of the glyphosate degradation may contain carbon dioxide, phosphate, ammonia, and methanol (Araujo et al., 2003). In the second pathway, sarcosine (*N*-methyl-glycine) is yielded through direct degradation of glyphosate by the action of C–P lyase enzyme (Schuette, 1998; Kafarski et al., 2000). The sarcosine can further be degraded into amino acids such as glycine, serine, cysteine, methionine, and histidine (Pipke et al., 1987).

A large number of soil microorganisms such as bacteria, fungi, actinomycetes, and some unidentified microbes are reported to be involved in glyphosate degradation (Borgard and Gimsing, 2008). Furthermore, as reported by Gimsing et al. (2004) for *Pseudomonas* sp., the degradation rate is strongly correlated with the population size of soil microbes. With regard to degradation in aqueous mediums, Lipok et al. (2007) concluding their findings with mixed culture of *Spirulina* spp., reported that the species exhibited a remarkable ability to degrade glyphosate, where the rate of glyphosate disappearance from the medium was independent of its initial concentration. They suggested that the degradative pathway for glyphosate in *Spirulina* spp. might differ from those exhibited in other bacteria. According to them, occurrence of herbicide metabolism in *Spirulina* is evident, because the species can grow in a medium containing phosphonate as the only source of phosphorus, where the rate of herbicide transformation was found to be dependent upon the cells' phosphorus status. Lipok et al. (2009) reconfirmed the ability of the cyanobacterium *Streptomyces platensis* and bacterium *Streptomyces lusitanus* to catalyze glyphosate metabolism. According to Forlani et al. (2008), four cyanobacterial strains (*Anabaena* sp., *Lentinula boryana*, *Microcystis aeruginosa*, and *Nostoc punctiforme*) out of the six strains studied were able to use the glyphosate as the only source of phosphorus. Dyhrman et al. (2006) too stated the existence of phosphorous-dependent glyphosate transformation with marine cyanobacterium *Trichodesmium erythraeum*. Glyphosate as a source of nitrogen for microorganism was also reported (Klimek et al., 2001) with *Penicillium chrysogenum* and then with *Alternaria alternate* (Lipok et al., 2003). However, reports on the utilization of glyphosate as a source of nitrogen by cyanobacteria are not yet available in the literature. Elaborating their findings with cyanobacterium *Anabaena variabilis*, Ravi and Balakumar (1998) reported that extracellular phosphatases are able to hydrolyze the C–P bond of glyphosate; however, this claim has not been reiterated so far by the other authors. Forlani et al. (2008) based on their results, stated that extracellular phosphatases seems unlikely to contribute any substantial scale to glyphosate degradation. As described earlier, different steps are reported to be involved in the process of glyphosate degradation in different strains of microorganisms. Some of them in fact can utilize glyphosate as a source of nutrient. In this regard, cyanobacterial strains that possess the ability to use this phosphonate as a source of phosphorus is of practical significance, because such strains could effectively be employed in treating the problematic waters.

3.6.1 Mechanism of Chlorpyrifos Degradation by Cyanobacterium *Synechocystis* sp.

Chlorpyrifos is applied on a large scale in rice fields as a broad spectrum organophosphate insecticide for the control of foliar insects. Singh et al. (2011) studied the degradation of chlorpyrifos by a rice field cyanobacterium *Synechocystis* sp. strain PUPCCC 64 so that the organism is able to reduce insecticide pollution *in situ*. The organism tolerated chlorpyrifos up to 15 mg/L. Major fraction of chlorpyrifos was removed by the organism during the first day followed by slow uptake. Biomass, pH, and temperature influenced the insecticide removal and the organism exhibited maximum chlorpyrifos removal at 100 mg protein/L biomass, pH 7.0, and 30°C. The cyanobacterium metabolized chlorpyrifos producing a number of degradation products as evidenced by GC–MS chromatogram. One of the degradation products was identified as 3,5,6-trichloro-2-pyridinol. Six major peaks with the retention time 11.11, 12.29, 13.56, 14.52, 15.00, and 16.20 min were observed in the GC–MS chromatogram of cell extracts (Figure 3.3). The peak with retention time 16.20 min matched with standard chlorpyrifos, whereas peak with retention time 11.11 min corresponded to standard 3,5,6-trichloro-2-pyridinol (TCP) (Figure 3.3). Other unidentified peaks appear to be the products of further degradation of TCP (Figure 3.3). Analysis of supernatant and biomass wash for chlorpyrifos degradation products also showed the presence of TCP. In the experimental set up for the chlorpyrifos degradation study, 250 mL

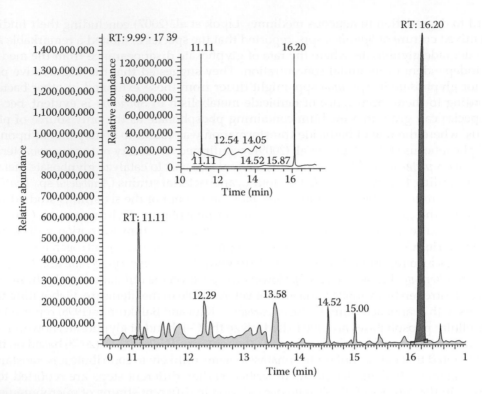

FIGURE 3.3
GC–MS chromatogram of cell extract of the *Synechocystis* sp. strain PUPCCC 64. Peaks with retention time (RT) 11.11 and 16.20 min correspond to TCP and chlorpyrifos. Inset: GC–MS chromatogram of standard TCP (RT 11.11 min) and chlorpyrifos (RT 16.20 min).

cultures contained a total of 1250 µg chlorpyrifos. After 60 h, chlorpyrifos recovery from cell extract, biomass wash, and medium was 113.12, 9.92, and 138.33 µg, respectively. Along with chlorpyrifos, the amount of TCP detected was 75.42, 38.16, and 613.33 µg, respectively. The presence of TCP in all extracts indicated that the organism degraded the insecticide intracellularly as well as extracellularly. Rapid depletion of chlorpyrifos from the culture medium may be due to extracellular degradation of chlorpyrifos. Detailed mechanism of chlorpyrifos degradation in cyanobacteria is not known. The mechanism of chlorpyrifos degradation in bacteria and fungi is fairly understood and a number of degradation products such as diethylthiophosphoric acid, TCP, chlorodihydro-2-pyridone, dihydroxypyridine, tetrahydro-2-pyridone, and maleamide semialdehyde have been identified (Singh and Walker, 2006). Detection of TCP as one of the chlorpyrifos byproduct during the current study indicates that chlorpyrifos degradation mechanism in cyanobacterium may be similar as in bacteria.

3.6.2 Mechanism of Anilofos Degradation by Cyanobacterium *Synechocystis* sp. Strain PUPCCC 64

Anilofos (*S*-[2-[(4-chlorophenyl)(1-methylethyl)amino]-2-oxoethyl]O,O-dimethylphosphorodithioate) belongs to the organophosphate group and is widely used as a preemergence and early postemergence herbicide for the control of annual grasses, sedges, and some broad-leaved weeds in transplanted and direct-seeded rice crops. Singh et al. (2013)

studied anilofos tolerance and its mineralization by the common rice field cyanobacterium *Synechocystis* sp. strain PUPCCC 64. The organism tolerated anilofos up to 25 mg/L. The herbicide caused inhibitory effects on photosynthetic pigments of the test organism in a dose-dependent manner. The test organism showed intracellular uptake and metabolized the herbicide. Uptake of herbicide by test organism was fast during the initial 6 h followed by slow uptake until 120 h. The organism exhibited maximum anilofos removal at 100 mg protein/L, pH 8.0, and 30°C. Its growth in phosphate-deficient basal medium in the presence of anilofos (2.5 mg/L) indicated that herbicide was used by the strain PUPCCC 64 as a source of phosphate.

The results indicated that neither did *Synechocystis* sp. strain PUPCCC 64 form any degradation products nor released into the medium. Of the total 1000 µg anilofos (peak area 7.96105 AU [absorbance units]; Figure 3.4a) present in culture medium at the start of the experiment, only 69.7 µg (peak area 0.556105 AU; Figure 3.4b) and 91.6 µg (peak area 0.736105 AU; Figure 3.4c) herbicide was recovered from biomass and supernatant, respectively, after 4 days. From these results, it follows that nearly 839 µg anilofos seems to have been mineralized by the organism into CO_2 and H_2O. It is possible that *Synechocystis* sp. strain PUPCCC 64 uses the released phosphate component of anilofos as a nutrient. To test this hypothesis, the PUPCCC 64 strain was grown for 12 days in anilofos (2.5 mg/L) phosphate-deficient basal medium. Whereas strain PUPCCC 64 did not grow in phosphate-deficient medium after day 4, it survived at a slow growth rate in the presence of anilofos until day 12. This confirmed that strain PUPCCC 64, in the absence of phosphate in the medium, used the phosphate component of anilofos as a nutrient.

No reports indicating route(s) of anilofos degradation in microorganisms are available. Since *Synechocystis* sp. strain PUPCCC 64 grew in phosphate-deficient anilofos supplemented medium, it is possible that this strain causes hydrolytic cleavage of P–S aryl linkage in anilofos releasing phosphate to be utilized as a nutrient. From the above discussion, it is clear that many environmental cyanobacteria and microalgae have developed a tolerance to otherwise toxic pesticides including herbicides and even are able to utilize the degradation products to sustain growth.

3.7 Biodegradation of Pesticides Using Recombinant Cyanobacteria

Pesticidal degradative genes in microbes have been found to be located on plasmids, transposons, and/or on chromosomes. Recent studies have provided clues to the evolution of degradative pathways and the organization of catabolic genes, thus making it much easier to develop genetically engineered microbes for the purpose of decontamination. Genetic manipulation offers a way of engineering microorganisms to deal with a pollutant, including pesticides that may be present in the contaminated sites. The simplest approach is to extend the degradative capabilities of existing metabolic pathways within an organism either by introducing additional enzymes from other organisms or by modifying the specificity of the catabolic genes already present. Continuous efforts are required in this direction, and at present several bacteria capable of degrading pesticides have been isolated from the natural environment. Catabolic genes responsible for the degradation of several xenobiotics, including pesticides, have been identified, isolated, and cloned into various other organisms such as algae, fungi, and so on. In addition, recombinant DNA studies have made it possible to develop DNA probes that are being

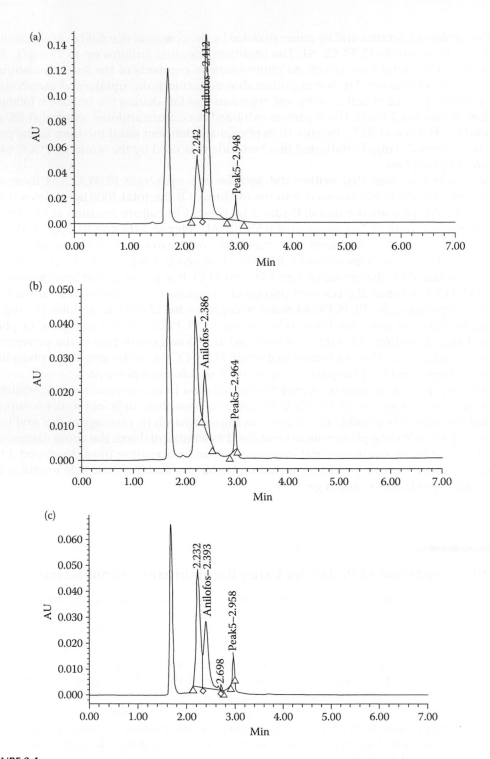

FIGURE 3.4
HPLC chromatogram showing the peaks of (a) standard anilofos, (b) cell extract, and (c) supernatant. First large peak in the HPLC chromatogram represents solvent used to dissolve extracted anilofos.

used to identify microbes from diverse environmental communities with a unique ability to degrade pesticides.

Genetic engineering in filamentous N_2-fixing cyanobacteria usually involves *Anabaena* sp. PCC 7120 and several other nonaggregating species. Mass culture and harvest of such species are more energy consuming relative to aggregating species. To establish a gene transfer system for aggregating species, Qiong et al. (2010) tested many species of *Anabaena* and *Nostoc*, and identified *Nostoc muscorum* FACHB244 as a species that can be genetically manipulated using the conjugative gene transfer system. To promote biodegradation of organophosphorus pollutants in aquatic environments, they introduced a plasmid containing the organophosphorus-degradation gene (*opd*) into *Anabaena* sp. PCC 7120 and *N. muscorum* FACHB244 by conjugation. The *opd* gene was driven by a strong promoter, P*psbA*. From both species, they obtained transgenic strains having organophosphorus-degradation activities. At 25°C, the whole-cell activities of the transgenic *Anabaena* and *Nostoc* strains were 0.163 ± 0.001 and 0.289 ± 0.042 unit/µg Chl *a*, respectively. However, most colonies resulting from the gene transfer showed no activity. Polymerase chain reaction (PCR) and DNA sequencing revealed deletions or rearrangements in the plasmid in some of the colonies (Figure 3.5). Expression of the green fluorescent protein gene from the same promoter in *Anabaena* sp. PCC 7120 showed similar results (Figure 3.6). These results suggested that there is the potential to promote the degradation of organophosphorus pollutants with transgenic cyanobacteria and that selection of high-expression transgenic colonies is important for genetic engineering of *Anabaena* and *Nostoc* sp. For the first time, they established a gene transfer and expression system in an aggregating filamentous N_2-fixing cyanobacterium. The genetic manipulation system of *N. muscorum* FACHB244 could be utilized in the elimination of pollutants and large-scale production of valuable proteins or metabolites.

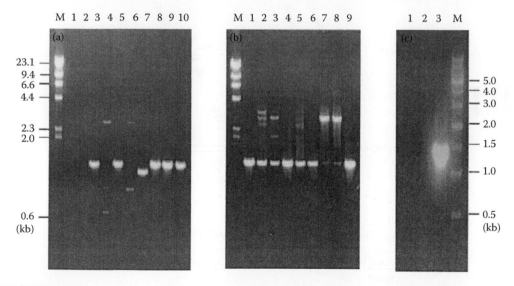

FIGURE 3.5
PCR examination of P*psbA-opd* in *Anabaena* 7120 and *Nostoc* 244 transformed with pHB303. PCR primers used were pRL25-1 and pDU1-3; the positive control was the PCR product generated using pHB303 as the template; molecular weight markers (M) were fragments of lambda DNA cut with Hind(III) (a and b) or 0.5-kb ladders (c); electrophoresis was performed with 0.7% agarose gel and 0.5× TAE buffer. (a) *Anabaena* 7120 transformed with pHB303. Lanes 1–4 and 6–9, Opd – strains; lane 5, an Opd + strain; lane 10, positive control. (b) *Nostoc* 244 transformed with pHB303. Lanes 1–3 and 5–8, Opd – strains; lane 4, an Opd + strain; lane 9, the positive control. (c) Lane 1, *Anabaena* 7120; lane 2, *Nostoc* 244; lane 3, positive control.

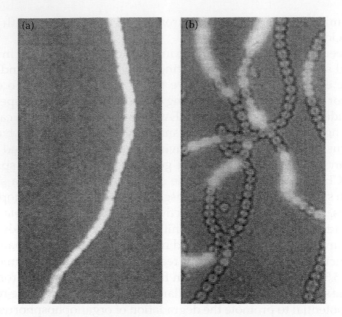

FIGURE 3.6
GFP fluorescence in two cell lines of *Anabaena* 7120 (pHB1153). An image of GFP-based fluorescence is superimposed upon the bright-field image. (a) Cell line showing GFP fluorescence in all cells. (b) One of seven cell lines that showed GFP fluorescence in a small proportion of cells.

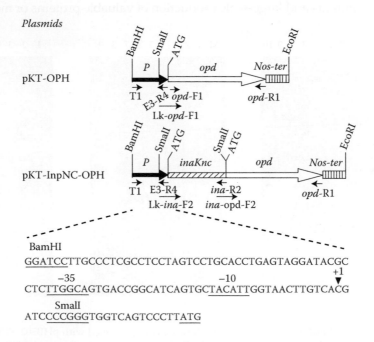

FIGURE 3.7
Plasmids harboring the *opd* gene. The gene cassettes harboring the PtRNA promoter (P), *inaKnc* gene encoding the truncated ice nucleation protein, *opd* gene encoding organophosphorus hydrolase, and nopaline synthase terminator (Noster) are shown. The gene cassettes were cloned into pKGT plasmid. Locations of primers used in this study are indicated. The PtRNA promoter sequence including the –35 and –10 regions, starting site of mature transcript (+1), and start codon (ATG) are indicated.

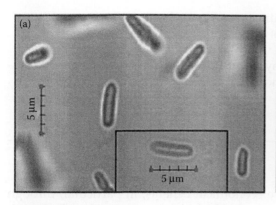
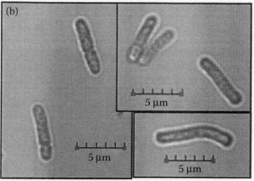

FIGURE 3.8
CLSM images of *Synechococcus* cells expressing OPH. *Synechococcus* cells harboring pKT-OPH (a) or pKT-InpNC-OPH (b).

The *opd* gene, encoding organophosphorus hydrolase (OPH) from *Flavobacterium* sp. capable of degrading a wide range of OPs, was surface and intracellular expressed in *Synechococcus* PCC 7942, a prime example of photoautotrophic cyanobacteria (Figure 3.7). OPH was displayed on the cyanobacterial cell surface using the truncated ice nucleation protein as an anchoring motif. A minor fraction of OPH was displayed onto the outermost surface of cyanobacterial cells, as verified by immune staining visualized under confocal laser scanning microscopy (Figure 3.8) and OPH activity analysis; however, a substantial fraction of OPH was buried in the cell wall, as demonstrated by proteinase K and lysozyme treatments. The cyanobacterial outer membrane acts as a substrate (paraoxon) diffusion barrier affecting whole-cell biodegradation efficiency. After freeze-thaw treatment, permeabilized whole cells with intracellular-expressed OPH exhibited 14-fold higher bioconversion efficiency (V_{max}/K_m) than that of cells with surface-expressed OPH (Wipa and Fa-Aroonsawat, 2008). As cyanobacteria have simple growth requirements and are inexpensive to maintain, expression of OPH in cyanobacteria may lead to the development of a low-cost and low-maintenance biocatalyst that is useful for detoxification of OPs.

3.8 Conclusion

Cyanobacteria serve as agents for the intrinsic or engineered processes of remediation from pesticides. Advantages of the use of cyanobacteria include the following:

1. Lower cost compared with the use of heterotrophic microorganisms, since cyanobacteria use carbon dioxide the source of carbon and sunlight the source of energy; some species also can fix atmospheric nitrogen.

2. Degradation of target pollutants to very low levels due to the cometabolic nature of degradation processes.

3. The possibility of combining aerobic and anaerobic processes within one microorganism, since filamentous, nitrogen-fixing cyanobacteria maintain aerobic metabolism in their vegetative cells and anaerobic conditions in heterocysts.

4. The possibility of reprocessing and utilization of cyanobacterial biomass, generated during remediation, by a variety of biotechnologies.

It is also possible that further research may lead to the discovery of novel genetic systems and enzymes of biodegradation in cyanobacteria.

3.9 Perspective

The use of pesticides has brought impressive economical benefits, in particular, through enhanced agricultural productivity. However, due to the high percentage of applied pesticides often reported to be deposited on nontarget areas, contamination of soil and water is inevitable. Under this background, frequent assessments on the impacts to nontarget organisms are of prime importance. Biological degradation by organisms (fungi, bacteria, cyanobacteria, viruses, and protozoa) can efficiently remove pesticides from the environment, especially organochlorines, organophosphates, and carbamates used in agriculture. The enzymatic degradation of synthetic pesticides with microorganisms represents the most important strategy for pollutant removal, in comparison with nonenzymatic processes.

However, the vast majority of the current investigations dealing with the impact of pesticides on aquatic organisms and mechanism involved in degradation are based on laboratory bioassays. Furthermore, toxicity studies are often targeted on individual strains, because such findings assist in assessing the direct impacts of pesticides on the organisms of concern. However, generation of remediation strategies for contaminated waters, merely based on such findings would not be advisable. Thus, field studies conducted in natural environments are encouraged as they could come up with much broader understanding as to how cyanobacteria cope with pesticide toxicity.

References

Abdullah AR, Bajet CM, Matin, MA, Khan, DD, and Sulaiman AH. 1997. Ecotoxicology of pesticides in the tropical paddy field system. *Environment Toxicology and Chemistry* 16: 59–70.

Abraham WR, Nogales B, Golyshin PN, Pieper DH, and Timmis KN. 2002. Polychlorinated biphenyl-degrading microbial communities and sediments. *Current Opinion in Microbiology* 5: 246–253.

Alexander M. 1994. *Biodegradation and Bioremediation*. Academic Press, San Diego, CA.

Alves SA, Albergaria JT, Fernandes DV, Alvim-Ferraz MC, and Delerue-Matos C. 2010. Remediation of soils combining soil vapor extraction and bioremediation. *Chemosphere* 80: 823–828.

Araujo ASF, Monteiro RTR, and Abarkeli RB. 2003. Effect of glyphosate on the microbial activity of two Brazilian soils. *Chemosphere* 52: 799–804.

Arunakumara KKIU, Buddhi CW, and Min-Ho Y. 2013. Metabolism and degradation of glyphosate in aquatic cyanobacteria: A review. *African Journal of Microbiology Research* 7: 4084–4090.

Badii M and Landeros J. 2007. Plaguicidas que afectan la salud humana y la sustentabilidad. *CULCYT/ Toxicología de Plaguicidas* 4(19): 21–34.

Bansal OP. 2012. Degradation of pesticides. In: Hamir S and Nollet LML (Eds.), *Pesticides Evaluation of Environmental Pollution*. CRC Press, New York, pp. 47–77.

Barton JWT, Kuritz LEO, Connor CYM, Maskarinec MP, and Davison BH. 2004. Reductive transformation of methyl parathion by the cyanobacterium *Anabaena* sp. strain PCC7120. *Applied Microbiology and Biotechnology* 65: 330–335.

Becker JG and Seagren EA. 2010. Bioremediation of hazardous organics. In: Mitchell R and Ji-Dong G. (Eds.), *Environmental Microbiology*. Wiley-Blackwell, New Jersey, pp. 177–212.

Bhashkar M, Sreenivasulu C, and Venkatesvarlu K. 1992. Interactions of monocrotophos and quinalphos with *Anabaena torulosu* isolated from rice soil. *Biochemistry International* 28: 767–773.

Borggaard OK and Gimsing AL. 2008. Fate of glyphosate in soil and the possibility of leaching to ground and surface waters. *Pest Management Science* 64: 441–456.

Carr NG and Whitton BA. 1982. *The Biology of Cyanobacteria*. Blackwell Scientific Publications Ltd., Oxford.

Castenholz RW and JB Waterbury 1989. In: Staley JT, Bryant MP, Pfennig N, and Holt JG (Eds.), *Oxygenic Photosynthetic Bacteria. Group I. Cyanobacteria. Bergey's Manual of Systematic Bacteriology*. The Williams & Wilkins Co., Baltimore, pp. 1710–1727.

Chaillan F, Gugger M, Saliot A, Coute A, and Oudot J. 2006. Role of cyanobacteria in the biodegradation of crude oil by a tropical cyanobacterial mat. *Chemosphere* 62: 1574–1582.

Chino-Flores C, Dantán-González E, Vázquez-Ramos A, Tinoco-Valencia R, Díaz-Méndez R, Sánchez-Salinas E, Castrejón-Godínez ML et al. 2012. Isolation of the *opdE* gene that encodes for a new hydrolase of *Enterobacter* sp. capable of degrading organophosphorus pesticides. *Biodegradation* 23: 387–397.

Damalas CA. 2009. Understanding benefits and risks of pesticide. *Scientific Research and Essays* 4: 945–949.

Damalas CA and Eleftherohorinos IG. 2011. Pesticide exposure, safety issues, and risk assessment indicators. *International Journal of Environmental Research and Public Health* 8: 1402–1419.

Debarati P, Gunjan P, Janmejay P, and Rakesh VJK. 2005. Accessing microbial diversity for bioremediation and environmental restoration. *Trends in Biotechnology* 23: 135–142.

Diaz E. 2004. Bacterial degradation of aromatic pollutants: A paradigm of metabolic versatility. *International Journal of Microbiology* 7:173–180.

Doggett SM and Rhodes RG. 1991. Effects of a diazinon formulations on unialgal growth rates and phytoplankton diversity. *Bulletin of Environmental Contamination and Toxicology* 47: 36–42.

Duke SO. 2011. Glyphosate degradation in glyphosate-resistant and -susceptible crops and weeds. *Journal of Agricultural Food Chemistry* 59: 5835–5841.

Dyhrman ST, Chapell PD, Haley ST, Moffett JW, Orchard ED, and Waterbury JB. 2006. Phosphonate utilization by the globally important marine diazotroph *Trichodesmium*. *Nature* 439: 68–71.

El-Enany AE and Issa AA. 2000. Cyanobacteria as a biosorbent of heavy metals in sewage water. *Environment Toxicology and Pharmacology* 8: 95–101.

El Sebae AH. 1993. Special problems experienced with pesticide use in developing countries. *Regulatory Toxicology and Pharmacology* 17: 287–291.

Eskew DL, Catano-Anolles G, Bassam BJ, and Gresshoff PM. 1993. DNA amplification fingerprinting of *Azolla-Anabaena* symbiosis. *Plant Molecular Biology* 21: 363–373.

Finley SD, Broadbelt LJ, and Hatzimanikatis V. 2010. In silico feasibility of novel biodegradation pathways for 1,2,4-trichlorobenzene. *BMC Systems Biology* 4: 4–14.

Fleming H and Haselkorn R. 1974. The program of protein synthesis during heterocyst differentiation in nitrogen-fixing blue-green algae. *Cell* 3 159–170.

Fogg GE. 1987. Marine planktonic cyanobacteria. In: Fay P and Van Baalen C. (Eds.), *The Cyanobacteria*. Elsevier Biomedical Press, Amsterdam, pp. 393–413.

Forlani G, Pavan M, Gramek M, Kafarski P, and Lipok J. 2008. Biochemical bases for a widespread tolerance of cyanobacteria to the phosphonate herbicide glyphosate. *Plant Cell Physiology* 49: 443–456.

Gavrilescu M. 2005. Fate of pesticides in the environment and its bioremediation. *Engineer in Life Science* 5: 497–526.

Gebhardt JS and Nierzmicki-Bauer SA. 1991. Identification of a common cyanobacterial symbiont associated with *Azolla* spp. through molecular and morphological characterization of free living and symbiotic cyanobacteria. *Applied and Environmental Microbiology* 57: 2141–2146.

Gimsing AL, Borggaard OK, Jacobsen OS, and Sørensen AJ. 2004. Chemical and microbiological soil characteristics controlling glyphosate mineralization in Danish surface soils. *Applied Soil Ecology* 27: 233–242.

Hayes VEA, Ternan NG, and McMullan G. 2000. Organophosphate metabolism by a moderately halophilic bacterial isolate. *FEMS Microbiology Letters* 186: 171–175.

Hussain S, Siddique T, Arshad M, and Saleem M. 2009. Bioremediation and phytoremediation of pesticides: Recent advances. *Critical Review in Environmental Science and Technology* 39: 843–907.

Iranzo M, Sain-Pardo I, Boluda R, Sanchez J, and Mormeneo S. 2001. The use of microorganisms in environmental remediation. *Annals of Microbiology* 51: 135–143.

Jain RK, Kapur M, Labana S, Lal B, Sarma PM, Bhattacharya D, and Thakur IS. 2005. Microbial diversity: Application of microorganisms for the biodegradation of xenobiotics. *Current Science* 89: 101–112.

Kafarski P, Lejczak B, and Forlani G. 2000. Biodegradation of pesticides containing carbon to phosphorus bond. In: Hall JC, Hoagland RE, and Zablotowicz RM (Eds.), *Pesticide Biotransformation in Plants and Microorganisms. Similarities and Divergencies*. ACS, Washington, DC, pp. 145–163.

Kannaiyan S, Aruna SJ, Kumari SMP, and Hall DO. 1997. Immobilized cyanobacteria as a biofertilizer for rice crops. *Journal of Applied Phycology* 9: 167–174.

Klimek M, Lejck B, Kafarski P, and Forlani G. 2001. Metabolism of the phosphonate herbicide glyphosate by a non-nitrate utilizing strain of *Penicillium chrysogenum*. *Pesticide Management Science* 57: 815–821.

Kuritz T, Bocanera LV, and Rivera NS. 1997. Dechlorination of lindane by the cyanobacterium *Anabaena* sp. strain PCC 7120 depends on the function of the *nir* operon. *Journal of Bacteriology* 179: 3368–3370.

Kuritz T and Wolk CP. 1995. Use of filamentous cyanobacteria for biodegradation of organic pollutants. *Applied Environmental Microbiology* 61: 234–238.

Latifi AM, Khodi S, Mirzaei M, Miresmaeili M, and Babavalian H. 2012. Isolation and characterization of five chlorpyrifos degrading bacteria. *African Journal of Biotechnology* 11: 3140–3146.

Lefebvre DD, Kelly D, and Budd K. 2007. Biotransformation of Hg(II) by cyanobacteria. *Applied Environmental Microbiology* 73: 243–249.

Lerbs W, Stock M, and Parthier B. 1990. Physiological aspects of glyphosate degradation in *Alcaligenes* sp. strain GL. *Archives of Microbiology* 153: 146–150.

Lipok J, Dombrovska L, Wieczorek P, and Kafarski P. 2003. The ability of fungi isolated from stored carrot seeds to degrade organo-phosphonate herbicides. In: Del Re AAM, Capri E, Padovani L, and Trevisan M. (Eds.), *Pesticide in Air, Plant, Soil and Water System. Proceedings of the XII Symposium Pesticide Chemistry*, Piacenza, Italy.

Lipok J, Owsiak T, Młynarz P, Forlani G, and Kafarski P. 2007. Phosphorus NMR as a tool to study mineralization of organophosphonates—The ability of *Spirulina* spp. to degrade glyphosate. *Enzyme Microbial Technology* 41: 286–291.

Lipok J, Studnik H, and Gruyaert S. 2010. The toxicity of roundups 360 SL formulation and its main constituents: Glyphosate and isopropylamine towards non-target water photoautotrophs. *Ecotoxicology Environmental Safety* 73: 1681–1688.

Lipok J, Wieczorek D, Jewgiński M, and Kafarski P. 2009. Prospects of *in vivo* ^{31}P NMR method in glyphosate degradation studies in whole cell system. *Enzyme Microbial Technology* 44: 11–16.

Majewski MS and Capel PD. 1995. *Pesticides in the Atmosphere*. Ann Arbor Press, Ann Arbor, MI.

Megharaj M, Madhavi DR, Sreeinvasaulu C, Umamaheswari A, and Venkateswarlu K. 1994. Biodegradation of methylparathion by soil isolates of microalgae and cyanobacteria. *Bulletin of Environmental Contamination and Toxicology* 53: 292–297.

Megharaj M, Ramakrishnan B, Venkateswarlu K, Sethunathan N, and Naidu R. 2011. Bioremediation approaches for organic pollutants: A critical perspective. *Environment International* 37: 1362–1375.

Megharaj M, Venkateswarlu K, and Rao AS. 1987. Metabolism of monocrotophos and quinalphos by algae isolated from soil. *Bulletin of Environmental Contamination and Toxicology* 39: 251–256.

Mishra V, Lal R, and Srinivasan S. 2001. Enzymes and operons mediating xenobiotic degradation in bacteria. *Critical Reviews in Microbiology* 27: 133–166.

Narro ML, Cerniglia CE, Van Baalen C, and Gibson DT. 1992. Metabolism of phenanthrene by the marine cyanobacterium *Agmenellum quadruplicatum* PR-6. *Applied Environmental Microbiology* 58: 1351–1359.

Nirmal Kumar JI, Amb MK, Kumar RN, Bora A, and Khan SR. 2013. Studies on biodegradation and molecular characterization of 2,4-D ethyl ester and pencycuron induced cyanobacteria by using GC–MS and 16S rDNA sequencing. *Proceedings of the International Academy of Ecology and Environmental Sciences* 3: 1–24.

Ortiz-Hernández ML, Quintero-Ramírez R, Nava-Ocampo AA, and Bello-Ramírez AM. 2003. Study of the mechanism of *Flavobacterium* sp. for hydrolyzing organophosphate pesticides. *Fundamental Clinical Pharmacology* 17: 717–723.

Ortiz-Hernández ML, Sánchez-Salinas E, Olvera-Velona A, and Folch-Mallol JL. 2011. Pesticides in the environment: Impacts and its biodegradation as a strategy for residues treatment. In: Stoytcheva M. (Ed.), *Pesticides-Formulations, Effects, Fate*. In Tech, Croatia, pp. 551–574.

Orus MI and Marco E. 1991. Disappearance of trichlorophon from cultures with different cyanobacteria. *Bulletin of Environmental Contamination and Toxicology* 47: 392–397.

Paul D, Pandey G, Pandey J, and Jain RK. 2005. Accessing microbial diversity for bioremediation and environmental restoration. *Trends in Biotechnology* 23: 135–142.

Pipke R, Amrhein N, Jacob GS, Kishore GM, and Schaefer J. 1987. Metabolism of glyphosate in an *Arthrobacter* sp. GLP-1. *European Journal of Biochemistry* 165: 267–273.

Porzio E, Merone L, Mandricha L, Rossia M, and Manco G. 2007. A new phosphotriesterase from *Sulfolobus acidocaldarius* and its comparison with the homologue from *Sulfolobus solfataricus*. *Biochemistry* 89: 625–636.

Qiong LI, Qing T, Xudong XU, and Hong GAO. 2010. Expression of organophosphorus-degradation gene (*opd*) in aggregating and non-aggregating filamentous nitrogen-fixing cyanobacteria. *Chinese Journal of Oceanology and Limnology* 28: 1248–1253.

Qiu X, Zhong Q, Li M, Bai W, and Li B. 2007. Biodegradation of p-nitrophenol by methyl parathion-degrading *Ochrobactrum* sp. B2. *International Biodeterioration and Biodegradation* 59: 297–301.

Quesada A and Valiente EF. 1996. Relationship between abundance of N2-fixing cyanobacteria and environmental features of Spanish rice fields. *Microbial Ecology* 31: 59–71.

Ramakrishnan B, Megharaj M, Venkateswarlu K, Sethunathan N, and Naidu R. 2011. Mixtures of environmental pollutants: Effects on microorganisms and their activities in soils. *Reviews of Environmental Contamination and Toxicology* 211: 63–120.

Ravi V and Balakumar H. 1998. Biodegradation of the C–P bond in glyphosate by the cyanobacterium *Anabaena variabilis*. *Journal of Scientific and Industrial Research* 57: 790–794.

Rippka RJ, Deruelles JB, Waterbury M, Herdman RY, and Stanier RY. 1979. Generic assignments, strain histories and properties of pure cultures of cyanobacteria. *Journal of General Microbiology* 111: 1–61.

Riya P and Jagatpati T. 2012. Biodegradation and bioremediation of pesticides in soil: Its objectives, classification of pesticides, factors and recent developments. *World Journal of Science and Technology* 2: 36–41.

Rojano-Delgado AM, Ruiz-Jiménez J, Castro MDL, and De Prado R. 2010. Determination of glyphosate and its metabolites in plant material by reversed-polarity CE with indirect absorptiometric detection. *Electrophoresis* 31: 1423–1430.

Schoefs O, Perrier M, and Samson R. 2004. Estimation of contaminant depletion in unsaturated soils using a reduced-order biodegradation model and carbon dioxide measurement. *Applied Microbiology and Biotechnology* 64: 53–61.

Schroll R, Brahushi R, Dorfler U, Kuhn S, Fekete J, and Munch JC. 2004. Biomineralisation of 1,2,4-trichlorobenzene in soils by an adapted microbial population. *Environmental Pollution* 127: 395–401.

Schuette J. 1998. *Environmental Fate of Glyphosate*. DPR, Sacramento, CA, p. 13.

Scott C, Pandey G, Hartley CJ, Jackson CJ, Cheesman MJ, Taylor MC, Pandey R et al. 2008. The enzymatic basis for pesticide bioremediation. *Indian Journal of Microbiology* 48: 65–79.

Singh PK. 1973. Effect of pesticides on blue-green algae. *Archives of Microbiology* 89: 317–320.

Singh BK and Walker A. 2006. Microbial degradation of organophosphorus compounds. *FEMS Microbiological Review* 30: 428–471.

Singh DP, Khattar JIS, Kaur M, Kaur G, Gupta M, and Singh Y. 2013. Anilofos tolerance and its mineralization by the cyanobacterium *Synechocystis* sp. strain PUPCCC 64. *PLoS One* 8(1): e53445.

Singh DP, Khattar JIS, Nadda J, Singh Y, Garg A, Kaur N, and Gulati A. 2011. Chlorpyrifos degradation by the cyanobacterium *Synechocystis* sp. strain PUPCCC. *Environmental Science and Pollution Research* 18: 1351–1359.

Sorkhoh N, Al-Hasan R, Radwan S, and Hopner T. 1992. Self-cleaning of the Gulf. *Nature (London)* 359: 109.

Subramanian G, Sekar S, and Sampoornam S. 1994. Biodegradation and utilization of organophosphorus pesticides by cyanobacteria. *International Biodegradation and Biodeterioration* 33: 129–143.

Theriot CM and Grunden AM. 2010. Hydrolysis of organophosphorus compounds by microbial enzymes. *Applied Microbiology and Biotechnology* 89: 35–43.

Torres RD. 2003. El papel de los microorganismos en la biodegradación de compuestos tóxicos. *Ecosistemas* 2: 1–5.

Trigo A, Valencia A, and Cases I. 2009. Systemic approaches to biodegradation. *FEMS Microbiological Reviews* 33: 98–108.

Wipa C and Fa-Aroonsawat S. 2008. Biodegradation of organophosphate pesticide using recombinant cyanobacteria with surface- and intracellular-expressed organophosphorus hydrolase. *Journal of Microbial Biotechnology* 18: 946–951.

Wolk CP, Ernst A, and Elhai J. 1994. Heterocyst metabolism and development. In: Bryant D. (Ed.), *Molecular Biology of Cyanobacteria*. Kluwer Academic Publishers, Dordrecht, The Netherlands, pp. 769–823.

4

Biomedical Waste: Its Effects and Safe Disposal

Bamidele T. Odumosu

CONTENTS

4.1 Introduction

Waste generated by health-care activities accounts for more than half the deaths associated with waste-related diseases. Public health is meant to be protected by the health sector through clinics and hospitals which provide health services to the general public, help in the management of infections, and disseminate relevant information regarding health issues. The health sectors' role in the risk associated with communal disease epidemics, which are a direct result of infectious and hazardous waste from medical facilities, is a pressing concern globally. The waste generated in the course of health-care activities includes various biomedical materials from used needles and syringes to blood, pharmaceuticals, human body parts, toxic chemicals, and so on. These and many others constitute a high risk for human infection due to their highly infectious and hazardous content.

FIGURE 4.1
Rag pickers scavenging for items among waste by the roadside in Nigeria.

Efficient handling and management of medical waste is a major problem globally because of growing populations and increasing medical attention at various health facilities that are generating more and more waste in the environment. More than half of the nations of the world, at the least, are experiencing environmental challenges associated with waste disposal. Unlike hazardous industrial waste which is normally encountered outside urban settlements, biomedical waste is relatively closer to residential zones due to the location of clinics, pharmacies, and hospitals in the heart of many cities.

It has been shown that this waste can be washed down and thus find its way into rivers, streams and underground water, which could lead to epidemics of water-borne diseases and influence the increase of multiple drug-resistant pathogens in the communal water supply. Direct contact of contaminated waste with sharp edges such as needles and broken vials capable of damaging the skin and introducing infectious agents are some of the risks associated with improper disposal of waste from medical sources. Waste workers and rag pickers who rummage through all kinds of waste material while trying to salvage items for sale are often the worst affected in the process (Figure 4.1 shows a rag picker scavenging for items in waste by the roadside). They may be exposed to poisonous materials and infectious objects and so are often in danger of serious infections and diseases such as hepatitis, plague, cholera, etc., which pose grave health risks.

Some of the major reasons for the ineffective management and control of biomedical waste in many countries are financial and technological constraints, and lack of education and proper training of personnel on duty. Many countries still facing biomedical waste challenges today are largely the developing ones due to their slow pace in technological advancement and civic development. Although developed countries are not often left out of this dire situation, the frequency of occurrence is far lower than in developing countries.

4.2 Definition of Terms

Several studies and reports on health-care-related waste have emerged on the issue of hospital waste and its management; as a result, many conflicting opinions have come up regarding the acceptance of definitions for specific terms associated with this topic, for example, regulated medical waste, infectious waste, biohazardous waste, and biological waste. Unfortunately, there are no universally agreed definitions for "hospital waste" and "medical waste." According to the Environmental Protection Agency (EPA) and Center for Disease Control and Prevention (CDC, 2003), hospital waste refers to all waste, biological or nonbiological, that is discarded and not intended for further use. Medical waste refers to materials generated as a result of patient diagnosis, treatment, or immunization of human beings or animals. Given the diversity of scientific opinions and interests of individuals, groups and agencies involved in the medical waste issue (e.g., physicians, health departments, hospitals, trade unions, state, and federal legislators), these differences are not surprising. However, in this chapter, we will be considering medical waste as a subset of hospital waste and continue to use biomedical waste in place of these descriptive terms.

4.3 Sources of Biomedical Waste

There are different sources of waste generated from the biomedical point of view. It is important to stress that the hospital is not the only source of biomedical waste; there are other health-care facilities of significant concern (Figure 4.2 shows various sources of biomedical waste). As said earlier, hospital waste refers to biological and nonbiological waste that is disposed of and is no longer intended for further use in the hospital. Although hospitals are considered the primary generators of waste by volume, there are also several

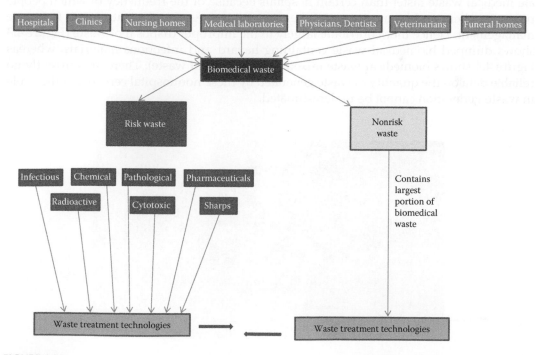

FIGURE 4.2
Flowchart showing sources of biomedical waste.

FIGURE 4.3
Dumped medical waste containing hazardous and infectious matter.

nonhospital facilities that generate waste. There are numerous private dentists' and private physicians' offices, medical clinics, veterinarian centers, and medical laboratories that generate hazardous and infectious medical waste across the globe. Some of these sites generate medical waste faster than certain hospitals because of the frequency of which people patronize them and because most of them do not possess an efficient system for waste management. Hence they contribute largely to the improper disposal of waste (Figure 4.3 shows dumped biomedical waste containing hazard and infectious materials, whereas Figure 4.4 shows biomedical waste mixed with municipal waste). There are currently no reliable data on the quantity of waste generated by these nonhospital centers but their role in waste generation cannot be underestimated.

FIGURE 4.4
Medical waste mixed with municipal waste poses a threat to the public.

4.4 Types of Hospital Waste

Normally, waste is mostly categorized based on the risk it holds. Different authors, scientists, observers, and regulatory bodies have classified medical waste in various ways and from different perspectives. The classification is (i) based on their form or state, that is, solid or liquid form, (ii) based on their sources, and (iii) some focused on their potential risk to the public. The Ministry of Environment and Forests in India and other regulatory bodies have also described 10 categories of biomedical waste (MoEF, 1998; Eigenheer and Zanon, 1991). These categories cover all the types of waste that are generated in all medical facilities including hospitals, which are known to generate the highest amount of waste. All these classifications are mostly based on what is usually obtained in each country or geographical zone. However, the definitions and classifications of the regulatory authorities of how a particular waste should be classified or handled, treated and disposed are usually adopted because of the serious ramifications involved. All categories and classifications of medical waste can be seen to generally fall into two main broad types: risk and nonrisk waste.

4.4.1 Risk Waste

Risk waste consists of several aspects of hospital and other biomedical wastes considered dangerous to the public because of the presence of certain components with the intrinsic ability of spreading infections or causing harm to the community. Risk waste may be subdivided into many categories, but it is important to note that each category is unique in its components, sources, and the potential hazard it may cause. We shall be discussing below the subdivisions of risk waste.

4.4.1.1 Infectious Waste

Infectious waste is biomedical waste that has been previously contaminated with any type of pathogen with sufficient virulence capable of infecting a susceptible host thereby resulting in an infectious disease. This includes bacteria, virus, parasite, or fungi that are capable of causing infection or spread of infectious diseases in the community. Some believe that infectious waste could just be classified as any waste with the presence of microorganisms. However, this approach needs careful consideration because certain factors are necessary for waste to be actually termed infectious. It is important for waste to harbor pathogenic microorganisms with virulence capabilities and in sufficient amount of dosage required for infection in a susceptible host. Such waste can be termed infectious.

4.4.1.2 Pathological Waste

This includes anatomical waste from human and animals usually generated from hospitals and veterinary hospitals. Examples include blood, body parts such as amputated legs or arms, placenta, and tissues. They are potentially dangerous with a high risk of disease and infection in a susceptible individual who has direct contact with them.

4.4.1.3 Microbiological/Clinical Waste

This is closely related to pathological waste in terms of high risk but the contents are quite different from pathological waste. It is usually generated from laboratories during the course of experiments and clinical trials. Examples include human and animal cell cultures from clinical experiments, microorganisms, blood collection tubes, body fluid, drainage bags, vials, used culture dishes, and other materials that were previously in contact with infectious agents.

4.4.1.4 Sharps

Sharps consist of both used and unused sharp objects such as hypodermic needles, syringes, scalpels, broken ampoules, and glassware. They are considered highly dangerous because they are not only capable of causing punctures and cuts on the skin but they also harbor dangerous pathogenic bacteria, which may be introduced into the body via wounds or punctures. Sharps are usually generated in the operation theatre where surgical activities are carried out. Other areas where they are generated include all wards and laboratories.

4.4.1.5 Pharmaceutical Waste

Under normal circumstances, pharmaceutical waste constitutes a very low percentage (3%) of medical waste. However, the unsafe disposal of such items may pose a significant threat. Pharmaceutical waste is generated by various means but not limited to the following: chemicals, substandard and expired drugs, preparations of drugs added to an intravenous solution, pharmaceutical products such as drugs. It is very important to note that pharmaceutical waste does not include empty glass ampoules, drugs, and other metabolic products excreted by patient undergoing therapy, empty pills bottles or strip packages from where the drug/capsules have been previously removed.

4.4.1.6 Chemical Waste

Chemical waste materials include those that are generated during the production of biological preparations such as disinfectants and insecticides, in medical, dental, and veterinary laboratories.

4.4.1.7 Radioactive Waste

These are solid, liquid, or gaseous forms of waste generated from the medical or research use of radiology, radionuclide (e.g., during radioimmunoassay), and other radiological procedures which are capable of emission of radiation at above the level set by regulatory authorities as exempt. Examples are found in nuclear medicine treatments, cancer related therapies, and medical devices that use radioactive isotopes.

4.4.1.8 Cytotoxic and Genotoxic Wastes

These include unused cytotoxic drugs, solid materials such as sharp objects, tissues, IV bags, and any other items which may have come into contact with a cytotoxic drug and/or carcinogenic matter.

4.4.2 Nonrisk Waste

Nonrisk waste is general waste that is generated within hospitals and they are similar to municipal waste from a normal home. They are considered noninfectious and nonhazardous and are generated by almost everyone in the hospital. Nonrisk waste includes office waste, food waste, that is, leftover food, fruits, vegetables, etc., aerosols, as well as noninfectious, nonanatomical waste from patient care areas (i.e., disposable diapers, pads, papers, cartons, gloves, trays, catheters/bags (empty), and casts).

4.4.3 Human Health and Environmental Concerns

Ensuring safe public health and a clean environment in the face of improper medical waste disposal is of great significance. While there are studies indicating unavailability of data to show the direct evidence of the harm done by the mishandling of medical waste, the dangers associated with these occurrences are obvious. Inefficient handling and management of medical waste may lead to problems such as spread of disease and infections, environmental pollution, etc., and the groups of people at the highest risk of such encounters are health-care staff, rag pickers and scavengers, municipal workers, and the public. Several ways in which medical waste pose dangers to the public shall be discussed below.

4.4.3.1 Spread of Infection and Disease

As discussed earlier in the previous section, medical waste often contains certain materials that are termed "infectious." Some of these materials have been discussed in the previous sections. There is a strong evidence of transmission of certain infectious agents such as blood-borne pathogens via medical waste (Collins and Kennedy, 1992; Simonsen et al., 1999). It has been estimated that the chances of infection after a needle-stick injury from a contaminated syringe is 0.3% for HIV, 1.8% for hepatitis C but 30% for hepatitis B (Batterman, 2004). The hepatitis B virus is very persistent in dry air and can survive for several weeks on a surface, brief exposure to boiling water and to some antiseptics, including 70% ethanol. An infective dose of hepatitis B or C virus can survive for up to a week in a blood droplet trapped inside a hypodermic needle (Thompson et al., 2003). Due to unsafe medical waste management, there is risk of the spread of such an infection, which may affect those in-house as well as the surrounding population. There are other notorious bacteria living as opportunistic pathogens such as *Pseudomonas aeruginosa* and *Acinetobacter baumannii* that are commonly found adhering to medical devices, for example, hypodermic needles, blades, etc., and capable of causing serious infections in immunocompromised individuals. The presence of these and many more pathogens can pose a serious threat to the public especially during cases of injury from sharps such as needles, blades, etc., that are easily found littering places where medical waste is dumped (Altaf et al., 2004).

Waste recycling is an important activity with many benefits. It helps in the protection of the environment against pollution, production of biogas through anaerobic digestion processes, composting, and so on. Normally, medical waste is often excluded from municipal waste that is meant for recycling. Cases whereby recycling facilities becomes hazardous usually occur whenever medical waste enters the mix. Unauthorized sweepers and rag pickers who scavenge waste sites for disposable items such as bottles, hypodermic needles, blades, etc., often cause such unfortunate occurrences. They process these untreated items either by granulation or making them into other products, or by washing and repackaging them for resale. Rag pickers and sweepers with such practices are common in many parts of the world with particular reports in Asian countries with large populations, for example, India, Pakistan, Bangladesh, as well as some African countries, for example, South Africa (Altaf and Mujeeb, 2002; Mujeeb et al., 2003). These activities are dangerous and can constitute a major threat to the public and cause the spread of infection among susceptible individuals who may by chance get into contact with these recycled items.

The possibility of transference of disease and infection from animals to humans (zoonosis) is well established. The spread of infection and disease through vectors such flies, mosquitos, insects, as well as mammals, such as monkeys and bats, has been well documented. According to the UK Health and Safety Executive (2000), there are approximately 40 zoonotic diseases in the United Kingdom of which salmonella and influenza (avian flu and H1N1) are among the notable ones. Animals, such as dogs, cats, rats, cattle, goats, birds, and even elephants, are usually found roaming waste sites looking for something to eat. These animals face a grave danger of contracting diseases such as tuberculosis, anthrax, and other infections which may possibly be transferred to humans who consume their meat. However, the extent to which they can spread infection has not been extensively reported.

4.4.3.2 Spread of Resistant Pathogen

Antibiotic-resistant bacteria pose an enormous risk to global health because they are responsible for the failure of treatment in many hospitals. Inadequate waste management from the hospital, especially in waste that contains pathogenic organisms, can affect the environment in a number of ways. Firstly, they affect the flora and fauna by increasing the pathogenic strains of hospital origin and decreasing the commensals that contribute beneficially to the ecosystem. Secondly, they pose great risk to the public because hospital waste harboring resistant nosocomial bacteria that are found at waste dumping sites can contaminate the groundwater, streams, and rivers via flooding thereby making such water unfit for drinking. There have been reports of groundwater contaminated with drug resistant Gram-negative bacteria such as *Escherichia coli*, *Klebsiella pneumoniae*, *Enterobacter* spp., etc. Most of these isolated bacteria are commonly encountered in hospitals (Lin et al., 2004; Soge et al., 2009). Another significant way by which the environment and public health is affected is through the spread of resistance genes. Certain bacteria are known to easily acquire resistance genes via plasmids and integrons through their interaction with other resistant bacteria of the same or different genera in the environment, this is called horizontal gene transfer. Many bacterial resistant genes that are widely disseminated across the globe are mostly of hospital origin (Mooij et al., 2009; Potron et al., 2009; Odumosu et al., 2013). Antibiotic-resistant bacteria are a pressing global health concern. Over the last three decades several types of bacteria have emerged with different resistance genes causing therapeutic failure among individuals that are infected. Recently, other bacteria with a new gene called New Delhi Metallo β-lactamase have emerged from India and have been detected in more than 13 countries in Europe, Asia, North America, the Middle East, and Africa. The gene was initially isolated from a hospital patient from India and has since been predominantly found among people who had previously been to India or Pakistan.

4.4.3.3 Dangers to the Public

The public is also at risk of activities associated with inefficient medical waste management in society. Developing countries where people do not have effective water supply are in great danger of consuming water that is already contaminated with bacteria washed from waste sites. These can result to development of sickness and ailments such as tuberculosis, typhoid, and other water borne diseases. The role of rag pickers and scavengers who assist unscrupulous recyclers in the repackaging of medical equipment such as syringes, hypodermic needles, expired drug, and other waste medical items for resale, also contribute immensely to the dangers to the public. They end up selling such material to small pharmacies patronized by members of the public who may be asked to provide their own syringes, needles, and even drugs. Anyone involved in these unfortunate scenarios are at high risk of contracting disease and infection. Many of such cases of outbreak of diseases have been reported in countries such as India where doctors are blamed for reusing syringes and also transacting in second-hand syringes (Solberg, 2009).

4.4.4 Waste Disposal Systems and the Challenges

Waste treatment may be defined as any method, technique, or process designed to change the biological character or composition of any medical waste to reduce and/or eliminate its potential for causing disease. Effective hospital waste disposal systems are essential in ensuring the safety of health workers and the public. The rationale for proper hospital waste management is because improper waste management is a direct violation of the right to a healthy environment. The public has a right to a safe and healthy environment that is free from infectious diseases and hazardous materials arising from hospital waste. The need to spend so much money on waste disposal is to ensure public safety especially from sharps that may cause injuries leading to serious infections (Brown, 1993). The risk of contracting diseases from nosocomial pathogens, exposure to hazardous chemicals, and many more have been previously established through research studies. Hence, a proper channel for disposing hospital waste is essential to safeguard the life of humans and animals.

The percentage of hospital waste that is in fact harmful is low compared with other waste that is generated within hospitals. The first step toward an effective waste management process is to segregate the different types of waste at the initial source of generation (Philips, 1991). Through segregation different types of waste can be identified, labeled, and placed in different waste containers and can be treated differently. The aim of this process is to avoid mixing together the risk and nonrisk waste, especially the infectious part of waste which represents a lower percentage, which is to be carefully separated and treated differently since it contains dangerous waste matter.

In collection of waste, certain procedures are usually followed. According to several regulatory bodies, waste collected at room temperature should not be allowed to stay at room temperature for more than 24 h. This is important to prevent the incubation of pathogens and the deterioration of such waste matter. Generally, different wastes are categorized based on the level of risk they carry and are assigned specific containers with designated colors. For instance human anatomical waste are collected in yellow plastic containers, microbiological waste in yellow/red, waste sharps in plastic bag puncture-proof in translucent blue/white containers, domestic hospital waste in green bags, etc. Hospital waste disposal methods are broadly classified under incineration and nonincineration technologies (Figure 4.5a,b shows municipal waste workers sorting and transferring waste containing both municipal and biomedical waste).

4.4.4.1 *Incineration Technology*

This is a high-temperature thermal process (between 900°C and 1200°C) employing the combusting of waste under controlled conditions to convert it into inert material and gases (Pruss et al., 1999). Incineration has been the most widely used method of treating hospital waste in many parts of the world. According to Chauhan et al. (2002), in the Biomedical Waste (Management and Handling) Rules, incineration has been recommended for human anatomical waste, animal waste, cytotoxic drugs, discarded medicines, and soiled waste.

FIGURE 4.5
(a) Waste workers transferring waste including biomedical waste awaiting further treatment. (b) Waste workers sorting waste before transporting.

The design of the biohazardous waste incinerator requires sufficient residence time to ensure the destruction of disease-causing pathogens and to convert combustible materials into noncombustible residue or ash. There are three major types of incinerators used for hospital waste: multiple hearth type, rotary kiln, and controlled air types. All the types of incinerators have both primary and secondary combustion; in each case, a secondary combustion chamber is used to completely combust all noncombusted gases leaving the primary incinerator chamber to ensure optimal combustion.

The public health concern related to incinerator technology is the emission of persistent organic pollutants particularly dioxins and furans. Dioxin and furans are chemical products that are produced when products are burned. They get distributed into the air after burning of certain hospital waste products. According to the US EPA, dioxins and furans are cancer-causing substances in humans. In addition, people exposed to dioxins and furans have experienced changes in hormone levels. Other nondioxin pollutants of concern from incinerators include mercury and other heavy metal pollutants, halogenated hydrocarbons, acid gases that are precursors of acid rain, particulates which damage lung function, and greenhouse gases. Of all the above, mercury pollution is one of the dangerous visible effects because it is a powerful neurotoxin that impairs motor sensory and cognitive functions. Production of ash and slag as a product of incinerators is also considered a source of pollution. Both ash and slag are defined as hazardous wastes under international law because they often contain heavy metals that may leach out.

4.4.4.2 Nonincinerator Technology

Nonincineration technology treatment is an alternative method to burning of medical waste and it includes four basic processes: (i) thermal, (ii) chemical, (iii) irradiative, and (iv) biological.

4.4.4.2.1 Thermal Processes

Thermal processes are methods that rely on thermal energy (heat) for the decontamination of pathogens in waste. This process is further subdivided into low-heat, medium-heat, and high-heat thermal processes. The importance of the subclassification is due to the physical and chemical mechanisms that distinctly take place at medium and high temperatures.

4.4.4.2.2 Low-Heat Thermal Processes

This process makes use of thermal energy to destroy pathogens in waste at temperatures insufficient to cause chemical breakdown or to support combustion or pyrolysis. Low-heat thermal process heat at the range of 93–177°C (200–350°F). The low-heat processes consist of two basic categories: wet heat (steam) and dry heat (hot air) treatment method. Autoclaving is an example of a wet heat treatment method. It involves the use of steam to decontaminate waste and allows for the treatment of only limited quantities of waste hence is only recommended for highly infectious waste such as microbial cultures and other infectious matter. Another example of wet heat is microwave treatment that uses the action of moist heat and steam generated by microwave energy to disinfect the waste (Vela and Wu, 1979). Dry heat disinfection does not make use of water or steam but rather heats the waste by conduction, natural or forced convection, and/or thermal radiation using infrared heaters.

4.4.4.2.3 Medium-Heat Thermal Processes

The temperature range for medium-heat thermal processes is approximately 177–370°C (350–700°F) and it involves the chemical breakdown of organic material. This processes operates on a technology been referred to as reverse polymerization using high-intensity microwave energy and thermal depolymerization using heat and high pressure.

4.4.4.2.4 High-Heat Thermal Processes

High-heat thermal processes operate at higher temperature ranging from around 540°C to 8300°C (1000–15,000°F) or more. The high heat produced involves both chemical and physical changes resulting in total destruction of the waste. There is also a significant reduction in the total mass and volume of the waste compared with low-heat thermal processes. High-heat thermal process can cause a size reduction up to 90%–95%, whereas the low-heat thermal processes can only reduce size about 60%–70%.

4.4.4.2.5 Chemical Processes

Chemical processes involves the use of disinfectants such as sodium hypochlorite, peracetic acid, hydrogen peroxide, or dry inorganic chemicals to treat liquid waste such as blood, urine, stools, or hospital sewage. This type of technology has been in existence since the mid-1980s. According to the EPA, this type of method is most appropriate for liquid hospital waste although it can be used to treat solid wastes as well. Chemical processes often involve the shredding or grinding of waste in order to enhance the exposure of the waste to chemical agents. A study carried out by Barek et al. (1998) described the efficacy of three tested chemical methods viz. oxidation with sodium hypochlorite (NaClO, 5%) hydrogen peroxide (H_2O_2, 30%), and Fenton reagent ($FeCl_2 \cdot 2H_2O$; 0.3 g in 10 mL H_2O_2, 30%), for the degradation of four anticancer drugs amsacrine, azathioprine, asparaginase, and thiotepa. The results showed a high degree of degradation with the chemical methods without any mutagenic residue. Other methods available in this process involve the use of encapsulating compounds that can solidify sharps, blood, or other body fluids within a solid matrix prior to disposal.

4.4.4.2.6 Irradiation Processes

Unlike the chemical and thermal processes that reduce or physically alters waste, irradiation-based treatment involves the use of electron beams such as gamma radiation, UV irradiation, or Cobalt 60 that uses a shower high-energy electron to destroy microorganisms in the waste by causing chemical dissociation and rupturing of their cell walls. The radiation sterilizing effect is by penetration and inactivation of microbial contaminants. This technology requires certain safety precautions such as the use of protective covering, etc., to prevent occupational hazards. The pathogen destruction efficacy of this method relies solely on the amount of the dose absorbed by the mass of waste, which, in turn, is related to waste density and electron energy. Although the EPA did not recommend this method for pathological waste, it is an alternative method of decontamination of all types of waste except pathological wastes.

4.4.4.2.7 Biological Processes

This involves the use of enzymes to degrade organic waste. It is a rarely used method in hospital waste management.

4.5 Conclusion and Recommendation

Biomedical waste management is a serious concern especially in developing countries. It poses a significant impact on human health and the environment. The challenge of medical waste needs urgent attention if any good is expected from the imminent threat it carries. It is very important to know that medical waste and its management requires a great deal of expertise, resources, and human cooperation. One of the problems facing most countries in the developing world is lack of education and the proper knowledge of the risks involved. The situation is not totally out of control; however, the possibility for improvement exists if proper steps are taken to ensure effective waste disposal and management. There are recommendations believed to be of significance in effective waste management.

Programs raising awareness on the effect of improper disposal of medical waste and other related issues will go a long way in educating the staff and personnel who are directly involved in the generation of waste. Training of personnel on duty is very important. By organizing training programs on the safe handling of medical waste, medical staff will begin to use their initiative in handling issues associated with waste and can pass on the knowledge to others as a form of continued in-house training of the sanitary staff. The role of government in this matter is also of high significance by implementing laws and regulations regarding the disposal and management of medical waste. Certain studies recommended incentives and penalty for staff with good sanitary practice and defaulters. This can also serve as an additional method of improving the practice.

References

Altaf A, Janjua NZ, Aamir JK, Mujeeb SA, and Samad L. 2004. An assessment of the quality of syringes in Pakistan. Study conducted by safe injection network, Karachi, Pakistan funded by World Health Organization, 30 p.

Altaf A and Mujeeb SA. 2002. Unsafe disposal of medical waste: A threat to the community and environment. *Journal of Pakistan Medical Association* 56(2): 232–233.

Barek J, Cvaka J, Zima J, Meo MD, Laget M, Michelon J, and Castegnaro M. 1998. Chemical degradation of wastes of antineoplastic agents amsacrine, azathioprine, asparaginase and thiotepa. *The Annals of Occupational Hygiene* 42(4): 259–266.

Batterman S. 2004. *Assessment of Small-Scale Incinerators for Healthcare Waste*. WHO, Geneva, 69 p.

Brown, J. 1993. Hospital waste management that saves money--and helps the environment and improves safety. *Regulatory analyst. Medical waste*, 1(10): 1–3.

Centers for Disease Control and Prevention. 2003. Guidelines for environmental infection control in health-care facilities: Recommendations of CDC and the Healthcare Infection Control Practices Advisory Committee (HICPAC). *MMWR* 52: 1–4.

Chauhan MS and Kishore M. 2002. Existing solid waste management in hospitals of Indore city. *Indian Journal of Environment and Science* 6: 43–49.

Collins CH and Kennedy DA. 1992. The microbiological hazards of municipal and clinical wastes. *Journal of Applied Bacteriology* 73: 1–6.

Eigenheer E and Zanon U. 1991. O quefazer com osresiduoshospitaleres. Propostaparaclassificacao, embalagem, coleta e destinacao final. *Arquivos Brasileiros de Medicina* 65(3): 233–237.

Health and Safety Executive (HSE). 2000. *Health and Safety Statistics 1999–2000.* HSE, London.

Lin J, Biyela PT, and Puckree T. 2004. Antibiotic resistance profiles of environmental isolates from Mhlathuze River, KwaZulu-Natal (RSA). *Water SA* 30: 23–28.

MoEF. 1998. *GoI, The Gazette of India: Extraordinary, Notification on the Bio-medical Waste (Management and Handling) Rules, [Part II—Sec.3(ii)], No. 460.* Ministry of Environment and Forests, New Delhi, pp. 1–20.

Mooij MJ, Willemsen I, Lobbrecht M, Vandenbroucke-Grauls C, Kluytmans J, and Savelkoul PH. 2009. Integron class 1 reservoir among highly resistant gram-negative microorganisms recovered at a Dutch teaching hospital. *Infection Control and Hospital Epidemiology* 30: 1015–1018.

Mujeeb SA, Adil MM, Altaf A, Hutin Y, and Luby S. 2003. Recycling of injection equipment in Pakistan. *Infection Control and Hospital Epidemiology* 24(92): 145–146.

Odumosu BT, Adeniyi BA, and Chandra R. 2013. Analysis of integrons and associated gene cassettes in clinical isolates of multidrug resistant *Pseudomonas aeruginosa* from Southwest Nigeria. *Annals of Clinical Microbiology and Antimicrobials* 12: 29–36.

Phillips G. 1999. Microbiological aspects of clinical waste. *Journal of Hospital Infection* 41(1): 1–6.

Potron A, Poirel L, Bernabeu S, Monnet X, Richard C, and Nordmann P. 2009. Nosocomial spread of ESBL-positive *Enterobacter cloacae* co-expressing plasmid-mediated quinolone resistance Qnr determinants in one hospital in France. *Journal of Antimicrobial Chemotherapy* 64: 653–654.

Pruss A, Giroult E, and Rushbrook P. 1999. *Safe Management of Wastes from Health—Care Activities.* World Health Organisation, Geneva, pp. 77–128.

Simonsen L, Kane A, Lloyd J, Zaffran M, and Kane M. 1999. Unsafe infections in the developing world and transmission of bloodborne pathogens: A review. *Bulletin of the World Health Organization* 77(10): 789–800.

Soge OO, Giardino MA, Ivanova IC, Pearson AL, Meschke JS, and Roberts MC. 2009. Low prevalence of antibiotic-resistant gram-negative bacteria isolated from rural south-western Ugandan groundwater. *Water SA* 35: 343–347.

Solberg KE. 2009. Trade in medical waste causes death in India. *Lancet* 373: 1067.

Thompson SC, Boughton CR, and Dore GJ. 2003. Blood-borne viruses and their survival in the environment: Is public concern about community needle stick exposures justified? *Australian and New Zealand Journal of Public Health* 27(6): 602–607.

US Environmental Protection Agency. *Priority PBTs: Dioxins and Furans Fact Sheet.* Office of Pollution Prevention and Toxics, Washington, DC. Available at www.epa.gov/epawaste/hazard/wastemin/minimize/factshts/dioxfura.pdf (accessed 7/2/2014).

Vela GR and Wu JF. 1979. Mechanism of lethal action of 2,450-MHz radiation on microorganisms. *Applied and Environmental Microbiology* 37(3): 550–553.

Waste Incineration: A Dying Technology. Global Anti-Incinerator Alliance Global Alliance for Incinerator Alternatives. Available at http://www.no-burn.org (accessed 8/2/2014).

5

Biological Nitrogen Removal in Wastewater Treatment

Rima Biswas and Tapas Nandy

CONTENTS

5.1 Introduction

Wastewater discharges containing nitrogen can be toxic to aquatic life, cause oxygen depletion and eutrophication in receiving water, and affect chlorine disinfection efficiency (US EPA, 1993). Hence reducing nitrogen levels from wastewater discharges is necessary. Nitrogen compounds can be removed from wastewater by a variety of physicochemical and biological processes. As biological nitrogen removal is more effective and relatively inexpensive, it has been widely adopted over physicochemical processes. In most modern wastewater treatment plants (WWTPs) nitrogen, which is generally in the form of ammonium or organic nitrogen, is removed by biological nitrification/denitrification (Equation 5.3). In the first step, ammonium is converted to nitrate (nitrification, Equation 5.1) which is then, in a second step, converted to nitrogen gas (denitrification, Equation 5.2). Benefits of the process are high removal efficiency, high process stability and reliability, relatively easy process control, low area requirement, and moderate cost.

$$NH_4^+ + 2O_2 + 2HCO_3^- \rightarrow NO_3^- + 3H_2O + 2CO_2 \qquad (5.1)$$

$$5C + 4NO_3^- + 2H_2O \rightarrow CO_2 + 4HCO_3^- + 2N_2 \qquad (5.2)$$

(a)

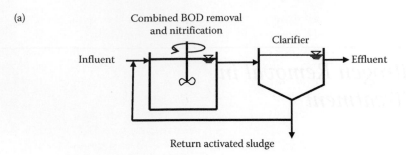

(b)

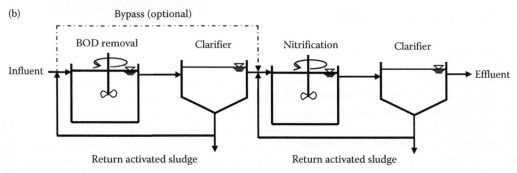

FIGURE 5.1

Process configuration used for biological nitrification. (a) Single-sludge suspended growth system, (b) two-sludge suspended growth system. (Adapted from Metcalf & Eddy, Inc., 2003. *Wastewater Engineering: Treatment and Reuse.* 4th Edition. New York: McGraw-Hill.)

$$4NH_4^+ + 8O_2 + 5C + 4HCO_3 \rightarrow 2N_2 + 10H_2O + 9CO_2 \tag{5.3}$$

Generally, the conventional biological nitrogen removal process is used for treating wastewater with relatively low nitrogen concentrations (total nitrogen concentration less than 100 mg N/L). Some wastewater streams such as anaerobic digester effluents, landfill leachate, and industrial wastewater contain high concentrations of nitrogen.

The rise in the demand for environmental protection has forced the development of more efficient nitrogen removing technologies in wastewater treatment. The new technologies that have developed are more efficient than the conventional nitrification-denitrification processes in terms of time and space. Based on the knowledge available on nitrogen-removing microorganisms, a number of new cost-effective technologies have been developed for efficient removal of ammonia from wastewater (Figure 5.1). The state-of-the-art of these newly developed processes is presented along with their current status in field applications.

5.2 Conventional Nitrogen Removing Processes

Conventionally, autotrophic nitrification followed by heterotrophic denitrification is a well-established process of removing a high concentration of ammonia as well as nitrate from wastewaters.

Nitrification is the term used to describe the two-step biological process in which ammonia (NH_4–N) is oxidized to nitrite (NO_2–N) and nitrite is oxidized to nitrate (NO_3–N). The first step is carried out by a group of autotrophic ammonia-oxidizing bacteria (AOB) belonging to the genera *Nitrosomonas, Nitrosococcus, Nitrosospira, Nitrosolobus,* and *Nitrosovibrio.* The second step is carried out by autotrophic nitrite-oxidizing bacteria (NOB) belonging to the genera *Nitrobacter, Nitrococcus, Nitrospira, Nitrospina,* and *Nitroeystis.*

Generally, nitrification can be accomplished in both suspended growth and attached growth biological processes (Metcalf & Eddy, Inc., 2003). For the suspended growth process, the more common approach is to achieve nitrification along with biochemical oxygen demand (BOD) removal in the same single-sludge process, while in the case of complex wastewater where there is significant potential of the presence of toxic compounds, a two-sludge suspended growth system is considered (Figure 5.1).

Stoichiometrically, for oxidizing each gram of ammonia to nitrate, 4.25 g of O_2 and 7.07 g of alkalinity as $CaCO_3$ is required. Besides, dissolved oxygen (DO) and alkalinity, nitrification in wastewater process must consider various issues such as wastewater nitrogen concentration, BOD concentration, alkalinity, temperature, toxic compounds in wastewater, etc. Nitrification is pH sensitive and the optimal pH range is between 7.5 and 8.0. The rates decline significantly near pH values 5.8–6.0. In the activated sludge process (ASP), a neutral pH of 7.0–7.2 is usually maintained. In single-sludge systems nitrification can take place at a reasonable rate at a neutral pH, but can suffer competition from fast growing heterotrophs that have their broad optimal pH range at 6.5–8.5. And also nitrification consumes alkalinity, and sufficient alkalinity is essential to buffer the pH drop during nitrification. Nitrifying organisms are also sensitive to a wide range of toxic compounds commonly present in wastewater at concentrations that are much lower than the concentrations that would affect other aerobic heterotrophic bacteria. Hence in many cases, nitrifiers continue to grow and nitrify in the presence of these toxic compounds but at rates much lower than their optimal rate. Metals are also a concern for nitrification. Nickel (0.25 mg/L), chromium (0.25 mg/L), and copper (0.1 mg/L) are also known to inhibit nitrification completely. Nitrification is also affected by free ammonia (NH_3) and free nitric acid (HNO_2). The inhibition effects are depending upon the concentration of the nitrogen species, pH, and temperature. At 20°C and pH 7.0, the NH_4–N and NO_2–N concentrations of 100 and 20 mg/L may initiate inhibition of ammonia and nitrite oxidation, respectively (US EPA, 1993).

Denitrification is the biological reduction of nitrate to nitric oxide, nitrous oxide, and nitrogen gas under anaerobic conditions. It is an integral part of the biological nitrogen removal process (Figure 5.2). Denitrification takes place after ammonia is autotrophically nitrified to nitrate in the presence of an external carbon source such as methanol. A variety of electron donors and carbon sources such as methanol, acetate, glucose, ethanol, acetic acid, lactic acid, and at times wastewater have been used for denitrification. Methanol (CH_3OH), being relatively inexpensive, has gained widespread application. For complete denitrification, 2.47 g of methanol is required per gram of nitrate nitrogen (McCarty et al., 1969). Carbon sources can also be derived from the wastewater. In this arrangement, a denitrification (anoxic) tank is followed by a nitrification (aeration) tank. Nitrate produced in the aeration tank is recycled back to the anoxic tank. This process is called substrate driven or preanoxic denitrification. In another arrangement, denitrification occurs after nitrification and the electron donor source is from an endogenous decay. This arrangement is termed as postanoxic denitrification and here the denitrification depends on endogenous respiration for energy and sometimes an external carbon source is added. Postanoxic processes include both suspended and attached growth systems.

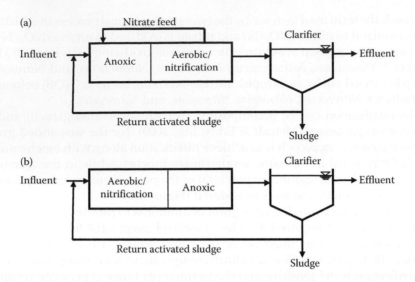

FIGURE 5.2
Denitrification process. (a) Pre-anoxic denitrification, (b) post-anoxic denitrification. (Adapted from Metcalf & Eddy, Inc., 2003. *Wastewater Engineering: Treatment and Reuse.* 4th Edition. New York: McGraw-Hill.)

The differences between autotrophic nitrifiers and the heterotrophic denitrifiers underscore the complexity of the design and operation of nitrification/denitrification systems. The autotrophic nitrifiers are slow growing, sensitive to organic carbon concentrations, and require high DO concentration. On the contrary, heterotrophic denitrifiers are fast growing bacteria, resilient to inhibitory compounds, and require high organic carbon concentrations with a low to absent DO level. Providing heterogeneous environmental conditions to nitrifying and denitrifying organisms is the greatest challenge in successful biological nitrogen removal (Bernet et al., 2001). Moreover, the nitrification reaction consumes a high oxygen concentration, requiring 4.2 g of oxygen for each gram of ammonium nitrogen nitrified (Gujer and Jenkins, 1975). Theoretical oxygen consumption for *Nitrosomonas* and *Nitrobacter* are 3.16 mg O_2 per mg NH_4–N oxidized and 1.11 mg O_2 per mg NO_2–N oxidized, respectively. Moreover, 7.07 mg alkalinity as $CaCO_3$ is required per milligram ammonia nitrogen oxidized (Khin and Annachhatre, 2004). Severe pH depletion can occur through acid production during the nitrification process, when the alkalinity in the wastewater is low. Therefore, in cases where the alkalinity of the wastewater is depleted by the acid produced by nitrification, the proper alkalinity must be supplemented by chemical addition, such as lime, for denitrification to occur. In addition, during denitrification, the requirement of organic carbon is significant.

Nitrogen removal from high chemical oxygen demand (COD) wastewater through conventional nitrification-denitrification has many inherent operational problems. The autotrophic nature of nitrifiers makes them vulnerable to vivacious competition by fast growing heterotrophic bacteria. Besides, stiff competitions by heterotrophs, autotrophic AOB are inhibited by many compounds such as cyanide, thiocyanate, and acetyline that are commonly present in wastewater. Still, nitrifying activity has always been reported in conventional activated sludge systems. Such activity is mostly due to the presence of heterotrophic nitrifiers such as *Bacillus* sp., *Pseudomonas stutzeri*, *Thiosphaera pantotropha*, *Alcaligenes faecalis*, and partly due to some autotrophic AOB (Gupta, 1997; Kim et al., 2005; Lin et al., 2006). Nitrification by heterotrophic bacteria is an alternative mechanism under stressful conditions and not a normal energy generating pathway. Moreover, heterotrophic nitrification is much slower (generally 10^4–10^5 times) than autotrophic nitrification and in wastewater with high C/N ratio the nitrifying activity of heterotrophs is usually not reported. In the chemostat process treating nitrogenous wastewater, *T. pantotropha* could achieve a nitrification rate equivalent to *Nitrosomonas europaea* only when the heterotrophs outnumber the autotrophs by a 250:1 ratio (van Niel et al., 1993). Thus, the major ammonia-oxidizing activity is undertaken by the autotrophic bacteria.

To avoid the inhibition of nitrifiers in the presence of heterotrophs and organic compounds, a few modifications in the operational process have been developed. These include an extended aeration system, anoxic followed by aerobic treatment, etc. The advantages in the "anoxic followed by aerobic treatment" process are the utilization of available COD in wastewater for denitrification, and the inhibition of the denitrification process can be avoided by excessive pH drop by the preceding nitrification step. However, denitrification in the first place could deplete the DO from the wastewater. Hence, the cost of aeration for nitrification may be more than in the conventional process. Furthermore, in wastewater containing ammonia, denitrification as the first step is not practically possible due to the absence of nitrate as an electron acceptor. Hence, under such conditions nitrification, at least partial, if not complete, is an essential prerequisite. The effluent from first stage partial nitrification can be combined with raw effluent and then be subjected to the denitrification unit. Such a process calls for third stage nitrification to complete the nitrification of untreated ammonia. No doubt, the process becomes more complex to accommodate strikingly different groups of microorganisms for complete removal of ammonia in wastewater.

5.3 ANAMMOX

Generally unusual observations made during biological processes lead to investigation to understand the underlying microbial phenomenon. Conversely, sometimes ideas on certain biological processes come first followed by the actual search for their existence in nature. This happened when an Austrian chemist, Broda (1977) wondered about the existence of certain nitrifying microorganisms which could catalyze a hitherto unknown reaction based on thermodynamic calculations.

$$NH_4^+ + NO_2^- \rightarrow N_2 + 2H_2O \tag{5.4}$$

Unaware of such predictions, Mulder detected the unusual disappearance of nitrogen in his denitrifying pilot plant in 1999 at the Gist Brocades Fermentation Company. The reactor received two effluent streams, one from the methanogenic pilot plant containing ammonium, sulfide, and organic compounds and another from a nitrifying plant. Unexpectedly after 100 days of operation, ammonia was decreasing with no buildup of NO_2^- or NO_3^-. He coined the term ANAMMOX for anaerobic ammonium oxidation in the presence of NO_3^-. However, Mulder was not sure of the biological nature of the reaction. Research carried out in collaboration with J.G. Kuenen and his group, comprising Astrid van de Graaf, Mike Jetten, and Marc Strous, not only established the biological nature of the reaction and its mechanism, but also established the identity of the responsible microorganism that belonged to the group *Planctomycetes*. The first ANAMMOX bacteria identified was *Canditatus Brocardia, ANAMMOXidans* (Strous et al., 1999). Further research through physiological and genomic analysis revealed many unexpected facts. CO_2 can be used as a carbon source to produce biomass and NO_2^- functions not only as an electron acceptor for ammonium oxidation, but also an electron donor for the reduction of CO_2.

$$NH_4^+ + 1.32NO_2 + 0.066HCO_3^- + 0.13H^+ \rightarrow 1.02N_2 + 0.26NO_3^-$$
$$+ 0.066CH_2O_{0.5}N_{0.15} + 2.03H_2O \qquad (5.5)$$

Besides NO_2^-, NO_3^- can also act as an electron acceptor. The autotrophic ANAMMOX bacteria can oxidize formate, acetate, and propionate to CO_2 through a dissimilatory pathway. An assimilatory pathway for the utilization of an organic carbon source does not exist. The organism can generate NO_2^- and NH_3 form NO_3^- using formate, acetate, and propionate as an electron donor; and is thereby capable of producing its electron donors and acceptors.

The ANAMMOX conversion is an elegant shortcut in the natural nitrogen cycle. The reaction needs NH_4^+/NO_2^- in a ratio of 1:1.3. Generally, wastewater does not always have both ammonia and nitrite in the required ratio of 1:1.3. In combination with nitrification ANAMMOX bacteria convert ammonium (NH_4^+) directly into dinitrogen gas (Figure 5.3). Hence, ANAMMOX process has been always applied in association with other low-cost technologies that could partially convert ammonia in the wastewater into nitrite.

The ANAMMOX® process is an innovative biological process that represents a major breakthrough in nitrogen removal. It is a cost-effective, robust and sustainable way of removing ammonium from wastewater and from waste gas. Compared with conventional nitrifation/denitrification, operational costs are reduced by up to 90% as are CO_2 emission levels. This brings the plant's carbon footprint down to a minimum.

Working closely with the Delft University of Technology, Paques developed the patented ANAMMOX® process for commercial purposes. The first full-scale ANAMMOX® plant started up in the Netherlands in 2002. Ten years later, there are 11 full-scale ANAMMOX® reactors operational. The ANAMMOX process is widely accepted as it achieved 60% savings in aeration cost and 100% savings in organic carbon dosing (Figure 5.4).

5.4 SHARON

The single reactor high-activity ammonium removal over nitrite (SHARON) process was originally developed for the removal of ammonia via the so called "nitrite route" as represented in Figure 5.4 (Hellinga et al., 1997; Jetten et al., 1997). In principle, this process involves the removal of ammonia to nitrite stage through partial nitrification followed by denitritation of nitrite to nitrogen. Both autotrophic nitrification and heterotrophic denitritation take place under high temperature (35°C) and pH above 7.0 with no sludge retention (Hellinga et al., 1997). Partial nitrification of ammonium to nitrite reduces the requirement of aeration up to 25% and denitrification of nitrite to nitrogen also reduces the cost of the addition of the electron donor by 40%. Basically the SHARON process involves a single stage completely mixed tank, which is intermittently aerated to accommodate sequential nitrification and denitrification. The process can be carried out in a chemostat such as a continuous stirred tank reactor or in a sequential batch reactor (SBR) with suspended biomass. Alternatively, two separate tanks, one for nitrification and another for denitrification,

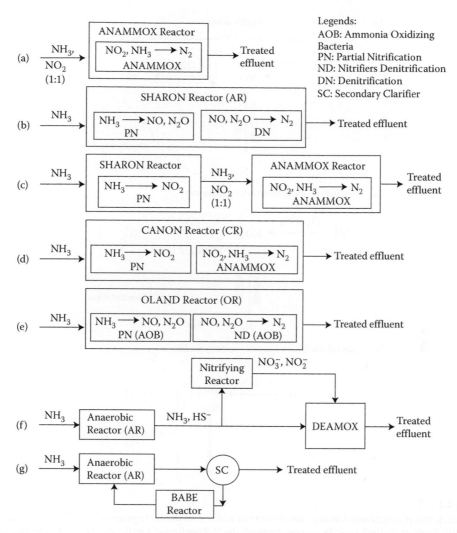

FIGURE 5.3
Schematic figures of (a) ANAMMOX process, (b) SHARON process, (c) SHARON-ANAMMOX process, (d) CANON process, (e) OLAND process, (f) DEAMOX process, and (g) BABE® process. (Adapted from Kalyuzhnyi S and Gladchenko M. 2009. *Desalination*, 248(1): 783–793; Paredes D et al. 2007. *Engineering in Life Sciences*, 7(1): 13–25; Bagchi S et al. *Critical Reviews in Environmental Science and Technology*, 42(13): 1353–1418.)

have also been used to lower the requirement of aerator capacity. Depending upon the site-specific conditions at the WWTP, the selection of the proper configuration is made.

For achieving stable performance, the SHARON process requires partial nitrification of ammonia at nitrite level. Stable partial nitrification can be achieved by eliminating the growth of NOB without harming the activity of AOB in the nitrifying culture (Bagchi et al., 2009). NOB can be washed out of a nitrifying culture through controlling temperature, pH, hydraulic retention time, substrate concentration, and DO. The mechanisms to obtain stable partial nitrification are discussed elsewhere. In the SHARON process, NOB are washed out from the system by maintaining a higher operating temperature (35°C) and a lower sludge retention time (SRT) or to be more specific, lower aeration retention

(a)   Nitrification

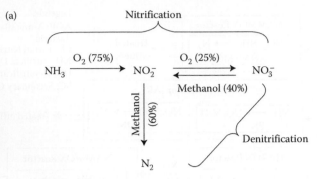

(b)   Nitritation (exclusive nitrite generation)

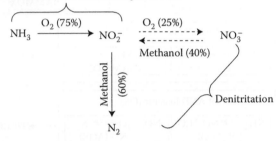

(c)   Partial nitritation (partial nitrite generation)

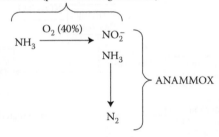

FIGURE 5.4
Savings in terms of cost of aeration and cost of electron addition in the three processes. (a) Conventional nitrification/denitrification, (b) the SAHRON process (through the "Nitrite Route"), (c) the partial nitritation/ANAMMOX. (Adapted from Bagchi S et al. *Critical Reviews in Environmental Science and Technology*, 42(13): 1353–1418.)

time (ART) of 1.5–2 days (Hellinga et al., 1997). Temperature and ART form the two major process parameters in the SHARON process. In a single stage configuration, cyclic phases of "aeration" and "no aeration" are provided for nitrification and denitrification to occur intermittently. The accumulation of nitrite, which is toxic to AOB, is prevented by the denitrification step. Denitrification also restores alkalinity consumed during the nitrification process. For denitrification to proceed, an organic carbon source has to be additionally supplied. Besides temperature (between 30°C and 40°C) and ART, the inlet ammonia concentration, system pH, DO concentrations, and system nitrite concentrations are also to be critically controlled for successful operation of the SHARON process.

In wastewater treatment, the SHARON process can substantially reduce high ammonia concentration along with organics. The first full-scale demonstration of the SHARON® process was carried out at WWTPs in Utrecht and Rotterdam Dokhaven (both in the Netherlands) (Mulder et al., 2001; van Kempen et al., 2001). At present, six full-scale plants

TABLE 5.1

Full-Scale SHARON Plants in the Netherlands

Site	Since in Operation	System Configuration	Wastewater	Load (kg N/day)	Efficiency (%)
Utrecht	1997	Two (3000 m^3 and 1500 m^3)	Sludge dewatering	900	90–95
Rotterdam-Dokhaven	1999	Single (1800 m^3)	Sludge dewatering	850	85–98
Zwolle	2003	Two (900 m^3 and 450 m^3)	Sludge dewatering	410	85–95
Beverwijk	2003	Two (1500 m^3 and 750 m^3)	Sludge dewatering	1200	85–95
The Hague-Houtrust	2005	Single (2000 m^3)	Sludge dewatering	1300	85–98
Groningen-Garmerwolde	2005	Two (4900 m^3 and 2450 m^3)	Sludge dewatering	2400	≥95

Source: Adapted from Bagchi S et al. *Critical Reviews in Environmental Science and Technology*, 42(13): 1353–1418.

have been developed in the Netherlands and one in New York, USA. The applied nitrogen loads in these plants are in the range of 400–2500 kg N/day (Table 5.1). Because of the high temperature involved, it also has high potential for application in tropical regions.

5.5 SHARON–ANAMMOX® Process

In the SHARON–ANAMMOX® process, partial nitrification in the SHARON process is used to deliver feed having an NH_4^+/NO_2^- ratio close to 1 for the ANAMMOX process. The combination of the SHARON and ANAMMOX processes (SHARON–ANAMMOX® process) offers several advantages over the conventional nitrification–denitrification process namely 60% lower oxygen requirement, no need for organic carbon and low production of the excess biomass. This process has been successfully demonstrated on laboratory scale on various types of wastewater such as anaerobic sludge digesters (Vazquez-Padin et al., 2009), sludge-rejected water from domestic WWTPs, food and agro industry, coke oven effluent, and animal wastewater such as slaughter house wastewater and piggery wastewater. The first full-scale application of this process was demonstrated in Rotterdam (the Netherlands) in 2002 (van der Star et al., 2007).

5.6 CANON

The completely autotrophic nitrogen removal over nitrite (CANON) process is a combination of partial nitrification and ANAMMOX in a single aerated reactor, employing two groups of bacteria namely, *Nitrosomonas*-like aerobic microorganisms and *Planctomycetes*-like anaerobic bacteria (Sliekers et al., 2002). This process is critically controlled under a limited oxygen condition. AOB consume all available oxygen while partially oxidizing ammonia to nitrite. This creates an oxygen-free environment needed for the ANAMMOX process. Thus, the cooperation and coexistence of aerobic and anaerobic AOB (AnAOB) is necessary for the successful operation of this process (Figure 5.1). The fluorescent *in situ* hybridization (FISH) assay indicated 40% of AOB and 60% of ANAMMOX bacteria in CANON biomass with the absence of aerobic NOB (Sliekers et al., 2002). However, prolonged exposure to ammonia-limited conditions result in the development of NOB in CANON biomass.

Virtually, CANON is an integration of the SHARON–ANAMMOX process into one single reactor. Although similar operational controls are needed for both processes, CANON provides more sensitive operational controls in terms of DO, nitrogen-surface load, biofilm thickness, temperature, etc. The ANAMMOX bacteria are reversible inhibited by low concentration of oxygen, above 0.5% air saturation. So an appropriate DO concentration enables the consumption of oxygen to an extent in which the DO concentration is not over the toxic threshold for ANAMMOX bacteria and inadequate for the growth of NOB. The oxygen-limited condition (<0.5% air saturation) provides an adequate environment for stable interaction between *Nitrosomonas*-like aerobic microorganisms and *Planctomycetes*-like anaerobic bacteria (Schmidt et al., 2002).

The CANON process has been tested in different lab-scale reactors. The SBR has been considered as an effective reactor system for CANON due to efficient biomass retention (Zhang et al., 2008). However, in a gas-lift reactor, higher ammonia removal rate of 1.5 kg N/m^3 day has also been achieved as against SBR (0.3 kg N/m^3 day).

A few pilot- and full-scale applications of the CANON process are reported in the literature (Joss et al., 2009; Vazquez-Padin et al., 2009). In a separate study, urea was used as feed for the CANON system instead of conventional wastewater. The system adapted very quickly and reached to full capacity within two weeks of adaptation. This has widened the scope of application of this process to urine waste along with other low C:N ratio wastewaters.

5.7 OLAND

The Oxygen-Limited Autotrophic Nitrification Denitrification (OLAND) process refers to the removal of ammonium-N under strict oxygen limitation by autotrophic organisms (Kuai and Verstraete, 1998). Initially, the process was considered to be dependent mainly on aerobic ammonium-oxidizing bacteria (AerAOB), but recently it has been reported that under conditions of low oxygen supply and absence of organic electron donors, both AerAOB and AnAOB co-exist in the OLAND process. The β-subclass of Proteobacteria includes most of the AerAOB, with Nitroso-species (*Nitrosomonas, Nitrosococcus, Nitrosospira*, etc.) as the principal members, while the AnAOB belong to a deep-branching lineage within the order of the *Planctomycetales*. Thus, in principle the process mechanism of OLAND is similar to CANON. Table 5.2 shows the comparative performances of the OLAND and CANON processes. An application of the OLAND process on a pilot scale is reported for the treatment of digested black water using a single-stage SBR (Vlaeminck et al., 2009). Similar to CANON, another process known as the deammoniafication process or DEMON has been successfully implemented at WWTP Strass (Austria) since 2006 (Wett, 2007).

TABLE 5.2

Comparison of Maximum Nitrogen Loading Rate, Nitrogen Removal Rate in Different CANON, and OLAND Processes

Process	Reactor Type	NLR_{max} (kg N/m^3 · day)	NRR_{max} (kg N/m^3 · day)
CANON	SBR	0.131	0.075
	SBR	NR	0.08
	SBR	0.22	0.12
	SBR	0.46	0.36
	SBR	NR	0.5
	MABR	0.87	0.77
	Gas lift	5.5	1.5
OLAND	SBR	0.13	0.05
	RBC	0.08307[a]	0.083[a]
	RBC	0.716	0.7

Source: Adapted from Bagchi S et al. *Critical Reviews in Environmental Science and Technology*, 42(13): 1353–1418.

Note: NLR_{max}, maximum nitrogen-loading rate; NRR_{max}, maximum nitrogen removal rate; SBR, sequential batch reactor; MABR, membrane-aerated bio-reactor; RBC, rotator biological contractor; CSTR, continuous stirred tank reactor.

[a] Values are in kg N/m^2 day.

5.8 DEAMOX

The Denitrifying AMmonia OXidation (DEAMOX) process was proposed by Mulder in 2004 for the ANAMMOX reaction occurring in the autotrophic denitrifying plant (Kalyuzhnyi et al., 2006). DEAMOX eliminates the requirement of partial nitritation by introducing denitritation of nitrate using sulphide as an electron donor (Equation 5.6).

$$NO_3^- + 0.25HS^- \rightarrow NO_2^- + 0.25SO_4^- + 0.25H^+ \tag{5.6}$$

The process is applicable for sulphur-bearing nitrogenous wastewater such as Baker's yeast effluent. In treatment of Baker's yeast effluent, the whole process operates in combination of the three reactor system; anaerobic reactor, DEAMOX reactor and nitrifying reactor. In the anaerobic reactor ammonia and sulphide are generated which is partially fed to the nitrifying reactor and the remaining is fed to the DEAMOX reactor (Figure 5.1). In the nitrifying reactor, ammonia is nitrified into nitrate and nitrite. The third reactor, i.e., the DEAMOX reactor, receives effluent from both the anaerobic and nitrifying reactors. The denitrification of nitrate takes place in the presence of sulphide (Equation 5.6) and the ANAMMOX bacteria coexisting in the DEAMOX reactor along with the autotrophic denitrifying bacteria oxidize ammonia using nitrite as an electron acceptor (Equation 5.5). Basically the scheme in the DEAMOX process is similar to the autotrophic denitrifying reactor plant in which ANAMMOX was first detected and then discovered by Mulder (1992).

Since, the standard application of this process is restricted to sulphur-bearing wastewater, the incorporation of heterotrophic denitratation using volatile fatty acids (VFA) as a more widespread electron donor has been investigated to implement the process for commonly found wastewater. However, the feasibility study showed a severe competition for nitrite between autotrophic ANAMMOX and heterotrophic denitritation resulting in poor removal efficiency.

The typical sulfide driven DEAMOX processes have maximum nitrogen loading rate of 1 kg N/m^3 day with more than 90% removal efficiency (Kalyuzhnyi et al., 2006). The organics-driven DEAMOX process has similar loading capacity (1.24 kg N/m^3 · day) with relatively lower ammonia removal (40%), mainly due to the competition of heterotrophic denitrifiers. Although the process is 10 times lower than the ANAMMOX process, the simple process operation is claimed to favor the DEAMOX process over the ANAMMOX process. So far, this process has been tested only in bench-scale reactors.

5.9 SNAD

Low-cost biological nitrogen-removing (BNR) technologies such as ANAMMOX, SHARON, SHARON–ANAMMOX, CANON, OLAND, and DEAMOX have proved to be very effective in removing ammonia from industrial effluents. However, most ammonium rich wastewater is produced with a certain concentration range of organics (such as old landfill leachates) and the presence of organics could upset the autotrophic bacteria which are mostly responsible in all low-cost BNR technologies except in the SHARON process. The application of the combined SHARON–ANAMMOX process has been also limited to wastewater with low total organic carbon to nitrogen ratio (C:N) ≤0.15. Increase of C:N ratio to more than 0.3 can lead to competition between heterotrophic bacteria and autotrophic bacteria in the SHARON process (Wang and Kang, 2005).

Studies have demonstrated that ANAMMOX and denitrification processes can coexist in the same environment (Chen et al., 2009). Recently, partial nitrification, ANAMMOX, and denitrification have been simultaneously investigated on bench scale in a single-stage nonwoven-rotating biological contractor (NRBC) bioreactor (Chen et al., 2009). The simultaneous partial nitrification, ANAMMOX, and denitrification (SNAD) process was found to potentially remove ammonium and organics from wastewater in a single NRBC reactor with total nitrogen removal efficiency of 70% and COD removal efficiency of 94% at nitrogen and COD loading rates of 0.69 kg N/m^3 day and 0.34 kg/m^3 day, respectively. Recently, the SNAD process has been demonstrated at full scale for the treatment of landfill leachate (Wang et al., 2010; Xu et al., 2010).

5.10 AS/PN–ANAMMOX

The AS/PN–ANAMMOX is a two-stage process comprising combined ASP-partial nitrification (AS/PN) followed by the ANAMMOX process. The concept promotes the coexistence of heterotrophic and autotrophic microorganisms in the conventional extended ASP for organic carbon removal and nitrification. Lamsam et al. (2008) have demonstrated that partial nitrification can be achieved in ASP by controlling three key parameters, namely DO (0.3–0.5 mg/L), pH (around 8), and temperature (35°C). It differs from the SHARON process by the inclusion of an aeration tank and a settler with sludge recirculation. For accomplishing partial nitrification in ASP, SRT of 12–15 days was required (Lamsam et al., 2008). The AS/PN is reported to have several advantages over the conventional extended ASP for organic removal along with nitrification. First, the process consumes 66% less oxygen, as calculated from the theoretical equations 5.8 through 5.10 based on the Henze et al. (2000) assumption of $C_{18}H_{19}O_9N$ as a chemical formula for organic matter. Second, the effluent from AS/PN contains effluent ammonia and nitrite in the ratio of 1:1, which is desirable for the downstream ANAMMOX process. In addition, a conventional ASP can be upgraded into AS/PN by controlling pH, temperature, and DO (Lamsam et al., 2008).

ASP with nitrification:

$$C_{18}H_{19}O_9N + 19.5O_2 \rightarrow 18CO_2 + 9H_2O + H^+ + NO_3^- \tag{5.7}$$

ASP without nitrification:

$$C_{18}H_{19}O_9N + 17.5O_2 + H^+ \rightarrow 18CO_2 + 8H_2O + NH_4^+ \tag{5.8}$$

AS/PN:

$$C_{18}H_{19}O_9N + 18O_2 \rightarrow 18CO_2 + 8H_2O + H^+ + 0.5\,NH_4^+ + 0.5NO_2^- \tag{5.9}$$

The operating conditions of the AS/PN process for coupling with the ANAMMOX process were identified as pH 7.7–8.2 and DO 0.5–0.9 mg/L to achieve over 85% COD removal as well as partial nitrification. The maximum nitrogen removal rate for the ANAMMOX process was found to be 0.6 kg N/m³ day. The process was applied for treating seafood processing wastewater at bench scale.

5.11 BABE® Process

The bio-augmentation batch–enhanced (BABE)® process was developed to bio-augment nitrification in the ASP that operated at suboptimal solid retention times (Salem et al., 2004). The principle was to implement a nitrification reactor (BABE reactor) in the sludge return line (Figure 5.1). The BABE reactor would be fed with an internal N-rich flow (such as effluent from the sludge treatment). Nitrifiers grown in the BABE reactor would ultimately reach the activated sludge reactor. Hence, the nitrification capacity of the ASP can be augmented with nitrifiers grown in the BABE reactor. Thus, the BABE reactor functions basically as a culture tank supplementing the ASP process with the specially cultivated microorganisms in the culture tank. The BABE technology has been tested in full scale at the WWTP Garmerwolde in Groningen (the Netherlands).

The concept of bio-augmentation may be of special interest in a partial nitritation-ANAMMOX system for bio-augmenting both the cultures in a similar manner and for seeding into the main reactor.

5.12 Future Perspective

These applications indicate that the partial nitrification- and ANAMMOX-based processes are more effective in removing ammonia present in wastewater.

Nitrite is a common intermediate in at least three different oxidative or reductive biochemical pathways that occur in nature (nitrification, denitrification, and dissimilatory or assimilatory nitrate reduction). Nitrite accumulation or partial nitrification has been reported in the literature for decades. In engineered systems, partial nitrification is of interest as it offers cost savings in aeration as well as in the form of a lesser need for the addition of organic carbon when compared with conventional denitrification. However, nitrite accumulation is not a rule of nature. Nitrite is an unstable intermediate of the nitrification process. Hence, it is very essential for process engineers to understand the process control for achieving successful nitrite accumulation in a reactor.

The ANAMMOX reaction, on the contrary, is limited by the slow growth rate of ANAMMOX bacteria. The practical application of the ANAMMOX process is still limited by its long start-up periods (up to 1 year) due to the very low growth rate and low cell yield of ANAMMOX organisms. Loss of a fraction of the sludge due to wash out with the effluent could further augment the start-up period. Hence, it is essential to use a reactor with high biomass retention. So far a large range of bioreactor types have been evaluated for the enrichment of ANAMMOX bacteria: fixed bed reactors, fluidized bed reactors, UASB reactors, SBR, gas-lift reactors. Among them, the SBR was accepted for ANAMMOX enrichment for its simplicity, efficient biomass retention, homogeneity of mixture in the reactor, stability and reliability for a long period of operation, stability under substrate-limiting conditions, and high nitrogen conversions.

Thus, the recent BNR technologies have the potential to remove high concentrations of ammonia in wastewater. In spite of their high removal efficiency, the quantum of full-scale applications of these processes is trivial. The issues that create bottlenecks are to be dealt with in detail before these processes can claim commercial success.

References

Bagchi S, Biswas R, and Nandy T. 2012. Autotrophic ammonia removal processes: Ecology to technology. *Critical Reviews in Environmental Science and Technology* 42(13): 1353–1418.

Bagchi S, Biswas R, Roychoudhury K and Nandy T. 2009. Stable partial nitrification in an up-flow fixed bed bioreactor under oxygen limiting environment. *Environmental Engineering Science* 26(8): 1309–1318.

Bernet N, Dangcong P, Delgenes J and Molletta R. 2001. Nitrification at low oxygen concentration in biofilm reactor. *Journal of Environmental Engineering* 127: 266–271.

Broda E. 1977. Two kinds of lithotrophs missing in nature. *Zeitschrift Fur Allgemeine Mikrobiologie* 17: 491–493.

Chen HH, Liu ST, Yang FL, Yuan X and Wang T. 2009. The development of simultaneous partial nitrification, ANAMMOX and denitrification (SNAD) process in a single reactor for nitrogen removal. *Bioresource Technology* 100: 1548–1554.

Gujer W and Jenkins D. 1975. The contact stabilization activated sludge process-oxygen utilization, sludge production and efficiency. *Water Research* 9: 553–560.

Gupta AB. 1997. *Thiosphaera pantotropha*: A sulphur bacterium capable of simultaneous heterotrophic nitrification and aerobic denitrification. *Enzyme and Microbial Technology* 21: 589–595.

Hellinga C, van Loosdrecht MCM and Heijnen JJ. 1997. The Sharon process for nitrogen removal in ammonium rich waste water. *Mededelingen Faculteit Landbouwwetenschappen, Universiteit Gent* 62(4b): 1743–1750.

Henze M, Gujer W, Matsuo T and van Loosdrecht M. 2000. *Activated Sludge Models ASM1, ASM2, ASM2d and ASM3*. Scientific and Technical Reports. IWA Publishing, London, UK.

Jetten MSM, Horn SJ and van Loosdrecht MCM. 1997. Towards a more sustainable municipal wastewater treatment system. *Water Science and Technology* 35: 171–180.

Joss A, Salzgeber D, Eugster J, König R, Rottermann K, Burger S, Fabijan P, Leumann S, Mohn J, and Siegrist H. 2009. Full-scale nitrogen removal from digester liquid with partial nitritation and anammox in one SBR. *Environmental Science & Technology* 43(14): 5301–5306.

Kalyuzhnyi S and Gladchenko M. 2009. DEAMOX–New microbiological process of nitrogen removal from strong nitrogenous wastewater. *Desalination*, 248(1): 783–793.

Kalyuzhnyi SV, Gladchenko M, Mulder A and Versprille B. 2006. DEAMOX—New biological nitrogen removal process based on anaerobic ammonia oxidation coupled to sulphide-driven conversion of nitrate into nitrite. *Water Research* 40(19): 3637–3645.

Khin T and Annachhatre AP. 2004. Novel microbial nitrogen removal processes. *Biotechnology Advances* 22: 519–532.

Kim KJ, Park KJ, Cho KS, Nam SW, Park TJ and Bajpai R. 2005. Aerobic nitrification–denitrification by heterotrophic *Bacillus* strains. *Bioresource Technology* 96: 1897–1906.

Kuai LP and Verstraete W. 1998. Ammonium removal by the oxygen limited autotrophic nitrification–denitrification system. *Applied Environmental Microbiology* 64: 4500–4506.

Lamsam A, Laohaprapanon S and Annachhatre AP. 2008. Combined activated sludge with partial nitrification (AS/PN) and ANAMMOX processes for treatment of seafood processing wastewater. *Journal of Environmental Science and Health, Part A. Toxic/Hazardous Substances and Environmental Engineering* 43(10): 1198–1208.

Lin Y, Tanaka S and Kong H. 2006. Characterization of a newly isolated heterotrophic nitrifying bacterium. *Water Practice and Technology* 1(3): 1–8.

McCarty PL, Beck L and St Amant P. 1969. Biological denitrification of wastewaters by addition of organic materials. *Proceedings of the 24th Industrial Waste Conference*, West Lafayette, IN, USA, pp. 1271–1285.

Metcalf & Eddy. Inc., 2003. *Wastewater Engineering: Treatment and Reuse*. 4th Edition. New York: McGraw-Hill.

Mulder A. 1992. Anoxic ammonium oxidation. US Patent 427849(5078884).

Mulder JW, van Loosdrecht MCM, Hellinga C and van Kempen R. 2001. Full scale application of the SHARON process for treatment of rejection water of digested sludge dewatering. *Water Science and Technology* 43(11): 127–134.

Paredes D, Kuschk P, Mbwette TSA, Stange F, Müller RA, and Köser H. 2007. New aspects of microbial nitrogen transformations in the context of wastewater treatment—A review. *Engineering in Life Sciences* 7(1): 13–25.

Salem S, Berends DHJG, van der Roest HF, van der Kuij RJ and van Loosdrecht MCM. 2004. Full scale application of BABE® technology. *Water Science and Technology* 50(7): 87–96.

Schmidt I, Sliekers O, Schmid M, Cirpus I, Strous M, Bock E, Kuenen JG and Jetten MSM. 2002. Aerobic and anaerobic ammonia oxidizing bacteria—Competitors or natural partners? *FEMS Microbiology and Ecology* 39: 175–181.

Sliekers AO, Derwort N, Gomez JL, Strous M, Kuenen JG and Jetten MSM. 2002. Completely autotrophic nitrogen removal over nitrite in one single reactor. *Water Research* 36: 2475–2482.

Strous M, Kuenen JG and Jetten MSM. 1999. Key physiology of anaerobic ammonium oxidation. *Applied Environmental Microbiology* 65: 3248–3250.

US EPA. 1993. Manual for nitrogen control. US Environmental Protection Agency Technical Report, EPA/625/R-93/010 Office of Research and Development and Office of Water, Washington, DC.

van der Star WRL, Wiebe R, Abma WR, Blommers D, Mulder JW, Tokutomi T, Strous M, Picioreanu C and van Loosdrech MCM. 2007. Startup of reactors for anoxic ammonium oxidation: Experiences from the first full-scale ANAMMOX reactor in Rotterdam. *Water Research* 41(18): 4149–4163.

van Kempen R, Mulder JW, Uijterlinde CA and van Loosdrecht MCM. 2001. Overview: Full scale experience of the SHARON® process for treatment of rejection water of digested sludge dewatering. *Water Science and Technology* 44(1): 145–152.

van Niel EWJ, Arts PAM, Wesselink BJ, Robertson LA and Kuenen JG. 1993. Competition between heterotrophioc and autotrophic nitrifiers for ammonia in chemostat cultures. *FEMS Microbiology and Ecology* 102: 109–118.

Vazquez-Padin JR, Pozo MJ, Jarpa M, Figueroa M, Franco A, Mosquera-Corral A, Campos JL and Mendez R. 2009. Treatment of anaerobic sludge digester effluents by the CANON process in an air pulsing SBR. *Journal of Hazardous Materials* 166: 336–341.

Vlaeminck SE, Terada A, Smets BF, van der Linden D, Boon N, Verstraete W and Carballa M. 2009. Nitrogen removal from digested black water by one-stage partial nitritation and ANAMMOX. *Environmental Science Technology* 43: 5035–5041.

Wang JL and Kang J. 2005. The characteristic of anaerobic ammonium oxidation (ANAMMOX) by granular sludge from an EGSB reactor. *Process Biochemistry* 40: 1973–1978.

Wang CC, Lee PH, Kumar M, Huang YT, Sung S and Lin JG. 2010. Simultaneous partial nitrification, anaerobic ammonium oxidation and denitrification (SNAD) in a full-scale landfill-leachate treatment plant. *Journal of Hazardous Materials* 175: 622–628.

Wett B. 2007. Development and implementation of a robust deammonification process. *Water Science and Technology* 56(7): 81–88.

Xu ZY, Zeng GM, Yang ZH, Xiao Y, Cao M, Sun HS, Ji LL and Chen Y. 2010. Biological treatment of landfill leachate with the integration of partial nitrification, anaerobic ammonium oxidation and heterotrophic denitrification. *Bioresource Technology* 101: 79–86.

Zhang L, Zheng P, Tang C-J and Jin R-C. 2008. Anaerobic ammonium oxidation for treatment of ammonium-rich wastewaters. *Journal of Zhejiang University Science B* 9(5): 416–426.

6

Bioconversion of Industrial CO$_2$ Emissions into Utilizable Products

Shazia Faridi and Tulasi Satyanarayana

CONTENTS

6.1 Introduction

Global warming as a result of greenhouse effect is changing the Earth's climate in dramatic ways, and human activities are primarily responsible for the rate at which this change is taking place. Carbon dioxide is the major greenhouse gas and we have been overfilling our atmosphere with it as a result of industrialization. We burn fossil fuels (coal, natural gas, and oil) for obtaining energy for almost everything from fuels in cars and aero planes to heating our homes that releases CO_2 into the atmosphere. As a result, the atmospheric CO_2 concentration has increased from preindustrial level of 280–390 ppm at present. The elevated CO_2 is strengthening the greenhouse effect as a result of which Earth is getting hotter, increasing the chances of weather disasters, drought and floods, and thus, adversely affecting our health. We need to take immediate steps to reduce our carbon emissions and slow down the speed of global warming. Several attempts are, therefore, being made to reduce carbon emissions by shifting to renewable energy sources, but fossil fuels will continue to be the mainstay of energy generation around the world through at least the next few decades. Consequently, in order to stabilize and ultimately reduce carbon emissions, various carbon capture and sequestration (CCS) technologies are being developed and assessed for mitigating carbon emissions from power plants, industrial processes, and other stationary sources of CO_2 while improving the efficiencies of the power plants.

Carbon capture and storage or CCS is the capture of CO_2 from fossil fuel-based power plants, transportation, and storage of it to a site from where it will not escape into the atmosphere. The search for safe and beneficial ways to utilize the captured CO_2 has received global attention in order to reduce CO_2 emissions, thus expanding CCS to carbon capture, utilization, and storage (CCUS).

Carbon sequestration techniques are broadly categorized into physical and biological. Physical sequestration methods (geological sequestration, sequestration in depleted oil and gas reservoirs, unmineable coal seams, saline aquifers, and ocean sequestration) have high costs associated with them as CO_2 has to be separated (captured) from the remainder of the exhaust gases from fossil-fuel combustion, concentrated and transported, so that it could be disposed of for instance, in depleted oil and gas wells, in deep saline aquifers, in the deep ocean, or deposition through conversion into minerals such as peridotites or serpentinites. Moreover, there is a leakage risk associated with the sequestered CO_2. Hence, improvements in existing CO_2 capture systems and revolutionary new capture and sequestration concepts are needed to provide economical, greener, and safer sequestration techniques.

Biological sequestration (bioconversion) methods mimic the CO_2 conversion methods used by biological systems such as microbes and their enzymes for converting CO_2 into value-added products that are inert and stable. Such methods are economical as they do not need pure CO_2 and also do not require separation and compression.

Bioconversion of CO_2 to industrially important by-products may be grouped under two major categories:

- Bioconversion of CO_2 into mineral carbonates via biomineralization.

Biomineralization refers to the conversion of CO_2 into value-added mineral carbonates employing microbes or their enzyme carbonic anhydrase (CA). Mineral carbonates thus formed find application in many industries as discussed in Section 6.1.1.

- Biofuel production from CO_2

CO$_2$ is converted into biofuel via indirect pathway by cultivation of microbes proficient in photosynthesis. Cyanobacteria and microalgae are, therefore, gaining more attention in research for combating the increasing level of CO$_2$. Algae convert CO$_2$ to biomass utilizing solar energy and water by the process of photosynthesis. Algal biomass thus generated can be used for the production of biofuels (methane, methanol, ethanol, hydrogen, or biodiesel). Their carbohydrate contents can specifically be fermented into bioethanol and biobutanol. Bacteria such as methanogens and acetogens are also capable of carrying out CO$_2$ conversion into fuels such as methane, methanol, ethanol, and butanol. Research is also being focused on the conversion of CO$_2$ into carbon monoxide (CO), which serves as a building block for production of fuels. In combination of other processes, CO can also be used in generating other useful marketable chemicals.

This chapter aims at reviewing the information on the possible CO$_2$ sequestration routes using microbes and their enzymes, feasibility and achievements of the processes, recent developments and uses of the by-products. A major part of this review focuses on the biomineralization of CO$_2$ using CA and development of suitable carbon capture processes utilizing CA. The utility of microbes in the production of biofuels using CO$_2$ from flue gas and other industrial emissions has also been discussed.

6.1.1 Biomineralization of CO$_2$

6.1.1.1 Origin of the Concept

Mineralization of CO$_2$ refers to the conversion of CO$_2$ present in flue gases and industrial emissions into mineral carbonates such as magnesium and calcium carbonates (MgCO$_3$ and CaCO$_3$) using industrial wastes or mineral oxides (e.g., oxides of magnesium and calcium). The concept arises from the natural CO$_2$ sequestration process termed silicate weathering, which is responsible for locking up vast amounts of the earth's carbon in the form of limestone rocks and uses atmospheric CO$_2$ and naturally occurring mineral rocks such as wollastonite (CaSiO$_3$), serpentine (Mg$_3$Si$_2$O$_5$(OH)$_4$), and olivine (Mg$_2$SiO$_4$). These rocks are weathered slowly by the action of wind and rain; react with CO$_2$ and water forming silica and carbonates (Huijgen et al., 2007; Santos et al., 2007).

Weathering of silicates occur in both fresh and salt waters as CO$_2$ is soluble in water and exists in equilibrium with dissolved CO$_2$, HCO$_3^-$, and CO$_3^{2-}$. The series of reactions involved in CO$_2$ mineralization is described as follows (Farrell, 2011):

Gaseous CO$_2$ dissolves rapidly in water to produce a loosely hydrated aqueous form.

$$CO_2(g) \rightarrow CO_2(aq) \tag{6.1}$$

Aqueous CO$_2$ then reacts with water to form carbonic acid (2).

$$CO_2(aq) + OH^- \rightarrow H_2CO_3 \tag{6.2}$$

In the next step, carbonic acid dissociates into bicarbonate and carbonate ions.

$$H_2CO_3 \rightarrow HCO_3^- + H^+ \tag{6.3}$$

$$HCO_3^- + OH^- \rightarrow CO_3^{2-} + H_2O \tag{6.4}$$

Finally, carbonate ions in the presence of divalent metal cations such as Ca, Mg, and Fe get precipitated as mineral carbonate

$$CO_3^{2-} + Ca^{2+} \rightarrow CaCO_3 \downarrow \quad \text{(calcite)}$$
$$CO_3^{2-} + Mg^{2+} \rightarrow MgCO_3 \downarrow \quad \text{(magnesite)}$$
$$CO_3^{2-} + Ca^{2+} + Mg^{2+} \rightarrow CaMg(CO_3)_2 \downarrow \quad \text{(dolomite)}$$
$$CO_3^{2-} + Fe^{2+} \rightarrow FeCO_3 \downarrow \quad \text{(siderite)}$$

(6.5)

It is the pH that controls the outcome of the carbonation reaction. At pH below 8, the reaction (6.2) is insignificant due to the absence of OH^- ions. At mid-pH value between 8 and 10, both reaction (6.2) and reaction (6.3) occur, while at pH greater than 10, reaction (6.2) predominates. HCO_3^- (bicarbonate) and CO_3^- dominate at high pH as there is an abundant supply of OH^- that leads to the precipitation of $CaCO_3$. At acidic pH carbonate solubility increases, and thus, carbonate precipitation is favored under basic conditions, while acidic conditions favor carbonate dissolution. In order to increase calcite precipitation with increasing CO_2, basic conditions are required and the pH needs to be buffered at higher values. Mineral carbonation has gained interest and is being extensively studied for direct capture and mineralization of CO_2 from the exhausts of power plants (flue gas). Some pilot scale studies have also been conducted to demonstrate the feasibility of the process (Reddy et al., 2010).

Sequestration of CO_2 in the form of mineral carbonates offers several advantages over other sequestration processes as outlined below:

- Mineral carbonation is an environmentally benign and effective means of carbon sequestration with CO_2 locked permanently in them. Carbonates and silica produced naturally via this mineralization of CO_2 are evidently stable over geological time scale. Thus, carbon sequestration via mineralization avoids the need for long-term monitoring of sequestered product. The conversion of anthropogenic CO_2 into solid carbonates was first proposed by Seifritz in 1990 to be a stable method for long-term sequestration of CO_2 and later many researchers studied extensively the mineralization of CO_2 (Haywood et al., 2001; Druckenmiller and Maroto-Valer, 2005; Liu et al., 2005; Stolaroff et al., 2005; Mirjafari et al., 2007; Favre et al., 2009; Ramanan et al., 2009a,b).

- Raw material for mineralization of CO_2 is abundant. Huge amounts of appropriate and easily accessible mineral silicates are present many times more than is required to sequester all anthropogenic CO_2 emissions. These minerals constitute a huge CO_2 reservoir, estimated to comprise an amount of carbon equivalent to $150,000 \times 10$ metric tons of CO_2 (Wright et al., 1995). Interestingly, industries such as metal refining and combustion produce huge amounts of wastes such as stainless steel slags, ashes, and brines (rich in MgO and CaO) (Soong et al., 2006), which are hazardous wastes as they are unstable and risk leaching of heavy metals into the atmosphere, and therefore, have an economic penalty associated with them owing to their storage as hazardous waste. Mineral carbonation using such wastes can serve dual purpose by offering a cost effective way to remediate these hazardous waste while also sequestering CO_2 (Stolaroff et al., 2005). For demonstrating the use of fly ash in mineralization of CO_2 from flue gas, a pilot scale study has been carried out in the United States of America at one of the largest coal fired power plants. Flue gas was introduced in a fluidized bed reactor containing fly ash particles and it was shown that during the first 2 min of reaction, CO_2 and SO_2 concentrations decreased from 13.0% to 9.6% and from 107.8 to 15.1 ppm, respectively (Reddy et al., 2010).

- Mineral carbonation is economically feasible, and thus, provides business models to make large-scale CCS commercially attractive. Mineral carbonates formed after sequestration not only can be stored in silicate mines but also provides industrially valuable and useful by-products such as chemicals, cements, and construction materials, white pigment in paints, cement component, a therapeutic source in antacids and calcium supplements, and tableting excipient as well as remediation of waste feedstocks (Ciullo, 1996). Calcium carbonate has its main application in construction either as a building material (cement) or limestone aggregate for building of roads, as an ingredient of cement or as the starting material for the preparation of builder's lime by burning in a kiln and also in the purification of iron in a blast furnace and in oil industry where it is added to drilling fluids as a formation-bridging and filtercake-sealing agent.

- Mineralization process parameters can be designed and optimized to produce high-purity valuable metals, silica, and carbonate mineral powders (www.cacaca. co.uk). The pure carbonates are valuable and are used in applications such as white pigments or fillers, for example, paper making and are required in large quantities. If silica powders are generated with sufficient purity and desirable particle size, they are worth thousands of pounds per tonne and are used as a chemical in the electronics, glass, construction, and plastics industries. Research is also underway to extract more value-added products from sequestration of carbon, for example, Cambridge Carbon Capture Ltd. is investigating the recovery of valuable trace metals such as platinum, rare-earths, and stainless steel alloys as by-products of carbon sequestration using a variety of industrial slags and mining wastes.

- Mineral carbonation provides a cost-effective way for carbon sequestration as flue gas can directly be used for CO_2 capture and does not require cost and energy-intensive large-scale solvent capture of CO_2 from industrial wastes (flue gas); thus serving as an onsite scrubber to provide a plant-by-plant solution to reducing CO_2 emissions. On the other hand, it avoids the cost and public acceptability issues of transportation of supercritical CO_2 in pipelines and storage in underground structures.

In spite of being a safer and a potential CO_2 sequestration method, it suffers limitation of being an extremely slow process under ambient temperature and pressure. Dreybodt et al. (1997) studied the kinetics of calcite precipitation in the reaction converting CO_2 to $CaCO_3$ and concluded that Equation 6.2, that is, the formation of HCO_3^- (bicarbonate) is the rate limiting step except at high pH. Mirjafari et al. (2007) reported that the equilibrium constants for reactions (6.2) and (6.3) are 2.6×10^{-3} and 1.7×10^{-4}, respectively. Reactions (6.3) and (6.4) are very rapid, the rate being virtually diffusion controlled (Gutknecht et al., 1977). Thus, if a feasible means to accelerate the hydration of CO_2 (6.2) could be found, it should be possible to sequester large quantities of anthropogenic CO_2 into mineral carbonates. Therefore, to be used as a viable method for CO_2 sequestration from anthropogenic sources, the hydration of CO_2 (6.2) must be significantly accelerated.

6.1.1.2 Role of Carbonic Anhydrase in Biomineralization

Biominerals existing in nature hold the key to accelerate the mineralization of CO_2. Similar mineralization of CO_2 is carried out by a variety of marine organisms, resulting in the formation of $CaCO_3$ in the shells of snails and eggs, exoskeleton in invertebrates, bivalves, silicates in algae and diatoms, and extensive oolitic limestone bed formation (Miyamoto et al., 1996).

A large number of microbial strains (various cyanobacteria, eukaryotic microalgae, *Bacillus, Pseudomonas, Vibrio,* and sulfate-reducing bacteria) capable of carrying out calcification, the conversion of CO_2 into $CaCO_3$, have been reported (Barabesi et al., 2007; Ercole et al., 2007; Pomar and Hallock, 2008). All these organisms harbor a biocatalyst called CA, which actually is responsible for acceleration of CO_2 hydration (6.2), thereby speeding up the entire carbonate formation. This enzyme is an effective biocatalyst and nature's way of efficiently converting carbon dioxide to bicarbonate.

6.1.1.3 Carbonic Anhydrase: A Brief Overview of Physiological Functions

Carbonic anhydrases (EC 4.2.1.1) are metalloenzymes containing Zn^{2+} in its active site and catalyzes the reversible and rapid hydration of CO_2-releasing protons.

$$CO_{2(aq)} + H_2O \rightarrow HCO_3^- + H^+$$

This enzyme accelerates the reaction dramatically, provided the pH is above the pKa of CO_2/HCO_3^- equilibrium. The activity of CA and its role in mineralization of CO_2 in molluscs has been studied in detail. Initiation of shell formation occurs by conversion of CO_2 to HCO_3^- and CO_3^{2-} by the action of CA. First, carbonates and bicarbonates that are formed mend into epithelial tissues along with Ca^{2+}. The proteins and mucopolysaccharides are synthesized by mantle zone. Then, $CaCO_3$ is synthesized in extrapallial space around the boundary of living tissue and grows epitaxially forming the exoskeleton (Wilbur and Anderson, 1950; Nielsen and Frieden, 1972; Wilbur and Saleuddin, 1983; Duvail et al., 1998; Medakovic, 2000; Henry et al., 2003; Yu et al., 2006; Lee et al., 2010).

Carbonic anhydrases are one of the fastest enzymes known and among them human isozyme HCAII is the fastest known CA with a k_{cat} ranging between 10^4 and 10^6/s depending on the class and the organism (Steiner et al., 1975; Tripp et al., 2001). The rate of this reaction is limited by the physical processes such as diffusion and proton transfer. The CAs are ubiquitous in nature present in all types of living organisms, including animals, plants, algae, and bacteria. Presently, they belong to at least five distinct gene families (designated as α, β, γ, δ, and ϵ) encoding them, and these have no significant amino acid sequence similarity indicative of convergent evolution of catalytic function (Tashian, 1989; Kaur et al., 2012). The first CA was discovered by Meldrum and Roughton (1933) while looking for the factors responsible for a rapid transition of bicarbonate anions HCO_3^- from erythrocytes toward the lung capillary. In 1939, CAs present in plants have been shown to be different from the previously known CAs (Neish, 1939). In 1940, first CA was purified by Keilin and Mann from bovine erythrocytes and showed that CA contains a Zn atom in their active site. In 1963, Veitch and Blankenship (1963) discovered CAs in prokaryotes and in 1972 first α-CA was purified from *Neisseria sicca* (Adler et al., 1972). The sequencing of the first CA of *Escherichia coli* was done in the 1990s (Guilloton et al., 1992), which was classified as the first β-CA having a molecular weight of 24 kDa.

Carbonic anhydrases play an essential role in various biological processes such as CO_2 and ion transport across biological membranes, respiration, supply of bicarbonate to various metabolic pathways, acid–base regulation, and biomineralization. They are also an important component of carbon-concentrating mechanism (CCM) in various photosynthetic organisms, plants, algae, and cyanobacteria, enabling them to fix CO_2 for carrying out photosynthesis. These are also involved in cyanate degradation in prokaryotes (Medakovic et al., 1997; Smith and Ferry, 1999; Badger et al., 2003; Badger, 2000). In some species, more than one class of CA is present.

6.1.1.4 Catalytic Mechanism of CA

Zn^{2+} present in the active site of the enzyme has a central role in catalysis (Figure 6.1). Zn^{2+} brings the two reactants water and CO_2 in close proximity as the positive charge on zinc attracts the oxygen of water molecule (Figure 6.2). This in turn brings down the pKa of water molecule from 15.7 to 7.0 and facilitates the removal of proton, leaving zinc bound to a hydroxide ion, which is a strong nucleophile (Campbell and Reece, 2005; Berg et al., 2007). Bicarbonate is thus formed by nucleophillic attack of hydroxide ion on CO_2.

Steps involved in catalysis by CA.

$$E\text{-}ZnH_2O \rightarrow E\text{-}ZnOH^- + H^+ \qquad \text{(A)}$$

$$E\text{-}ZnOH^- + CO_{2(aq)} \rightarrow E\text{-}ZnHCO_3^- \qquad \text{(B)}$$

$$E\text{-}ZnHCO_3^- + H_2O \rightarrow E\text{-}ZnH_2O + HCO_3^- \qquad \text{(C)}$$

At the end of the reaction, the water molecule bound to zinc loses a proton regenerating the active site of CA. Thus, CA has the inherent potential for accelerated mineralization of CO_2, and therefore, can be applied to sequester CO_2 from industrial flue gases. In addition to allowing fast, safe, and long-term sequestration of CO_2, biomineralization via CA enables the process to occur at ambient or near-ambient conditions.

6.1.1.5 Proposed Technical Designs for Carbon Capture
Using CA before Biomineralization

The capturing of CO_2 from remainder of the gases is the biggest contributor to the cost associated with the CCS process. The three widely used methods (postcombustion, precombustion, and oxyfuel) for CO_2 scrubbing requires large input of energy consuming up to 40% of a power station's energy. Carbon capture assisted by CA can offset this cost

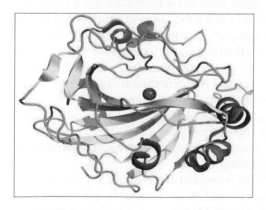

FIGURE 6.1
(**See color insert.**) Structure of carbonic anhydrase (http://en.wikipedia.org/wiki/File:Carbonic_anhydrase.png).

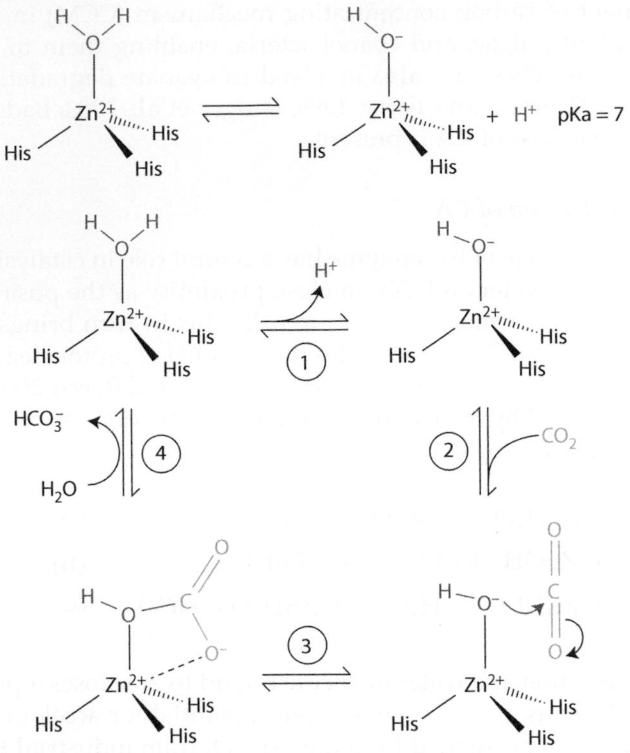

FIGURE 6.2
(See color insert.) CO_2 hydration at the active site of CA. (Adapted from Berg JM, Tymoczko JL, and Stryer L. 2007. *Biochemistry*, 6th ed. Sara Tenney, New York.)

and when combined with further mineralization of scrubbed CO_2 brings about significant reduction in overall cost of the CCUS process.

CO_2 Solution Company has proposed large-scale reactors for CO_2 capture for using CA as an onsite scrubber of CO_2 from coal-fired power generation, oil sands, and other CO_2–intensive industries (www.co$_2$ solutions.com/en/the-process). CA is immobilized onto a solid matrix in a bioreactor. The capture of CO_2 directly from flue gases by CA and its separation from other flue gases requires a bioreactor containing a matrix onto which CA is immobilized and which acts as a solid support for the enzyme. Water is sprinkled into the bioreactor to maintain aqueous condition from the top with the help of a pump and flue gas is made to enter from the bottom of the bioreactor, which gets converted into aqueous CO_2 as it bubbles up through the cylinder. Immobilized CA converts this CO_{2aq} to bicarbonate ions, which is extracted for mineralization or other uses. CO_2-free air leaves from the top. Goff and Rochelle (2004) showed that using CA to capture CO_2 makes the process economical when compared with other nonbiological capture methods, such as an amine solution. The process is shown in Figure 6.3.

Different variations of such basic bioreactors have been studied since then for increasing the efficiency of carbon capture by CA.

Bhattacharya et al. (2003) increased the rate of CO_2 dissolution from flue gas by using a water sprayer and immobilized CA by covalently grafting it on silica-coated porous steel. The group has shown that best CO_2 capture is possible when there is a horizontal inflow and outflow of the CO_2-carrying gas (at 60°C) with continuous water spraying from the top of the

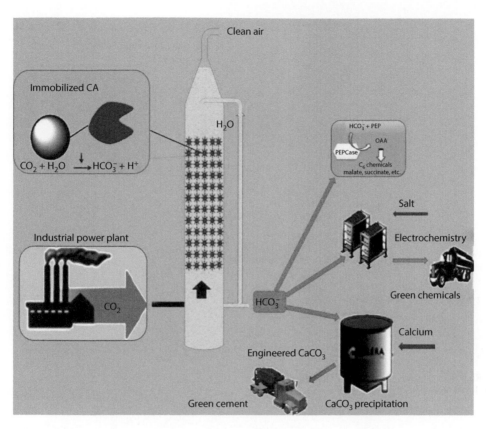

FIGURE 6.3
(See color insert.) Schematic representation of CO_2 sequestration processes using CA.

reactor. Larachi et al. (2012) have recently shown that by combining the immobilized enzyme absorption technique with ion-exchange resin results in the removal of excess enzyme, which may cause inhibition by bicarbonate ions. Iliuta and Larachi (2012) analyzed the mass transfer exchange mechanism of gas–liquid and liquid–solid packed-bed reactors and a similar mass exchange and CO_2 hydration kinetics were studied by Lacroix and Larachi (2008) who reported that for packed bed reactors gas–liquid and liquid–solid mass transfer exchange mechanisms could considerably affect the CO_2-hydration kinetics. The time required for mass transfer from gas to liquid phases is more than the characteristic time required for the reactions to occur in the liquid phase. Transport phenomena are not quick enough for enabling CO_2 to diffuse across the liquid phase before reaching the enzyme containing solid surface. Therefore, the reactor configurations and gas–liquid–solid contact patterns must be carefully selected to maximally exploit the catalytic potential of CA in capturing CO_2. Based on the above two studies, Iliuta and Larachi (2012) developed a novel a three-phase monolith reactor having HCAII enzyme immobilized onto the longitudinal channels wash coat of a postcombustion column. Aqueous slurry (containing resin exchange beads) is flown that enables *in situ* removal of inhibitory bicarbonate ions (Figure 6.4).

National Aeronautics and Space Administration (NASA) has designed another bioreactor (Figure 6.5) for capturing CO_2 directly from the ambient atmosphere of confined inhabited cabins, which have CO_2 concentration of 0.1% or less, by making use of thin aqueous films carrying dissolved CA (Ge et al., 2002; Cowan et al., 2003a,b). The enzyme

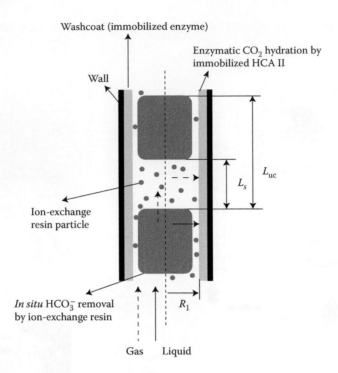

FIGURE 6.4
A model of the 3 phase reactor. (Adapted from Iliuta I and Larachi F. 2012. *Separation and Purification Technology* 86: 199–214.)

selectively allowed CO_2 to diffuse inside in a ratio of 1400:1 by comparison with N_2 and 866:1 by comparison with O_2. A schematic illustration of the design involved is shown in Figure 6.3. The core of the liquid membrane is composed of a very thin (e.g., 330 μm thick) layer of CA solution in phosphate buffer, sandwiched between two microporous and hydrophobic polypropylene membranes. The integrity of the membrane and rigidity is provided by shielding it with thin metal grids. The CO_2 from the atmosphere gets purified spontaneously and gets dissolved inside the liquid membrane on one face of the membrane, diffuses through it and evaporates out the liquid membrane on the opposite face both in vacuum and in a carrier gas. The liquid membrane, however, tends to get dried up during prolonged operations.

Ward and Robb (1967) successfully applied this technique of simple diffusion liquid enzyme membranes by using CA dissolved in caesium or potassium bicarbonate solutions to capture CO_2 from gases containing up to 5% CO_2. This work further instigated Matsuyama and colleagues to develop similar enzymatic liquid films to capture CO_2 from gases containing up to 15% CO_2, which closely represents CO_2 concentration in flue gas (Matsuyama et al., 1994, 1999).

Another CO_2-capture technology employing CA has been developed by Carbozyme Company as illustrated in Figure 6.6. The Carbozyme reactor is composed of two hollow fiber, microporous propylene microfiber membranes separated by a thin-liquid membrane (CLM). CA is immobilized onto the hollow fiber wall to ensure the contact between incoming CO_2 and CA at the gas–liquid interface to achieve maximum catalytic efficiency. Microporous membrane offers low resistance to the flow of gas, and therefore, to the CO_2 diffusion before getting converted to bicarbonates by CA. Thus, CO_2 faces little diffusion resistance before

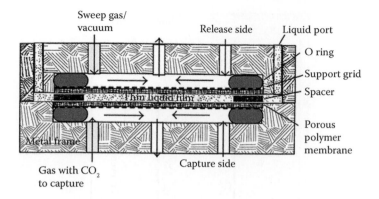

FIGURE 6.5
A design of thin liquid membrane system for CO$_2$ capture developed by NASA.

being captured selectively by CA and converted to bicarbonate at the feed side. This design offers significant cost reduction and performance advantages over conventional amine or ammonia-scrubbing processes for capture of CO$_2$ from flue gases and is applicable for flue gases having CO$_2$ concentration ranging from 1% to approximately 20% or higher.

Several improvements in designs have been made since then, relating to the geometry of microfiber network, the nature of hollow microfibers, and by the use of different types of CA. This technique was found to be efficient for a flue gas carrying 0.05%–40% CO$_2$ at 15–85°C with a particular γ-CA isozyme (Trachtenberg et al., 2007). Zhang and co-workers had also successfully designed such hollow fiber membrane reactors with some CA immobilized in nanocomposite hydrogel/hydrotalcite thin films, which were used as thin layers to separate the fibers (Zhang et al., 2009, 2010, 2012a,b).

Another CA-mediated CO$_2$ capture system has been developed by Trachtenberg et al. (2007) for continuous CO$_2$ capture suitable for broad spectrum of applications

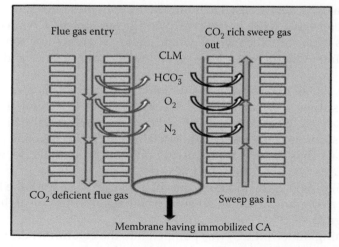

FIGURE 6.6
A modified representation of hollow microporous membrane-based CO$_2$ capture designed by Carbozyme, Inc.

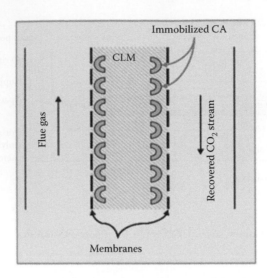

FIGURE 6.7
A contained liquid system designed by Trachenberg et al. (2009).

(e.g., submarine and postcombustion CCS). They employed two gas-permeable membranes operated at different pressures having a solvent layer in between the two (Figure 6.7).

When flue gas is applied, CO_2 diffuses through one membrane and gets absorbed into the solvent, which is then desorbed at the other membrane operated at pressures slightly lower than that of the first membrane. The CA can be used as free enzyme in solution (Cowan et al., 2003a,b) or it can be immobilized onto the membrane, which significantly enhances mass transfer just at the gas–liquid interface (Trachtenberg et al., 2009). This process suffers from disadvantages such as there is a relatively high concentration of oxygen in the recovered CO_2 rich stream, which exceeds the admissible limit for further CO_2 processing and storage. And there is an issue of extensive solvent evaporation. A solution was proposed to overcome the above problem for using the contained liquid membrane (CLM) system wherein a hydrogel was used to immobilize CA in the hollow fiber contactor for air purification (Cheng et al., 2008). In the case of flue gas treatment, CA was trapped within membrane pores which are filled with an ionic liquid providing low solvent evaporation, and thereby, maintaining the water level needed for enzyme activity and stability (Neves et al., 2012).

Dziedzic et al. (2006) developed a gas-to-liquid membrane CO_2 sequestration system (Figure 6.8) consisting of a hollow fiber membrane module, which allows the contact between the CO_2-bearing gas stream and the absorbing liquid stream. Bicarbonate formation occurs in a separate conversion module having controlled alkaline pH and a supply of CA. Bicarbonate thus formed is pumped to mineralization module, which is supplemented with a Ca^{2+} source to form solid $CaCO_3$.

Liu et al. (2005) provided the proof-of-concept at the lab scale with the help of a contactor composed of a column packed with CA immobilized in chitosan–alginate beads and a liquid mixer (Figure 6.9).

Liquid containing dissolved CO_2 was pumped into the column where CA catalyzed the conversion of CO_2 into bicarbonate. The effluent from the column was mixed with an aqueous solution containing metal ions having the composition similar to that of water streams produced by oil/gas wells. Calcium carbonate formation was then monitored by varying

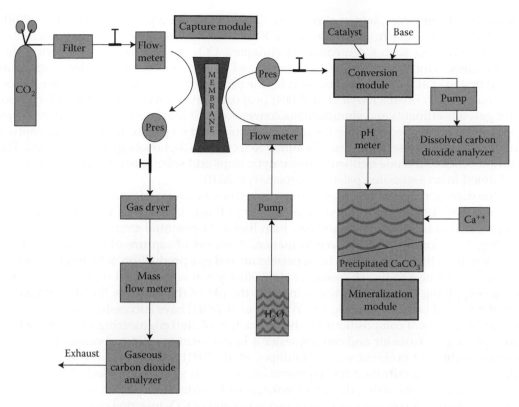

FIGURE 6.8
An illustration of gas-to-liquid hollow fiber membrane system developed by Dziedzic et al. (2006).

the concentration of CO$_2$ being pumped into the packed column. The results showed that the rate of carbonate precipitation increases if the enzyme activity is constantly maintained in the presence of a buffer.

Novozyme Company has patented some of the research where it proposed to combine various CO$_2$ capture and release units similar to those developed by the CO$_2$ Solution or

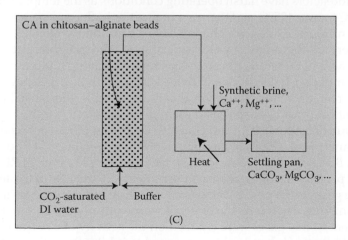

FIGURE 6.9
Design of lab scale model by Liu et al. (2005).

Carbozyme Companies, interconnected by fluid circulation pipes (Saunders et al., 2010). They have used α-CA from *Bacillus clausii* KSM-K16, for studies with hollow microfiber units. By utilizing such CA-mediated techniques, CO_2 is captured and converted into bicarbonates, which may further be mineralized in the presence of divalent cations as shown by Liu et al. (2005), or can be used for the production of chemicals and other important compounds. Bhattacharya et al. (2004) proposed an innovative unit for CO_2 capture from gaseous effluents and simultaneous conversion of it into useful compound by developing a spray fixed bed absorber having CA immobilized on porous particles. The capture unit is included in a series of reactors aimed to fix CO_2 in 3-phosphoglycerate (3PGA). The multiple reactors include multienzyme catalytic steps and solar panels for ATP production as claimed in an associated patent (Bhattacharya, 2001).

In the direction of utilizing wastes from industries as source of cations, Liu et al. (2005) used synthetic brines with compositions similar to those of produced waters from oil and gas production in the Permian and San Juan Basins as potential sources for cations (Ca^{2+} and Mg^{2+}) in biomimetic carbonate formation. A means of capture of CO_2 has also been designed, with the use of wastes from petroleum and gas production, which is rich in Ca^{2+} ions such as brine and fly ash (Soong et al., 2006). Fly ash when mixed with brine, in addition to supplying Ca^{2+} ions, serves to increase the pH of the mixture, thereby enhancing the efficiency of carbonation reaction. Walenta et al. (2011) have successfully attempted the use of CA in cement compositions for the production of civil engineering materials, which can capture CO_2 from air and can sequester it in the form of solid carbonates within the porous coatings of building walls. Mahinpey et al. (2011) investigated the application of CA catalyzed mineralization for sequestration of CO_2 in saline formations. He proposed the use of CA in minimizing the risk of leakage of CO_2 from the injection wellbore. A sufficient volume of CA solution can be pumped at the end of CO_2 injection period through the injection wellbore. When the CO_2 come in contact with the enzyme solution, solid particles would precipitate speeding up pore occlusion resulting in reduced permeability of porous media near the wellbore.

6.1.1.6 Screening of Efficient CA for CO_2 Capture from Industrial Flue Gas

There are many practical problems associated with the use of CAs in actual carbon sequestration. These bioreactors have harsh operating conditions as the temperature of flue gases ranges from 50°C to over 125°C, and there are impurities in flue gas such as organic amines and sulfur and nitrogen oxides, which may have an inhibitory effect on the functioning of CA. Moreover, for the efficient CO_2 sequestration, the pH should be maintained at higher values. There is, therefore, a need to look for CAs having high catalytic activity and remains stable under such conditions without losing catalytic efficiency. The research is in progress to screen and characterize CA that is alkalistable, thermostable, and is able to tolerate the effects of other toxic species present in flue gas.

The CA was purified, sequenced, and characterized from *Citrobacter freundii* and *Bacillus subtilis* by Ramanan et al. (2009a,b) having a molecular weight of 24 and 37 kDa, respectively. CA from *B. subtilis* was found to be stable over the pH range of 7.0–11.0 with optimum activity at pH 8.3. One of the main concerns related to the use of CA for sequestration of CO_2 from flue gases is the possible effects of other components present in the flue gas on the activity of enzyme. Therefore, the effect of various inhibitors and metal ions particularly those present in flue gases were checked on the activity of CA from *B. subtilis* and it was found that Pb^{2+} and Hg^{2+} at 1 mM were found to inhibit the activity of CA, and metal ions such as Co^{2+}, Cu^{2+}, and Fe^{3+} stimulated enzyme activity. SO_4^{2-}, the main component of flue gas, which is generally known to inhibit CA, surprisingly was found to activate CA (Ramanan et al., 2009a,b).

Sharma and Bhattacharya (2010) characterized and compared CAs isolated and purified from *Pseudomonas fragi* (PCA), *Micrococcus luteus* 2 (MTCA), and *Micrococcus lylae* (MLCA). Among them, CA from *P. fragi* exhibited the highest specific activity next to BCA. CA from *Mi. luteus* 2 was found to be highly stable in a temperature range of 35–45°C followed by CAs from *P. fragi*, *Mi. lylae,* and BCA. The effect of other metal ions and anions on the activity of these CAs was checked and it was found that Cd^{2+}, Zn^{2+}, Co^{2+}, and Fe^{2+} stimulated CA activity and PCA and MLCA were able to retain 80% and 90% of their activity, respectively, in the presence of 100 mM sulfate ion and 50 mM nitrate ion.

Bond et al. (2001) have investigated the effects of SO_4^{2-} and NO_3^- ions on CA by directly supplementing these to the aqueous solution at 0–5°C. They showed that with less than 0.1 M ion concentration, the effect was negligible. A γ-class CA has been purified and characterized from *Methanosarcina thermophila* (CAM), which is optimally active at 55°C, with a k_{cat} of 10^5/s and exhibits 50% residual activity after 15 min of incubation at 70°C and is inactivated at 75°C (Alber and Ferry, 1994). A β-class CA has been studied from *Methanobacterium* (CAB), which is comparatively more thermostable, retaining 50% residual activity after 15 min at 85°C and gets inactivated at 90°C, but has a lower k_{cat} value than that of CAB (104/s) (Smith and Ferry, 1999).

One more γ-class CA has been the focus of study, which is from *Pyrococcus horikoshii* (Jeyakanthan et al., 2008). This CA has not been fully characterized but this organism shows optimum growth at 98°C (Ferry, 2010). Several other heat-stable CAs for the extraction of CO$_2$ from CO$_2$-containing media have been reported (Borchert and Saunders, 2010, 2011). A detailed study on a thermophilic α-class CA from *B. clausii* has also been done. This CA exhibits higher thermostability than CAM with 17% residual activity after 15 min at 80°C in 1.0 M sodium bicarbonate at pH 8.05. A 0.6 g/L of this enzyme is capable of extracting >99% of CO$_2$ from a 15% CO$_2$ gas stream against only 33% removal without CA (Borchert and Saunders, 2010). A highly stable CA from the thermophilic organism *Caminibacterium ediatlanticus* DSM 16658 has T_m of 109°C at pH 9.0 and has a residual activity of 40% after 15 min in 1.0 M sodium bicarbonate at 100°C, and also shows increased CO$_2$ extraction with 1.0 M sodium bicarbonate at pH 9.0 (Borchert and Saunders, 2010).

Recently, a β-class CA CahB1 from the cyanobacterium *Coleofasciculus chthonoplaste* has been characterized in terms of its catalytic activity and effect of various inorganic anions and small molecules on the activity. The enzyme has a good hydration activity having a k_{cat} of 2.4×10^5/s and a k_{cat}/K_m of 6.3×10^7 M^{-1} s^{-1}. Perchlorate and tetrafluoroborate did not affect the enzyme, whereas selenate and selenocyanide were found to be weak inhibitors with KIs of 29.9–48.61 mM. Various halides, pseudohalides, carbonate, bicarbonate, trithiocarbonate, and a range of heavy metal ions containing anions were found to inhibit the enzyme activity at submillimolar–millimolar range with KIs ranging from 0.15 to 0.90 mM. *N,N*-diethyldithiocarbamate, sulfamate, sulfamide, phenylboronic acid, and phenylarsonic acid were the best inhibitors of CahB1 with KIs in the range of 8–75 µM (Vullo et al., 2014).

The extraction and purification of enzyme from wild-type microbes is a costly affair. Cloning of the gene encoding the particular enzyme provides an economic way of producing a large quantity of that protein. All CA-encoding genes (can, cynT, caiE, pay, and yrdA) from *E. coli* have been cloned and overexpressed and their expression profiles have been studied in response to different growth conditions (Merlin et al., 2003). Premkumar et al. (2003) cloned α-type CA (Dca), which is associated with the plasma membrane of the extremely salt-tolerant, unicellular, green alga *Dunaliella salina*. The CA is active over broad salinity of 0–4 M NaCl. Kaur et al. (2010) cloned a putative γ-CA encoding gene of

Azospirillum brasilense and overexpressed it in *E. coli*. Its expression was induced in the stationary phase and in the presence of high CO_2. This CA is cotranscribed with the *N*-acetyl-γ-glutamate-phosphate reductase suggesting a possible link between arginine metabolism and an unknown CO_2-dependent metabolic process that utilizes this CA. An α-type CA of *Neisseria gonorrhoeae* was cloned and overexpressed in *E. coli*. The recombinant CA, in purified as well as nonpurified crude form, shows comparable CO_2 hydration activity to commercial BCA and considerably promoted the formation of solid $CaCO_3$ (Kim et al., 2012). Recently, a full-length open-reading frame of marine diatom encoding a CA gene from *Thalassiosira weissflogii* has been cloned and functionally expressed in *E. coli* using pTWIN2 expression vector (Lee et al., 2013).

Carbozyme Company has patented the use of CA from *Myceliophthora thermophila*, which is active at temperatures as high as 85°C. While Kanth et al. (2012) expressed a codon optimized α-type CA from *Dunaliella* sp. (Dsp-aCAopt) and characterized its kinetic properties to use it in the conversion of CO_2 to calcite, and concluded that Dsp-aCAopt is an effective enzyme catalyzing the mineralization of CO_2 to $CaCO_3$ as the calcite form in the presence of Ca^{2+} ions producing 8.9 mg of calcite per 100 μg (172 U/mg enzyme) in the presence of 10 mM of Ca^{2+}.

Capasso et al. (2012) cloned and expressed an α-CA (SspCA) from *Sulfurihydrogenibium yellowstonense* YO3AOP1, which thrives in hot springs at temperatures up to 110°C in *E. coli*. The recombinant CA exhibits high thermostability at 70°C for several hours. The effects of oxides of nitrogen and sulfur (typically contained in flue gas) were studied on SspCA by De Luca et al. (2012), and the inhibition constants were in the range of 0.58–0.86 mM for the anions NO_2^-, NO_3^-, and SO_4^{2-}. The stability of SspCA at high temperatures was optimal, which is about 53 days at 40°C and about 8 days at 70°C making this CA suitable for industrial use.

Another approach to improve the activity and stability of CA under nonnatural conditions could be by employing protein engineering to design functional CA having high activity, thermostability, pH optimum, and tolerance to toxic impurities (Daigle and Desrochers, 2009; Newman et al., 2010). Based on the available information on the structure and function relationship of CAs, protein engineering research is being carried out for obtaining CA variants displaying desirable functions in activity and stability. Cordexis company has done four rounds of site-directed mutagenesis on a CA from a mesophillic organism that resulted in increased enzyme stability with a residual activity of >40% at 82°C in 5 M MDEA at pH 11.8. The half-life ($t_{1/2}$) increased by five times to 20 h in 4.2 M MDEA at 75°C (Savile et al., 2011).

6.1.1.7 Immobilization of CA for Use in Carbon Capture

Immobilization is a technique in which an enzyme is attached to an inert, insoluble material restricting its movement in space and provides increased resistance to changes in conditions such as pH or temperature. It also possesses other benefits such as it makes it possible to recycle the enzyme for reusing it, less amount of immobilized enzyme is needed to carry out a reaction, easy recovery of product from the reaction mixture and thus, ultimately reducing the expenses associated with the process (http://en.wikipedia.org/wiki/Immobilized_enzyme).

Immobilization of CA on various matrices is the most effective method to stabilize them in order to increase its reusability, operational stability, and to limit exposure to denaturing conditions for the accelerated CO_2 capture under relevant process conditions. It enables enzyme bioreactors to operate under continuous conditions without losing the enzyme. Various techniques for immobilizing CA are now being investigated, such as adsorption and entrapment, covalent attachment, cross-linking, and affinity. Prabhu et al. (2009) checked different chitosan and sodium alginate-based materials for entrapment of whole cells of *Bacillus pumilus* (extracellular CA producer), and these were found to enhance the associated CA activity when compared with the free cells. Immobilization of purified CAs from different bacterial species (*P. fragi*, *Mi. lylae*, *Mi. luteus* 2, and *B. pumilus*) has also been studied on chitosan and alginate beads and it was reported that immobilized CAs exhibit improved storage stability with a retention of 50% of its initial activity after 30 days (Sharma and Bhattacharya, 2010; Prabhu et al., 2011). Wanjari et al. (2011) had also studied immobilization of CA of *B. pumilus* and its effect on kinetic parameters and also on the mineralization efficiency of the enzyme. They reported an increase in the precipitation of CaCO$_3$ compared with the free enzyme and K_m and V_{max} for immobilized enzymes were 2.36 mM and 0.54 µmol/min/mL, respectively, and for the free enzymes were 0.87 mM and 0.93 µmol/min/mL, respectively, by using esterase assay for measuring the activity. Prabhu et al. (2011) studied the kinetic constants of the CA from *B. pumilus* after immobilization on chitosan-activated alumina-carbon composite beads and the K_m and v_{max} values were 10.35 mM and 0.99 µmol/min/mL.

Silylated chitosan beads were used to immobilize CA by Yadav et al. (2010). The immobilized CA possessed increased storage stability over free enzyme and with a residual activity of 50% up to 30 days. They also determined the kinetic constant of immobilized CA for mineralization of CO$_2$ by quantification of the CaCO$_3$ precipitate using gas chromatography after dissolving the precipitate by the addition of HCl to evolve the captured CO$_2$(g). While V_{max} remained unchanged, K_m was higher for the immobilized CA when compared with the free CA. The best CO$_2$ sequestration capacity and improved stability was achieved with CA immobilized on core–shell CA-chitosan nanoparticles (SEN-CA), obtained by covering the surface of CA by applying a thin layer of chitosan (Yadav et al., 2011). The ordered mesoporous aluminosilicate has also been used for immobilizing CA, on which CA exhibited K_m, V_{max}, and k_{cat} values of 0.158 mM, 2.307 µmol/min/mL, and 1.9/s, respectively (Wanjari et al., 2012). Vinoba et al. (2012a) studied immobilization of BCA by using octa(aminophenyl)silsesquioxane-functionalized Fe$_3$O$_4$/SiO$_2$ nanoparticles and showed 26-fold higher enhancement in activity than the free enzyme, and the immobilized enzyme could be used up to 30 cycles with the retention of 82% of its activity after 30 days. Vinoba et al. (2011) have also compared the immobilization of BCA on SBA-15 by employing different techniques comprising covalent attachment, adsorption, and cross-linked enzyme aggregation, and reported similar CaCO$_3$ precipitation abilities by the three immobilization techniques. The k_{cat} values for CO$_2$ hydration for BCA immobilized via cross linking were 0.58, 0.36, 0.78/s in comparison with 0.79/s for free enzyme, respectively, indicating that the crossed-linked BCA exhibited fairly higher hydration rate compared with the rest of immobilization techniques used above, although it was lower than the free BCA (Vinoba et al., 2012b).

A CA from *Rhodobacter sphaeroides* has been cross linked to electrospun polystyrene/poly(styrene-comaleic anhydride) nanofibers (CLEA), which retained more than 94.7% of activity for 60 days of storage at 4°C and retained more than 45.0% activity after 60 cycles. The immobilized CA also maintained continuous bicarbonate supply, which induced enhancement in the growth of *R. sphaeroides* in the presence of 5% CO$_2$, which in turn increased the production of organic substances including carotenoid, bacterio-chlorophyll, porphyrin, and coenzyme from it (Park et al., 2012).

Extracellular CA from *B. subtilis* VSG-4 has been purified followed by its immobilization and characterization by Oviya et al. (2012). Immobilization was achieved by entrapment within a chitosan–alginate polyelectrolyte complex (CA PEC) hydrogel. The optimum pH and temperature for this form of CA were found to be 8.2°C and 37°C. The maximum CO_2 mineralization was obtained with immobilized CA (480 mg $CaCO_3$/mg protein). The immobilized enzyme had much higher storage stability than the free enzyme. In another attempt by Oviya et al. (2012) *E. coli* MO1-derived CA was purified and immobilized in chitosan–alginate polyelectrolyte complex (CA-PEC) with an immobilization potential of 94.5%. The K_m and V_{max} of the immobilized CA came out to be 19.12 mM and 416.66 μmol/min/mg, whereas the K_m and V_{max} of the free counterpart were 18.26 mM and 434.78 μmol/min/mg. Both the forms of the CA were most active and stable at pH 8.2 and at 37°C. The CO_2 sequestration potential in terms of $CaCO_3$ precipitation of both immobilized and free CA was 267 and 253 mg/mg of enzyme, respectively.

The major disadvantage of using immobilization via adsorption is the leakage (leaching) of enzyme from the solid support during repeated use. The covalent attachment of enzyme onto a solid support can solve the problem of enzyme leaching with repeated cycle of reuse. Bhattacharya et al. (2003) immobilized CA on γ amino-propyltriethoxysilane-coated iron particles, by grafting via dicarbocarbodiimide (DCC) bonds or dicarboxy bonds. Immobilization was also attempted by copolymerization of CA with gluteraldehyde in methacrylic acid polymer beads. These methods (particularly the DCC and dicarboxy coupling) proved to be efficient in minimizing leaching with 98% activity retention. While Zhang et al. (2009, 2010) studied the CA immobilization by covalently grafting it on a hybrid poly(acrylic acid-coacrylamide)/hydrotalcite anocomposite termed "PAA-AAm/HT," they reported an immobilization of up to 4.6 mg of CA per gram of support with retention of 76.8% of the initial activity. In another effort to immobilize CA, Zhang et al. (2012a,b) grafted CA by covalent coupling mechanism onto silica nanoparticles formed by spray pyrolysis and reported a considerably enhanced thermal stability in comparison with the free CA.

Besides covalent linking and adsorption, enzymes can be effectively immobilized within porous supports such as polyurethane foam (Kanbar and Ozdemir, 2010). CA immobilized on polyurethane foam can be used for seven continuous cycles for capturing CO_2 in aqueous media in a temperature range from 35°C to 45°C. Recently, Migliardini et al. (2013) studied the immobilization and CO_2 capture potential of recombinant SspCA, which has been earlier cloned and purified by Capasso et al. (2012). Polyurethane foam was used as a support for immobilizing SspCA. The immobilization conferred enhanced stability with retention of catalytic activity. Also, immobilized CA performed efficiently in capturing CO_2 when used in a bioreactor, supplied with CO_2-resembling conditions close to those existing in power plant emissions.

Membranes have also recently been used as immobilization supports for using CA in CO_2 sequestration. Cheng et al. (2008) used hollow fiber membrane to immobilize CA, and performed laboratory and pilot scale experiments to check its feasibility for industrial applications. Based on their experiments they reported a successful CO_2 capture with 80% sequestration of CO_2. The successful immobilization of CA in the pores of a permeable polymeric membrane was carried out by Favre and Pierre (2011) using BCA by simply immersing the membrane in an enzyme solution. Recently, a recombinant enzyme SspCA, from the thermophile bacterium *Sulfurhydrogenibium* sp. has been characterized and immobilized by silanization of a siliceous support (Sipernat®) with subsequent activation with glutaraldehyde. It emerged as a potential biocatalyst for CO_2 capture processes based on regenerative absorption into alkaline solutions (Russoa et al., 2013).

The kinetics of $CaCO_3$ precipitation in a $CaCl_2$ solution and other earth cation carbonates has been investigated by Pocker and Bjorkquist (1977), Bond et al. (2001), Liu et al. (2005), and

Druckenmiller and Maroto-Valer (2005). For efficient CO$_2$ mineralization, enzyme could be directly added in the precipitation medium, or it could be first used to catalyze the formation of HCO$_3^-$ in a pH ranging from 8.55 to 8.7, then the Ca^{2+}-containing brine would be added. Mirjafari et al. (2007) measured the amount of CaCO$_3$ precipitated in a buffered aqueous CaCl$_2$·2H$_2$O solution, after the addition of CO$_2$ saturated water.

Favre et al. (2009) examined the deposition kinetics of CaCO$_3$ in buffers of varying pH and different enzyme concentration and showed that the presence of CA could hugely increase the apparent CaCO$_3$(s) precipitation rate during the first minute with an enzyme mass ≤0.3 mg/mL at high pH. If the pH was not set to high value (e.g., 10), the amount of CaCO$_3$(s) precipitated was independent of the presence of enzyme. CA only decreases the time needed to achieve equilibrium where CaCO$_3$(s) precipitation stops. When the concentration of enzyme increases, it results in faster increase in the formation rate of HCO$_3^-$ and H$^+$ during the first deprotonation step. This leads to a more rapid decrease of pH to a low value, undesirable for CO$_3^{2-}$ ions formation. This in turn stopped CaCO$_3$ precipitation early. Therefore, the final amount of CaCO$_3$(s) precipitated depends only on the initial buffer nature, pH, and concentration of enzyme. When the precipitates were analyzed by x-ray diffraction, it was observed that in absence of enzyme at initial pH 8.4 or 9.4, and temperature of 20°C, calcite was the only phase present. Vaterite was predominant at an initial pH 10.5. While at pH 10.5, the vaterite to calcite conversion was favored by the enzyme.

Ramanan et al. (2009a) compared CA from *Ci. freundii* and *B. subtilis* for their CO$_2$ mineralization potential and observed that purified CA is more active than the crude. Acceleration in CaCO$_3$ precipitation was observed by Kim et al. (2012) using a recombinant α-CA of *N. gonorrhoeae*.

Li et al. (2010) observed similar CO$_2$ sequestration efficiency in calcite precipitation with CA from microbial origin and BCA. Sharma et al. (2011) reported a greater than two-fold improvement in CaCO$_3$ precipitation over a period of 5 min using purified CA from *P. fragi* immobilized on chitosan. A significant enhancement in precipitation rates were also achieved with CA from *B. pumilus* adsorbed on chitosan beads (Wanjari et al., 2011).

6.1.1.8 Using Whole Microbial Cells for Biomineralization

Whole cells of microorganisms have the potential to be used for expediting the process of carbon capture and storage (CCS) and this is being investigated by a number of scientists. The whole cell approach offers significant economical advantage because the purification of enzyme is no longer required.

Cyanobacteria are known to carry out calcification, magnificent example being the presence of stromatolites and whitings, which are extensive precipitations of fine-grained CaCO$_3$ along with organic compounds that can turn whole water bodies such as Lake Michigan and the Great Bahama Bank into milky state. They generally flourish in high levels of CO$_2$ and also many cyanobacteria are halophilic, and therefore, are potential systems for CO$_2$ sequestration from flue gas (Ono and Cuello, 2007). Cyanobacteria utilize solar energy for the conversion of carbon dioxide into calcium carbonate and this calcification is a nonobligate process that is intimately associated with photosynthetic activities. The CCM present in autotrophs enables the cells to raise the concentration of CO$_2$ at the vicinity of the carboxylating enzyme ribulose 1,5-bisphosphate carboxylase/oxygenase (Rubisco) up to 1000-fold higher than in the surrounding medium by the action of CA, extracellular surface properties, and environmental conditions (Riding, 2006). These can be cultured in marine waters, saline drainage water, or brine from petroleum refining industry or CO$_2$ injection sites while producing biomass that can be converted into biofuel. Yates and

Robbins (1998) performed laboratory experiments to show that a single bloom (3.2×10^9 L) of the eukaryotic microalga *Nannochloris attomus* is able to precipitate 1.6×10^3 T of $CaCO_3$ in 12 h. Another research by the same authors while studying marine whiting events on the Bahama Bank concluded that blooms of cyanobacteria and green algae can sequester 5700 T CO_2/year as $CaCO_3$.

Ramanan et al. (2010) has shown enhanced sequestration of CO_2 via calcite deposition using algae *Chlorella* sp. and *Spirulina platensis*. The sequestration efficiencies achieved were 46% with *Chlorella* sp. and 39% with *Sp. platensis* with the application of 10% CO_2. Furthermore, to affirm the role played by CA in calcite formation, acetazolamide, a well-known inhibitor of CA was used.

In addition, several heterotrophic bacterial strains are also capable of mineralization of CO_2 via production of extracellular CA. Zhang et al. (2011) showed that whole cells of *Bacillus mucilaginous* can sequester CO_2 directly from the atmosphere with the help of extracellular CA produced by it. Yoshida et al. (2010) have isolated a moderately thermophilic bacterial strain *Geobacillus thermoglucosidasius*, which catalyzes the formation of fluorescent calcite crystals. These calcite crystals are excited by a wavelength interval of 260–400 nm, and have wide emission wavelengths ranging from 350 to 600 nm, which is a novel fluorescence property of calcite crystals formed by *G. thermoglucosidasius*. These crystals find many applications such as filler in rubber and plastics, fluorescent particles in stationery ink, and fluorescent marker. Interestingly, Jo et al. (2013) have developed a whole cell catalyst by expressing a γ-class CA from *N. gonorrhoeae* into the periplasmic space of *E. coli* having comparable catalytic activity to HCA II and studied the role of this periplasmic whole-cell CA system in bringing down the cost associated with the CO_2 sequestration process. The periplasmic whole-cell CA system can easily achieve the cost reduction by approximately 14%. Patel et al. (2013) achieved similar localization of CA in periplasmic space of *E. coli* by using two different signal sequences. With periplasmic CAs they reported a significant enhancement in CO_2-hydration activity with approximately 50%–70% increase in $CaCO_3$ precipitation relative to unanalyzed reactions. Their operational stabilities were found to be more than satisfactory (100% retention after 24 h of use for all four whole-cell biocatalysts). The periplasmic CAs have also exhibited improved thermal stabilities with more than 88% retention of activity up to 95°C for three of four whole cell biocatalysts but with compromised k_{cat} and catalytic efficiencies (k_{cat}/K_m).

Fan et al. (2011) had successfully cloned and expressed α-CA on the outer membrane of *E. coli* from *Helicobacter pylori* by means of a surface-anchoring system derived from ice nucleation protein from *Pseudomonas syringae*. CO_2 sequestration using whole cells of *E. coli* displaying CA on its surface was checked for its effectiveness in a CLM device and showed that CA-displaying *E. coli* cells can be efficiently used in CLM for CO_2 sequestration with an improved stability for at least 3 days.

6.1.2 Photosynthetic Bioconversion of Flue Gas CO_2 into Fuels

Biofuel produced using CO_2 from waste gases not only helps in CCUS but also is an attractive option for providing alternate fuels. In addition to plants, photosynthetic microbes such as algae and autotrophic bacteria (cyanobacteria and anoxyphotoautotrophs) also play an important role in the fixation of carbon dioxide. These are autotrophs which use carbon dioxide (CO_2) as the source of carbon for growth and transport them across their plasma membrane to be stored as inorganic carbon reservoir for photosynthesis (Zak et al., 2001; Badger and Price, 2003). Both cyanobacteria and algae play an important role in carbon and oxygen cycle by converting atmospheric CO_2 into organic matter by fixing CO_2 and producing O$_2$ during photosynthesis. Both can be developed as a microbial cell factory for converting atmospheric CO$_2$ into beneficial products such as biofuel by using solar energy.

Cyanobacteria are photosynthetic nitrogen fixing prokaryotic bacteria that thrive in wide variety of habitats such as oceans, fresh water, damp soil, temporarily moistened rocks in deserts, bare rock and soil, and even Antarctic rocks. Algae are very large and diverse group of autotrophic organisms ranging from single-celled microalgae to large seaweeds such as the giant kelps that grow up to 65 m in length. These are very simple organisms as they lack many distinct cells and organ types present in other land plants and highly abundant forms present on earth, and these are responsible for producing more than half of the world's primary production. Algae are widespread in water bodies in terrestrial environments and are found in unusual environments such as on snow and on ice too (http://en.wikipedia.org/wiki/Biosequestration). As in plants, RuBisCO catalyzed CO$_2$ fixation is the main pathway for CO$_2$ incorporation in microalgae and cyanobacteria under optimal conditions. Uptake of CO$_2$ into cells as bicarbonate is the rate-limiting step of biomass production and this limitation is overcome by the presence of CCM. CCM contains the very efficient enzyme CA that enables them to capture and deliver inorganic carbon to RuBisCo, and thus, enhancing photosynthesis.

In addition, other pathways also exist for CO$_2$ fixation in microalgae. Microalgae also carry out carboxylation of phosphenol pyruvate catalyzed by phosphenol pyruvate carboxylase. Oxaloacetate thus formed gets converted into C4 dicarboxylic acids; for example, into malate or citrate, and subsequently into amino acids such as aspartate or glutamate. Therefore, under limited light conditions, microalgae are capable of carrying out carbon assimilation that is preferentially channeled in the direction of synthesis of amino acids and other necessary cell constituents, while in the presence of adequate light, pentose phosphate pathway predominates (Fay, 1983; Campbell et al., 1998; Zak et al., 2001).

With the rising global warming issue, microalgae and cyanobacteria are gaining more attention for use in carbon sequestration. The concept is to convert CO$_2$ from flue gases directly into biomass using intensive photosynthesis. Algae are known to thrive at high concentrations of carbon dioxide and nitrogen dioxide (NO$_2$). In addition, the fast growth rates of algae and their capacity to grow practically in any kind of environment make them attractive for capturing CO$_2$ emitted by power plants, steel, and cement plants as well as transport vehicles exhaust and other industrial sources worldwide.

The use of microalgae for sequestration of CO$_2$ has several merits over other CO$_2$ sequestration methods. The microalgae strains enable direct sequestration of CO$_2$ from flue gas without the need of separation of CO$_2$. Thus, flue gases can be directly fed to algal production facilities for significantly increasing productivity resulting in the capture and storage of CO$_2$ in their biomass. One kilogram of algal dry cell weight utilizes around 1.83 kg of CO$_2$. Annually, around 54.9–67.7 tons of CO$_2$ can be sequestered from raceway ponds equivalent to yearly dry weight biomass production of 30–37 tons per hectare (Brennan and Owend, 2010).

Conceptually, cultivation of microalgae near power plants is fairly simple. Flue gas emissions from the power plants can be piped to the open (ponds) or closed (photobioreactors) algal cultivation systems. The closed systems results in higher photosynthetic efficiency due to accurate control of the operational variables, showing decreased risk of contamination and water loss by evaporation. But these are costly, complicated to operate, and hard to scale up (Borowitzka, 1999). A number of designs have been projected for closed photobioreactors such as bubble column, air-lift, tubular (loop), and stirred tank reactors (Jacob-Lopes et al., 2009). Open pond systems are proposed to be oval (raceway), circular, or rectangular (Borowitzka, 1999).

For algal-based CO_2 sequestration to be useful in this milieu, the cost associated with the cultivation of algae should be at least comparable to other conventional nonbiological CO_2 sequestration methods. The offset for operational costs for CO_2 sequestration can be provided by the commercial profit from energy rich algal biomass, which can be utilized for several applications such as production of biofuels (Figure 6.10). In addition, algae can secrete many extracellular proteins and sugars into culture media when grown in photobioreactors. Such compounds have application in the production of food supplements for humans and in pharmaceuticals, and production of secondary metabolite. Some of the compounds secreted by them possess unique properties for special applications, for example, they can be used as bioemulsifier, bioflocculant, agar–agar substitute or cosmetic material. Microalgae are also known to produce nonmethane hydrocarbon (ethane, ethylene, propane, propylene, butane, isobutane, pentane, hexane, isoprene, and ethylene) (Schobert and Elstner, 1980; Shaw et al., 2003), organohalogens (chloroform, trichloroethylene, bromomethane, chloromethane, and iodomethane) (Scarratt and Moore, 1996), and aldehydes (propanal, hexanal, *n*-heptanal, formaldehyde, acetaldehyde, furfural, and valeraldehyde) (Schobert and Elstner, 1980; Nuccio et al., 1995). These compounds are constantly produced by algae and are secreted in the culture media of photobioreactors. Their biomass can be used as an animal feed such as some species are frequently used as feed for fish and shellfish (Novoa et al., 1999) and can be used for electricity generation upon combustion directly or by fermentation of the algal biomass to methane (Yen and Brune, 2007).

Since microalgae and cyanobacteria can import both CO_2 and HCO_3^- through the cell membrane, the process of carbon capture by CA can also be integrated with the cultivation of algae or cyanobacteria. Bicarbonate produced after CA-mediated carbon process can also be used for other applications. Chi et al. (2011) successfully utilized the bicarbonates

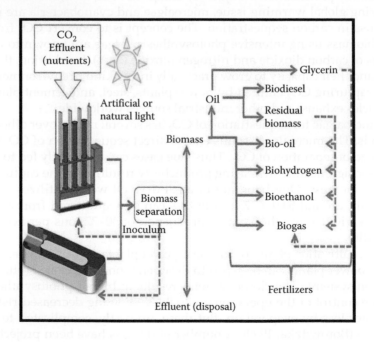

FIGURE 6.10
Scheme for the utilization of microalgae biomass produced in wastewater. (Adapted from Schneider RCS et al. 2012. *Potential Production of Biofuel from Microalgae Biomass Produced in Wastewater, Biodiesel—Feedstocks, Production and Applications.* InTech, Croatia, Europe.)

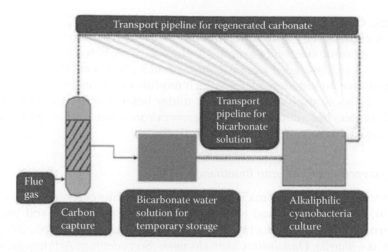

FIGURE 6.11
(See color insert.) Closed-loop bicarbonate/carbonate recirculation of bicarbonate produced from carbon capture by algal culture. (Adapted from Chi Z, Fallon JVO, and Shulin Chen. 2011. *Trends in Biotechnology* 29: 537–541.)

produced from the capture of carbon as a feed for cultivating algae or cyanobacteria culture (Figure 6.11). The carbonate regenerated by the culture process can again be reused as an absorbent to capture more CO_2 avoiding the requirement of additional energy for regeneration of carbonate. This process would appreciably bring down the cost associated with the carbon capture process as it is much economical and safe to transport aqueous bicarbonate solution than to transport compressed CO_2 and also bicarbonate serves as a superior alternative to raise an algal culture system.

6.1.2.1 Bioconversion of CO₂ into Biodiesel

Some strains of algae biomass are naturally rich in lipids. These can be extracted and converted into biodiesel. Additionally, the chemical composition of microalgae can be manipulated to enhance the production and accumulation of lipids by growing them under stressful conditions (e.g., low nitrogen) or in the presence of supplemental reductants (sugar, glycerol). Many unicellular algae can facultatively produce up to 60% of neutral lipids such as triacylglycerol per gram of dry weight, thus making them one of the most efficient biofuel producers (Weyer et al., 2010). Notably, biofuel production from algal systems can be maximized by tightly controlling and optimizing temperature, pH, nutrient, and CO_2 concentrations. Moreover, light quantity and quality (wavelength) may also be controlled by altering pond depth or using frequency-shifting fluorophores to increase photosynthetically active radiation. Therefore, by optimizing light harvesting efficiency and enhancing metabolic flux in the direction of increased oil or biomass accumulation efficiency of biomass, oil production from algae can be increased at least two- to threefold (Stephens et al., 2010). Biodiesel produced using microalgae is similar to biodiesel produced from vegetable oil in fatty acid content (14–22 carbon atoms) (Qin, 2005; Mata et al., 2010). Bjerk (2012) produced biodiesel using a combination of microalgae belonging to different genera: *Chlorella* sp., *Euglena* sp., *Spirogyra* sp., *Scenedesmus* sp., *Desmodesmus* sp., *Pseudokirchneriella* sp., *Phormidium* sp. (cyanobacteria), and *Nitzschia* sp.

Another candidate microorganism for biodiesel production is *Ralstonia eutropha,* which is a facultatively chemolithoautotrophic bacterium and is able to utilize CO_2 as a sole carbon source. Genetic engineering of *R. eutropha* was carried out by Muller et al. (2013) by cobbling together genes from different organisms such as integration and overexpression of the cytoplasmic version of gene TesA (thioesterase), heterologous expression of the acyl coenzyme A oxidase gene from *Mi. luteus*, and *fadB* and *fadM* from *E. coli* and deletion of two putative β-oxidation operons. Such modifications resulted in the production of 50–65 mg/L of diesel-range methyl ketones under heterotrophic growth conditions and 50–180 mg/L under chemolithoautotrophic growth conditions utilizing CO_2 and H_2 as the sole carbon source and electron donor, respectively.

6.1.2.2 Bioconversion of CO_2 into Bioethanol

After oil extraction, the left over microalgae biomass contains carbohydrates such as starch and glycogen, which can be used for the production of bioethanol and provides a sustainable alternative for the production of renewable biofuels (John et al., 2011). Microalgal species such as *Chlorella, Dunaliella, Chlamydomonas, Scenedesmus, Arthrospira*, and *Spirulina* have a carbohydrate content of approximately 70% of the biomass, and therefore, are suitable for bioethanol production (Harun and Danquah, 2011). The process of bioethanol production vary depending on the type of microbial biomass used starting with pretreatment of the biomass with an acid hydrolysis step for conversion carbohydrates from the cell wall into simple sugars, followed by saccharification, fermentation, and finally recovery of the product. In place of acid hydrolysis, other means such as enzymatic digestion or gamma radiation or use of fungi are good alternatives to render the process sustainable (Bjerk, 2012; Chen et al., 2012; Yoon et al., 2012).

Cyanobacteria also produce ethanol as a by-product of natural fermentation but the yields are low. An impressive approach for direct conversion of CO_2 into biofuel using cyanobacteria is genetic engineering of the existing strains by the introduction of the genes encoding the key enzymes of ethanol biosynthesis from pyruvate.

For instance, Deng and Coleman (1999) created a genetically engineered strain of *Synechococcus* sp. strain PCC7942, which is capable of producing ethanol from CO_2 under oxygenic photosynthesis without the need for special conditions, such as an anaerobic environment. A novel pathway has been created as shown in Figure 6.12, by integrating the genes encoding pyruvate decarboxylase (pdc) and alcohol dehydrogenase II (adh) from the bacterium *Zymomonas mobilis* under the control of promoter from the rbcLS operon encoding the cyanobacterial ribulose-1,5-bisphosphate carboxylase/oxygenase. The genes were first cloned into a shuttle vector and then transformed into *Synechococcus* sp., resulting in integration and overexpression of the integrated gene with concomitant production of ethanol, which diffused from the cells into the culture medium.

Genetic engineering was again used to modify *Synechococcus elongatus* for ethanol production from CO_2 and water in an attempt by Deng and Coleman (1999). A rate of 0.18 µg/L/h of ethanol production by the engineered strain was reported. The latest cyanobacterial strain engineered has been patented by Joule Unlimited (www.jouleunlimited.com) and is reported to produce bioethanol at a rate of approx. 1 mg/h.

Acetogenic bacteria are also gaining attention in CCUS owing to their potential to convert CO_2 into biofuels. Ramachandriya et al. (2013) have proposed and tested an exceptional hybrid technology by employing acetogenic bacteria to sequester CO_2 from large point sources into fuels and chemicals using renewable hydrogen (H_2). They have successfully demonstrated the use of *Clostridium carboxidivorans* and *Clostridium ragsdalei* in bioconversion of CO_2 and H_2 to ethanol, along with production of acetic acid, n-butanol, and n-hexanol. Furthermore, by varying the nutrient composition, it is possible to promote

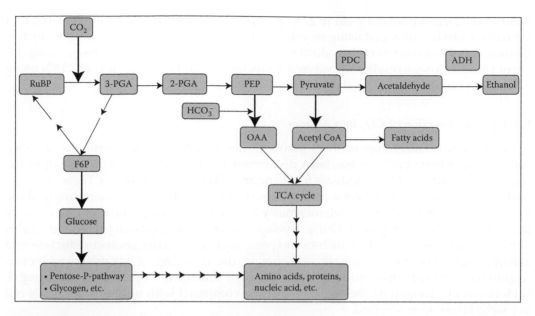

FIGURE 6.12
Photosynthesis and photoassimilate metabolism in cyanobacteria. Abbreviations: 2-PGA, 2-phosphoglyceric acid; 3-PGA, 3-phosphoglyceric acid; F6P, fructose-6-phosphate; OAA, oxaloacetate; PEP, phosphoenolpyruvic acid; RuBP: ribulose 1,5- bisphosphate; TCA cycle, tricarboxylic acid cycle; acetyl CoA, acetyl coenzyme A. The pathway at the upper right is the added pathway for ethanol synthesis. (Adapted from Deng MD and Coleman JR. 1999. *Applied and Environmental Microbiology* 65: 523–528.)

a particular metabolic pathway in these bacteria relative to other. Out of the two bacteria, *Cl. carboxidivorans* has a high ethanol production potential with a yield of 66.5% by consumption of 23% of CO$_2$ when it was supplied with 75% H$_2$ and 25% CO$_2$ feed gas.

6.1.2.3 Production of Biogas, Biohydrogen, and Bio-Oil

The residual algal biomass can also be subjected to anaerobic digestion (fermentation) and produce biogas in the same way as other agro-residues are used. Furthermore, the by-products of fermentation are rich in nutrients and can be recycled and used as fertilizers (Chae et al., 2006).

In addition to biogas using enzymatic and chemical processes, the residual biomass can also be converted to biohydrogen and bio-oils. Hydrogen from them can be generated by chemical processes such as gasification, partial oxidation of oil and water electrolysis. Cyanobacteria are the preferred organisms used for the production of biohydrogen reaction being catalyzed by nitrogenases and hydrogenases (Tamagnini et al., 2007). The studies on *Anabaena* sp. has showed that its biomass has the potential for the production of biohydrogen in the presence of sufficient levels of air, water, minerals, and light as the process can be photosynthetic (Marques et al., 2011; Ferreira et al., 2012).

It is possible to convert any biomass to bio-oil, and in this direction a number of micro-algal species have been considered for bio-oil production such as *Chlamydomonas*, *Chlorella*, *Scenedesmus*, *Scenedesmus dimorphus*, *Sp. platensis*, *Chlorogloeopsis fritschiiwer*, and *Dunaliella tertiolecta*.

The process involves the use of hydrothermal liquefaction in the temperature range of 200–350°C and pressures of 15–20 MPa to convert biomass into bio-oil. In one such attempt,

Biller et al. (2012) reported a yield of 27%–47% using microalgae, taking into account that microalgae can be cultivated using recycled nutrients. Shuping et al. (2010) studied hydrothermal liquefaction for bio-oil production from *Dunaliella tertiolecta* biomass by using different catalyst dosage conditions, varying temperatures and time. A yield of 25.8% using 5% sodium carbonate as catalyst at 360°C has been reported.

6.1.2.4 Bioconversion of CO_2 into Alkanes

Alkanes are the main components of gasoline, diesel, and jet fuels. Recently, a two-step alkane biosynthetic pathway has been discovered in cyanobacteria, which catalyze the conversion of fatty acid intermediates to alkane and alkene (www.ls9.com). These enzymes could be exploited to photosynthetically convert CO_2 into alka(e)nes. Interestingly, these enzymes can also be expressed heterologously in *E. coli* bestowing upon them the capacity to produce alcohol. Wang et al. (2013) developed a series of genetically modified strains of *Synechocystis* sp. PCC6803, which overexpress acyl–acyl carrier protein reductase and aldehyde-deformylating oxygenase resulting in the doubling of alka(e)ne production (Figure 6.13). An enhancement in alka(e)ne by eight times with a production of 26 mg/L (1.1% of cell dry weight) by the engineered strain compared with wild type (0.13% of cell dry weight) has been reported.

6.1.2.5 Bioconversion of CO_2 into 1-Butanol

Butanol serves as a chemical feedstock and is an efficient fuel having 90% of the energy of petrol and is better than ethanol in many aspects. It is likely that in near future, ethanol will be replaced by butanol. Significant research is underway to find ways to produce butanol from biomass (http://www.energy-without-carbon.org/Butanol).

Lan and Liao (2011) were successful in producing 1-butanol using CO_2 by genetic engineering in cyanobacterial strain *Sy. elongatus* PCC 7942. They have introduced an entire modified

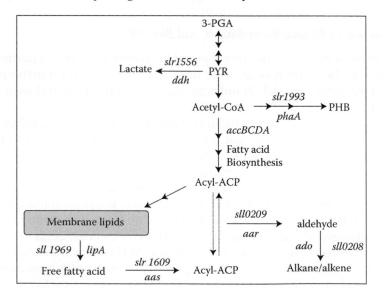

FIGURE 6.13
Schematic overview of fatty acid, alkane (alkene), and main competing metabolic pathways in *Synechocystis* sp. PCC6803. (Adapted from Wang W, Liu X, and Lu X. 2013. *Biotechnology for Biofuels* 6: 69.)

CoA-dependent 1-butanol production pathway by chromosomal integration into the genome of *Sy. elongatus*. Genes for butanol biosynthesis were acquired from different bacterial strains hbd (3-hydroxybutyryl-CoA dehydrogenase), crt (Crotonase), and adhE2 (aldehyde/alcohol dehydrogenase) genes were taken from *Clostridium acetobutylicum*, the ter (trans-2-enoyl-CoA reductase) gene was from *Treponema denticola*, while the atoB (thiolase) gene was from *E. coli*. The butanol production from the genetically modified strain is, however, oxygen-sensitive and required inhibition of photosynthesis by shifting to dark anoxic conditions.

6.1.2.6 Bioconversion of CO_2 into Isobutaraldehyde

A breakthrough research published in nature biotechnology reported direct conversion of CO_2 into isobutyraldehyde and isobutanol by genetic engineering of a cyanobacterium *Sy. elongatus* PCC7942. Isobutyraldehyde is a raw material to synthesize several other chemicals such as the production of neopentylglycol, which is used as polyester, and isobutanediol, which is used in personal care industry. It can be converted into isobutanol, which after processing, can be used in place of petrol. An IPTG inducible expression cassette containing gene for ketoacid decarboxylase from *Lactococcus lactis*13 was integrated into the genome and to increase the flux to the keto acid precursor, 2ketoisovalerate (KIV)16, the gene for alsS from *B. subtilis* and the ilvC and ilvD genes from *E. coli* were also integrated into its genome. Furthermore, to overcome the limitation of low turnover rates of Rubisco in CO_2 fixation and competition between O_2 and CO_2 for the active site of the enzyme, additional genes for Rubisco rbcLS genes from the related *Sy. elongatus* strain PCC6301 were integrated downstream of the original gene. The biosynthetic pathway involves photosynthetic CO_2 fixation into pyruvic acid, which is further converted into isobutaraldehyde by the newly added enzymes. Furthermore, separation of isobutaraldehyde from the product is fairly easy as it can be vaporized into the gas phase easily. The new genetically engineered strain now exhibits higher productivity owing to enhanced CO_2 fixation rates and direct conversion of CO_2 to isobutyraldehyde with a production rate of 6230 µg/L/h and produced 1.1 g/L of isobutyraldehyde over 8 days with a steady state (Atsumi et al., 2009).

Several reactor and photobioreactor designs are being investigated for successful sequestration of CO_2 using flue gas. In a recent study by Arata et al. (2013) an air-lift reactor for sequestration of CO_2 with the simultaneous abatement of NO_x from flue gas using cyanobacterial strain *Sp. platensis* has been investigated. They reported the process to be feasible both in terms of productivity (86.8 mg/L/day) and CO_2 sequestration (229 mg/day). Using fed-batch for feeding of flue gas resulted in achievement of a high CO_2 sequestration (407 mg/day), 90.0% removal of NO_x, and a biomass production of 188.7 mg/L/day.

Thus, the key challenges for using algae for CO_2 sequestration and subsequent biodiesel production lies in screening and isolation of microalgal strains capable of high CO_2 tolerance and growth rate and high oil content, possessing tolerance to toxic impurities present in flue gases and thermophillic. Much research has been focused on studying tolerance of algae to high concentration of CO_2 such as that present in flue gases.

Euglena gracilis is one of the high CO_2-tolerant species. The growth of this species was increased under 5%–45% concentration of CO_2, although the best growth was at 5% CO_2 (Nakano et al., 1996). *Chlorococcum littorale* can successfully grow at 60% CO_2 using the stepwise adaptation technique (Kodama et al., 1993). Hanagata et al. (1992) reported an alga *Scenedesmus* sp., which could grow at 80% CO_2 but the maximum cell mass was observed in the presence of 10%–20% CO_2. Kubler et al. (1999) studied the effect of elevated CO_2 levels on the seaweed *Lomentaria articulata* and found that twice the ambient CO_2 concentration had significantly affected daily net carbon gain and total wet biomass production rates; 52% and 314% greater than those were at the ambient CO_2 conditions, respectively. The microalgae such as *Chlorella* grow much better at 100,000 ppm of CO_2 than in the ambient air (Yue and Chen, 2005). At still higher concentration of CO_2, growth rates decline, but it is still higher than that at the baseline. Similar observations have been reported by Watanabe et al. (1992) for *Chlorella*. It can thus be concluded that a pollutant from power plants (flue gases containing CO_2) can act as a nutrient for the algae.

Furthermore, a strain of alga, *Chlorella* sp. T-1 discovered by Maeda et al. (1995) is capable of growing at 100% CO_2, although the highest growth rate was achieved at 10%. Yun et al. (1996) successfully sequestrated 0.624 g CO_2 per liter per day (0.026 g/L/h) using algal culture. Doucha et al. (2005) also reported 10%–50% reduction in CO_2 using *Chlorella*. The highest fixation rate has been reported by Morais and Costa (2007) using *Spirulina* (53.29% at 6% CO_2 v/v). Some other researchers concentrated on studying the effect of trace impurities such as NO_x and SO_2 on CO_2 sequestration by microalgae. Yoshihara et al. (1996) observed *Nannochloris* sp. to grow at 100 ppm of nitric oxide (NO). While *Dunaliella tertiolecta* at 1000 ppm of NO and 15% CO_2 could remove 51%–96% of nitric oxide depending on the growth conditions (Nagase et al., 1998). Matsumoto et al. (1995) showed *Tetraselmis* sp. to grow in the presence of actual flue gas with 185 ppm of SO_x and 125 ppm of NO_x in addition to 14.1% CO_2. Maeda et al. (1995) studied the tolerance of a strain of *Chlorella* and reported that the strain can grow at various concentrations and combinations of trace elements.

Since the temperature of flue gas is around 120°C, the use of thermophilic algae will reduce the operational cost of the process by eliminating the need to cool down the flue gas. In addition to this, some thermophiles are known to produce unique secondary metabolites (Edwards, 1990), which may further reduce overall cost of CO_2 sequestration. A disadvantage of using thermophilic algae is the increased loss of water due to evaporation. *Cyanidium caldarium* is a thermophilic species but requires pure CO_2 (Seckbach et al., 1971). Miyairi (1995) studied the growth characteristics of *Sy. elongatus* at high CO_2 concentrations and recorded good growth at 60% CO_2 and 60°C.

Apart from this, the use of marine microalgae for CO_2 sequestration is also being evaluated as they can allow the use of seawater directly as a growing medium thereby reducing the maintenance costs of microalgae cultures. Many power plants are situated along the coastal areas, and therefore, algae can be cultivated nearby using flue gas and sea water. In this direction, a number of marine algal species have been studied for CO_2 sequestration abilities that include *Tetraselmis* sp. (Laws and Berning, 1991; Matsumoto et al., 1995), *Synechococcus* sp. (Takano et al., 1992), *Ch. littorale* (Pesheva et al., 1994), *Chlamydomonas* sp. (Miura et al., 1993), *Nannochloropsis salina* (Matsumoto et al., 1995), and *Phaeodactylum tricornutum* (Matsumoto et al., 1995).

CO_2 assimilation ability is a crucial criterion in selecting algal species for CO_2 sequestration. Laws and Berning (1991) conducted an experiment to compare bubbling CO_2 gas versus adding carbonated water as a means of introducing CO_2 into the microalgal cultures, and reported that bubbling CO_2 led to 96% ± 11% utilization efficiencies, while using carbonated water supported 81% ± 11% efficiencies.

6.1.3 Nonphotosynthetic CO_2 Bioconversion to Fuel

6.1.3.1 Production of Methane Using Methanogens

The conversion of CO_2 to organic compounds such as methane by using microbes is one of the most effective and attractive process for the sequestration of CO_2, which is supported by the presence of natural process where methane is produced on a vast scale by the microbes present in the earth's crust. Methanogens belong to the Euryarchaeota kingdom of Archaea, which very efficiently convert CO_2 to methane as they utilize CO_2 as a carbon source and reduce it to methane using hydrogen gas or formate as an electron-donating reducing agent. Methanogens dwell in diverse anaerobic environments containing carbon dioxide, acetate, and low sulfate concentrations such as landfills, digestive systems of animals, in deep ocean vents, and in coal seams.

Methanogens generate methane by several pathways; the best-described pathways involve the fermentation of carbon dioxide and acetic acid as terminal electron acceptors.

$$CO_2 + 4H_2 \rightarrow CH_4 + 2H_2O$$
$$CH_3COOH \rightarrow CH_4 + CO_2$$

The pathway of methanogenesis is relatively complex, involving cofactors such as F420, coenzyme B, coenzyme M, methanofuran, and methanopterin.

Generally, methanogenesis involves a consortium of chemosynthetic communities or food chain of microbial organisms, living in close association with cold hydrocarbon seeps, and display complex relationships such as mineralization of CO_2 and methanogenesis (Sassen et al., 1993). Both work together leading to a series of biochemical reactions that result in the production of methane in energy-yielding cellular processes.

$$CH_3COO^- + H^+ \rightarrow CH_4 + CO_2 \tag{6.5}$$

$$CO_2 + 4H_2 \rightarrow CH_4 + 2H_2O \tag{6.6}$$

Methanogens are dependent upon hydrogen-producing acetogenic microbes as well as fermentatives to provide them growth substrates, as shown in Figure 6.14.

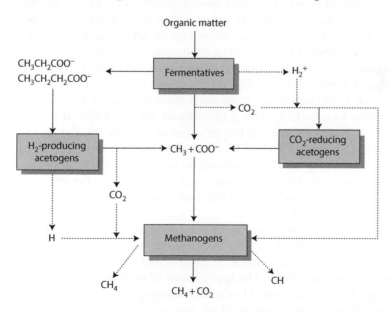

FIGURE 6.14

Anaerobic microbial food chain (consortium) converting complex organic matter to methane in freshwater environments. Solid lines represent the major routes of methanogenesis. (Adapted from Stevens SH, Ferry JG, and Schoell M. 2012. Methanogenic conversion of CO_2 into CH_4. doi: 10.2172/1041046.)

There is an interspecies electron transfer mechanism operating in methanogenic consortium. Since the production of hydrogen gas is thermodynamically unfavorable under standard conditions of equimolar reactants and products, the thermodynamically favorable utilization of hydrogen gas by methanogens is essential for growth of acetogenic microbes. Thus, the hydrogen producers and hydrogen consumers are interdependent and exist as a consortium of metabolic groups.

The bioconversion of CO_2 to methane offers several advantages. First, it requires much lower energy for reduction of CO_2. Burning the methane produced by these microbes would be a carbon neutral process, which does not contribute to atmospheric carbon dioxide as these microbes utilize CO_2 from the atmosphere to form methane, and hence, burning methane will just return the extracted CO_2. Growth conditions of most methanogens are simple, requiring only inorganic nutrients, CO_2, hydrogen (or formate), and the absence of oxygen. Methanogens can flourish under extreme environmental conditions, such as total darkness, can tolerate a broad range of pressures, temperatures (ranging from 9°C and 110°C), and other environmental extremes. Methane produced from microbes is more environment friendly than biofuels produced from feedstocks such as corn ethanol. Production of methane from microbes eliminates the competition with food production and other agricultural resources such as land, irrigation, and fertilizers (http://www.greenoptimistic.com/2012/07/31/methane-producing-microbes-as-new-source-of-renewable-energy/).

Hydrogenotrophic methanogens are known to occupy oil reservoirs universally. These methanogens can be exploited for *in situ* conversion of CO_2 into methane. The sequestration of CO_2 in the form of methane has the potential to reproduce natural gas deposits in reservoirs, thereby helping to meet rapidly increasing demand for clean fuels. A preliminary investigation of the US reservoirs indicates that environmental conditions prevailing there are ideal for 84 Gt of CO_2 sequestration using methanogen technology. This is equivalent to about 40 years of current power generation emissions. Sugai et al. (2010) had shown the existence of a symbiotic relationship in oil reservoirs between oil-degrading and hydrogen-producing bacteria *Thermotoga naphthophila* and *Thermoanaerobacter* sp. and *Methanobacterium* sp. and *Methanothermcococcus* sp. which are known as hydrogenotrophic methanogens, enabling *in situ* conversion of CO_2 to methane. Lee et al. (2012) studied biological conversion of CO_2 to CH_4 using hydrogenotrophic methanogens in a fixed bed reactor. Methanogens have been shown to have the potential to be effective in converting CO_2 to CH_4 with a conversion rate of 100% at 3.8 h retention time. In a research conducted by the US Department of Energy 98, the species of methanogens were studied for their CO_2 to CH_4 conversion potential under the same conditions that typically prevail in geologic sequestration sites. Among these, *Methanosarcina vacuolata* and *Methanosarcina barkeri*, *Methanococcus vannielii*, *Methanobacterium uliginosum*, *Methanobacterium ivanovii*, and *Methanobacterium subterraneum* showed optimum methanogenesis (Stevens et al., 2012).

The potential of methanogens can also be utilized in combination with geological sequestration where CO_2 is proposed to be stored in deep subsurfaces such as coal beds. The conditions prevailing there are suitable to use methanogens for simultaneous conversion of sequestered CO_2 into methane. The applications of methanogens in geologic sequestration sites is expected to make use of existing commercial injection technologies that are in use for water flooding and enhanced oil recovery at mature oil fields. Mayumi et al. (2013) simulated conditions prevailing in oil reservoirs by constructing microcosms mimicking reservoir conditions (55°C, 5 MPa) using samples from high-temperature oil reservoir. A high CO_2 injection into deep subsurface high temperature oil reservoirs induced drastic changes in microbial communities and methanogenic functions that enhanced methanogenesis.

A conceptualized diagram of methanogen injection scheme is shown in Figure 6.13, where methanogenic archaeal cultures, nutrients, and cofactors are injected into the CO_2 storage reservoir using low-cost water-injection wells. The horizontal pinnate drilling has been shown to be an economical approach and could be employed to enhance the distribution of agents, especially in low-permeability reservoirs.

National Thermal Power Corporation of India has recently proposed a bioreactor for sequestration of CO_2 from flue gas using consortia of methogens as shown in Figure 6.14.

The successful application of naturally occurring methanogens to produce methane using CO_2 would allow the conversion of subeconomic (high CO_2) natural gas deposits into pure and economical methane deposits and would also open up this CCS technology to widespread applications in industries such as power generation, chemical, petroleum, and other industries. The methanogenic technology is still in a budding precommercial stage of research and development. It is an area of high-potential research that can be explored and possibly exploited. Industry must only recognize candidate reservoirs and research should be done to screen and engineer archaea to efficiently catalyze conversion of CO_2 to methane.

6.1.4 Other Heterotrophic Bacteria and Their Enzymes in CO_2 Bioconversion

Wood et al. (1941) proposed that reduction of CO_2 occurs during the fermentation of glycerol by certain genera of heterotrophic bacteria such as propionic acid bacteria, *Propionibacterium*, *Escherichia*, and *Citrobacter* and demonstrated that oxaloacetate (OAA) is produced when CO_2 and pyruvate combines. In the bacteria that he studied, the assimilated CO_2 has been shown to get incorporated into acetate, lactate, and succinate in *Aerobacter indologenes*, lactate and succinate in *Proteus vulgaris*, *Streptococcus paracitrovorus*, and *Staphylococcus candidus*, acetate and lactate in *Clostridium welchii*, lactate and succinate in *St. paracitrovorus*, and lactate in *Cl. acetobutylicum*. This pathway has the potential to be used for carbon sequestration combined with oxaloacetate production. The pathway requires the presence of CA, which converts CO_2 into bicarbonate which is further utilized by carboxylases such as phosphoenolpyruvate (PEP) carboxylase and pyruvate carboxylase to form oxaloacetate (Norici et al., 2002). Such anapleurotic pathway exists in organisms to compensate the loss of oxaloacetate siphoned off for the synthesis of amino acids of the aspartate family.

Thus, heterotrophic bacteria can be used to sequester carbon and in turn produce oxaloacetate and amino acids from CO_2. *Corynebacterium glutamicum* is well-known producer of amino acids such as glutamic acid and lysine. This bacterium contains the enzymes, PEP carboxylase and pyruvate carboxylase and PEP pyruvate-oxaloacetate node (Sauer and Eikmanns, 2005), and thus, fixes carbon in the form of amino acids. At elevated CO_2, increased supply of bicarbonate to PEP carboxylase and pyruvate carboxylase with the help of CAs makes the conditions favorable for lysine production (Gubler et al., 1994; Jetten et al., 1994).

$$CO_2 + H_2O \xrightarrow{\text{CA}} HCO_3 + PEP \xrightarrow{\text{PEPcase}} OAA \rightarrow \underset{\text{(malate, succinate, etc.)}}{\text{C4 Chemicals}}$$

To demonstrate the possibility of CO_2 sequestration and increased production of value-added four carbon compound, oxaloacetate, an integrated system using CA and PEPCase was developed. *Escherichia coli* was used for cloning and expression of a gene for PEPCase

1 from marine diatom *Phaeodactylum tricornutum* and a recombinant C-terminal codon optimized αCA protein from *Dunaliella* sp. were tested in a cell free system, which were composed of the two recombinant enzymes along with a substrate (PEP) for PEPCase 1. The cell free biocatalytic system successfully produced OAA thus illustrating the potential of this system in capturing and sequestration of CO_2 from stationary sources coupled to OAA production (Chang et al., 2013).

6.1.4.1 Enzymatic Conversion of CO_2 into Methane

In a recent research work carried out by Yang et al. (2012), the potential of enzymes is further exploited by remodeling a nitrogenase MoFe protein by site-directed mutagenesis (α-70Val → Ala, α-195His → Gln) endowing it the ability to catalyze reduction of CO_2 to yield methane (CH_4). One nanomolar of the remodeled MoFe nitrogenase catalyzes the formation of 21 nmol of CH_4 within 20 min under optimized conditions. The doubly substituted nitrogenase MoFe protein can accomplish the eight-electron reduction of CO_2 to CH_4 and this CO_2 reduction can also be coupled to the reduction of other substrates resulting in the formation of longer chain, value-added hydrocarbons (e.g., propylene $=H_2C=CH–CH_3$) formation through the reductive coupling of CO_2 and acetylene ($HC\equiv CH$). An insight into the mechanistic details of the reaction catalyzed by nitrogenase can make it possible to employ this enzyme directly for sequestration of CO_2 into methane or other hydrocarbons.

6.1.4.2 Enzymatic Capture of Carbon into Methanol

To cut down the emissions from power plants, scientists have been working on finding alternative sources of renewable energy. Methanol is the simplest alcohols and one of the oldest and most versatile of all renewable sources of energy. Thus, conversion of CO_2 to methanol offers a potential new technology not only for sequestration of CO_2 but also the efficient production of fuel. This would be a carbon neutral approach, which means that producing and burning of this fuel will not augment the carbon dioxide in the atmosphere. Methanol generation by partial hydrogenation of carbon dioxide has been accomplished by electrocatalysis and photocatalysis. Oxide-based catalysts are mainly used for the industrial fixation of carbon dioxide (Azuma et al., 1990; Fan and Fujimoto, 1994; Heleg and Willner, 1994). A distinctive approach for the direct conversion of gaseous CO_2 to methanol involves the use of three dehydrogenases. The series of reactions starts with CO_2 reduction by formate dehydrogenase (FateDH), resulting in the formation of formic acid followed by its reduction to formaldehyde by the action of formaldehyde dehydrogenase (FaldDH) and finally formaldehyde conversion to methanol catalyzed by alcohol dehydrogenase (ADH).

$$CO_2 \xrightarrow{\text{FatePD}} HCOOH \xrightarrow{\text{FaldDH}} HCHO \xrightarrow{\text{ADH}} CH_3OH$$

The process is made thermodynamically favorable by using excess of NADH. The reduced nicotinamide adenine dinucleotide (NADH) acts as a terminal electron donor for each dehydrogenase-catalyzed reduction. Obert and Dave (1999) have successfully encapsulated these dehydrogenases in sol–gel matrix using the biocompatible synthesis method. The pores of the sol–gels act as nanoreactors for biocatalytic reactions. This immobilization resulted in significantly higher productivity of methanol compared with free enzymes. Similar encapsulation studies have been carried out by Xu et al. (2006) and Jiang et al. (2009) by immobilization of enzymes into alginate–silica nanoparticles as well as by coimmobilization in multicompartmented titanate systems. The coimmobilized multienzymes displayed superior catalytic activity and stability as hybrid titania-protamine acts as a structural scaffold with three entrapped dehydrogenases acting as biomolecular machines working in an enzymatic assembly line resulting in enhanced conversion of CO_2 to methanol owing to more constructive interactions among enzymes in a nanoscale environment. Such integrated multienzyme systems will provide a novel solution for green and efficient utilization of carbon dioxide (Sun et al., 2009).

Furthermore, Xin et al. (2004) have investigated the production of methanol from CO_2 by methanotrophic bacteria. Methanotrophs are known to oxidize methane to CO_2 in a multistep reaction pathway catalyzed by a series of enzymes including methane monooxygenase, methanol dehydrogenase, formaldehyde dehydrogenase, and formate dehydrogenase. The conversion of CO_2 to methane is an opposite reaction catalyzed by the same enzyme that requires considerable amount of energy. Xin and his coworkers (2004) have successfully reported the bioconversion of CO_2 to methanol using the resting cells of methanotrophic bacterium *Methylosinus trichosporium* IMV 3011 using methane as a substrate for the regeneration of reducing equivalent. By this new route, it is possible to produce methanol while sequestering CO_2.

Cazelles et al. (2013) constructed a very efficient artificial cascade reaction for the production of methanol from CO_2 by combining enzymes of the methanol biosynthetic pathway. The enzymes were extracted from different organisms and their amount to be used in the cascade was optimized so that every subsequent enzyme has a higher conversion rate than the previous one for preventing the inhibition of enzyme by substrate accumulation. The three enzymes along with a NAD⁺ regenerating system (phosphite dehydrogenase, PTDH) PTDH were immobilized in silica nanocapsules nanostructured by phospholipids (NPS) (Figure 6.15), which resulted in a 55 times higher activity of the hybrid nanobioreactor when compared with the free enzymes in solution and production of 4.3 mmol/g commercial enzymatic powder of methanol in 3 h at room temperature and pressure of 5 bar.

At present, bioconversion of CO_2 to methanol is in the preliminary stage. This needs to be investigated further for understanding the reaction kinetics and its applicability and feasibility in the sequestration of CO_2 from flue gas.

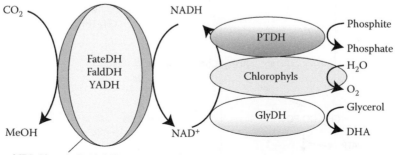

NPS: Phospholipids/silica nanocapsules

FIGURE 6.15
Schematic representation of CO_2 valorization into methanol using different NADH regenerating systems. (Adapted from Cazelles R et al. 2013. *New Journal of Chemistry* 37: 3721–3730.)

6.1.4.3 Bioconversion of CO₂ into Acetic Acid

CO_2 can also be converted to acetate anaerobically by acetogenic bacteria catalyzed by multienzyme system by the Wood–Ljungdahl pathway. CO dehydrogenase is one of the enzymes involved in such pathway, which catalyzes the reversible interconversion of CO_2 and CO and contributes to the carboxyl group in acetate.

$$4H_2 + 2CO_2 \rightarrow CH_3COOH + 2H_2O$$

Methyl group is produced by first conversion of CO_2 into formate by formate dehydrogenase, followed by a series of enzymatic reaction as shown in the scheme Figure 6.16. This methyl group is then converted into acetyl-CoA by the action of carbon monoxide dehydrogenase utilizing CO. *Azotobacter vinelandii* contains vanadium nitrogenase that has been shown to reduce CO to ethylene, ethane, propane, and butane (Lee et al., 2010) (Figure 6.16).

Acetogens such as *Clostridium ljungdahlii* and *Cl. ragsdalei* are known to produce acetic acid along with ethanol utilizing CO_2 as the source of energy. Sakai et al. (2005) studied acetic acid production using an acetogen *Moorella thermoacetica*. HUC22-1 by supplying a mixture of CO_2 and H_2 and reported a production of 20 g/L of acetic acid (339 mM) after 220 h of fermentation. Unlike other acetogens, *Mo. thermoacetica* produces only acetic acid as the end-product. Using acetogens for carbon sequestration is at an early stage. It

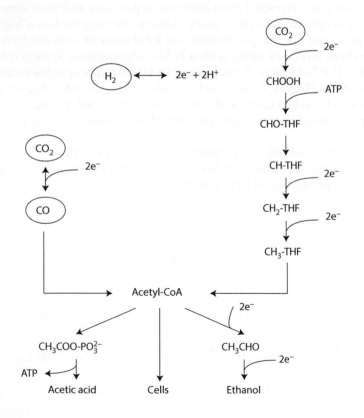

FIGURE 6.16
Wood–Ljungdahl (W–L) metabolic pathway for syngas fermentation. (Adapted from Ljungdahl LG. 1986. *Annual Review of Microbiology* 40: 415–50.)

requires considerable research for screening strains capable of tolerating high CO_2 as well as impurities present in flue gas along with developing a well-designed and effective system for cultivation.

Another genetic engineering feat has been carried out recently on extreme thermophile *Pyrococcus furiosus*, which confers it the ability to convert gaseous CO_2 and hydrogen into 3-hydroxypropionate which is an industrially important chemical and serves as a building block for the synthesis of other chemicals. This has been accomplished by integration and expression of five genes from the carbon fixation pathway of the *Metallosphaera sedula*, an autotrophic archaeon with T_{opt} of 73°C. Both cell free extracts and whole cells of the engineered strain are able to produce 3-hydroxypropionate at approximately 30°C below the optimum growth temperature of the organism. This bioconversion approach has a high potential to be used in CO_2 sequestration from industrial flue gas. An extensive research is called for studying its applicability for this purpose (Keller et al., 2013).

6.2 Future Perspectives

Bioconversion of CO_2 from industrial emissions is an important and challenging task. There are microbes and pathways existing in nature to carry out such bioconversions some of which are already known. Serious and concerted efforts are needed to widen the scope of biological CCUS by discovering novel microbes and their enzymes, which may open up unexplored opportunities to sequester and convert CO_2 directly from power plants into stable and useful products. Biomineralization and CO_2 conversion to biofuels by various photosynthetic and heterotrophic bacteria is a promising alternative approach to meet the increasing energy demands. For example, characterization of the performance and stability of the potential enzymes, cyanobacteria and algae and other heterotrophic systems under realistic operating conditions is currently lacking, and therefore, deserves to be examined and assess the effectiveness of the processes. Genetic manipulation of the enzymes and microbes to further improve their efficiencies and properties desirable in carbon sequestration may prove beneficial. Further research efforts are called for improving and optimizing the process designs for large-scale industrial biomimetic CCUS. An individual bioconversion process is not adequate enough to sequester all the power sector emissions no matter how efficient it is. The present day need is, therefore, to integrate all the existing biomimetic processes along with physical methods in order to address the CCUS of industrial CO_2 emissions.

6.3 Conclusions

The ongoing and unavoidable increase in anthropogenic CO_2 emissions is leading to enhanced global warming, thereby heating up our planet in an unprecedented way. The temperature of Earth has increased by about 0.8°C. Scientists have warned that an increase in global temperature by more than 2°C will result in disastrous changes in the environment. Almost 200 nations agreed in 2010 to circumvent hazardous effects of climate change by limiting the rise of global Earth's temperature to below 2°C (3.6°F). We have to act now to slow down the rate of global warming by reducing our carbon emissions, adopting an ecofriendly and green lifestyle, and shifting to renewable energy sources. There is also an urgent need to explore ways to sequester carbon emitted into the atmosphere by the continued use of fossil fuels.

Most of the research on CCS is concentrating on the capturing of CO_2 from the point sources of emissions. Abiotic processes for carbon sequestration include storage of CO_2 in deep oceans, underground geological structures, unmineable coal mines, and oil wells. These have high potential for storage but there is a cost penalty associated with them along with leakage risk. The present day need is not just to store CO_2 but to recycle it into useful products. To achieve this goal we need to look for cheaper ways to convert CO_2 chemically or biologically into industrially important products.

Nature is equipped with living organisms having diverse mechanisms to fix atmospheric CO_2. Mimicking such biological processes would provide natural, cost-effective and secure ways to mitigate CO_2 emissions, and at the same time to convert it into value-added products. Microbes thriving at every possible niche on Earth have important roles to play in global carbon and other mineral cycles. These are equipped with highly efficient enzymes, some of which are remarkably excellent at capturing CO_2 and converting it into either biomass or other products. Whole microbial cells as well as their purified enzymes can be employed for CCUS. A significant benefit of microbial systems is that they do not require separation and compression of CO_2 from flue gases, and thus, decrease the costs of biomimetic CCUS. Carbon capture by the action of CA is one such process, which has the potential to convert CO_2 into bicarbonate and other value-added minerals. Mineral carbonates thus formed have diverse industrial applications. CA is an efficient enzyme that converts CO_2 into bicarbonate at a faster rate, which can be further mineralized. Extensive global efforts are underway to find appropriate CA, which can be used under harsh operational conditions. Several companies and researchers are attempting to develop technologies using CA in a biomimetic approach for simultaneous capture and mineralization of CO_2 directly from industrial flue gases and industrial emissions. The development of enzyme immobilization technique for use in biomimetic CCS is also receiving attention. The use of CA as a biocatalyst in this process greatly offsets the costs normally associated with carbon capture.

The sequestration of CO_2 from industrial flue gases via bioconversion into fuels represents a novel approach for carbon mitigation. Biofuel serves dual purpose in carbon sequestration by providing an alternative to fossil fuels with simultaneous sequestration of carbon from coal-based power plants. The photosynthetic bacteria and algae proficient in photosynthesis can be cultivated near the stationary sources of flue gas emissions that results in the incorporation of CO_2 into biomass that may finance the cost associated with the process. Biomass can be harvested and converted into biodiesel, bioethanol, human feed, pharmaceuticals, and several others. In addition, the exploitation of natural molecular pathways existing in the heterotrophic bacteria coupled with genetic engineering techniques for efficiently converting CO_2 into useful biochemicals and biofuels such as methane, butanol, and isobutaraldehyde represents breakthrough technologies for cost-effective carbon sequestration.

The present-day need is to further explore such beneficial microbes, molecular pathways, and the enzymes involved and understand the most suitable conditions for using them to safely capture and convert industrial CO_2 emissions into commercially important by-products.

References

Adler L, Brundell J, Falkbring SO, and Nyman PO. 1972. Carbonic anhydrase from *Neisseria sicca*, strain 6021 I. Bacterial growth and purification of the enzyme. *Biochimica Biophysica Acta Enzymology* 284: 298–310.

Alber BE and Ferry JG. 1994. Carbonic anhydrase from the archaeon *Methanosarcina thermophila*. *Proceedings of the National Academy of Science of the United States of America* 91: 6909–6913.

Arata S, Strazza C, Lodi A, and Del Borghi A. 2013. *Spirulina platensis* culture with flue gas feeding as a cyanobacteria based carbon sequestration option. *Chemical Engineering and Technology* 36: 91–97.

Atsumi S, Higashide W, and Liao JC. 2009. Direct photosynthetic recycling of carbon dioxide to isobutyraldehyde. *Nature Biotechnology* 27: 1177–1180.

Azuma H, Hashimoto K, Hiramoto M, and Sakata TJ. 1990. Electrochemical reduction of carbon dioxide on various metal electrodes in low-temperature aqueous $KHCO_3$ media. *Electrochemical Society* 137: 1772–1778.

Badger M. 2000. The roles of carbonic anhydrases in photosynthetic CO_2 concentrating mechanisms. *Photosynthesis Research* 77: 83–94.

Badger MR and Price GD. 2003. CO_2 concentration mechanisms in cyanobacteria: Molecular components, their diversity and evolution. *Journal of Experimental Botany* 54: 609–622.

Barabesi C, Galizzi A, Mastromei G, Rossi M, Tamburini E, and Perito B. 2007. *Bacillus subtilis* gene cluster involved in calcium carbonate biomineralization. *Journal of Bacteriology* 189: 228–235.

Berg JM, Tymoczko JL, and Stryer L. 2007. *Biochemistry*, 6th ed. Sara Tenney, New York.

Bhattacharya S, Nayak A, Schiavone M, and Bhattacharya SK. 2004. Solubilization and concentration of carbon dioxide: Novel spray reactors with immobilized carbonic anhydrase. *Biotechnology and Bioengineering* 86: 37–46.

Bhattacharya S, Schiavone M, Chakrabarti S, and Bhattacharya SK. 2003. CO_2 hydration by immobilized carbonic anhydrase. *Biotechnology and Applied Biochemistry* 38: 111–117.

Bhattacharya SK. 2001. Patent number US20016258335B1.

Biller P, Ross AB, Skill SC, Lea-Langton A, Balasundaram B, Hall C, Riley R et al. 2012. Nutrient recycling of aqueous phase for microalgae cultivation from the hydrothermal liquefaction process. *Algal Research* 1: 70–76.

Bjerk TR. 2012. Microalgae cultivated in photobiorreator and joint reactor for bioremediation and production of biofuels. In: *Portuguese: Cultivo de microalgasemfotobiorreator e reatormistovisando a biorremediação e produção de biocombustíveis).* Universidade de Santa Cruz do Sul, Santa Cruz do Sul-Rio Grande do Sul.

Bond GM, Stringer J, Brandvold DK, Simsek AM, and Margaret-Gail EG. 2001. Development of integrated system for biomimetic CO_2 sequestration using the enzyme carbonic anhydrase. *Energy and Fuel* 15: 309–316.

Borchert M and Saunders P. 2010. *Heat-Stable Carbonic Anhydrases and Their Use*. USPTO, Editor. Novozymes A/S.

Borchert M and Saunders P. 2011. Heat stable carbonic anhydrases and their use. US Patent US7892814.

Borowitzka MA. 1999. Commercial production of microalgae: Ponds, tanks, tubes and fermenters. *Journal of Biotechnology* 70: 313–321.

Brennan L and Owend P. 2010. Biofuels from Microalgae—A review of technologies for production, processing, and extractions of biofuels and co-products. *Renewable and Sustainable Energy Reviews* 14: 557–577.

Campbell D, Hurry V, Clarke AK, Gustafsson P, and Oquist G. 1998. Chlorophyll fluorescence analysis of cyanobacterial photosynthesis and acclimation. *Microbiology and Molecular Biology Reviews* 30: 667–680.

Campbell NA and Reece JB. 2005. *Biology*, 7th ed. Benjamin/Cummings Publishing Co., NY.

Capasso C, De LV, Carginale V, Cannio R, and Rossi M. 2012. Biochemical properties of a novel and highly thermostable bacterialα-carbonic anhydrase from *Sulfurihydrogenibium yellowstonense* YO3AOP1. *Journal of Enzyme Inhibition and Medicinal Chemistry* 27: 892–897.

Cazelles R, Drone J, Fajula F, Ersen O, Moldovan S, and Galarneau A. 2013. Reduction of CO_2 to methanol by a polyenzymatic system encapsulated in phospholipids–silica nanocapsules. *New Journal of Chemistry* 37: 3721–3730.

Chae SR, Hwang EJ, and Shin HS. 2006. Single cell protein production of *Euglena gracilis* and carbon dioxide fixation in an innovative photo-bioreactor. *Bioresource Technology* 97: 322–329.

Chang KS, Jeon H, Gu MB, Pack SP, and Jin ES. 2013. Conversion of carbon dioxide to oxaloacetate using integrated carbonic anhydrase and phosphoenol-pyruvate carboxylase. *Bioprocess and Biosystems Engineering* 36: 1923–1928.

Chen R, Yue Z, Deitz L, Liu Y, Mulbry W, and Liao W. 2012. Use of an algal hydrolysate to improve enzymatic hydrolysis of lignocellulose. *Bioresource Technology* 108: 149–154.

Cheng LH, Zhang L, Chen HL, and Gao CJ. 2008. Hollow fiber contained hydrogel-CA membrane contactor for carbon dioxide removal from the enclosed spaces. *Journal of Membrane Science* 324: 33–43.

Chi Z, Fallon JVO, and Shulin Chen. 2011. Bicarbonate produced from carbon capture for algae culture. *Trends in Biotechnology* 29: 537–541.

Ciullo PA (ed.). 1996. Industrial minerals and their uses. In: *A Handbook and Formulatory*. William Andrew Publishing, Westwood, NJ, pp. 26–28.

Cowan RM, Ge JJ, Qin YJ, and McGregor ML. 2003a. CO_2 capture by means of an enzyme-based reactor. *Annals of the New York Academy of Science* 984: 453–469.

Cowan RM, Ge JJ, Qin YJ, McGregor ML, and Trachtenberg MC. 2003b. CO_2 capture by means of an enzyme-based reactor. *Annals of the New York Academy of Science* 984: 453–469.

Daigle R and Desrochers M. 2009. Carbonic anhydrase having increased stability under high temperature conditions. US Patent 7521217.

De Luca V, Vullo D, Scozzafava A, Carginale V, Rossi M, Supuran CT, and Capasso C. 2012. Anion inhibition studies of an α-carbonic anhydrase from the thermophilic bacterium *Sulfurihydrogenibium yellowstonense* YO3AOP1. *Bioorganic and Medicinal Chemistry Letters* 22: 5630–5634.

Deng MD and Coleman JR. 1999. Ethanol synthesis by genetic engineering in cyanobacteria. *Applied and Environmental Microbiology* 65: 523–528.

Doucha J, Straka F, and Livansky K. 2005. Utilization of flue gas for cultivation of microalgae (*Chlorella* sp.) in an outdoor open thin-layer photobioreactor. *Journal of Applied Phycology* 17: 403–412.

Dreybodt W, Eisenlohr L, Madry B, and Ringer S. 1997. Precipitation kinetics of calcite in the system $CaCO_3$–H_2O–CO_2. *Geochimica et Cosmochimica Acta* 61: 3987.

Druckenmiller ML and Maroto-Valer MM. 2005. Carbon sequestration using brine of adjusted pH to form mineral carbonates. *Fuel Processing Technology* 86: 1599–1614.

Duvail L, Moal J, and Peron MF. 1998. CGRP-like molecules and carbonic anhydrase activity during the growth of *Pecten maximus*. *Comparitive Biochemistry and Physiology* 120: 475–480.

Dziedzic D, Gross KB, Gorski RA, and Johnson JT. 2006. Feasibility study of using brine for carbon dioxide capture and storage from fixed sources. *Journal of Air Waste Management and Assessment* 56: 1631–1641.

Edwards C. 1990. Microbiology of extreme environments. In: Graham C. (Ed.), *Thermophiles*. Open University Press, Milton Keynes, Great Britain, p. 218.

Ercole C, Cacchio P, Botta AL, Centi V, and Lepidi A. 2007. Bacterially induced mineralization of calcium carbonate: The role of exopolysaccharide sand capsular polysaccharides. *Microscopy and Microanalysis* 13: 42–50.

Fan L and Fujimoto K. 1994. Hydrogenation of carbon dioxide to methanol by lanthana-supported palladium catalysts. *Chemistry Letters* 105–108.

Fan LH, Liu N, Yu MR, Yang ST, and Chen HL. 2011. Cell surface display of carbonic anhydrase on *Escherichia coli* using ice nucleation protein for CO_2 sequestration. *Biotechnology and Bioengineering* 108: 2853–2864.

Farrell A. 2011. Carbon dioxide storage in stable carbonate minerals. Basalt laboratory studies of interest to carbon capture and storage. *Geology* 394: 1–24.

Favre N, Christ ML, and Pierre AC. 2009. Biocatalytic capture of CO_2 with carbonic anhydrase and its transformation to solid carbonate. *Journal of Molecular Catalysis B: Enzymatic* 60: 163–170.

Favre N and Pierre AC. 2011. Synthesis and behaviour of hybrid polymer-silica membranes made by sol gel process with adsorbed carbonic anhydrase enzyme, in the capture of CO_2. *Journal of Sol–Gel Science and Technology* 60: 177–188.

Fay P. 1983. *The Blue Greens: (Cyanophyta-cyanobacteria)*. Edward Arnold, London, p. 88.

Ferreira AF, Marques AC, Batista AP, Pass M, Gouveia L, and Silva CM. 2012. Biological hydrogen production by *Anabaena* sp.: Yield, energy and CO_2 analysis including fermentative biomass recovery. *International Journal of Hydrogen Energy* 37: 179–190.

Ferry JG. 2010. The γ class of carbonic anhydrases. *Biochimica et Biophysica Acta (BBA): Proteins Proteomics* 1804: 374–381.

Ge J, Cowan RM, McGregor ML, and Trachtenberg MC. 2002. Enzyme-based CO_2 capture for advanced life support. *Life Support and Biosphere Science* 8: 181–189.

Goff GS and Rochelle GT. 2004. Monoethanolamine degradation: O_2 mass transfer effects under CO_2 capture conditions. *Industrial and Engineering Chemistry Research* 43: 6400–6408.

Gubler M, Park SM, Jetten M, Stephanopoulos G, and Sinskey AJ. 1994. Effects of phosphoenol pyruvate carboxylase deficiency on metabolism and lysine production in *Corynebacterium glutamicum*. *Applied Microbiology and Biotechnology* 40: 857–863.

Guilloton MB, Korte JJ, Lamblin AF, Fuchs JA, and Anderson PM. 1992. Carbonic anhydrase in *Escherichia coli*, a product of the cyn operon. *Journal of Biological Chemistry* 267: 3731–3734.

Gutknecht J, Bisson MA, and Tosteson FC. 1977. Diffusion of carbon dioxide through lipid bilayer membrane. *Journal of General Physiology* 69: 779–794.

Hanagata N, Takeuchi T, Fukuju Y, Barnes DJ, and Karube I. 1992. Tolerance of microalgae to high CO_2 and high temperature. *Phytochemistry* 31: 3345–3348.

Harun R and Danquah MK. 2011. Enzymatic hydrolysis of microalgae biomass for bioethanol production. *Chemical Engineering Journal* 168: 1079–1084.

Haywood HM, Eyre JM, and Scholes H. 2001. Carbon dioxide sequestration as stable carbonate minerals environmental barriers. *Environmental Geology* 41: 11–16.

Heleg V and Willner I. 1994. Photocatalysed CO_2-fixation to formate and H_2 evolution by eosin modified Pd–TiO_2 powders. *Journal of the Chemical Society, Chemical Communications* 32: 2113–2114.

Henry RP, Gehnrich S, Wihrauch D, and Towle DW. 2003. Salinity-mediated carbonic anhydrase induction in the gills of the euryhaline green crab, *Carcinus maenas*. *Comparitive Biochemistry and Physiology A* 136: 243–258.

Huijgen WJJ, Comans RNJ, and Witkamp GJ. 2007. Cost evaluation of CO_2 sequestration by aqueous mineral carbonation. *Energy Conversion and Management* 48: 1923–1935.

Iliuta I and Larachi F. 2012. New scrubber concept for catalytic CO_2 hydration by immobilized carbonic anhydrase II and *in situ* inhibitor removal in three phase monolith slurry reactor. *Separation and Purification Technology* 86: 199–214.

Jacob-Lopes E, Revah S, Hernández S, Shirai K, Franco TT. 2009. Development of operational strategies to remove carbon dioxide in photobioreactors. *Chemical Engineering Journal* 153: 120–126.

Jetten MSM, Pitoc GA, Follettie MT, and Sinskey AJ. 1994. Regulation of phospho(enol)-pyruvate and oxaloacetate converting enzymes in *Corynebacterium glutamicum*. *Applied Microbiology and Biotechnology* 41: 47–52.

Jeyakanthan J, Rangarajan S, Mridula P, Kanaujia SP, Shiro Y, Kuramitsu SSY, and Sekar K. 2008. Observation of a calcium-binding site in the γ-class carbonic anhydrase from *Pyrococcus horikoshii*. *Acta Crystallographica Section D: Biological Crystallograhy D* 64: 1012–1019.

Jiang Y, Sun Q, Jiang Z, Zhang L, Li J, and Li L. 2009. The improved stability of enzyme encapsulated in biomimetic titania particles. *Material Science Engineering C* 29: 328.

Jo BH, Kim IG, Seo JH, Kang DG, and Cha HG. 2013. Engineered *Escherichia coli* with periplasmic carbonic anhydrase as a biocatalyst for CO_2 sequestration. *Applied and Environmental Microbiology* 79: 6697–6705.

John RP, Anisha GS, Nampoothiri KM, and Pandey A. 2011. Micro and macroalgal biomass: A renewable source for bioethanol. *Bioresource Technology* 102: 186–193.

Kanbar B and Ozdemir E. 2010. Thermal stability of carbonic anhydrase immobilized within poly-urethane foam. *Biotechnology Progress* 26: 1474–1480.

Kanth BK, Min K, Kumari S, Jeon H, Jin ES, Lee J, and Pack SP. 2012. Expression and characterization of codon-optimized carbonic anhydrase from *Dunaliella species* for CO_2 sequestration application. *Applied Biochemistry and Biotechnology* 167: 2341–2356.

Kaur S, Bhattacharya A, Sharma A, and Tripathi AK. 2012. Diversity of microbial carbonic anhy-drases, their physiological role and applications. In: Satyanarayana T, Johri BN, and Prakash A. (Eds.), *Microorganisms in Environmental Management*. Springer, Heidelberg, Germany, pp. 151–173.

Kaur S, Mishra MN, and Tripathi AK. 2010. Gene encoding gamma-carbonic anhydrase is cotran-scribed with argC and induced in response to stationary phase and high CO_2 in *Azospirillum brasilense* Sp7. *BMC Microbiology* 10: 184.

Keilin D and Mann T. 1940. Carbonic anhydrase: Purification and nature of the enzyme. *Biochemical Journal* 34: 1163–1176.

Keller MW, Schut GJ, Lipscomb GL, Menon AL, Iwuchukwu IJ, Leuko TT, Thorgersen MP et al. 2013. Exploiting microbial hyperthermophilicity to produce an industrial chemical, using hydrogen and carbon dioxide. *Proceedings of the National Academy of Science of the United States of America* 110: 5840–5845.

Kim IG, Jo BH, Kang DG, Kim CS, Choi YS, and Cha HJ. 2012. Biomineralization-based conversion of carbon dioxide to calcium carbonate using recombinant carbonic anhydrase. *Chemosphere* 87: 1091–1096.

Kodama M, Ikemoto H, and Miyachi S. 1993. A new species of highly CO_2-tolerant fast growing marine microalga suitable for high-density culture. *Journal of Marine Biotechnology* 1: 21–25.

Kubler JA, Johnston AM, and Raven JA. 1999. The effects of reduced and elevated CO_2 and O_2 on the seaweed *Lomentaria articulata*. *Plant, Cell and Environment* 22: 1303–1310.

Lacroix O and Larachi F. 2008. Scrubber designs for enzyme-mediated capture of CO_2. *Recent Patents on Chemical Engineering* 1: 93.

Lan EI and Liao JC. 2011. Metabolic engineering of cyanobacteria for 1-butanol production from carbon dioxide. *Metabolic Engineering* 13(4): 353–63.

Larachi F, Lacroix O, and Grandjean BPA. 2012. Carbon dioxide hydration by mobilized carbonic anhydrase in Robinson-Mahoney and packed-bed scrubbers-role of mass transfer and inhibitor removal. *Chemical Engineering Science* 73: 99–115.

Laws EA and Berning JL. 1991. A study of the energetics and economics of microalgal mass culture with the marine chlorophyte *Tetraselmis suecica*: Implications for use of power plant stack gases. *Biotechnology and Bioengineering* 37: 936–947.

Lee CC, Hu Y, and Ribbe MW. 2010. Vanadium nitrogenase reduces CO. *Science* 329: 642.

Lee JC, Kim JH, Chang WS, and Pak D. 2012. Biological conversion of CO_2 to CH_4 using hydrogeno-trophic methanogen in a fixed bed reactor. *Journal of Chemical Technology and Biotechnology* 87: 844–847.

Lee RBY, Smith JAC, and Rickaby REM. 2013. Cloning, expression and characterization of the δ-carbonic anhydrase of *Thalassiosira weissflogii* (Bacilli-riophyceae). *Journal of Phycology* 49: 170–177.

Lee SW, Park SB, Jeong SK, and Lim KS. 2010. On carbon dioxide storage based on biomineralization strategies. *Micron* 41: 273–282.

Li W, Liu L, Chen W, Yu L, Li W, and Yu H. 2010. Calcium carbonate precipitation and crystal morphology induced by microbial carbonic anhydrase and other biological factors. *Process Biochemistry* 45: 1017–1021.

Liu N, Bond GM, Abel TA, McPherson BJ, and Stringer J. 2005. Biomimetic sequestration of CO_2 in carbonate form: Role of produced waters and other brines. *Fuel Processing Technology* 86: 1615–1625.

Ljungdahl LG. 1986. The autotrophic pathway of acetate synthesis in acetogenic bacteria. *Annual Review of Microbiology* 40: 415–50.

Maeda K, Owada M, Kimura N, Omata L, and Karube I. 1995. CO_2 fixation from the flue gas on coal fired thermal power plant by microalgae. *Energy Conversion Management* 36: 717–720.

Mahinpey N, Asghari K, and Mirjafari K. 2011. Biological sequestration of carbon dioxide in geological formations. *Chemical Engineering Research and Design* 89: 1873–1878.

Marques AE, Barbosa AT, Jotta J, Coelho MC, Tamangini P, and Gouveia L. 2011. Biohydrogen production by *Anabaena* sp. PCC 7120 wild-type and mutants under different conditions: Light, nickel, propane, carbon dioxide and nitrogen. *Biomass and Bioenergy* 35: 4426–4434.

Mata TM, Martins AA, and Caetano NS. 2010. Microalgae for biodiesel production and other applications: A review. *Renewable and Sustainable Energy Reviews* 14: 217–232.

Matsumoto H, Shioji N, Hamasaki A, Ikuta Y, Fukuda Y, Sato M, Endo N, and Tsukamoto T. 1995. Carbon dioxide fixation by microalgae photosynthesis using actual flue gas discharged from a boiler. *Applied Biochemistry and Biotechnology* 51: 681–692.

Matsuyama H, Terada A, Nakagawara T, Kitamura Y, and Teramoto M. 1999. Facilitated transport of CO_2 through polyethylenimine/poly(vinyl alcohol) blend membrane. *Journal of Membrane Science* 163: 221–227.

Matsuyama H, Teramoto M, and Iwai K. 1994. Development of a new functional cation-exchange membrane and its application to facilitated transport of CO_2. *Journal of Membrane Science*. 93: 237–244.

Mayumi D, Dolfing J, Sakata S, Maeda H, Miyagawa Y, Ikarashi M, Tamaki H et al. 2013. Carbon dioxide concentration dictates alternative methanogenic pathways in oil reservoirs. *Nature Communications* 4: 1–6.

Medakovic D. 2000. Carbonic anhydrase activity and biomineralization process in embryos, larvae and adult blue mussels *Mytilus edulis*. *Helgoland Marine Research* 54: 1–6.

Medakovic D, Popovic S, Grazeta B, Plazonic M, and Brenko M. 1997. X-ray diffraction study of calcification processes in embryos and larvae of the brooding oyster *Ostreaedulis*. *Marine Biology* 129: 615–623.

Meldrum NU and Roughton FJW. 1933. Carbonic anhydrase: Its preparation and properties. *Journal of Physiology* 80: 113–142.

Merlin C, Masters M, McAteer S, and Coulson A. 2003. Why is carbonic anhydrase essential to *Escherichia coli. Journal of Bacteriology* 185: 6415–6424.

Migliardini F, Luca VD, Carginale V, Rossi M, Corbo P, Supuran CT, and Capasso C. 2013. Biomimetic CO_2 capture using a highly thermostable bacterial α-carbonic anhydrase immobilized on a polyurethane foam. *Journal of Enzyme Inhibition and Medicinal Chemistry* 29: 146–150.

Mirjafari P, Asghari K, and Mahinpey N. 2007. Investigating the application of enzyme carbonic anhydrase for CO_2 sequestration purposes. *Industrial and Engineering Chemistry Research* 46: 921–926.

Miura Y, Yamada W, Hirata K, Miyamoto K, and Kiyohara M. 1993. Stimulation of hydrogen production in algal cells grown under high CO_2 concentration and low temperature. *Applied Biochemistry and Biotechnology* 39: 753–761.

Miyairi S. 1995. CO_2 assimilation in a thermophilic cyanobacterium. *Energy Conversion and Management* 36: 763–766.

Miyamoto H, Miyashita T, Okushima M, Nakano S, Morita T, and Matsushiro A. 1996. A carbonic anhydrase from the nacreous layer in oyster pearls. *Proceedings of the National Academy of Science of the United States of America* 93: 9657–9660.

Morais MGD and Costa JAV. 2007. Biofixation of carbon dioxide by *Spirulina* sp. and *Scenedesmus obliquus* cultivated in a three-stage serial tubular photobioreactor. *Journal of Biotechnology* 129: 439–445.

Muller J, MacEachran D, Burd H, Sathitsuksanoh N, Bi C, Yeh YC, Lee TS et al. 2013. Engineering of *Ralstonia eutropha* H16 for autotrophic and heterotrophic production of methyl ketones. *Applied and Environmental Microbiology* 79: 4433–4439.

Nagase H, Eguchi K, Yoshihara K, Hirata K, and Miyamoto K. 1998. Improvement of microalgal NOx removal in bubble column and airlift reactors. *Journal of Fermentation and Bioengineering* 86: 421–423.

Nakano YY, Miyatake K, Okuno H, Hamazaki K, Takenaka S, Honami N, Kiyota M et al. 1996. Growth of photosynthetic algae *Euglena* in high CO_2 conditions and its photosynthetic characteristics. *Acta Horticulture* 440: 49–54.

Neish AC. 1939. Studies on chloroplasts. Their chemical composition and the distribution of certain metabolites between the chloroplasts and remainder of the leaf. *Biochemical Journal* 33: 300–308.

Neves LA, Afonso C, Coelhoso IM, and Crespo JG. 2012. Integrated CO_2 capture and enzymatic bioconversion in supported ionic liquid membranes. *Separation and Purification Technology* 97: 34–41.

Newman LM, Clark L, Ching C, and Zimmerman S. 2010. Carbonic anhydrase polypeptides and uses thereof. US Patent WO10081007.

Nielsen SA and Frieden E. 1972. Carbonic anhydrase activity in molluscs. *Comparitive Biochemistry and Physiology A* 135: 271–278.

Norici A, Dalsass A, and Giordano M. 2002. Role of phosphoenolpyruvate carboxylase in anaplerosis in the green microalga *Dunaliella salina* cultured under different nitrogen regimes. *Physiologia Plantarum* 116: 186–191.

Novoa OMA, Domýnguez LJ, Castillo OL, and Palacios MCA. 1999. Effect of the use of the micro-algae *Spirulina maxima* as fish meal replacement in diets for tilapia, *Oreochromis mossambicus* (Peters) fry. *Aquaculture Research* 71: 219–225.

Nuccio J, Seaton PJ, and Kieber RJ. 1995. Biological production of formaldehyde in the marine environmental. *Limnology and Oceanography* 40: 521–527.

Obert R and Dave BC. 1999. Enzymatic conversion of carbon dioxide to methanol: Enhanced methanol production in silica sol gel matrices. *Journal of American Chemical Society* 121: 2192–2193.

Ono E and Cuello JL. 2007. Carbon dioxide mitigation using thermophilic cyanobacteria. *Biosystems Engineering* 96: 129–134.

Oviya M, Giri SS, Sukumaran V, and Natarajan P. 2012. VSG-4 and its application as CO_2 sequesterer. *Preparative Biochemistry and Biotechnology* 42: 462–475.

Park JM, Kim M, Lee HJ, and Jang A. 2012. Enhancing the production of *Rhodobacter sphaeroides* derived physiologically active substances using carbonic anhydrase-immobilized electrospun nanofibers. *Biomacromolecules* 13: 3780–3786.

Patel TN, Park AHA, and Banta S. 2013. Periplasmic expression of carbonic anhydrase in *Escherichia coli*: A new biocatalyst for CO_2 hydration. *Biotechnology and Bioengineering* 110: 1865–1873.

Pesheva I, Kodama M, Dionisio-Sese ML, and Miyachi S. 1994. Changes in photosynthetic characteristics induced by transferring air grown cells of *Chlorococcum littorale* to high CO_2 conditions. *Plant and Cell Physiology* 35: 379–387.

Pocker Y and Bjorkquist DW. 1977. Stopped-flow studies of carbon dioxide hydration and bicarbonate dehydration in H_2O and D_2O. Acid–base and metal ion catalysis. *Journal of American Chemical Society* 99: 6537–6543.

Pomar L and Hallock P. 2008. Carbonate factories: A conundrum in sedimentary geology. *Earth-Science Reviews* 87: 134–169.

Prabhu C, Valechha A, Wanjari S, Labhsetwar N, Kotwal S, Satyanarayana T, and Rayalu S. 2011. Carbon composite beads for immobilization of carbonic anhydrase. *Journal of Molecular Catalysis B: Enzymatic* 71: 71–78.

Prabhu C, Wanjari S, Gawande S, Das S, Labhsetwar N, Kotwal S, Puri AK et al. 2009. Immobilization of carbonic anhydrase enriched microorganism on biopolymer based materials. *Journal of Molecular Catalysis B: Enzymatic* 60: 13–21.

Prabhu, C, Wanjari S, Puri A, Bhattacharya A, Pujari R, Yadav R, Das S et al. 2011. Region specific bacterial carbonic anhydrase for biomimetic sequestration of carbon dioxide. *Energy and Fuels* 25: 1327–1332.

Premkumar L, Bhageshwar UK, Irena Gokhman I, Zamir A, and Sussmanb JL. 2003. An unusual halotolerant α-type carbonic anhydrase from the alga *Dunaliella salina* functionally expressed in *Escherichia coli*. *Protein Expression and Purification* 28: 151–157.

Qin J. 2005. Bio-hydrocarbons from algae: Impacts of temperature, light and salinity on algae growth. Rural industries research and development corporation report. RIRDC, Adelaide, Australia.

Ramachandriya KD, Kundiyana DK, Wilkins MR, Terrill JB, Atiyeh HK, and Huhnke RL. 2013. Carbon dioxide conversion to fuels and chemicals using a hybrid green process. *Applied Energy* 112: 289–299.

Ramanan R, Kannan K, Deshkar A, Yadav R, and Chakrabarti T. 2010. Enhanced algal CO_2 sequestration through calcite deposition by *Chlorella* sp. and *Spirulina platensis* in a mini-raceway pond. *Bioresource Technology* 101: 2616–2622.

Ramanan R, Kannan K, and Sivanesan SD. 2009a. Biosequestration of carbon dioxide using carbonic anhydrase enzyme purified from *Citrobacter freundii*. *World Journal of Microbiology and Biotechnology* 25: 981–987.

Ramanan R, Kannan K, Vinayagamoorthy N, Ramkumar KM, Sivanesan SD, and Chakrabarti T. 2009b. Purification and characterization of a novel plant-type carbonic anhydrase from *Bacillus subtilis*. *Biotechnology and Bioprocess Engineering* 14: 32–37.

Reddy KJ, Weber H, Bhattacharyya P, Morris A, Taylor D, Christensen M, Foulke T et al. 2010. Instantaneous capture and mineralization of flue gas carbon dioxide: Pilot scale study. *Nature Proceedings* http://dx.doi.org/10.1038/npre.2010.5404.1.

Riding R. 2006. Cyanobacterial calcification cyanobacterial calcification, carbon dioxide concentrating mechanisms, and Proterozoic–Cambrian changes in atmospheric composition. *Geobiology* 4: 299–316.

Russoa ME, Sciallaa S, Lucac VD, Capasso C, Olivierib G, and Marzocchellab A. 2013. Immobilization of carbonic anhydrase for biomimetic CO_2 capture. *Chemical Engineering and Transactions* 32: 1867–1872.

Sakai S, Nakashimada Y, Inokuma K, Kita M, Okada H, and Nishio N. 2005. Acetate and ethanol production from H_2 and CO_2 by *Moorella* sp. using a repeated batch culture. *Journal of Bioscience and Bioengineering* 99: 252–258.

Santos A, Fernandez TJA, and Serna MA. 2007. Chemically active silica aerogel wollastonite composites for CO_2 fixation by carbonation reactions. *Industrial and Engineering Chemistry Research* 46: 103–107.

Sassen R, Roberts HH, Aharon P, Larkin J, Chinn EW, and Carney R. 1993. Chemosynthetic bacterial mats on cold hydrocarbon seeps, Gulf of Mexico continental slope. *Organic Geochemistry* 20(1): 77–89.

Sauer U and Eikmanns BJ. 2005. The PEP-pyruvate-oxaloacetate node as the switch point for carbon flux distribution in bacteria. *FEMS Microbiology Reviews* 29: 765–794.

Saunders P, Salmon S, Borchert M, and Lessard LP. 2010. Modular membrane reactor and process for carbon dioxide extraction. Novozymes, International Patent WO2010/014774 A2.

Savile C, Nguyen L, Alvizo O, Balatskaya S, Choi G, Benoit M, Fusman I et al. 2011. Low cost biocatalyst for the acceleration of energy efficient CO_2 capture. *ARPA-E Energy Innovation Summit*. June 6–8, Washington, DC.

Scarratt MG and Moore RM. 1996. Production of methyl chloride and methyl bromide in laboratory cultures of marine phytoplankton. *Marine Chemistry* 54: 263–272.

Schneider RCS, Bjerk TR, Gressler PD, Souza MP, Corbellini VA, and Lobo EA. 2012. In: Zhen Fang (Ed.), *Potential Production of Biofuel from Microalgae Biomass Produced in Wastewater, Biodiesel—Feedstocks, Production and Applications*. InTech, Croatia, Europe.

Schobert SB and Elstner EF. 1980. Production of hexanal and ethane by *Phaeodactylum triconutum* and its correlation to fatty acid oxidation and bleaching of photosynthetic pigments. *Plant Physiology* 66: 215–219.

Seckbach J, Gross H, and Nathan MB. 1971. Growth and photosynthesis of *Canidium caldarium* cultured under Pure CO_2. *Israel Journal of Botany* 20: 84–90.

Seifritz W. 1990. CO_2 disposal by means of silicates. *Nature* 345: 486.

Sharma A and Bhattacharya A. 2010. Enhanced biomimetic sequestration of CO_2 into $CaCO_3$ using purified carbonic anhydrase from indigenous bacterial strains. *Journal of Molecular Catalysis B: Enzymatic* 67: 122–128.

Sharma A, Bhattacharya A, and Shrivastava A. 2011. Biomimetic CO_2 sequestration using purified carbonic anhydrase from indigenous bacterial strains immobilized on biopolymeric materials. *Enzyme and Microbial Technology* 48: 416–426.

Shaw SL, Chisholm SW, and Prinn RG. 2003. Isoprene production by *Prochlorococcus*, a marine cyanobacterium and other phytoplankton. *Marine Chemistry* 80: 227–245.

Shuping Z, Kaleem I, Chun L, and Tong J. 2010. Production and characterization of bio-oil from hydrothermal liquefaction of microalgae *Dunaliella tertiolecta* cake. *Energy* 35: 5406–5411.

Smith KS and Ferry JG. 1999. A plant-type (β-class) carbonic anhydrase in the thermophilic methanoarchaeon *Methanobacterium thermoautotrophicum*. *Journal of Bacteriology* 181: 6247–6253.

Soong Y, Fauth DL, and Howard BH. 2006. CO_2 sequestration with brine solution and fly ashes. *Energy Conversion and Management* 47: 1676–1685.

Steiner H, Jonsson BH, and Lindskog S. 1975. The catalytic mechanism of carbonic anhydrase. Hydrogen-isotope effects on the kinetic parameters of the human C isoenzyme. *European Journal of Biochemistry* 259: 253–259.

Stephens E, Ross IL, King Z, Mussgnug JH, Kruse O, Posten C, Borowitzka MA et al. 2010. An economic and technical evaluation of microalgal biofuels. *Nature Biotechnology* 28: 126–128.

Stevens SH, Ferry JG, and Schoell M. 2012. Methanogenic conversion of CO_2 into CH_4. doi: 10.2172/1041046.

Stolaroff JK, Lowry GV, and Keith DW. 2005. Using CaO and MgO rich industrial waste streams for carbon sequestration. *Energy Conversion and Management* 46: 687–699.

Sugai Y, Purwasena IA, Sasaki K, and Fujiwara K. 2010. Evaluation of the potential of microbial conversion process of CO_2 into CH_4 by investigating the microorganisms in high CO_2 content oil field. In: *Canadian Unconventional Resources and International Petroleum Conference*, October 9–21, Calgary, Alberta, Canada, pp. 19–21.

Sun Q, Jiang Y, Jiang Z, Zhang L, Sun X, and Li J. 2009. Green and efficient conversion of CO_2 to methanol by biomimetic coimmobilization of three dehydrogenases in protamine-templated titania. *Industrial and Engineering Chemistry Research* 48: 4210–4215.

Takano H, Takeyama H, Nakamura N, Sode K, Burgess JG, Manabe E, Hirano M et al. 1992. CO_2 removal by high-density culutre of a marine cyanobacterium *Synechococcus* sp. using an improved photobioreactor employing light-diffusing optical fibers. *Applied Biochemistry and Biotechnology* 34: 449–458.

Tamagnini P, Leitao E, Oliveira P, Ferreira D, Pinto F, and Harris D. 2007. Cyanobacterial hydrogenases. *FEMS Microbiology Reviews* 31: 692–720.

Tashian RE. 1989. The carbonic anhydrases: Widening perspectives on their evolution, expression and function. *BioEssays* 10: 186–192.

Trachtenberg MC, Cowan M, Smith DA, Horazak DA, Jensen MD, Laum JD, Vucelic AP et al. 2009. Membrane-based, enzyme facilitated, efficient carbon dioxide capture. *Energy Procedia* 1: 353–360.

Trachtenberg MC, Smith DA, Cowan RM, and Wang X. 2007. Flue gas CO_2 capture by means of a biomimetic facilitated transport membrane. In: *Proceedings of the AIChE Spring Annual Meeting*, Houston, Tex, USA (http://www.carbozyme.us/pub abstracts.shtml).

Tripp BC, Smith K, and Ferry JG. 2001. Carbonic anhydrase: New insights for an ancient enzyme. *Journal of Biological Chemistry* 276: 48615–48618.

Veitch FP and Blankenship LC. 1963. Carbonic anhydrases in bacteria. *Nature* 197: 76–77.

Vinoba M, Bhagiyalakshmi M, Jeong SK, and Nam SC. 2012b. Carbonic anhydrase immobilized on encapsulated magnetic nanoparticles for CO_2 sequestration. *Chemistry* 18: 12028–12034.

Vinoba M, Bhagiyalakshmi M, Jeong SK, Yoon YI, and Nam SC. 2012a. Immobilization of carbonic anhydrase on spherical SBA-15 for hydration and sequestration of CO_2. *Colloids and Surfaces B* 90: 91–96.

Vinoba M, Kim DH, Lim KS, Jeong SK, Lee SW, and Alagar M. 2011. Biomimetic sequestration of CO_2 and reformation to $CaCO_3$ using bovine carbonic anhydrase immobilized on SBA-15. *Energy and Fuels* 25: 438–445.

Vullo D, Kupriyanova EV, Scozzafava A, Capasso C, and Supuran CT. 2014. Anion inhibition study of the β-carbonic anhydrase (cahB1) from the cyanobacterium *Coleofasciculus chthonoplastes* (ex-*Microcoleus chthonoplastes*). *Bioorganic and Medicinal Chemistry* 22: 1667–1671.

Walenta G, Morin V, Pierre AC, and Christ L. 2011. Cementitious compositions containing enzymes for trapping CO_2 into carbonates and/or bicarbonates. PCT International Application. WO 2011048335 A1 2011042.

Wang W, Liu X, and Lu X. 2013. Engineering cyanobacteria to improve photosynthetic production of alka(e)nes. *Biotechnology for Biofuels* 6: 69.

Wanjari S, Prabhu C, Satyanarayana T, Vinu A, and Rayalu S. 2012. Immobilization of carbonic anhydrase on mesoporous aluminosilicate for carbonation reaction. *Microporous and Mesoporous Materials* 160: 151–158.

Wanjari S, Prabhu C, Yadav R, Satyanarayana T, Labhsetwar N, and Rayalu S. 2011. Immobilization of carbonic anhydrase on chitosan beads for enhanced carbonation reaction. *Process Biochemistry* 46: 1010–1018.

Ward WJ and Robb WL. 1967. Carbon dioxide-oxygen separation: Facilitated transport of carbon dioxide across a liquid film. *Science* 156: 1481–1484.

Watanabe Y, Ohmura N, and Saiki H. 1992. Isolation and determination of cultural characteristics of microalgae which functions under CO$_2$ enriched atmosphere. *Energy Conversion and Management* 33: 545–552.

Weyer KM, Bush DR, Darzins A, and Willson BD. 2010. Theoretical maximum algal oil production. *Bioenergy Research* 3: 204–213.

Wilbur KM and Anderson NG. 1950. Carbonic anhydrase and growth in the oyster and busycon. *Biological Bulletin* 98: 19–24.

Wilbur KM and Saleuddin ASM. 1983. Shell formation. In: *The Mollusca*, Vol. 4.. Academic Press, New York, London, pp. 235–287.

Wood HG, Werkman CH, Hemingway A, and Nier AO. 1941. The position of carbon dioxide in succinic acid synthesized by heterotrophic bacteria. *Journal of Biological Chemistry* 139: 377–381.

Wright J, Colling A, and Open University Course Team. 1995. *Seawater: Its Composition, Properties and Behaviour*. 3rd ed. Pergamon-Elsevier, Oxford.

Xin JY, Cui JR, Niu JZ, Hua SF, Xia CG, Li SB, and Zhu LM. 2004. Biosynthesis of methanol from CO$_2$ and CH$_4$ by methanotrophic bacteria. *Biotechnology* 3: 67–71.

Xu S, Lu Y, Li J, and Jiang Z. 2006. Efficient conversion of CO$_2$ to methanol catalyzed by three dehydrogenases co-encapsulated in an alginate-silica (ALG- SiO$_2$) hybrid gel. *Industrial Engineering and Chemistry Research* 45: 4567.

Yadav R, Satyanarayana T, Kotwal S, and Rayalu S. 2011. Enhanced carbonation reaction using chitosan based carbonic anhydrase nanoparticles. *Current Science* 100: 520–524.

Yadav R, Wanjari S, Prabhu C, Vivek K, Labhsetwar N, Satyanarayana T, Kotwal S et al. 2010. Immobilized carbonic anhydrase for the biomimetic carbonation reaction. *Energy and Fuels* 24: 6198–6207.

Yang ZY, Moure VR. Dean DR, and Seefeldt LC. 2012. Carbon dioxide reduction to methane and coupling with acetylene to form propylene catalysed by remodeled nitrogenase. *Proceedings of National Academy of Science of the United States of America* 109: 19644–19648.

Yates KK and Robbins LL. 1998. Production of carbonate sediments by a unicellular green alga. *American Mineralogist* 83: 1503–1509.

Yen HW and Brune DE. 2007. Anaerobic co-digestion of algal sludge and waste paper to produce methane. *Bioresource Technology* 98: 130–134.

Yoon M. Choi J, Lee JW, and Park DH. 2012. Improvement of saccharification process for bioethanol production from *Undaria* sp. by gamma irradiation. *Radiation Physics and Chemistry* 81: 999–1002.

Yoshida N, Higashimura E, and Saeki Y. 2010. Catalytic biomineralization of fluorescent calcite by the thermophilic bacterium *Geobacillus thermoglucosidasius*. *Applied and Environmental Microbiology* 76: 7322–7327.

Yoshihara K, Nagase H, Eguchi K, Hirata K, and Miyamoto K. 1996. Biological elimination of nitric oxide and carbon dioxide from flue gas by marine microalga NOA-113 cultivation in a long tubular photobioreactor. *Journal of Fermentation and Bioengineering* 82: 351–354.

Yu Z, Xie L, Lee S, and Zhang R. 2006. A novel carbonic anhydrase from the mantle of the pearl oyster (*Pinctada fucata*). *Comparative Biochemistry and Physiology* 143: 190–194.

Yue L and Chen W. 2005. Isolation and determination of cultural characteristics of a new highly CO$_2$ tolerant fresh water microalgae. *Energy Conversion and Management* 46: 1868–1876.

Yun YS, Park JM, and Yang JW. 1996. Enhancement of CO_2 tolerance of *Chlorella vulgaris* by gradual increase of CO_2 concentration. *Biotechnology Techniques* 10: 713–716.

Zak E, Norling B, Maintra R, Huang F, Andersson B, and Pakrasi B. 2001. The initial steps of biogenesis of cyanobacterial photosystems occurs in plasma membranes. *Plant Biology* 32: 13443–13448.

Zhang S, Lu Y, Rostab-Abadi M, and Jones A. 2012a. Immobilization of a carbonic anhydrase enzyme onto flame spray pyrolysis based silica nanoparticles for promoting CO_2 absorption into a carbonate solution for post-combustion CO_2 capture. In: *Proceedings of the 243rd ACS National Meeting and Exposition*, San Diego, CA, USA March.

Zhang YT, Dai XG, and Xu GH. 2012b. Modelling of CO_2 mass transport across a hollow fiber membrane reactor filled with immobilized enzyme. *AIChE Journal* 58: 2069–2077.

Zhang YT, Zhang L, Chen HL, and Zhang HM. 2010. Selective separation of low concentration CO_2 using hydrogel immobilized CA enzyme based hollow fiber membrane reactors. *Chemical Engineering Science* 65: 3199–3207.

Zhang YT, Zhi TT, Zhang L, Huang H, and Chen HL. 2009. Immobilization of carbonic anhydrase by embedding and covalent coupling into nanocomposite hydrogel containing hydrotalcite. *Polymer* 50: 5693–5700.

Zhang Z1, Lian B, Hou W, Chen M, Li X, and Li Y. 2011. *Bacillus mucilaginosus* can capture atmospheric CO_2 by carbonic anhydrase. *African Journal of Microbiology Research* 5: 106–112.

7

The Role of Bioreactors in Industrial Wastewater Treatment

Ahmed ElMekawy, Gunda Mohanakrishna, Sandipam Srikanth, and Deepak Pant

CONTENTS

7.1 Introduction

Industrialization across the globe has resulted in the contamination of soils, groundwater, sediments, surface water, and air with hazardous and toxic chemicals, which is one of the major problems to be resolved by the research presently being carried out globally. Providing clean and affordable water to meet human needs is another grand challenge of the twenty-first century. The more the world industrializes, the more are the waste generation and contamination problems. In general, groundwater represents about 98% of the available fresh water on the planet and thus, protecting and restoring groundwater quality is of high importance. Water supply across the globe struggles to keep up with the fast growing demand, which is exacerbated by population growth, global climate change, and water quality deterioration. This widespread problem represents a significant technical and economic challenge. Globally, a huge amount of capital and resources is being spent for treating trillions of litres of wastewater annually, consuming significant amounts of energy (ElMekawy et al., 2013, 2014a,b). Therefore, there is a need for developing a wider application of cost-effective, *in situ* remediation approaches that take advantage of natural phenomena, such as bioremediation. Biological treatment is an important and integral part of any wastewater treatment plant that treats wastewater that has soluble organic impurities or a mix of the two sources from either municipality or industry (Pant and Adholeya, 2007). The economic advantage of biological treatment over other treatment processes such as chemical oxidation, thermal oxidation, etc., in terms of capital investment and operating costs has established its place in any integrated wastewater treatment plant (Mittal, 2011). The current chapter describes the major existing challenges of industrial wastewater treatment and how advanced biological processes are dealing with them. The role of different bioreactors in treating industrial effluents has also been discussed in detail including recent advances.

7.1.1 Major Challenges in Industrial Wastewater Treatment

The principle of bioremediation lies in the implementation of the necessary processes and actions for the transformation of a contaminated environment to its original status (Thassitou and Arvanitoyannis, 2001). Bioventing, bioreactor operation, composting, bioaugmentation and biostimulation, are some of the examples of biodegradation approaches. Bioremediation mainly involves chemical transformations mediated by microorganisms that satisfy their nutritional and energy requirements with simultaneous detoxification of the immediate environment. Application of microbial treatment has been extensively studied for domestic, agricultural and industrial wastes and subsurface pollution in soils, sediments and marine environments. However, the ability of each microorganism to degrade the waste fully depends on the nature of each contaminant in that particular waste (Pant and Adholeya, 2009). Since most of the waste sources contain a wide range of components and will not be similar in any two cases, this makes the degradation process more difficult. The most effective approach to solve this problem is to use a mixed culture (mixture of different bacterial or fungal species and strains, each specific to the biodegradation of one or more types of contaminants) (Pant et al., 2010).

The presence of excess nutrients such as sulfur, nitrogen, phosphates, ammonia, etc., makes the wastewater more complex and needs specific groups of bacterial population to fulfill the treatment process. Similarly, presence of xenobiotics, pharmaceutical products, micropollutants, endocrine disrupting compounds, etc., makes the process more complicated and the survival of most of the bacteria is hindered in this condition. These components need special attention and a separate treatment approach. Similarly, the bioremediation approach may also be used to break down heavy metals such as arsenic, lead, and mercury. However, some of the heavy metals such as cadmium and lead are not readily absorbed or captured by microorganisms. They need a special unit of treatment process to be included in the wastewater treatment system. Likewise, when considering any industrial wastewater, the treatment process is completely dependent on the nature of the waste and its constituents. A single process can never fulfill all the required criteria for industrial or complex wastewater treatment and multiple integrated approaches were needed to obtain good treatment efficiency (Mohanakrishna et al., 2010; Mohanakrishna and Venkata Mohan, 2013). In the following sections, we have tried to focus on the possible bioremediation approaches, bioreactors used for the treatment of industrial wastewater and the recent advancements in wastewater treatment technologies.

7.2 Conventional Bioprocesses

Conventionally, biological wastewater treatment was considered majorly for the secondary treatment process. Primary treatment through a grit chamber and sedimentation tank removes grit (large particles) and suspended solid particles. The effluent of primary treatment mainly contains dissolved organic matter and nitrogen and phosphorus based nutrients. The selection of a secondary treatment process is based on the nature of wastewater and the concentration of organic matter (either chemical oxygen demand (COD) or biochemical oxygen demand (BOD)). Low strength/lower organic loading condition and higher biodegradable wastewater can be treated very efficiently by aerobic processes. Wastewater with higher organic loading concentrations can be treated effectively by anaerobic processes. Indeed, other factors such as energy consumption, treatment time, type of pollutants present in wastewater also influence the selection of process for the treatment. In the aerobic treatment process, the bacterial biomass oxidizes the organic materials present in wastewater into carbon dioxide and water. Various configurations of aerobic processes, such as activated sludge process (ASP), trickling filter, and rotating biological contactor (RBC), were well-known aerobic processes. ASP is a suspension configuration and the other two have biofilm configurations.

The ASP contains aeration tanks which allows the suspended growth of bacterial biomass. Supply of oxygen can be through diffused aerators or suspended aerators. Hydraulic retention time for the domestic effluents usually ranges from 6 to 12 h, whereas for industrial effluents it will be more than 24 h. An aeration step is followed by sedimentation step for the separation of bacterial biomass and clarified wastewater/treated water as supernatant. The separated biomass will be recycled to maintain required mixed-liquor-suspended solids (MLSS)/biomass in the aeration tank.

The trickling filter, also called a biofilter (BF) contains a column of supporting media or substratum that supports the growth of aerobic bacterial biofilm. Varied kinds of materials have been used as supporting media such as stones, plastic discs, wooden chips etc. The wastewater is periodically applied over the column, during which the organic matter present in the wastewater diffuses into the bacterial film and gets oxidized. The oxygen required for the oxidation process is normally supplied by the natural flow of air in the column. The effluents of this column are collected in a secondary clarifier. Complete details about BFs for the treatment of emissions and effluents are discussed in a later section. The RBC is another type of biofilm configured wastewater treatment system, which contains a media disc or panel that rotates at a slower rate. The biofilm attached on the rotating disc is simultaneously exposed to the wastewater and oxygen, facilitating oxidation of organic matter present in the wastewater. In both the configurations the biomass present on the supporting material grows with time. Periodically, portions of the biofilm removed from the media and the sloughed off biomass is separated in the secondary clarifier (Gernaey et al., 2004; Grady et al., 2011).

The above mentioned aerobic wastewater treatment processes are energy intensive. Anaerobic digestion (AD) process requires less energy than aerobic process. Moreover, some amount of energy can be generated in the form of methane gas from the treatment of wastewater. Compared with aerobic treatment processes, anaerobic processes generate less sludge (Lettinga, 1995; Juretschko et al., 2002). AD is a sequential combination of four different processes that involve four distinct groups of microorganisms. The first stage is hydrolysis, during which the complex or insoluble organic matter present in wastewater is broken down into soluble and simple molecules such as sugars and amino acids. Acidogenesis is the second stage, during which sugars are converted to organic acids. During the third stage (acetogenesis), the organic acids are converted to acetic acid and carbon dioxide along with hydrogen and water. Finally, the methanogenesis (fourth stage) takes place resulting in methane production and wastewater treatment (Speece, 1983; Ueno et al., 1996; Venkata Mohan et al., 2005, 2008). For the treatment of industrial wastewater in large scale applications, AD was found to be the more efficient process than the aerobic process. Since this process is an energy source, the wastewater is considered as the renewable substrate (Speece, 1983, Ueno et al., 1996; Rajeshwari et al., 2000; Mohanakrishna and Venkata Mohan, 2013). All the configurations and operating modes related to AD are discussed in the next sections.w

7.3 Advanced Bioprocesses

7.3.1 Biohydrogen Production

Biological wastewater treatment systems can also be operated for hydrogen production from the treatment. Dark fermentation and photo fermentation are found to be suitable processes for hydrogen production. Acidogenic fermentation is also called dark fermentation. It is a truncated version of AD, in which the AD does not proceed to methanogenesis resulting in H_2 as the end product. As the hydrogen is found to have a heating value higher than methane, the wastewater treatment systems can operate for hydrogen production. It can be achieved by eliminating the methanogenic group of bacteria from anaerobic sludge. As the final step is eliminated from AD, the organic acids are retained in the effluents of ·acidogneic process, thus limiting the treatment efficiency. Although theoretically wastewater treatment is limited to only 33%, several studies exhibited more than 70% treatment efficiency (Mohanakrishna et al., 2012). The effluents of acidogenic fermentation that are rich in organic acids can be used as suitable substrate for other energy generating processes such as microbial fuel cells (MFCs) for bioelectricity generation (Mohanakrishna et al., 2010a; Pant et al., 2013), polyhydroxyalkanoates (Reddy and Venkata Mohan, 2012), hydrogen production by photo fermenting bacteria (Srikanth et al., 2009).

7.3.2 Photo Fermentation

Photo fermentation is a process that utilizes the organic acids present in wastewater as substrate and converts them to hydrogen and carbon dioxide. Due to the operational limitations of AD and dark fermentation, their effluents are often found to have a high amount of organic acids. These can be further treated in photo fermentation to get additional energy from the wastewater. Although, the theoretical possibility of photo biological hydrogen production is found to be effective from organic acids, several researchers reported complex wastewater and glucose as substrate using mixed bacterial communities (Srikanth et al., 2009). This is due to the complex microbial metabolisms that take place in the system. This process is found to be very efficient as it extends the wastewater treatment efficiency. The process of photo biological hydrogen production and the bacteria responsible for this are more sensitive to diverse environmental conditions, high organic loading rates, etc. The process is readily inhibited as the turbidity or opacity of the reactor contents increases. As this process has high stoichiometric efficiencies, extensive research is being conducted to adapt it for commercial scale operations.

7.3.3 Anammox Process

Besides organic matter, nitrogenous compounds are also present in wastewater. Leachates from landfills and different types of effluents such as from the petrochemical, pharmaceutical, fertilizer and food industries, contains large quantities of ammonium. By the end of the AD process, most of the biologically amenable nitrogen compounds are converted to ammonium. As this ammonia causes serious environmental problems, it is also important to treat it before disposal (Carrera et al., 2003). Conventionally, biological nitrogen removal takes place with nitrifying and denitrifying bacteria. Several distinct groups of bacteria act on nitrogenous compounds and converts them to molecular nitrogen. Among the nitrification and denitrification processes, nitrification requires oxygen. Commercial scale bioreactors such as sequencing batch reactors (SBRs), BFs, etc., are specially designed for nitrogen removal from wastewater. Anammox is a newly discovered microbial metabolism discovered in 1995 in a fluidized bed bioreactor in which ammonium oxidation is possible under anoxic or anaerobic conditions (Mulder et al., 1995). This process converts ammonium and nitrite directly to dinitrogen (Figure 7.1). This process is estimated to be responsible for more than 30% of the global nitrogen production. The anammox process is being integrated with the wastewater treatment plants which proceed with two distinct processes. In the first stage, ammonium undergoes partial oxidation forming nitrite. In the second stage, ammonia and nitrite is converted to dinitrogen in the anammox process. Various designs were proposed to integrate both the processes in one reactor for practical application in wastewater treatment plants (van der Star et al., 2007).

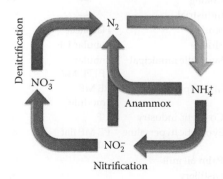

FIGURE 7.1
Biological nitrogen cycle with an emphasis on differentiation between the conventional nitrification-denitrification process and the anammox process.

7.4 Types of Bioreactors

Several types of bioreactors are employed in wastewater treatment, in which polluted water is either recycled in an aerobic or anaerobic tank where free suspended microbes or immobilized cells on a matrix are used to metabolize organic materials, forming a sludge that is recycled or discharged. The efficiencies of both bioreactor technologies are illustrated in Table 7.1 in terms of chemical oxygen demand (COD).

TABLE 7.1

Overview of the Wastewater Treatment Using Different Bioreactor Technologies and Their Performances in Terms of COD Removal Efficiency

Bioreactor Type	Aeration	Wastewater Source	Membrane	COD Removal (%)	References
SBR	Anoxic/aerobic	Olive mill	—	90	Chiavola et al. (2014)
	Anoxic/aerobic	Domestic	—	83	Fernandes et al. (2013)
	Anoxic/aerobic	Municipal	—	94	Bagheri et al. (2014)
	Anaerobic/ anoxic/aerobic	Synthetic	—	98	Puay et al. (2015)
	Anaerobic	Alcohol	—	76	Intanoo et al. (2014)
	Aerobic	Leachate + domestic	—	73	Mojiri et al. (2014)
	Aerobic	Swine	—	76	Daverey et al. (2013)
AD	Anaerobic	Sewage	—	43	Bajón Fernández et al. (2014)
	Anaerobic	Rice straw	—	57	Mussoline et al. (2012)
BF	Aerobic	Leachate	—	80	Ferraz et al. (2014)
	Aerobic	Domestic	—	90	Luo et al. (2014)
CSTR	Aerobic	Hydrocarbon-rich industrial	—	95	Gargouri et al. (2011)
	Anaerobic	Swine	—	65	Kim et al. (2013)
PBB	Aerobic	Synthetic	—	98	Dizge et al. (2011)
	Anaerobic/ aerobic	Slaughterhouse	—	93	Del Pozo and Diez (2005)
MBBR	Aerobic	Synthetic	—	44	Shore et al. (2012)
	Aerobic	Coking	—	89	Gu et al. (2014)
	Aerobic	Industrial	—	90	Dvořák et al. (2014)
MBR	Anaerobic	Sludge	Flat sheet	99	Ersahin et al. (2014)
	Anaerobic	Sludge	Tubular UF	99	Dereli et al. (2014)
	Anaerobic	Synthetic municipal	Tubular PTFE MF	95	Ho and Sung (2010)
	Aerobic	Municipal	PE MF module	99	Mohammed et al. (2008)
	Aerobic	Cosmetic industry	UF	83	Friha et al. (2014)
	Aerobic	Synthetic hypersaline	PE MF flat sheet	81	Sharghi et al. (2014)
UASB	Anaerobic	Palm oil mill	—	62	Singh et al. (2013)
	Anaerobic	Distillery	—	87	Sridevi et al. (2014)
	Anaerobic	Heavy oil	—	74	G. Liu et al. (2013)
	Anaerobic	Berberine antibiotic	—	98	Qiu et al. (2013)

Note: UF: ultrafiltration; PTFE: poly-tetrafluoroethylene; MF: microfiltration; PE: polyethylene; COD: chemical oxygen demand.

7.4.1 Free Suspended Cell Technology

7.4.1.1 Anaerobic Digesters (AD)

The AD is a multifaceted technology that requires strict anaerobic conditions with oxidation reduction potential less than −200 mV to be carried out, and relies on the synchronized metabolism of a mixed microbial culture to mainly convert organic material to methane (CH_4) and carbon dioxide (CO_2). The produced biogas with great calorific outcome is considered as a source of renewable energy. Normally, AD of sludge employs airtight reactors to perform the main four steps of organic material (Figure 7.2a). Several organic materials can be degraded anaerobically by microorganisms, apart from cellulosic materials as anaerobic microbes are unable to digest lignin (Appels et al., 2008; ElMekawy et al., 2013).

Mainly, AD uses three modes of operation. The first one is the standard rate digestion, in which there is no heating or mixing process of the reactor sludge, and therefore it is considered the simplest mode using an extended degradation period (1–2 months) (Qasim, 1999; Tchobanoglous et al., 2003; Turovskiy and Mathai, 2006). The standard rate digestion was upgraded to the high rate digesters, in which the reactor sludge content is mixed and heated, in order to obtain thick and uniform sludge. The processing of the reactor content results in a homogeneous environment, which positively reflects on the reactor volume and the efficiency of the whole process stability (Turovskiy and Mathai, 2006). The high-rate digester was further improved to obtain the two stage digester, by coupling the high rate digester with a secondary digester, which is used for the digested component storage and supernatant draw off without heating or mixing (Qasim, 1999; Tchobanoglous et al., 2003; Turovskiy and Mathai, 2006).

The majority of high rate digesters run in the mesophilic temperature range (30–38°C) (Qasim, 1999). AD can also be carried in the thermophilic temperature range, where the digestion process by thermophilic bacteria, takes place at 50–57°C. The digestion rate of the thermophilic process is faster compared with that of the mesophilic one, due to the increased rate of biochemical reaction at elevated temperatures (Qasim, 1999; Tchobanoglous et al., 2003; Turovskiy and Mathai, 2006).

7.4.1.2 Continuous Stirred Tank Reactors (CSTR)

Activated sludge represents the most common aerobic process used for the treatment of agitated and aerated suspension of a mixed culture of bacterial growth that metabolizes wastewater (Sheng et al., 2008). It has high productivity with flexible operation and potential nutrient elimination. The ASP could be operated in several operational modes, that is, continuously stirred tank reactor (CSTR), plug flow reactor, or SBR. CSTRs consist of a constant volume tank equipped with agitation system to mix all the components together (Figure 7.2b). They are also equipped with influent and effluent ports for the inflow of reactants and the harvest of products, respectively (Chan et al., 2009).

7.4.1.3 Sequencing Batch Reactors (SBR)

The SBR is one of the wastewater treatment processes established on the basis of ASP principles, but differ from it in combining all of the process steps into a single tank, while traditional processes take place in several tanks (Bagheri et al., 2014). SBR has been effectively used in industrial and municipal wastewater treatment (Mace and Mata-Alvarez, 2002; Mohanakrishna et al., 2011; Çınar et al., 2008), and was applied in the biological treatment

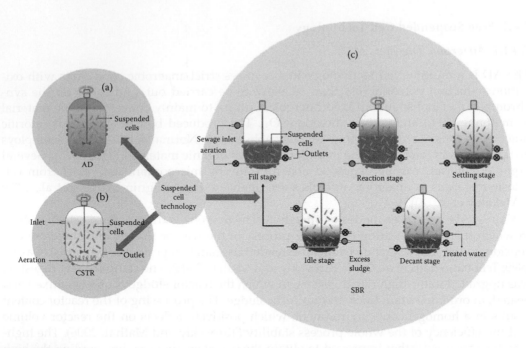

FIGURE 7.2
(See color insert.) Schematic of different types of bioreactors utilizing free suspended microbial cells for waste treatment.

of wastewater, with the order of anaerobic, anoxic and aerobic settings for phosphorus, nitrogen, and carbon elimination (Littleton et al., 2014).

The standard SBR technique for wastewater treatment process has five steps: fill, reaction, settlement, draw and idle, which run consecutively in a batch reactor (Figure 7.2c) (Lee and Park, 1999; Gali et al., 2008). Wastewater, with phosphate, ammonia, and carbon content, is pumped to the bioreactor tank during the filling step, and then mixed with the other pre-existing biomass (Lee and Park, 1999). The reactor is then shifted to the aerobic reaction step, in which the existing wastewater is mechanically agitated and aerated in order to provide several biological metabolisms to take place, such as phosphorus uptake/release, nitrification, denitrification and carbonaceous BOD removal (Lee and Park, 1999; Mahvi 2008; Rodríguez et al., 2011). The activated sludge is then settled under anoxic conditions in the settling stage, without flow, mixing or aeration in the reactor, and the sludge mass (sludge blanket) is separated from the clarified treated wastewater (Lee and Park 1999; Casellas, Dagot, and Baudu 2006), which is then removed from the bioreactor in the drawing stage (Lee and Park, 1999; Aziz et al., 2011; Wu and Chen, 2011). The idle step takes place between the drawing and the fill steps, in which a slight volume of the activated sludge at the SBR reactor bottom is pumped out (wasting process) (Mahvi, 2008). The time of this step differs based on the flow rate of the influent and the process scheme (Bagheri et al., 2014).

7.4.2 Immobilized Cell Technology

7.4.2.1 Upflow Anaerobic Sludge Blanket (UASB) Reactors

Upflow anaerobic sludge blanket (UASB) reactors are a stable technology which has operated effectively in the field of wastewater treatment for several decades (Tchobanoglous et al., 2003). They are commonly applied as a pretreatment process prior to AD of different types of municipal and industrial wastewater. It was observed that it is an effective technology to overcome some of the problems accompanying the automatic aerobic setups, by lowering the consumption of energy and production of sludge (Chan et al., 2009). The UASB reactor involves the upflow of wastewater across a condensed sludge bed with a high population of active microorganisms (Sperling and Chernicharo, 2005; Vlyssides et al., 2009). This bioreactor depends on the formation of dense particles with minute diameter (1–4 mm) developed via the spontaneous immobilization of the anaerobic consortium, which is a vital condition for the efficient process of UASB bioreactor (Figure 7.3a). In general, this type of bioreactor can eliminate over 60% of the COD from several wastewater types (Chan et al., 2009).

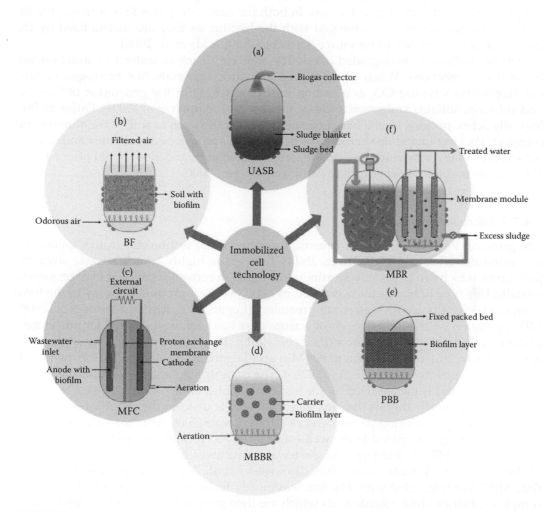

FIGURE 7.3
Schematic diagram overviews the six different types of bioreactors utilizing immobilized bacterial cells for waste/wastewater treatment. (a) Upflow anaerobic sludge blanket (UASB) bioreactor; (b) biofilter (BF); (c) microbial fuel cell (MFC); (d) moving bed biofilm reactor (MBBR); (e) packed bed bioreactor (PBB); (f) membrane bioreactors (MBR).

7.4.2.2 Biofilters

Biofilters are one of the main biological gaseous waste treatment and odor elimination processes, with efficacious applications in the field of petrochemicals, tobacco industry. These BFs also used for wastewater treatment like trickling filters (McNevin and Barford, 2000; Duan et al., 2006). BFs were found to have economic capital and lower operating costs, high in efficient treatment capacity and simple maintenance (Busca and Pistarino, 2003; Kikuchi, 2006). A BF is a tank of organic material cultured with microorganisms, in which polluted air commonly flows upwards (Figure 7.3b). The stream either flows in a counter-current or concurrent mode and contacts the liquid solution which supplies nutrients and confers the suitable conditions to preserve the activity and viability of the developing biofilm. Counter-current passes the polluted gas from the bottom and liquids were sprayed from the top of the bed. In the case of concurrent sprays of polluted air and liquid are pumped through the top of the bed. In both the cases, the gases flow through the BF, in which contaminants are exchanged with the biofilm, as they are metabolized by the bacteria as an energy or carbon source (Ortiz et al., 2003; Ma et al., 2006).

Organic pollutants are degraded to odorless materials such as water, CO_2 and biomass by oxidative reactions. When metabolizing inorganic materials like hydrogen sulfide, autotrophic bacteria use CO_2 as a carbon source, resulting in the generation of biomass and sulfur or sulfate (Andersson and Grennberg, 2001; Barona et al., 2004). Different biologically active packing media have been employed in BFs such as soil, horse manure and compost. Packing materials should have high liquid and gas permeabilities and surface area as they play a vital role in water and gaseous exchange, and also should offer a favorable surface for biofilm development (Song and Kinney, 2000).

7.4.2.3 Microbial Fuel Cells (MFC)

The growing interest in bioelectrochemical systems (BES) technology in the application of wastewater treatment is due to its ability to degrade highly complex waste streams/ pollutants, which can be effectively mineralized than in conventional treatment processes. Initially, this type of bioreactor was called as MFC, where it produced energy in the form of bioelectricity along with wastewater treatment (Logan, 2008; Rismani-Yazdi et al., 2008). MFC operation with different kinds of wastewaters has been well studied for power generation and bioremediation aspects (Pant et al., 2010). The MFC types usually employed in wastewater treatment are the flat plate (Min and Logan, 2004), stacked (Aelterman et al., 2006a,b), and upflow MFCs (He et al., 2005). The organic substrate is fed into the anodic compartment and is metabolized by the bacterial biofilm on the anode. The bacterial biocatalyst anaerobically digests the organic substrate to transform the chemical energy into electrical energy, producing protons and electrons (Logan, 2008; Rismani-Yazdi et al., 2008; Zhou et al., 2011). Compared to the well-established AD technology, restricted industrial applications of MFC technology and wastewater treatment is rendering it an initial stage (Aelterman et al., 2006a,b; Logan 2008; Chiranjeevi et al., 2012). To overcome this restriction, MFC was integrated with different wastewater technologies in order to degrade the complex substrates into volatile acids which are then metabolized to several value-added products (ElMekawy et al., 2014a,b). Later, when the potential of this system in treating complex pollutants and synthesizing a diverse range of biocommodities were observed and renamed it as BES that covers all kinds of biologically catalyzed electrochemical systems. BES typically contains two compartments, anode and cathode, separated by a proton exchange membrane (PEM) (Figure 7.3c). In this chapter, we try to summarize the role of

BES in waste/wastewater remediation aspects. A detailed discussion on the role of MFC as a BES for electroremediation has been discussed in Section 7.6.1.

7.4.2.4 Moving Bed Biofilm Reactors (MBBR)

Moving bed biofilm reactor (MBBR) technology is based on the merger between the advantages of both the BF process and those of the ASP. The MBBR is similar to the activated sludge bioreactor in terms of the whole tank volume utilization for microbial growth, which is normally not the feature of most biofilm reactors. Alternatively, it does not require any sludge recycle as the case of the activated sludge bioreactor. This is obtained by allowing the growth of the microbial biomass on carriers which move freely in the liquid volume of the bioreactor (Figure 7.3d) (Odegaard, 2006). Only the leftover biomass has to be separated as there is no sludge recirculation, which is considered as an advantage compared with the activated sludge technology. The bioreactor may be operated in aerobic, anaerobic, or anoxic conditions, where the movement of the biofilm carrier in the aerobic condition is carried out by the agitation effect of the air, while in anaerobic and anoxic conditions the carriers are continuously moved by a mixing shaft. The biofilm carrier commonly used in MBBR, that is, K1, is fabricated from polyethylene with high density (0.95 g/cm^3) and has the shape of a small pipe with an internal cross and external limbs (Odegaard, 2006).

7.4.2.5 Packed Bed Bioreactors (PBR)

Fixed bed bioreactors are progressively employed in wastewater treatment process due to the flexible and compact features of this technology (Iliuta, 1997; Deront et al., 1998; Benthack et al., 2001). Packed bed bioreactors (PBRs) are tubular tanks packed with solid particles of catalyst (Fogler, 2006), which has a high nutrient exchange per weight (Figure 7.3e). This exchange capacity depends on the solid catalyst content rather than the size of the reactor.

The packing material is beneficial for bacterial immobilization to form biofilms for the treatment at an industrial scale. The catalyst is packed in the column and the nutrients are fed either from bottom or top of the reactor (Martinov et al., 2010; Venkata Mohan et al., 2008b). There are several traditional granular packing materials such as ceramic pieces, volcanic rocks, and clay balls. Also, the fibrous packing materials, that is, polyethylenevinylacetate, were introduced with their comprehensive specific surface and plastic nature, described to improve the cells adhesion to form a biofilm (Hadjiev et al., 2007).

7.4.2.6 Membrane Bioreactors (MBR)

Membrane bioreactors (MBRs) are an integration of the traditional ASP with suspended biomass growth and ultrafiltration (UF) or microfiltration (MF) membrane systems (Judd, 2006). The biological part of the reactor is employed in the waste biodegradation and the membrane compartment is responsible for the physical filtration of the biologically treated water from the mixture. The conventional gravitational sedimentation clarifier in the ASP is replaced by the membrane filtration system with a pore diameter in the range of 0.01–0.1 μm, which allows bacteria and particles to be filtered out of a permeate (Figure 7.3f) (Hoinkis et al., 2012). The anaerobic MBR is a combined setup of the anaerobic bioreactor and UF or MF membrane filtration with low pressure, in which membranes can physically filtrate suspended solids, such as biological and inert materials (Chang, 2014).

7.5 The Main Challenges Associated with Bioreactors

Microorganisms can grow in aerobic or anaerobic conditions in several types of bioreactors, either in the form of suspended cells or immobilized cells on the surface of a solid support. AD is considered to be one of the most applied technologies for wastewater treatment with free suspended cells. Even with the advantages of AD, some obstacles are expected (Appels et al., 2008). Among several limitations of bioreactors for industrial wastewater treatment, incomplete degradation of the organic portion and the relatively slow rate can be considered as critical challenges. Furthermore, the AD has a high cost with the process exposure to different inhibitors (ElMekawy et al., 2013). Moreover, the produced supernatant has a poor quality with the presence of residual COD and by products such as organic acids (Appels et al., 2008). Also, some other components, such as the volatile siloxanes, can result in serious destruction in the end user machines such as generators and boilers, because of microcrystalline silica formation. The leftover sludge has an amplified content of heavy metals and different industrial pollutants as the organic fraction is reduced during the digestion process, while the nonmetabolized fraction is left without digestion. The AD is incapable of attaining complete degradation of several contaminants. Phenolic components and their by-products accumulates in the reactors and then discharged in the effluent or inhibit the biomass (Beccari et al., 1996; ElMekawy et al., 2013).

Additionally, the problem of foaming is considered as one of the most challenging for AD technology (Subramanian and Pagilla, 2014). Foaming is a multi-phase (solid, gas, and liquid) phenomenon resulting from the interaction of surfactants or materials with active surfaces and also the biogas generated inside the digester. Also, foam could be affected by the unbalanced ratio between waste activated sludge and primary sludge solids in the reactor feed. Vagarous mixing is likely to develop the entrapment of air bubbles in the liquid, producing foam (Campbell, 1999). Keeping extended retention time for the sludge is one of the problems of AD technology because of the slow rate of the anaerobic biomass growth (Ho and Sung, 2010).

The UASB reactor is one of the most well-established efficient AD technologies for wastewater treatment with more than thousands of working UASB bioreactors worldwide (Tiwari et al., 2006). The stability and efficiency of the UASB bioreactor relies greatly on the early start-up, which is influenced by several chemical, biological, and physical factors (Ghangrekar et al., 1996). These factors include the operating conditions, type of wastewater, and growth of active biomass in the sludge (Chong et al., 2012). The colloidal and suspended constituents of wastewater in the form of protein, cellulose, and fat have negative effect on the efficiency of the UASB bioreactor, preventing the operation of the reactor at higher organic loads causing the decline of microbial activity and washing out of biomass (Torkian et al., 2003). The higher loads of organic materials require a long adaptation period for the inoculated seed sludge (Mohanakrishna et al., 2011). This delay represents the major drawback of the industrial scale application of UASB bioreactors (Vlyssides et al., 2008). Also, temperature has a crucial effect on the start-up digestion rate of sludge, where the main methanogenic bacterial component of UASB particles (35°C) are likely to be reproducing in 72 h, while at a low temperature (10°C), the generation time can reach 50 days (Bhuptawat et al., 2007). This is due to the direct positive effect of higher temperatures on methanogenesis, usually accompanied with low effluent quality because of sludge wash out (Liu and Tay, 2004).

On the contrary, there are several operational problems with bioreactor technologies that utilize microbial biofilms (immobilized cells). One of the main challenges is the loss of the biofilm integrity as it grows, which requires the continuous removal of biosolids either by the start over of the system or the detachment of the biomass from the packing material with a backwash step. In spite of the recent progress in MBRs, there are several challenges that restrict their widespread industrial applications (Dizge et al., 2011). Membrane fouling is considered the most important challenge as it can raise the relevant operational costs. The fouling problem could be caused by the accumulation of the microorganisms themselves or their products, as well as the influent particles into the membrane pores and on the membrane surface (Lin et al., 2009). Membrane fouling is influenced by several variables such as influent properties, operating conditions, effluent nature, membrane characteristics, and cake layer (He et al., 2005; Le-Clech et al., 2006). The extracellular polymeric substances and soluble microbial products are considered the most important source of organic fouling, where they can be attached to the flocculated or suspended materials, to form microbial agglomeration or to bind directly to the membrane (Choo and Lee, 1996; Brockmann and Seyfried, 1996; He et al., 2005). Membrane fouling can lead to the regular washing of the membranes, in which the operation is stopped and further costs are added for the chemicals normally used in this process (Mohammed et al., 2008).

Moreover, the drawbacks related to the MBR reflect on its overall costs, in which the membrane units (UF and MF) and the consumed energy for the pressure gradient lead to high capital cost (Mohammed et al., 2008). Another disadvantage is that the activated sludge waste could be poorly filtered and when processed at high retention time, the inorganic materials accumulate in the bioreactor leading to an adverse effect on the microbial growth or membrane composition (Cicek et al., 1999a,b).

7.6 Technology Integrations and Advanced Bioprocesses

7.6.1 Bioelectrochemical Systems (BES)

Energy generation from wastewater treatment through microbial metabolism is due to the fermentation (substrate oxidation) of organic pollutants. Substrate falls in the metabolic flux of the microbe generating energy rich reducing equivalents (protons [H^+] and electrons [e^-]), which get reduced in presence of an electron acceptor to complete the electron transport chain. Separating these two processes (oxidation and reduction) by an ion permeable membrane in a system equipped with electrodes, helps us in harnessing the energy (Venkata Mohan et al., 2013a,b). Protons are transported to the cathode through the solution electrode interface across the ion selective membrane generating a potential difference between anode and cathode against which the electrons will flow through the circuit across the external load (current) (Pant et al., 2012). The reducing equivalents generated during the BES operation have multiple applications in energy generation as well as in waste remediation areas. Broadly the BES application can be classified as a wastewater treatment unit and system for the recovery of value added products, apart from power generation. When the waste/wastewater functions as an electron donor or acceptor, its remediation gets manifested either through anodic oxidation or cathodic reduction under defined conditions (Pant et al., 2010). Very recently, reduction of some substrates or carbon dioxide (CO_2) as electron acceptors during the BES operation has also been reported, increasing its commercial viability (Rabaey and Rozendal, 2010; Pant et al., 2012).

Both the anode and cathode chambers can be an option for treating waste based on the nature of the waste selected. Treatment of wastewater in the anode chamber is mainly through microbial metabolism and partly due to the induced electrochemical oxidation (EO) mechanism. The presence of oxidizing agents (which gain electrons) such as chlorine, bromine, and ozone increases the potential differences between electrodes and thus the redox potential of the system which in turn favours EO resulting in both pollutant as well as carbon removal. Apart from EO, direct and indirect anodic oxidation (DAO and IAO) mechanisms are two possible ways described for the pollutant treatment at the anode of the BES (Venkata Mohan and Srikanth, 2011). When the simple organic fraction of waste gets oxidized at the anode through microbial metabolism, the reducing equivalents [e^- and H^+] formed during metabolism help in generation of primary oxidation species under *in situ* biopotential (Wilk et al., 1987; Israilides et al., 1997; Mohanakrishna et al., 2010b; Venkata Mohan and Srikanth, 2011). The primary oxidants formed during DAO further react on the anode yielding secondary oxidants such as chlorine dioxide and ozone, which will have significant positive impact on the treatment, especially for color removal efficiency. The reactions between water and free radicals near the anode also yield secondary oxidants, viz., nascent oxygen, free chlorine and hydrogen peroxide, hypochloric acid, etc. (Venkata Mohan and Srikanth, 2011). Generally, pollutants present in waste are adsorbed on the anode surface and get destroyed by the anodic electron transfer reactions during DAO, while during the IAO, these pollutants will be oxidized by the oxidants (primary and secondary) formed electrochemically on the anode surface.

The cathode can also participate in effective remediation of waste streams and pollutants such as azo dyes, nitrobenzene, nitrates, sulfates, etc., likewise the anode. The major route of the wastewater treatment at the cathode is considering these pollutants as terminal electron acceptors to make the electrical circuit closed in the absence of oxygen. However, their function as electron acceptors is based on the thermodynamic hierarchy. Unlike anode, the cathode chamber can be maintained under different microenvironments (aerobic, anaerobic, and microaerophilic) to increase the treatment efficiency based on the nature of the pollutant (Venkata Mohan and Srikanth, 2011; Srikanth et al., 2012). Generally, oxygen is considered as the TEA at cathodes (abiotic) but in biocathodes, microorganisms will be used it as the catalyst for the terminal reduction reaction. Depending on the terminal electron acceptors adopted at the cathode, they can be classified as aerobic and anaerobic biocathodes (He and Angenent, 2006). However, the efficiency of treatment as well as energy output varies among the microenvironments studied. Apart from the electron acceptor conditions, the treatment in the cathode chamber is also through direct or indirect oxidation processes (similar to the anode) based on the primary and secondary oxidizing species (Aulenta et al., 2010; Srikanth and Venkata Mohan, 2012). The oxidizing species also react with the primary cationic species viz., Na^+ and K^+, under biopotential leading to their removal as salt. Bicarbonates will be formed from the reaction between carbon dioxide (from air sparging or aerobic metabolism) and water which further reacts with the cationic species forming respective salts. Maintenance of cathodic pH is very crucial to sustain the microbial activity at the cathode, in spite of continuous reduction reactions. The *in situ* bicarbonate buffering mechanism formed at the cathode helps to overcome this drop in cathodic pH which is essential in continuing the reduction reaction as well as in maintaining the metabolic activities of microbes. Physiologically favorable redox conditions in the cathode chamber support the rapid metabolic activities of aerobic consortia thus resulting in higher substrate removal (Mahmoud et al., 2014; Torres, 2014).

The anaerobic microenvironment at the cathode supports the removal of specific pollutants and toxic components of wastewater, especially when they act as electron acceptors. Instead of oxygen, other substances such as nutrients viz., nitrogen and sulfur and metal ions viz., iron, manganese, and chromium will act as terminal electron acceptors in the case of the anaerobic biocathode. This helps in the removal of these toxic substances from wastewater along with power generation (Clauwaert et al., 2007; Hamelers et al., 2010; Huang et al., 2011). Both the anode and cathode chambers function as anaerobic treatment units in this case except for the variation of the presence of electrodes in each chamber and the connection in the circuit across an external resistance/load. On the contrary, the microaerophilic environment at the cathode switches between aerobic and anaerobic microenvironments. This has an advantage over aerobic and anaerobic biocathode operations, especially in the wastewater treatment sector. Some pollutants such as azo dyes need both the environments for complete mineralization. The anaerobic condition helps in splitting the azo bond, while the aerobic condition helps in mineralization of dye metabolites (Venkata Mohan et al., 2013a,b). The lower DO levels maintained at cathode during this operation helps in initiating electrochemical oxidation reactions as well as maintaining strong reduction reactions. The survival of facultative microbes which can carry out both metabolic functions will increase the treatment efficiency (Srikanth et al., 2012).

7.6.2 Nanotechnology

Nanotechnology holds great potential in advancing industrial wastewater treatment to improve process efficiency. The unique properties of nanomaterials and their convergence with current treatment technologies present great opportunities to revolutionize wastewater treatment. Our current water treatment, distribution, and discharge practices, generally based on conveyance and centralized systems, are not very sustainable. However, recent advancements in nanotechnology offer tremendous opportunities to develop efficient water supply systems. The highly efficient, multifunctional processes enabled by nanotechnology are envisaged to provide high performance and affordable wastewater treatment solutions, generally less reliant on large infrastructures (Qu et al., 2013). Nanotechnology based wastewater treatment helps to overcome some of the major challenges of existing treatment technologies and also allows us the economic utilization of unconventional water sources.

Qu and his co-workers (2013) recently reported the role of nanotechnology in wastewater treatment in their detailed review. Various applications for the integration of nanotechnology have been integrated into wastewater treatment but most of them are still in the laboratory research stage (Table 7.2). However, some have made their way to pilot testing and

TABLE 7.2

Consolidated Table of the Most Used Nano Materials in Bioremediation, Their Specific Properties, and Respective Applications

Material	Specific Properties	Applications
Silver (Ag) nano particles	Broad-spectrum antimicrobial activity, low toxicity to humans, wide range of applications with ease of use	Disinfection/decontamination of wastewater; anti-biofouling agents; solar disinfectants
Carbon nanotubes		
TiO_2 nano particles	High chemical stability and selectivity; photocatalytic activity; low cost and human toxicity	
Fullerene derivatives		
Carbon nanotubes	High-specific surface area with accessible adsorption sites; short intra-particle diffusion distance; interaction with wide range of contaminant; tunable surface chemistry; high rate re-usability	Detection of recalcitrant contaminants; adsorptive removal of contaminants; adsorptive media filters; reactive nano-adsorbents

commercialization among which, nanoadsorbents, nanotechnology enabled membranes, and nanophotocatalysts are the most promising ones. These three categories could be interesting in full scale application based on their current research state, cost of nanomaterials involved, and compatibility with the existing infrastructure. Different types of carbon/metal based nanomaterials were developed as adsorbents or membranes or photocatalysts depending on the pollutant to be removed. The potential applications of nanomaterials in wastewater treatment were briefed by Qu et al. (2013). Although this technology enabled great promise in wastewater treatment processes, there are still several limitations which have to be overcome. The challenges faced by wastewater treatment nanotechnologies are important, but many of these challenges are perhaps only temporary, technical hurdles, high cost, and potential environmental and human risk. The two major research hurdles for full-scale applications of nanotechnology in wastewater treatment include (i) studies with real wastewater under more realistic conditions and (ii) long-term efficacy of these nanotechnologies in wastewater treatment processes.

7.6.3 Bioaugmentation

Problems associated with the degradation of the organics under complex system conditions by the indigenous population may be overcome by an augmentation strategy resulting in enhancement of the process efficiency. Bioaugmentation can be explained as a process by which the application of indigenous or wild-type or genetically modified organisms to the bioreactor or to the polluted sites improves the performance of the ongoing biological processes (Venkata Mohan et al., 2009). Bioaugmentation is the application/introduction of indigenous or allochthonous wild-type or genetically modified organisms to polluted hazardous waste sites or bioreactors in order to accelerate the removal of undesired compounds (van Limbergen et al., 1998). Bioaugmentation has several advantages such as to improve the start-up of a reactor, to enhance reactor performance, to protect the existing microbial community against adverse effects, to accelerate the onset of degradation and/or to compensate for organic or hydraulic overloading (Wilderer et al., 1991; Stephenson and Stephenson, 1992; Hajji et al., 2000; Jianlong et al., 2002; Venkata Mohan et al., 2007, 2009). Bioaugmentation has been also used as a treatment strategy for contaminated soils (Newby et al., 2000; Rojas-Avelizapa et al., 2003; Shailaja et al., 2007; Rodrigo et al., 2008). Bioaugmentation with genetically modified organisms carrying plasmid-encoded catabolic genes has the potential to enhance the breakdown of xenobiotic compounds by increasing the degradation potential of an indigenous microbial population via horizontal gene transfer. Bioaugmentation mediated enhanced degradation of xenobiotics in a sequencing mode activated sludge system has been shown previously (Bathe, 2004; Bathe et al., 2005). However, bioaugmented species often fail to compete with the indigenous population and it may also cause process inhibition (Bouchez et al., 2000). This approach involves increasing the metabolic capabilities of the native microflora present in either soil or wastewater by the addition of organisms having the specific required biological activity.

7.7 Constructed Wetlands

Constructed wetlands are emerging to be highly promising for the treatment of industrial wastewaters. The process is known as an economically feasible and environmentally friendly approach. The design of constructed wetlands consists of various common plants such as *Eichornia, Typha, Schoenoplectus, Phragmites,* and *Cyperus* which are emergent plants. The bed of the system is constructed with permeable substrata such as gravel (Mbuligwe, 2005; Davies et al., 2009; Vymazal, 2009; Kumar et al., 2011). The wastewater infiltrates through the gravel that provides more contact to the plant roots that develop a good rhizosphere zone (Haberl et al., 2003). In case of floating plants such as *Eichornia,* the fluffy roots harbor rhizosphere biota. Such developed rhizosphere biota is involved in the treatment of wastewater. In the presence of toxic effluents, the growth and treatment efficiency of plants is affected. The combination of microorganisms and fungi helps in the sustainability of the treatment process (Afzal et al., 2013). In the case of low toxic effluents algae is considered to improve the dissolved oxygen concentration of the system. This also helps for the effective biological nitrogen removal process (Venkata Mohan et al., 2010, 2011). Constructed wetlands are also designed to treat specific pollutants present in different types of wastewater. Particular biological components that specifically function for specific pollutant removal are used in constructed wetlands for wastewater treatment (Table 7.3).

Major research was also focused on enhanced nitrogen removal from industrial wastewaters. Hybrid constructed wetlands were developed for wastewater containing higher nitrogen concentrations. Varied redox conditions prevailing in the constructed wetlands are more suitable for nitrification and denitrification processes. Single vertical flow and horizontal vertical flow patterns were simple designs that provide anoxic/anaerobic microenvironments due to the permanent saturation of the filtration bed of the constructed wetlands, hence, providing suitable conditions for denitrification. Free-drain vertical flow constructed wetlands are aerobic due to the intermittent feeding that allows oxygen diffusion into the filtration bed (Vymazal, 2007). Photosynthesis of algae, cyanobacteria, and submerged plants also helps in the nitrification process by improving the dissolved oxygen levels (Venkata Mohan et al., 2010). Denitrification may take place at the bottom layer or sediments of decaying litter material. In addition, high pH values prevail in the system due to the photosynthetic activity of microphytes and submerged plants, which may be involved in the volatilization of ammonia (Kadlec and Wallace, 2009). The average removal of ammoniacal nitrogen by vertical wetlands was found to be 56%, whereas horizontal constructed wetlands registered at 29% (Vymazal, 2013). The recirculation also enhances the removal of total nitrogen, especially in constructed wetlands integrated with horizontal flow and vertical flow (Brix et al., 2003). A study by Ayaz et al. (2012) compared removal efficiency with and without recirculation that exhibited 100% recirculation, resulted in about 20% increase of total nitrogen removal efficiency. In the case of Kantawanichkul and Neamkam (2003), in their hybrid system, 100% recirculation was found to be superior to 50% and 200% recirculation in removal of total nitrogen.

7.8 Conclusions

Over the years, different types of bioreactors have played an important role in the treatment of wastewater from both domestic and industrial sources. While the established bioreactor types such as AD, CSTR, UASB, and SBR continue to function at a large scale globally for treatment for various types of wastewater, newly emergent ones such as MFCs and MBBRs are also making their presence felt. With new types of pollutants emerging and more often with recalcitrant natures, it becomes imperative to tune the existing types

TABLE 7.3

Typical Examples of the Constructed Wetlands That are Reported for Treating Different Wastewaters and the Biological Components Used in the System

Design of CW	Major Biological Components of CW	Type of Wastewater	Country	Specific/Major Pollutants Removed	Other Pollutants	References
Electrolysis-integrated tidal flow constructed wetland (CW)	Juncus effusus[a]	Synthetic wastewater	China	Ammonia, phosphorous	Sulfide	Ju et al. (2014)
Vertical flow CW reactor	Typha domingensis, Microbacterium arborescens TYSI04, Bacillus pumilus PIRI30	Textile effluents	Pakistan	COD, BOD	TDS and TSS	Shehzadi et al. (2014)
Lagoon wetland	Hydrocottle spp., Phragmites australis	Municipal wastewater	USA	Pharmaceutically active compounds		Conkle et al. (2008)
Subsurface horizontal flow Constructed wetland	Canna indica, Typha latifolia, Phragmites australis, Stenotaphrum secundatum, and Iris pseudacorus	Tannery wastewater	Portugal	COD, BOD	Ammonia	Calheiros et al. (2007)
Ecologically engineered treatment system (EETS)	Eichhorniia crassipes, Cyphoma gibbosum, Asparagus, Lycopersicum esculentum, Hydrilla verticillata, Oriza sativa, Tagetes erecta, Cyprinus carpio	Effluents of acidogenic fermentation and domestic sewage	India	COD, color (melanoidic pigment of distillery wastewater)	Nitrates	Venkata Mohan et al. (2010)
EEETS	Eichhorniia crassipes, Cyphoma gibbosum, Asparagus, Lycopersicum esculentum, Hydrilla verticillata, Oriza sativa, Tagetes erecta, Cyprinus carpio	Estrogenic endocrine disrupting compounds present in domestic sewage	India	Estriol (E3, natural), 17α-ethinylestradiol (EE2, synthetic)	COD, nitrates turbidity	Kumar et al. (2011)
Surface flow CW	Phragmites australis and Typha latifoli[b]	Antibiotics in groundwater	Sweden	Variable for different antibiotics between 53% and 99%	—	Berglund et al. (2014)
flow, flow CW system	Macrophytes[c]	Wide range of wastewater	Canada	COD, TKN	—	Brisson and Chazarenc (2009)
Vertical flow CW	Acoruscalamus L	Heavily polluted river water	China	COD, total nitrogen	Total phosphorous	Dong et al. (2012)

[a] Electrodes integrated in the system to supply electricity that triggered nitrogen transformations and phosphorous removal.

[b] 12 different types of antibiotics were treated and the removal efficiency of the antibiotics was ranged between 53% and 99%.

[c] Comparative study with various macrophytes used in constructed wetlands treating using wide range of wastewater.

of bioreactors to these pollutants and also upgrade the bioreactors for improved functionality. The only thing which can be said with surety is that bioreactors will remain important and continue to play a crucial role in any wastewater treatment facility.

References

Aelterman P, Rabaey K, Clauwaert P, and Verstraete W. 2006a. Microbial fuel cells for wastewater treatment. *Water Science and Technology* 54(8): 9–15.

Aelterman P, Rabaey K, Pham HT, Boon N, and Verstraete W. 2006b. Continuous electricity generation at high voltages and currents using stacked microbial fuel cells. *Environmental Science and Technology* 40(10): 3388–3394.

Afzal M, Khan S, Iqbal S, Mirza MS, and Khan QM. 2013. Inoculation method affects colonization and activity of *Burkholderia phytofirmans* PsJN during phytoremediation of diesel-contaminated soil. *International Biodeterioration and Biodegradation* 85: 331–336.

Andersson A and Grennberg K. 2001. Isolation and characterization of a bacterial population aimed for a biofilter treating waste-gases from a restaurant. *Environmental Engineering Science* 18(4): 237–248.

Appels L, Baeyens J, Degrève J, and Dewil R. 2008. Principles and potential of the anaerobic digestion of waste-activated sludge. *Progress in Energy and Combustion Science* 34(6): 755–781.

Aulenta F, Reale P, Canosa A, Rossetti S, Panero S, and Majone M. 2010. Characterization of an electro-active biocathode capable of dechlorinating trichloroethene and cis-dichloroethene to ethene. *Biosensors and Bioelectronics* 25(7): 1796–1802.

Ayaz SC, Aktas Ö, Findik N, Akca L, and Kinaci C. 2012. Effect of recirculation on nitrogen removal in a hybrid constructed wetland system. *Ecological Engineering* 40: 1–5.

Aziz SQ, Aziz HA, and Yusoff MS. 2011. Powdered activated carbon augmented double react-settle sequencing batch reactor process for treatment of landfill leachate. *Desalination* 277(1–3): 313–320.

Bagheri M, Mirbagheri SA, Ehteshami M, and Bagheri Z. 2014. Modeling of a sequencing batch reactor treating municipal wastewater using multi-layer perceptron and radial basis function artificial neural networks. *Process Safety and Environmental Protection* 93: 111–123.

Bajón Fernández Y, Soares A, Villa R, Vale P, and Cartmell E. 2014. Carbon capture and biogas enhancement by carbon dioxide enrichment of anaerobic digesters treating sewage sludge or food waste. *Bioresource Technology* 159: 1–7.

Barona A, Elías A, Arias R, Cano I, and González R. 2004. Biofilter response to gradual and sudden variations in operating conditions. *Biochemical Engineering Journal* 22(1): 25–31.

Bathe S. 2004. Conjugal transfer of plasmid pNB2 to activated sludge bacteria leads to 3-chloroaniline degradation in enrichment cultures. *Letters in Applied Microbiology* 38: 527–531.

Bathe S, Schwarzenbeck N, and Hausner M. 2005. Plasmid-mediated bioaugmentation of activated sludge bacteria in a sequencing batch moving bed reactor using pNB2. *Letters in Applied Microbiology* 41: 242–247.

Beccari M, Bonemazzi F, Majone M, and Riccardi C. 1996. Interaction between acidogenesis and methanogenesis in the anaerobic treatment of olive oil mill effluents. *Water Research* 30(1): 183–189.

Benthack C, Srinivasan B, and Bonvin D. 2001. An optimal operating strategy for fixed-bed bioreactors used in wastewater treatment. *Biotechnology and Bioengineering* 72(1): 34–40.

Berglund B, Khan GA, Weisner SE, Ehde PM, Fick J, and Lindgren PE. 2014. Efficient removal of antibiotics in surface-flow constructed wetlands, with no observed impact on antibiotic resistance genes. *Science of the Total Environment* 476: 29–37.

Bhuptawat H, Folkard GK, and Chaudhari S. 2007. Innovative physico-chemical treatment of wastewater incorporating Moringaoleifera seed coagulant. *Journal of Hazardous Materials* 142(1–2): 477–482.

Boon N, Goris J, De Vos P, Verstraete W, and Top E. 2000. Bioaugumentation of activated sludge by an indigenous 3-chloroaniline-degrading comamonas testosterone strain 12gfp. *Applied Environmental Microbiology* 66: 2906–2913.

Bouchez T, Patureau D, Dabert P, Wagner M, Deigenes P, and Moletta R. 2000. Successful and unsuccessful bioaugumentation experiments monitored by fluorescent *in situ* hybridization. *Water Science and Technology* 41: 61–68.

Brisson J and Chazarenc F. 2009. Maximizing pollutant removal in constructed wetlands: Should we pay more attention to macrophyte species selection? *Science of the Total Environment* 407(13): 3923–3930.

Brix H, Arias CA, and Johansen N-H. 2003. Experiments in a two-stage constructed wetland system: Nitrification capacity and effects of recycling on nitrogen removal. In: Vymazal J. (Ed.), *Wetlands—Nutrients, Metals and Mass Cycling*. Backhuys Publishers, Leiden, The Netherlands, pp. 237–258.

Brockmann M and Seyfried C. 1996. Sludge activity and cross-flow microfiltration—A non-beneficial relationship. *Water Science and Technology* 34(9): 205–213.

Busca G and Pistarino C. 2003. Technologies for the abatement of sulphide compounds from gaseous streams: A comparative overview. *Journal of Loss Prevention in the Process Industries* 16(5): 363–371.

Campbell G. 1999. Creation and characterization of aerated food products. *Trends in Food Science and Technology* 10(9): 283–296.

Carrera J, Baeza J A, Vicent T, and Lafuente J. 2003. Biological nitrogen removal of high-strength ammonium industrial wastewater with two-sludge system. *Water Research* 37(17): 4211–4221.

Casellas M, Dagot C, and Baudu M. 2006. Set up and assessment of a control strategy in a SBR in order to enhance nitrogen and phosphorus removal. *Process Biochemistry* 41(9): 1994–2001.

Chan YJ, Chong MF, Law CL, and Hassell DG. 2009. A review on anaerobic–aerobic treatment of industrial and municipal wastewater. *Chemical Engineering Journal* 155(1–2): 1–18.

Chang S. 2014. Anaerobic Membrane Bioreactors (AnMBR) for Wastewater Treatment. *Advances in Chemical Engineering and Science* 4: 56–61.

Chiavola A, Farabegoli G, and Antonetti F. 2014. Biological treatment of olive mill wastewater in a sequencing batch reactor. *Biochemical Engineering Journal* 85: 71–78.

Chiranjeevi P, Mohanakrishna G, and Venkata Mohan S. 2012. Rhizosphere mediated electrogenesis with the function of anode placement for harnessing bioenergy through CO_2 sequestration. *Bioresource Technology* 124: 364–370.

Chong S, Sen TK, Kayaalp A, and Ang HM. 2012. The performance enhancements of upflow anaerobic sludge blanket (UASB) reactors for domestic sludge treatment—A state-of-the-art review. *Water Research* 46 (11): 3434–3470.

Choo K-H and Lee C-H. 1996. Membrane fouling mechanisms in the membrane-coupled anaerobic bioreactor. *Water Research* 30(8): 1771–1780.

Cicek N, Dionysiou D, Suidan M, Ginestet P, and Audic J. 1999a. Performance deterioration and structural changes of a ceramic membrane bioreactor due to inorganic abrasion. *Journal of Membrane Science* 163(1): 19–28.

Cicek N, Franco JP, Suidan MT, Urbain V, and Manem J. 1999b. Characterization and comparison of a membrane bioreactor and a conventional activated-sludge system in the treatment of wastewater containing high-molecular-weight compounds. *Water Environmental Research* 71(1): 64–70.

Çınar Ö, Yaşar S, Kertmen M, Demiröz K, Yigit NÖ, and Kitis M. 2008. Effect of cycle time on biodegradation of azo dye in sequencing batch reactor. *Process Safety and Environmental Protection* 86(6): 455–460.

Clauwaert P, Rabaey K, Aelterman P, De Schamphelaire L, Pham TH, Boeckx P, and Verstraete W. 2007. Biological denitrification in microbial fuel cells. *Environmental Science and Technology* 41(9): 3354–3360.

Conkle JL, White JR, and Metcalfe CD. 2008. Reduction of pharmaceutically active compounds by a lagoon wetland wastewater treatment system in Southeast Louisiana. *Chemosphere* 73(11): 1741–1748.

Daverey A, Hung N-T, Dutta K, and Lin J-G. 2013. Ambient temperature SNAD process treating anaerobic digester liquor of swine wastewater. *Bioresource Technology* 141: 191–198.

Davies LC, Cabrita G, Ferreira R, Carias C, Novais J, and Martins-Dias S. 2009. Integrated study of the role of *Phragmites australis* in azo-dye treatment in a constructed wetland: From pilot to molecular scale. *Ecological Engineering* 35: 961–970.

Del Pozo R and Diez V. 2005. Integrated anaerobic-aerobic fixed-film reactor for slaughterhouse wastewater treatment. *Water Research* 39(6): 1114–1122.

Dereli RK, Grelot A, Heffernan B, van der Zee FP, and van Lier JB. 2014. Implications of changes in solids retention time on long term evolution of sludge filterability in anaerobic membrane bioreactors treating high strength industrial wastewater. *Water Research* 59: 11–22.

Deront M, Samb F, Adler N, and Peringer P. 1998. Biomass growth monitoring using pressure drop in a cocurrent biofilter. *Biotechnology and Bioengineering* 60(1): 97–104.

Dizge N, Tansel B, and Sizirici B. 2011. Process intensification with a hybrid system: A tubular packed bed bioreactor with immobilized activated sludge culture coupled with membrane filtration. *Chemical Engineering and Processing: Process Intensification* 50(8): 766–772.

Dong H, Qiang Z, Li T, Jin H, and Chen W. 2012. Effect of artificial aeration on the performance of vertical-flow constructed wetland treating heavily polluted river water. *Journal of Environmental Sciences* 24(4): 596–601.

Duan H, Koe LCC, Yan R, and Chen X. 2006. Biological treatment of H(2)S using pellet activated carbon as a carrier of microorganisms in a biofilter. *Water Research* 40(14): 2629–2636.

Dvořák L, Lederer T, Jirků V, Masák J, and Novák L. 2014. Removal of aniline, cyanides and diphenylguanidine from industrial wastewater using a full-scale moving bed biofilm reactor. *Process Biochemistry* 49(1): 102–109.

ElMekawy A, Diels L, Bertin L, De Wever H, and Pant D. 2014a. Potential biovalorization techniques for olive mill biorefinery wastewater. *Biofuels, Bioproducts and Biorefining* 8: 283–293.

ElMekawy A, Diels L, De Wever H, and Pant D. 2013. Valorization of cereal based biorefinery byproducts: Reality and expectations. *Environmental Science and Technology* 47(16): 9014–9027.

ElMekawy A, Sandipam S, Vanbroekhoven K, De Wever H, and Pant D. 2014b. Bioelectro-catalytic valorization of dark fermentation effluents by acetate oxidizing bacteria in bioelectrochemical system (BES). *Journal of Power Sources* 262: 183–191.

Ersahin ME, Ozgun H, Tao Y, and van Lier JB. 2014. Applicability of dynamic membrane technology in anaerobic membrane bioreactors. *Water Research* 48: 420–429.

Fernandes H, Jungles MK, Hoffmann H, Antonio RV, and Costa RHR. 2013. Full-scale sequencing batch reactor (SBR) for domestic wastewater: Performance and diversity of microbial communities. *Bioresource Technology* 132: 262–268.

Ferraz FM, Povinelli J, Pozzi E, Vieira EM, and Trofino JC. 2014. Co-treatment of landfill leachate and domestic wastewater using a submerged aerobic biofilter. *Journal of Environmental Management* 141C: 9–15.

Fogler HS. 2006. *Elements of Chemical Reaction Engineering*. Prentice Hall, Upper Saddle River.

Friha I, Karray F, Feki F, Jlaiel L, and Sayadi S. 2014. Treatment of cosmetic industry wastewater by submerged membrane bioreactor with consideration of microbial community dynamics. *International Biodeterioration and Biodegradation* 88:125–133.

Gali A, Dosta J, Lopez-Palau S, and Mata-Alvarez J. 2008. SBR technology for high ammonium loading rates. *Water Science and Technology* 58: 467–472.

Gargouri B, Karray F, Mhiri N, Aloui F, and Sayadi S. 2011. Application of a continuously stirred tank bioreactor (CSTR) for bioremediation of hydrocarbon-rich industrial wastewater effluents. *Journal of Hazardous Materials* 189(1–2): 427–434.

Gernaey KV, van Loosdrecht M, Henze M, Lind M, and Jørgensen SB. 2004. Activated sludge wastewater treatment plant modelling and simulation: State of the art. *Environmental Modelling and Software* 19(9): 763–783.

Ghangrekar M, Asolekar S, Ranganathan K, and Joshi S. 1996. Experience with UASB reactor start-up under different operating conditions. *Water Science and Technology* 34(5–6): 421–428.

Grady Jr CL, Daigger GT, Love NG, Filipe CD, and Leslie Grady CP. 2011. *Biological Wastewater Treatment* (No. Ed. 3). IWA Publishing.

Gu Q, Sun T, Wu G, Li M, and Qiu W. 2014. Influence of carrier filling ratio on the performance of moving bed biofilm reactor in treating coking wastewater. *Bioresource Technology* 166C: 72–78.

Haberl R, Grego S, Langergraber G, Kadlec RH, Cicalini A-R, Dias SM, Novais JM et al. 2003. Constructed wetlands for the treatment of organic pollutants. *Journal of Soils and Sediments* 3: 109–124.

Hadjiev D, Dimitrov D, Martinov M, and Sire O. 2007. Enhancement of the biofilm formation on polymeric supports by surface conditioning. *Enzyme and Microbial Technology* 40(4): 840–848.

Hajji KT, Lépine F, Bisaillon JG, Beaudet R, Hawari J, and Guiot SR. 2000. Effects of bioaugmentation strategies in UASB reactors with a methanogenic consortium for removal of phenolic compounds. *Biotechnology Bioengineering* 67: 417–423.

Hamelers HVM, TerHeijne A, Sleutels TH, Jeremiasse AW, Strik DP, and Buisman CJN. 2010. New applications and performance of bioelectrochemical systems. *Applied Microbiology and Biotechnology* 85(6): 1673–1685.

He Y, Xu P, Li C, and Zhang B. 2005. High-concentration food wastewater treatment by an anaerobic membrane bioreactor. *Water Research* 39(17): 4110–4118.

He Z and Angenent LT. 2006. Application of bacterial biocathodes in microbial fuel cells. *Electroanalysis* 18(19–20): 2009–2015.

He Z, Minteer SD, and Angenent LT. 2005. Electricity generation from artificial wastewater using an upflow microbial fuel cell. *Environmental Science and Technology* 39(14): 5262–5267.

Ho J and Sung S. 2010. Methanogenic activities in anaerobic membrane bioreactors (AnMBR) treating synthetic municipal wastewater. *Bioresource Technology* 101(7): 2191–2196.

Hoinkis J, Deowan SA, Panten V, Figoli A, Huang RR, and Drioli E. 2012. Membrane bioreactor (MBR) technology—A promising approach for industrial water reuse. *Procedia Engineering* 33: 234–241.

Huang L, Regan JM and Quan X. 2011. Electron transfer mechanisms, new applications, and performance of biocathode microbial fuel cells. *Bioresource Technology* 102(1): 316–323.

Iliuta I. 1997. Performance of fixed bed reactors with two-phase upflow and downflow. *Journal of Chemical Technology and Biotechnology* 68(1): 47–56.

Intanoo P, Suttikul T, Leethochawalit M, Gulari E, and Chavadej S. 2014. Hydrogen production from alcohol wastewater with added fermentation residue by an anaerobic sequencing batch reactor (ASBR) under thermophilic operation. *International Journal of Hydrogen Energy* 39(18): 9611–9620.

Israilides CJ, Vlyssides AG, Mourafeti VN, and Karvouni G. 1997. Olive oil wastewater treatment with the use of an electrolysis system. *Bioresource Technology* 61: 163–170.

Jianlong W, Xianghun Q, Libo W, Yi Q, and Hegemann W. 2002. Bioaugmentation as a tool to enhance the removal refractory compound in coke plant wastewater. *Proceedings of Biochemistry* 38: 777–781.

Ju X, Wu S, Zhang Y, and Dong R. 2014. Intensified nitrogen and phosphorus removal in a novel electrolysis-integrated tidal flow constructed wetland system. *Water Research* 59: 37–45.

Judd, S. 2006. In: S. Judd, (Ed.), *The MBR Book: Principles and Applications of Membrane Bioreactors for Water and Wastewater Treatment*. Elsevier, Amsterdam.

Juretschko S, Loy A, Lehner A, and Wagner M. 2002. The microbial community composition of a nitrifying-denitrifying activated sludge from an industrial sewage treatment plant analyzed by the full-cycle rRNA approach. *Systematic and Applied Microbiology* 25(1): 84–99.

Kadlec RH and Wallace SD. 2009. *Treatment Wetlands*, 2nd ed. CRC Press, Boca Raton, FL.

Kantawanichkul S and Neamkam P. 2003. Optimum recirculation ratio for nitrogen removal in a combined system: Vertical flow vegetated bed over horizontal flow sand bed. In: Vymazal J. (Ed.), *Wetlands: Nutrients, Metals and Mass Cycling*. Backhuys Publishers, Leiden, The Netherlands, pp. 75–86.

Kikuchi R. 2006. Pilot-scale test of a soil filter for treatment of malodorous gas. *Soil Use and Management* 16(3): 211–214.

Kim W, Cho K, Lee S, and Hwang S. 2013. Comparison of methanogenic community structure and anaerobic process performance treating swine wastewater between pilot and optimized lab scale bioreactors. *Bioresource Technology* 145: 48–56.

Kumar AK, Chiranjeevi P, Mohanakrishna G, and Mohan SV. 2011. Natural attenuation of endocrine-disrupting estrogens in an ecologically engineered treatment system (EETS) designed with floating, submerged and emergent macrophytes. *Ecological Engineering* 37(10): 1555–1562.

Le-Clech P, Chen V, and Fane TAG. 2006. Fouling in membrane bioreactors used in wastewater treatment. *Journal of Membrane Science* 284(1–2): 17–53.

Lee DS and Park JM. 1999. Neural network modeling for on-line estimation of nutrient dynamics in a sequentially-operated batch reactor. *Journal of Biotechnology* 75(2–3): 229–239.

Lettinga G. 1995. Anaerobic digestion and wastewater treatment systems. *Antonie van Leeuwenhoek* 67(1): 3–28.

Lin HJ, Xie K, Mahendran B, Bagley DM, Leung KT, Liss SN, and Liao BQ. 2009. Sludge properties and their effects on membrane fouling in submerged anaerobic membrane bioreactors (SAnMBRs). *Water Research* 43(15): 3827–3837.

Littleton HX, Daigger GT, Strom PF, and Cowan RA. 2014. Simultaneous biological nutrient removal: Evaluation of autotrophic denitrification, heterotrophic nitrification, and biological phosphorus removal in full-scale systems. *Water Environment Research: A Research Publication of the Water Environment Federation* 75(2): 138–150.

Liu G, Ye Z, Tong K, and Zhang Y. 2013. Biotreatment of heavy oil wastewater by combined upflow anaerobic sludge blanket and immobilized biological aerated filter in a pilot-scale test. *Biochemical Engineering Journal* 72: 48–53.

Liu Y and Tay J-H. 2004. State of the art of biogranulation technology for wastewater treatment. *Biotechnology Advances* 22(7): 533–563.

Logan BE. 2008. *Microbial Fuel Cells*. John Wiley and Sons, New York.

Luo W, Yang C, He H, Zeng G, Yan S, and Cheng Y. 2014. Novel two-stage vertical flow biofilter system for efficient treatment of decentralized domestic wastewater. *Ecological Engineering* 64: 415–423.

Ma Y, Zhao J, and Yang B. 2006: Removal of H_2S in waste gases by an activated carbon bioreactor. *International Biodeterioration and Biodegradation* 57(2): 93–98.

Mace S and Mata-Alvarez J. 2002. Utilization of SBR technology for wastewater treatment: An overview. *Industrial and Engineering Chemistry Research* 41(23): 5539–5553.

Mahmoud M, Parameswaran P, Torres CI, and Rittmann BE. 2014. Fermentation pre-treatment of landfill leachate for enhanced electron recovery in a microbial electrolysis cell. *Bioresource Technology* 151: 151–158.

Mahvi AH. 2008. Sequencing batch reactor: A promising technology in wastewater treatment. *Iranian Journal of Environmental Health Science and Engineering* 5(2): 79–90.

Martinov M, Hadjiev D, and Vlaev S. 2010. Gas–liquid dispersion in a fibrous fixed bed biofilm reactor at growth and non-growth conditions. *Process Biochemistry* 45(7): 1023–1029.

Mbuligwe SE. 2005. Comparative treatment of dye-rich wastewater in engineered wetland systems (EWSs) vegetated with different plants. *Water Research* 39: 271–280.

McNevin D and Barford J. 2000. Biofiltration as an odour abatement strategy. *Biochemical Engineering Journal* 5(3): 231–242.

Min B and Logan BE. 2004. Continuous electricity generation from domestic wastewater and organic substrates in a flat plate microbial fuel cell. *Environmental Science and Technology* 38(21): 5809–5814.

Mittal A. 2011. Biological wastewater treatment. *Water Toda* 8: 32–44.

Mohammed TA, Birima AH, Noor MJMM, Muyibi SA, and Idris A. 2008. Evaluation of using membrane bioreactor for treating municipal wastewater at different operating conditions. *Desalination* 221(1–3): 502–510.

Mohanakrishna G, Krishna Mohan S, and Venkata Mohan S. 2012. Carbon based nanotubes and nanopowder as impregnated electrode structures for enhanced power generation: Evaluation with real field wastewater. *Applied Energy* 95: 31–37.

Mohanakrishna G, Venkata Mohan S, and Sarma PN. 2010a. Utilizing acid-rich effluents of fermentative hydrogen production process as substrate for harnessing bioelectricity: An integrative approach. *International Journal of Hydrogen Energy* 35(8): 3440–3449.

Mohanakrishna G, Venkata Mohan S, and Sarma PN. 2010b. Bio-electrochemical treatment of distillery wastewater in microbial fuel cell facilitating decolorization and desalination along with power generation. *Journal of Hazardous Materials* 177: 487–494.

Mohanakrishna G, Subhash GV, and Venkata Mohan S. 2011. Adaptation of biohydrogen reactor to higher substrate load: Redox controlled process integration strategy to overcome limitations. *International Journal of Hydrogen Energy* 36: 8943–8952.

Mohanakrishna G, and Venkata Mohan S. 2013. Multiple process integrations for broad perspective analysis of fermentative H_2 production from wastewater treatment: Technical and environmental considerations. *Applied Energy* 107: 244–254.

Mojiri A, Aziz HA, Zaman NQ, Aziz SQ, and Zahed MA. 2014. Powdered ZELIAC augmented sequencing batch reactors (SBR) process for co-treatment of landfill leachate and domestic wastewater. *Journal of Environmental Management* 139: 1–14.

Mulder A, Van De Graaf AA, Robertson LA, and Kuenen JG. 1995. Anaerobic ammonium oxidation discovered in a denitrifying fluidized bed reactor. *FEMS Microbiology Ecology* 16(3): 177–184.

Mussoline W, Esposito G, Lens P, Garuti G, and Giordano A. 2012. Design considerations for a farm-scale biogas plant based on pilot-scale anaerobic digesters loaded with rice straw and piggery wastewater. *Biomass and Bioenergy* 46: 469–478.

Newby DT, Gentry TJ, and Pepper IL. 2000. Comparison of 2,4-dichlorophenoxyacetic acid degradation and plasmid transfer in soil resulting from bioaugmentation with two different pJP4 donors. *Applied and Environmental Microbiology* 66(8): 3399–3407.

Odegaard H. 2006. Innovations in wastewater treatment: The moving bed biofilm process. *Water Science and Technology* 53(9): 17–33.

Ortiz I, Revah S, and Auria R. 2003. Effects of packing material on the biofiltration of benzene, toluene and xylene vapours. *Environmental Technology* 24(3): 265–275.

Pant D and Adholeya A. 2007. Biological approaches for treatment of distillery wastewater: A review. *Bioresource Technology* 98: 2321–2334.

Pant D and Adholeya A. 2009. Concentration of fungal ligninolytic enzymes produced during solid-state fermentation by ultrafiltration. *World Journal of Microbiology and Biotechnology* 25: 1793–1800.

Pant D and Adholeya A. 2010. Development of a novel fungal consortium for the treatment of molasses distillery wastewater. *The Environmentalist* 30: 178–182.

Pant D, Arslan D, Van Bogaert G, Alvarez Gallego Y, De Wever H, Diels L, and Vanbroekhoven K. 2013. Integrated conversion of food waste diluted with sewage into volatile fatty acids through fermentation and electricity through a fuel cell. *Environmental Technology* 34(13–14): 1935–1945.

Pant D, Singh A, Van Bogaert G, Olsen SI, Nigam PS, Diels L, and Vanbroekhoven K. 2012. Bioelectrochemical systems (BES) for sustainable energy production and product recovery from organic wastes and industrial wastewaters. *RSC Advances* 2:1248–1263.

Pant D, Van Bogaert G, Diels L, and Vanbroekhoven K. 2010. A review of the substrates used in microbial fuel cells (MFCs) for sustainable energy production. *Bioresource Technology* 101(6): 1533–1543.

Puay N-Q, Qiu G, and Ting Y-P. 2015. Effect of ZnO nanoparticles on biological wastewater treatment in a sequencing batch reactor (SBR). *Journal of Cleaner Production* 88: 139–145.

Qasim SR. 1999. *Wastewater Treatment Plants: Planning, Design and Operation.* CRC Press, Boca Raton.

Qiu G, Song Y-H, Zeng P, Duan L, and Xiao S. 2013. Characterization of bacterial communities in hybrid upflow anaerobic sludge blanket (UASB)-membrane bioreactor (MBR) process for berberine antibiotic wastewater treatment. *Bioresource Technology* 142: 52–62.

Qu X, Alvarez PJJ, and Li Q. 2013. Applications of nanotechnology in water and wastewater treatment. *Water Research* 47: 3931–3946.

Rabaey K and Rozendal R. 2010. Microbial electrosynthesis—Revisiting the electrical route for microbial production. *Nature Reviews Microbiology* 8: 706–716.

Rajeshwari KV, Balakrishnan M, Kansal A, Lata K, and Kishore VVN. 2000. State-of-the-art of anaerobic digestion technology for industrial wastewater treatment. *Renewable and Sustainable Energy Reviews* 4(2): 135–156.

Reddy MV and Venkata Mohan S. 2012. Influence of aerobic and anoxic microenvironments on polyhydroxyalkanoates (PHA) production from food waste and acidogenic effluents using aerobic consortia. *Bioresource Technology* 103(1): 313–321.

Rismani-Yazdi H, Carver SM, Christy AD, and Tuovinen OH. 2008. Cathodic limitations in microbial fuel cells: An overview. *Journal of Power Sources* 180(2): 683–694.

Rodrigo JS, Benedict JC, Okeke BFM, Teixeira AS, Peralba MCR, and Flavio AOC. 2008. Microbial consortium bioaugmentation of a polycyclic aromatic hydrocarbons contaminated soil. *Bioresource Technology* 99: 2637–2643.

Rodríguez DC, Pino N, and Peñuela G. 2011. Monitoring the removal of nitrogen by applying a nitrification-denitrification process in a Sequencing Batch Reactor (SBR). *Bioresource Technology* 102(3): 2316–2321.

Rojas-Avelizapa NG, Martinez-Cruz J, Zermeno-Eguia Lis JA, and Rodriguez-Vazquez R. 2003. Levels of polychlorinated biphenyls in Mexican soils and their biodegradation using bioaugmentation. *Bulletin of environmental contamination and toxicology* 70(1): 0063–0070.

Shailaja S, Ramakrishna M, Venkata Mohan S, and Sarma PN. 2007. Biodegradation of di-*n*-butyl phthalate (DnBP) in bioaugmented bioslurry phase reactor. *Bioresource Technology* 98: 1561–1566.

Sharghi EA, Bonakdarpour B, and Pakzadeh M. 2014. Treatment of hypersaline produced water employing a moderately halophilic bacterial consortium in a membrane bioreactor: Effect of salt concentration on organic removal performance, mixed liquor characteristics and membrane fouling. *Bioresource Technology* 164: 203–213.

Shehzadi M, Afzal M, Khan MU, Islam E, Mobin A, Anwar S, and Khan QM. 2014. Enhanced degradation of textile effluent in constructed wetland system using *Typha domingensis* and textile effluent-degrading endophytic bacteria. *Water Research* 58: 152–159.

Sheng G-P, Yu H-Q, and Cui H. 2008. Model-evaluation of the erosion behavior of activated sludge under shear conditions using a chemical-equilibrium-based model. *Chemical Engineering Journal* 140(1–3): 241–246.

Shore JL, M'Coy WS, Gunsch CK, and Deshusses MA. 2012. Application of a moving bed biofilm reactor for tertiary ammonia treatment in high temperature industrial wastewater. *Bioresource Technology* 112: 51–60.

Singh L, Siddiqui MF, Ahmad A, Rahim MHA, Sakinah M, and Wahid ZA. 2013. Application of polyethylene glycol immobilized *Clostridium* sp. LS2 for continuous hydrogen production from palm oil mill effluent in upflow anaerobic sludge blanket reactor. *Biochemical Engineering Journal* 70: 158–165.

Song J and Kinney KA. 2000. Effect of vapor-phase bioreactor operation on biomass accumulation, distribution, and activity: Linking biofilm properties to bioreactor performance. *Biotechnology and Bioengineering* 68(5): 508–516.

Speece RE. 1983. Anaerobic biotechnology for industrial wastewater treatment. *Environmental Science and Technology* 17(9): 416A–427A.

Sperling M and Chernicharo CA. 2005. *Biological Wastewater Treatment in Warm Climate Regions*. IWA Publishing, London.

Sridevi K, Sivaraman E, and Mullai P. 2014. Back propagation neural network modelling of biodegradation and fermentative biohydrogen production using distillery wastewater in a hybrid upflow anaerobic sludge blanket reactor. *Bioresource Technology* 165: 233–240.

Srikanth S, Reddy MV, and Venkata Mohan S. 2012. Microaerophilic microenvironment at biocathode enhances electrogenesis with simultaneous synthesis of polyhydroxyalkanoates (PHA) in bioelectrochemical system (BES). *Bioresource Technology* 125: 291–299.

Srikanth S and Venkata Mohan S. 2012. Change in electrogenic activity of the microbial fuel cell (MFC) with the function of biocathode microenvironment as terminal electron accepting condition: Influence on overpotentials and bio-electro kinetics. *Bioresource Technology* 119: 241–251.

Srikanth S, Venkata Mohan S, Prathima Devi M, Peri D, and Sarma PN. 2009. Acetate and butyrate as substrates for hydrogen production through photo-fermentation: Process optimization and combined performance evaluation. *International Journal of Hydrogen Energy* 34(17): 7513–7522.

Stephenson D and Stephenson T. 1992. Bioaugmentation for enhancing biological wastewater treatment. *Biotechnology Advances* 10: 549–559.

Subramanian B and Pagilla KR. 2014. Anaerobic digester foaming in full-scale cylindrical digesters—Effects of organic loading rate, feed characteristics, and mixing. *Bioresource Technology* 159: 182–192.

Tchobanoglous G, Burton FL, and Stensel HD. 2003. *Wastewater Engineering: Treatment and Reuse.* McGraw-Hill Education, New York.

Thassitou PK and Arvanitoyannis IS. 2001. Bioremediation: A novel approach to food waste management. *Trends in Food Science and Technology* 12: 185–196.

Tiwari MK, Guha S, Harendranath CS, and Tripathi S. 2006. Influence of extrinsic factors on granulation in UASB reactor. *Applied Microbiology and Biotechnology* 71(2): 145–154.

Torkian A, Eqbali A, and Hashemian S. 2003. The effect of organic loading rate on the performance of UASB reactor treating slaughterhouse effluent. *Resources Conservation and Recycling* 40(1): 1–11.

Torres CI. 2014. On the importance of identifying, characterizing, and predicting fundamental phenomena towards microbial electrochemistry applications. *Current Opinion in Biotechnology* 27: 107–114.

Turovskiy IS and Mathai PK. 2006. *Wastewater Sludge Processing.*Wiley, New York.

Ueno Y, Otsuka S, and Morimoto M. 1996. Hydrogen production from industrial wastewater by anaerobic microflora in chemostat culture. *Journal of Fermentation and Bioengineering* 82(2): 194–197.

van der Star WRL, Abma WR, Blommers D, Mulder JW, Tokutomi T, Strous M, Picioreanu C, and Van Loosdrecht MCM. 2007. Startup of reactors for anoxic ammonium oxidation: Experiences from the first full-scale anammox reactor in Rotterdam. *Water Research* 41: 4149–4163.

Van Limbergen H, Top EM, and Verstreete W. 1998. Bioaugmentation in activated sludge: Current features and future perspectives. *Applied Microbiology Biotechnology* 50:16–23.

Venkata Mohan S, Mohanakrishna G, and Sarma PN. 2008a. Integration of acidogenic and methanogenic processes for simultaneous production of biohydrogen and methane from wastewater treatment. *International Journal of Hydrogen Energy* 33: 2156–2166.

Venkata Mohan S, Mohanakrishna G, Ramanaiah SV, and Sarma PN. 2008b. Simultaneous biohydrogen production and wastewater treatment in biofilm configured anaerobic periodic discontinuous batch reactor using distillery wastewater. *International Journal of Hydrogen Energy* 33(2): 550–558.

Venkata Mohan S, Mohanakrishna G, and Chiranjeevi P. 2011. Sustainable power generation from floating macrophytes based ecological microenvironment through embedded fuel cell along with simultaneous wastewater treatment. *Bioresource Technology* 102: 7036–7042.

Venkata Mohan S and Srikanth S. 2011. Enhanced wastewater treatment efficiency through microbial catalyzed oxidation and reduction: Synergistic effect of biocathode microenvironment. *Bioresource Technology* 102: 10210–10220.

Venkata Mohan S, Babu PS, and Srikanth S. 2013a. Azo dye remediation in periodic discontinuous batch mode operation: Evaluation of metabolic shifts of the biocatalyst under aerobic, anaerobic and anoxic conditions. *Separation and Purification Technology* 118: 196–208.

Venkata Mohan S, Chandrasekhara Rao N, Prasad KK, and Sarma PN. 2005. Bioaugmentation of anaerobic sequencing batch biofilm reactor (ASBBR) with immobilized sulphate reducing bacteria (SRB) for treating sulphate bearing chemical wastewater. *Process Biochemistry* 40: 2849–2857.

Venkata Mohan S, Falkentoft CY, Nancharaiah V, Belinda S, Sturm MS, Wattiau P, Wilderer PA et al. 2009. Bioaugmentation of microbial communities in laboratory and pilot scale sequencing batch biofilm reactors using the TOL plasmid. *Bioresource Technology* 100: 1746–1753.

Venkata Mohan S, Mohanakrishna G, Chiranjeevi P, Peri D, and Sarma PN. 2010. Ecologically engineered system (EES) designed to integrate floating, emergent and submerged macrophytes for the treatment of domestic sewage and acid rich fermented-distillery wastewater: Evaluation of long term performance. *Bioresource Technology* 101(10): 3363–3370.

Venkata Mohan S, Mohanakrishna G, Veer Raghavulu S, and Sarma PN. 2007. Enhancing biohydrogen production from chemical wastewater treatment in anaerobic sequencing batch biofilm reactor (AnSBBR) by bioaugmenting with selectively enriched kanamycin resistant anaerobic mixed consortia. *International Journal of Hydrogen Energy* 32: 3284–3292.

Venkata Mohan S, Srikanth S, Velvizhi G, and Babu ML. 2013b. Microbial fuel cells for sustainable bioenergy generation: Principles and perspective applications (Chapter 11). In: Gupta VK and Tuohy MG (Eds.), *Biofuel Technologies: Recent Developments*. Springer, Berlin Heidelberg.

Vlyssides A, Barampouti E, and Mai S. 2009. Influence of ferrous iron on the granularity of a UASB reactor. *Chemical Engineering Journal* 146(1): 49–56.

Vlyssides A, Barampouti EM, and Mai S. 2008. Granulation mechanism of a UASB reactor supplemented with iron. *Anaerobe* 14(5), 275–279.

Vogel TM. 1996. Bioaugmentation as a soil bioremediation approach. *Current Opinion in Biotechnology* 7:311–316.

Vymazal J. 2007. Removal of nutrients in various types of constructed wetlands. *Science of the Total Environment* 380: 48–65.

Vymazal J. 2009. The use constructed wetlands with horizontal sub-surface flow for various types of wastewater. *Ecological Engineering* 35: 1–17.

Vymazal J. 2013. The use of hybrid constructed wetlands for wastewater treatment with special attention to nitrogen removal: A review of a recent development. *Water Research* 47(14): 4795–4811.

Wilderer PA, Rubio MA, and Davids L. 1991. Impact of the addition of pure cultures on the performance of mixed culture reactor. *Water Research* 25: 1307–1313.

Wilk IJ, Altmann RS, and Berg JD. 1987. Antimicrobial activity of electrolyzed saline solutions. *Science of Total Environment* 63: 191–197.

Wu YC and Chen JS. 2011. Toward the identification of EMG-signal and its bio-feedback application. *International Journal of Mechatronics and Automation* 1(2): 112–120.

Zhou M, Chi M, Luo J, He H, and Jin T. 2011. An overview of electrode materials in microbial fuel cells. *Journal of Power Sources* 196(10): 4427–4435.

8

Microbial Genomics and Bioremediation of Industrial Wastewater

Atya Kapley, Niti B. Jadeja, Vasundhara Paliwal,
Trilok C. Yadav, and Hemant J. Purohit

CONTENTS

8.1 Wastewater Treatment Process

Wastewater can be defined as any water that has been affected by man-made activities and contains substances arising from anthropogenic activities. Depending on its source of origin, it can be broadly classified into municipal wastewater or industrial wastewater.

The contaminants could be organic compounds, inorganic, and heavy metals arising from industrial activities and pathogens such as bacteria or viruses. Hence, the wastewater needs to be remediated before the water can be released back into the environment. This chapter will deal with the bioremediation of industrial wastewater.

Conventionally, wastewaters were treated using physicochemical approaches, but recently microbial degradation is emerging as a low-cost and efficient alternative as they provide the option of complete mineralization of organic compounds (Shah et al., 2014). The most popular "green route" for wastewater treatment is the activated sludge (AS) process. This is an aerobic route of degradation, carried out by the activated biomass comprising biological flocs containing microbes that are responsible for degradation of organics in the wastewater. The microbial population is dominated by bacteria that contain the genetic information required for degradation (More et al., 2014). The AS process was developed in 1914 by Arden and Lockett (Arden and Lockett, 1914). This system works on naturally occurring bacteria in bioreactors, which survive at the cost of the pollutants as carbon source. These bacteria with other life forms such as protozoa and fungi collectively create a balanced microbial community, which is referred to as AS. The bacteria utilize organic carbon for their growth, and thereby make wastewater safe for its disposal. The overall conceptual process of a wastewater treatment plant (WWTP) fairly remains the same with the wastewater being first pretreated and then pooled into an equalization tank. The wastewater is then channeled into the biological unit, the aerators that provide the necessary aerobic conditions that are required for biological degradation. To achieve a required nutritional balance, the supplementation of nutrients such as nitrogen and phosphate is carried out, which helps in sustaining the biomass in the treatment plant by supporting bacterial growth. Sometimes the influent wastewater has sufficient nutrients and may not need any supplements. The hydraulic retention time (HRT) or the time the wastewater remains in the aeration chamber depends on the source of wastewater, can range from a few hours to a few days, which is decided by the degradation rates. Mostly, to maintain the desired degradation kinetics the microbial biomass has to be recycled; therefore, it is separated from the treated wastewater to produce clarified secondary effluent. The treated wastewater duly assessed can be released into the environment. A diagrammatic representation is provided in Figure 8.1.

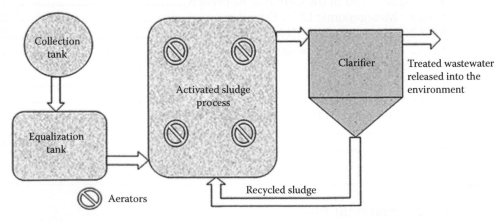

Diagram is a respresentation of the process and not made to scale

FIGURE 8.1
An overview of activated sludge process in WWTPs.

There are two main methods of biological treatment of wastewater, aerobic or anaerobic methods. The treatment output is higher in the aerobic method but the energy required for treatment is also higher than anaerobic method. Both methods have a different set of merits and demerits and are being used depending on target wastewater, and have been compared in a report by Eckenfelder et al. (1988). The ultimate goal of biological treatment is to reduce or mineralize organic compounds to carbon dioxide and water.

Most industries have their own effluent treatment plants (ETPs) to decontaminate their wastes. Large-scale industries such as oil refineries, petrochemical plants, and chemical plants use a combination of chemical and biological routes for wastewater treatment. However, developing countries, with their small-scale industry (SSI) set ups, majorly comprising of wastewater from dyes and dye-intermediates, bulk drugs and pharmaceuticals, and fine chemicals, are unable to afford the cost of operating an ETP. Hence, a common treatment facility is created, namely, the common ETP (CETP) in order to make a cooperative movement of pollution control especially to treat the effluent, emanating from the clusters of compatible SSIs. The major objective of the CETP is therefore, to reduce the treatment cost to be borne by an individual member unit to a maximum while protecting the environment to a maximum.

Although the principle of wastewater treatment is very simple, control of the treatment process is very complex, because of the large number of variables governing bacterial performance, such as nutrient availability, pH, temperature, flow rate, and so on. Domestic wastewater mainly includes organic matter, either in solution or in particle form. During treatment process 1 nm to 100 µm of particle size adsorb on the flocs surface of AS. The organic matter includes proteins, amino acids, peptides, carbohydrate, fats, and fatty acids, which are easily biodegradable. The C:N:P ratio of domestic wastewater is near to ideal for the growth of AS bacteria that is 100:10:1. However, industrial wastewaters are more variable in composition.

8.1.1 Physicochemical Parameters Affecting WWTPs

Wastewater treatment plants involve either one or a combination of physical, chemical, and biological means. Most of the WWTPs include combination of these methods depending on organic loading to treat wastewater (Narmadha and Kavitha, 2012). Chemical oxygen demand (COD) is the parameter most commonly used to measure the amount of organic compounds in water and to evaluate the treatment efficiency. It is the most practicable parameter applied at the treatment plants. COD provides the total organic loading data; however, to discriminate from biologically degaradative organics another parameter, that is, biochemical oxygen demand or BOD is applied. BOD is the difference in oxygen concentration in a test wastewater sample before and after incubation with active biomass for a specific length of time under controlled degaradative conditions. During this time, microorganisms in the sample oxidize the organic matter, using the available dissolved oxygen.

The physical parameters have strong effects on the chemical and biological parameters and govern the ability of a system to receive and assimilate pollution. To evaluate these parameters, it is necessary to know the physical status of wastewater in terms of temperature, weather, and stream flow. These variables have profound collective effect on efficiency of treatment; for example, it has been reported that on the rise of temperature from 15°C to 35°C, the COD removal decreased while the sludge volume index (SVI) and suspended solids in effluent increased (Ghanizadeh and Sarrafpour, 2001). Considering chemical parameters, pH is an important parameter in the biological and chemical systems of natural waters. On the rise of pH from 5.7 to 9.0, SVI and suspended solids in effluent decreased while percent of COD removal increased (Ghanizadeh and Sarrafpour, 2001). Alkalinity or buffering capacity of water is the capacity to neutralize acid. Alkalinity is expressed as the amount of milligrams per liter (mg/L) of calcium carbonate. Some commonly occurring materials such as carbonates, bicarbonates, phosphates, and hydroxides, in natural waters increase the alkalinity. High calcium and magnesium carbonate contain compounds such as limestone, and dolomite can be used as a neutralizer. The effect of acid precipitation is determined by buffering capacity of natural water; thus, it is also important. With high acidity, the greater is the sensitivity of surface water to acid precipitation owing to decrease in species diversity.

Dissolved oxygen is essential for all aerobic processes. Its concentration in wastewater varies with the composition. The types and concentrations of dissolved and suspended solids, temperature, also affect the amount of dissolved oxygen in a lake or water stream. Low-dissolved oxygen concentration is damaging to aquatic life. During summer, critical conditions can occur for these two reasons; first, at high temperature solubility of oxygen is lower and second, oxygen demand increases because of high metabolic rate of microorganisms. However, increased oxygen saturation levels can also be harmful. Different level of dissolved oxygen not only affects the treatment efficiency of industrial wastewater but also the nitrification, denitrification, and the behavior of the microbes in aETP/CETP (Yadav et al., 2014). Oxygen functions as the electron acceptor for microorganisms over nitrate, so it can inhibit denitrification reaction and aerobic conditions repress enzymes involved in denitrification (Zumft, 1997). However, high DO concentrations enhance the activity of nitrifying bacteria in the biofilm reactor (Hagedorn-Olsen et al., 1994; Lie and Welander, 1994). The oxygen concentration is a very important limiting factor in partial nitrification (Bae et al., 2001; Bernet et al., 2001; Ruiz et al., 2003). The specific growth rate of both ammonium-oxidizing bacteria (AOB) and nitrite-oxidizing bacteria (NOB) is affected by low DO concentrations, depending on its oxygen saturation constant; however, its influence on the NOB (1.1 mg O_2/L) is significantly greater than on the AOB (0.3 mg O_2/L) (Wiesmann, 1994). Park and Noguera (2004) observed changes in the population structure at low oxygen concentrations, which could affect nitrite accumulation rates. Oxygen level for nitrite accumulation is in the range of 0.5–1.5 mg O_2/L for suspended cultures (Bernet et al., 2001; Botrous et al., 2004; Ciudad et al., 2005).

Nitrate is a required nutrient for plants and microbes. Excess nitrates may lead to eutrophication by promoting algae and plant growth resulting in lower oxygen levels. Large amounts of nitrates in drinking water supplies (public and wells) can cause blue babies syndrome in infants. Organic wastes from some sewage treatment plants, municipal wastewater, septic tanks, and feed lot discharges are the major sources of nitrogen pollution. Indirect sources include fertilizers from lawns and farms, which leach out of soil and enter water through runoff, animal wastes, leachate from waste disposal in dumps or sanitary landfills. Phosphorus is an essential nutrient to microbial growth and present as phosphate. Phosphorus present as an element is toxic and can bioaccumulate up the food web in much the same way as mercury and other toxic chemicals.

8.1.2 Biological Parameters Affecting WWTPs

Biological parameters detect water quality problems that other methods may miss or underestimate. Resident microflora are continual monitors of environmental quality, increasing the detection of episodic events (spills, dumping, treatment plant malfunctions, nutrient enrichment), nonpoint source pollution (agricultural pesticides), cumulative pollution (multiple impacts over time or continuous low-level stress), or other impacts that chemical sampling is unlikely to detect. Impacts on the physical habitat such as sedimentation from storm water runoff and the effects of physical or structural habitat changes (dredging, filling, and channelization) can also be detected (USEPA, OW., 1990) 1990). The last decade has seen researchers monitoring and studying the microflora of WWTPs in an attempt to improve degradative efficiency. The application of genomics tools in environmental impact assessment can detect a perturbation in an ecosystem (Purohit et al., 2003) and this principle could be used in WWTPs to monitor the changes in degradation over time in WWTPs.

8.1.3 Operational Parameters in WWTPs

The main operational parameters at a WWTP are described below. Table 8.1 defines their calculation methods.

8.1.3.1 SVI

Sludge volume index is the volume in milliliters of settled sludge occupied by 1 g of dry sludge solids after 30 min of settling in a 1000 mL graduated cylinder. It is used to monitor the sludge-settling characteristics and quality in the final clarifier (Dick and Vesilind, 1969).

8.1.3.2 Sludge Density Index

Sludge density index (SDI) determines settling characteristics of sludge and return sludge pumping rates. A low SVI and high SDI sludge have good settling and compaction characteristics.

TABLE 8.1

Parameter Associated with Operation and Control of Wastewater at CETP

Parameter	Calculation Formula	Average Range
Sludge volume index (SVI)	$\text{SVI} = \dfrac{\text{Settled sludge volume (in mL/L) after 30 min} \times 1000}{\text{MLSS (mg/L)}} = \dfrac{\text{mL}}{\text{g}}$	50–150 mL/g
Sludge density index (SDI)	$\text{SDI} = \dfrac{\text{MLSS\%} \times 100}{\text{\%volume occupied by MLSS after 30 min, settling}}$	1.0–2.5
Sludge age (days)	$\text{Sludge age} = \dfrac{\text{MLSS in aerator, in lbs LSS\%} \times 100}{\text{Primary effluent suspended solids, lbs/days}}$	3–15 days
Mean cell residence time (MCRT) or solids retention time (SRT)	$\text{MCRT or SRT} = \dfrac{\text{Suspended solids (SS) in total secondary system, lbs}}{(\text{SS wasted, lbs/day} + \text{SS lost in effluent, lbs/day})}$	
F/M ratio	$\text{F/M} = \dfrac{\text{BOD added to aerator (lbs/day)}}{\text{MLSS under aeration}}$	0.15–0.5 based on BOD value

8.1.3.3 Sludge Age

Sludge age is defined as the average time in days that the suspended solids remain in the entire system. Sludge age considers solids entering the aerator; measured as primary effluent suspended solids in mg/L, and solids or organisms available to degrade the wastes; measured as mixed liquor-suspended solids (MLSSs, mg/L). Higher sludge ages are required to maintain a sufficient biological mass during winter season while, biological activity increases in the summer time so lower sludge ages normally produce a higher quality effluent. Thus, to accommodate seasonal variations, the sludge age should be adjusted at least twice a year. This is due to the fact that the waste characteristics, process design, flexibility in operation, and process control equipment are different for all facilities. A low sludge age produce a light, fluffy, buoyant type of sludge particle commonly referred to as straggler floc, which settles slowly in a final clarifier while, high sludge age or too many solids in the system tends to produce a darker, more granular type of sludge particle, commonly called pin floc, which settles too fast in a final clarifier.

8.1.3.4 Mean Cell Residence Time or Solids Retention Time

Mean cell residence time (MCRT) and solids retention time (SRT) are refinement of the sludge age and takes into consideration the total solids inventory in the secondary or biological system.

8.1.3.5 The Food/Mass or the Food/Microorganism Ratio

It is commonly referred to as food/mass or the food/microorganism (F/M) ratio and is based upon the ratio of food fed to the microorganisms per day to the mass of microorganisms held under aeration. It is a simple calculation, using the results from the influent BOD test to the aerator and the MLSS test. The optimum F/M varies from plant to plant and can be determined by trial and error. F/M ratio provides information on the biological behavior of AS.

8.1.3.6 Mixed Liquor-Suspended Solids

Constant MLSSs select a certain MLSS concentration or range of mix liquor concentrations that produces the best effluent and the highest removal efficiencies. This specific value or range must be determined experimentally. One rule of thumb for AS systems is that for every pound of BOD removed in the secondary system half a pound of new solids is generated through reproduction of the organisms and addition of new organisms from the influent wastes.

8.1.3.7 Return Activated Sludge Control

Mixed liquor-suspended solid settled in a clarifier are returned to the aeration tank as the return AS (RAS). The RAS makes it possible for the microorganisms to be in the treatment system longer than the flowing wastewater. For conventional AS operations, the RAS flow is generally about 20%–40% of the incoming wastewater flow. It is obtained by two methods: constant RAS flow rate control and constant percentage RAS flow rate control.

Industrial wastewater is loaded with toxic chemicals discharged from various treatment plants. Due to the presence of different types of toxic compounds present in the wastewater, the complexity for achieving efficient treatment increases. For effective treatment of industrial wastewater, bio-augmentation strategies have emerged as a powerful tool to accelerate the mineralization of toxic compounds (Yu et al., 2010; Li et al., 2013).

8.2 Microbes in Bioremediation

Microbial entities are copious and rampant in natural as well as anthropogenic ecologies of existing environment, which include polluted soils, industrial wastewaters, and the like. Ever-evolving microbial communities coexist in these extreme habitats owing to the fact that their genetic machinery keeps adapting, making it possible for the microbes to survive harsh conditions. Strategies based on bioremediation employ these evolved microbial communities that degrade the organic pollutants, detoxify the inorganic pollutants, and thus the study of these microbial communities yields insights of interrelated pathways involved in biodegradation processes. Microbial detoxification of heavy metals (Ow, 1996; Khan et al., 2000) diesel, oil, and petroleum compounds have been extensively studied (Chhatre et al., 1996; Margesin and Schinner, 2001; Yergeau et al., 2012; Fuentes et al., 2014; Shah et al., 2014). Several marine bacteria have also been reported for their potential in bioremediation owing to the advantage that marine organisms can survive adverse conditions without genetic manipulations (Geiselbrecht et al., 1996; Yakimov et al., 2007; Dash et al., 2013). Figure 8.2 gives an overview of the techniques used in analyzing the AS of a WWTP and generating information that can be used to improve treatment efficiency.

8.2.1 Bioremediation Strategies at WWTPs

Wastewater Treatment Plants can be termed as man-made anthropogenic ecologies, wherein an array of organic pollutants such as amines, aromatics, PCBs, PAHs, pesticides are cotreated. These compounds are often recalcitrant to natural degradation, thus pose a threat to natural ecosystems. The concept of biological treatment of wastewaters based on AS processes (ASP) is one of the oldest biological routes to treatment of wastes (Bramucci et al., 2003; Purohit et al., 2003; Sheik et al., 2014).

8.2.1.1 Bioaugmentation Studies in Industrial Wastewaters

Bioaugmentation is the introduction of designed microbial consortia harboring the necessary catabolic potential required to degrade the pollutants of interest present on a particular contaminated site (El and Agathos 2005; Paliwal et al., 2012). The introduction of these designed consortia increases the biological diversity thereby broadening the existing gene pool (Kapley et al., 1999; Qureshi and Purohit, 2002). Bioaugmentation has mainly been focused on treatment of contaminated soils, but now research is being carried out to develop bioaugmentation strategies for wastewater treatment as well (Li et al., 2013). Analyses of the microbial community in WWTPs have aided understanding, the interplay of several degradation pathways (Khardenavis et al., 2008; Kapley and Purohit, 2009). The inhabiting microbiota thrives on toxic compounds and thus brings about bioremediation of the anthropogenic site (Moharikar et al., 2005; Chandra et al., 2012).

The samples from WWTP are used as a source of isolating catabolically potential bacteria, which bear genetic machinery to degrade toxic compounds (Chandra et al., 2009).

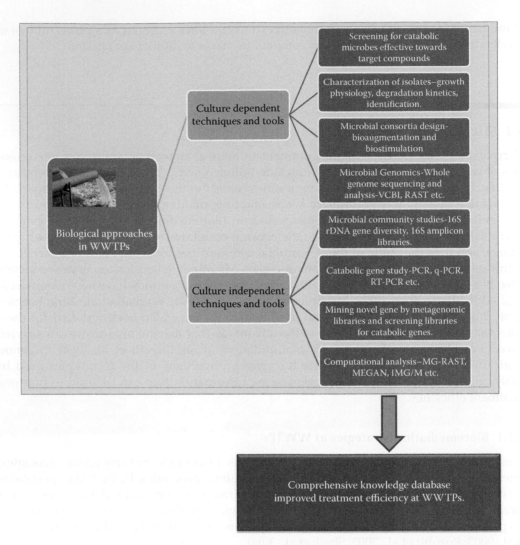

FIGURE 8.2
An overview of techniques used in analysing the activated biomass of wastewater treatment plants.

Bacterial cultures are screened for their degradation ability toward toxic compounds. Lab-scale experiments to understand growth physiology of microbes, degradation kinetics, molecular biology studies, and so on constitute the basic steps in characterization of bacterial isolates (Kapley et al., 2001; Purohit, 2003; Khardenavis et al., 2010; Selvakumaran et al., 2011). The efficient microbes are selected to design consortia for further application onsite.

Bioaugmentation of AS has been proposed as a powerful strategy to boost efficiency of WWTP (Zheng et al., 2013). Bioaugmentation scores over conventional techniques as it accelerates the rate of degradation thereby removing undesired toxic compounds faster. The efficiency of the bioaugmentation process depends on many factors, such as the chemical property and the concentration of pollutant, strains being used, and so on. The strain selected for bioaugmentation should be persistent or tolerating high concentration of contaminant, possessing the genetic capability (catabolic potential), and compatibility or ability to survive in harmony with the indigenous microflora (McLaughlin et al., 2006;

Wang et al., 2009). The introduced bacteria are sometimes not able to cope up with the new environmental conditions, and consequently decline rapidly (El and Agathos et al., 2005). To address this problem, plasmid-mediated bioaugmentation has been reported as a new bioaugmentation technique in some studies. In this method, horizontal gene transfer (HGT) via conjugation between the introduced plasmids and the indigenous bacteria results in the formation of transconjugants harboring the encoded functions (Bathe et al., 2009; Król et al., 2012; Tsutsui et al., 2010, 2013). However, for the exchange of information through plasmids, the compatibility of the donor (bacteria having the required plasmid) and recipient strains (i.e., the existing microbial community in the AS) is important. The microbial community in the AS is affected by many operational parameters, such as the sludge retention time, the ratio of return sludge, and the BOD-MLSS loading rate. These parameters, therefore, are important while studying the fate of introduced plasmids in the AS produced in the industrial wastewaters. In a study, conjugative transfer of a self-transmissible, broad host range, mercury resistance, and partial 2,4-dichlorophenoxyacetic acid (2,4-D)-degrading plasmid pJP4 from *Escherichia coli* HB101 or *Pseudomonas putida* KT2440 to AS bacteria was examined by filter mating. The AS samples were collected from a WWTP and laboratory reactors were set up with different sludge retention time. It was seen that the transfer frequency of plasmid Pjp4 depended on the donor strain and the recipient bacteria. Also, more transconjugants were isolated from AS sampled from the reactors with longer sludge retention time operations (Tsutsui et al., 2010). Therefore, plasmid-mediated bioaugmentation may provide a better option than conventional; however, further studies are required to understand the stability and feasibility of these plasmids in various operational conditions of the ASP.

The biological functioning of these plasmids along with the complete nucleotide sequence can further help us in designing bioaugmentation strategies. A study demonstrated the use of genomic information in enhancing the degradative potential. The complete nucleotide sequencing of an IncP-1 chloroaniline degradation plasmid, pWDL7::*rfp* and its close relative pNB8c was carried out along with the expression pattern, function, and bioaugmentation potential of the putative 3-chloroaniline (3-CA) oxidation genes. From the nucleotide sequence it was found that the plasmid pWDL7::*rfp* carries two copies of a fully functional *dca* gene cluster suggesting a potential usage for bioaugmentation to enhance removal of 3-CA by transferring the genetic information to indigenous population. To verify the ability of pWDL7::*rfp* to transfer 3-CA degradation capacity from *P. putida* UWC3 to a mixed bacterial population residing in the AS of a WWTP, plate matings was carried out. Little degradation was seen during the initial period with 0.16 mM or 0.39 mM3-CA in minimal medium without additional carbon (MMO); however, when pyruvate (0.227 mM) was added as an additional carbon source 3-CA was rapidly removed. The authors suggested that plasmid pWDL7::*rfp* might successfully accelerate bioaugmentation of WWTP by removing recalcitrant chloroanilines, thereby saving both time and operation cost (Król et al., 2012).

8.3 Role of Genomics and Metagenomics in Wastewater Treatment

The genomic revolution created by next generation sequencing has created a huge opportunity for scientific community to make progress in wastewater microbiology (Daims et al., 2006; Kapley and Purohit, 2009). Recent advances in DNA sequencing has brought

TABLE 8.2

A Comparison of Sequencing Technologies

Sequencing Type	Technique	Advantages	Disadvantages
First-generation sequencing	Sanger sequencing	Long reads (~900 bps), suitable for small projects	Low throughput, expensive
Next-generation sequencing	454	Low error rate, medium read length (~400–600 bps)	Relatively high cost per base, must run at large scale, medium/high startup costs
	Illumina	Low error rate, lowest cost per base, tons of data	Must run at very large scale, short read length (50–75 bp), runs take multiple days, high startup costs, *de novo* assembly difficult
	SOLiD	Low startup costs, scalable (10–1000 Mb of data per run), medium/low cost per base, low error rate, fast runs (<3 h)	New, developing technology, cost not as low as illumina, read lengths only ~100–200 bps so far
	HeliScope (single molecule sequencing)	No amplification stage, DNA fragments are attached to array, higher throughput	High error rates, time consuming

down the cost of genome sequencing, creating a possibility for single research groups to generate large amount of sequence data (Kircher et al., 2010).

Next generation sequencing (NGS) technologies has revolutionized fields of genetics, clinical research, environmental genomics, agriculture, pharmaceuticals, and so on by decreasing the cost of DNA sequencing and increasing the output data. The inexpensive production of large volumes of sequence data is the primary advantage over conventional methods. The most common platforms for massively parallel DNA sequencing read production are in reasonably widespread use at present: the Roche/454 (Margulies et al., 2005), the Illumina/Solexa Genome Analyzer (Bentley et al., 2008), and the Applied BiosystemsSOLiD TM System (http://marketing.appliedbiosystems.com/images/Product/Solid Knowledge/flash/102207/solid.html). In recent time, two more systems present include the HelicosHeliscope TM (www.helicosbio.com), and Pacific Biosciences SMRT (www.pacificbiosciences.com) instruments. Table 8.2 compares the features of these sequencing technologies.

8.3.1 Microbial Genomics of Industrial Wastewater

Genome sequencing of a large number of bacteria has provided a quantum leap in the understanding of metabolic gene pool available in these bacteria. New degradation pathways that have not been previously reported can be identified and exploited for biotechnological applications. An improved understanding of the catabolic gene pool available in the ASP will help in developing strategies for optimizing industrial wastewater treatment. Various tools associated with molecular biology and microbial genomics are now being applied to understand the microbial population and their interrelationship with each another. Only a small portion of this population can be isolated in the laboratory as pure microbial culture; the rest comprises unculturable microbial population (Bramucci et al., 2006). Studying both the populations is important for further improvement in environmental genomic approaches prerequisite for wastewater treatment.

Information obtained by genome sequencing of potential bacteria isolated from industrial wastewater can be used for bioremediation of contaminated sites. Table 8.3 provides a list of bacteria isolated from various contaminated niches, for which many genome sequencing projects are reported. For example, studies have been carried out by sequencing genome of some *Pseudomonas* species isolated from different wastewater. Members of the genus *Pseudomonas* are known to show high metabolic versatility enabling them to colonize diverse niches such as sites containing toxic compounds (Silby et al., 2011). One such *Pseudomonas* species is *Pseudomonas aeruginosa* strain PFK10, which was isolated from a CETP at Ankleshwar industrial area of Gujarat, India. The strain can decolorize as well as degrade azo dyes. Many genes for oxidases as well as laccase-like multicopper oxidases were identified in the genome. From the in vitro plate studies it was seen that the strain could hydrolyze casein, starch, gelatin, and produce important enzymes such as catalase and nitrate reductase. From the genetic information about the metabolic pathways, the strain can be used for treatment of sites contaminated with textile effluents (Faldu et al., 2014). Even bacteria isolated from pollutant-contaminated sites may prove effective for industrial wastewater treatment. *Pseudomonas putida* CSV86 is a bacterial strain isolated from petroleum contaminated site. It has been shown both functionally and chromosomally that the strain can utilize aromatic compounds such as naphthalene, 1- and 2-methylnaphthalene, phenylacetic acid (PA) and *p*-hydroxyphenylacetic acid (4-HPA), salicylate, benzylalcohol, benzoate, and *p*-hydroxybenzoate. In addition it can grow on vanillin, veratraldehyde and ferulic acid, homogentisate, as well as grow on aromatic compounds in the presence of heavy metals such as copper cadmium, cobalt, and arsenic as the carbon source. The strain is plasmid free and has the capability of transferring naphthalene degrading property to other bacteria. The degradative capacity along with the conjugation capability makes this bacterium a possible candidate or donor for dissemination of catabolic potential for the process of genetic bioaugmentation (Phale et al., 2013; Paliwal et al., 2014; Shah, 2014).

8.3.2 Metagenomics

The term metagenomics implies a combination of various analyses for the study of multiple genomes, as coined by Handelsman et al. (1998). The study of the total genetic material recovered from any sample niche is termed as metagenomics. This analysis is very useful in analyzing the microbial population, since it is well known that only 2%–5% of the bacterial population can be cultured. To actually harness the potential of the microbiota in wastewater degradation, the catabolic potential of the uncultivable 95% cannot be ignored. Hence, metagenomics is a very promising tool. Several processes in wastewater treatment are generally regarded as being a "black box" due to the very limited knowledge available regarding composition, dynamics, and stability of the microbial communities (Siezen and Galardini, 2008). But with the advent of genome sequencing of both culturable species our understanding of the role played by each bacteria or communities in WWTPs is increasing. The integration of knowledge derived from genomics combined with the engineering parameters, may help in solving the "black box," thereby tremendously aiding in development of efficient techniques for industrial wastewater treatments. Jadeja et al. (2014) have demonstrated the use of metagenomic tools in pointing out the genes absent in AS of a treatment plant. This information carries great potential for improving treatment efficiency by bioaugmentation of the missing catabolic potential.

TABLE 8.3

List of Bacteria Deposited in NCBI That Have Been Isolated from Various Contaminated Sites
(Genome Sequencing Projects are Completed/On-Going)

Organisms along with Accession no.	Source	References	Features of the Study
Novosphingobium nitrogenifigens Y88T or DSM 19370 (AEWJ00000000)	Isolated from nickel-enriched pulp and paper mill wastewater	Strabala et al. (2012)	Genes encoding for polyhydroxyalkanoate (phbA, phbB, phaC) nitrogen fixation proteins were identified. CDS encoding putative metal resistance/efflux proteins, including Ni, Cu, Hg, Co, and As resistance proteins were also identified in the genome
Cellulosilyticum lentocellum DSM 5427 (CP002582)	Isolated from estuarine, sediment of the River Don, Aberdeenshire, Scotland (the river received both domestic and paper mill effluent)	Miller et al. (2011)	Thirteen predicted open reading frames (ORFs) are annotated as cellulose degradation enzymes in the genome
Shewanella decolorationis S12 (AXZL00000000)	Isolated from activated sludge of a textile-printing waste water treatment plant in Guangzhou, China	Xu et al. (2013)	Genome analysis unpublished. May have potential use in dye degradation
Pseudomonas putida MO2 (PRJNA217511)	Isolated from wastewater can produce medium chain length polyhydroxyalkanoates from waste oil	—	Genome analysis unpublished
Ochrobacterum intermedium CCUG 57381 (PRJNA198781)	Isolated from a wastewater treatment plant receiving industrial effluent from pharmaceutical production contaminated with high levels of quinolones	Johnning et al. (2013)	Genome analysis unpublished
Serratia marcescens WW4 (CP003959)	Isolated from a paper machine in Taiwan	Chung et al. (2013)	One complete *pig* gene cluster for the biosynthesis of prodigiosin (i.e., a red pigment with antibiotic activities) Additionally, several putative bacteriocin genes were also identified
Pseudomonas putida LS46 (ALPV00000000)	Isolated by enrichment from wastewater	Sharma et al. (2013)	Can synthesize medium-chain-length polyhydroxyalkanoates (mcl-PHAs). Genome analysis unpublished
Lactococcus chungangensis DSM 22330 (PRJNA243309)	Isolated from a foam sample taken from an activated sludge	—	Genome analysis unpublished
Pseudomonas otitidis LNU-E-001 (PRJNA241481)	Isolated from activated sludge of a wastewater treatment plant in Northeast China	—	Phenanthrene-degrading bacteria. Genome analysis unpublished

(Continued)

TABLE 8.3 (*Continued*)

List of Bacteria Deposited in NCBI That Have Been Isolated from Various Contaminated Sites (Genome Sequencing Projects are Completed/On-Going)

Organisms along with Accession No.	Source	References	Features of the Study
Comamonas testosteroni I2 (PRJNA232439)	Isolated from activated sludge	Boon et al. (2000)	Found to be able to mineralize 3-chloroaniline (3-CA). Genome analysis unpublished
Virgibacillus sp. CM-4 (AUQA00000000)	Isolated from activated sludge treating saline sewage in Hong Kong	—	Halophilic bacterial species. Genome analysis unpublished
Corynebacterium doosanense CAU 212 = DSM 45436 (PRJNA219342)	Isolated from activated sludge	—	Genome analysis unpublished
Sphingobium sp. YL23 (ASTG00000000)	Isolated from the sewage sludge of a full-scale domestic wastewater treatment plant in Fujian Province, China	Hu et al. (2013)	Can degrade Bisphenol A (BPA). The cytochrome P450 monooxygenase system was found in the YL23 genome which showed high similarity to the previously reported P450s of *Sphingomonas bisphenolicum* AO1 involved in BPA degradation
Sphingomonas sp. YL-JM2C (ASTM00000000)	Isolated from activated sludge	—	Can degrade triclosan under aerobic conditions. Genome analysis unpublished
Pseudomonas putida TRO1 (APBQ00000000)	Isolated from activated sludge, Aalborg West WTTP	—	Can utilize triclosan as a sole source of carbon and energy. Genome analysis unpublished
Dyella ginsengisoli strain LA-4 (AMSF00000000)	Isolated from activated sludge	Kong et al. (2013)	The continuous biphenyl metabolism clusters *bphA1A2*(orf1)*A3A4BCX0* and *bphX1*(orf2)*X2X3D* were found in a small contig
Uncultured bacterium plasmid PSP21 (NC_019021)	Plasmid isolated from activated sludge bacteria of a municipal wastewater treatment plant	Szczepanowski et al. (2011)	IncP-1α antibiotic resistance plasmids sequenced for comparing conserved regions in plasmids
Uncultured bacterium plasmid PB11 (NC_019022)	Plasmid isolated from activated sludge bacteria of a municipal wastewater treatment plant	Szczepanowski et al. (2011)	IncP-1α antibiotic resistance plasmids sequenced for comparing conserved regions in plasmids
Uncultured bacterium plasmid PB5 (NC_019020)	Plasmid isolated from activated sludge bacteria of a municipal wastewater treatment plant	Szczepanowski et al. (2011)	IncP-1α antibiotic resistance plasmids sequenced for comparing conserved regions in plasmids
Kineosphaera limosa NBRC 100340 (BAHD00000000)	Isolated from phosphorus removal activated sludge	—	This strain accumulates polyhydroxyalkanoate (PHA) in anaerobic condition. The genome sequence of this strain would give clue for understanding of phosphorus removal process in activated sludge. Genome analysis unpublished

(*Continued*)

TABLE 8.3 (*Continued*)

List of Bacteria Deposited in NCBI That Have Been Isolated from Various Contaminated Sites
(Genome Sequencing Projects are Completed/On-Going)

Organisms along with Accession No.	Source	References	Features of the Study
Sulfurospirillum multivorans DSM 12446 (CP007201)	Isolated from activated sludge of a sewage treatment plant in Stuttgart, Germany	—	It is capable to reductive dehalogenation of tetrachloroethene (PCE) and trichloroethene (TCE) to cis-dichloroethene (DCE). Genome analysis unpublished
Patulibacter medicamentivorans strain I11 (AGUD00000000)	Isolated from activated sludge with ibuprofen degradation capacity.	Almeida et al. (2013)	Genetic analysis revealed a cluster of genes for ibuprofen degradation (ipfABDEF)
Paracoccus sp. TRP (AEPN00000000)	Isolated from activated sludge from the HuaYang pesticide plant in Shandong, China	Li et al. (2011)	Can completely biodegrade chlorpyrifos and 3,5,6-trichloro-2-pyridinol. Genome analysis unpublished
Comamonas testosteroni strain CNB-2 (NC_013446)	Isolated from CNB-contaminated activated sludge	Ma et al. (2009)	Gene clusters for benzoate, gentisate, phenol, 3-hydroxybenzoate, 4-hydroxybenzoate, protocatechuate, and vanillate were identified in the genome
Nakamurella multipartita DSM 44233 (CP007201)	Isolated from activated sludge acclimated with sugar-containing synthetic wastewater	Tice et al. (2010)	Capable of accumulating large amounts of polysaccharides in its cells. Genome analysis unpublished
Methylobacillus flagellatus KT (CP000284)	Isolated from activated sludge found at the wastewater treatment plant in Moscow, Russia	Chistoserdova et al. (2007)	Organism contains multiple formaldehyde degradation pathways including a ribulose monophosphate (RMP) cycle and a linear pathway utilizing tetrahydromethanopterin-dependent enzymes
Comamonas testosteroni KF-1 (AAUJ00000000)	Isolated from activated sludge	Schleheck et al. (2004)	Able to degrade 3-(4-sulfophenyl)butyrate, a component of many synthetic surfactants. Genome analysis unpublished
Acidovorax temperans CB2 (PRJNA39849)	Isolated from an activated sludge wastewater treatment plant in northern New Zealand	—	Genome analysis unpublished
Acinetobacter calcoaceticus PHEA-2 (NC_016603)	Isolated from industrial wastewater in China	Zhan et al. (2011)	Pathways predicted for the catabolism of catechol (*cat*), protocatechuate (*pca*), and phenylacetate (*pha*) in the genome
Methylobacterium extorquens DM4 (NC_012988)	Isolated from industrial wastewater sludge in Switzerland	Vuilleumier et al. (2011)	Dichloromethane utilization (*dcm*) gene cluster identified in the genome

(*Continued*)

TABLE 8.3 (*Continued*)

List of Bacteria Deposited in NCBI That Have Been Isolated from Various Contaminated Sites
(Genome Sequencing Projects are Completed/On-Going)

Organisms along with Accession No.	Source	References	Features of the Study
Bacillus cereus SJ1 (ADFM00000000)	Isolated from chromium contaminated wastewater of a metal electroplating factory	He et al. (2010)	A putative chromate transport operon, chrIA1, and two additional *chrA* genes encoding putative chromate transporters that likely confer chromate resistance were identified
Lysinibacillus fusiformis ZC1 (ADJR00000000)	Isolated from chromium (Cr) contaminated wastewater of a metal electroplating factory	He et al. (2011)	Large numbers of metal(loid) resistance genes specifically, a *chrA* gene encoding a putative chromate transporter conferring chromate resistance was identified
Alcaligenes sp. Strain HPC1271 (AMXV00000000)	Isolated from activated sludge of a common effluent treatment plant that treats industrial wastewater	Kapley et al. (2013)	Gene for type I nonribosomal polyketide synthetase (enterobactin synthase) along with clusters for type I polyketide synthetase (PKS) and colicin synthase were identified in the genome
Klebsiella pneumoniae strain: EGD-HP22 (PRJNA242503)	A bacterium performing denitrification	—	Genome analysis unpublished
Morganella sp. EGD-HP17 (AZRH00000000)	A *Morganella* strain isolated from mobile toilets	—	Genome analysis unpublished
Ochrobactrum sp. EGD-AQ16 (AWEU00000000)	Desert bacterial isolate with combined potential for biofilm formation and Cr(VI) reduction under salinity.	—	Genome analysis unpublished
Acinetobacter baumannii EGD-HP18 (AVST00000000)	Study of ammonical-nitrogen removal by nitrification denitrification process using laboratory isolates	—	Genome analysis unpublished
Pantoea dispersa EGD-AAK13 (AVSS00000000)	Demonstration of hydrolytic capacities of enzymes from bacteria isolated from vegetable market soil	—	Genome analysis unpublished
Serratia marcescens EGD-HP20 (AVSR00000000)	Demonstration of feather waste management by keratin hydrolyzing bacteria	—	Genome analysis unpublished
Klebsiella pneumoniae EGD-HP19 (NZ_AUTW00000000)	Study of ammonical-nitrogen removal by nitrification denitrification process using laboratory isolates	—	Genome analysis unpublished

(Continued)

TABLE 8.3 (*Continued*)

List of Bacteria Deposited in NCBI That Have Been Isolated from Various Contaminated Sites
(Genome Sequencing Projects are Completed/On-Going)

Organisms along with Accession No.	Source	References	Features of the Study
Bacillus amyloliquefaciens EGD-AQ14 (AVQH00000000)	Plant growth-promoting bacteria isolated from saline desert plant rhizosphere of Kuchch, Gujarat	—	Genome analysis unpublished
Pseudomonas fluorescens EGD-AQ6 (AVQG00000000)	Biodegradability of aromatic compounds in biofilm forming *Pseudomonas* isolated from sewage sludge	—	Genome analysis unpublished
Pseudomonas mendocina EGD-AQ5 (AVQF00000000)	Biodegradation of aromatic compounds in biofilm forming *Pseudomonas mendocina*	—	Genome analysis unpublished
Bacillus sp. EGD-AK10 (AVPM00000000)	Atrazine degrading bacterium isolated from Indian agricultural soil	—	Genome analysis unpublished
Pseudogulbenkiania ferrooxidans EGD-HP2 (AVPH00000000)	Color producing isolate from Loktak lake	—	Genome analysis unpublished
Staphylococcus sp. EGD-HP3 (AVOQ00000000)	Salt tolerant isolate from ETP	—	Genome analysis unpublished
Alcaligenes sp. EGD-AK7 (AVOG00000000)	Atrazine degrading *Alcaligenes* sp.	Sagarkar et al. (2014)	Genome analysis unpublished
Pseudomonas sp. EGD-AK9 (AVOF00000000)	Atrazine degrading bacteria isolated from Indian agricultural soil	—	Genome analysis unpublished
Lactobacillus plantarum EGD-AQ4 (AVAQ00000000)	Isolation of probiotic bacteria from nonalcoholic fermented bamboo shoot products of Assam, North-East India	—	Genome analysis unpublished
Enterobacter sp. EGD-HP1 (PRJNA211614)	Isolate overproducing gases from Loktak lake	—	Genome analysis unpublished
Enterococcus gallinarum EGD-AAK12 (AVPC00000000)	Demonstration of hydrolytic capacities of enzymes from bacteria isolated from vegetable market soil	—	Genome analysis unpublished

8.3.3 Techniques Used in the Genomic Approach

8.3.3.1 *Metagenomic DNA Extraction*

Since the metagenome is a representative of all the DNA contained in a sample, extraction procedures are usually optimized separately for each niche. Several research groups have worked on optimizing protocols for the isolation of high-quality DNA from environmental samples such as sludge, soil, wastewaster, and so on. It involves troubleshooting problems such as organism lysis, integrity of high molecular weight DNA, humic acid contamination, color removal, and so forth. (Purohit et al., 2003; Bertrand et al., 2005; Desai and Madamwar, 2007). A large number of commercial kits are also available for this purpose.

8.3.3.2 *16S rDNA Gene Diversity Studies*

Polymerase chain reactions (PCRs) and cloning of microbial genes are the techniques being practiced since 1980s, but in recent times, the decreasing cost of DNA sequencing, and emerging high throughput methods to clone, store, and screen environmental libraries have enlarged possibilities. PCR remains the most powerful and popular tool to understand 16S diversity in any ecology as it unravels microorganisms underlying community assembly with the aid of cost-effective sequencing platforms. Microbial community structures in wastewaters have been previously studied to measure biodegradation potential relying in the process (Roest et al., 2005; Liu et al., 2007). Many uncultivable bacteria have been observed by the use of TA clone libraries, for example, the presence of uncultured gammaproteobacterium and *Kleibsella* in anoxic bioreactor treating azo dyes was demonstrated by Dafale et al. (2010). Several studies support the combination of culture-based and culture-independent techniques to cover the microbial diversity of complex ecosystems efficiently. Microbial diversity studies for two different ETPS of pesticide and pharmaceutical industries revealed dominance of genus such as *Alcaligenes, Bacillus,* and *Pseudomonas, Brevundimonas, Citrobacter, Pandoraea,* and *Stenotrophomonas* in pesticide ETP and dominance of *Agrobacterium, Brevibacterium, Micrococcus, Microbacterium, Paracoccus,* and *Rhodococcus* specifically to pharmaceutical ETP (Rani et al., 2008). In another study, diversity analyses of activated biomass from a CETP using culture as well as culture-independent techniques revealed 26 different genera belonging to phyla *Proteobacteria, Actinobacteria,* and *Firmicutes* using culture-based techniques and 21 different genera from the phyla *Proteobacteria, Firmicutes, Planctomycetes,* and *Bacteroidetes* by applying 16S rDNA gene cloning (Kapley et al., 2007a,b).Variations in microbial diversity can also be monitored as demonstrated in fed-batch reactor operation with wastewater containing nitroaromatic residues (Kapley et al., 2007b). In another study, the diversity of *Citrobacter* genus, from activated biomass of ETPs was studied using 16S rDNA analysis and applied in salicylate utilization (Selvakumaran et al., 2008). In studying biodegradation of a tannery wastewater treating CETP, bacterial diversity studies revealed the presence of *Escherichia* sp., *Stenotrophomonas* sp., *Bacillus* sp., and *Cronobacter* sp. and *Burkholderiales* bacterium (Chandra et al., 2011). The catabolic genes such as catechol 1,2-dioxygenase, catechol 2,3-dioxygenase, toluene dioxygenase-iron-sulfur protein component, benzene dioxygenase, and naphthalene dioxygenase were studied using PCR primers in combination with time-dependent change in bacterial population by the amplification of 16S rDNA product, followed by restriction digestion in AS metagenome (Moharikar et al., 2003). nxrB gene coding for nitrate oxidoreductase proved to be a phylogenetic and functional marker in the detection of *Nitrospira* sp. in AS as well as soil metagenomes (Pester et al., 2013). Ammonia monooxygenase gene is also a very commonly used gene marker for ammonia removal processes. Specificity of various primers coding for ammonia monooxygenase was checked where pros and cons of primers reported were discussed (Junier et al., 2008). Fluorescence *in situ* hybridization (FISH) analysis of an anaerobic sludge digester led to proposing a new bacterial candidate division in the study (Guermazi et al., 2008). Genetic diversity and quantification of functional microbes was studied using FISH and denaturing gradient gel electrophoresis (DGGE) in completely autotrophic nitrogen removal over nitrite (CANON) process for two membrane biological reactors in low and high ammonia conditions for nitrogen removal (Zhang et al., 2014). In current times, amplicon libraries prove to be a common alternative to conventional TA cloning methods. In a recent study, microbial community changes in response to various dissolved oxygen levels was studied (Yadav et al., 2014).

8.3.3.3 Gene Quantification Studies

Real-time PCR (q-PCR) to monitor catabolic potential in sludge samples.

Monitoring genes involved in degradation of target compounds can also be studied by analyzing the copy numbers of the target gene. Real-time PCR enables the quantification of marker genes, which allows measuring functional gene levels and thus biodegradation efficiency with respect to time, or other varying parameters of the study. For analyzing biodegradation pathways in WWTPs, q-PCR can be used as biomonitor for marker genes of the process. A number of examples demonstrate the importance of this tool. Some of them are aniline degrading bacterial strains in AS in a recent study (Kayashima et al., 2013), comparison of the effect of silver and zinc oxide nanoparticles on functional bacterial community (Chen et al., 2013), comparison of bacterial chromosomal tbpA gene in *E. coli*, chromosomal phlA gene on *Cupravidisnecator,* and plasmid pJP4-encoded *tfdB* gene where plasmid-mediated bioaugmentation enhanced degradation of 2,4-D (Tsutsui et al., 2013). Applying q-PCR, reliable detection and quantification of Microthrix sp. for sludge bulking and foaming problems across 29 samples from different WWTPs was demonstrated (Vanysacker et al., 2014).

q-PCR was used to quantify AOB and ammonia-oxidizing archae levels in lab-scale and full-scale WWTP (Ye et al., 2012). q-PCR for 16S rDNA gene revealed *Kuenia* genus in annamox bacteria in municipal sewage waste in comparison with landfill leachate treatment and monosodium glutamate WWTP (Shen et al., 2012). Effect of various growth patterns that is, attached, suspended, or both; on nitrifying bacterial groups (ammonia oxidizing and nitrite oxidizing) over a period of 120 days was monitored using q-PCR (Wang et al., 2012). In another study, mesophilic anaerobic digester and thermophilic aerobic digestion in sequential digestion for sludge reduction and methane production was studied. The former maintained more diverse bacteria and the latter had dominant archael population as monitored by tools such as DGGE and q-PCR (Jang et al., 2014). Sequential aeration caused decrease in NOB and tenfold increase in anaerobic AOB enhancing autotrophic nitrogen removal (Pellicer-Nàcher et al., 2014). Long-term low dissolved oxygen-enriched AOB and NOB in AS increased nitrification efficiency, and *N. europaea* was dominant in AOBs and *Nitrospira* was dominant species among NOBs. Nitrifier community shifts helped balance the adverse effect of long-term low dissolved oxygen conditions (Liu et al., 2013). Real-time reverse transcription-PCR (RT-PCR) and terminal restriction fragment length polymorphism (t-RFLP) were used to study transcriptional response of AOBs to dimethyl sulfide (Fukushima et al., 2014).

Antibiotic-resistant bacteria (ARBs) and antibiotic-resistant genes (ARGs) are drawing attention worldwide. ARGs travel among bacteria via plasmids, integrons, transposons, and so on (Schwarz and Chaslus-Dancla, 2001; Pruden et al., 2006). Class 1 integrons and

ARGs in sewage treatment plant were investigated where q-PCR was used to study levels of integrons and other mobile genetic elements in influent, AS, and effluent. Sewage treatment plants proved to be a hotspot for HGT of ARGs.9 tetracycline resistance genes and two mobile elements were quantified by q-PCR in oxytetracycline treatment plant (Liu et al., 2012). In another similar study, nine tetracycline resistance genes and four mobile elements were quantified in sludge exposed to oxytetracycline (Liu et al., 2013). PCR and q-PCR were employed to investigate nine beta lactam resistance genes abundance in AS from 15 geographically different sewage treatment plants of East Asia and North America (Yang et al., 2012).

8.3.4 Metagenomic Libraries and Screening Strategies

Application of molecular biology techniques to metagenome samples, allows access to microbial diversity, identification of novel protein coding ORFs, understanding reactions of biochemical pathways, and reconstruction of genomes of uncultivable diversity from metagenomic sequences. In this direction, construction of metagenomic library has proved to be a powerful tool to access the meta-genetic wealth resting in anthropogenic ecologies such as WWTPs. Thus, a metagenomic library from samples of such niches will reveal "who is present, what are they doing and how they are working." Several studies have exploited metagenomic libraries for mining novel genes and hence derived metabolic pathways undertaken by the unculturable diversity of ETPs and AS. Many studies based on AS metagenomes employing metagenomic library constructions have yielded novel catalysts in organic compound degradation. Metagenomic libraries that used AS metagenomic DNA have resulted in novel catabolic enzymes as reported in several studies. A novel arsenic resistant gene *arsN* was reported from a metagenomic library from industrial ETP metagenome (Chauhan et al., 2009). BAC and PUC vector libraries resulted in cellulose metabolizing clones and lipolytic clones (Sharma et al., 2012). 4-Nitrotoluene catalyzing enzyme was isolated from fosmid metagenomic library using AS metagenome (Kimura et al., 2009). In another similar study, fosmid-based metagenomic library of forest soil and AS metagenomes lead to novel genes involved in quorum sensing phenomena in microbes (Nasuno et al., 2012). In similar fosmid-based study, 43 extradioloxygenases (EDOs) could be categorized into four new subfamilies, thus shedding light on genetic evolution of EDO genes (Suenaga et al., 2007). This study used 96-well plates for screening EDO-positive clones'. The use of such microtiter plate enables screening of a large number of clones thus increasing chances of securing positive clones from metagenomic libraries (Suenaga et al., 2009). In another study, the use of robotics in applying clones onto macroarray coupled with probe labeling, membrane hybridization, and signal detection methodologies succeeded in screening 384-well plates (Pham et al., 2007). The operon trap GFP expression vector in combination with FACs is another approach popularly known as SIGEX (substrate induced gene expression screening), which screens clones of catabolic genes with respect to substrate of choice (Uchiyama et al., 2008). Thirty-three positive clones for benzoate and two for naphthalene were found in a metagenomic library from groundwater metagenome using SIGEX (Uchiyama et al., 2005). This approach will give interesting results in response to substrate expected in WWTPs. A modified methodology of SIGEX works on product-induced gene expression by a transcription activator and sensory cells responsive to product hence termed product-induced gene expression (PIGEX) (Uchiyama and Miyazaki, 2010).

Any metagenomic library is a store of metabolic genes and hence can be screened for various activities. In this regard, making such metagenomic libraries accessible publically may encourage different research groups to screen these libraries for respective activity of interest. Such an approach has been initiated where a publicly accessible collection of fosmid libraries of environmental soil metagenome was made by Neufeld and colleagues (2011). Similar databases for WWTP metagenomic libraries can allow us to take complete advantage of the technique.

8.3.5 Designing Consortia with Respect to WWTP

Understanding the composition of microbial community is the key for efficient removal of organic and inorganic load in a WWTP, which can be achieved by genome and metagenome analysis (Valentín-Vargas et al., 2012). From the information obtained by these analyses, the crucial players governing the community dynamics can be identified. This method also allows us to estimate the missing catabolic potential required to treat target waste that can be remediated by bioaugmentation. A microbial consortia can be designed to fill the lacking genes or pathways thereby improving the performance of the plant. Domde et al. (2007) demonstrated this technique in the treatment of wastewater contaminated with hydrocarbons. They reported an improvement in the degradation efficiency by 37%, with the inclusion of a bacterium containing the *alkB* gene in AS that did not have the potential to degrade lower alkanes. The addition of a new gene pool improved the degradative efficiency. Many bacteria with diverse catabolic potential have isolated from ETPs for use in bioaugmentation, as consortium members (Narde et al., 2004; Khardenavis et al., 2007; Raj et al., 2007a,b; Chandra et al., 2007, 2008; Selvakumaran et al., 2008; Thangaraj et al., 2008; Verma et al., 2010; Kapley et al., 2013).

8.4 Computational Tools in Metagenomics

Next-generation sequencing methods yield large amounts of sequence data that need to be analyzed using computational tools. This need gave rise to the development of bioinformatics that can help analyze large sequence data. Figure 8.3 represents steps involved in metagenome sequence analysis and tools explored. Tools such as MOTHUR (Schloss et al., 2009) and QIIME (Caporaso et al., 2010) allow analysis and comparison of 16S rDNA gene amplicon data of metagenomes. MEtaGenomeANalyzer (MEGAN) is another similar tool to analyze taxonomy and functional genes in metageomes (Huson et al., 2007). Previous version of MEGAN allowed analysis of only random shotgun sequences but recent version MEGAN four allows analysis of 16S sequences also (Huson and Mitra, 2012) and genes involved in degradative pathways (More et al., 2014). Integrated microbial genomes-metagenomes (IMG/M) (Markowitz et al., 2012) is another such computational server, which stores genomes as well as metagenomes and provides an analysis pipeline. Here the data are categorized into plasmid, viruses, and genome fragments. More than 160 metatranscriptomic data across 16S RNA sequence projects are also submitted in IMG (data of January 2014). Metagenomics rapid annotation using subsystem technology (MG-RAST) (Meyer et al., 2008) is one of the most commonly used metagenome analysis tool where diverse sample data are submitted. It allows phylogenetic analysis of 16S rDNA amplicon data as well as whole genome-sequencing projects too. Functional annotations of metagenomic sequences are based on databases of KEGG, COG, and the like. Statistical analysis such as principle component

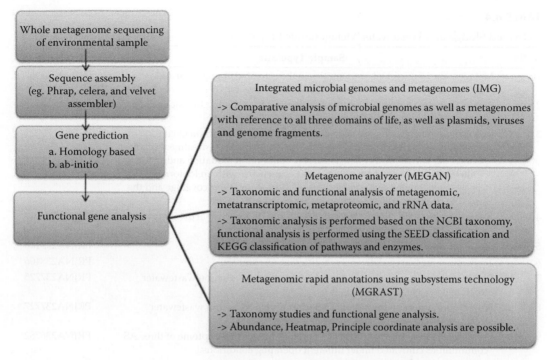

FIGURE 8.3
Steps involved in metagenome sequence analysis and frequently used computational tools.

analysis (PCA), heat map, and so on can also be performed using MG-RAST. Various WWTP and AS metagenomes are studied using either a combination, or any one of these computational tools. Table 8.4 lists the AS-whole metagenome sequencing projects submitted in National Centre for Biotechnology Information (NCBI) (till March 2014).

Bioinformatic analysis of metagenomic data reveals information, which contributes toward an understanding of uncultivable diversity. Several research groups have carried out whole metagenome sequencing of sludge samples and studied taxonomic as well as functional potential of the biomass. Some examples are highlighted here: whole metagenome sequence of an EBPR process was analyzed by MG-RAST and q-FISH, focusing on accumulibacter microdiversity. This study employed tools such as MEGAN, MG-RAST, and q-FISH, for analysis of the metagenomic sequences (Albertsen et al., 2012). In another study, metagenomic fosmid library of a refinery WWTP of petroleum was analyzed for phenol hydroxylases and catechol dioxygenases using MG-RAST (Silva et al., 2013). AS from two WWTP revealed p450 gene abundance; and *Microbacterium* sp. as the dominant population using MG-RAST (Fang et al., 2013). In yet another study, aerobic and anaerobic sludge of tannery sludge wastewaters in China were studied using MGRAST wherein *Proteobacteria, Firmicutes, Bacteriodes,* and *Actinomycetes* phyla were shown to dominate the aerobic sludge, whereas archeal population dominated the anaerobic sludge (Wang et al., 2013). More et al. (2013) demonstrated the use of metagenome analysis in improving wastewater treatment. The gene profile obtained by metagenome analysis was used to activate the sludge and pilot scale studies demonstrated increased efficiency of treatment.

TABLE 8.4

Activated Sludge and Wastewater-Metagenomic Projects Submitted in NCBI (as of March 2014)

S. No.	Sample Type/Site	Accession No.
1	Metagenome of microbial communities in the activated sludge (AS) and biofouling layers on membrane in membrane bioreactors.	PRJNA242367
2	Bacterial diversity of SBRs Genome sequencing for analyzing the bacterial community structure of AC sludge.	PRJNA239647
3	Metagenome sequencing of *Kuenenia stuttgartiensis* TB1 enrichment culture. An anammox culture enriched from a Taiwanese wastewater treatment plant performing partial nitrification, anaerobic ammonium oxidation and denitrification was used for metagenome sequencing with an Iontorrent PGM. Contigs were binned to *K. stuttgartiensis* TB1 based on their coverage and the resulting draft genome was annotated using prokka.	PRJNA239221
4	Anaerobic digestion sludge targeted locus (Loci), 16S microbial diversity research in anaerobic digestion sludge.	PRJNA239749
5	Study the microbial biodiversity of AS.	PRJNA239186
6	To examine the bacterial diversity in different AS samples.	PRJNA238406
7	Bacterial community composition of AS from sludge reduction wastewater treatment system.	PRJNA237725
8	Fungal community composition of AS from sludge reduction wastewater treatment system.	PRJNA237727
9	Preliminary study comparing metagenome and metatranscriptome of three AS communities cultivated under different operating conditions.	PRJNA236782
10	Metagenomic data sets of AS samples: (1) to compare the MBR microbial diversity and functional potentials with those of influent sewage and AS from the CAS system; (2) to evaluate the variation of inter- and intra-population patterns in the metagenome; (3) to investigate whether the microbial community in MBRs present higher nitrification or denitrification potentials; and (4) to find out the genes that contribute to membrane biofouling.	PRJNA236430
11	Microbial community composition of AS from sludge reduction wastewater treatment system.	PRJNA236147
12	Microbial community functional structures in wastewater treatment plants as characterized by GeoChip. To understand microbial community functional structures of AS in wastewater treatment plants (WWTPs) and the effects of environmental factors on their structure, 12 AS samples were collected from 4 WWTPs in Beijing. GeoChip 4.2 was used to determine the microbial functional genes involved in a variety of biogeochemical processes. The results showed that, for each gene category, such as egl, amyA, nir, ppx, dsrAsox, and benAB, there were a number of microorganisms shared by all 12 samples, suggestive of the presence of a core microbial community in the AS of four WWTPs. Variance partitioning analyses (VPA) showed that a total of 53% of microbial community variation can be explained by wastewater characteristics (25%) and operational parameters (23%), respectively. This study provided an overall picture of microbial community functional structures of AS in WWTPs and discerned the linkages between microbial communities and environmental variables in WWTPs. Overall design: Four full-scale wastewater treatment systems located in Beijing were investigated. Triplicate samples were collected in each site.	PRJNA234489
13	AS Metagenome	PRJNA233634
14	This study analyzed the taxonomic compositions of AS microbial communities residing within the aeration basins of Swiss wastewater treatment plants. The goal was to generate hypotheses about relationships between community composition, operational metrics, and performance metrics of wastewater treatment plants.	PRJNA232662

(Continued)

TABLE 8.4 *(Continued)*

Activated Sludge and Wastewater-Metagenomic Projects Submitted in NCBI (as of March 2014)

S. No.	Sample Type/Site	Accession No.
15	AS of SBR	PRJNA227484
16	Profiling filamentous bacteria in foaming AS using high throughput sequencing in an anaerobic-anoxic-oxic (A2/O) wastewater treatment plant in Zhengzhou, China.	PRJNA219426
17	DNA from AS targeted locus (Loci)	PRJNA219408
18	AS microbial communities from Denmark Metagenome	PRJNA219165
19	Waste AS Metagenome	PRJNA218566
20	Metagenomic analysis of AS	PRJNA61401
21	PHA-producing WAS 16S rRNA transcripts	PRJNA218058
22	The effect of triclosan on AS	PRJEB1049
23	PHA-accumulating microorganims Metagenome	PRJNA215014
24	AS samples Metagenome	PRJNA214923
25	16S rRNA gene–454 pyrosequencing. Microbiome (eubacteria) study of AS from wastewater treatment plants and from microbial enrichments promoting anammox process	PRJNA210433
26	16S rRNA gene–454 pyrosequencing. Microbiome (eubacteria) study of AS from wastewater treatment plants and from microbial enrichments promoting anammox process	PRJNA209121
27	Foam and foaming AS metagenome	PRJNA194095
28	AS Metagenome targeted locus (Loci)	PRJNA193182
29	Seasonal dynamics of microbial communities in the AS over 4 years	PRJNA192924
30	Stanley aerobic AS sample metagenome	PRJNA189660
31	AS 16S rRNA gene	PRJNA182716
32	Bacterial survey on industrial AS through 454 pyrosequencing	PRJNA182181
33	AS Metagenome	PRJNA178295
34	Sewage sludge metagenome	PRJNA175357
35	Waste AS metagenome	PRJNA174730
36	AS (acetate acclimated)	PRJNA174280
37	AS Targeted locus	PRJNA168603
38	AS Targeted locus	PRJNA86117
39	AS Metagenome	PRJNA242292

8.5 Omics: A Way Forward to Efficient Biological Wastewater Treatment

Taking metagenomics further, comparative metagenomics can be used to understand gene pools from different treatment plants or time series analysis for the same plant can help in designing the bioaugmentation strategy, wherever necessary (Jadega et al., 2014). Wastewater treatment processes have been in practice for the last few decades. To keep up with the increasing levels of pollution, new strategies involving the biological component of the activated biomass need to be incorporated if we aim to improve treatment efficiency and bring about a sustained effort in minimization of pollutants reaching the environment. Rising environmental concerns in the recent times have triggered the design and development of novel means for wastewater treatment. In this regard, understanding each and every aspect of the wastewater treatment process is mandatory. However, in order to have this come to culmination, it is essential that we first understand the individual community members and functional potential residing in them. Once we gain these insights, wastewater treatment processes can be reoriented, aiming at degradation of recalcitrant compounds in the wastewater using microbial community intelligence. The biological composition of the process should be regarded as equivalent parameter as other physical operational parameters of ETPs/CETPs. A combination of omics (genomics, metagenomics, proteomics, transcriptomics) approach will help comprehend the process physiology of AS in combination with other physical parameters being monitored at ETPs/CETPs.

The wastewater treatment environment creates a scenario, which brings the genetically different species under stressed environment. This leads to evolution and genetic breeding, which omics approach can unveil. The metagenomics approach will excavate the newly emerged biochemical pathways that can generate new molecules via transformations or synthesis.

References

Albertsen M, Hansen LBS, Saunders AM, Nielsen PH, and Nielsen KL. 2012. A metagenome of a full-scale microbial community carrying out enhanced biological phosphorus removal. *The ISME Journal* 6(6): 1094–1106.

Almeida B, Vaz-Moreira I, Schumann P, Nunes OC, Carvalho G, and Barreto Crespo MT. 2013. *Patulibacter medicamentivorans* sp. nov., isolated from activated sludge of a wastewater treatment plant. *International Journal of Systematic and Evolutionary Microbiology* 63(Pt 7): 2588–2593.

Ardern E and Lockett WT. 1914. Experiments on the oxidation of sewage without the aid of filters. *Journal of the Society of Chemical Industry* 33(10): 523–539.

Bae W, Baek S, Chung J, and Lee Y. 2001. Optimal operational factors for nitrite accumulation in batch reactors. *Biodegradation* 12(5): 359–366.

Bathe S, Schwarzenbeck N, and Hausner M. 2009. Bioaugmentation of activated sludge towards 3-chloroaniline removal with a mixed bacterial population carrying a degradative plasmid. *Bioresource Technology* 100(12): 2902–2909.

Bentley DR, Balasubramanian S, Swerdlow HP, Smith GP, Milton J, Brown CG, Hall KP et al. 2008. Accurate whole human genome sequencing using reversible terminator chemistry. *Nature* 456(7218): 53–59.

Bernet N, Dangcong P, Delgenès J-P, and Moletta R. 2001. Nitrification at low oxygen concentration in biofilm reactor. *Journal of Environmental Engineering* 127(3): 266–271.

Bertrand H, Poly F, Van TV, Lombard N, Nalin R, Vogel TM, and Simonet P. 2005. High molecular weight DNA recovery from soils prerequisite for biotechnological metagenomic library construction. *Journal of Microbiological Methods* 62(1): 1–11.

Boon N, Goris J, De Vos P, Verstraete W, and Top EM. 2000. Bioaugmentation of activated sludge by an indigenous 3-chloroaniline-degrading *Comamonas testosteroni* strain, I2gfp. *Applied and Environmental Microbiology* 66(7): 2906–2913.

Botrous A, Dahab M, and Mihaltz P. 2004. Nitrification of high-strength ammonium wastewater by a fluidized-bed reactor. *Water Science and Technology* 49(5–6): 65–71.

Bramucci MG, Chen MW, and Nagarajan V. Bacterial plasmid having genes encoding enzymes for the degradation of aromatic compounds. US Patent 6,548,292, issued April 15, 2003.

Bramucci M and Nagarajan V. 2006. Bacterial communities in industrial wastewater bioreactors. *Current Opinion in Microbiology* 9(3): 275–278.

Caporaso JG, Kuczynski J, Stombaugh J, Bittinger K, Bushman FD, Costello EK, Fierer N et al. 2010. QIIME allows analysis of high-throughput community sequencing data. *Nature Methods* 7(5): 335–336.

Chandra R, Naresh Bharagava R, Kapley A, and Purohit HJ. 2009. Isolation and characterization of potential aerobic bacteria capable for pyridine degradation in presence of picoline, phenol and formaldehyde as co-pollutants. *World Journal of Microbiology and Biotechnology* 25(12): 2113–2119.

Chandra R, Naresh Bharagava R, Kapley A, and Purohit HJ. 2011. Bacterial diversity, organic pollutants and their metabolites in two aeration lagoons of common effluent treatment plant (CETP) during the degradation and detoxification of tannery wastewater. *Bioresource Technology* 102(3): 2333–2341.

Chandra R, Naresh Bharagava R, Kapley A, and Purohit HJ. 2012. Characterization of *Phragmites cummunis* rhizosphere bacterial communities and metabolic products during the two stage sequential treatment of post methanated distillery effluent by bacteria and wetland plants. *Bioresource Technology* 103(1): 78–86.

Chandra R, Raj A, Purohit HJ, and Kapley A. 2007. Characterisation and optimisation of three potential aerobic bacterial strains for kraft lignin degradation from pulp paper waste. *Chemosphere* 67(4): 839–846.

Chandra R, Singh S, Reddy MMK, Patel DK, Purohit HJ, and Kapley A. 2008. Isolation and characterization of bacterial strains *Paenibacillus* sp. and *Bacillus* sp. for kraft lignin decolorization from pulp paper mill waste. *The Journal of General and Applied Microbiology* 54(6): 399–407.

Chauhan NS, Ranjan R, Purohit HJ, Kalia VC, and Sharma R. 2009. Identification of genes conferring arsenic resistance to *Escherichia coli* from an effluent treatment plant sludge metagenomic library. *FEMS Microbiology Ecology* 67(1): 130–139.

Chen J, Tang Y-Q, Li Y, Nie Y, Hou L, Li XQ, and Wu XL. 2013. Impacts of different nanoparticles on functional bacterial community in activated sludge. *Chemosphere* 104: 141–148.

Chhatre S, Purohit H, Shanker R, and Khanna P. 1996. Bacterial consortia for crude oil spill remediation. *Water Science and Technology* 34(10): 187–193.

Chistoserdova L, Lapidus A, Han C, Goodwin L, Saunders L, Brettin T, Tapia R et al. 2007. Genome of *Methylobacillus flagellatus*, molecular basis for obligate methylotrophy, and polyphyletic origin of methylotrophy. *Journal of Bacteriology* 189(11): 4020–4027.

Chung W-C, Chen L-L, Lo W-S, Kuo P-A, Tu J, and Kuo C-H. 2013. Complete genome sequence of *Serratia marcescens* WW4. *Genome Announcements* 1(2): e00126-13.

Ciudad G, Rubilar O, Munoz P, Ruiz G, Chamy R, Vergara C, and Jeison D. 2005. Partial nitrification of high ammonia concentration wastewater as a part of a shortcut biological nitrogen removal process. *Process Biochemistry* 40(5): 1715–1719.

Dafale N, Agrawal L, Kapley A, Meshram S, Purohit H, and Wate S. 2010. Selection of indicator bacteria based on screening of 16S rDNA metagenomic library from a two-stage anoxic–oxic bioreactor system degrading azo dyes. *Bioresource Technology* 101(2): 476–484.

Daims H, Taylor MW, and Wagner M. 2006. Wastewater treatment: A model system for microbial ecology. *Trends in Biotechnology* 24(11): 483–489.

Dash HR, Mangwani N, Chakraborty J, Kumari S, and Das S. 2013. Marine bacteria: Potential candidates for enhanced bioremediation. *Applied Microbiology and Biotechnology* 97(2): 561–571.

David WO. 1996. Heavy metal tolerance genes: Prospective tools for bioremediation. *Resources, Conservation and Recycling* 18: 135–149.

Desai C and Madamwar D. 2007. Extraction of inhibitor-free metagenomic DNA from polluted sediments, compatible with molecular diversity analysis using adsorption and ion-exchange treatments. *Bioresource Technology* 98(4): 761–768.

Dick RI and Vesilind PA. 1969. The sludge volume index: What is it? *Water Pollution Control Federation* 41(7): 1285–1291.

Domde P, Kapley A, and Purohit HJ. 2007. Impact of bioaugmentation with a consortium of bacteria on the remediation of wastewater-containing hydrocarbons. *Environmental Science and Pollution Research International* 14(1): 7–11.

Eckenfelder WW, Patoczka JB, and Pulliam GW. 1988. Anaerobic versus aerobic treatment in the USA. In: Wang LK, Tay J, Tay STL, and Hung Y. (Eds.), *Anaerobic Digestion 1988, Proceedings of the 5th International Symposium on Anaerobic Digestion*, Pergamon Press, Oxford, UK, pp. 105–114.

El Fantroussi S and Agathos SN. 2005. Is bioaugmentation a feasible strategy for pollutant removal and site remediation? *Current Opinion in Microbiology* 8(3): 268–275.

Faldu PR, Kothari VV, Kothari CR, Rawal CM, Domadia KK, Patel PA, Bhimani HD et al. 2014. Draft genome sequence of textile azo dye-decolorizing and -degrading *Pseudomonas aeruginosa* strain PFK10, isolated from the common effluent treatment plant of the Ankleshwar industrial area of Gujarat, India. *Genome Announcements* 2(1): e00019-14.

Fang H, Cai L, Yu YL, and Zhang T. 2013. Metagenomic analysis reveals the prevalence of biodegradation genes for organic pollutants in activated sludge. *Bioresource Technology* 129: 209–218.

Fuentes S, Méndez V, Aguila P, and Seeger M. 2014. Bioremediation of petroleum hydrocarbons: Catabolic genes, microbial communities, and applications. *Applied Microbiology and Biotechnology* 98(11): 4781–4794.

Fukushima T, Whang L-M, Lee Y-C, Putri DW, Chen P-C, and Wu Y-J. 2014. Transcriptional responses of bacterial amoA gene to dimethyl sulfide inhibition in complex microbial communities. *Bioresource Technology* 165: 137–144.

Geiselbrecht AD, Herwig RP, Deming JW, and Staley JT. 1996. Enumeration and phylogenetic analysis of polycyclic aromatic hydrocarbon-degrading marine bacteria from Puget sound sediments. *Applied and Environmental Microbiology* 62(9): 3344–3349.

Ghanizadeh GH and Sarrafpour R. 2001. The effects of temperature and pH on settlability of activated sludge flocs. *Iranian Journal of Public Health* 30(3–4): 139–142.

Guermazi S, Daegelen P, Dauga C, Rivière D, Bouchez T, Godon JJ, Gyapay G et al. 2008. Discovery and characterization of a new bacterial candidate division by an anaerobic sludge digester metagenomic approach. *Environmental Microbiology* 10(8): 2111–2123.

Hagedorn-Olsen C, Mller I, Tttrup H, and Harremoës P. 1994. Oxygen reduces denitrification in biofilm reactors. *Water Science and Technology* 29(10–11): 83–91.

Handelsman J, Rondon MR, Brady SF, Clardy J, and Goodman RM. 1998. Molecular biological access to the chemistry of unknown soil microbes: A new frontier for natural products. *Chemistry and Biology* 5(10): R245–R249.

He M, Li X, Guo L, Miller SJ, Rensing C, and Wang G. 2010. Characterization and genomic analysis of chromate resistant and reducing *Bacillus cereus* strain SJ1. *BMC Microbiology* 10(1): 221.

He M, Li X, Hongliang Liu, Miller SJ, Wang G, and Rensing C. 2011. Characterization and genomic analysis of a highly chromate resistant and reducing bacterial strain *Lysinibacillus fusiformis* ZC1. *Journal of Hazardous Materials* 185(2): 682–688.

Hu A, Lv M, and Yu C-P. 2013. Draft genome sequence of the bisphenol A-degrading bacterium *Sphingobium* sp. strain YL23. *Genome Announcements* 1(4): e00549-13.

Huson DH, Auch AF, Qi J, and Schuster SC. 2007. MEGAN analysis of metagenomic data. *Genome Research* 17(3): 377–386.

Huson DH and Mitra S. 2012. Introduction to the analysis of environmental sequences: Metagenomics with MEGAN. In: Anisimova M. (Ed.), *Evolutionary Genomics*. Humana Press, New York, pp. 415–429.

Jadeja NB, More RP, Purohit HJ, and Kapley A. 2014. Metagenomic analysis of oxygenases from activated sludge. *Bioresource Technology* 165: 250–256.

Jang HM, Cho HU, Park SK, Ha JH, and Park JM. 2014. Influence of thermophilic aerobic digestion as a sludge pre-treatment and solids retention time of mesophilic anaerobic digestion on the methane production, sludge digestion and microbial communities in a sequential digestion process. *Water Research* 48: 1–14.

Johnning A, Moore ERB, Svensson-Stadler L, Shouche YS, Larsson DGJ, and Kristiansson E. 2013. Acquired genetic mechanisms of a multiresistant bacterium isolated from a treatment plant receiving wastewater from antibiotic production. *Applied and Environmental Microbiology* 79(23): 7256–7263.

Junier P, Kim O-S, Molina V, Limburg P, Junier T, Imhoff JF, and Witzel K-P. 2008. Comparative in silico analysis of PCR primers suited for diagnostics and cloning of ammonia monooxygenase genes from ammonia-oxidizing bacteria. *FEMS Microbiology Ecology* 64(1): 141–152.

Kapley A, De Baere T, and Purohit HJ. 2007a. Eubacterial diversity of activated biomass from a common effluent treatment plant. *Research in Microbiology* 158(6): 494–500.

Kapley A, Prasad S, and Purohit HJ. 2007b. Changes in microbial diversity in fed-batch reactor operation with wastewater containing nitroaromatic residues. *Bioresource Technology* 98(13): 2479–2484.

Kapley A and Purohit HJ. 2009. Diagnosis of treatment efficiency in industrial wastewater treatment plants: A case study at a refinery ETP. *Environmental Science and Technology* 43(10): 3789–3795.

Kapley A, Purohit HJ, Chhatre S, Shanker R, Chakrabarti T, and Khanna P. 1999. Osmotolerance and hydrocarbon degradation by a genetically engineered microbial consortium. *Bioresource Technology* 67(3): 241–245.

Kapley A, Sagarkar S, Tanksale H, Sharma N, Qureshi A, Khardenavis A, and Purohit HJ. 2013. Genome sequence of *Alcaligenes* sp. strain HPC1271. *Genome Announcements* 1(1): e00235-12.

Kapley A, Tolmare A, and Purohit HJ. 2001. Role of oxygen in the utilization of phenol by *Pseudomonas* CF600 in continuous culture. *World Journal of Microbiology and Biotechnology* 17(8): 801–804.

Kayashima T, Suzuki H, Maeda T, and Ogawa HI. 2013. Real-time PCR for rapidly detecting aniline-degrading bacteria in activated sludge. *Chemosphere* 91(9): 1338–1343.

Khan AG, Kuek C, Chaudhry TM, Khoo CS, and Hayes WJ. 2000. Role of plants, mycorrhizae and phytochelators in heavy metal contaminated land remediation. *Chemosphere* 41(1): 197–207.

Khardenavis AA, Kapley A, and Purohit HJ. 2007. Simultaneous nitrification and denitrification by diverse *Diaphorobacter* sp. *Applied Microbiology and Biotechnology* 77(2): 403–409.

Khardenavis AA, Kapley A, and Purohit HJ. 2008. Phenol-mediated improved performance of active biomass for treatment of distillery wastewater. *International Biodeterioration and Biodegradation* 62(1): 38–45.

Khardenavis AA, Kapley A, and Purohit HJ. 2010. Salicylic-acid-mediated enhanced biological treatment of wastewater. *Applied Biochemistry and Biotechnology* 160(3): 704–718.

Kimura N, Sakai K, and Nakamura K. 2009. Isolation and characterization of a 4-nitrotoluene-oxidizing enzyme from activated sludge by a metagenomic approach. *Microbes and Environments/ JSME* 25(2): 133–139.

Kircher M and Kelso J. 2010. High-throughput DNA sequencing—Concepts and limitations. *Bioessays* 32(6): 524–536.

Kong C, Wang L, Li P, Qu Y, Tang H, Wang J, Zhou H et al. 2013. Genome sequence of *Dyella ginsengisoli* strain LA-4, an efficient degrader of aromatic compounds. *Genome Announcements* 1(6): e00961-13.

Król JE, Penrod JT, McCaslin H, Rogers LM, Yano H, Stancik AD, Dejonghe W et al. 2012. Role of IncP-1β plasmids pWDL7::rfp and pNB8c in chloroaniline catabolism as determined by genomic and functional analyses. *Applied and Environmental Microbiology* 78(3): 828–838.

Li K, Wang S, Shi Y, Qu J, Zhai Y, Xu L, Xu Y et al. 2011. Genome sequence of *Paracoccus* sp. strain TRP, a chlorpyrifos biodegrader. *Journal of Bacteriology* 193(7): 1786–1787.

Li Q, Wang M, Feng J, Zhang W, Wang Y, Gu Y, Song C et al. 2013. Treatment of high-salinity chemical wastewater by indigenous bacteria-bioaugmented contact oxidation. *Bioresource Technology* 144: 380–386.

Lie E and Welander T. 1994. Influence of dissolved oxygen and oxidation-reduction potential on the denitrification rate of activated sludge. *Water Science and Technology* 30(6): 91–100.

Liu G and Wang J. 2013. Long-term low DO enriches and shifts nitrifier community in activated sludge. *Environmental Science and Technology* 47(10): 5109–5117.

Liu M, Zhang Y, Ding R, Gao Y, and Yang M. 2013. Response of activated sludge to the treatment of oxytetracycline production waste stream. *Applied Microbiology and Biotechnology* 97(19): 8805–8812.

Liu M, Zhang Y, Yang M, Tian Z, Ren L, and Zhang S. 2012. Abundance and distribution of tetracycline resistance genes and mobile elements in an oxytetracycline production wastewater treatment system. *Environmental Science and Technology* 46(14): 7551–7557.

Liu X-C, Zhang Y, Yang M, Wang Z-Y, and Lv W-Z. 2007. Analysis of bacterial community structures in two sewage treatment plants with different sludge properties and treatment performance by nested PCR-DGGE method. *Journal of Environmental Sciences* 19(1): 60–66.

Ma Y-F, Zhang Y, Zhang J-Y, Chen D-W, Zhu Y, Zheng H, Wang S-Y et al. 2009. The complete genome of *Comamonas testosteroni* reveals its genetic adaptations to changing environments. *Applied and Environmental Microbiology* 75(21): 6812–6819.

Margesin R and Schinner F. 2001. Biodegradation and bioremediation of hydrocarbons in extreme environments. *Applied Microbiology and Biotechnology* 56(5–6): 650–663.

Margulies M, Egholm M, Altman WE, Attiya S, Bader JS, Bemben LA, Berka J et al. 2005. Genome sequencing in microfabricated high-density picolitre reactors. *Nature* 437(7057): 376–380.

Markowitz VM, Chen IM, Chu K, Szeto E, Palaniappan K, Grechkin Y, Ratner A et al. 2012. IMG/M: The integrated metagenome data management and comparative analysis system. *Nucleic Acids Research* 40(D1): D123–D129.

McLaughlin H, Farrell A, and Quilty B. 2006. Bioaugmentation of activated sludge with two *Pseudomonas putida* strains for the degradation of 4-chlorophenol. *Journal of Environmental Science and Health part* A 41(5): 763–777.

Meyer F, Paarmann D, D'Souza M, Olson R, Glass EM, Kubal M, Paczian T et al. 2008. The metagenomics RAST server—A public resource for the automatic phylogenetic and functional analysis of metagenomes. *BMC Bioinformatics* 9(1): 386.

Miller DA, Suen G, Bruce D, Copeland A, Cheng JF, Detter C, Goodwin LA et al. 2011. Complete genome sequence of the cellulose-degrading bacterium *Cellulosilyticum lentocellum*. *Journal of Bacteriology* 193(9): 2357–2358.

Moharikar A, Kapley A, and Purohit HJ. 2003. Detection of dioxygenase genes present in various activated sludge. *Environmental Science and Pollution Research* 10(6): 373–378.

Moharikar A and Purohit HJ. 2003. Specific ratio and survival of *Pseudomonas* CF600 as co-culture for phenol degradation in continuous cultivation. *International Biodeterioration and Biodegradation* 52(4): 255–260.

Moharikar A, Purohit HJ, and Kumar R. 2005. Microbial population dynamics at effluent treatment plants. *Journal of Environmental Monitoring* 7(6): 552–558.

More RP, Mitra S, Raju SC, Kapley A, and Purohit HJ. 2014. Mining and assessment of catabolic pathways in the metagenome of a common effluent treatment plant to induce the degradative capacity of biomass. *Bioresource Technology* 153: 137–146.

Narde GK, Kapley A, and Purohit HJ. 2004. Isolation and characterization of *Citrobacter* strain HPC255 for broad-range substrate specificity for chlorophenols. *Current Microbiology* 48(6): 419–423.

Narmadha D and Mary Selvam Kavitha VJ. 2012. Treatment of domestic waste water using natural flocculants. *International Journal of LifeSciences Biotechnology and Pharma Research* 1(3): 206–213.

Nasuno E, Kimura N, Fujita MJ, Nakatsu CH, Kamagata Y, and Hanada S. 2012. Phylogenetically novel LuxI/LuxR-type quorum sensing systems isolated using a metagenomic approach. *Applied and Environmental Microbiology* 78(22): 8067–8074.

Neufeld J, Engel K, Jiujun Cheng, Moreno-Hagelsieb G, Rose D, and Charles T. 2011. Open resource metagenomics: A model for sharing metagenomic libraries. *Standards in Genomic Sciences* 5(2): 203.

Paliwal V, Puranik S, and Purohit HJ. 2012. Integrated perspective for effective bioremediation. *Applied Biochemistry and Biotechnology* 166(4): 903–924.

Paliwal V, Raju SC, Modak A, Phale PS, and Purohit HJ. 2014. *Pseudomonas putida* CSV86: A candidate genome for genetic bioaugmentation. *PLoS One* 9(1): e84000.

Park H-D and Noguera DR. 2004. Evaluating the effect of dissolved oxygen on ammonia-oxidizing bacterial communities in activated sludge. *Water Research* 38(14): 3275–3286.

Pellicer-Nàcher C, Franck S, Gülay A, Ruscalleda M, Terada A, Abu Al-Soud W, Hansen MA et al. 2014. Sequentially aerated membrane biofilm reactors for autotrophic nitrogen removal: Microbial community composition and dynamics. *Microbial Biotechnology* 7(1): 32–43.

Pester M, Maixner F, Berry D, Rattei T, Koch H, Lücker H, Nowka B et al. 2013. NxrB encoding the beta subunit of nitrite oxidoreductase as functional and phylogenetic marker for nitrite-oxidizing Nitrospira. *Environmental Microbiology* 16: 3055–3071.

Phale PS, Paliwal V, Raju SC, Modak A, and Purohit HJ. 2013. Genome sequence of naphthalene-degrading soil bacterium *Pseudomonas putida* CSV86. *Genome Announcements* 1(1): e00234-12.

Pham VD, Palden T, and DeLong EF. 2007. Large-scale screens of metagenomic libraries. *Journal of Visualised Experiments* e(4): 201.

Pruden A, Pei R, Storteboom H, and Carlson KH. 2006. Antibiotic resistance genes as emerging contaminants: Studies in northern Colorado. *Environmental Science & Technology* 40(23): 7445–7450.

Purohit HJ. 2003. Biosensors as molecular tools for use in bioremediation. *Journal of Cleaner Production* 11(3): 293–301.

Purohit HJ, Kapley A, Moharikar AA, and Narde G. 2003. A novel approach for extraction of PCR-compatible DNA from activated sludge samples collected from different biological effluent treatment plants. *Journal of Microbiological Methods* 52(3): 315–323.

Qureshi AA and Purohit HJ. 2002. Isolation of bacterial consortia for degradation of p-nitrophenol from agricultural soil. *Annals of Applied Biology* 140(2): 159–162.

Raj A, Chandra R, Reddy MMK, Purohit HJ, and Kapley A. 2007a. Biodegradation of kraft lignin by a newly isolated bacterial strain, *Aneurinibacillus aneurinilyticus* from the sludge of a pulp paper mill. *World Journal of Microbiology and Biotechnology* 23(6): 793–799.

Raj A, Reddy MK, Chandra R, Purohit HJ, and Kapley A. 2007b. Biodegradation of kraft-lignin by *Bacillus* sp. isolated from sludge of pulp and paper mill. *Biodegradation* 18(6): 783–792.

Rani A, Porwal S, Sharma R, Kapley A, Purohit HJ, and Kalia VC. 2008. Assessment of microbial diversity in effluent treatment plants by culture dependent and culture independent approaches. *Bioresource Technology* 99(15): 7098–7107.

Roest K, Heilig HGHJ, Smidt H, de Vos WM, Stams AJM, and Akkermans ADL. 2005. Community analysis of a full-scale anaerobic bioreactor treating paper mill wastewater. *Systematic and Applied Microbiology* 28(2): 175–185.

Ruiz G, Jeison D, and Chamy R. 2003. Nitrification with high nitrite accumulation for the treatment of wastewater with high ammonia concentration. *Water Research* 37(6): 1371–1377.

Sagarkar S, Bhardwaj P, Yadav TC, Qureshi A, Khardenavis A, Purohit HJ, and Kapley A. 2014. Draft genome sequence of atrazine-utilizing bacteria isolated from Indian agricultural soil. *Genome Announcements* 2(1): e01149-13.

Schleheck D, Knepper TP, Fischer K, and Cook AM. 2004. Mineralization of individual congeners of linear alkylbenzenesulfonate by defined pairs of heterotrophic bacteria. *Applied and Environmental Microbiology* 70(7): 4053–4063.

Schloss PD, Westcott SL, Ryabin T, Hall JR, Hartmann M, Hollister EB, Lesniewski RA et al. 2009. Introducing mothur: Open-source, platform-independent, community-supported software for describing and comparing microbial communities. *Applied and Environmental Microbiology* 75(23): 7537–7541.

Schwarz S and Chaslus-Dancla E. 2001. Use of antimicrobials in veterinary medicine and mechanisms of resistance. *Veterinary Research* 32(3–4): 201–225.

Selvakumaran S, Kapley A, Kalia VC, and Purohit HJ. 2008. Phenotypic and phylogenic groups to evaluate the diversity of *Citrobacter* isolates from activated biomass of effluent treatment plants. *Bioresource Technology* 99(5): 1189–1195.

Selvakumaran S, Kapley A, Kashyap SM, Daginawala HF, Kalia VC, and Purohit HJ. 2011. Diversity of aromatic ring-hydroxylating dioxygenase gene in Citrobacter. *Bioresource Technology* 102(7): 4600–4609.

Shah A, Kato AS, Shintani N, Kamini NR, and Nakajima-Kambe T. 2014. Microbial degradation of aliphatic and aliphatic-aromatic co-polyesters. *Applied Microbiology and Biotechnology* 98(8): 3437–3447.

Shah MP. 2014. Microbiological removal of phenol by an application of *Pseudomonas* spp. ETL: An innovative biotechnological approach providing answers to the problems of FETP. *Journal of Applied and Environmental Microbiology* 2(1): 6–11.

Sharma N, Tanksale H, Kapley A, and Purohit HJ. 2012. Mining the metagenome of activated biomass of an industrial wastewater treatment plant by a novel method. *Indian Journal of Microbiology* 52(4): 538–543.

Sharma PK, Fu J, Zhang X, Fristensky BW, Davenport K, Chain PSA, Sparling R et al. 2013. Draft genome sequence of medium-chain-length polyhydroxyalkanoate-producing *Pseudomonas putida* strain LS46. *Genome Announcements* 1(2): e00151-13.

Sheik AR, Muller EE, and Wilmes P. 2014. A hundred years of activated sludge: Time for a rethink. *Frontiers in Microbiology* (5): 47.

Shen L-D, Hu A-H, Jin RC, Cheng D-Q, Zheng P, Xu X-Y, Hu B-L et al. 2012. Enrichment of anammox bacteria from three sludge sources for the startup of monosodium glutamate industrial wastewater treatment system. *Journal of Hazardous Materials* 199: 193–199.

Siezen RJ and Galardini M. 2008. Genomics of biological wastewater treatment. *Microbial Biotechnology* 1(5): 333–340.

Silby MW, Winstanley C, Godfrey SAC, Levy SB, and Jackson RW. 2011. *Pseudomonas genomes*: Diverse and adaptable. *FEMS Microbiology Reviews* 35(4): 652–680.

Silva CC, Hayden H, Sawbridge T, Mele P, De Paula SO, Silva LC, Vidigal PM et al. 2013. Identification of genes and pathways related to phenol degradation in metagenomic libraries from petroleum refinery wastewater. *PLoS One* 8(4): e61811.

Strabala TJ, Macdonald L, Liu V, and Smit A-M. 2012. Draft genome sequence of *Novosphingobium nitrogenifigens* Y88T. *Journal of Bacteriology* 194(1): 201.

Suenaga H, Koyama Y, Miyakoshi M, Miyazaki R, Yano H, Sota M, Ohtsubo Y et al. 2009. Novel organization of aromatic degradation pathway genes in a microbial community as revealed by metagenomic analysis. *The ISME Journal* 3(12): 1335–1348.

Suenaga H, Tsutomu Ohnuki, and Kentaro Miyazaki. 2007. Functional screening of a metagenomic library for genes involved in microbial degradation of aromatic compounds. *Environmental Microbiology* 9(9): 2289–2297.

Szczepanowski R, Eikmeyer F, Harfmann J, Blom J, Rogers LM, Top EM, and Schlüter A. 2011. Sequencing and comparative analysis of IncP-1α antibiotic resistance plasmids reveal a highly conserved backbone and differences within accessory regions. *Journal of Biotechnology* 155(1): 95–103.

Thangaraj K, Kapley A, and Purohit HJ. 2008. Characterization of diverse *Acinetobacter* isolates for utilization of multiple aromatic compounds. *Bioresource Technology* 99(7): 2488–2494.

Tice H, Mayilraj S, Sims D, Lapidus A, Nolan M, Lucas S, Glavina Del Rio T et al. 2010. Complete genome sequence of *Nakamurella multipartita* type strain (Y-104T). *Standards in Genomic Sciences* 2(2): 168.

Tsutsui H, Anami Y, Matsuda M, Hashimoto K, Inoue D, Sei K, Soda S et al. 2013. Plasmid-mediated bioaugmentation of sequencing batch reactors for enhancement of 2, 4-dichlorophenoxyacetic acid removal in wastewater using plasmid pJP4. *Biodegradation* 24(3): 343–352.

Tsutsui H, Anami Y, Matsuda M, Inoue D, Sei K, Soda S, Ike M. 2010. Transfer of plasmid pJP4 from *Escherichia coli* and *Pseudomonas putida* to bacteria in activated sludge developed under different sludge retention times. *Journal of Bioscience and Bioengineering* 110(6): 684–689.

Uchiyama T, Abe T, Ikemura T, and Watanabe K. 2005. Substrate-induced gene-expression screening of environmental metagenome libraries for isolation of catabolic genes. *Nature Biotechnology* 23(1): 88–93.

Uchiyama T and Miyazaki K. 2010. Product-induced gene expression, a product-responsive reporter assay used to screen metagenomic libraries for enzyme-encoding genes. *Applied and Environmental Microbiology* 76(21): 7029–7035.

Uchiyama T and Watanabe K. 2008. Substrate-induced gene expression (SIGEX) screening of metagenome libraries. *Nature Protocols* 3(7): 1202–1212.

USEPA, OW. 1990. Biological Criteria, National Program Guidance for Surface Waters, Washington, DC: USEPA Office of Water, Regulations and Standards. EPA-448/5–90-00.

Valentín-Vargas A, Toro-Labrador G, and Massol-Deya AA. 2012. Bacterial community dynamics in full-scale activated sludge bioreactors: Operational and ecological factors driving community assembly and performance. *PLoS One* 7(8): e42524.

Vanysacker L, Denis C, Roels J, Verhaeghe K, and Vankelecom IFJ. 2014. Development and evaluation of a TaqMan duplex real-time PCR quantification method for reliable enumeration of *Candidatus microthrix*. *Journal of Microbiological Methods* 97: 6–14.

Verma V, Raju SC, Kapley A, Kalia VC, Daginawala HF, and Purohit HJ. 2010. Evaluation of genetic and functional diversity of *Stenotrophomonas* isolates from diverse effluent treatment plants. *Bioresource Technology* 101(20): 7744–7753.

Vuilleumier S, Nadalig T, Ul Haque MF, Magdelenat G, Lajus A, Roselli S, Muller EE et al. 2011. Complete genome sequence of the chloromethane-degrading *Hyphomicrobium* sp. strain MC1. *Journal of Bacteriology* 193(18): 5035–5036.

Wang F, Liu Y, Wang J, Zhang Y, and Yang H. 2012. Influence of growth manner on nitrifying bacterial communities and nitrification kinetics in three lab-scale bioreactors. *Journal of Industrial Microbiology and Biotechnology* 39(4): 595–604.

Wang M, Yang G, Min H, Lv Z, and Jia X. 2009. Bioaugmentation with the nicotine-degrading bacterium *Pseudomonas* sp. HF-1 in a sequencing batch reactor treating tobacco wastewater: Degradation study and analysis of its mechanisms. *Water Research* 43(17): 4187–4196.

Wang Z, Zhang X-X, Huang K, Miao Y, Shi P, Liu B, Long C et al. 2013. Metagenomic profiling of antibiotic resistance genes and mobile genetic elements in a tannery wastewater treatment plant. *PLoS One* 8(10): e76079.

Wiesmann U. 1994. Biological nitrogen removal from wastewater. In: *Biotechnics/Wastewater*. Springer, Berlin, pp. 113–154.

Xu M, Fang Y, Liu J, Chen X, Sun G, Guo J, Hua Z et al. 2013. Draft genome sequence of *Shewanella decolorationis* S12, a dye-degrading bacterium isolated from a wastewater treatment plant. *Genome Announcements* 1(6): e00993-13.

Yadav TC, Khardenavis AA, and Kapley A. 2014. Shifts in microbial community in response to dissolved oxygen levels in activated sludge. *Bioresource Technology* 165: 257–264.

Yakimov MM, Timmis KN, and Golyshin PN. 2007. Obligate oil-degrading marine bacteria. *Current Opinion in Biotechnology* 18(3): 257–266.

Yang Ying, Tong Zhang, Xu-Xiang Zhang et al. 2012. Quantification and characterization of β-lactam resistance genes in 15 sewage treatment plants from East Asia and North America. *Applied Microbiology and Biotechnology* 95(5): 1351–1358.

Ye Lin, Tong Zhang, Taitao Wang, and Zhiwei Fang. 2012. Microbial structures, functions, and metabolic pathways in wastewater treatment bioreactors revealed using high-throughput sequencing. *Environmental Science and Technology* 46(24): 13244–13252.

Yergeau E, Sanschagrin S, Beaumier D, and Greer CW. 2012. Metagenomic analysis of the bioremediation of diesel-contaminated Canadian high Arctic soils. *PLoS One* 7(1): e30058.

Yu F-B, Ali SW, Guan L-B, Li S-P, and Zhou S. 2010. Bioaugmentation of a sequencing batch reactor with *Pseudomonas putida* ONBA-17, and its impact on reactor bacterial communities. *Journal of Hazardous Materials* 176(1): 20–26.

Zhan Y, Yan Y, Zhang W, Yu H, Chen M, Lu W, Ping S et al. 2011. Genome sequence of *Acinetobacter calcoaceticus* PHEA-2, isolated from industry wastewater. *Journal of Bacteriology* 193(10): 2672–2673.

Zhang X, Li D, Liang Y, Zeng H, He Y, Zhang Y, Zhang J. 2014. Performance and microbial community of completely autotrophic nitrogen removal over nitrite (CANON) process in two membrane bioreactors (MBR) fed with different substrate levels. *Bioresource Technology* 152: 185–191.

Zheng Y, Chai L-Y, Yang Z-H, Tang C-J, Chen Y-H, and Shi Y. 2013. Enhanced remediation of black liquor by activated sludge bioaugmented with a novel exogenous microorganism culture. *Applied Microbiology and Biotechnology* 97(14): 6525–6535.

Zumft WG. 1997. *Cell biology and molecular basis of denitrification*. Microbiology and Molecular Biology Reviews 61(4): 533–616.

Web References

http://marketing.appliedbiosystems.com/images/Product/SolidKnowledge/flash/102207/solid.
   html (accessed, March 2014).
www.helicosbio.com (accessed, March 2014).
www.pacificbiosciences.com (accessed, March 2014).

FIGURE 1.1
View of byproducts of the sugar industries, pressmud (a) molasses (b), and bagasses (c and d).

Effluent   Pressmud

FIGURE 1.2
Mixing of effluent and pressmud with the help of an aerotiller.

FIGURE 1.5
View of vermicomposting.

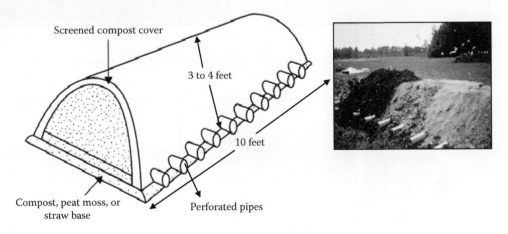

FIGURE 1.9
Passively aerated windrows.

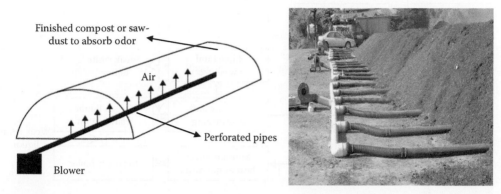

FIGURE 1.10
Aerated static pile.

FIGURE 1.16
Some fungi-like *Trichoderma* (yellow color) and *Aspergillus* (white color) growing on compost of distillery effluent and pressmud.

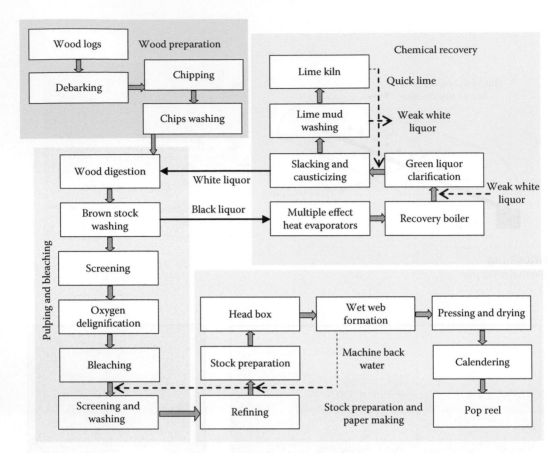

FIGURE 2.1
Process flow diagram for integrated pulp and paper manufacturing.

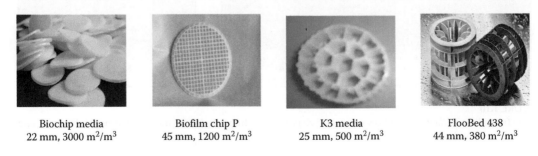

Biochip media	Biofilm chip P	K3 media	FlooBed 438
22 mm, 3000 m^2/m^3	45 mm, 1200 m^2/m^3	25 mm, 500 m^2/m^3	44 mm, 380 m^2/m^3

FIGURE 2.6
Commonly used carriers (with dia and effective surface areas) in MBBR.

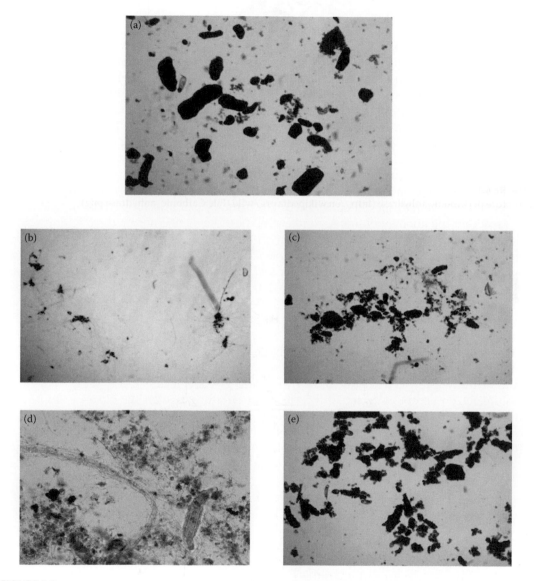

FIGURE 2.7
Microphotograph of biosludge in different bioreactors. (a) Conventional activated-sludge process (ASP); (b) low-sludge bioprocess (LSB); (c) LSB followed by ASP; (d) moving bed biofilm reactor (MBBR); (e) MBBR followed by ASP.

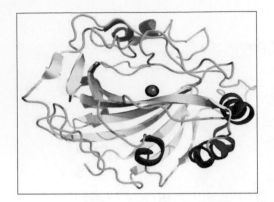

FIGURE 6.1
Structure of carbonic anhydrase (http://en.wikipedia.org/wiki/File:Carbonic_anhydrase.png).

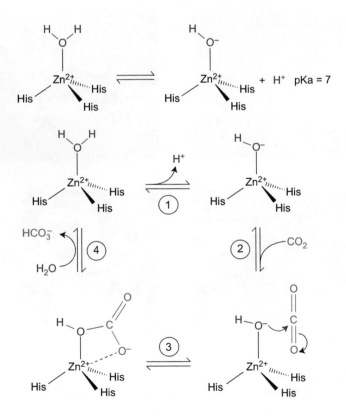

FIGURE 6.2
CO_2 hydration at the active site of CA. (Adapted from Berg JM, Tymoczko JL, and Stryer L. 2007. *Biochemistry*, 6th ed. Sara Tenney, New York.)

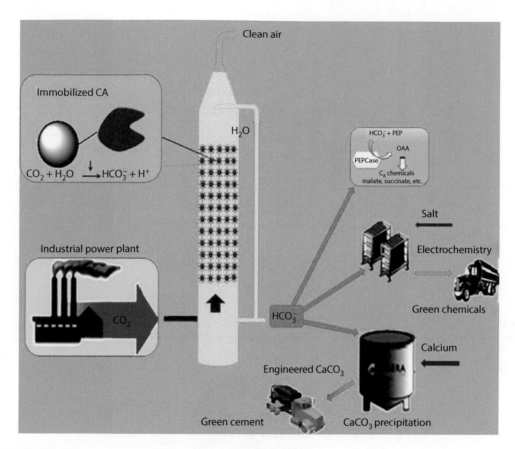

FIGURE 6.3
Schematic representation of CO_2 sequestration processes using CA.

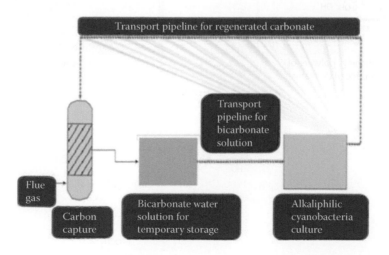

FIGURE 6.11
Closed-loop bicarbonate/carbonate recirculation of bicarbonate produced from carbon capture by algal culture. (Adapted from Chi Z, Fallon JVO, and Shulin Chen. 2011. *Trends in Biotechnology* 29: 537–541.)

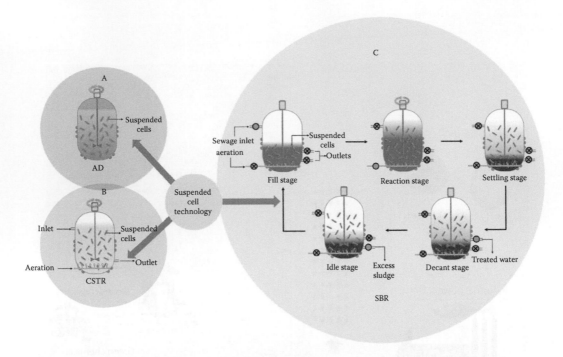

FIGURE 7.2
Schematic of different types of bioreactors utilizing free suspended microbial cells for waste treatment.

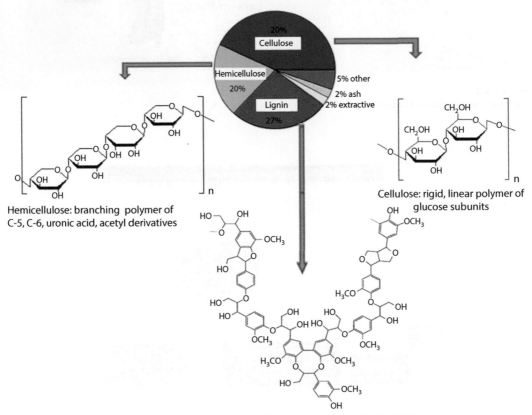

FIGURE 10.1
Lignocellulosic component of terrestrial plant cell wall.

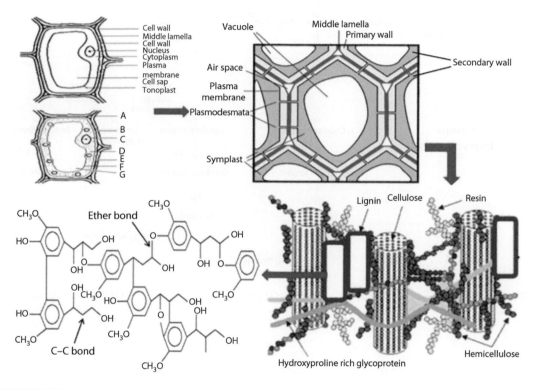

FIGURE 10.2
General structure of lignin with phenylpropanoid structure and methoxy substitute.

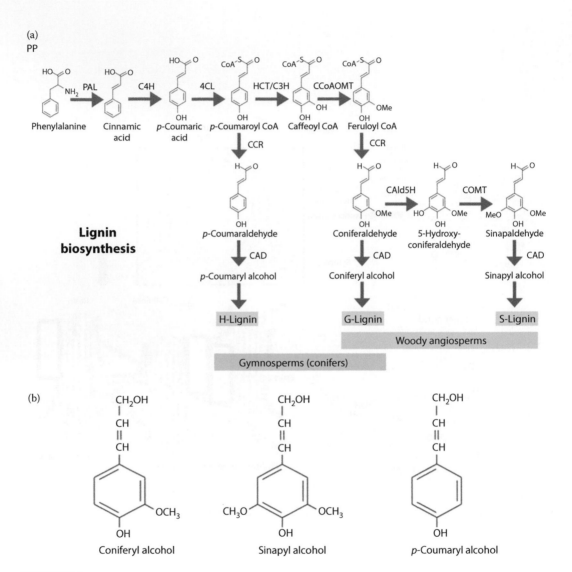

FIGURE 10.3
(a) Biosynthesis of lignin and (b) its precursor.

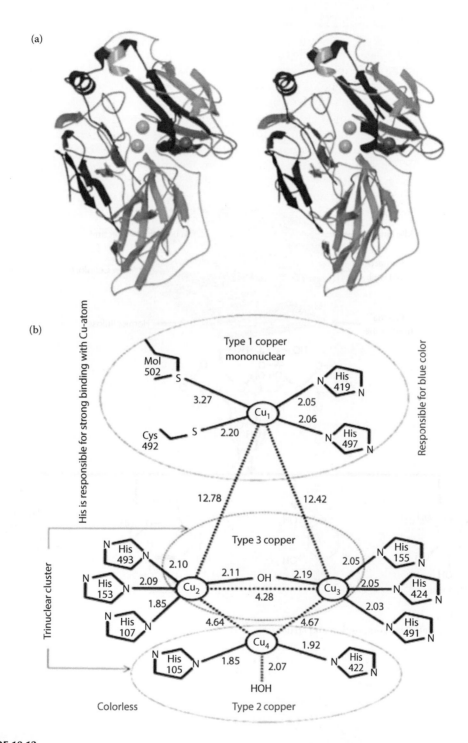

FIGURE 10.13

Cartoon representation of the three-dimensional structure of the *C. cinereuslaccase*. The figure is color-ramped from the N-terminus (blue) to the C-terminus (red). The Cu-atoms are shown as shaded spheres, with the T1 site in blue and the T3 pair in yellow (a). The laccase active site showing the relative orientation of the copper atoms including interatomic distances among all relevant ligands (b).

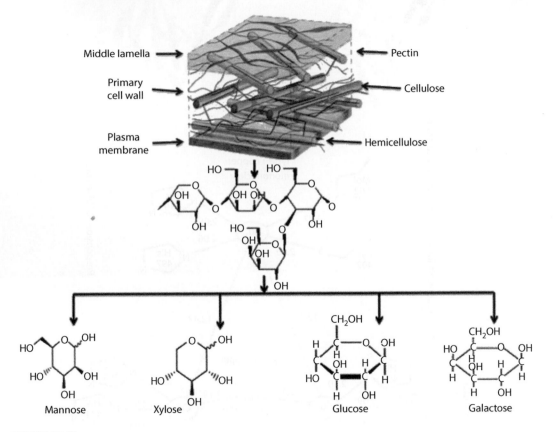

FIGURE 10.15
Showing hemicellulose composition and structure.

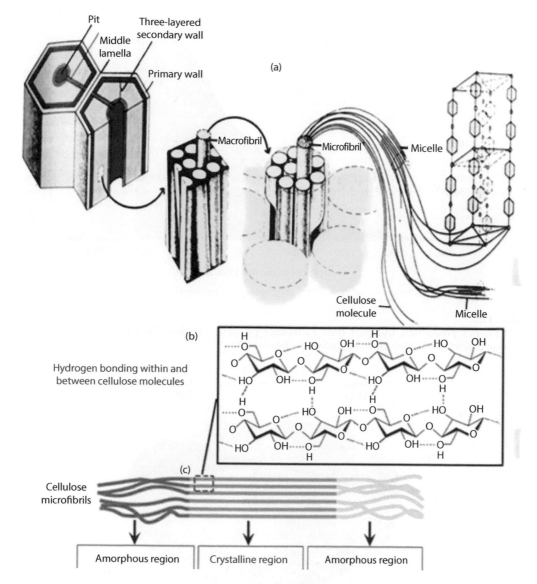

FIGURE 10.21
(a) Structural composition of cellulose and arrangement of macro and microfibrils. (b) Chemical structure and bonding arrangement of cellulose molecule. (c) Showing cellulose microfibirils and amorphous and crystalline regions.

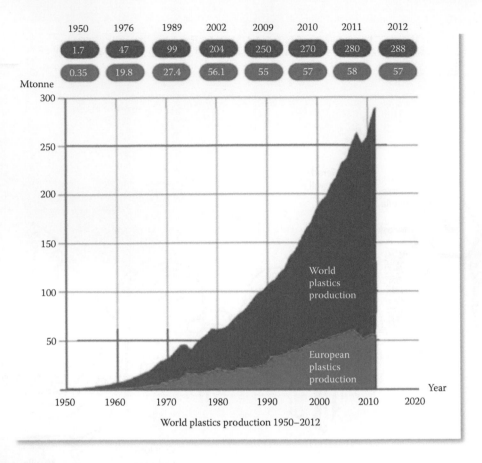

1950	1976	1989	2002	2009	2010	2011	2012
1.7	47	99	204	250	270	280	288
0.35	19.8	27.4	56.1	55	57	58	57

World plastics production 1950–2012

FIGURE 12.1
World plastics production. (Adapted from PlasticsEurope, 2012. Plastics—The Facts 2012. An analysis of European plastics production, demand, and waste data for 2011.)

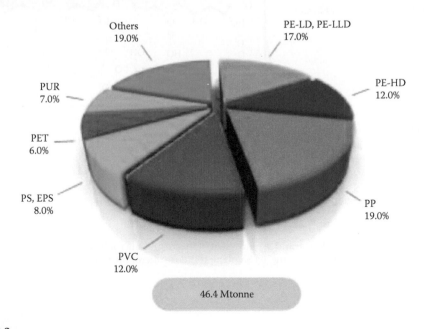

FIGURE 12.2
Plastics demand by resin type. (Adapted from PlasticsEurope, 2010. Plastics—The Facts 2010. An analysis of European plastics production, demand, and recovery for 2009. http://www.plasticseurope.org/documents/document/20101006091310-final_plasticsthefacts_28092010_lr.pdf.)

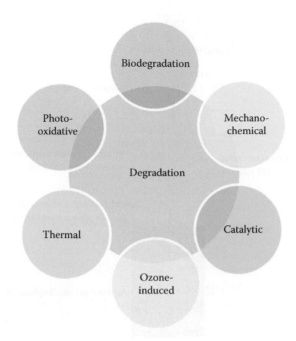

FIGURE 12.3
Types of plastic degradation.

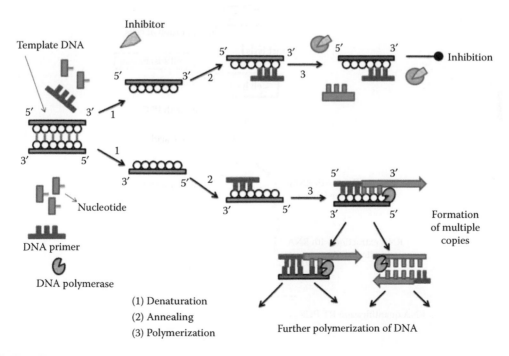

(1) Denaturation
(2) Annealing
(3) Polymerization

Further polymerization of DNA

FIGURE 15.3
Mechanism of inhibition in PCR.

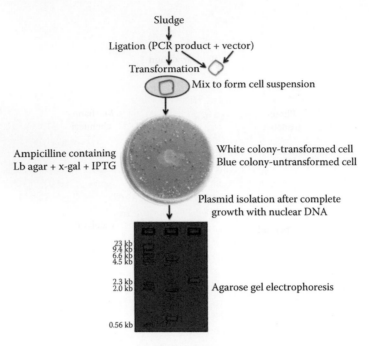

FIGURE 15.6
Schematic diagram of RFLP.

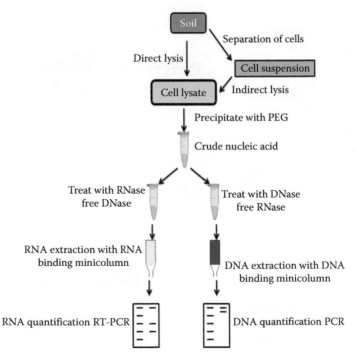

FIGURE 15.7
Isolation and purification of nucleic acid (DNA and RNA) from soil bacteria.

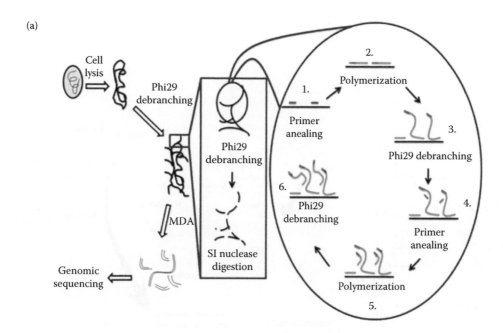

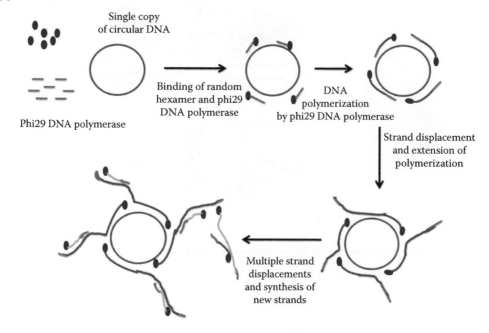

FIGURE 15.8
(a) MDA of a linear DNA fragment, (b) MDA of a circular single-stranded DNA.

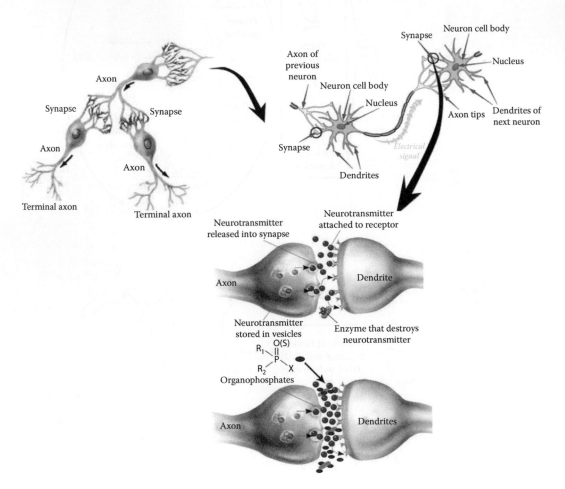

FIGURE 16.12
Mode of action of organophosphates on the mammalian nervous system.

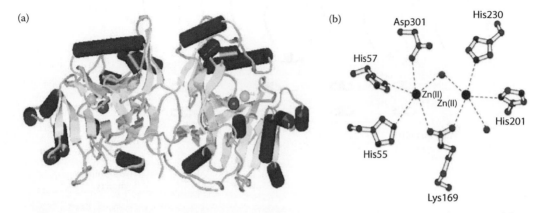

FIGURE 16.16
(a) Ribbon diagram of organophosphorus hydrolase (OPH). (Adapted from Zheng Y et al. 2008. In: *The 2nd International Conference on Bioinformatics and Biomedical Engineering*, Vol. 1, IEEE, pp. 13–16.) (b) Representation of the structure of the binuclear metal center within the active site of OPH. (Adapted from Benning MM et al. 2001. *Biochemistry* 40: 2712–2722.)

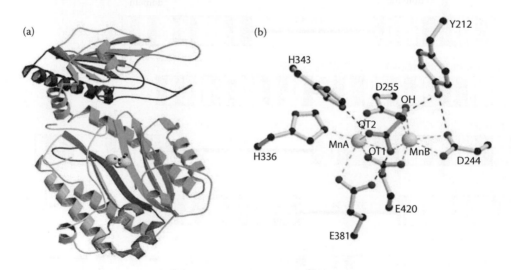

FIGURE 16.18
Ribbon diagram of organophosphorus acid anhydrolase (OPAA) determined from *Alteromonas* sp. strain JD6.5. (a) Ball-and-stick model of OPAA active site residues (yellow bonds) that are involved in the coordination of the two metal ions (yellow spheres) and in the interaction with the glycolate molecule (green bonds). Metal coordinations are shown as green dashed lines, while hydrogen bonds and other contacts with the glycolate are shown as red dashed lines (b). (Adapted from Vyas NK et al. 2010. *Biochemistry* 49(3): 547–559.)

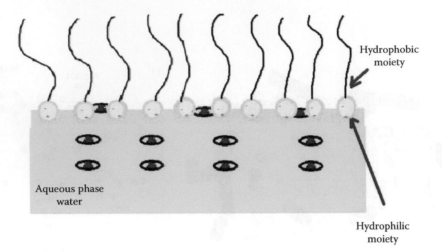

FIGURE 17.4
Accumulation of biosurfactants at the interface between liquid and air. (Modified from Pacwa-Płociniczak M et al. 2011. *International Journal of Molecular Science* 12: 633–654.)

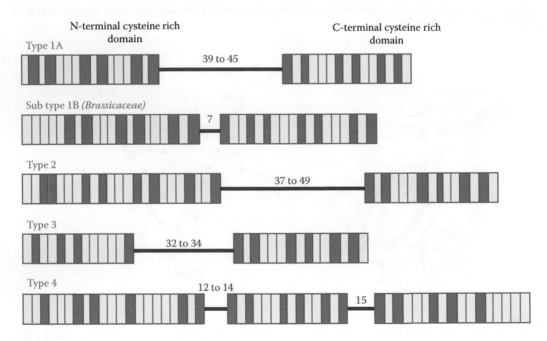

FIGURE 18.1
Classification of plant MT (Type 1, Type 2, Type 3, and Type 4) according to the cysteine motifs in their N- and C-terminal regions and the length of the C corresponding spacer regions. Blue and yellow boxes represent cysteine (Cys) and other amino acids (X), respectively. Red line and number indicate the spacer region containing the amino acid number. Type 1 contains six Cys-X-Cys motifs that are equally distributed among two domains. Type 2 contains Cys–Cys motif in amino acid positions 3 and 4 of these types and Cys-X-X-Cys motif is present at the end of the N-terminal cysteine-rich domain. Overall, the N-terminal domain of types 2 MTs are highly conserved and the C-terminal domain contains three Cys-X-Cys motifs. Type 3 contains only four Cys residues in the N-terminal domain and six Cys residues in the C-terminal are arranged in the Cys-X-Cys motif. Type 4 has three cysteine rich domains, each containing 5 or 6 Cys residues. Most of the cysteines present as Cys-X-Cys motif.

9

Persistent Organic Pollutants and Bacterial Communities Present during the Treatment of Tannery Wastewater

Gaurav Saxena and Ram Naresh Bharagava

CONTENTS

9.1 Introduction

Tannery industries (place where raw hides/skins are processed into leather), often called as leather industry is a widespread and common industry all over the world. Tannery industries play an important role in the economy of many developing countries as in the cases of Turkey, China, India, Pakistan, Brazil, and Ethiopia (Lofrano et al., 2013). However, it is major polluters worldwide due to its complex nature of wastewater. In developing countries, most of the tanneries are at the small-scale level and cannot afford the expensive treatment plant at their own cost because the effluent treatment plants (ETPs) are excessively costly to construct and operate, and also produce a large quantity of sludge. Therefore, a common ETP (CETP) is used for the treatment of wastewater generated from a cluster of tannery industries. The CETP is an activated sludge process (ASP)-based treatment plant that receives wastewater from cluster of tanneries through pipelines or tanks and after treatment, the treated wastewater from CETP is finally discharged into the water body.

Tanning (processing of raw hide/skin into leather) is a chemical process. There are several methods that are used for tanning of raw hide/skin, but the most common are vegetable and chrome tanning methods. Chrome tanning is the most common type of tanning method in which a large quantity of chemical compounds or a mixture of chemical compounds used concerned chemical compounds that can be released into the environment where they negatively affect all life forms (Kumar et al., 2008; Siqueira et al., 2011) and the environment too (Meric et al., 2005; Oral et al., 2007; Tigini et al., 2011; Shakir et al., 2012) or may inhibit nitrification process (Jochimsen and Jekel, 1997; Szpyrkowicz et al., 2001). The outline of leather production and type of pollutants generated is presented in the Figure 9.1.

However, the chemical compounds used during the leather tanning process are also summarized in Table 9.1.

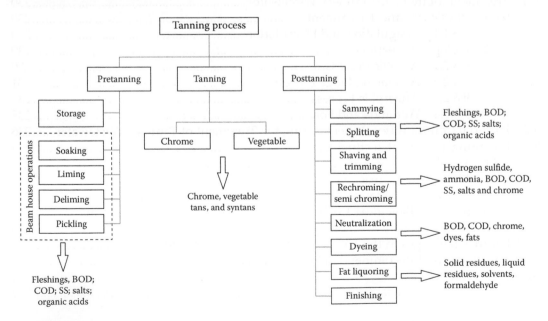

FIGURE 9.1
Schematic representation of leather production and type of pollutants released.

TABLE 9.1

Chemical Compounds Used at Different Steps in Leather Tanning Process

Tanning Operation	Chemical Used	Application
Beam house and tanyard	Biocides	Prevent bacterial growth that can damage the hides or skins during soaking
	Surfactants	Help in the wetting back of the hides or skins
	Degreasers	Remove natural fats and greases from the hides or skins
	Swell regulating agents	Prevent uneven swelling of hides or skins during liming
	Sodium sulfide	Destroys hairs on hides or skins
	Lime	Swelling of hides or skins
	Sodium hydrosulfide	Destroys hairs on hides or skins
	Low sulfide unhairing agents	Reduce the amount of sulfides used thus reducing the environmental impact of tanneries
	Caustic soda	Help in swelling of hides or skins during soaking or liming
	Soda ash	Raise the pH of hides or skins
	Ammonium sulfate	Helps to remove lime from hides or skins
	Ammonium chloride	Helps to remove lime from hides or skins during the deliming
	Sodium metabisulfite	Prevent formation of toxic hydrogen sulfide gas during deliming. It also acts as a bleaching agent
	Formic acid	Lower the pH of hides or skins
	Sulfuric acid	Lower the pH of hides or skins during the pickling
	Salt	Prevent acid swelling of hides or skins during the pickling
	Sodium formate	Assist penetration of chrome tanning salts into hides or skins
	Chromium sulfate	Act as tanning agent used to make wet blue
	Aldehyde tanning agents	Tanning agents used to make wet white
	Magnesium oxide	Raises the pH of hide or skin to allow chromium or aldehyde to chemically bind to the skin protein
	Fungicide	Prevent moulds or fungal growth on tanned hides or skins
Dye house	Surfactants/wetting agents	Help in wetting back of wet blue in dye house
	Degreasers	Remove grease or fats that may be present on the wet blue as a result of the wet blue coming into contact with machinery
	Sodium formate	Raise the pH during neutralization process
	Sodium bicarbonate	Raise the pH during neutralization process
	Formic acid	Reduces the pH for rechroming process
	Chrome syntans	Improve softness of final leather during rechroming process
	Chromium sulfate	Improve softness of final leather during rechroming process
	Syntans	Give properties such as softness, fullness, roundness to leather
	Resins	Give fullness and a tight grain to final leather
	Polymers	Give fullness and a tight leather grain to final leather

(Continued)

TABLE 9.1 (*Continued*)

Chemical Compounds Used at Different Steps in Leather Tanning Process

Tanning Operation	Chemical Used	Application
	Dyes	Give a color to final leather desired by customer
	Dyeing auxiliaries	Disperse dyes evenly
	Fat liquors	Give softness to final leather
Finishing	Acrylic resins	Give properties such as adhesion, water resistance to final leather
	Butadiene resins	Give specific properties such as good coverage to leather finish
	Polyurethane resins	Give properties such as toughness and light fastness to leather finish
	Fillers	Fill small blemishes on the leather surface
	Dullers	Reduces the gloss of the finish
	Cross linkers	Toughen the leather finish and improve water resistance
	Handle modifiers	Give the leather surface a waxy or slippery feel
	Nitrocellulose lacquers	Used in the top coat of a leather finish
	Acrylic lacquers	Used in the top coat of a leather finish
	Polyurethane lacquers	Used in the top coat of a leather finish
	Viscosity modifiers	Used to increase the viscosity of a finish mixture
	Pigments	Coloring agents that help hide defects on the leather surface
	Dyes	Coloring agents that slightly change the color of the leather finish or give the leather finish a more natural look
	Defoamers	Prevent bubbles from forming in the finish mixture

The inherent nature of the tanning process is that large volume of water is consumed (Khwaja et al., 2001) and an average of 30–35 m^3 of wastewater is produced per ton of raw hide processed (Lofrano et al., 2013). Tannery wastewater (TWW) is ranked as the highest polluted wastewater among all the industrial wastes (Camargo et al., 2003), which contain a complex mixture of both inorganic and organic pollutants (OPs). TWW also contains a variety of persistent OPs (POPs), which do not degrade much during the secondary (biological) treatment process in industries, and goes into the environment.

The amount of OPs in TWW is large and is represented mainly by high COD value, which is usually several thousand mg/L O$_2$ (Lofrano et al., 2013). The OPs (proteic and lipidic components) are originated from raw hides/skins or introduced during the tanning processes (Lofrano et al. 2013). Proteins, mainly collagen and their hydrolysis products that is, amino acids derived from the skin are predominant, while others such as fats are in low concentration (Alam et al., 2010). The most important organics used in tanning of skin are natural and synthetic tannins, fatty aldehydes, and quinones (Alam et al., 2010). Tanneries also use compounds such as aliphatic amines, nonionic surfactants, oils, and pigments. Most of these pollutants are in a soluble form, but a lot of them exist in suspension and only a few as colloids (Ates et al., 1997; Cassano et al., 2001; Di Iaconi et al., 2002).

Other OPs that remain in TWW include chlorinated phenols, polychlorinated biphenyls (PCBs), polycyclic aromatic hydrocarbons (PAHs), alkylated benzenes, aliphatic hydrocarbons, formaldehyde resins, phthalate esters, and pesticides residues (Mwinyihija, 2010; Lofrano et al., 2013). The United States Environmental Protection Agency (USEPA) and some of its international counterparts have classified most of the phthalic acid esters, such as diethyl phthalate, benzyl butyl phthalate, di-*n*-butyl phthalate, and di-(2-ethyl hexyl) phthalate in TWW, as priority pollutants and endocrine-disrupting compounds (Moore, 2000; Alam et al., 2010). The high concentrations of pollutants with low biodegradability in TWW represent a serious and actual technological and environmental challenge (Di Iaconi et al., 2002). Hence, the treatment of TWW is only an option to minimize the pollutants load from wastewater for its safe disposal into the environment. Therefore, this chapter will present the state of the art of the chemical compounds used during tanning process in leather industry, POPs that remains in TWW even after the conventional treatment process, treatment methods including emerging treatment technologies (ETTs) for TWW, and bacterial communities present during the degradation and detoxification of TWW at CETP.

9.2 Tannery Wastewater

9.2.1 Wastewater Generation and Its Characteristics

In tannery industries, an average of 30–35 m^3 of wastewater is produced per ton of raw hide processed (Lofrano et al., 2013). However, the wastewater production varies greatly ranging from 10 to 100 m^3/ton hide processed, depending on the raw material, the finishing product, and the production processes adopted (Tunay et al., 1995). The characteristics of TWW vary considerably from industry to industry depending upon the size of the tannery industries, chemicals used, as well as process adopted by the tannery industries.

Tannery wastewater is characterized mainly by the measurement of biochemical oxygen demand (BOD), chemical oxygen demand (COD), suspended solids (SSs), and total dissolved solids (TDS) and specific pollutants such as chromium and sulfides, which imparts high bacterial activity (Durai and Rajasimman, 2011). OPs (proteic and lipidic components) originate either from skins (it is calculated that the raw skin has 30% loss of organic material during the working cycle) or introduced during the tanning process (Lofrano et al., 2013). High concentrations of these constituents make the possible discharge of TWW into the water bodies and pose serious threats in the environment (Leta et al., 2004).

In general, TWW is basic, having a dark brown color, and a high content of organic substances that vary according to the chemicals used and process adopted by the tannery industries (Kongjao et al., 2008). The beamhouse and tanning processes are the most contaminating steps; the former mainly adds high organic load and sulfide content, while the latter adds inorganic salts of chloride, ammonia, chromium, and sulfate to TWW. The beamhouse wastewater is characterized by an alkaline pH and the TWW by a very high acidic pH as well as high COD. The physicochemical characteristics of TWW are given in Table 9.2.

Tunay et al. (1995) have reported COD value of 27,600 mg/L for beamhouse wastewater, which indicates the very highest amount of the salt load in the beamhouse area. About 40%–45% (w/w) of common salts are used to preserve the animal skins, which are removed during the soaking process (Sundarapandiyan et al., 2010). Lime and sodium sulfide or sulfydrate are normally used during the liming-unhairing operation. It is also observed that TWW are highly rich in nitrogen especially organic nitrogen, but very poor in phosphorous (Durai and Rajasimman, 2011). Organic solvents are the most widely

TABLE 9.2

Physico-Chemical Characteristics of Treated Tannery Wastewater

Parameters	Wastewater[a]	Permissible Limit
pH	8.85	6.0–8.0
Conductivity (moles/cm)	11,000	850
Total solids (TS; mg/L)	2477	2200
Alkalinity (mg/L)	729	500
Total dissolved solids (TDS; mg/L)	2219	2100
Total suspended solids (TSS; mg/L)	258	100
DO (mg/L)	2.8	4.0–6.0
BOD (mg/L)	267	30
COD (mg/L)	458	250
Sulfate (mg/L)	2400	1000
Chloride (mg/L)	354	600
Magnesium (mg/L)	234	200
Phosphate (mg/L)	5.5	5.0
Nitrate (mg/L)	12.08	10
Total nitrogen (mg/L)	229.04	780
Fluoride (mg/L)	3.5	2.0l
Phenol (mg/L)	10.5	1.0
Total chromium (mg/L)	19.57	2.0
Oil and grease	17	10
Cr^{6+} (mg/L)	3.5	0.1
Cu^{2+} (mg/L)	1.9	3.0
Mn^{2+} (mg/L)	1.6	2.0
Zn^{2+} (mg/L)	3.8	5.0
As^{3+} (mg/L)	0.30	0.2
Pb^{2+} (mg/L)	0.08	0.1
Cd^{2+} (mg/L)	ND[b]	2.0
Ni^{2+} (mg/L)	3.2	2.5
Co^{2+} (mg/L)	0.22	1.5
Fe^{2+} (mg/L)	3.5	3.0

Source: Adapted from Verma T, Ramteke PW and Garg SK. 2008. *Environmental Monitoring Assessment* 145: 243–249; Chandra R et al. 2011. *Bioresource Technology* 102: 2333–2341.
[a] Average of triplicate samples.
[b] Not detected.

used chemicals in degreasing process, which produces a considerable amount of volatile compounds (Cassano et al., 2001). If a tannery is processing salted hides then the major salt component in wastewater is the sodium chloride that comes from the hide and skin preservation. The wet-end re-tanning, dyeing, and fat liquoring processes contribute only a minor fraction of total salt load, which dominantly originated from the hides during the initial presoak and main soak process. TWW contains very high amount of total chromium, that is, up to 4950 mg/L (Cooman et al., 2003). However, the coloring of hides and skins usually involves the combination of dyes with the tanned skin fibers to form insoluble compounds. Retanning and wet finishing streams have relatively low BOD and TSS, but high COD and contain trivalent chromium, tannins, sulfonated oils, and spent dyes (USEPA, 1986). Although COD is the most common parameter used for TWW characterization the total organic carbon (TOC) is more reliable than COD.

9.2.2 Biodegradation Characteristics

The BOD_5/COD or BOD_5/TOC ratio continues to be used for biodegradability information of TWW, but BOD_5 is a controversial parameter because the metal inhibitors like chromium negatively affects the biodegradation of TWW (Ates et al., 1997). Regarding the COD measurement, it can give more reliable information about the composition of this kind of wastewater. TWW requires COD fractionation for the identification of biodegradable and inert COD components (Orhon et al., 1999). Lofrano et al. (2013) have reported the COD fractionation of TWW and also correlated with the inhibitors effect of the particulate portion of TWW. The integration of physical particle size distribution of COD with COD fractionation has been recently studied for a better interpretation of COD fractions and biodegradability of TWW (Karahan et al., 2008).

9.3 Persistent Organic Pollutants

During the end of the 20th century, the global environment became polluted with a number of persistent, fat-soluble chemical pollutants, commonly referred to as the POPs. The pollution of global environment with a complex mixture of POPs has resulted from industrial discharges and anthropogenic activities, applications, as well as the inadvertent formation of by-products of incomplete combustion or industrial processes. POPs are a group of chemical compounds that originated from different anthropogenic activity, but have some common characteristics such as semivolatility, lipophilicity, bioaccumulation, which make the POPs resistant to photolytic, biological, as well as chemical degradation over a reasonable period of time (Chandra and Chaudhary, 2013) and these persist in the environment for a long period of time and have potential significant impacts on human health and the environment (Samaranda and Gavrilescu, 2008) due to their very high toxicity, their prolonged persistence in polluted ecosystems, and their extremely limited biodegradability (Chandra and Chaudhary, 2013).

The importance of persistent environmental chemicals such as pesticides was first identified by Rachel Carson an American and a courageous woman who was a scientist in the early 1960s with the publication of her seminal work in her classic book *Silent Spring* (Prest et al., 1970). The POPs have three common characteristics: (i) one or more cyclical ring structures of either aromatic or aliphatic nature, (ii) a lack of polar functional groups, and (iii) a variable amount of halogen usually chlorine. If some key properties of POPs are known, then the environmental chemists can make predictions about their fate and behavior in natural environment. These properties include aqueous solubility, vapor pressure, partition coefficients between water:solid and air:solid or liquid, and half-live in air, water, and soil (Chandra and Chaudhary, 2013).

A large amount of waste discharged from eight largest industries such as oil, cement, leather, textile, steel, pulp paper, tannery, and distillery industries includes a variety of gaseous, liquid, and solid waste into the environment and persist for long period of time. These POPs containing waste causes serious threats to the receiving water bodies (Chandra and Chaudhary, 2013).

9.3.1 Categories of Persistent Organic Pollutants

9.3.1.1 Intentionally Produced Persistent Organic Pollutants

These compounds are produced intentionally as commercial products through different chemical reactions with chlorine. These organic molecules are called as organochlorine compounds (OCs) and have high lipophilicity and neurotoxicity. On the basis of their applications, these OCs are categorized into two types, that is, pesticides and industrial products.

9.3.1.1.1 Pesticides

Pesticides contain a large part (70%) of the POPs and their persistence in the environment depends largely on the physical and chemical properties of the soil and its toxicity, moisture content, temperature, composition of soil micro flora, and the plant species growing in soil. Various types of pesticides are summarized in Table 9.3.

TABLE 9.3

List of Pesticides as Intentionally Produced POPs

Organophosphate	Carbamates	Organochlorine Insecticides	Pyrethroides
Parathion	Aldicarb	DDT	Allethrin
Malathion	Carbofuran	Chlordane	Bifenthrin
Methyl	Carbaryl	Mirex	Cyfluthrin
Chlorpyrifos	Ethienocarb	HCB	Cypermethrin
Diazinon	Fenobucarb	Endrin	Cyphenothrin
Dichlorvos	Oxamyl	Aldrin	Deltamethrin
Phosmet	Methomyl	Dieldrin	Esfenvalerate
Fenitrothion		Toxaphene	Etofenprox
Tetrachlorvinphos		Heptachlor	Fenpropathrin
Azinphos methyl		Hexachlorobene (HCB)	Fenvalerate
			Flucythrinate
			Imiprothrin
			Metofluthrin
			Permethrin
			Prallethrin
			Resmethrin
			Silafluofen
			Sumithrin
			τ-Fluvalinate
			Tefluthrin
			Tetramethrin
			Tralomethrin

Source: Adapted from Chandra, R. and Chaudhary, S. 2013. *International Journal of Bioassays* 2(09): 1232–1238.

TABLE 9.4

List of Industrial POPs Product from Different Sources

S. No.	Industrial Product	Source
1.	Polychlorinated biphenyls (PCBs)	Transformer, capacitor, electric motors
2.	Hexachlorobenzene (HCB)	Agriculture
3.	Chlordecone	Agricultural insecticide
4.	Pentachlorobenzene	Industrial
5.	Hexabromobiphenyl	Flame retardant
6.	Hexabromodiphenyl ether	Flame retardant
7.	Heptabromodiphenyl ether	Flame retardant
8.	Tetrabromodiphenyl ether	Flame retardant
9.	Pentabromodiphenyl ether	Flame retardant
10.	Polychlorinated biphenyls (PCBs)	Industrial
11.	Alpha hexachlorocyclohexane	Insecticide
12.	Beta hexachlorocyclohexane	Insecticide
13.	Lindane	Agricultural insecticide
14.	Perfluorooctane sulfonic acid	Industrial
15.	Perfluorooctane sulfonyl fluoride	Industrial

Source: Adapted from Chandra, R. and Chaudhary, S. 2013. *International Journal of Bioassays* 2(09): 1232–1238.

9.3.1.1.2 Industrial Products

Persistent OPs are also produced for their various uses in industrial processes, such as coolants for electrical transformers. This group includes PCBs and hexachlorobenzene (HCB), which are summarized in Table 9.4.

9.3.1.2 Unintended By-Products

These are produced as unwanted by-products resulting from the combustion or burning of chlorine-containing compounds. These are mainly three types, that is, PAHs, dioxin, and furan compounds.

9.3.2 Toxicological Profiles of POPs and Related Health Hazards

The toxicological properties of POPs are mainly because of their very low solubility and high lipid solubility leading to their bioaccumulation in tissues (Guzzella et al., 2005). They enter the body through food and are transferred to all the trophic levels of the ecosystem. POPs are highly toxic in nature and have a wide range of chronic effects such as endocrine disruption, mutagenicity, and carcinogenicity (Sultan et al., 2001; Adeola, 2004; Lee et al., 2006). These have also been able to persist in the environment for decades causing serious health hazards such as cancer, birth defects, and learning disabilities, immunological, behavioral, neurological, and reproductive discrepancies in human and animals, developmental abnormalities, neurological impairment, and tumor (Bolt and Degan, 2002; Sweetman et al., 2005; Chandra and Chaudhary, 2013).

In wild life, it causes egg shell aberration in birds to extinction of certain bird species (WWF, 1999); other serious effects including cancer, twisted spines, skeletal deformations, and death of beluga whales. In Florida's Lake Apoka, stunted penis, hormone disruption, and reproductive failure have been found among alligators with disrupted reproductive development, deformity, immune toxicity, hormonal deficiencies, to overall population decimation have been reported (Abelsohn et al., 2002).

9.3.3 Microbial Degradation of Persistent Organic Pollutants

A large number of bacterial and fungal species possess the capability to degrade POPs. Biodegradation is the catalyzed reduction of complex chemical compounds based on two processes, that is, growth and co-metabolism. In case of growth, OPs are used as a sole source of carbon and energy. The second possibility is co-metabolism in which the following events occur: cellular uptake of compounds, manipulation of substrate for ring fission, ring cleavage, conversion of cleaved product into standard metabolites, and utilization of metabolites.

Biodegradation generally depends on many factors. These factors include the structure of the compound, substituent's position in the molecule, and solubility of the compound and concentration of the pollutant (Chandra and Chaudhary, 2013). Environmental and soil condition such as organic carbon content also influence degradation rate of POPs. Addition of organic substrate and nutrients to contaminated soil could enhance microbial and degradation activity (Borja et al., 2005). Microorganisms have very large enzymatic facilities due to their genetic flexibility, short reproduction cycle, and high adaptive potential. The degradation of the POPs has been monitored by HPLC/DAD/FLD (Chandra and Chaudhary, 2013).

Biodegradation of chlorinated compounds is highly affected by the degree of chlorination and presence of other functional groups. Higher chlorinated congeners are only susceptible to anaerobic biotransformation, whereas lower congeners are only susceptible to aerobic biodegradation. Complete biodegradation of the higher chlorinated congeners requires a sequence of anaerobic and aerobic conditions.

A majority of microorganisms are able to metabolize aldrin to dieldrin by epoxidation (Tu et al., 1968). Dieldrin is highly persistent (persist for 5 years) in soil and residues (Ritter et al., 2007). Dieldrin could be further hydrolyzed by several bacteria and fungi strains, which convert dieldrin to water-soluble and solvent-soluble metabolites and to CO_2 (Lal and Saxena, 1982). It has shown that soil microorganisms metabolize heptachlor to many different products by many independent metabolic pathways (Miles and Moy, 1979). The insecticide endosulfan is structurally similar to chlordane and dieldrin. The degradation time of endosulfan and its metabolites could exceed 6 months in some acidic soil (Herman and Gisela, 2002). Beta-endosulfan has a longer half-life and is slowly converted to alpha-endosulfan. It was reported that mixed soil microbial culture was able to interconvert α and β isomers and consequently further metabolism.

Hexachlorobenzene is a fungicide that was first introduced in 1945 for seed treatment and is also a by-product of the industrial chemicals. Volatilization is an important process of HCB losses from soil. Estimated half-lives in soil under aerobic and anaerobic degradation range from 2.7 to 22.9 years (Ritter et al., 2007).

Chlorophenols have been introduced into the environment through their use as biocides in wood preservation. Wide spectrum of indigenous soil microorganisms is able to metabolize chlorophenols and utilize it as a carbon source (Mahmood et al., 2005). Chlorophenols are degradable under both aerobic and anaerobic conditions.

Other POP compounds are releases from the world's largest polluting industries such as pulp paper and distillery industrial waste, which are recalcitrant to degradation. These recalcitrant wastes include phenolic substances, fatty acids, resin acids, chlorolignin, molasses, dyes, pesticides, explosives, heavy metals, poly alcohols, dioxin, and furan derivatives.

There is not much information about soil microbial degradation of other POPs. These candidate and new POPs chemicals are often very stable and no longer used. In case of brominated diphenyl ethers (BDE), OctaBDE is not readily biodegradable in standard tests and is not expected to degrade rapidly under anaerobic conditions (POPRC, 2007). More highly brominated congeners have been found to degrade anaerobically in sewage sludge, although at a very slow rate (Gerecke et al., 2005). Lower brominated diphenyl ethers are usually more toxic and much more bioaccumulative.

Perfluorochemicals are nonpolar, highly fluorinated compounds that are chemically and biologically inert. They are used in surface treatments to provide soil and stain resistant coatings, in paper treatments to provide oil, grease, and water resistance. Perfluorooctane sulfonate (PFOS) and its salts are highly persistent in the environment and do not appear to degrade (WWF, 2005).

In pulp paper industry, various types of POPs are released in which PAH is one of them. PAH-degrading populations in soil are probably mostly not growing and degradation of PAHs is a complex process involving assimilation as well as co-metabolisms by many bacterial species. It has been observed that PAH degradation in soil is dominated by bacterial strains belonging to a very limited number of taxonomic groups such as *Sphingomonas, Burkholderia, Pseudomonas,* and *Mycobacterium* (Bouchez et al., 1999). PAHs degrading bacteria are able to produce bio-surfactants that increase the dissolution of PAHS and facilitate their bioavailability and biodegradation (Johnsen et al., 2005). Biodegradation of PAHs by isolated strains of white-rot fungi was also reported (Field and Sierra-Alvarez, 2007).

Polychlorinated dibenzo-*p*-dioxins and polychlorinated dibenzofurans are produced unintentionally due to incomplete combustion, during manufacturing of other chlorinated compounds, and in the chlorine bleaching of wood pulp. They are also toxic, persistent, and bioaccumulating man-made compounds.

9.4 Persistent Organic Pollutants (POPs) in Tannery Wastewater

In case of leather tanning industry, there has been an increasing environmental concern regarding the release of various POPs in TWW, which do not degrade much during the secondary treatment process in industries and goes in the environment. The treatment of TWW is very complex due to the presence of a variety of chemical compounds, which are used during the tanning process (Schrank et al., 2009). It has been shown that biological treatment processes involving microbes are known as the most environmental friendly and cost-effective but inefficient for effective removal of POPs from TWW, which goes into the environment causing serious threats to soil and water ecosystem as well as to humans and animal life. However, the various types of POPs are reported by few researchers (Table 9.5), which were detected and identified by gas chromatography–mass spectroscopy (GC–MS) analysis of TWW extracted with different solvent systems.

But, there is no much information available about POPs that remains in TWW after secondary treatment and their biodegradation as well as their toxicological effects in the environment.

9.5 Toxicity of TWW in the Environment

Tannery industries are considered as one of the major source of pollution because the wastewater generated from these industries poses serious environmental impacts on soil, land, and water, terrestrial and atmospheric system (Mwinyihija, 2010). Usually tanning industries discharge their wastewater into nearby rivers and is indirectly being used for irrigation of crops and vegetables. This practice has ultimately led to the movement of potentially toxic metals from water to plants and ultimately to human beings (Sinha et al., 2008). It is well reported that TWW is the major source of pollution in the environment as it has very high concentration of chromium, which goes in soil and water and create serious health hazards in the environment. The Cr^{6+} is a potent carcinogen to humans and animals as it enters the cells *via* surface-transport system and gets reduced to Cr^{3+} inducing genotoxicity (Matsumoto et al., 2006). Thus, the use of Cr-loaded wastewater used for irrigation practices disrupts the several physiological and cytological processes in cells (Shanker et al., 2005; Chidambaram et al., 2009) leading to reduced root growth, biomass, seed germination, early seedling development (Irfan and Akinici, 2010), and induces chlorosis, photosynthetic impairment, and finally leading to plant death (Akini and Akini, 2010). In the environment, chromium (Cr^{6+}) contamination alters the structure of soil microbial communities as well as reduce their growth retarding the bioremediation process and if Cr^{6+} enters in food chain, it causes skin irritation, eardrum perforation, nasal irritation, ulceration, and lung carcinoma in humans and animals along with accumulating in placenta impairing the fetal development in mammals (Chandra et al., 2011). Extensive use of chromium salts in tanning industries has resulted in chromium contamination in soil and ground water at production sites, which pose a serious threat to human health, fish, and other aquatic biodiversity (Turick et al., 1996).

There are a number of studies that have highlighted the number of health hazards in the tanning industry including occupational exposures (Battista et al., 1995), water and land contamination affecting seed germination in various crop plants (Asfaw et al., 2012; Kohli and Malaviya, 2013), aquatic and terrestrial biota, and humans (Barnhart, 1997), as well as acute toxicity in *Vibrio fischeri* (Jochimsen and Jekel, 1997) and the microcrustacean *Daphnia magna* (Tisler et al., 2004), sea urchin (De Nicola et al., 2007), and marine micro algae (Meric et al., 2005). Verma et al. (2008) have conducted the quality assessment of treated TWW with special emphasis on pathogenic *Escherichia coli* detection through serotyping.

However, Alam et al. (2009) have studied the genotoxic and mutagenic effects of dichloromethane, methanol, acetonitrile, and acetone extracts of agricultural soil irrigated with TWW by using different mutants of *Salmonella typhimurium*, such as TA97a, TA98, TA100, TA102, and TA104 and DNA repair-defective/SOS-defective *recA*, *lexA*, and *polA* mutants of *E. coli* k-12 in presence and absence of S9 liver homogenate suspension. They found that all the soil extracts have significant mutagenicity on Ames tester strains, but the dichloromethane soil extract exhibited the maximum mutagenic potential of 17.3 (–S9) and 20.0 (+S9) revertants/mg soil equivalent in TA100. A significant decline in the survival of DNA repair-defective *E. coli* K-12 mutants was observed compared to their isogenic wild-type counterparts when treated with different soil extracts. Out of all the tested mutants of *E. coli* K-12, *PolA* mutant was found to be the most sensitive strain toward all the four soil extracts. In addition, the similar study was also performed by Alam and his co-workers in 2010.

TABLE 9.5

Persistent Organic Pollutants (POPs) Identified in Tannery Wastewater Using Different Extraction Solvents through GC-MS Analysis

Extraction Solvent	Persistent Organic Pollutants	References
Dichloromethane (DCM)	2,4-bis(1,1-dimethyl)phenol	Alam et al. (2009)
	10-Methylnonadecane	
	Docosane	
	bis(2-ethylhexyl)phthalate	
	2,6,10-Dodecatrien-1-ol-3,7,11-trimethyl acetate	
Methanol	1, 3-Hexadien-5-yn	Alam et al. (2009)
	1, 2-Benzenedicarboxylic acid, di-isooctyl ester	
Acetonitrile + acetone	2,2,3-Trimethyl oxepane	Alam et al. (2009)
	Benzene	
	3-Nitropthalic acid	
Chloroform + hexane	2-(2-hydroxy)-2 propyl cyclohexanol	Alam et al. (2010)
	Dibutyl phthalate	
	Tetratetracontane	
	Bis(2-methoxyethyl)phthalate	
	Hexatriacontane	
	Heneicosane	
	Docosane	
	Tricosane	
	1,2-Benzenedicarboxylic acid, di-isooctyl ester (di-isooctyl phthalate)	
Dichloromethane (DCM)	Phenyl *N*-methylcarbamate	Alam et al. (2010)
	Caprolactam	
	Octacosane	
	2,6,10,15-Tetramethylheptadecane	
	Nonadecane	
	2,6,10,14-Tetramethylhexadecane	
	Triacontane	
	Heptadecane	
	Tetracosane	
	Eicosane	
	9-Methylnonadecane	
	Heptadecane	
	1,2-benzenedicarboxylic acid, di-isooctyl ester (di-isooctyl phthalate)	
	Dotriaconatn	
Ethyl acetate	L-(+)-Lactic acid	Chandra et al. (2011)
	Acetic acid	
	Benzene	
	2-Hydroxy-3-methyl-butanoic acid	

9.6 Treatment Methods for Tannery Wastewater

9.6.1 Physicochemical Treatment

9.6.1.1 Coagulation and Flocculation

The coagulation and flocculation of TWW has been investigated by using different types of inorganic coagulants such as aluminum sulfate (AlSO$_4$), ferric chloride (FeCl$_3$), ferrous sulfate (FeSO$_4$) to reduce the total organic load (BOD and COD) and total solids (TDS and TSS) as well as to remove the toxic metals such as chromium before the biological treatment processes (Ates et al., 1997; Song et al., 2004; Lofrano et al., 2006). However, each coagulant operates most effectively at specific pH and extent of pH range largely depends on the nature of coagulants, characteristics of the wastewater to be treated, and dosage of the coagulant (Song et al., 2004). There are several studies that have been conducted to investigate the effectiveness of different coagulants used for the treatment of TWW in terms of COD and chromium removal (Ates et al., 1997; Song et al., 2004).

9.6.2 Biological Treatment

9.6.2.1 Aerobic Processes

Biological treatment processes are generally used for treatment of industrial wastewater to reduce the organic content as these processes have many economic advantages over the physicochemical treatment oxidation (Dogruel et al., 2006). The high concentration of tannins and other poorly biodegradable compounds as well as toxic metals can inhibit the biological treatment processes (Lofrano et al., 2013). Stasinakis et al. (2002) have observed a significant inhibition in heterotrophic growth in the presence of 10 mg/L Cr^{6+}. Farabegoli et al. (2004) reported that chromium concentration had less influence on denitrifying bacteria than on nitrification bacteria.

A typical sequencing batch reactor (SBR) has been proved to be more capable of carrying out the biological processes such as nitrification and denitrification in the presence of inhibitors (Farabegoli et al., 2004; Ganesh et al., 2006) due to the selection and enrichment of particular microbial species. The performance of SBR for nitrogen removal in TWW, with a wide range of temperature (7–30°C), was studied and full nitrification and denitrification was achieved by adjustment of the sludge age for each temperature range (Murat et al., 2006).

The biodegradation of naphthalene-2-sulfonic acid, which is a main component of the naphthalenesulfonate, by *Arthrobacter* sp. 2AC and *Comamonas* sp. 4BC was reported and these two bacterial strains were isolated from tannery activated sludge (Song et al., 2005). Song and Burns (2005) described the degradation of all components of the condensation product of 2-naphthalenesulfonic acid and formaldehyde (CNSF) by fungus *Cunninghamella polymorpha* and suggested that the combination of *C. polymorpha* and *Arthrobacter* sp. 2 AC or *Comamonas* sp. 4BC was effective for the treatment of TWW. Conventional cultures could not treat saline wastewaters of values higher than 3%–5% (w/v) and shift in salt concentration causes with significant reactor failures in the system performance. Senthilkumar et al. (2008) have isolated *Pseudomonas aeruginosa, Bacillus flexus, Exiguobacterium homiense,* and *Staphylococcus aureus* from soak liquor, marine soil, salt lake saline liquor, and seawater, respectively, and studied the biodegradation of tannery soak liquor by these halotolerant bacterial consortia. An appreciable COD removal (80%) was observed at 8% (w/v) salinity for mixed salt tolerant consortia, but increase in salt concentration to 10% (w/v) resulted in a decrease in COD removal efficiency.

The presence of sulfide, chromium, chloride, and fluctuations in temperature also has adverse effects on the nitrification process. The impact of temperature on organic carbon and nitrogen removal had been studied for a full-scale industrial-activated sludge plant during the treatment of TWW (Gorgun et al., 2007). It was observed that temperature changes had a minor influence on COD removal efficiency (4%–5%), while the total nitrogen removal was affected significantly by the temperature. Insel et al. (2009) investigated the performance of intermittent aeration type of operation when temperature was fluctuated between 21°C and 35°C and they found that an increase in the aeration intensity improved nitrification performance and the application of intermittent aeration also improved the total nitrogen removal up to 60%.

9.6.2.2 Activated Sludge Process

9.6.2.2.1 Introduction

The ASP is the most generally applied suspended growth biological (aerobic) wastewater treatment method that primarily removes the dissolved organic solids as well as settleable and non-settleable SSs. In ASP, a suspension of bacterial biomass (the activated sludge) is mainly responsible for the removal of organics. These organisms are cultivated in aeration tanks, where they are provided with dissolved oxygen (DO) and food from the wastewater. Depending on the design and the specific applications, an activated sludge treatment plant can remove organic nitrogen (N) removal and phosphorus (P), besides the removal of organic carbon substances.

The processes used most frequently for the biological treatments of TWW in CETPs in India are the ASP and the upflow anaerobic sludge blanket (UASB) process (Kadam, 1990; Rajamani et al., 1995; Jawahar et al., 1998). However, the biological treatment of TWW using ASP has been reported by many workers (Murugesan and Elangoan, 1994; Jawahar et al., 1998; Eckenpfelder, 2002; Tare et al., 2003; Hayder et al., 2007; Ramteke et al., 2010).

9.6.2.2.2 Principle

The ASP uses microorganisms, which feed on OPs in wastewater and produce high quality wastewater. The basic principle behind all the ASPs is that as microorganisms grow, they form particles that clump together. These particles (floc) get settled to the bottom of the tank, leaving a relatively clear liquid, free of organic material, and SSs.

9.6.2.2.3 Terminology

1. Mixed liquor SSs (MLSS) and mixed liquor volatile SSs (MLVSS): The microbial biomass that is responsible for removal of BOD are the "active" part of activated sludge. The solids under aeration are referred to as the MLSS. The portion of MLSS that is actually eaten by microorganisms is referred to as the MLVSS. The inventory of biomass is calculated as pounds of microorganisms based on the volume of the tank and the concentration of the MLVSS.

2. Return activated sludge (RAS) and waste activated sludge (WAS): As the mixed liquor moves to the secondary clarifier, the activated sludge get settled to the bottom of tank and is removed. One of two things happens to the settled sludge. Most of it is returned to the aeration basins to keep enough activated solids in the tanks to handle the incoming BOD. This is known as the RAS. A small portion of sludge

that is removed from the system as the MLSS inventory grows is referred to as WAS.

3. Detention time (DT): DT is the length of time during which the MLSS remains under aeration, differs with each type of ASP. RAS flows can be used to manipulate the DT in the aeration tanks and increase the RAS flow at night maintains the proper DT as influent flows drop.

4. Food-to-microorganism ratio (F:M ratio): It is a baseline established to determine how much food a single pound of microorganisms will eat every day. It is used to control activated sludge solids inventory. A pound of microorganisms eat around 0.05–0.6 pounds of food per day depending on the process.

5. Mean cell residence time (MCRT) or sludge age: Another controlling parameter is the length of time for which the microbial cells stay in the process. If a system wastes 5% of solids every day, then MLSS would only remain for an average of about 20 days (100%/5% per day = 20 days). This value is known as the MCRT or sludge age.

6. Sludge volume index (SVI): SVI is the measurement of how well the activated sludge settles in the clarifier. Sludge settleability largely depends on the conditions of the microorganisms. Good settling sludge will have an SVI between 80 and 120 as the sludge becomes lighter and the settled volume increases with an increase in SVI.

9.6.2.2.4 Design

A conventional ASP (CASP) is shown in Figure 9.2 and includes the following ones:

1. *Aeration tank*: Aerobic oxidation of organic matter is carried out in this tank. Primary effluent is introduced and mixed with RAS to form the mixed liquor, which contains approximately 1500–2500 mg/L of SSs. Aeration is provided by mechanical means. An important characteristic of the ASP is the recycling of a large portion of the biomass. This makes the mean cell residence time (i.e., sludge age) much greater than the hydraulic retention time. This practice helps to

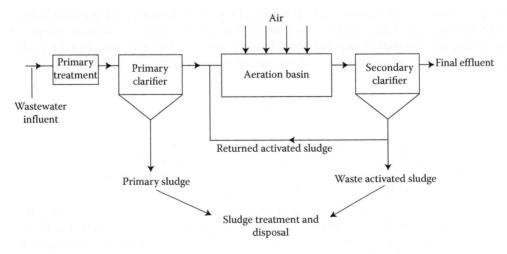

FIGURE 9.2
Flow diagram of conventional activated sludge process.

maintain a large number of microorganisms that effectively oxidize the organic compounds in a relatively short time. In the aeration basin, detention time varies between 4 and 8 h.

2. *Sedimentation tank*: This tank is used for the sedimentation of microbial flocs (sludge) produced during the oxidation phase in aeration tank. A portion of sludge present in clarifier is recycled back to aeration basin and the remaining is wasted to maintain a proper F/M ratio.

9.6.2.2.5 Equation

Activated sludge process is an aerobic, biological process, which uses the microbial metabolism to remove substances causing oxygen demand. The qualitative biochemical reaction taking place in the organic matter stabilization process can be summarized as

$$\text{Inert matter + Organic matter + Oxygen + Nutrients + Microorganisms}$$
$$\rightarrow \text{New microorganisms} + CO_2 + H_2O + \text{Additional inert matter}$$

9.6.2.2.6 Activated Sludge Treatment Process

There are basically three types of ASP. These accomplish the biochemical reduction of organics using the aeration basin return and waste sludge systems. The detention time, MLSS, and F:M ratio vary in each case. An important controlling parameter shared by all types of ASP has the DO requirement of 2.0–4.0 mg/L. The fact that aerobic conditions exist in aeration basin means that the mixed liquor should have a light earthy odor that is not objectionable. This DO level is maintained by aeration equipment such as blowers and diffusers or mechanical aerators.

9.6.2.2.6.1 Conventional Activated Sludge Process

Conventional ASP has an aeration basin detention time of 4–6 h. During this time, the microorganisms stabilize the BOD before the mixed liquor leaves the basin. The MLSS concentration usually ranges from 2000 to 3500 mg/L, F:M ratios should be between 0.2 and 0.5, whereas the MCRT or sludge age varies from 5 to 15 days.

9.6.2.2.6.2 Contact Stabilization Process

Contact stabilization uses two separate aeration processes in which the primary effluent enters the contact chamber where microorganism break down the BOD and increase the settleability of the organics that are not yet oxidized. The raw organics and MLSS get settled down just such as CASP. But, instead of returning the RAS to the contact basin, it is pumped to another aeration basin called stabilization basin, whenever it is aerated until the organics have been used or stabilized by the microorganisms. The wastewater from the stabilization basin is returned to the contact basin, to maintain the MLSS, and the process begins for next step.

The main advantage of the contact stabilization process is that most of the solids and BOD reduction occurs off-line from the main flow. This prevents massive solid loss during the hydraulic shocks on the system and reduces recovery time since the bulk of the biomass is present in the stabilization basin.

The detention time in contact basin ranges from 0.5 to 2.0 h, whereas in stabilization basin, it ranges from 4 to 8 h, about the same as in conventional processes. The MLSS concentration ranges from 1200 to 2000 mg/L in contact chamber, whereas 4000–6000 mg/L in stabilization chamber, same as in the RAS. Contact stabilization has the highest F:M ratios of 0.6.

9.6.2.2.6.3 Extended Aeration Process   An extended aeration system is designed to completely stabilize all of the organic matter present in the aeration basins. The DT ranges from 16 to 24 h and MLSS ranges from 3000 to 5000 mg/L. It has lowest F:M ratio of any of the ASPs, usually 0.05–0.2. Extended aeration plants normally have pretreatment but not primary clarifiers. RAS is returned to the head works and waste sludge is sent to an aerobic digester.

9.6.2.2.6.4 Oxidation Ditches   An oxidation ditch is a form of extended aeration activated sludge. The aeration basin is a large oval-shaped tank that resembles a racetrack and in this process, wastewater enters the ditch and is circulated around the track by means of a large horizontal brush/rotor. The rotor assembly is partly submerged in the ditch. As it rotates, it pushes the mixed liquor around the track. The rotor also provides the required aeration to maintain a DO level of 2 mg/L in the basin. The BOD load for oxidation ditches may vary from 10 to 50 lbs/1000 cu ft/day. The oxidation ditch effluent passes to the secondary clarifier and RAS is returned to the ditch.

The velocity and DO level can be adjusted by changing the rotor speed and operating depth. The effluent weir is a slide gate that can be raised and lowered to change the water level in the ditch. This also changes how deep the rotor is submerged in the mixed liquor. When it runs deeper, it aerates better. The proper velocity for an oxidation ditch should be about 1 ft/s and if the velocity falls below 1 ft/s, there is a possibility that sludge will settle in the corners of the ditch resulting in the septic conditions and odors.

Some ditches are designed with a concrete wedge at the exit of each bend. As the flow comes "out of the turn," the wedge forces the water at the outside to the inside as it comes down the "straightaway" (Lofrano et al., 2013). This helps in mixing the flow and creates turbulence where settling is most likely to occur. Oxidation ditches, as with other extended air systems, do not have primary treatment. Pretreatment may be limited to bar screens. This means that grit will not be removed until it settles in the oxidation ditch. The grit buildup in the ditch can result in odor and loss of detention time. It should be removed anytime the unit is drained for service.

9.6.2.2.6.5 Sequenced Batch Reactor   A SBR is a process, which is used in small-scale plants of industries. In this process, the reactor basin is filled and aerated for a certain period of time, usually for 1–3 h and after the aeration process is complete, the reactor is allowed to settle and effluent is decanted from the top of the unit.

When the decanting cycle is completed, the reactor is again filled with raw sewage and the process is repeated. This process is popular because the entire process uses only one tank. Most plants do not have clarifiers or RAS systems. Large equalization basin is required in this process, since the influent flow must be contained while the reactor is in the aeration cycle.

9.6.2.2.7 Applications

Activated sludge process is widely used by large cities and communities where the large volume of wastewater is treated economically. The ASP plants are too good choices for the isolated facilities, such as hospitals or hotels, cluster situations, subdivisions, and small communities to adapt other alternation process.

9.6.2.2.8 Limitations

9.6.2.2.8.1 Sludge Bulking   Sludge blanket problems manifest themselves in several different ways. Straggler floc is common in young sludge. It can be eliminated by reducing wasting so the sludge ages. Pin floc occurs as the sludge becomes too old. It can be remedied by increasing the wasting rates to reduce the sludge age.

Filamentous bacteria are long stringy bacteria that are present in sludge of all ages. These are beneficial in small numbers and help to stick sludge particles together but in large numbers they bridge the particles and reduce the settleability of the sludge and when this occurs sludge "bulking" problems take place. Bulking sludge is dark brown in color and can also be accompanied by dark, oily foam accumulation in the aeration basin. *Nocardia* is a type of filamentous bacteria that is known to cause bulking and foaming problems in ASP. They are long cylindrical organisms that cluster in groups such as broom bristles. They appear in older sludge and thrive when pH and oxygen levels drops below 6.0–1.0 mg/L and addition of 1.0–1.5 mg/L of chlorine to the RAS flow is often used in wastewater systems to control the growth of filamentous *Nocardia*.

Hydraulic shock loading can cause a "wash out" of the sludge blanket. When the upward velocity in clarifier exceeds the settling velocity of sludge, the solids are blown out of the tank and over the weirs. Flow equalization basins are the only way to avoid hydraulic shock to the system when flow spike. Rising sludge occurs when the DO is depleted and a portion of the sludge blanket becomes septic. Chunks of sludge will float to the surface as anaerobic gases are produced and trapped in sludge blanket. Besides losing solids from the system, the septic conditions mean the bugs are also very sick and may not perform well when they are returned to aeration.

9.6.2.2.8.2 Foaming The bioactivity in aeration basin will always result in some foam buildup in basin. The color of foam is an indicator of sludge age and condition. A crisp, white foam is indicative of young sludge while rich medium tan foam is associated with good settling sludge. Dark brown, oily foam can be found in older sludge. It is common in aerobic digesters, but not a good sign in an aeration basin.

9.6.2.3 Anaerobic Process

Anaerobic treatment processes are the subset of processes in which microorganisms break down biodegradable material in the absence of oxygen, used for industrial or domestic purposes to manage waste and/or to release energy.

There has recently been growing interest within the scientific community regarding the anaerobic treatment of TWW due to the several drawbacks of its application (Lofrano et al., 2013): (i) the implementation of adequate technology for H_2S desorption and treatment is required due to the consistent production of sulfide as a result of the reduction of sulfate, which occurs in the absence of alternative electron acceptors such as oxygen and nitrate; (ii) high protein component affects selection of biomass, slow kinetics of hydrolysis, and also inhibits the granular sludge formation.

The anaerobic treatment of TWW is mainly performed by using the anaerobic filters (AF) composed of both upflow anaerobic filters (UAF) and downflow anaerobic filters (DAF) and UASB reactors (Lefebvre et al., 2006; El-Sheikh et al., 2011). Only a few experiments have referred to the expanded granular sludge bed and anaerobic baffled reactor (ABR) (Zupancic and Jemec, 2010).

9.6.2.4 Wetlands and Ponds

Constructed wetlands (CWs) may be an interesting treatment option for leather TWW. Several efforts have been undertaken in the last decades for selecting plant species tolerant to this peculiar wastewater (Calheiros et al., 2007, 2008; Calheiros et al., 2012), selecting suitable supporting media or substrate (Calheiros et al., 2008) and for approaching the bacterial dynamics (Aguilar et al., 2008; Calheiros et al., 2009a,b).

The choice of plants plays an important role in CWs, as they must survive the potential toxic effects of wastewater and its variability. The potential of CWs for the phytoremediation of chromium (10 and 20 mg Cr/dm^3) from primary treated leather TWW by *Pennisetum purpureum, Brachiaria decumbens,* and *Phragmites australis* has been investigated by Mant et al. (2004). Calheiros et al. (2008) evaluated the use of *Canna indica, Typha latifolia, Ph. australis, Stenotaphrum secundatum,* and *Iris pseudacorus* in CWs receiving wastewater from a tannery industry, under two different hydraulic loading rates, 3 and 6 cm/day. *Phragmites australis* and *T. latifolia* were only the plants that were able to establish successfully. Calheiros et al. (2012) tested *Arundo donax* and *Sarcocornia fruticosa* in two series of horizontal subsurface flow CWs used to treat the effluent of conventional biological treatment system operating at a tannery.

Tadesse et al. (2004) also reported the influence of pH, temperature, and dissolved oxygen in a pilot-scale advanced integrated wastewater pond system treating TWW. According to their studies, a combination of advanced facultative pond, secondary facultative pond, and maturation pond all arranged in series, preceded with simple pretreatment can adequately treat combined raw TWW.

9.7 Bacterial Communities Present during the Treatment of Tannery Wastewater

Environmental pollution due to wastewater discharged from tannery industries causes serious problems in environment as well as health threats to both humans and animal. Since it contains a complex mixture of both inorganic and OPs, TWW also contains a variety of POPs, which do not degrade much during the secondary treatment process in industries and goes into the environment. Microbial communities were also reported in the degradation of TWW; especially bacterial communities reported by many workers in literature (Table 9.6).

However, Camargo et al. (2003) have reported a number of bacteria such as *Pseudomonas* sp., *Microbacterium* sp., *Desulfovibrio* sp., *Enterobacter* sp., *E. coli, Shewanella alga,* and *Bacillus* sp., which play an important role in detoxification of hexavalent chromium (Cr^{6+}) by the periplasmic biosorption, intracellular bioaccumulation, and biotransformation through direct enzymatic reaction or indirectly with metabolites. Megharaj et al. (2003) have isolated an *Arthrobacter* sp. and a *Bacillus* sp., from a long-term TWW contaminated soil, and examined for their capability to tolerate and reduce Cr^{6+} to Cr^{3+}. Both the bacterial isolates tolerate Cr^{6+} at 100 mg/mL on a minimal salts agar medium supplemented with 0.5% glucose, but only Arthrobacter was able to grow in liquid medium at this concentration. *Arthrobacter* sp. could reduce Cr^{6+} up to 50 μg/mL, while *Bacillus* sp. was not able to reduce Cr^{6+} beyond 20 μg/mL. Further, the *Arthrobacter* sp. was distinctly superior over *Bacillus* sp. in terms of their Cr^{6+}-reducing ability and resistance to Cr^{6+}. Assays with permeabilized (treated with toluene or Triton X 100) cells and crude extracts demonstrated that the Cr^{6+} reduction was mainly associated with that soluble protein fraction of the cell. *Arthrobacter* sp. has a great potential for bioremediation of Cr^{6+}-containing waste. Bacterial

TABLE 9.6

Bacterial Communities Reported in the Degradation of Tannery Wastewater

Bacteria	Reduction in COD (%)	Reduction in BOD (%)	Chromium	References
Halophiles	95	—	—	Lefebvre et al. (2005)
Thiobacillus ferrooxidans	69	72	5	Mandal et al. (2010)
S. condensate, Rhizoclonium hieroglyphicum	—	—	>75	Onyancha et al. (2008)
Escherichia coli	98.46	90	—	Ramteke et al. (2010)
Vibrio sp.	87.5			
Pseudomonas sp.	96.15			
Bacterial strain	—	—	87	Shakoori et al. (2000)
Pseudomonas aeruginosa, Bacillus flexus, Exiguobacterium homiense, Staphylococcus aureus	80	—	—	Sivaprakasam et al. (2008)
Acidithiobacillus thiooxidans	—	—	99.7	Wang et al. (2007)
Acinetobacter sp.				Srivastava et al. (2007)
E. coli sp., *Stenotrophomonas* sp., *Cronobacter* sp., *Burkholderia* sp.	86.87	92	87.96	Chandra et al. (2011)

strains (CrT-11, CrT-12, *Brevibacterium* sp. CrT-13, CrT-14) were isolated from the TWW. All strains could resist very high concentration of $K_2Cr_2O_7$ that is up to 40 mg/mL on nutrient agar and 25 mg/mL in nutrient broth. They have wide pH (5–9) and temperature (24–42°C) growth range. They exhibited multiple metals (Ni, Zn, Mn, Cu, Co, and Pb) and antibiotics (streptomycin, ampicillin, Cr^{6+} in to Cr^{3+} aerobically). *Brevibacterium* sp. CrT-13 accumulated and reduced more Cr^{6+} at all the concentrations applied in comparison to the other strains. These bacterial strains also took up and reduced Cr^{6+} present in industrial effluents and their reduction potential was not significantly affected in the presence of different metallic salts (Costa, 2003). Sinha et al. (2011) isolated tem bacteria from TWW that were found to be efficient in reducing 70% Cr^{6+} under anoxic condition. This included five isolates of *Pseudomonas*, three isolates of *Micrococcus* and two isolates of *Aeromonas*. They suggested the isolated bacteria for possible use in Cr^{6+} detoxification from TWW. Farag and Zaki (2010) isolated four chromium-resistant bacterial strains (S1, S2, S3, and S4) from TWW collected from Burgelarab, Alexandria, and Egypt. Among these isolates, S3 and S4 were identified as *Acinetobacter* and *Pseudomonas*, respectively, based on the 16S rDNA gene sequence analysis. These isolates displayed different degrees of chromate reduction under aerobic conditions. They found that the strain S4 was able to reduce a wide range of Cr^{6+} concentrations from 20 to 200 mg/L, while it was reducing 64.4% of Cr^{6+} at 160 mg L^{-1} within 72 h. Singh et al. (2013) reported the detoxification of hexavalent chromium by an indigenous facultative anaerobic *Bacillus cereus* strain (FA-3) isolated from TWW. FA-3 was tolerant to 1400 µg/mL of Cr^{6+} and reduced a maximum of 72% Cr^{6+} at 1000 µg/mL chromate concentration and therefore was suggested for the bioremediation of chromate containing wastes. Chaturvedi (2011) isolated a chromate-removing strain from spent chrome effluent and identified as *Bacillus circulans* strain MN1. The isolated strain was studied for resistance to Cr^{6+} and its ability to remove Cr^{6+}. He found that the strain was capable of tolerating Cr^{6+} concentration as high as 4500 mg/L, but the cells growth was heavily influenced when initial Cr^{6+} concentration was increased between 1110 and 4500 mg/L while Cr^{6+} at 500–1110 mg/L did not suppress the cells growth. He also demonstrated that the cells removed toxic Cr^{6+} more efficiently at 30°C compared with that at 25 and 35°C. The optimum initial pH for Cr^{6+} removal was 5.6 and final pH values of 5.1–5.6 were observed for initial pH 5.2–5.7. Sen et al. (2014) isolated a total of 38 bacterial strains from the TWW-enriched soil that were able to grow at 200 mg/L Cr^{6+} with minimum inhibitory concentration (MIC) of Cr^{6+} ranging from 50 to 750 mg/L. The potent Cr^{6+}-resistant isolates showed a very high tolerance level to 750 mg/L and were able to show 100% Cr^{6+} reduction up to 200 mg/L within 48 h. Benazir et al. (2010) selected a consortia of *Bacillus subtilis*, *P. aeruginosa*, and *Saccharomyces cerevisiae* for chromium remediation ability in their immobilized forms. They found that the chromium content of the effluent was around 770 mg/L before remediation, after which it reduced to 5.2–5.7 mg/L. The best activity was observed by *Sa. cerevisiae-P. aeruginosa* consortia, followed by immobilized beads of *Sa. Cerevisiae*, and *Sa. cerevisiae-B. subtilis consortia*. Tewari et al. (2012) selected a consortium of pentachlorophenol (PCP) degrading bacterial strains isolated from TWW in Jajmau, Kanpur, India, which were enriched in a lab-scale chemostat for enhancement of their degradation potential and used in a lab-scale bioreactor for the degradation of PCP. The isolated bacterial strains were identified as *Flavobacterium* species. The enrichment potential for PCP degradation by this consortium was enhanced during growth in the chemostat 91.0% reduction in PCP levels was observed within 80 h. Revealing possible applications of this consortium for bioremediation of sites contaminated with PCP. Bacterial consortiums were able to degrade the PCP concentration of 30 mg/L.

9.8 Emerging Treatment Technologies

9.8.1 Membrane Processes

In recent years, membrane technologies have been focused and their cost is continuing to reduce while the application possibilities are ever extending. The use of membrane technologies applied to the leather industry represents an economic advantage, especially in the recovery of chromium from residual waters of leather tanning process. Several studies have shown that cross flow microfiltration, ultrafiltration, nanofiltration, reverse osmosis (RO), and supported liquid membranes can be applied in leather industry for the recovery of chromium from spent liquors (Ashraf et al., 1997; Cassano et al., 2001; Labanda et al., 2009), reuse of wastewater and chemicals of deliming/bating liquor (Gallego-Molina et al., 2013), reduction in polluting load of unhairing and degreasing (Catarino et al., 2009), removal of salts, and in biological treatment of TWW in the light of their reuse.

Reverse osmosis with a plane membrane has been used as posttreatment process to remove refractory organic compounds (chloride and sulfate) by De Gisi et al. (2009). The high quality of permeate produced by the RO system with a plane membrane allowed the reuse of TWW within the production cycle, thus reducing groundwater consumption.

9.8.2 Membrane Bioreactors

Membrane bioreactor (MBR) has been attracting much attention from scientists and engineers for TWW treatment due to the numerous advantages over CASP, such as elimination of settling basins, independence of process performance from filamentous bulking, or other phenomena affecting settleability (Suganthi et al., 2013). MBR systems essentially consists of a combination of membrane and biological reactor systems. The separation of biomass from wastewater by membranes also allows the concentration of MLSS in bioreactor to be increased significantly. However, from the studies of Munz et al. (2009), it is possible to infer how the kinetics of nitrification are effectively reduced by the presence of tannins, without large differences between biomass selected with either the CASP and the MBR. One of the main drawbacks of membrane application is a significant fouling due to the clogging, adsorption, and cake layer formation by the pollutants onto the membrane. In recent years, extensive work is in progress to reduce biofouling.

9.8.3 Advanced Oxidation Processes

There has been increasing number of studies on advanced oxidation processes (AOPs) to treat TWW and chemicals. AOPs refers to the set of chemical treatment processes that uses strong oxidizing agents (O_3, hydrogen peroxide [H_2O_2]) and/or catalysts (Fe, Mn, TiO_2) sometimes supported in activity by high-energy radiation, for example, UV light (Schrank et al., 2009; Di Iaconi et al., 2012). These processes are all based on the production and utilization of hydroxyl radicals, which are very powerful oxidants that quickly and unselectively oxidize a broad range of organic compounds. The scientific interest toward AOPs application to high strength wastewater has increased remarkably in the past 20 years. AOPs can reduce the concentration of pollutants several hundred ppm to less than 5 ppb and therefore significantly brings the level of COD and TOC down, which earned it the credit of "wastewater treatment processes of the 21st century" (Munter, 2001). Generally, AOPs can be used to treat wastewater generated after secondary treatment processes, which is then called tertiary treatment (Audenaert et al., 2011). The pollutants are converted to a large extent into stable inorganic compounds such as water, carbon dioxide, and salts, that is, they undergo mineralization. However, most studies evaluated the efficiency of the treatment by COD removal; TOC remains a more suitable parameter to investigate the state of mineralization occurring in the process (Schrank et al., 2009, 2005). A goal of wastewater purification by means of AOP procedures is the reduction of the chemical pollutants and the toxicity to such an extent that the cleaned wastewater may be reintroduced into receiving streams or, at least, into a conventional sewage treatment. AOPs still have not been put into commercial use on a large scale (especially in developing countries) even up to today mostly because of the relatively high costs. Nevertheless, its high oxidative capability and efficiency make AOPs a popular technique in tertiary treatment in which the most recalcitrant organic and inorganic pollutants are to be eliminated. The increasing interest in water reuse and more stringent regulations regarding water pollution are currently accelerating the implementation of AOPs at full scale.

9.9 Combination of Biological Treatment Processes with Physical/Chemical Treatment Process

Due to complexity of TWW, the biological pretreatment could not be stabilized all the time and nitrification process sometime gets inhibited. To overcome these impediments, the combined application of physical/chemical treatment methods with biological treatment methods is preferred. The oxidative treatment distinctly improved the aerobic biodegradation of refractory organic compounds and found to be optimum at ozone consumption in range of about 2 g O_3/g DOC for both batch and continuous operating conditions. Moreover, nitrification could be established during subsequent aerobic degradation and the remaining ammonia is completely removed.

Jochimsen et al. (1997) investigated the combined biological treatment of TWW substreams, that is, beamhouse and tanyard wastewater and the application of an oxidative treatment by ozone, followed by a second aerobic treatment process. Jochimsen and his team found that the combined oxidative and biological treatment of beamhouse and tanyard wastewater was effective and meets the maximum permissible limits for COD and ammonia for their direct discharge of TWW into rivers. Mandal et al. (2010) have also studied the combined treatment of TWW by aerobic treatment process incorporating the *Thiobacillus ferrooxidans* and Fenton's reagents. The sole treatment by Fenton's oxidation (FO) involving the introduction of 6 g $FeSO_4$ and 266 g H_2O_2 in a liter of wastewater at pH of 3.5 and 30°C for 30 min under batch condition reduced the COD, BOD, sulfide, total chromium, and color up to 69%, 72%, 88%, 5%, and 100% and *Th. ferrooxidans* alone showed the maximum reduction to an extent of 77%, 80%, 85%, 52%, and 89%, respectively, in 21 days treatment at pH 2.5, temperature 30°C and 16 g/L of $FeSO_4$. The combined treatment under batch conditions involving 30 min chemical treatment by FO followed by 72 h biochemical treatment by *Th. ferrooxidans* under batch condition gave rise to 93%, 98%, 72%, 62%, and 100% removal efficiency of COD, BOD, sulfide, chromium, and color at pH 2.5 and temperature 30°C. They observed a decrease in photoabsorption of the Fenton's reagent treated samples, as compared to the blanks at 280, 350, and 470 nm wavelengths. This may be the key factor for stimulating the biodegradation by *Th. ferrooxidans*.

Biological treatment of TWW carried out in a sequencing batch biofilm reactor with chemical oxidation by ozone is an innovative treatment process. Moreover, it has been proved that the combined treatment processes produce a very low amount of sludge as compared to the value reported for conventional biological treatment (Di Iaconi et al., 2002).

9.10 Future Challenges in the Treatment of Tannery Wastewater

The major problems with the TWW is its complex nature due to the POPs such as tannins and other poorly biodegradable compounds as well as metals that can inhibit the biological treatment. Besides POPs, TWW also contains some toxic metals ions such as Cu^{2+}, Cr^{3+}, Fe^{3+}, Zn^{2+}, Pb^{2+}, and so on, present in TWW also have a high inhibitory and antimicrobial activity reducing the anaerobic digestion of TWW. Stasinakis et al. (2002) observed a significant inhibition of heterotrophic growth in the presence of 10 mg/L Cr^{6+}. Recently, the sequential applications of bacteria and wetland plants have been reported to be very promising for the degradation and detoxification of TWW, but this has to be optimized yet with the detailed microbiology of wetland plants, plants rhizosphere, and detoxification mechanisms. Moreover, the nature of POPs in TWW and extent of their toxicity need to be explained in detail for their complete degradation and detoxification during the treatment process at CETPs.

9.11 Conclusion and Recommendations

1. TWW is ranked as one of the major source of pollution in the environment.

2. TWW characteristics mainly depend on the chemical compounds used in tanning process.

3. TWW contains a complex mixture of both inorganic and OPs, which create serious health hazards to humans and animals.

4. TWW also contain a variety of POPs, which do not degrade much during the secondary (biological) treatment and go into the environment causing toxicity to the environment.

5. There are many treatment options available for TWW *viz.* physico-chemical, chemical, and biological treatment methods.

6. The combined application of physical or chemical treatment process with biological treatment process to treat the TWW would give the satisfactory results as compared to the individual treatment process.

7. The ETTs-like membrane filtration and oxidation processes are also currently using/under investigation for the treatment of TWW.

8. It appears that ETTs are promising to remove POPs completely. However, there is still a need to optimize these treatment technologies for economic reasons.

9. The organization of tanneries in industrial cities is another common approach, which helps much to abate the pollution in parallel to strengthen the discharge limits for TWW.

10. The integrated pollution prevention strategy of EU and greening economy that includes the shifting chemicals with the natural ones, water minimization technologies and water recycling will continue to spend the efforts for solving the environmental problems that occurs due to TWW.

However, it is not yet possible to distinguish which treatment methods is the best to treat the TWW. It is clear that continuous efforts are required to attract the minds of scientific communities to increase more and more knowledge about the treatment processes, so that treatment of TWW can be made feasible for environmental safety in near future. Hence, the aim of this chapter has been to arouse curiosity among those working in this area. This chapter has therefore mainly focused on the POPs in TWW that remain, even after the secondary (biological) treatment process, conventional and advanced ETTs for TWW and bacterial communities present during the degradation and detoxification of TWW for its safe disposal into the environment.

Acknowledgments

Authors are thankful to Prof. Ram Chandra, Head, Department of Environmental Microbiology, School for Environmental Sciences, Babasaheb Bhimrao Ambedkar University (A Central University), Vidya Vihar, Raebareli Road, Lucknow 226 025 (UP), India, for valuable suggestions accorded and discussions during the preparation of this chapter. Authors are also highly grateful to the Department of Science and Technology (DST), Government of India (GOI), New Delhi, for providing the financial support as "Major Research Project" (Grant No.: SB/EMEQ-357/2013) for this work and the University Grant Commission (UGC) Fellowship received by Gaurav Saxena is also duly acknowledged.

References

Abelsohn A, Gibson BL, Sanborn MD, and Weir E. 2002. Identifying and managing adverse environmental health effects: Persistent organic pollutants. *Canadian Medical Association Journal* 166(12): 1549–1554.

Adeola FO. 2004. Boon or Bane? The environmental and health impacts of persistent organic pollutants (POPs). *Human Ecology Review* 11(1): 27–35.

Aguilar JRP, Cabriales JJP, and Vega MM. 2008. Identification and characterization of sulfur-oxidizing bacteria in an artificial wetland that treats wastewater from a tannery. *International Journal of Phytoremediation* 10: 359–370.

Akinici IE and Akinci S. 2010. Effect of chromium toxicity on germination and early seedling growth in melon (*Cucumis melo* L.). *African Journal of Biotechnology* 9: 4589–4594.

Alam MZ, Ahmad S, and Malik A. 2009. Genotoxic and Mutagenic potential of Agricultural soil irrigated with tannery effluents at Jajmau (Kanpur), India. *Achieves of Environmental Contamination and Toxicology* 57: 463–476.

Alam MZ, Ahmad S, Malik A, and Ahmad M. 2010. Mutagenicity and genotoxicity of tannery effluents used for irrigation at Kanpur, India. *Ecotoxicology and Environmental Safety* 73: 1620–1628.

Asfaw A, Sime M, and Itanna F. 2012. Determining the effect of tannery effluent on seeds germination of some vegetable in Ejersa areas of east shoa, Ethiopia. *International Journal of Scientific and Research Publications* 2(12): 1–10.

Ashraf CM, Ahmad S, and Malik MT. 1997. Supported liquid membrane technique applicability for removal of chromium from tannery wastes. *Waste Managemant* 17: 211.

Ates E, Orhon D, and Tunay O. 1997. Characterization of tannery wastewater for pretreatment-selected case studies. *Water Science and Technology* 36: 217–223.

Audenaer WTM, Vermeersch Y, Van Hulle SWH, Dejans P, Dumouilin A, and Nopens I. 2011. Application of a mechanistic UV/hydrogenperoxide model at full-scale: Sensitivity analysis, calibration and performance evaluation. *Chemical Engineering Journal* 171(1): 113–126.

Barnhart J. 1997. Chromium chemistry and implications for environmental fate and toxicity. *Journal of Soil Contamination* 6: 561–568.

Battista G, Comba P, Orsi D, Norpoth K, and Maier A. 1995. Nasal cancer in leather workers: An occupational disease. *Journal of Cancer Research and Clinical Oncology* 121: 1–6.

Benazir JF, Suganthi D, Rajvel M, Padmini P, and Mathithumilan B. 2010. Bioremediation of chromium in tannery effluent by microbial consortia. *African Journal of Biotechnology* 9(21): 3140–3143.

Bolt HM and Degen GH. 2002. Comparative assessment of endocrine modulators with oestrogenic activity II. Persistent organochlorine pollutants. *Archieves of Toxicology* 76: 187–103.

Borja J, Taleon DM, Auresenia J, and Gallardo S. 2005. Polychlorinated biphenyls and their biodegradation. *Process Biochemistry* 40(6): 1999–2013.

Bouchez M, Blanchet D, Bardin V, Haeseler F, and Vandecasteele JP. 1999. Efficiency of defined strains and of soil consortia in the biodegradation of polycyclic aromatic hydrocarbon (PAH) mixtures. *Biodegradation* 10: 429–435.

Calheiros CSC, Duque AF, Moura A, Henriques IS, Correia A, Rangel AOSS, and Castro PML. (2009a). Changes in the bacterial community structure in two-stage constructed wetlands with different plants for industrial wastewater treatment. *Bioresource Technology* 100: 3228–3235.

Calheiros CSC, Quiterio PVB, Silva G, Crispim LFC, Brix H, Moura SC, and Castro PML. 2012. Use of constructed wetland systems with Arundo and Sarcocornia for polishing high salinity tannery wastewater. *Journal of Environmental Managemant* 95: 66–71.

Calheiros CSC, Rangel AOSS, and Castro PML. 2007. Constructed wetland systems vegetated with different plants applied to the treatment of tannery wastewater. *Water Research* 41: 1790–1798.

Calheiros CSC, Rangel AOSS, and Castro PML. 2008. Evaluation of different substrates to support the growth of *Typha latifolia* in constructed wetlands treating tannery wastewater over long-term operation. *Bioresource Technology* 99: 6866–6877.

Calheiros CSC, Rangel AOSS, and Castro PML. 2009b. Treatment of industrial wastewater with two-stage constructed wetlands planted with *Typha latifolia* and *Phragmites australis*. *Bioresource Technology* 100: 3205–3213.

Camargo, FAO, Bento FM, Okeke BC, and Frankenberger WT. 2003. Chromate reduction by chromium resistant bacteria isolated from soils contaminated with dichromate. *Journal of Environmental Quality* 32: 1228–1233.

Cassano A, Molinari R, Romano M, and Drioli E. 2001. Treatment of aqueous effluent of the leather industry by membrane processes. A review. *Journal of Membrane Science* 181: 111–126.

Catarino J, Mendonca E, Picado A, Lanca A, Silva L, and DePinho MN. 2009. Membrane-based treatment for tanning wastewaters. *Canadian Journal of Civil Engineering* 36: 356–362.

Chandra R, Bharagava RN, Kapley A, and Purohit HJ. 2011. Bacterial diversity, organic pollutants and their metabolites in two aeration lagoons of common effluent treatment plant (CETP) during the degradation and detoxification of tannery wastewater. *Bioresource Technology* 102: 2333–2341.

Chandra R and Chaudhary S. 2013. Persistent organic pollutants in environment and health hazards. *International Journal of Bioassays* 2(09): 1232–1238.

Chaturvedi MK. 2011. Studies on chromate removal by chromium-resistant *Bacillus* sp. isolated from tannery effluent. *Journal of Environmental Protection* 2: 76–82.

Chidambaram AP, Sundaramoorthy A, Murugan K, and Baskaran SGL. 2009. Chromium induced cytotoxicity in black gram (*Vigna mungo* L). *Iranian Journal of Environmental Health, Science and Engineering* 6: 17–22.

Cooman K, Gajardo M, Nieto J, Bornhardt C, and Vidal G. 2003. Tannery wastewater characterization and toxicity effects on *Daphnia* spp. *Environmental Toxicology* 18: 45–51.

Costa M. 2003. Potential hazards of hexavalent chromate in our drinking water. *Oxicology and Applied Pharmacology* 188(1): 1–5.

De Gisi S, Galasso M, and De Feo G. 2009. Treatment of tannery wastewater through the combination of a conventional activated sludge process and reverse osmosis with a plane membrane. *Desalination* 249: 337–342.

De Nicola E, Meric S, Gallo M, Iaccarino M, Della Rocca C, Lofrano G, Russo T, et al. 2007. Vegetable and synthetic tannins induce hormesis/toxicity in sea urchin early development and in algal growth. *Environmental Pollution* 146: 46–54.

Di Iaconi C. 2012. Biological treatment and ozone oxidation: Integration or coupling? *Bioresource Technology* 106: 63–68.

Di Iaconi C, Lopez R, Ramadori AC, Di P, and Passino R. 2002. Combined chemical and biological degradation of tannery wastewater by a periodic submerged filter. *Water Research* 36: 2205–2214.

Dogruel S, Genceli EA, Babuna FG, and Orhon D. 2006. An investigation on the optimal location of ozonation within biological treatment for a tannery wastewater. *Journal of Chemical Technology and Biotechnology* 81: 1877–1885.

Durai, G and Rajasimmam. 2011. Biological treatment of tannery wastewater—A review. *Journal of Environmental Science and Technology* 4(1): 1–17.

Eckenfelder WW. 2002. Industrial Water Pollution Control. McGraw-Hill, Singapore.

El-Sheikh Mahmoud A, Hazem I, Saleh J, Flora R, and AbdEl-Ghany MR. 2011. Biological tannery wastewater treatment using two stage UASB reactors. *Desalination* 276: 253–259.

Farag S and Zaki S. 2010. Identification of bacterial strains from tannery effluent and reduction of hexavalent chromium. *Journal of Environmental Biology* 31(5): 877–882.

Field JA and Sierra-Alvarez R. 2007. Biodegradability of chlorinated aromatic compounds. Science dossiers of EuroChlor. Available online at http//www.eurochlor.org/sciencedossiers

Gallego-Molina A, Mendoza-Roca JA, Aguado D, and Galiana-Aleixandre MV. 2013. Reducing pollution from the deliming-bating operation in a tannery. Wastewater reuse by microfiltration membranes. *Chemical Engineering Research and Desalination* 91: 369–376.

Ganesh R, Balaji G, and Ramanujam RA. 2006. Biodegradation of tannery wastewater using sequencing batch reactor-respirometric assessment. *Bioresource Technology* 97(15): 1815–1821.

Gerecke AC, Hartmann PC, Heeb NV, Kohler HPE, Giger W, Schmid P, Zennegg M, and Kohler M. 2005. Anaerobic degradation of decabromodiphenyl ether. *Environmental Science and Technology* 39(4): 1078–1083.

Gorgun E, Insel G, Artan N, and Orhon D. 2007. Model evaluation of temperature dependency for carbon and nitrogen removal in a full-scale activated sludge plant treating leather-tanning wastewater. *Journal of Environmetal Sciences and Health. A Toxic Hazardous Substances Environmental Engineering* 42: 747–756.

Guzzella L, Roscioli C, Vigano L, Saha M, Sarkar SK, and Bhattacharya A. 2005. Evaluation of the concentration of HCH, DDT, HCB, PCB and PAH in the sediments along the lower stretch of Hugli estuary, West Bengal, northeast India. *Environment International* 31(4): 523–534.

Hayder S, Aziz JA, and Ahmad MS. 2007. Biological treatment of tannery wastewater using activated sludge process, Pakistan. *Journal of Engineering and Applied Sciences* 1: 61–66.

Hermann MB and Gisela HD. 2002. Comparative assessment of endocrine modulators with oestrogenic activity. II. Persistent organochlorine pollutants. *Archives of Toxicology* 76 (4): 187–193.

Insel GH, Gorgun E, Artan N, and Orhon D. 2009. Model based optimization of nitrogen removal in a full scale activated sludge plant. *Environmental Engineering and Sciences* 26: 471–480.

Irfan EA and Akinici S. 2010. Effect of chromium toxicity on germination and early seedling growth in melon (*Cucumismelo* L.). *African Journal of Biotechnology* 9: 4589–4594.

Jawahar AJ, Chinnadurai M, Ponselvan JKS, and Annadurai G. 1998. Pollution from tanneries and options for treatment of effluent. *Indian Journal of Environmental Protection* 18: 672–672.

Jochimsen J and Jekel MP. 1997. Partial oxidation effects during the combined oxidative and biological treatment of separated stream of tannery wastewater. *Water Science and Technology* 35: 337–345.

Jochimsen JC, Schenk H, Jekel MR, and Hegemann W. 1997. Combined oxidative and biological treatment for separated streams of tannery wastewater. *Water Science and Technology* 36: 209–216.

Johnsen AR, Wick LY, and Harms H. 2005. Principles of microbial PAH-degradation in soil. *Environmental Pollution* 133(1): 71–84.

Kadam RV. 1990. Treatment of tannery wastes. *Indian Journal of Environmental Protections* 10: 212–212.

Karahan O, Dogruel S, Dulekgurgen E, and Orhon D. 2008. COD fractionation of tannery wastewater-particle size distribution, biodegradability and modeling. *Water Research* 42: 1083–1092.

Khwaja AR, Singh R, and Tandon SN. 2001. Monitoring of Ganga water and sediments *vis-a-vis* tannery pollution at Kanpur (India): A case study. *Environmental Monitoring and Assessment* 68: 19–35.

Kohli R and Malaviya P. 2013. Impact of tannery effluent on germination of various varieties of wheat (*Triticum aestivum* L). *Journal of Applied and Natural Science* 5(2): 302–305.

Kongjao S, Damronglerd S, and Hunsom M. 2008. Simultaneous removal of organic and inorganic pollutants in tannery wastewater using electrocoagulation technique. *Korean Journal of Chemical Engineering* 25(4): 703–709.

Kumar V, Majumdar C, and Roy P. 2008. Effects of endocrine disrupting chemicals from leather industry effluents on male reproductive system. *Journal of Steroid Biochemistry and Molecular Biology* 111: 208–216.

Labanda J, Khaidar MS, and Llorens J. 2009. Feasibility study on the recovery of chromium (III) by polymer enhanced ultrafiltration. *Desalination* 249: 577–581.

Lal R and Saxena DM. 1982. Accumulation, metabolism and effects of organochlorine insecticides on microorganisms. *Microbiological Reviews* 46(1): 95–127.

Lee DH, Lee IK, Song K, Steffes M, Toscano W, Baker BA, and Jacobs Jr. DR. 2006. A strong dose-response relation between serum concentrations of persistent organic pollutants and diabetes. *Diabetes Care* 29: 1638–1644.

Lefebvre O, Vasudevan N, Torrijosa M, Thanasekaran K, and Moletta R. 2006. Anaerobic digestion of tannery soak liquor with an aerobic post-treatment. *Water Research* 40: 1492–500.

Leta S, Assefa F, Gumaelius L, and Dalhammar G. 2004. Biological nitrogen and organic matter removal from tannery wastewater in pilot plant operations in Ethiopia. *Applied Microbiology and Biotechnology* 66: 333–339.

Lofrano G, Belgiorno V, Gallo M, Raimo A, and Meric S. 2006. Toxicity reduction in leather tanning wastewater by improved coagulation flocculation process. *Global Nest Journal* 8: 151–158.

Lofrano G, Meric S, Zengin GE, and Orhon D. 2013. Chemical and biological treatment technologies for leather tannery chemicals and wastewaters: A review. *Science of the Total Environment* 461: 265–281.

Mahmood S, Paton GI, and Prosser JI. 2005. Cultivation independent in-situ molecular analysis of bacteria involved in degradation of pentachlorophenol in soil. *Environmental Microbiology* 7(9): 1349–1360.

Mandal T, Maity S, Dasgupta D, and Datta S. 2010. Advanced-oxidation process and biotreatment: Their roles in combined industrial wastewater treatment. *Desalination* 250: 87–94.

Mant C, Costa S, Williams J, and Tambourgi E. 2004. Phytoremediation of chromium by model constructed wetland. *Bioresource Technology* 97(15): 1767–1772.

Matsumoto ST, Mnlovani SM, Malaguttii MIA, Dias AL, Fonseca IC, and Morales MAM. 2006. Genotoxicity and mutagenicity of water contaminated with tannery effluent, as evaluated by the micronucleus test and comet assay using the fish *Oreochromis niloticus* and chromosome aberrations in onion root tips. *Genetics and Molecular Biology* 29: 148–158.

Megharaj M, Avudainayagam S, and Naidu R. 2003. Toxicity of hexavalent chromium and its reduction by bacteria isolated from soil contaminated with tannery waste. *Current Microbiology* 47: 51–54.

Meric S, De Nicola E, Iaccarino M, Gallo M, Di Gennaro A, Morrone G, Warnau M, et al. 2005. Toxicity of leather tanning wastewater effluents in sea urchin early development and in marine microalgae. *Chemosphere* 61: 208–217.

Miles JRW and Moy P. 1979. Degradation of endosulfan and its metabolites by a mixed culture of soil- microorganisms. *Bulletin of Environmental Contamination and Toxicology* 23(1–2): 13–19.

Moore NP. 2000. The estrogenic potential of the phthalate esters. *Reproductive Toxicology* 14: 183–192.

Munter R. 2001. Advanced oxidation processes—current status and prospects. *Proceedings of the Estonian Academy of Sciences, Chemistry* 50(2): 59–80.

Munz G, De Angelis D, Gori R, Mori G, Casarci M, and Lubello C 2009. The role of tannins in conventional angogated membrane treatment of tannery wastewater. *Journal of Hazard Material* 164: 733–739.

Murat S, Insel G, Artan N, and Orhon D. 2006. Performance evaluation of SBR treatment for nitrogen removal from tannery wastewater. *Water Science and Technology* 53: 275–284.

Murugesan V and Elangoan R. 1994. Biokinetic parameters for activated sludge process treating vegetable tannery waste, *Indian Journal of Environment Protection* 14: 511–515.

Mwinyihija M. 2010. Main pollutants and environmental impacts of the tanning industry. In: Mwinyihija M. (Ed.), *Ecotoxicological Diagnosis in the Tanning Industry*. Springer-Verlag, Berlin, Heidelberg, Germany, pp. 17–35.

Oral R, Meric S, De Nicola E, Petruzzelli D, Della Rocca C, and Pagano G. 2007. Multi-species toxicity evaluation of a chromium-based leather tannery wastewater. *Desalination* 211: 48–57.

Orhon D, Ates GE, and Ubay CE. 1999. Characterization and modeling of activate sludge for tannery wastewater. *Water Environment Resources* 71: 50–63.

POPRC 2007. Draft Risk Profile for Commercial Octabromo-diphenyl Ether. Stockholm Convention on POPs, Persistent Organic Pollutants Review Committee, Available online at http://www.pops.int/documents/meetings/poprc/drprofile/default.html

Prest I, Jefferies DJ, and Moore NW 1970. Polychlorinated biphenyls in wild birds in Britain and their avian toxicity. *Environmental Pollution* 1: 3–26.

Rajamani S, Ramasami T, Langerwerf JSA, and Shappman JE. 1995. Environmental managemant in tanneries-feasible chromium recovery and reuse system. In: *Proceedings of the 3rd International Conference on Appropriate Waster Managemant Technologies for Developing Countries (AWMTDC95)*, Nagpur, India, pp. 965–969.

Ramteke PW, Awasthi S, Srinath T, and Joseph B. 2010. Efficiency assessment of common effluent treatment plant (CETP) treating tannery effluents. *Environmental Monitoring and Assessment* 169(1–4): 125–131.

Ritter L, Solomon KR, Forget J, Stemeroff M, and O'Leary C. 2007. *Persistent Organic Pollutants*. United Nations Environment Programme, Parent organization of United Nations, Nairobi, Kenya, Africa.

Samaranda C and Gavrilescu M. 2008. Migration and fate of persistent organic pollutants in the atmosphere—A modeling approach. *Environmental Engineering and Management Journal* 7: 743–761.

Schrank SG, Bieling U, Jose HJ, Moreira RFPM, and Schroder HFR. 2009. Generation of endocrine disruptor compounds during ozone treatment of tannery wastewater confirmed by biological effect analysis and substance specific analysis. *Water Science and Technology* 59: 31–38.

Sen P, Pal A, Chattopadhyay B, and Pal D. 2014. Chromium-tolerant bacteria in diversified soil microbial community in the bank of tannery waste water discharging canal of East Calcutta, West Bengal. *Journal of Biology and Environmental Sciences* 4: 233–238.

Senthilkumar S, Surianarayanan M, Sudharshan S, and Susheela R. 2008. Biological treatment of tannery wastewater by using salt-tolerant bacterial strains. *Microbial Cell Factories* 7: 15.

Shakir L, Ejaz S, Ashraf M, Aziz QN, Ahmad AA, Iltaf I, and Javeed A. 2012. Ecotoxicological risks associated with tannery effluent wastewater. *Environmental Toxicology and Pharmacology* 34: 180–191.

Shanker AK, Cervantes C, Loza-Tavera H, and Avudainayagam S. 2005. Chromium toxicity in plants. *Environment International* 31: 739–753.

Singh N, Verma T and Gaur R. 2013. Detoxification of hexavalent chromium by an indigenous facultative anaerobic Bacillus cereus strain isolated from tannery effluent. *African Journal of Biotechnology* 12(10): 1091–1103.

Sinha S, Singh S, and Mallick S. 2008. Comparative growth response of two varieties of *Vignaradiata* L. (var. PDM 54 and var. NM 1) grown on different tannery sludge applications: Effects of treated wastewater and ground water used for irrigation, *Environmental Geochemistry and Health* 30: 407–422.

Sinha SN, Biswas M, Paul D, and Rahaman S. 2011. Biodegradation potential of bacterial isolates from tannery effluent with special reference to hexavalent chromium. *Biotechnology, Bioinformatics and Bioengineering* 1(3): 381–386.

Siqueira IR, Vanzella C, Bianchetti P, Siqueira Rodrigues MA, and Stulp S. 2011. Anxiety-like behavior in mice exposed to tannery wastewater: The effect of photo-electro-oxidation treatment. *Neurotoxicology and Teratology* 33: 481–484.

Song Z and Burns RG. 2005. Depolymerisation and biodegradation of a synthetic tanning agent by activated sludges, the bacteria *Arthrobacter globiformis* and *Comamonas testosterone* and the fungus *Cunninghamella polymorpha*. *Biodegradation* 16: 305–318.

Song Z, Edwards SR, and Burns RG. 2005. Biodegradation of naphthalene-2-sulfonic acid present in tannery wastewater by bacterial isolates *Arthrobacter* sp. 2AC and *Comamonas* sp. 4BC. *Biodegradation* 16: 237–252.

Stasinakis AS, Mamais D, Thomaidis NS, and Lekkas TD 2002. Effect of chromium (VI) on bacterial kinetics of heterotrophic biomass of activated sludge. *Water Research* 36: 3342–3350.

Suganthi KV, Mahalaksmi M, and Balasubramanian N. 2013. Development of hybrid membrane bioreactor for tannery effluent treatment. *Desalination* 309: 231–236.

Sultan C, Balaguer P, Terouanne B, Georget V, Paris F, Jeandel C, Lumbroso S, and Nicolas JC. 2001. Environmental xenoestrogens, anti-androgens and disorders of male sexual differentiation. *Molecular and Cellular Endocrinology* 178: 99–105.

Sundarapandiyan S, Chandrasekar R, Ramanaiah B, Krishnan S, and Saravanan P. 2010. Electrochemical oxidation and reuse of tannery saline wastewater. *Journal of Hazardous Material* 180: 197–203.

Sweetman AJ, Valle MD, Prevedouros K, and Jones KC. 2005. The role of soil organic carbon in the global cycling of persistent organic pollutants (POPs): Interpreting and modeling field data. *Chemosphere* 60(7): 959–972.

Szpyrkowicz L, Kelsall GH, Kaoul SN, and De Faveri M. 2001. Performance of electrochemical reactor for treatment of tannery wastewaters. *Chemical Engineering and Science* 56: 157–186.

Tadesse I, Green FB, and Puhakka JA. 2004. Seasonal and diurnal variations of temperature, pH and dissolved oxygen in advanced integrated wastewater pond system® treating tannery effluent. *Water Research* 38: 645–654.

Tare V, Gupta S, and Bose P. 2003. Case studies on biological treatment of tannery wastewater in India. *Journal of Air and Waste Managemant Association* 53: 976–982.

Tewari CP, Shukla S, and Pandey P. 2012. Biodegradation of pentachlorophenol (PCP) by consortium of *Flavobacterium* sp. in tannery effluent. *Journal of Environmental Research and Development* 2A(October–December): 876–882.

Tigini V, Giansanti P, Mangiavillano A, Pannocchia A, and Varese GC. 2011. Evaluation of toxicity, genotoxicity and environmental risk of simulated textile and tannery wastewaters with a battery of biotests. *Ecotoxicology and Environmental Safety* 74: 866–873.

Tisler T, Zagorc-Koncan J, Cotman M, and Drolc, A. 2004. Toxicity potential of disinfection agent in tannery wastewater. *Water Research* 38: 3503–3510.

Tu CM, Miles JRW, and Harris CR. 1968. Soil microbial degradation of aldrin. Life Sciences Part 1 physiology and pharmacology and Part 2 biochemistry. *General and Molecular Biology* 7: 311–322.

Tunay O, Kabdasli I, Orhon D, and Ates E. 1995. Characterization and pollution profile of leather tanning industry in Turkey. *Water Science and Technology* 32: 1–9.

Turick CE, Apel WA, and Carmiol NS. 1996. Isolation of hexavalent chromium reducing anaerobes from hexavalent chromium contaminated and non-contaminated environments. *Applied Microbiology and Biotechnology* 44: 683–688.

US EPA. 1986. Guidelines for the health risk assessment of chemical mixtures (PDF). EPA/630/R-98/002

Verma T, Ramteke PW, and Garg SK. 2008. Quality assessment of treated tannery wastewater with special emphasis on pathogenic *E. coli* detection through serotyping. *Environmental Monitoring Assessment* 145: 243–249.

World Wildlife Fund. 1999. *Persistent Organic Pollutants: Hand-Me-Down Poisons that Threaten Wildlife and People*. WWF, Washington, DC.

WWF. 2005. Stockholm Convention "New POPs" Screening Additional POPs Candidates, World Wildlife Foundation, Stockholm, Sweden.

Zupancic GD and Jemec A. 2010. Anaerobic digestion of tannery waste: Semi-continuous and anaerobic sequencing batch reactor processes. *Bioresource Technology* 101: 26–33.

10

Microbial Degradation of Lignocellulosic Waste and Its Metabolic Products

Ram Chandra, Sheelu Yadav, and Vineet Kumar

CONTENTS

10.1 Introduction

Lignocellulose is the structural material used to make plant cell walls, and is therefore the main component of plant biomass. Lignocellulose consists of three main components: cellulose, hemicellulose, and lignin. Biomass in general consists of 40%–50% cellulose, 25%–30% hemicellulose, and 15%–20% lignin. Cellulose is the primary constituent of lignocellulosic biomass, which accounts for 30%–50% dry weight of lignocellulose, and is a polysaccharide composed of β-1,4-linked D-glucose units. Cellulose is used for the manufacture of paper and cardboard, and can be converted via the action of cellulase enzymes into glucose, for bioethanol production (Ahmad et al., 2010). The carbon cycle is closed primarily as a result of the action of cellulose-utilizing microorganisms present in soil and the guts of animals. Thus, microbial cellulose (MC) utilization is responsible for one of the largest material flows in the biosphere and is of interest in relation to the analysis of carbon flux at both local and global scales. The importance of MC utilization in natural environments is further enhanced by the status of ruminants as a major source of dietary protein. Finally, MC utilization is also an integral component of widely used processes such as anaerobic digestion and composting. MC, a polysaccharide synthesized in abundance by *Acetobacter xylinum*, has already been used quite successfully in wound-healing applications, proving that it could become a high-value product in the field of biotechnology.

Hemicelluloses are branched polysaccharides, which are associated with lignin and cellulose in plant cell walls, consist of other polysaccharides, principally xylans and mannans, which are closely associated with the cellulose filaments, and chemically linked with lignin. The major hemicellulose in hardwoods is xylan (15%–30% dry weight), a polysaccharide composed of β-1,4-linked D-xylose units, which can be substituted with other monosaccharide units; whereas softwood hemicellulose contains mainly galactoglucomannan (15%–20% dry weight), a polysaccharide composed of β-1,4-linked D-glucose, and D-galactose units (Bugg et al., 2011). They have a higher hydrolysis rate than cellulose in hydrothermal pretreatment. During hydrothermal pretreatment, water itself and organic acid liberated from side chain of hemicelluloses catalyzes the breakdown of long hemicelluloses chains to form shorter chain oligomers and/or sugar monomers (Roos et al., 2009; Tunc and Heiningen van, 2011). Indeed, in addition to hemicellulose, hydrothermal pretreatment of lignocellulosic materials also involves solubilization of extracts, small portions of cellulose, and lignin (Vázquez et al., 2005). Moreover, hemicellulose is absolutely the primary substance in the hydrolysate liquor, therefore hemicellulose utilization and value-added product conversion would largely benefit the effectiveness and competitiveness of the entire biofuel production process.

Enzymatic hydrolysis of hemicellulose requires endo-1,4-xylanase, β-xylosidase, α-glucuronidase, α-L-arabinofuranosidase, and acetylxylan esterase, which act on xylan degradation and saccharification (Saha, 2004; Carvalheiro et al., 2008), and -mannanase and -mannosidase, which cleave the glucomannan polymer backbone (Kumar et al., 2008).

Lignin is a complex aromatic heteropolymer, composed of phenylpropanoid aryl-C3 units, linked together via a variety of ether and C–C bonds. Lignin accounts for 15%–30% dry weight of lignocellulose, in which it forms a matrix that is closely associated with the cellulose filaments, and is covalently attached to hemicelluloses. Lignin occurs by polymerization of guaiacyl (G) units from precursor coniferyl alcohol, syringyl (S) units from precursor sinapyl alcohol, and *p*-hydroxyphenyl (H) units from precursor *p*-coumaryl alcohol. The ratio of G:S:H units varies from species to species, but softwoods are generally G-type lignins, containing mainly G units, while hardwoods are generally GS-type lignins, containing mixtures of G and S units, and grass lignins are G type lignins containing a higher proportion of H units (Rodrigues et al., 1999). The white-rot fungi degrade lignins not only to use them as carbon sources but also to remove a physical barrier against cellulose utilization. Due to their powerful degrading capabilities toward various recalcitrant chemicals, white-rot fungi and their lignin degrading enzymes have long been studied for biotechnical applications such as biobleaching (Takano et al., 2001), biodecolorization (Dias et al., 2003), and bioremediation (Beltz et al., 2001; Cheong et al., 2006). The lignin-degrading enzymes consist of laccase, lignin peroxidase, manganese peroxidase, and H_2O_2-supplying glucose oxidase for the peroxidase reactions. *Phanerochaete chrysosporium* is one of the most widely studied white-rot fungi with regards to lignin-degrading enzymes. But several potential bacterial species have been reported for degradation of kraft lignin that is, *Bacillus* sp., *Paenibacillus* sp., and *Aneurinibacillus aneurinilyticus* (Raj et al., 2007a,b; Chandra et al., 2008). Further the identification of low molecular weight aromatic compounds by gas chromatography–mass spectroscopy (GC–MS) from kraft lignin degradation by three *Bacillus* sp. is also reported by Raj et al. (2007a,b). In addition, the degradation of potential bacteria for chlorophenol is also reported by Singh et al. (2008), which is also a major by-product from pulp paper industry.

Although there are many different types of industries across the world pulp and paper industries are the major source of environmental pollution that produces large amount of lignocellulosic waste. The pulp manufacturing process includes wood digestion and bleaching. In the former one, the wood chips are cooked at elevated temperature and pressure in sodium hydroxide and sodium sulfate solution to break chips into fiber mass. The chemical reaction with wood fibers, dissolve all the depository materials that are hard to degrade. To overcome this problem, many studies had been done on microorganisms that are capable of degrading the lignocellulosic waste (Chandra and Abhishek, 2011). The decolorization and detoxification of rayon grade pulp paper mill effluent by mixed bacterial culture isolated from pulp paper mill effluent site is reported by Chandra and Singh (2012). Besides, the metabolic product characterization during the degradation is also reported by Chandra et al. (2011). In recent studies it has been also reported that the biodegradation of residual pollutants of pulp paper mill waste is also influenced by the ratio of bacterial community during decolorization and detoxification of pulp paper mill effluent (Raj et al., 2007a,b; Chandra et al., 2012).

Many microorganisms such as bacteria, fungi, and actinomycetes are reported, which have the capability to degrade and utilize cellulose and hemicellulose as carbon and energy sources. However, a much smaller group of filamentous fungi has evolved with the ability to break down lignin, the most recalcitrant component of plant cell walls. These are known as white-rot fungi, which possess the unique ability of efficiently degrading lignin to CO_2 (Sánchez, 2009). This degradative ability of white-rot fungi is due to the strong oxidative activity and low substrate specificity of their ligninolytic enzymes. Other lignocellulose degrading fungi are brown-rot fungi that rapidly depolymerize cellulosic materials while only modifying lignin. Some soft-rot fungi can degrade lignin, because they erode the secondary cell wall and decrease the content of acid-insoluble material (Klason lignin) in angiosperm wood. Soft-rot fungi typically attack higher moisture, and lower lignin content materials (Shary et al., 2007).

Due to increase in environmental contamination of lignocellulosic material as persistent organic pollutants (POPs) the ligninolytic enzyme provides a breakthrough in the microbial field of environmental biotechnology. Further, there is a need to detect and purify these enzymes. Presently, there are several techniques through which the enzymes could be extracted after the screening process. The knowledge of contaminants should be adequate, for this the contaminants after extraction are concentrated and characterized by available techniques such as thin layer chromatography, high-performance liquid chromatography (HPLC), and GC–MS (Chandra et al., 2013).

Worldwide lignocellulosic residue generation every year results in pollution of the environment and loss of valuable materials that can be bioconverted to several added-value products. Lignin can be removed by chemical or physical pretreatment, which then permits efficient bioconversion. Pretreatment can also be carried out microbiologically. This has advantages over nonbiological procedures of producing potentially useful by-products and minimal waste (Zimbardi et al., 1999; Howard et al., 2004). However, the available knowledge of lignocellulosic material chemical composition and their microbial degradation in environment is scattered. Moreover, the detection and analysis of various POPs discharged from pulp paper industry is also not much known. Hence, detailed knowledge on the above issue will facilitate to manage the lignocellulosic waste in the environment for sustainable development. Therefore, this chapter provides insight on updated knowledge of the composition of lignocellulosic material and microbial degradation in the environment. Further, various available techniques for the detection and analysis of degraded lignocellulosic metabolic products in the environment have been mentioned.

10.2 General Structure and Composition of Lignocellulosic Wood Wall

The major component of lignocellulose material is cellulose, along with lignin and hemicellulose. Cellulose and hemicellulose are macromolecules from different sugars; whereas lignin is an aromatic polymer synthesized from phenylpropanoid precursors. The composition and percentages of these polymers vary from one plant species to another which is described in Table 10.1. Moreover, the composition within a single plant varies with age, stage of growth, and other conditions (Jeffries, 1994). Hemicellulose and lignin cover microfibrils (which are formed by elemental fibrils). The orientation of microfibrils is different in the different wall levels. The composition of lignocellulose material is shown in Figure 10.1.

In softwoods, coniferyl alcohol is the principal constituent; the lignin of hardwoods is composed of guaiacyl and syringyl units. Grass lignin contains guiacyl, syringyl, and *p*-hydroxyphenyl units.

10.3 Lignin

The term lignin is derived from the Latin term lignum, meaning "wood." It is present in the cell wall, conferring structural support, impermeability, and resistance against microbial attack and oxidative stress. Lignin content also varies depending on the developmental stage of a plant (Boerjan et al., 2003). For example, lignin composition reaches

TABLE 10.1

Lignocellulose Contents of Common Agricultural Residues and Wastes

Lignocellulosic Materials	Cellulose (%)	Hemicellulose (%)	Lignin (%)
Hardwood stems	40–55	24–40	18–25
Softwood stems	45–50	25–35	25–35
Nut shells	25–30	25–30	30–40
Corn cobs	45	35	15
Paper	85–99	0	0–15
Wheat straw	30	50	15
Rice straw	32.1	24	18
Sorted refuse	60	20	20
Leaves	15–20	80–85	0
Cotton seed hairs	80–95	5–20	0
Newspaper	40–55	25–40	18–30
Waste paper from chemical pulps	60–70	10–20	5–10
Primary wastewater solids	8–15	NA	24–29
Fresh bagasse	33.4	30	18.9
Swine waste	6	28	NA
Solid cattle manure	1.6–4.7	1.4–3.3	2.7–5.7
Coastal Bermuda grass	25	35.7	6.4
Switch grass	45	31.4	12.0

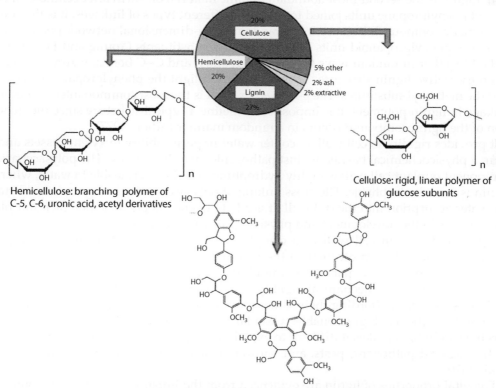

Hemicellulose: branching polymer of C-5, C-6, uronic acid, acetyl derivatives

Cellulose: rigid, linear polymer of glucose subunits

Lignin: complex, cross-linked polymer of aromatic rings (phenolic monomers)

FIGURE 10.1
(See color insert.) Lignocellulosic component of terrestrial plant cell wall.

higher syringyl/guaiacyl (S/G) ratios as plants attain maturity (Dixon et al., 2001). Lignin composition also changes in response to external and natural, environmental, or artificial stressors including, drought, low temperature, ultraviolet irradiation, mineral deficiency, mechanical wounding, and attack by pathogens and pests (Moura et al., 2010). For example, plants have been shown to produce lignin with enhanced levels of *p*-coumaryl alcohol at the sites of injury. Structurally, lignin is an amorphous heteropolymer, nonwater soluble and optically inactive molecule (Tuomela et al., 2000).

Lignin is frequently described as a random, complex, irregular, heterogenous, 3D, varyingly branched network of cross-linked, phenolic (aromatic) biopolymer. Lignin is composed of three main phenylpropane units (hydroxycinnamyl alcohols that vary in their degree of methoxylation: coniferyl, sinapyl, and *p*-coumaryl alcohols), in addition to several different minor phenolic compounds. Even the molecular mass of lignin is ranging from tens of thousands of Daltons to infinite (Janshekar et al., 1982). Lignins are racemic and therefore optically inactive. The racemic nature of lignins might arise from the fact that its polymerization is a nonenzymatic process (Ralph, 2006).

10.3.1 Structure and Properties of Lignin and Its Role in the Environment

Lignin has a varied, unique, and complicated chemical structure, which contains many aromatics. It is present in vascular plants not in bryophytes. Lignin exhibits a high degree of structural variability depending on the species, tissues, cells, and environmental condition. Lignin is the second-most-abundant organic material on earth after cellulose. It consists of phenylpropane units joined together by different types of linkages. It is the source of aromatic compounds because it is a random three-dimensional network polymer consisting of phenylpropanoid units rich in methoxyl substituents (Young and Frazer, 1987) linked together in random ways through ester, ether, and C—C bonds. Figure 10.2 shows a representative lignin structure. These figures highlight the phenylpropanoid structure and the methoxyl substitutes. The β-aryl-ether bond is the most common inter-monomeric linkage in lignin polymers. It is impossible to define a typical structure since the deposition of the lignin compound occurs in a random manner (Adler , 1977).

It provides rigidity to cell wall, to confer water impermeability to xylem vessels and to form a physicochemical barrier against pathogenic microbial attack. The polysaccharides component of plant cell wall is highly hydrophilic and thus permeable to water, whereas lignin is more hydrophobic. The cross linking of polysaccharides by lignin is an obstacle for water absorption into the cell wall. It is a heterogeneous phenolic polymer and plays crucial roles in the development and physiology of vascular plant.

Lignin is found in cell wall in a complex with cellulosic and hemicellulosic polysaccharides providing protection to these carbohydrates from biological degradation. Additionally, the lignin in paper is an inhibiting agent in paper degradation (in landfills). Lignin also provides mechanical rigidity enabling plants to defy gravitational forces and grow skyward, affording their sometimes immense size and volume (such as giant redwoods, sequoias). Lignin enables vascular integrity and protects the cell wall polysaccharides, which makes it difficult to exploit the sugars for biofuels. Lignin offer protection against pathogens, pests, and natural or mechanical wounding (Stutzenberger et al., 1970).

The vital properties of lignin are evidenced from the impaired growth and viability of plants with low or disorganized lignin (naturally or due to genetic engineering) (Martinez et al., 2009). Lignin enables the transport of water from the roots to the leaves through the xylem vasculature. The lignified cell walls are essentially dead cells linked together

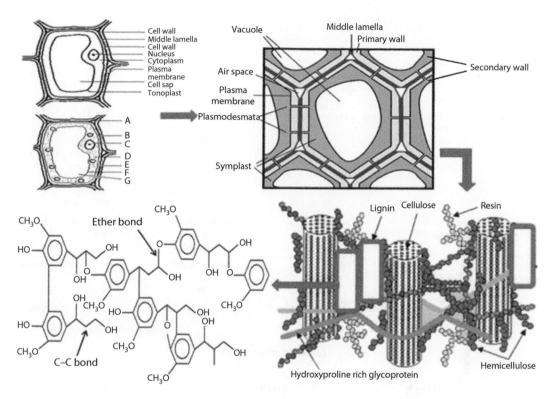

FIGURE 10.2
(See color insert.) General structure of lignin with phenylpropanoid structure and methoxy substitute.

to form long hollow tubes. The inside of these tubes are the sites for lignin deposition (i.e., lignification). Lignin is deposited in the secondary walls and among the fibers amid nonlignified tissues. Such polydisperse deposition also hinders the isolation and characterization of pure, intact lignin. It also plays a significant role in carbon cycle, formation of humus, and increases the photosynthetic productivity of plant by providing cation exchange capacity in the soil and expending the capacity of moisture retention between flood and drought condition by its hydrophilic nature (Ros Barceló et al., 2004).

10.3.2 Synthesis of Lignin and Its Precursor

Lignin formation is a complex process; it begins in the cytosol with the synthesis of glycosylated monolignols from the amino acid phenylalanine. Lignin is formed by radical polymerization of guaiacyl (G) units from precursor coniferyl alcohol, syringyl (S) units from precursor sinapyl alcohol, and *p*-hydroxyphenyl (H) units from precursor *p*-coumaryl alcohol (Figure 10.3b). Polymerization occurs in the cell walls after polysaccharide deposition and is initiated by peroxidase- or laccase-mediated one electron oxidation of phenylpropanoid precursor to phenoxy radicals, which undergoes coupling to form a complex, cross-linked lignin polymer. The final result of this polymerization is a heterogeneous structure whose basic units are linked by C–C and aryl–ether linkages, with aryl-glycerol β-arylether being the predominant structure (Bugg et al., 2011).

p-Coumaric acid undergoes the phenylpropanoid pathway in order to generate the three monolignols with varying methoxy substitutions (Ferrer et al., 2008). It begins

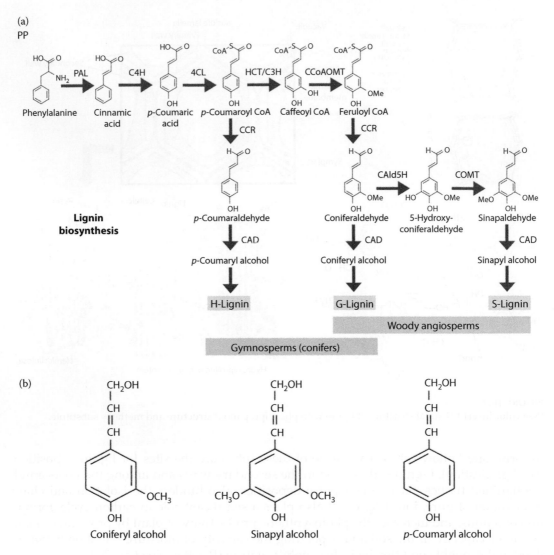

FIGURE 10.3
(See color insert.) (a) Biosynthesis of lignin and (b) its precursor.

with the enzymatic deamination of phenylalanine by phenylalanine ammonia lyase (Sticklen, 2008). After monolignol synthesis they are transported to lignification sites during wall development. Glucosyl transferases and β-glucosidases regulate the availability of monolignols during lignification (Samuels et al., 2002). Cell wall-associated β-glucosidases might release the aglycone (monolignols) for subsequent polymerization during lignin synthesis. Kaneda et al. (2008) suggested that "unknown membrane transporters, rather than Golgi vesicles, export monolignols." One reason for the uncertainty regarding the direct transport of monolignols is that these compounds are hydrophobic and must travel through the hydrophilic cellular matrix in order to reach the sites of lignin deposition.

Once the monolignols are transported to lignification sites, they undergo radicalization. Redox enzymes such as peroxidases, phenol oxidases, and laccases have been implicated in the oxidative radicalization of the monolignols (Vanwolme et al., 2010). Redox shuttle mediators involve monolignols in the radicalization of monolignols for the polymerization of lignin (Önnerud et al., 2002). Laccases utilize oxygen for oxidation, whereas peroxidases utilize hydrogen peroxide (H_2O_2) for oxidizing the monolignols (Boerjan et al., 2003). Cell-wall-associated peroxidases and laccases might additionally enable lignin polymerization at specific sites. The detailed structure of lignin biosynthetic cinnamic acid pathway is described in Figure 10.3a.

10.4 Microbial Degradation of Lignin

The degradation of lignin to enhance the release of fermentable sugars from plant cell walls presents a challenge for biofuel production from lignocellulosic biomass. The discovery of novel lignin-degrading enzymes from bacteria could provide advantages over fungal enzymes in terms of their production and relative ease of protein engineering. Lignin degradation is a slow process, which takes many years for degradation naturally. To remove lignin from the environment, many studies had been conducted on microorganisms by many researchers.

10.4.1 Bacterial Biodegradation

Although microbial degradation of lignin has been most intensively studied in white-rot and brown-rot fungi, there are a number of reports of bacteria that can break down lignin. In addition, these bacteria also showed the same enzymes, that is, peroxidise and laccases that were earlier reported into fungi. More recently, Chandra et al. (2013) have developed spectrophotometric analysis of lignin breakdown by different enzymes that are present in *Citrobacter freundii* and *Citrobacter* sp. It was found that peroxidise activity was low in kraft lignin when compared with synthetic lignin (Chandra and Bhargava, 2013). The reason might be due to the presence of different heavy metals and other pollution parameters, which have negative effect on bacteria cell. Further, the SDS-PAGE of purified enzyme has revealed its molecular weight 43 kDa, which lies between the range of MnP peroxidise family (Figure 10.4).

In another study on pulp paper mill effluent, the sample was analyzed for physicochemical parameters as per standard methods for wastewater analysis. Physicochemical parameters, that is, BOD was measured by 5-day method, COD was measured by open reflux method, while pH was measured by using pH meter (Water and Soil analysis kit ESICO Model 1160), and color was measured according to CPPA standard method at 465 nm by UV–vis spectrophotometer 2300 (Techcomp, Korea), and DO was measured by Clark type polarographic DO probe (835 A, Germany) detection level 0.1 mg/L. Lignin was measured according to Pearl and Benson (Chandra et al., 2007; Raj et al., 2007a,b; Chandra et al., 2011; Chandra et al., 2012; Chandra and Singh, 2012).

Degradation studies were also performed by inoculating 3% (v/v) of pre-grown seed culture of purified bacterial strains optical densiy (OD) 2.0. These cultures were inoculated axenically in separate flasks and mixed culture was prepared by inoculating 1:1:1 of each culture in 250 mL of Erlenmeyer flask containing 97 mL of autoclaved effluent samples. One percentage dextrose and 0.5% peptone was provided as additional carbon and nitrogen source. The inoculated broth cultures were incubated at 33 ± 2°C in temperature

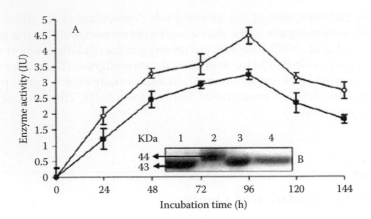

FIGURE 10.4
Showing manganese peroxidase activity and its molecular weight. A: Time course of peroxidise activity in synthetic ($\diamondsuit$) and kraft lignin (■). B: SDS PAGE of bacterial peroxidise. Lane 1: peroxidase standard; lane 2: protein ladder; lane 3: synthetic lignin; lane 4: kraft lignin.

control shaker (New Brunswick, Innova 4230, USA) at 130 rpm. The growth of axenic and mixed cultures was measured at 620 nm at every 24-h interval up to 216 h of incubation period. Autoclaved pulp paper mill effluent without bacterial inoculum was used as a control (Chandra et al., 2007; Raj et al., 2007; Chandra et al., 2011; Chandra et al., 2012; Chandra and Singh 2012).

In addition, the ligninolytic enzymes were analyzed where LiP assay was done by monitoring the oxidation of dye Azure B in the presence of H_2O_2. The reaction mixture contained sodium tartrate buffer (50 mM, pH 3.0), Azure B (32 M), 500 L of culture filtrate, and 500 L of H_2O_2 (2 M). OD was taken at 651 nm after 10 min. MnP assay was performed by the method as described by Oliveira et al. (2009), which is based on the oxidation of phenol red. Reaction mixture (4 mL) contained 1 mL of potassium phosphate buffer (pH 7.0), 1 mL of enzyme extract, 500 μL of MnSO4 (1 mM), 1 mL of phenol red (1 mM), and 500 μL H_2O_2 (50 M). Approximately, 1 mL sample was removed from reaction mixture and 40 μL of 5 M NaOH was added to stop the reaction. OD was taken at 610 nm at every 1 min interval. Moreover, laccase activity was detected by taking the absorbance at 450 nm. Reaction mixture was prepared by 3.8 mL of acetate buffer (10 mM, pH –5.0), 1 mL of guaiacol (2 mM), and 0.2 mL of enzyme extract. Then reaction mixture was incubated at 25°C for 2 h. One international unit of enzyme activity was defined as an activity of enzyme that catalyzed the conversion of 1 μM of substrate per min.

During the degradation of the pulp paper mill effluent, fast bacterial growth was observed at 48 h of incubation period followed by gradual increase up to 120 h and became almost constant at 216 h (Figure 10.5). During degradation, reduction in DO and pH was also recorded (Figure 10.5), where DO reduced up to 0.2 mg/L and pH up to 5 at the initial phase of growth due to fast bacterial growth and consumption of DO present in effluent, which resulted in the microaerophilic environment. The acidic pH might be due to the partial fermentation process and generation of acidic product. Besides, the growing bacterial biomass also releases H^+ in media due to metal uptake by active mode. At 216 h, 90.02% chlorophenol, 90% COD, 92.59% BOD, and 96.02% color removal was noted (Figure 10.6). These findings corroborated the earlier observation for the chlorophenol and lignin degradation. The results of this study were compared quantitatively with previous observations for white-rot fungi for removal of pollutants. Earlier studies on the treatment of pulp paper

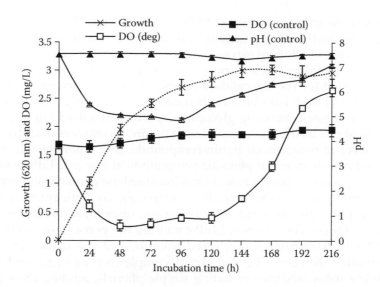

FIGURE 10.5
Bacterial growth curve, change in pH and dissolved oxygen (DO) by bacterial consortium during pulp paper mill effluent degradation.

effluent reported low COD (between 5000 and 8000) and reduction up to 83% COD, 62% lignin in the three step reactor. However, in another study average removal of color, COD, and BOD were 86.4%, 78.8%, and 70.5%, respectively, by fungus *P. chrysosporium*. Hence, this study revealed that there is a high reduction in BOD and COD when compared with previous observations.

To detect the role of extracellular enzymes during the degradation of pulp paper mill effluent, the enzymes secreted by bacteria were measured from the culture supernatant. LiP and MnP were recorded as dominating enzymes at an initial stage of degradation. Highest LiP activity was recorded at 48 h (5.56 IU/mL) and MnP at 72 h (3.58 IU/mL).

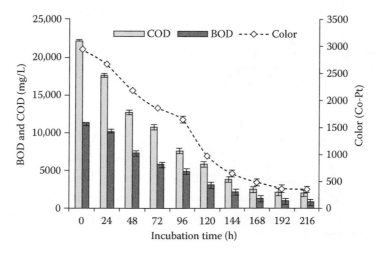

FIGURE 10.6
Reduction in BOD, COD, and color during degradation of pulp paper mill effluent by bacterial consortium.

However, the laccase activity was noted to be significantly higher in subsequent stages of bacterial growth, further higher activity was noted at 144 h of incubation (3.70 IU/mL) (Figure 10.7). Presence of all three extracellular ligininolytic enzymes in individual bacterial strains is a rare property. These enzymes function concurrently in the medium. These enzymes are reported for the degradation of phenolic and nonphenolic units of pollutants in which LiP and MnP have broad substrate range due to high redox potential. Both these enzymes are heme-containing glycoprotein, which required H_2O_2 for the activity. LiP plays a central role in the biodegradation of lignin. It catalyzes oxidative depolymerization of a variety of nonphenolic lignin compounds, β-O-4 nonphenolic lignin model compounds, and a wide range of phenolic compounds at low pH. LiPs oxidize the substrates in multistep electron transfers and form intermediate phenoxy and veratryl alcohol radical cations. These intermediate radicals undergo nonenzymatic reactions such as radical coupling and polymerization, side-chain cleavage, demethylation, and intramolecular addition and rearrangement, whereas MnP catalyzes the peroxide-dependent oxidation of Mn(II) to Mn(III). Thereafter, it is released from the enzyme surface in complex with oxalate or with other chelators. Chelated Mn(III) complex acts as a reactive low molecular weight, diffusible redox-mediator including simple phenols, amines, dyes, phenolic lignin substructures, and dimers. The oxidation potential of Mn(III) chelator is only limited to phenolic lignin structures. However, for the oxidation of nonphenolic substrates by Mn(III), reactive radicals must be formed in the presence of a second mediator. By measuring the enzyme activity it was justified that the loss of polymeric lignin can be attributed to the action of LiP and MnP because the production of extracellular enzyme requires H_2O_2, which was produced during glucose oxidation, available abundantly at initial phase of bacterial growth (Chandra and Singh, 2012) (Table 10.2).

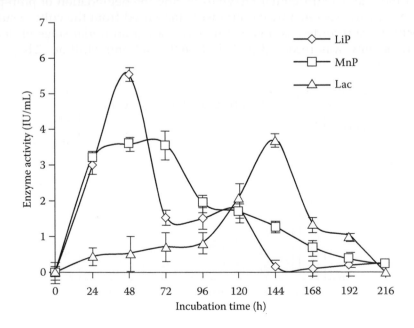

FIGURE 10.7
Extracellular ligninolytic enzyme activity during pulp paper effluent degradation by bacterial consortium.

TABLE 10.2

List of Microorganisms That Shows Ligninolytic Enzyme Activity

Microorganisms	Lignolytic Enzymes			References
	LiP	MnP	Laccase	
Bacteria				
Citrobacter freundii (F1581026)		+		Chandra and Bhargava (2013)
Camphylobacter (L04322)		+		
Geobacter (U28173)		+		
Nitrosomonas (Z46927)		+		
Bacillus (X77790)		+		
Methanobacterium (X99044)		+		
Pseudochrobactrum glaciale (FJ581024)	+	+	+	Chandra and Singh (2012)
Providencia rettgeri (GU193984)	+	+	+	
RCT2 *Pantoea* sp.	+	+	+	
Serratia marcescens	+	+	+	Chandra et al. (2012)
Serratia liquifaviens	+	+	+	
Bacillus cereus	+	+	+	
Grammothele subargentea LPSC no. 436	+	+	+	Saparrat et al. (2008)
Pseudomonas aeruginosa	+	+	+	Bholay et al. (2012)
Streptomyces viridosporus T7A	+	+	+	Ramachandra et al. (1988)
Sphingomonas paucimobilis SYK-6	+	+	+	Masai et al. (2007)
Pseudomonas putida mt-2		+	+	Ahmad et al. (2010)
Rhodococcus jostii RHA1		+	+	
Aneurinibacillus aneurinilyticus		+	+	Reid and Paice (1998), Bugg et al. (2011)
Fungi				
Phanerochaete chrysosporium	+	+	+	Bugg et al. (2011)
Phanerochaete flavido-alba	+	+	+	
Trametes trogii	+	+	+	
Bjerkandera sp. strain BOS55	+	+	+	
Phlebia ochraceofulva	+	+	+	
Phlebia tremellosa	+	+	+	
Trametes versicolor	+	+	+	Erden et al. (2009)
Bjerkandera adusta	+	+	+	
Phlebia radiata	+	+	+	
Pleurotus eryngii	+	+	+	
Ceriporiopsis subvermispora	+	+	+	
Marasmius quercophilus	+	+	+	
Coprinus cinereus	+	+	+	
Aspergillus oryzae				

Note: +, present.

10.4.2 Detection of Metabolic Product from Degraded Lignocellulosic Waste

10.4.2.1 *Extraction of Organic Pollutants*

Numerous products, developed to make everyday life easier, surround us in today's society. To a great extent, the basis for this production is the development of new chemical compounds with specific properties designed to meet certain physicochemical requirements. Unfortunately, this production has led to severe environmental pollution owing to the direct or indirect formation of various toxic compounds. One group of special interest is lignocellulosic waste that contains phenolic compounds, POPs, and the like. The hazardous nature of these wastes is a result of their toxicity, in combination with high chemical and biological stability, and a high degree of lipophilicity. In addition, these compounds bioaccumulate in food chain and cause detrimental health effects. Many compounds that are present in lignocellulosic biomass are extracted by liquid–liquid extraction in case of wastewater and by solid–liquid extraction in case of sludge.

10.4.2.1.1 *Effluent/Liquid Sample*

10.4.2.1.1.1 *Liquid–Liquid Extraction*   Solvent extraction methods use nonpolar solvents, which are miscible with water to extract the target compound from water by using the greater solubility of the target compound in the solvent than water. Ideally, one selectively extracts the target compound by using a solvent whose polarity is close to that of the target compound. Volatile solvents such as hexane, benzene, ether, ethyl acetate, and dichloromethane are usually used for the extraction of semivolatile compounds from water. Hexane is suitable for extraction of nonpolar compounds such as aliphatic hydrocarbons, benzene is suitable for aromatic compounds, and ether and ethyl acetate are suitable for relatively polar compounds containing oxygen. Dichloromethane has high extraction efficiency for a wide range of nonpolar to polar compounds. Dichloromethane is suitable for simultaneous analysis because of the following advantages: its boiling point is low and easy to reconcentrate after extraction, it is easy to separate from water because of its higher specific gravity, and it is nonflammable. However, dichloromethane, like benzene, is carcinogenic, and recent trends have been to refrain from using these solvents in liquid–liquid extractions. It is sometimes possible to selectively extract semi-volatile compounds from water by changing the character of samples, not changing solvents. For example, by changing the pH of samples, only acid or basic substances can be extracted. When pH of the water is less than 2, basic compounds become fully ionized and are not extracted by the solvent, allowing selective extraction of acidic and neutral compounds.

When extracting compounds are relatively soluble in water, salting-out techniques are used in order to increase extraction rates. Adding salt to an aqueous sample decreases the solvation power of the solution and the solubility of target compounds. This is useful not only for liquid–liquid extraction but also for headspace and solid-phase extraction (SPE) methods. Extraction is commonly achieved by shaking the water sample and solvent in a separating funnel. Therefore, large amounts of emulsion are formed, and it is difficult to separate the solvent from the aqueous phase. If this occurs, the emulsion is often efficiently dispersed (broken down) by adding either a small amount of ethanol, by sonicating the mixture in ultrasonic bath, or by adding anhydrous sodium sulfate, or continuous liquid–liquid extraction can be performed on samples that form emulsions. Continuous liquid–liquid extraction methods repeatedly circulate solvent in special glassware but, although this method has good extraction efficiency, it is not suitable for thermally unstable compounds because the extraction time is long.

10.4.2.1.1.2 *Solid-Phase Extraction*   Solid-phase extraction is a more rapid, modern alternative to liquid–liquid extraction. SPE is based on the principle that the components of interest are retained on a special sorbent contained in a disposable mini-column (cartridge). By using SPE one can remove matrix interferences (these either pass through the cartridge or are subsequently washed off) and then isolate with selective enrichment of one's target compounds. Solvent use is small. Common cartridges' packing materials (solid phases) are charcoal and XAD resin, silica gel chemically bonded with ozone depleting substances

(ODS), and high-polymer resin such as polystyrene and polyacrylate. Cartridges pre-packed with known quantities of adsorbent are on the market, and they are ready to use after simple conditioning.

10.4.2.1.1.3 Solid-Phase Microextraction   Solid-phase microextraction (SPME) is a method used to both extract and concentrate organic compounds in which a fiber needle attach-ment, which has been chemically coated with a fused silica equivalent to a GC liquid phase, is dipped directly into liquid samples, or exposed to the headspace vapors from liquid or solid samples. Because SPME has only recently been developed, there are few reports of its use with real environmental samples, and we must wait for the results of future investigation before recommending use. However, the fact that organic compounds can be analyzed easily and quickly without using any solvents suggests that this is the direction in which the next generation of analytical methods should proceed.

10.4.2.1.2 Sediments, Sludge Samples

10.4.2.1.2.1 Purge and Trap Method   This method, also known as the dynamic headspace method, removes (separates) volatile compounds from the sample matrix (in this case, water) by passing an inert gas such as helium or nitrogen through the matrix (purging). The target, volatile compounds are desorbed from the aqueous phase to the gas phase (purged) and are then separated from the stream of gas (trapped) by adsorbent filters. The adsorbent material is then heated in a stream of GC carrier gas (usually pure helium). This releases the trapped sub-stances into the carrier gas, the target analytes are introduced to GC, and analyzed. Typical trapping (adsorbent) materials are porous polymer beads, activated charcoal, silica gel, other GC column packing materials, or combinations of such materials. The purge and trap method can be used to extract volatile compounds from solid samples such as sediment. The sample (less than 1 g) is placed into the purge bottle, suspended in adding water, then treated and subsequently analyzed in the same manner as water samples. Samples that include high con-centration of volatile organic compounds (VOCs) are extracted with methanol, and then a part of extract is analyzed by purge and trap after being added to blank water.

10.4.2.1.2.2 Organic Solvent Extraction Method   There are three organic solvent extraction methods for semivolatile compounds from solid samples: (1) soxhlet extraction, (2) extrac-tion after mechanical mixing such as shaking, homogenization, or stirring, and (3) ultra-sonic extraction.

10.4.2.1.2.3 Soxhlet Extraction Method   In soxhlet extraction, organic components in solid samples are extracted from the matrix by continuously washing the solid with a volatile solvent in a specialized piece of glassware (soxhlet extraction apparatus). This is the most common method for extraction of organic compounds from solid samples, and is used as an extraction rate standard for the newly developed extraction method known as supercritical fluid extraction. Nonpolar solvents such as benzene or dichloromethane, polar solvents such as methanol, or mixtures of polar and nonpolar solvents whose boiling points are close to those of ethanol/benzene, or acetone/hexane are used. Benzene is known to be an especially efficient extraction solvent for polyaromatic hydrocarbons (PAHs), and acetone for sulfur-containing compounds. However, soxhlet extraction takes long time to get high extraction efficiency, and is not suitable for organic compounds, which are thermally unstable.

10.4.2.1.2.4 Ultrasonic Extraction Method   Ultrasonic extraction uses ultrasonic vibrations to extract samples with polar solvents in an ultrasonic bath. This is often used for chemical extraction from solid samples because it is simple.

10.4.2.1.2.5 Extraction by Mechanical Mixing Shaking and Stirring These methods are in essence derivations of liquid–liquid extraction. Sample extraction is achieved by simply placing solid samples in centrifuge containers with organic solvents and shaking. Subsequently, the sample matrix is separated after centrifugation.

Homogenization method is suitable for the extraction of nonpolar compounds in biological samples. Anhydrous sodium sulfate is added to the sample, the sample is then homogenized in the presence of a nonpolar solvent. Extracts usually include lipids, and so de-fatting processes are required before analysis.

10.4.2.2 Concentration of Extracted Organic Pollutants

Use a Kuderna-Danish (KD) concentrator or rotary evaporator to concentrate the extract or column chromatography eluant. Whether the KD method or rotary evaporation is chosen depends on the boiling point of the target compounds, their sublimation character, timeframe for analysis, and so on.

Kuderna-Danish concentration takes longer than rotary evaporation, but there is less loss of target chemical through evaporation, and this method is as applicable to low boiling compounds as to high boiling compounds. This method is able to concentrate samples down to a few milliliter. For further concentration, one must use a micro-Snyder column or evaporate under a stream of nitrogen. Evaporating under nitrogen gas may cause evaporative loss of low boiling point compounds. However, the micro-Snyder column method can concentrate samples containing low boiling point compounds to volumes of 0.5 mL. Rotary evaporation can concentrate large volumes of samples in a relatively short period of time. However, it has big evaporative losses and is not suitable for low boiling point compounds.

10.4.2.3 Characterization of Metabolite

10.4.2.3.1 Thin-Layer Chromatography

In thin-layer chromatography (TLC), the stationary phase is a polar absorbent, usually finely ground alumina or silica particles. This absorbent is coated on a glass slide or plastic sheet creating a thin layer of the particular stationary phase. Almost all mixtures of solvents can be used as the mobile phase. By manipulating the mobile phase, organic compounds can be separated.

In TLC, the sample can be spotted on precoated TLC plates. The solvent system can be used, that is, toluene: methanol: acetic acid (90:16:24 v/v) and the degradation product on TLC plates can be visualized under UV light. The T1 and T2 in Figure 10.8 shows the TLC analysis of the polymer compounds present in synthetic and kraft lignin, which has been broken down into smaller fractions during the bacterial degradation process because a clear fragmentation pattern was observed for bacterial degraded synthetic and kraft lignin samples compared with control samples.

10.4.2.3.2 High-Performance Liquid Chromatography

High-performance liquid chromatography (HPLC) works on the same principles as in thin layer chromatography and column chromatography. It has the ability to separate, identify, and quantitate the compounds that are present in any sample that can be dissolved in a liquid. Sample can be analyzed in HPLC by using a Waters 515 HPLC system equipped with 2487 UV/VIS detector. The samples will be injected followed by implementation of HPLC grade acetonitrile and water in 70:30 (V/V) ratio at the rate of 1 mL/min. The sample will

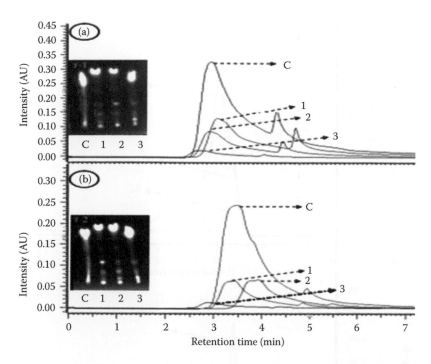

FIGURE 10.8
TLC and HPLC chromatogram of synthetic and kraft lignin. (a) TLC and HPLC chromatograph of synthetic lignin (b) TLC and HPLC chromatograph of kraft lignin.

be injected in reversed phase C-18 column at 27°C and the lignin content will be analyzed at 280 nm.

Figure 10.8 shows the HPLC analysis of degraded samples of lignin after 144 h of incubation period showed the clear cut reduction in peak area compared with control samples for axenic as well as mixed culture (Figure 10.8a), whereas kraft lignin degraded samples by axenic culture exhibited shifting of peaks compared with control (Figure 10.8b). Similar work has been done by Chandra et al. In this degradation study, the bacterial treated and untreated samples of pulp paper mill effluent were centrifuged at 5000 rpm for 10 min to separate bacterial biomass and particulate matter (Chandra and Singh, 2012). The obtained supernatant was acidified to pH 2 using 1 N HCl and extracted thrice with equal volume of ethyl acetate and dichloromethane for maximum extraction of lignin and chlorophenols, respectively. The collected organic layer was dewatered over anhydrous Na_2SO_4 and filtered through Whatman no. 54 filter paper (Whatman, UK). The filtered solvents were vacuum dried and obtained residues were dissolved in acetonitrile and samples were analyzed by using HPLC system (Hitachi, Japan) equipped with reverse phase column C-18 at 27°C and UV–vis diode array detector (L 2455), 20 µL of sample was injected followed by mobile phase containing acetonitrile and water in 70:30 (v/v) ratio at the flow rate of 1 mL/min. For lignin, the detection wave length was set at 280 nm and chlorophenol was detected at 320 nm.

The HPLC analysis of undegraded and bacterial degraded samples extracted in ethyl acetate and dichloromethane for lignin-related compounds and chlorophenol, respectively, showed the reduction in peak area. The reduction in peak area and generation of new peaks revealed the biodegradation and biotransformation of lignin-derived compounds

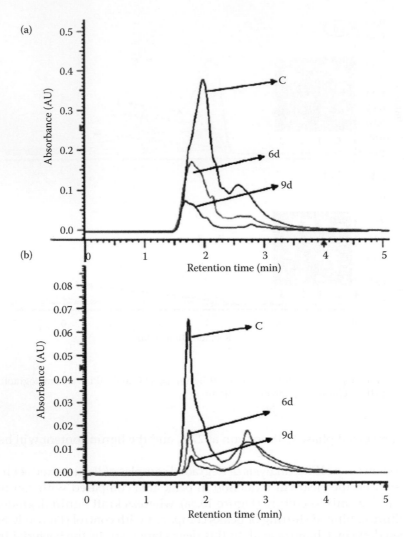

FIGURE 10.9
Hplc chromatogram of control and bacterial degraded sample of lignin related compounds (a) and chloro-phenol (b) of pulp paper effluent (C: control [untreated]; 6d: after 6th day bacterial treatment; 9d: after 9th day bacterial treatment).

and chlorophenol. The effluent treated by consortium showed the reduction in lignin-related compounds up to 90.69% at 9th day (Figure 10.9a), whereas chlorophenol reduction was noted to be 57.31% at 6th day with some new products. Further, it reduced up to 90.24% after 9th day of bacterial treatment (Figure 10.9b). This study revealed that consortium effectively reduced pollutants level in pulp paper mill effluent.

10.4.2.4 Gas Chromatography/Mass Spectrophotometer

Gas chromatography is undoubtedly one of the key techniques used for screening/iden-tification/quantification of many groups of nonpolar and/or semi-polar toxicants (or their GC amenable derivatization products). The high attainable separation power (potential number of theoretical plates) in combination with a wide range of the detectors employing various detection principles to which it can be coupled makes GC an important, often irreplaceable tool in the analysis of (ultra)trace levels of toxic components that may occur in environmental samples.

10.4.2.4.1 Derivatization

Derivatization involves chemically modifying the target compounds. It is common to convert hydroxyl group (phenols), carboxyl groups (fatty acids), amino groups (amines), and organometallic compounds (organotins) into their trimethylsilyl, ester, acyl, and alkyl derivatives. By derivatizing such chemical functional groups, one can make nonvolatile compounds volatile. Thus, one can analyze by GC compounds which normally cannot be analyzed by GC, including compounds that are normally adsorbed by separation columns. It is, however, possible to analyze some polar compounds without derivatization if one uses a fused-silica capillary column. One should not rush to derivatives target compounds because there are problems with derivatization that must be carefully considered, such as yield of derivatization, derivative recovery, stability of a derivative, identification, and confirmation.

10.4.2.4.1.1 Esterification
Methyl ester derivatives are generally stable; isolation is usually easy, and it is often possible to confirm the yield and stability of reaction products. After methylation, the increase in molecular weight of the methyl ester derivatives is small; some show molecular ions in mass spectra, and one may therefore use data reference systems. However, there are some problems, for example, methylating reagents are often not highly reactive, and many do not react with alcoholic hydroxyl groups and amine groups, although they are used for fatty acids.

10.4.2.4.1.2 Silylation
Trimethylsilylation (TMS) shows the highest reactivity of all derivatization methods for most compounds that have active hydrogens. However, this method should only be chosen when methylation and acetylation cannot be used, since there some disadvantages: (1) products of silylation are easily hydrolyzed and cannot be stored for long periods, (2) products of silylation often cannot be isolated, and therefore reaction solutions often have to be analyzed directly by GC, (3) injection of reaction solutions into GC tends to make GC "dirty" because inorganic silicone compounds are formed, and (4) polyethylene glycol columns react with silylating reagents.

10.4.2.4.1.3 Acetylation
This is one of the most-often used derivatization methods, because acylating agents react well with many functional groups, such as alcohols, thiols, and amines. Fluoroacetylation, particularly heptafluoropropionylation, is often used because highly sensitive analysis by electron capture detector (ECD) or MS is possible.

10.4.2.5 Sample Injection and Analysis

The acidified supernatants from control and treated sample were extracted with three volumes of ethyl acetate. The organic layer was collected, dewatered over anhydrous Na_2SO_4, and filtered through Whatman no. 54 filter paper. The residues were dried under a stream of nitrogen gas. The ethyl acetate extracts residues were analyzed as TMS derivatives as described. In this method, 100 mL dioxane and 10 mL pyridine were added to the residues and silylated with 50 mL trimethyl silyl (BSTFA [N,O-bis(trimethylsilyl) trifluoroacetamide] and trimethylchlorosilane [TMCS]). The mixture was heated at 60°C for 15 min with periodic shaking to dissolve the residues. An aliquot of 1 mL of silylated compounds was injected into the GC–MS equipped with a PE auto system XL gas chromatograph interfaced with a Turbomass mass spectrometric mass selective detector. The analytical column connected to the system was a PE-5MS capillary column 20 m × 0.18 mm internal diameter, 0.18 μm film thickness. Helium with flow rate of 1 mL/min was used as the carrier gas. The column temperature program was 50°C (5 min); 50–300°C (10°C/min, hold time: 5 min). The transfer line and ion source temperatures were maintained at 200°C and 250°C. A solvent time delay of 3.0 min was selected. In the full-scan mode, electron ionization (EI) mass spectra in the range of 30–550 (m/z) were recorded at an electron energy of 70 eV. All standard monomeric phenolic compounds (1 mg) were derivatized and chromatographed as above. The identification of low molecular weight lignin-related compounds derived from bacterial treatment was done by comparing their mass spectra with that of the National Institute of Standards and Technology (NIST) library available in the instrument and by comparing the retention time (RT) with those of authentic compounds available.

GC–MS is used in the present study because it has proved to be a suitable technique to analyze low molecular weight compounds released from lignin due to degradation. The total ion chromatographs (TICs) corresponding to the compounds extracted with ethyl acetate from the acidified supernatants obtained from un-inoculated (control) and inoculated alkali lignin sample with *A. aneurinilyticus* are shown in Figure 10.10a and 10.b and their peak identity is depicted in Table 10.3. Figure 10.10a, peak at RT 11.9 was identified as guaiacol, a single aromatic compound detected in ethyl acetate extract of control, which may be attributed to chemical oxidation of lignin due to aeration and agitation.

In addition to guaiacol, many acid-type compounds identified in the extract from the control (Table 10.3) revealed the partial degradation of lignin during anaerobic and aerobic treatment in industry. It has been reported in the literature that cupric oxide

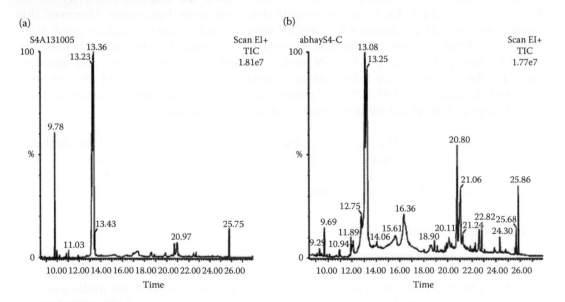

FIGURE 10.10

Total ion gas chromatogram of ethyl acetate extracts from control (a) and *Aneurinibacillus aneurinilyticus*-inoculated (b) kraft lignin (KL). The MS-identified compounds with respect to their retention time are summarized in Table 10.3.

TABLE 10.3

Compound Identified as Trimethylsilyl (TMS) Derivatives in Ethyl Acetate Extract from Control and *Aneurinibacillus aneurinilyticus* Degraded Kraft Lignin (KL) Samples as Given in Figure 10.10

S.No.	Peak Retention Time (mm)	Present in Control	Inoculated	Identified Compounds
1.	9.3	−	+	Acetic acid
2.	9.7	+	+	Propanoic acid
3.	10.5	+	−	Ethanedioic acid
4.	11.0	+	−	Butanoic acid
5.	11.9	+	+	Guaiacol[a]
6.	12.0	−	+	Valeric acid
7.	12.7	−	+	Phenylacetic acid
8.	15.6	−	+	Acetoguiacone
9.	16.3	−	+	Phenylpropionyl glycine
10.	18.9	−	+	Tetradecanoic acid
11.	20.6	+	−	Dibutyl phthalate
12.	20.7	+	−	Hexadecanoic acid
13.	20.8	+	+	Bis(2-methoxyethyl)phthalate
14.	21.1	−	+	Gallic acid[a]
15.	22.6	−	+	Ferulic acid[a]
16.	22.8	−	+	Octadecanoic acid
17.	24.3	−	+	Benzyl butyl phthalate
18.	24.9	−	+	1-Phenanthrenecarboxylic acid
19.	25.7	+	−	Bis(2-ethylhexyl) phthalate
20.	25.8	−	+	Di-otcyl phthalate

[a] Confirmed by match of retention time (RT) with known standards.

degradation of native lignin from different vegetal tissues showed that aldehyde-typem compounds were always more abundant than ketone or acid type. TIC of the sample degraded by *A. aneurinilyticus* (Figure 10.10) showed a significant increase in the number of peaks after 6 days of incubation when compared with the control (Chandra and Bhargava, 2013).

10.4.3 Fungal Degradation

Fungi are the only microorganisms studied extensively for the degradation of lignin. Lignin degradation by fungi has been discussed by a number of researchers. All fungi produced LiP, MnP, and laccase except *Daedalea flavida* and *Dichomitus squalens*, which lack LiP and MnP, respectively. White-rot fungi occur more commonly on wood species of Angiosperms than on Gymnosperms. Syringyl units are degraded, whereas guaiacyl units are more resistant to degradation. Transmission electron microscopy has revealed partial removal of the middle lamella by *Ceriporiopsis subvermispora* and *Pleurotus eryngii* and removal of lignin from secondary cell walls by *Phlebia radiate*. In recent years, more taxonomically diverse fungi have been studied for lignin degradation. In general, it has been found that the physiological process of lignin degradation is fungus-specific and is different from *Pha. chrysosporium*. Differences may be related to the taxonomic position and ecology of fungi. *Ganoderma lucidum* produces MnP in poplar wood but not in pine wood media.

10.4.3.1 Lignin-Degrading Enzymes

In nature, white-rot fungi are the predominant degraders of lignin, as these fungi contain specific genes for the enzymes necessary for depolymerization of lignin. These enzymes include lignin peroxidase (LiP), manganese peroxidase (MnP), laccase, and hydrogen peroxide-generating enzymes. Reactive oxygen species (ROSs) are also produced and are considered agents of wood decay by fungi. These enzymes are produced in different combinations, suggesting more than one strategy for biodegradation of lignin. The oxidation of phenolic compounds to phenoxy radicals takes place via lignin-degrading enzymes and oxidation of nonphenolic compounds via cation radicals. Based on the pattern of enzyme production, Hatakka, classified the white-rot fungi into three major categories: (1) the lignin–manganese peroxidase group (e.g., *P. chrysosporium* and *Phlebia radiata*), (2) the manganese peroxidase–laccase group (e.g., *Di. squalens* and *Rigidoporus lignosus*), and (3) the lignin peroxidase–laccase group (e.g., *Phlebia ochraceofulva* and *Junghuhnia separabilima*) (Hatakka, 1994). *P. chrysosporium* has been studied extensively as a model of lignin biodegradation research and production of LiP. Many researchers have recently discussed the oxidative mechanisms, molecular genetics, and applications of a variety of ligninolytic enzymes.

10.4.3.1.1 Lignin Peroxidases

Lignin peroxidase (LiP; EC 1.11.1.14) was discovered in *P. chrysosporium* (Glenn et al., 1983) and has become the most thoroughly studied peroxidase. LiPs are produced by most white-rot fungi, such as *Phanerochaete flavido-alba*, *Trametes trogii*, *Bjerkandera* sp. strain BOS55, *Phl. ochraceofulva*, and *Phlebia tremellosa*. Several isozyme forms have been detected in *P. chrysosporium* cultures and a number of other white-rot fungi (e.g., *Trametes versicolor*, *Bjerkandera adusta*, *Phlebia radiata*). The activity and number of LiP isozymes produced by *P. chrysosporium* vary from 2 to 15, depending on the strain, age of the culture, medium composition, and method of cultivation. LiP activity has not been detected in cultures of all three marine fungi (Raghukumar et al., 1996) or in one of the most active and widely studied biopulping fungus, *Ceriporiopsis subvermispora*. LiPs are glycosylated heme proteins secreted during secondary metabolism in nutrient-limited cultures. LiPs have a molecular weight of about 40 kDa, a low optimum pH of 2.5–3.0, acidic pIs, and a high redox potential. The three-dimensional crystal structure of LiP at 2–2.5 Å has also been established. The substrate-binding site of LiP has not been determined with certainty. The crystal structure indicates the heme access channel to be the substrate-binding site. The catalytic cycle of LiP is similar to that of other peroxidases. Figure 10.11 shows the catalytic cycle of LiP.

LiP is relatively nonspecific for reducing substrates, as it reacts with a wide range of lignin model compounds and even unrelated compounds. LiPs can oxidize the cleavage of β-O-4 linkages, Cα–Cβ linkages, and other bonds in lignin and its model compounds and are also involved in benzyl alcohol oxidations, side-chain cleavages, ring-opening reactions, demethoxylations, and oxidative dechlorination. LiP is unable to oxidize lignin to CO_2 in a cell-free system.

The mechanism of LiP action on lignin is still unclear, and it is also not certain whether the enzyme can oxidize lignin directly or requires certain radicals, such as varatryl alcohol (VA). LiP has the ability to oxidize ferrocytochrome c (12,300 Da), suggesting its action directly on polymeric lignin (Wariishi et al., 1994). Site-directed mutagenesis shows the presence of Trpin LiP isozyme H8 that is essential for activity toward VA, as two mutants with different amino acids do not exhibit this activity. Two distinct substrate interaction sites in LiP H8 seem to be present in both mutants. A resonance mirror biosensor indicates

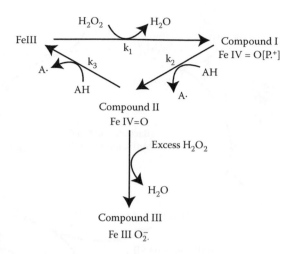

FIGURE 10.11
Catalytic cycle of lignin peroxidase.

the specific binding of LiP to synthetic lignin dehydrogenated polymerizate (DHP). It appears that a specific amino acid is responsible for lignin binding. VA and nonphenolic β-O-4 dimeric lignin compound can act as reductants for LiP L3 from *Phl. radiate*. VA plays various roles in the ligninolytic system of *P. chrysosporium*. VA is related to H_2O_2 production in the ligninolytic system of *Pl. eryngii*. Compounds such as 3,4-dimethoxytoluene, 1,4-dimethoxybenzene (1,4-DMB), 3,4,5-trimethoxybenzyl alcohol, and 2-Cl-1,4-DMB have been found to substitute the function of VA as a cofactor of LiP (Teunissen and Field, 1998).

10.4.3.1.2 *Manganese Peroxidases*

Manganese peroxidase (MnP; EC 1.11.1.13) is secreted in multiple forms in microenvironments by white-rot fungi and certain soil litter-decomposing fungi. A list of 56 fungi that produce MnP in liquid and/or solid-state fermentation has been compiled by Hofrichter (2002). MnP is secreted by a distinct group of Basidiomycetes, such as the families Coriolaceae, Meruliaceae, Polyporaceae, and the soil litter families Strophariaceae and Tricholomataceae. The molecular weight of MnP ranges between 38 and 62.5 kDa, and the molecular weight of the most purified MnP is 45 kDa. About 11 isozymes of MnP are known to be produced by *Ceriporiopsis subvermispora* (Lobos et al., 1994). Five isozymes in *P. chrysosporium* MP-1 have been detected to date (Kirk and Cullen, 1998). Like LiP, nitrogen-deficient conditions favor the production of MnP.

MnP is also glycosylated heme-containing extracellular peroxidase. Its catalytic cycle is similar to those of LiP and horseradish peroxidase (HRP), but it uses absolute Mn (II) as a substrate that is widespread in lignocellulose and soil. MnP has a lower redox potential than LiP and is different from other peroxidases, due to the structure of the binding site. Two calcium ions maintain the structure of the active site (Banci, 1997). A catalytic cycle is initiated by binding H_2O_2 or an organic peroxide to the native ferric enzyme, forming an iron–peroxide complex, as depicted in Figure 10.12. Peroxide oxygen–oxygen bond cleaves produce MnP compound I (Fe4+-oxo-porphyrin-radical complex). Subsequent reduction takes place through MnP compound II (Fe4+-oxo-porphyrin complex), and Mn (II) oxidizes to Mn (III), and this continues with the generation of native enzyme and release of the water molecule. MnP compound I is similar to LiP and HRP and can be reduced by electron donors such as ferrocyanide, phenolics, and Mn(II), whereas MnP compound II is

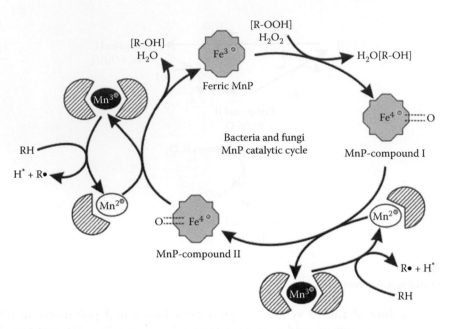

FIGURE 10.12
The catalytic cycle of manganese peroxidase (MnP).

very slowly reduced by other substrates and requires Mn(II) for completion of the catalytic cycle. A high concentration of H_2O_2 causes reversible inactivation of MnP by forming compound III, an inactive state (Hofrichter, 2002).

10.4.3.1.3 Laccases

Laccases (benzenediol:oxygen oxidoreductase; EC 1.10.3.2) are widely distributed in many white-rot Basidiomycetes. They are also found in higher plants and in several fungi belonging to Ascomycetes and Deuteromycetes. They are involved in lignin degradation and also serve other functions: in fungal pigmentation, pathogenicity, fructification formation, sporulation, and detoxification. Laccases are blue glycosylated multi-copper-containing oxidases that are larger than peroxidases and have a molecular weight of 60–80 kDa. Laccases contain four coppers per enzyme and are of three different types: type I, type II, and type III. Each type has a distinct role in the oxidation of laccase substrates. Laccases have an optimum pH between 3.0 and 5.7, but some laccases of soil-inhabiting Basidiomycetes have pH optima of 7.0. The optimum temperature is as high as 75°C for the laccase of the litter-decomposing fungus *Marasmius quercophilus* (Dedeyan et al., 2000). Laccases exhibit significant differences in redox potential. Three types of laccases are classified based on differing redox potential (Eggert et al., 1998). The crystal structure of laccase from *Coprinus cinereus* resolved at 2.2 Å. Laccase from *C. cinereus* expressed in *Aspergillus oryzae* has been crystallized and its cartoon-like 3D structure has been determined (Ducros et al., 2001) (Figure 10.13a).

Laccase can catalyze four single-electron oxidation of phenolic compounds, aromatic amines, and other compounds coinciding with reduction of molecular O_2 to H_2O. The use of molecular oxygen as the oxidant and the fact that water is the only by-product are very attractive catalytic features. Laccases catalysis mechanism (Figure 10.13b) occurs due to the reduction of the oxygen molecule including the oxidation of one electron with the wide range of aromatic compounds, which include polyphenol (Dedeyan et al., 2000) and also

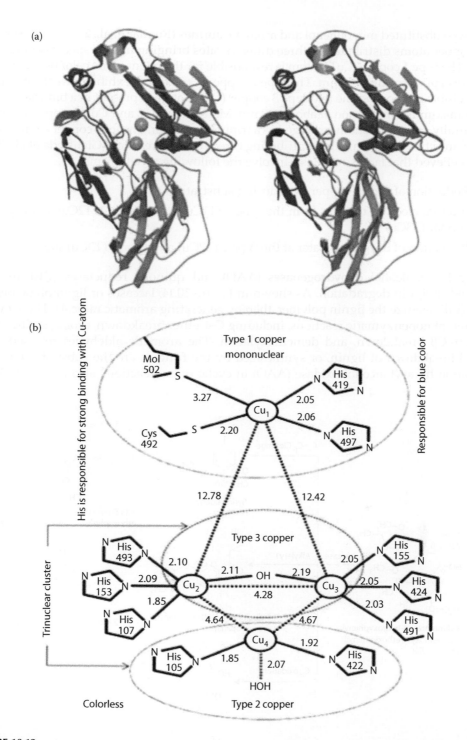

FIGURE 10.13
(See color insert.) Cartoon representation of the three-dimensional structure of the *C. cinereuslaccase*. The figure is color-ramped from the N-terminus (blue) to the C-terminus (red). The Cu-atoms are shown as shaded spheres, with the T1 site in blue and the T3 pair in yellow (a). The laccase active site showing the relative orientation of the copper atoms including interatomic distances among all relevant ligands (b).

methoxy-substituted monophenol and aromatic amines (Rubilar et al., 2008). Laccases have four copper atoms distributed in three different sites bringing unique spectroscopic properties. The type 1 copper (CuT1) atom is responsible for the intense blue color of enzymes by light adsorption around 610 nm. The type 2 copper (CuT2) atom exhibits a weak absorption in the visible region, and the two type 3 copper (CuT3) atom is present in a binuclear center with a maximum absorption of about 330 nm. Moreover, CuT2 and CuT3 copper atoms are structurally and functionally arranged as a trinuclear cluster. The four copper atoms form a part of an active site of enzyme contributing directly to the reaction (Gianfreda et al., 1999).

It is believed that laccase catalysis involve the following mechanisms:

a. Reduction of type 1 copper by reducing substrate.

b. Internal electron transfer from the type 1 (T1Cu) to the type 2 (T2Cu) and type 3 copper (T3Cu).

c. Reduction of oxygen to water at the type 2 (T2Cu) and type 3 (T3Cu) site.

Fungal aryl-alcohol dehydrogenases (AAD) and quinone reductases (QR) are also involved in lignin degradation. As shown in Figure 10.14, laccases or ligninolytic peroxidases (LiP) oxidize the lignin polymer, thereby generating aromatic radicals. These evolve in different nonenzymatic reactions, including C-4-ether breakdown, aromatic ring cleavage, Cα–Cβ breakdown, and demethoxylation. The aromatic aldehydes released from Cα–Cβ breakdown of lignin, or synthesized by the fungus, are the substrates for H_2O_2 generation by aryl alcohol oxidase (AAO) in cyclic redox reactions also involving AAD.

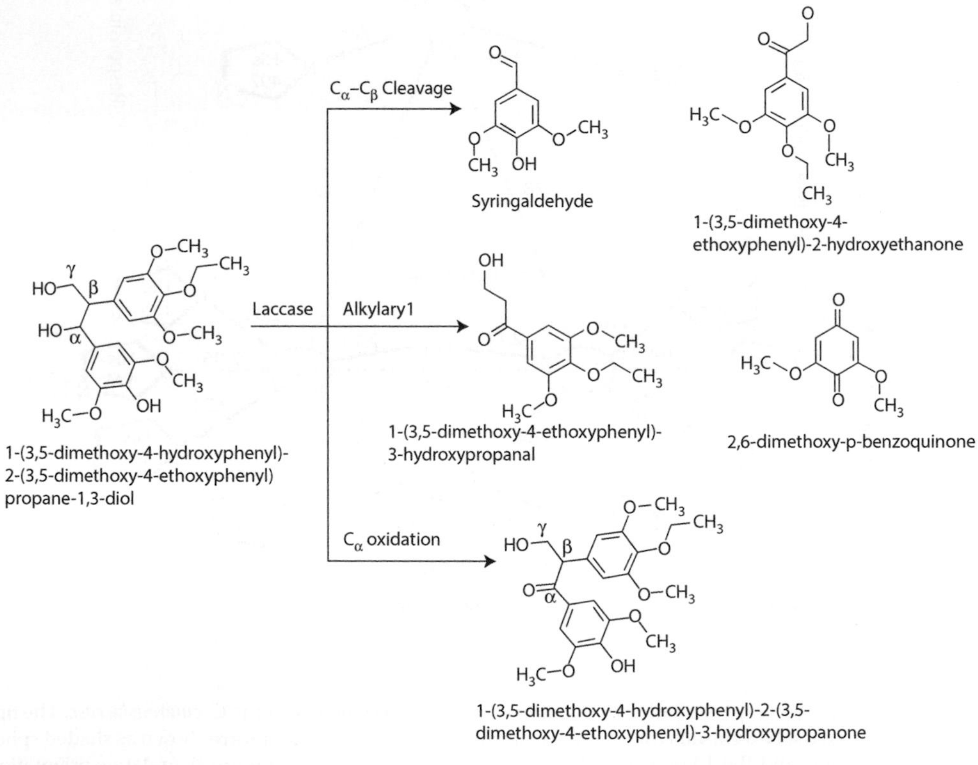

FIGURE 10.14
Laccase-catalyzed lignin model compound.

TABLE 10.4

Enzymes Involved in the Degradation of Lignin and Their Main Reactions

Enzyme Activity	Cofactor or Substrate	Main Effect or Reaction
Lignin peroxidase, LiP	H_2O_2, veratryl alcohol	Aromatic ring oxidized to cation radical
Manganese peroxidase, MnP	H_2O_2, Mn, organic acid as chelator, thiols, unsaturated lipids	Mn(II) oxidized to Mn(III); chelated Mn(III) oxidizes phenolic compounds to phenoxyl radicals; other reactions in the presence of additional compounds
Laccase, Lacc	O_2; mediators, for example, hydroxybenzotriazole or ABTS	Phenols are oxidized to phenoxyl radicals; other reactions in the presence of mediators
Glyoxal oxidase, GLOX	Glyoxal, methyl glyoxal	Glyoxal oxidized to glyoxylic acid; H_2O_2 production
Aryl alcohol oxidase, AAO	aromatic alcohols (anisyl, veratryl alcohol)	Aromatic alcohols oxidized to aldehydes; H_2O_2 production
Other H_2O_2 producing enzymes	Many organic compounds	O_2 reduced to H_2O_2

Phenoxy radicals from C4-ether breakdown can repolymerize on the lignin polymer if they are not first reduced by oxidases to phenolic compounds. The phenolic compounds formed can be again reoxidized by laccases or peroxidases. Phenoxy radicals can also be subjected to Cα–Cβ breakdown, yielding p-quinones (Table 10.4).

10.5 Hemicellulose

10.5.1 Structure, Composition, and Properties of Hemicelluloses

The word Hemicelluloses is made up of two words Hemi and Cellulose; hemi means peripheral, and therefore hemicelluloses means "peripheral cellulose." Hemicelluloses are wall polysaccharides, which are characterized by β-(1 → 4)-linked backbones of sugars in an equatorial configuration. It includes xyloglucans, xylans, mannans and glucomannans, and β-(1 → 3, 1 → 4)-glucans. This type of hemicellulose is present in the cell walls of all terrestrial plants, except for β-(1 → 3,1 → 4)-glucans, which are restricted to Poles and a few additional groups. The complete structure of the hemicelluloses and their abundance vary widely between different species and cell types. There are several types of hemicelluloses, which are xyloglucan, xylans, mannans and glucomannans, and β-(1 → 3, 1 → 4)-glucans (Figure 10.15).

Some polysaccharides, such as galactans, arabinans, and arabinogalactans are sometimes included in the hemicelluloses group, but since these appear to be a part of pectin molecules, at least in the initial synthesis, and do not share the equatorial β-(1 → 4)-linked backbone structure, we think they should not be included in the already heterogeneous group of hemicelluloses. We also consider that callose, which has a backbone entirely composed of β-(1 → 3)-linked glucose residues, should not be considered hemicelluloses (Scheller and Ulvskov, 2010).

All the hemicelluloses show differences in structural details between different species and in different cell types within plants. The main role of hemicelluloses is to bind

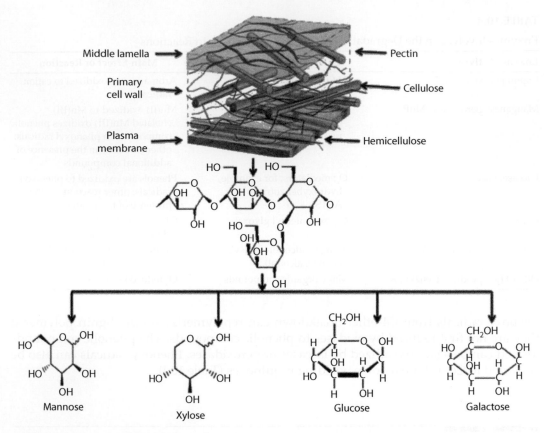

FIGURE 10.15
(See color insert.) Showing hemicellulose composition and structure.

cellulose microfibrils, thereby strengthening the cell wall. Hemicelluloses are synthesized by glycosyl transferases located in the Golgi membranes. Cell wall models that are shaped to demonstrate how wall polymers are organized in higher-order structures influence the idea about biological functions of the hemicelluloses.

The most excepted structural model has been proposed by McCann and Roberts as shown in the Figure 10.15 (McCann and Roberts, 1991). It is based on noncovalent interactions but it has a particular noncovalent interaction as the central tenet. Further, the tethering of glycans cross link cellulose microfibrils for the synthesis of cellulose.

A hemicellulose can be any of several heteropolymers (matrix polysaccharides) present in almost all plant cell walls along with cellulose. Although cellulose is in crystalline form, strong, and resistant to hydrolysis, hemicellulose has a random, shapeless arrangement with small strength. It is simply hydrolyzed by dilute acid or base as well as numerous hemicellulase enzymes. Dissimilar to cellulose, hemicellulose (also a polysaccharide) consists of shorter chains—500–3000 sugar units as opposite to 7000–15,000 glucose molecules per polymer seen in cellulose, and hemicellulose is a branched structure while cellulose is unbranched. Hemicelluloses are fixed in the cell walls of plants, sometimes in chains that form a "ground"— they bind with pectin to cellulose to form a network of cross-linked fibers (Figure 10.15).

Hemicelluloses are named according to their main sugar residues in the backbone. Xylans, consisting of D-xylose units, and glucomannans, consisting of D-glucose and D-mannose units contribute to the main hemicelluloses in hardwoods and softwoods. Trees are classified as softwoods (gymnosperms) and hardwoods (angiosperms). Globally, there are about 30,000 hardwoods and 520 softwood tree species known. In softwoods, O-acetyl-galactoglucomannan is the principal hemicellulose component and constitutes around 20% of the dry weight (Lundqvist et al., 2003). The glucose to mannose ratio is about 1:3, whereas the ratio of galactose to glucose can vary from 1:1 to 1:10 (Hakkila 1989). Softwood O-acetyl-galactoglucomannan has been reported to have an approximate degree of polymerization, DP, between 100 and 150, equivalent to a molecular weight around 16,000–24,000. The chief set of hemicelluloses found in hardwoods is the glucuronoxylans. Distinctively, this xylan is an O-acetyl-(4-O-methylglucurono)-b-D-xylan. The content of glucuronoxylan content of hardwood is, in general, between 15% and 30% by weight. Unlike softwoods, the 2,3 positions of the xylose backbone in hardwood may be partially acetylated with about seven acetyl residues per ten xylose units.

10.5.2 Various Types of Hemicellulose

Hemicellulose contains many different sugar monomers. In contrast, cellulose contains only glucose, but in hemicellulose besides glucose, sugar monomers can include xylose, mannose, galactose, arabinose, and rhamnose. Hemicelluloses contain most of the D-pentose sugars, and also small amounts of L-sugars occasionally. Xylose is the sugar monomer always present in the largest amount, but galacturonic acid and mannuronic acid also tend to be present. Hemicellulose is a heterogeneous polymer consisting of different kinds of neutral sugars such as pentoses, hexoses, and sugar acids. The most common hemicelluloses are xylan, xyloglucan, glucuronoxylan, arabinoxylan, and glucomannan.

10.5.2.1 Xylan and Its Derivatives

The term "xylan" is a catchall for polysaccharides that have α-(1–4)-D-xylopyranose backbone with a variety of side chains. Xylan is the predominant hemicellulose in most plant cell walls, generally comprising about 1/3 of the total plant biomass. The composition and linkages of the side chains determine the specific variety of xylan. Removal of these side chains generally enhances the rate of degradation by endoxylanase enzymes.

Xylans from grasses and annuals having an elevated level of α-L-arabinofuranoside substituents are referred to as arabinoxylans. Arabinoxylan is a hemicellulose found in both the primary and secondary cell walls of plants, including woods and cereal grains, consisting of copolymers of two pentose sugars—arabinose and xylose. Arabinoxylan chiefly serve a structural role in the plant cells. They are also the reservoirs of large amounts of ferulic acid and other phenolic acids, which are covalently linked to them. Arabinoxylan is one of the main components of soluble and insoluble dietary fibers, which are shown to exert various health benefits. In addition, arabinoxylans, owing to their bound phenolic acids, are shown to have antioxidant activity (Wyman et al., 2005).

Hardwood xylans, highly substituted with acetyl and 4-O-methyl lucuronic acid, are termed glucuronoxylans. Glucuronoxylan is the primary part of hemicellulose as found in hardwood trees, for example birch, containing glucuronic acid and xylose as its main constituents has α-(1–4)-xylan backbone, with 4-O-methylglucuronic acid side chains. 4-O-Methylglucuronic acid is linked to the xylan backbone by α-(1–2)-glycosidic bonds, and the acetic acid is esterified at the 2 and/or 3 carbon hydroxyl group. The molar ratio of xylose/glucuronic acid/acetyl residues is about 10:1:7 (Awano et al., 2002). The basic structure

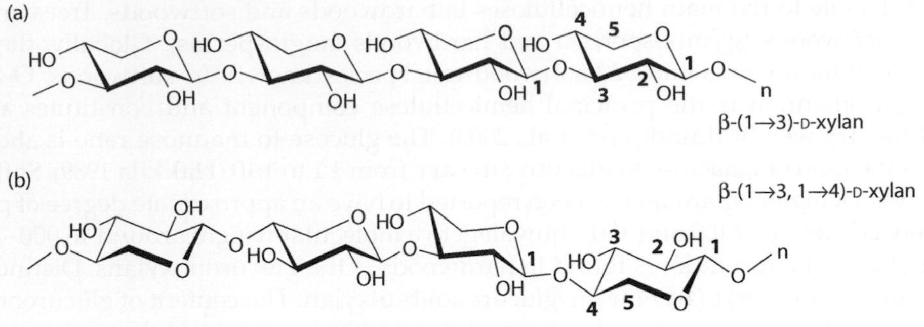

(a)

β-(1→3)-D-xylan

(b)   β-(1→3, 1→4)-D-xylan

FIGURE 10.16
Structure of xylan with linkage.

of XyG is shown in Figure 10.16, Xylans. XyGs are made of repetitive units, and a special one-letter code is used to denote the different XyG side chains. G denotes an unbranched Glc residue, while X denotes α-D-Xyl-(1 → 6)-Glc. The xylosyl residues can be substituted at O-2 with β-Gal (L side chain) or α-L-Ara*f* (S-side chain). A Gal residue substituted at O-2 with α-L-Fuc is designated as F. Other less common side chains have other designations.

10.5.2.2 Glucans and Its Derivatives

The dominant polysaccharide in plant biomass, glucans play important roles in plant cell wall structure and function. Glucan and xyloglucan are structurally similar to cellulose, being based on α-linked glucose backbone. β-Glucan consists of mixtures of β-(1–3)- and β-(1–4)-linked glucose residues.

Xyloglucans is the most abundant hemicelluloses in the primary wall of nongraminaceous plant, often comprising 20% of the dry mass of the wall (O'Neill and York, 2003). Xyloglucans are classified as XXXG type or XXGG type depending on the number of backbone residues that are branched. XXXG-type xyloglucans are present in the walls of numerous higher plants, whereas XXGG-type xyglucans occur in the walls of solanaceous plants. Xyloglucans dominate in primary walls of dicots and conifers, whereas (glucurono) arabinoxylans dominate in commelinid monocots. Xyloglucan (XyG) has been found in every land plant species that has been analyzed, including mosses, but has not been found in charophytes (Popper, 2008). XyG is the most abundant hemicellulose in primary walls of spermatophytes except for grasses.

β-Glucan is a glucopyranose polymer containing either β-(1–3) or mixed β-(1–3), β-(1–4) linkages. The ratio of (1–4) to (1–3) linkages varies by species and gives specific properties to individual h-glucan polymers. In maize coleoptiles, the mixed β-(1–3), β(1–4) glucan is deposited early in cell development, comprising greater than 70% of the cell wall material in the endosperm, and decreases rapidly after day 5. The second type of plant h-glucan consists of callose, a β-(1–3)-linked glucose polymer found primarily in rapidly growing structures such as pollen tubes and developing seeds.

10.5.2.3 Mannans and Its Derivatives

Mannan may refer to a plant polysaccharide that is a linear polymer of the sugar mannose. Plant mannans have β (1–4) linkages. It is a form of storage polysaccharide. Ivory nut is a source of mannan. Mannan may also refer to a cell wall polysaccharide found in yeasts.

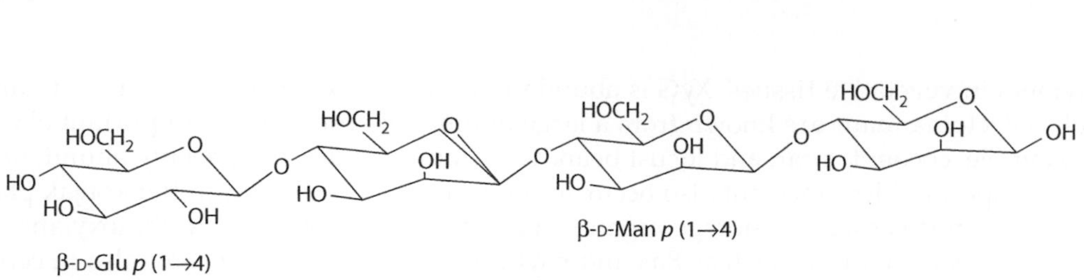

FIGURE 10.17
Glucomannan from angiosperms.

This type of mannan has a α (1–6)-linked backbone and α(1–2)- and α(1–3)-linked branches. It is serologically similar to structures found on mammalian glycoproteins. Mannose-containing polysaccharides including mannans, galactomannans, and galactoglucomannans are present in the walls of many plants. The glucomannoxylans (usually referred to as glucomannans and galactomannans) are made up of β-(1–4)-D-glucopyranose and β-D-mannopyranose residues in linear chains. Detection of mannan leads to lysis in the mannan-binding lectin pathway. This mannan is the source of mannan oligosaccharides (MOS) used as prebiotics in animal husbandry and nutritional supplements.

Galactomannans is abundantly found in cell wall of storage tissues, notably those from the endosperm of leguminous seeds (guar, locust bean, tara gum). The amount of galactose residues influences solubility, viscosity, and interaction with other polysaccharides. Galactomannan, believed to be extracellular carbohydrate storage, is found predominantly in secondary cell wall of softwood.

Glucomannan is a storage polysaccharide found mainly in the root of the Konjac plant (*Amorphophallus konjac*). It consists of α-(1–4)-linked mannopyranose and glucopyranose backbone in a ratio of 1.6:1. Hardwood glucomannans consist of β-(1–4)-linked glucose and mannose units forming chains that are slightly branched. The ratio of mannose:glucose is about 1.5:1 or 2:1 in most hardwoods. Softwood glucomannans have occasional galactose side branches linked α-(1-6) to the mannose main chain (Figure 10.17). The α-(1-6) linkage of galactose is very sensitive to acid and alkali and maybe cleaved during alkaline extraction. Softwood xylans and xylans from most graminaceous plants have single L-arabinofuranosyl units attached through alpha (α) linkages to some O-3 positions of the main chain.

10.5.3 Function of Hemicelluloses in Plants

The most important biological role of hemicelluloses is their contribution to strengthening the cell wall by interaction with cellulose and, in some walls, with lignin. XyG plays an additional role as a source of signal molecules. Breakdown products of XyG, most notably XXFG, were demonstrated to counteract auxin-induced cell expansion (York et al., 1984). Research in XyG oligomers as substrates for XET has received much attention, whereas studies of the oligosaccharin inhibitory activities dwindled at the end of the 1990s, and it is thus still not clear how important these responses are *in vivo*.

Hemicelluloses in the cell wall have the primary role of interacting with other polymers to ensure the proper physical properties of the wall. However, in a large number of cases, hemicelluloses have been recruited to the function of seed storage carbohydrate (Reid, 1985). This has happened independently many times in evolution, and it has been suggested that from a taxonomic viewpoint hemicelluloses are as important as starch is in the role as storage of carbohydrate in seeds (Meier and Reid, 1982). Much of our knowledge of hemicelluloses comes from the study of seed polysaccharides rather than the polymers in vegetative tissues. XyG is abundant in plants such as nasturtium and tamarind. Galactomannans are known from a large number of economically important plants, for example, coconut, guar, and locust bean. Galactomannans are especially abundant in the endosperm of legumes but also occur in other seeds. It is present in the konjak plant (*A. konjac*); in this case the storage organ is a corm and not the seed. Arabinoxylans are present in seeds of dicots such as flax and psyllium (Naran et al., 2008) and also in cereal endosperm. Cereal endosperm additionally contains β-(1 → 3, 1 → 4)-glucans.

10.6 Microbial Degradation of Hemicellulose

Hemicelluloses are biodegraded in monomeric sugars and acetic acid. Hemicellulases are enzymes that degrade hemicellulose and are frequently classified according to their function on different substrates. Xylan is the major carbohydrate found in hemicellulose. Its entire degradation requires the supportive action of a range of hydrolytic enzymes. The main feature should be made between endo-α,4-β-xylanase and xylan 1,4-β-xylosidase (Figure 10.18). The earlier generates oligosaccharides from the cleavage of xylan; the later works on xylan oligosaccharides, producing xylose (Jeffries, 1994).

In addition, hemicellulose biodegradation needs accessory enzymes such as xylan esterases, ferulic, and *p*-coumaric esterases, α-L-arabinofuranosidases, and α-4-O-methyl glucuronosidases acting synergistically to efficiently hydrolyze wood xylans and mannans. Hemicellulosic compounds are subject to degradation by various fungal and bacterial populations (Kirk and Cullen, 1998).

10.6.1 Bacterial Degradation of Hemicellulose

Bacteria play a role in both aerobic and anaerobic degradation of hemicelluloses. The decomposition of hemicellulose is similar to that of cellulose in that the initial depolymerization step takes place outside of the cell, and the sugars produced are then transported into the cell for catabolism or anabolism. Hemicellulose decomposition is much quicker than cellulose decomposition. Bacteria degrading xylan have been frequently

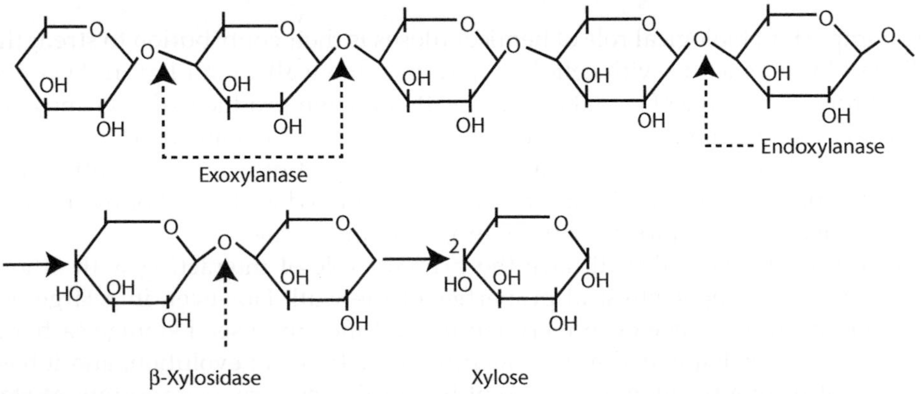

FIGURE 10.18
Enzymatic breakdown of hemicellulose.

isolated from compost, litter, soils, and sludges, the rumen system and other gastrointestinal tracts. Some plant pathogenic *Pseudomonas* sp., *Xanthomonas alfalfae*, and *Achromobacter* sp. have been isolated that also produce hemicellulases as a part of their enzyme system. Hemicellulolytic *Aeromonas, Bacteroides, Butyrivibrio, Ruminococcus,* and *Clostridium* species have been isolated from human and animal intestine. Plant cell walls are particularly resistant to microbial degradation. Physical access to the polymer can be restricted by the surrounding lignocellulosic components. Most xylanolytic microorganisms grown under natural conditions produce both cellulolytic and xylanolytic enzymes to efficiently degrade plant cell walls. In mannose-producing bacteria are *Aeromonas hydrophila, Cellulomonas* sp., *Streptomyces olivochromogenes, Polyporous versicolor, Trichoderma harzianum, Butyrivibrio* sp., *Bacteroides* sp., *Clostridium* sp. which also support for hemicellulose degradation. The bacterial degradation could be aerobic or anaerobic depending upon substrate and action of microorganism.

10.6.1.1 Aerobic Degradation

Hemicellulose is randomly acetylated, which reduces its enzymatic reactivity. Hemicellulose can be hydrolyzed by enzymatic hydrolysis by hemicellulase. When subjected to microbial decomposition, hemicelluloses degrade initially at faster rate and are first hydrolyzed to their component sugars. The hydrolysis is brought about by a number of hemicellulolytic enzymes known as "hemicellulases" excreted by the microorganisms. This enzyme system consists of three enzymes: exoxylanase, endoxylanase, and β-xylosidase (which split xylose and other short chain xylobioses). On hydrolysis hemicelluloses are converted into soluble monosaccharide/sugars (e.g., xylose, arabinose, galactose, and mannose), which are further convened to ethanol, CO_2 and H_2O are broken down to pentoses and CO_2 (Figure 10.19). Various microorganisms including fungi, bacteria, and actinomycetes both aerobic and anaerobic are involved in the decomposition of hemicelluloses. Bacterial xylanases have been described in several aerobic species and some ruminal genra (Blanco et al., 1999).

10.6.1.2 Anaerobic Degradation

Anaerobic cellulolytic and hemicellulolytic bacteria are responsible for the initial stage in the overall conversion of lignocellulosic material to methane and carbon dioxide. Hemicellulose in lignocellulose is efficiently used by converting into ethanol, methane, and recently it has also been reported to convert into hydrogen (Figure 10.20). The anaerobic digestion process may be thought in simple terms to be composed of three steps, which include

a. Hydrolysis of polymeric substrates to monomers

b. Fermentation of monomers to organic acids, hydrogen, and carbon dioxide

c. Conversion of organic acids, hydrogen, and carbon dioxide to methane

Butyrivibrio fibrisolvens are strictly anaerobic, curved rod-shaped bacteria. It has a Gram-positive cell wall structure, which stains Gram negative, because of its thin cell wall. Most strains ferment xylan, pectin, arabinose, glucose, fructose, and galactose. Genes encoding for an endoglucanase, cellodextrinase, β-glucosidase, two xylanases, and a dual β-xylosidase/α-L-arabinofuranosidase were cloned (Lin and Thomson, 1991). *Prevotella* (*Bacteroides*) *ruminicola* are strictly anaerobic, Gram negative, rod-shaped bacteria found in large amounts in the rumen. Depending on the strain they can ferment starch, xylan,

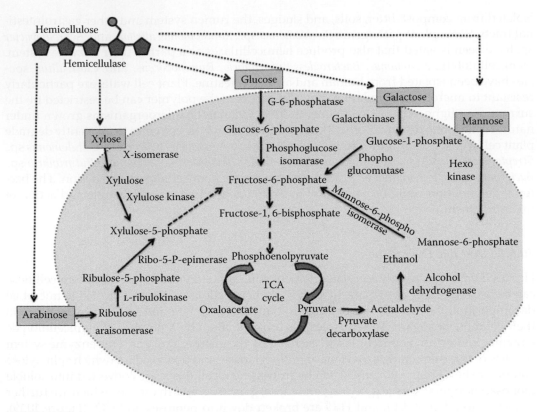

FIGURE 10.19
Enzymatic degradation of xylose, glucose, galactose, mannose, and arabinose.

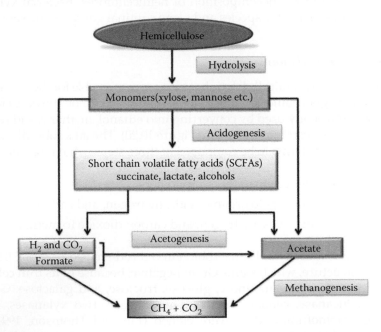

FIGURE 10.20
Anaerobic bacterial degradation of hemicellulose.

pectin, maltose, cellobiose, xylose, and arabinose. All ferment glucose, fructose, galactose, and lactose. The main fermentation products are acetate, propionate, succinate, and formate. Endoglucanase, β-xylosidase, α-L-arabinofuranosidase, and xylanase genes were cloned. *Clostridium thermolacticum* ferments xylan, starch, cellobiose, glucose, and xylose, but not arabinose.

Several xylanolytic enzymes and cellulolytic were isolated from cultures of the cellulolytic thermophilic anaerobe *C. stercorarium*. Three xylanases from *C. stercorarium* were purified and characterized. Two xylanase, six β-xylosidase, and α-L-arabinofuranosidase genes were sequenced and/or cloned. *Thermoanaerobacterium saccharolyticum* is a thermophilic anaerobic bacterium growing in the pH range 4–6.5 and up to temperatures of 66°C. From one *Thermoanaerobacterium* strain a xylanase, two xylosidases, two acetyl xylan esterases, and a α-glucuronidase were identified recently.

10.6.2 Fungal Degradation of Hemicellulose

Fungi play a central role in the degradation of plant biomass, producing an extensive array of carbohydrate-active enzymes responsible for polysaccharide degradation. The enzyme sets for plant cell wall degradation differ between many fungal species, and our understanding about fungal diversity with respect to degradation of plant matter is essential for the improvement of new strains and the development of enzymatic cocktails for industrial applications. In thermophilic fungus two species *Thermoascus aurantiacus* and *Talaromyces emersonii* are well known. Endo-1,4-β-xylanase is produced by *Tricoderma reesei, Aspergillus niger* that cleave endo-β-1,4-glycosidic bond. β-D-Xylosidase produced by *Aspergillus awamori* that break exo-β-1,4-glycosidic bond. The white-rod fungus *P. chrysosporium* has been shown to produce multiple endoxylanases. Optimum temperature for xylanases from fungal origin ranges from 40°C to 60°C. Fungal xylanases are generally less thermostable than bacterial xylanases.

10.6.3 Actinomycetes Degradation of Hemicellulose

In thermophilic actinomycetes, three species *Thermomonospora, Actinomadura*, and mannase-producing bacteria produce enzymes that help in the degradation of hemicellulose. Xylanases from actinobacteria are active at pH 6–7. However, alkaline activity of xylanases is described from *Bacillus* sp. or *Streptomyces viridosporus*.

Hemicellulose is made up of different sugars; hence, it works as a carbon source for microorganisms. Therefore, for the bioavailability of these carbons some enzymes that play a crucial role secreted by some microorganisms are given in Table 10.5.

10.7 Cellulose: Structure and Function

Cellulose is one of many polymers found in nature. It is a long chain of linked sugar molecules that gives wood its remarkable strength. Cellulose is a linear polysaccharide polymer with many glucose monosaccharide units. The acetal linkage is beta (β), which makes it different from starch. Wood, paper, and cotton all contain cellulose. Cellulose is an excellent fiber. Wood, cotton, and hemp rope are all made of fibrous cellulose. Cellulose is a polymer of glucose and is the most abundant organic material in nature (Brown and

TABLE 10.5

Various Enzymes Secreted by Different Microorganisms

S. No.	Microorganisms	Enzyme	Substrate	Mode of Action	References
Fungi					
1.	*Tricoderma reesei, Aspergillus niger*	Endo-1,4 β-xylanase	Xylan backbone	Endo-β-1,4-glycosidic bond	Poutanen (1988), Rombouts et al. (1988)
2.	*Aspergillus awamori*	β-D-xylosidase	Xylan oligomer	Exo-β-1,4-glycosidic bond	Thomson (1983)
3.	*Tricoderma reesei, Aspergillus niger*	α-L-*Arabino furanosida*	Arabinose group	α-1,2-glycosidic bond, α-1,3-glycosidic bond, α-1,5-glycosidic bond	Poutanen (1988), Rombouts et al. (1988)
4.	*Aspergillus niger*	Endo-1,5-α-arabinases	Arabinose group	α-1,5-glycosidic bond	Beldman et al. (1993)
5.	*Aspergillus awamori*	Arabino xylan-arabinofuranohydrolases	Arabinose group	α-1,4-glycosidic bond	Kormelink et al. (1991)
6.	*Aspergillus niger*	α-Glucuronidase	Glucuronic acid group	α-1,2-glycosidic bond	Yoshida et al. (1990)
7.	*Aspergillus niger and Tricoderma reesei*	Acetyl xylan esterase	Acetyl group	Ester bonds	Biely et al. (1986)
8.	*Aspergillus niger and Tricoderma reesei*	Acetyl esterases	Feruloyl group	Ester bonds	Tenkanen et al. (1991)
9.	*Aspergillus niger and Tricoderma reesei*	Feruloyl esterase	*p*-Coumaroyl group	Ester bonds	Tenkanen et al. (1991)
Bacteria					
10.	*Fibrobacter succinogenes*	β-Xylosidases	Xylan oligomer	Exo-β-1,4-glycosidic bond	Utt et al. (1991)
11.	*Ruminococcus albus*	β-Xylosidases	Xylan oligomer	Exo-β-1,4-glycosidic bond	Utt et al. (1991)
12.	*Prevotella(Bacteroides) ruminicola*	β-Xylosidases	Xylan oligomer	Exo-β-1,4-glycosidic bond	Utt et al. (1991)

Malcolm, 1996). It is the most common organic polymer, representing about 1.5×10^{12} tons of the total annual biomass production through photosynthesis. This is the same glucose that your body metabolizes in order to live, but you cannot digest it in the form of cellulose. Because cellulose is built out of a sugar monomer, it is called a polysaccharide. Cellulose is an insoluble molecule consisting of between 2000 and 14,000 residues with some preparations being somewhat shorter. It is found in plants as microfibrils (2–20 nm diameter and 100–40,000 nm long). Cellulose is also produced in a highly hydrated form by some bacteria (*Acetobacter xylinum*). Structural composition of cellulose, and arrangement of its macro and micofibrils, their arrangement of H-bonding in the cellulose molecule and its amorphous and crystalline region are showing in Figure 10.21.

Photosynthetic such as plant, algae, and some bacteria produce more than 100 million tons of organic matter each year from the fixation of carbon dioxide. Half of this biomass is made up of the biopolymer cellulose, which as a result is perhaps the most abundant molecule on the planet. The physical structure and morphology of native cellulose are complex, and fine structural details are difficult to determine experimentally (O'Sullivan, 1997). The chemical

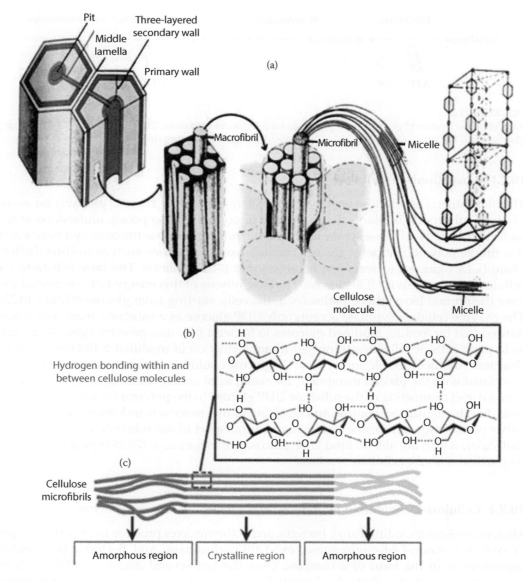

FIGURE 10.21
(See color insert.) (a) Structural composition of cellulose and arrangement of macro and microfibrils. (b) Chemical structure and bonding arrangement of cellulose molecule. (c) Showing cellulose microfibrils and amorphous and crystalline regions.

composition of cellulose is simple: the polysaccharide consists of D-glucose residues linked by β-1,4-glycosidic bonds to form linear polymeric chains of over 10,000 glucose residues. The individual chains adhere to each other along with their lengths by hydrogen bonding and Vander Waals forces, and crystallize shortly after biosynthesis. Although highly crystalline, the structure of cellulose is not uniform. Physical and chemical evidence indicates that cellulose contains both highly crystalline and less-ordered amorphous (Hon, 1994) (Figure 10.21). Although chemically simple, the extensive intermolecular bonding pattern of cellulose generates a crystalline structure that, together with other cell-wall components such as hemicellulose and lignin, results in very complex morphologies.

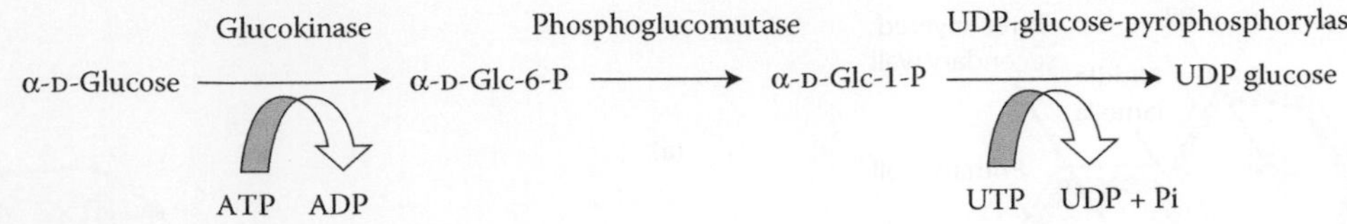

Glucokinase   Phosphoglucomutase   UDP-glucose-pyrophosphorylase

α-D-Glucose ⟶ α-D-Glc-6-P ⟶ α-D-Glc-1-P ⟶ UDP glucose

ATP   ADP   UTP   UDP + Pi

FIGURE 10.22
Intracellular activation of glucose as the precursor for cellulose biosynthesis. UDP_uridine-5-diphosphate; glc_glucose; glc-6-P_glucose-6-phosphate; glc-1-P_glucose-1-phosphate; Pi–inorganic phosphate.

10.7.1 Biosynthesis of Cellulose

The biosynthesis of cellulose is not yet completely elucidated. There is probably no major biochemical process in plants that is both so important and so poorly understood at the molecular level as cellulose synthesis. This is surprising, because the basic synthetic event is a simple polymerization of glucose residues from a substrate such as uridine diphosphate (UDP) glucose to form the homopolymer p-1,4-D-glucan. The only substrate for cellulose biosynthesis is UDP glucose. The biosynthesis of this energy-rich compound follows the normal biosynthetic pathways in the cells, starting from glucose (Figure 10.22). The enzyme cellulose synthase accepts only UDP glucose as a substrate; moreover, it was noticed that by feeding modified glucoses to bacteria (*Gluconacetobacter xylinum*), as well as to plant cells or cell extracts, no significant formation of modified celluloses could be detected. Another possible source for UDP glucose could be sucrose synthase, an enzyme associated with the plasma membrane, for example, of cotton fibers. Because of this location, a direct channeling of the substrate UDP glucose to the polymerizing enzyme is possible. But the regulation, control, and targeting of this process is unknown in wide areas. Other possible sources for the stabilizing and transport of the substrate are annexin-like molecules, which are able to bind UDP glucose, for example, a 170-kDa polypeptide was co-purified with the cellulose synthase.

10.7.2 Cellulose Degradation by Aerobic and Anaerobic Microorganisms

Microorganisms including fungi, bacteria, and actinomycetes produce mainly three types of cellulase—endo-1,4-β-D-glucanase, exo-1,4-β-D-glucanase, and β-glucosidase—either separately or in the form of a complex. Over the last several decades, cellulases have become better understood at a fundamental level; nevertheless, much remains to be learnt (Bhat and Bhat, 1997). Most of the cellulolytic microorganisms belong to eubacteria and fungi, although some anaerobic protozoa and slime molds are able to degrade cellulose. Cellulolytic microorganisms can establish synergistic relationships with noncellulolytic species in cellulosic wastes. The interactions between both populations lead to complete degradation of cellulose, releasing carbon dioxide and water under aerobic conditions, and carbon dioxide, methane, and water under anaerobic conditions (Leschine, 1995). Microorganisms capable of degrading cellulose produce a battery of enzymes with different specificities, working together. Cellulases hydrolyze the β-1,4-glycosidic linkages of cellulose. Traditionally, they are divided into two classes referred to as endoglucanases and cellobiohydrolases. Endoglucanases (endo-1,4-β-glucanases, EGs) can hydrolyze internal bonds (preferably in cellulose amorphous regions) releasing new terminal ends. Cellulose are degraded aerobically or anaerobically by extracellular microbial enzyme cellulose, which convert cellulose into cellobiose (two glucose unit), and it can be transported

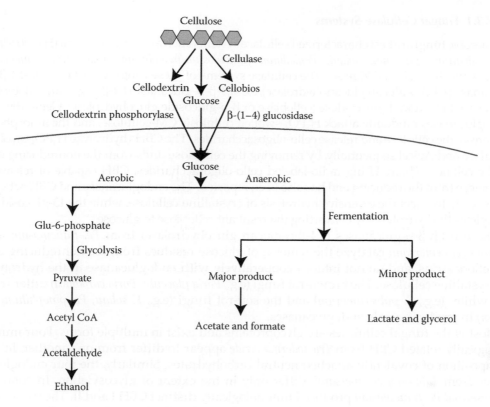

FIGURE 10.23
Degradation pathway of cellulose.

into the cell, β-1,4-glucosidase enzyme cleaves the cellobiose into glucose molecule. Finally glucose in aerobic condition enters into TCA cycle and in anaerobic condition enters into fermentation process (Figure 10.23).

10.7.3 Enzymes Involved in Cellulose Breakdown

Cellulases are enzymes that catalyze the hydrolysis of β-1,4-glycosidic bonds of cellulose, breaking insoluble cellulose into fermentable glucose. There are three general types of cellulases that collaborate to break down cellulosic biomass:

- Endo-β-1,-4-glucanase breaks internal bonds to disrupt the structure of cellulose and expose individual polysaccharide chains.
- Exo-β-1,4-D-glucanase can access the chains from the reducing end or the nonreducing end of the exposed chains and cleaves two to four units to produce tetrasaccharide or disaccharide such as cellobiose.
- Cellobiase or β-glucosidase hydrolyzes cellobiose to release D-glucose units.

These types of enzymes alone cannot hydrolyze complex crystalline cellulose structure efficiently but working together synergistically, they can increase the hydrolysis rates significantly. There are two main methods, which are acidic hydrolysis and enzymatic hydrolysis.

10.7.3.1 Fungal Cellulase Systems

The aerobic fungi are best characterized cellulase systems such as *P. chrysosporium* (*Sporotrichum pulverulentum*), *Fusarium solani*, *Penicillium funiculosum/pinophilum*, *Talaromyces emersonii*, *Trichoderma koningii*, and *T. reesei*. The cellulase systems of these fungi consist of endo-1,4-β-D-giucanase (1,4-β-D-glucan glucanohydrolase), exo-1,4-β-D-glucanase (1,4-β-D-glucan cellobiohydrolase [CBH]), and β-giucosidase (cellobiase or β-D-giucoside glucohydrolase). Generally, the endoglucanases randomly attack H_3PO_4-swollen cellulose, CM-cellulose, and the amorphous regions of the cellulose and release cello-oligosaccharides. The CBHs hydrolyze H_3PO_4-swollen cellulose and Avicel sequentially, by removing the cellobiose units from the nonreducing end of the cellulose chain. Using radio-labeled cello-oligosaccharides, CBH capable of releasing cellobiose from the reducing end has also been reported. The endoglucanase and CBH act synergistically to affect the extensive hydrolysis of crystalline cellulose, while the 13-glucosidase completes the hydrolysis by converting the resultant cellobiose to glucose.

The exo-1,β-β-D-giucanases (1,4-β-D-giucan giucohydrolase) from *Pe. funiculosum* and *Talaromyces emersonii* catalyze the removal of glucose residues from the nonreducing end of cellodextrins, and do not interact cooperatively with endoglucanase in the hydrolysis of crystalline cellulose. The brown-rot fungi (e.g., *Poria placenta*, *Poria carbonica*) differ from the white- (e.g., *S. pulverulentum*) and the soft-rot fungi (e.g., *F. solani*, *Pe. pinophilum*, *Tr. reesei*) in producing only endoglucanases.

Most of the fungal cellulases are glycoproteins and exist in multiple forms. Four immunologically related CBH from *Trichoderma viride* appear to differ from one another, in the composition of covalently attached neutral carbohydrates. Similarly, the four endoglucanases from *Talaromyces emersonii*, differ only in the extent of glycosylation. In contrast, *T. reesei and Pe. funiculosum* produce immunologically distinct CBH I and II. The two CBHs differing in amino acid composition have also been identified from *Fusarium lini*.

The extracellular cellulase components of most fungi are generally found to exist as individual entities. However, the presence of aggregates of extracellular enzyme in *Tr. reesei* culture filtrate has been reported. This complex consists of six proteins and exhibits cellulase, β-glucosidase, and xylanase activities. Also, the proteins appear to be aggregated with the help of the remnants of the fungal cell wall, where, Ca^{2+} may be involved. Although many fungi secrete separate cellulase components into culture medium, it is not yet clearly known how these components interact on the surface of crystalline cellulose and affect the extensive hydrolysis of cellulose.

Studies on rumen anaerobic fungi suggest that these microorganisms have a definite role in the initial colonization and degradation of lignocellulose in the rumen. Anaerobic fungi have been shown to produce extracellular cellulases and xylanases, which are important in the breakdown of lignocellulose. Several studies have been carried out on the production of cellulolytic enzymes from these microorganisms and indicated their importance in the degradation of cellulosic materials and subsequent fermentation in the rumen (Bauchop, 1979).

Anaerobic cellulolytic fungi belonging to genera *Neocallimastix*, *Cacomyces*, *Orpinomyces*, *Piromyces*, and *Rurainorayces* have been described and classified. At present, it is a well-established fact that obligately anaerobic fungi are found in the saliva, alimentary tracts, and feces of a number of animals such as ruminants and herbivorous nonruminant mammals. Among these species, *Neocallimastix frontalis* has been studied in most detail. Unlike that of aerobic fungi, the cellulase system of *N. frontalis* is a multicomponent enzyme complex termed crystalline cellulose-solubilizing factor (CCSF). It has a molecular mass of 700 kDa and consists of a number of subunits ranging in molecular mass from 68 to 135 kDa. So far, it has not been possible to dissociate the CCSF without losing much of its original activity toward crystalline cellulose. The presence of the multicomponent enzyme complex of this fungus suggests that it may hydrolyze the crystalline cellulose using a mechanism that is similar to that of the anaerobic cellulolytic bacterium *Clostridium thermocellum*. The *N. frontalis* (RK 21) is the most efficient degrader of crystalline cellulose so far known and plays an important role in the digestion of cellulose in the rumen. Therefore, it will be interesting to investigate the mechanism of cellulose degradation by the cellulase system of this fungus.

10.7.3.2 Bacterial Cellulase Systems

Cellulolytic bacteria may be aerobes or anaerobes. Most of the bacteria produce mainly endoglueanases. Only in case of actinomycete, *Microbispora bispora*, the presence of cellobiohydrolase in culture fluid has been convincingly demonstrated. However, the β-glucosidase in this actinomycete is cell bound. Also, the cellulase system of *M. bispora* was similar to those of aerobic cellulolytic fungi in effecting the hydrolysis of crystalline cellulose. In contrast, the anaerobic bacteria, *C. thermocellum*, *Clostridium cellulovorans*, *Acetivibrio cellulolyticus*, which degrade the highly crystalline cellulose such as cotton produce a high molecular-mass enzyme complex called cellulosome.

In case of *C. thermocellum*, the cellulosome consists of several endoglucanases and at least three exoglucanases. It has been suggested that these anaerobic bacteria must attach themselves to cellulose in order to affect cellulolysis. The activity of some of the anaerobic bacteria, *C. thermocellum*, *A. cellulolyticus*, and *Ruminococcus flavefaciens*, toward crystalline cellulose is dependent on either Ca^{2+} or Mg^{2+} and either DTT or other thiol reagents. The activities of *Ruminococcus albus* and *Fibrobacter (Bacteroides) succinogenes* against crystalline cellulose were diminished in the presence of air, but maintained under reducing conditions.

10.7.3.3 Bacterial Enzymes Involved in Cellulose Degradation

Although these anaerobic bacteria hydrolyze crystalline cellulose, there is no definite proof for the production of exoglucanase activity. *Acidothermus cellulolyticus*, anaerobic bacterium produces cell bound β-galactosidase enzyme which degrade cellulose. The cells of *Bacteroides cellulosolvens* were believed to comprise cellobiose phosphorylase, whereas the cells of *C. thermocellum* and *R. flavefaciens* produced both cellobiose phosphorylase and β-giucosidase. However, the β-giucosidase of *C. thermocellum* had a high K_m toward cellobiose and hence its role in the metabolism of cellobiose is difficult to understand. In contrast, cellobiose phosphorylase from *C. thermocellum* showed a high affinity for cellobiose. Also, *C. thermocellum* produced an intracellular cellodextrin phosphorylase, but its role is not well understood.

Among the bacterial cellulase systems, the most extensively studied system is that of *C. thermocellum*. This bacterium produces a very active cellulase in the form a complex termed cellulosome, which degrades the crystalline cellulose extensively in the presence of Ca^{2+} and a reducing agent, preferably dithiothreitol. The yellow affinity substance (YAS) which is produced by *C. thermocellum* during its growth on crystalline cellulose further facilitates the degradation of cellulose by binding the cellulosome to the substrate. Also, the genes coding for several endogiucanases of *C. thermocellum* have been cloned, expressed, and the corresponding proteins purified and characterized. Recently, recombinant EGA, EGC, and EGD from *C. thermocellum* have been crystallized and their 3D structures determined. Despite substantial progress, the mechanism of cellulose degradation by the cellulosome of *C. thermocellum* is not entirely clear. Attempts to dissociate and isolate the individual subunits of the *C. thermocellum* cellulosome in active form were unsuccessful for many years (Eriksson and Wood, 1985).

Studies reported the isolation of S_L (S_l) and Ss (S_8) subunits of *C. thermocellum* cellulosome in active form and their interaction in solubilizing Avicel, but the extent of Avicel solubilization was low. Recently, Bhat and Wood established conditions for the successful dissociation and reassociation of the cellulosome without losing much of its original activity. Using this novel method, four subunits of *C. thermocellum* cellulosome have been isolated, and their interaction in the solubilization of crystalline cellulose has been studied (Wood, 1985; Bhat et al., 1989). One major endoglucanase and two major exoglucanases corresponding to three different subunits of the cellulosome have been purified and characterized. It should now be possible to isolate all the subunits of *C. thermocellum* in active form and demonstrate the mechanism of cellulose degradation by this highly active cellulase system (Table 10.6).

TABLE 10.6

List of Cellulose Degrading Microorganism Their Enzyme and Initial Action

Microorganism	Enzyme	Initial Action
Aerobic bacteria		
Bacillus licheniformis 1,	Endoglucanase	--G--G--G--G--- ↑ ↑
Bacillus sp. (alkalophilic) 1139		Cleaves linkages at random
Bacillus sp. (alkalophilic) (cloned in *E.coli*) N-4	Endoglucanase cel A Endoglucanase cel B Endoglucanase cel C	
Bacillus sp. (neutrophilic) KSM-522	Endoglucanase	
Bacillus subtilis	Endoglucanase	
Microbispora bispora	Endoglucanase I Endoglucanase II Exoglucanase I Exoglucanase II	Hydrolysis of 1,4-β-D-glucosidic linkages in cellulose and cellotetraose, releasing cellobiose from the nonreducing ends of the chains
	β-Glucosidase	G-G, G-G-G-G ↑ ↑ Releases glucose from cellobiose and short chain cellooligosaccharides
Aerobic Bacteria		
Acetivibrio cellulolyticus ATCC33288	Exoglucanase C1	Hydrolysis of 1,4-beta-D-glucosidic linkages in cellulose and cellotetraose, releasing cellobiose from the nonreducing ends of the chains
	Endoglucanase C2 Endoglucanase C3	--G--G--G--G---; ↑ ↑ Cleaves linkages at random
	β-Glucosidase B1	G-G, G-G-G-G ↑ ↑ Releases glucose from cellobiose and short chain cellooligosaccharides
Mesophilic actinomycetes		
Streptomyces antibiotics *Streptomyces flavogriseus* *Streptomyces viridosporus* *Streptomyces nitrosporeus* *Streptomyces albogriseolus* *Micromonospora melanosporea*	Complete cellulase system	Catalyzes extensive hydrolysis of crystalline cellulose
Fungus		
Neocallimastix frotalis *Neocallimastix patriciarum* *Piromonas communis*	Endoglucanase or endocellulase	--G--G--G--G--- ↑ ↑ Cleaves linkages at random

Most cellulose is degraded aerobically, but 5%–10% is degraded anaerobically (Vogels, 1979). Thus, vast quantities of cellulose are degraded by cellulose-fermenting microorganisms in anaerobic environments. Anaerobic activity starts close to the surface in soils and composts, as well as in freshwater, marine, and estuarine sediments, indicating that aerobic conditions normally prevail only in a thin crust at the atmospheric boundary (Eriksson, 1978; Ljungdahl and Eriksson, 1985). The anaerobic degradation of soil organic matter plays an extremely important role in the global cycling of carbon. Soils are a huge reservoir of carbon: The top meter of soil contains about twice as much carbon as is found in the atmosphere (Post et al., 1982), and some recent studies suggest that the rates of accumulation and turnover of below ground carbon may be underestimated. Consequently, understanding the effects of predicted global warming on decomposition of organic matter in soils and sediments is essential for modeling future atmospheric and climatic changes (Jenkinson et al., 1991).

A community of physiologically diverse microorganisms is responsible for the anaerobic degradation of cellulose. This community structure contrasts with aerobic decomposition, which may be achieved through the activities of single species. For example, both carbohydrate and lignin components of wood are completely decomposed to CO_2 and H_2O by single species of white-rot fungi (Kirk, 1971). Anaerobic decomposition, on the other hand, requires mixed populations. The metabolic versatility of anaerobes arises largely because they can perform various fermentations and respirations, employing diverse electron acceptors (e.g., carbon dioxide, inorganic sulfur compounds in organic nitrogen compounds) in place of oxygen. In the absence of oxygen and certain other exogenous inorganic electron acceptors (for example, nitrate, Mn(IV) Fe(III), sulfate), cellulose is decomposed by the anaerobic community into CH_4, CO_2, and H_2O through a complex microbial food chain (Boone, 1991) shown diagrammatically in Figure 10.24.

The processes are similar in most anaerobic soils and sediments and in anaerobic digestors. Cellulolytic microbes produce enzymes that depolymerize cellulose, thereby producing cellobiose, cellodextrins, and some glucose. These sugars are fermented by cellulolytic and other saccharolytic microorganisms. By keeping cellobiose concentrations low, and

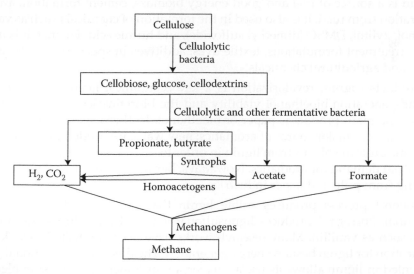

FIGURE 10.24

Anaerobic cellulose degradation by microbial communities in soil and sediments. Lactate, succinate, and ethanol are also produced by fermentative bacteria but usually do not accumulate. In environment where nitrate, Mn(IV), Fe(III), or sulfate are present, through final products of fermentation may differ.

thus preventing inhibition of the cellulase system by this product of cellulose hydrolysis, noncellulolytic cellobiose fermenters may play a very important role in this process. These fermentations yield CO_2, H_2, organic acids (e.g., acetate, propionate, butyrate), and alcohols. Very little H_2 escapes into the atmosphere because it is immediately consumed by methanogens or homoacetogens. Methanogens use H_2 to reduce CO_2 to CH_4, and homoacetogens use H_2 to reduce CO_2 to acetate. Some methanogenic species use acetate produced by fermenters or by homoacetogens through the acetoclastic cleavage to CH_4 and CO_2 (Mah, 1981). Syntrophic bacteria play a key role in the conversion of cellulose to CH_4 and CO_2. These organisms ferment fatty acids such as propionate and butyrate, or alcohols, and produce acetate, CO_2, and H_2. They grow only in the presence of H_2-consuming organisms through interspecies H_2 transfer. Syntrophic bacteria grow very slowly and thus the fermentation of fatty acids is usually the rate-limiting step in the anaerobic decomposition of cellulose (Wolin and Miller, 1987). Through the combined activities of several major physiological groups of microbes, cellulose is completely dissimilated to CO_2 and CH_4. Thus, as a source of CO_2 and CH_4, the anaerobic decomposition of cellulose plays a major role in carbon cycling on the planet. In marine environments, sulfate is plentiful and sulfate-reducing bacteria out-compete methanogens or H_2. Thus, H_2S is a major product of the anaerobic degradation of cellulose in marine systems.

10.8 Biotechnological Applications of Lignocellulosic Biomass

- Lignin is the most abundant source of aromatic compounds in nature and can generate a large amount of chemical reagents or adhesives to replace those derived from oil. This is the case of phenol-formaldehyde in which part of phenol can be replaced by lignin that results to be more available, less toxic, and less expensive than phenol.

- Lignin is a source of fuel and good energy biomass, cement formation, and dust separation from road. It is also used in the formation of chemicals such as vanillin, ethanol, xylitol, DMSO (dimethyl sulfoxide), and humic acid. Further, it is used in water treatment formulations, textile dyes, additives in specialty oilfield applications, and agricultural chemicals.

- By-products (lignin) revalorization is a key economical factor that will ensure second-generation bioethanol viability and the biorefineries implementation no matter the main product is paper or bioethanol. In this sense, lignin revalorization is nature's major source of aromatics; new ways are needed to produce small aromatic building blocks from lignin in order to satisfy the enormous and diverse industrial demand for aromatics. Nowadays, lignin is available as lignosulfonates coming from pulp and paper industries.

- Lignoboost process produces sulfite lignin that can be used as a dispersant (Inventia), Borregard produces lignosulfonates and also produces lignin derivatives, such as vanillin. Many research works have been conducted to look for an application for lignin besides energy generation. The diversity of functional groups presented in lignin allows its use as dispersant in cement and gypsum blends, as emulsifier or chelating agent for removing heavy metals from industrial effluents.

- Xylans.

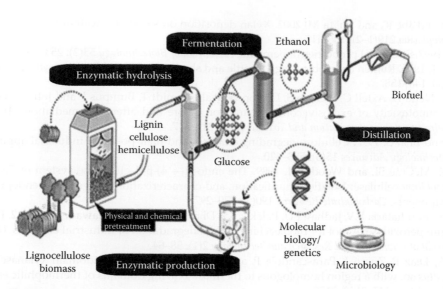

FIGURE 10.25
Conversion of lignocellulose into biofuels.

- Polymers: "super gel" for wound dressing, micro and nano particles for controlled drug delivery
- Oligosaccharides: novel functional food ingredients modifying food flavor
- Physiochemical characteristics: model component for enzymatic assay
- Adhesive properties: wet end additive in paper-making (**Xyloglucans)**
- β-glucans
- Accepted as functional bioactive ingredients
- Interaction with starch (reduced enthalpy of starch gelatinozation, reduced swelling of starch granule)
- Viscosity enhancing effect: thickening agent for gravies, salad dressing, or icecream formulation; stabilizing of emulsions and foams
- Change in the precipitation of mouth-feel for beverages

The two most common examples of biofuels are bio-ethanol and bio-diesel (Figure 10.25). Biofuels are considered the cleanest liquid fuel alternative to fossil fuel and it is estimated that replacing fossils fuels with biofuels could decrease CO_2 emissions by 60%–90%.

References

Adler E. 1977. Lignin chemistry—past, present and future. *Wood Science and Technology* 11(3): 169–218.
Ahmad M, Taylor CR, Pink D, Burton KS, Eastwood DC, and Bending GD. 2010. Development of novel assays for lignin degradation: Comparative analysis of bacterial and fungal lignin degraders. *Molecular Biosystems* 6(5): 815–821.

Awano T, Takabe K, and Fujita M. 2002. Xylan deposition on secondary wall of *Fagus crenata* fiber. *Protoplasma* 219(1–2): 106–115.

Banci L. 1997. Structural properties of peroxidases. *Journal of Biotechnology* 53(2): 253–263.

Bauchop T. 1979. Rumen anaerobic fungi of cattle and sheep. *Applied and Environmental Microbiology* 38(1): 148–158.

Beltz LA, Neira D, Axtell C, Iverson S, Deaton W, Waldschmidt T, Bumpus J, and Johnston C. 2001. Immunotoxicity of explosivecontaminated soil before and after bioremediation. *Archives of Environmental Contamination and Toxicology* 40: 311–317.

Bhat M and Bhat S. 1997. Cellulose degrading enzymes and their potential industrial applications. *Biotechnology Advances* 15(3): 583–620.

Bhat KM, McCrae SI, and Wood TM. 1989. The endo-(1→ 4)-β-d-glucanase system of *Penicillium pinophilum* cellulase: Isolation, purification, and characterization of five major endoglucanase components. *Carbohydrate Research* 190(2): 279–297.

Bholay A., Borkhataria B.V, Jadhav PU, Palekar KS, Dhalkari MV, and Nalawade PM. 2012. Bacterial lignin peroxidase: A tool for biobleaching and biodegradation of industrial effluents. *Universal Journal of Environmental Research and Technology* 2(1): 58–64.

Blanco A, Díaz P, Zueco J, Parascandola P, and Pastor JF. 1999 A multidomain xylanase from a *Bacillus* sp. with a region homologous to thermostabilizing domains of thermophilic enzymes. *Microbiology* 145: 2163–2170.

Boerjan W, Ralph J, and Baucher M, 2003. Lignin biosynthesis. *Annual Review of Plant Biology* 54(1): 519–546.

Boone DR. 1991. Ecology of methanogenesis. In: Roger JE and Whitman WB (Eds.), *Microbial Production and Consumption of Greenhouse Gases: Methane, Nitrogen Oxides, and Halomethanes*. American Society for Microbiology, pp. 57–70.

Brown Jr and Malcolm, R. 1996. The biosynthesis of cellulose. *Journal of Macromolecular Science, Part A: Pure and Applied Chemistry* 33(10): 1345–1373.

Bugg TDH, Ahmad M, Hardiman EM, and Rahmanpour R. 2011. Pathways for degradation of lignin in bacteria and fungi. *Natural Product Reports* 28(12): 1883–1896.

Carvalheiro F, Duarte LC, and Giri FM. 2008. Hemicellulose biorefineries: A review on biomass pre-treatments. *Journal of Scientific and Industrial Research* 67: 849–864.

Chandra R, Abhishek A, and Sankhwar M. 2011. Bacterial decolorization and detoxification of black liquor from rayon grade pulp manufacturing paper industry and detection of their metabolic products. *Bioresource Technology* 102(11): 6429–6436.

Chandra R and Bharagava RN 2013. Bacterial degradation of synthetic and kraft lignin by axenic and mixed culture and their metabolic products. *Journal of Environmental Biology* 34(6): 991–999.

Chandra R, Raj A, Purohit HJ, and Kapley, A. 2007. Characterisation and optimisation of three potential aerobic bacterial strains for kraft lignin degradation from pulp paper waste. *Chemosphere* 67: 839–846.

Chandra R and Singh R. 2012. Decolourisation and detoxification of rayon grade pulp paper mill effluent by mixed bacterial culture isolated from pulp paper mill effluent polluted site. *Biochemical Engineering Journal* 61:49–58.

Chandra R, Singh R, and Yadav S. 2012. Effect of bacterial inoculum ratio in mixed culture for decolourization and detoxification of pulp paper mill effluent. *Journal of Chemical Technology and Biotechnology* 87(3): 436–444.

Chandra R, Singh S, Reddy KMM, Patel DK, Purohit HJ, and Kapley A. 2008. Isolation and characterization of bacterial strains *Paenibacillus* sp. and *Bacillus* sp. for kraft lignin decolorization from pulp paper mill waste. *Journal of General and Applied Microbiology* 54: 399–407.

Cheong S, Yeo S, Song HG, and Choi H. 2006. Determination of laccase gene expression during degradation of 2,4,6-trinitrotoluene and its catabolic intermediates in *Trametes versicolo*. *Microbiology Research* 161:316–320.

Dedeyan B, Klonowska A, Tagger S, Tron T, Iacazio G, Gil G, and Le Petit J 2000. Biochemical and molecular characterization of a laccase from *Marasmius quercophilus*. *Applied Environmental Microbiology* 66(3): 925–929.

Dias A, Bezerra R, Lemos P, and Pereiram A. 2003. In vivo and laccase characterization of xenobiotic azo dyes by basidiomycetous fungus: Characterization of its lignolytic system. *World Journal of Microbiology and Biotechnology* 19:969–975.

Dixon RA, Chen F, Gua D, and Parvathi K. 2001. The biosynthesis of monolignols: A "metabolic grid", or independent pathways to guaiacyl and syringyl units? *Phytochemistry* 57(7): 1069–1084.

Ducros V, Brzozowski AM, Wilson KS. Østergaard P, Schneider P, Svendson A, and Davies GJ. 2001. Structure of the laccase from *Coprinus cinereus* at 1.68 A resolution: Evidence for different "type 2 Cu-depleted" isoforms. *Acta Crystallographica Section D: Biological Crystallography* 57(Pt 2): 333–336.

Eggert C, LaFayette PR, Temp U, Eriksson KL, and Dean JFD. 1998. Molecular analysis of a laccase gene from the white rot fungus Pycnoporus cinnabarinus. *Applied and Environmental Microbiology* 64(5): 1766–1772.

Erden E, Ucar MC, Gezer T, and Pazarlioglu, NK. 2009. Screening for ligninolytic enzymes from autochthonous fungi and applications for decolorization of Remazole Marine Blue. *Brazilian Journal of Microbiology* 40(2): 346–353.

Eriksson KE. 1978. Enzyme mechanisms involved in cellulose hydrolysis by the rot fungus *Sporotrichum pulverulentum*. *Biotechnology and Bioengineering* 20(3): 317–332.

Eriksson KE and Wood TM. 1985. Biodegradation of cellulose. In: Higuchi T (Ed.), *Biosynthesis and Biodegradation of Wood Component*, Academic Press, New York, pp. 469–503.

Ferrer JL, Austin MB, Stewart Jr C, and Noel JP. 2008. Structure and function of enzymes involved in the biosynthesis of phenylpropanoids. *Plant Physiology and Biochemistry* 46(3): 356–370.

Gianfreda L, Xu F, and Bollag JM. 1999. Laccases: A useful group of oxidoreductive enzymes. Bioremediation Journal 3(1): 1–26.

Glenn JK, Morgan MA, Mayfield MB, Kuwahara M, and Gold MH. 1983. An extracellular H_2O_2-requiring enzyme preparation involved in lignin biodegradation by the white rot basidiomycete *Phanerochaete chrysosporium*. *Biochemical and Biophysical Research Communications* 114(3): 1077–1083.

Hakkila P. 1989. *Utilization of Residual Forest Biomass*. Springer-Verlag, Berlin, Heidelberg.

Hatakka A. 1994. Lignin-modifying enzymes from selected white-rot fungi: Production and role from in lignin degradation. *FEMS Microbiology Reviews* 13(2): 125–135.

Hofrichter M. 2002. Review: Lignin conversion by manganese peroxidase (MnP). *Enzyme and Microbial Technology* 30(4): 454–466.

Hon DN-S. 1994. Cellulose: A random walk along its historical path. *Cellulose* 1(1): 1–25.

Howard R, Abotsi E, Van Rensburg ELJ and Howard S. 2004. Lignocellulose biotechnology: Issues of bioconversion and enzyme production. *African Journal of Biotechnology* 2(12): 602–619.

Janshekar H, Haltmeier T and Brown C. 1982. Fungal degradation of pine and straw alkali lignins. *European Journal of Applied Microbiology and Biotechnology* 14(3): 174–181.

Jeffries TW. 1994. Biodegradation of lignin and hemicelluloses. In: Ratledge C (Ed.), *Biochemistry of Microbial Degradation*. Springer, the Netherlands, pp. 233–277.

Jenkinson D, Adams DE, and Wild A. 1991. Model estimates of CO_2 emissions from soil in response to global warming. *Nature* 351(6324): 304–306.

Kaneda M, Rensing KH, Wong JC, Banno, B, Mansfield SD, and Samuels AL. 2008. Tracking monolignols during wood development in lodgepole pine. *Plant Physiology* 147(4): 1750–1760.

Kirk TK. 1971. Effects of microorganisms on lignin. *Annual review of Phytopathology* 9(1): 185–210.

Kirk TK and Cullen D. 1998. Enzymology and molecular genetics of wood degradation by white-rot fungi. In: Yong RA and Akhtar M (Ed.), *Environmentally Friendly Technologies for the Pulp and Paper Industry*. Wiley, New York, pp. 273–307.

Kormelink FJM, Leeuwen MJF, Wood TM, and Voragen AGJ. 1993. Purification and characterization of three endo-(1, 4)-β-xylanases and one β-xylosidase from *Aspergillus awamori*. *Journal of Biotechnology* 27(3): 249–265.

Kumar R, Singh S, and Singh OV. 2008. Bioconversion of lignocellulosic biomass: Biochemical and molecular perspectives. *Journal of Indian Microbiology and Biotechnology* 35:377–391.

Leschine SB. 1995. Cellulose degradation in anaerobic environments. *Annual Reviews in Microbiology* 49(1): 399–426.

Lin LL and Thomson JA. 1991. An analysis of the extracellular xylanases and cellulases of *Butyrivibrio fibrisolvens* H17c. *FEMS Microbiology Letters* 84(2): 197–204.

Ljungdahl LG and Eriksson KE. 1985. Ecology of microbial cellulose degradation. In: Marshall KC (Ed.), *Advances in Microbial Ecology*, Vol. 8, Springer, US, pp. 237–299.

Lobos S, Larraín J, Salas L, Cullen D, and Vicuña R. 1994. Isoenzymes of manganese-dependent peroxidase and laccase produced by the lignin-degrading basidiomycete *Ceriporiopsis subvermispora*. *Microbiology* 140(10): 2691–2698.

Lundqvist J, Jacobs A, Palm M, Zacchi G, Dahlman O, and Stålbrand H. 2003. Characterization of galactoglucomannan extracted from spruce (*Picea abies*) by heat-fractionation at different conditions. *Carbohydrate Polymers* 51 (2): 2003–2011.

Mah, R.A. 1981. The methanogenic bacteria, their ecology and physiology. In: Hollaender A, Robert R, Rogers P, San Pietro A, Valentine R, Wolfe R (Eds.), *Trends in the Biology of Fermentations for Fuels and Chemicals*. Springer, pp. 357–374.

Martinez AT, Ruiz-Dueñas FJ, Martínez MJ, del Río JC, and Gutiérrez A. 2009. Enzymatic delignification of plant cell wall: From nature to mill. *Current Opinion in Biotechnology* 20(3): 348–357.

Masai E, Katayama Y, and Fukuda M. 2007. Genetic and biochemical investigations on bacterial catabolic pathways for lignin-derived aromatic compounds. *Bioscience, Biotechnology, and Biochemistry* 71(1): 1–15.

McCann MC and Roberts K. 1991. Architecture of the primary cell wall. In: Lloyd CW (Ed.), *The Cytoskeletal Basis of Plant Growth and Form*. Academic Press, London, pp. 109–129.

Meier H and Reid JSG. 1982. Reserve polysaccharides other than starch in higher plants. In: Loewus FA and Tanner W (Eds.), *Plant carbohydrates I*. Springer, US, pp. 418–471.

Moura JCMS, Bonine CAV, De Oliveira Fernandes Viana J, Dornelas MC, and Mazzafera P. 2010. Abiotic and biotic stresses and changes in the lignin content and composition in plants. *Journal of Integrative Plant Biology* 52 (4): 360–376.

Naran R, Chen G, and Carpita NC. 2008. Novel rhamnogalacturonan I and arabinoxylan polysaccharides of flax seed mucilage. *Plant physiology* 148 (1): 132–141.

Oliveira PL, Duarte MCT, Ponezi, AN, and Durrant, LR. 2009. Purification and Partial characterization of manganese peroxidase from *Bacillus pumilus* and *Paenibacillus* sp. *Brazilian Journal of Microbiology* 40(4): 818–826.

O'Neill MA and York WS. 2003. The composition and structure of plant primary cell walls. In: Rose JKC (Ed.), *The Plant Cell Wall*. CRC Press, Boca Raton, pp. 1–54.

Önnerud H, Zhang L, Gellerstedt G, and Henriksson G. 2002. Polymerization of monolignols by redox shuttle–mediated enzymatic oxidation a new model in lignin biosynthesis I. *The Plant Cell Online* 14(8): 1953–1962.

O'sullivan AC. 1997. Cellulose: The structure slowly unravels. *Cellulose* 4(3): 173–207.

Popper ZA. 2008. Evolution and diversity of green plant cell walls. *Current Opinion in Plant Biology* 11(3): 286–292.

Post WM, Emanuel WR, Zinke PJ, and Stangenberger AG. 1982. Soil carbon pools and world life zones. *Nature* 298: 156–159.

Raghukumar C, Chandramohan D, Michel Jr FC, and Redd CA. 1996. Degradation of lignin and decolorization of paper mill bleach plant effluent (BPE) by marine fungi. *Biotechnology Letters* 18(1): 105–106.

Raj A, Reddy MMK, and Chandra R. 2007a. Identification of low molecular weight aromatic compounds by gas chromatography–mass spectrometry (GC–MS) from kraft lignin degradation by three Bacillus sp. *International Biodeterioration and Biodegradation* 59: 292–296.

Raj A, Chandra R, Reddy MMK, Hemant JP, and Kapley A. 2007b. Biodegradation of kraft lignin by a newly isolated bacterial strain, *Aneurinibacillus aneurinilyticus* from the sludge of a pulp paper mill. *World Journal of Microbiology and Biotechnology* 23: 793–799.

Ralph J. 2006. What makes a good monolignol substitute? In: Hayashi T (Ed.), *The Science and Lore of the Plant Cell Wall Biosynthesis, Structure and Function*. BrownWalker Press, USA, pp. 285–293.

Ramachandra M, Crawford DL, and Hertel G. 1988. Characterization of an extracellular lignin peroxidase of the lignocellulolytic actinomycete *Streptomyces viridosporus*. *Applied and Environmental Microbiology* 54(12): 3057–3063.

Reid ID and Paice MG. 1998. Effects of manganese peroxidase on residual lignin of softwood kraft pulp. *Applied and Environmental Microbiology* 64(6): 2273–2274.

Reid JSG. 1985. Cell wall storage carbohydrates in seeds-biochemistry. *Advances in Botany Research* 11: 125–155.

Rodrigues J, Meier D, Faix O, and Pereira H. 1999. Determination of tree to tree variation in syringyl/ guaiacyl ratio of *Eucalyptus globulus* wood lignin by analytical pyrolysis. *Journal of Analytical and Applied Pyrolysis* 48(2): 121–128.

Roos AA, Persson T, Krawczyk H, Zacchi G, and Ståbrand H. 2009. Extraction of water soluble hemicelluloses from barley husks. *Bioresource Technology* 100: 763–769.

Ros Barceló A, Gómez Ros LV, Gabaldón C, López-Serrano M, Pomar F, Carrión JS, and Pedreño MA. 2004. Basic peroxidases: The gateway for lignin evolution? *Phytochemistry Reviews* 3(1–2): 61–78.

Rubilar O, Diez MC, and Gianfreda L. 2008. Transformation of chlorinated phenolic compounds by white rot fungi. *Critical Reviews in Environmental Science and Technology* 38(4): 227–268.

Saha BC. 2004. Lignocellulose biodegradation and applications in biotechnology. *ACS Symposium Series* 889:2–34.

Samuels A, Rensing K, Douglas C, Mansfield S, Dharmawardhana D, and Ellis B. 2002. Cellular machinery of wood production: Differentiation of secondary xylem in *Pinus contorta* var. *latifolia*. *Planta* 216(1): 72–82.

Sánchez C. 2009. Lignocellulosic residues: Biodegradation and bioconversion by fungi. *Biotechnology Advances* 27(2): 185–194.

Saparrat MCN, Mocchiutti P, Liggieri CS, Aulicino MB, Caffini NO, Balatti PA, and Martínez MJ. 2008. Ligninolytic enzyme ability and potential biotechnology applications of the white-rot fungus *Grammothele subargentea* LPSC no. 436 strain. *Process Biochemistry* 43(4): 368–375.

Scheller HV and Ulvskov P. 2010. Hemicelluloses. *Plant Biology* 61(1): 263.

Schmidt A, Tenkanen M, Thomson AB, and Woidemann A. 1998. Hydrolysis of solubilized hemicellulose derived from wet oxidized wheat straw by a mixture of commercial fungal enzyme preparations. Riso National Laboratory, Roskilde, Denmark.

Shary S, Ralph SA, and Hammel KE. 2007. New insights into the ligninolytic capability of a wood decay ascomycete. *Applied and Environmental Microbiology* 73(20): 6691–6694.

Sticklen MB. 2008. Plant genetic engineering for biofuel production: Towards affordable cellulosic ethanol. *Nature Reviews Genetics* 9(6): 433–443.

Stutzenberger FJ, Kaufman, AJ, and Lossin RD. 1970. Cellulolytic activity in municipal solid waste composting. *Canadian Journal of Microbiology* 16(7): 553–560.

Takano M, Nishida A, and Nakamura M. 2001. Screening of wood-rotting fungi for kraft pulp bleaching by the poly R decolorization test and biobleaching of hardwood kraft pulp by *Phanerochaete crassa* WD1694. *Journal of Wood Science* 47: 63–68.

Tenkanen M, Schuseil PJ, and Poutanen K. 1991. Production, purification and characterization of an esterase liberating phenolic acid from lignocellulosics. *Journal of Biotechnology* 18: 69–84.

Teunissen PJM and Field JA. 1998. 2-Chloro-1,4-dimethoxybenzene as a mediator of lignin peroxidase catalyzed oxidations. *FEBS Letters* 439(3): 219–223.

Tunc MS and van Heiningen ARP. 2011. Characterization and molecular weight distribution of carbohydrates isolated from the autohydrolysis extract of mixed southern hardwoods. *Carbohydrate Polymers* 83, 8–13.

Tuomela M, Vikman M, Hatakka A, and Itävaara M. 2000. Biodegradation of lignin in a compost environment: A review. *Bioresource Technology* 72(2): 169–183.

Utt EA, Keshav KF, and Ingram LD. 1991, Sequencing and expression of the *Butyrivibrio fibrisolvens* xyl B gene encoding a novel bifunctional protein with β-D-xylosidase and α-L-arabinofuranosidase activities. *Applied Environmental Microbiology* 57, 1227–1234.

Vanholme R, Demedts B, Morreel K, Ralph J, and Boerjan W. 2010. Lignin biosynthesis and structure." *Plant Physiology* 153(3): 895–905.

Vázquez MJ, Garrote G, Alonso J, Domínguez H, and Parajó J. 2005. Refining of autohydroly-
    sis liquors for manufacturing xylooligosaccharides: Evaluation of operational strategies.
    Bioresource Technology 96: 889–896.

Vogels GD. 1979. The global cycle of methane. *Antonie van Leeuwenhoek* 45(3): 347–352.

Wariishi H, Sheng D, and Gold MH. 1994. Oxidation of ferrocytochrome c by lignin peroxidise.
    Biochemistry 33(18): 5545–5552.

Wolin MJ and Miller TL. 1987. Bioconversion of organic carbon to CH_4 and CO_2. *Geomicrobiology
    Journal* 5(3–4): 239–259.

Wood TM. 1985. Properties of cellulolytic enzyme systems. *Biochemical Society Transactions* 13(2):
    407–410.

Wyman CE, Decker SR, Himmel ME, Brady JW, Skopec CE, and Viikari L. 2005. Hydrolysis of cellu-
    lose and hemicelluloses. In: Dumitriu S (Ed.), *Polysaccharides: Structural Diversity and Functional
    Versatility*. CRC Press, USA, Vol. 1, pp. 1023–1062.

York WS, Darvill AG, and Albersheim P. 1984. Inhibition of 2,4-dichlorophenoxyacetic acid-stimu-
    lated elongation of pea stem segments by a xyloglucan oligosaccharide. *Plant Physiology* 75(2):
    295–297.

Yoshida N, Kaneoya M, Uchida M, and Morita H. 1990. Specific compounds having hydroxyl group
    reacted with esters in the presence of hydrolases. Patent no. 4,962,031.

Young LY and Frazer AC. 1987. The fate of lignin and lignin-derived compounds in anaerobic envi-
    ronments. *Geomicrobiology Journal* 5(3–4): 261–293.

Zimbardi F, Viggiano D, Nanna F, Demichele M, Cuna D, and Cardinale G. 1999. Steam explosion of
    straw in batch and continuous systems. *Applied Biochemistry and Biotechnology* 77(1–3): 117–125.

11

Advanced Oxidative Pretreatment of Complex Effluents for Biodegradability Enhancement and Color Reduction

Prachi Tembhekar, Kiran Padoley, Togarcheti Sarat Chandra, Sameena Malik, Abhinav Sharma, Suvidha Gupta, Ram Awatar Pandey, and Sandeep Mudliar

CONTENTS

11.1 Introduction

In recent years, complex industrial wastewater generated from distillery, heterocyclic, pharmaceutical, chemical manufacturing industries, and so on, has become a major area of concern. These complex effluents are characterized by high chemical oxygen demad (COD), color, presence of recalcitrant intermediates and poor biodegradability, which warrant the need for efficient and sustainable processes for their treatment. Conventional biological treatment processes have severe limitations due to the high recalcitrant nature of these effluents, which often results in ineffective treatment. Alternative physicochemical options are capital and energy intensive and also may generate secondary waste streams. The advanced oxidation processes (AOP) such as wet air oxidation (WAO), ozonation, UV, H_2O_2, and the like, are emerging techniques and have immense potential for effective treatment of complex effluents. AOPs are characterized by generation of hydroxyl radicals that are highly reactive and can easily degrade/alter/modify recalcitrant/toxic contaminants. They are, however, presently beset with problems of high capital and energy intensiveness and are unsustainable as standalone options. The synergistic combination of AOP pretreatment of complex effluent for biodegradability enhancement followed by biological treatment has potential for providing high treatment efficiency along with high overall sustainability. Therefore, the primary emphasis of any advanced oxidative pretreatment process should be on partial oxidation/minimal mineralization leading to the formation of biologically degradable intermediates. Minimum mineralization ensures the process feasibility and economics. This chapter discusses WAO and ozonation as pretreatment options for complex effluent to enhance biodegradability to facilitate subsequent biodegradation via anaerobic and aerobic biological processes. Discussions pertaining to the recent developments in this area along with future challenges and prospects exemplified through relevant case studies are presented.

11.2 Advanced Oxidative Pretreatment of Complex Industrial Effluent for Biodegradability Enhancement: Concept, Mechanism, and Methodology

11.2.1 The Concept

Based on the characteristics of the complex effluent (e.g., composition and concentration), the specific treatment objectives, and the oxidation potential of the oxidant in question; AOPs may be used either for the complete mineralization of all pollutants to carbon dioxide, water, and mineral salts or as pretreatment option for biodegradability enhancement via conversion of biorecalcitrant molecules to more biodegradable molecules (Karaca and Tasdemir, 2014; Merayo et al., 2014; Rizzo et al., 2014). The selection of appropriate AOP and its efficiency for pretreatment/complete mineralization of wastewater contaminants is a function of the concentration, oxidation potential, and wastewater characteristics. In general, an AOP process aiming at complete mineralization might become extremely cost-energy intensive and unsustainable is the long-run, since the highly oxidized end-products that are formed during chemical oxidation tend to be refractory to total oxidation by chemical means. The anaerobic–aerobic biological treatment is ineffective due to high retention times and lower biodegradation rates along with limited COD removal. It has been reported that single method may not offer a enable complete solution to treat the complex effluent like distillery spent wash or pharmaceutical wastewater and use of hybrid methods involving combination of AOPs and biological processes are likely to be more promising and efficient. The AOP-based pretreatment option provides an opportunity to enhance biodegradability and facilitate integration with biological process viz., anaerobic digestion for biogas generation or enhanced and efficient aerobic biodegradation and thereby exploit the synergistic combination of both the processes (Figure 11.1).

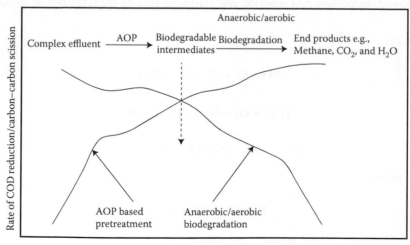

FIGURE 11.1

The concept of AOP-based pretreatment for biodegradability enhancement and its integration with biological treatment. (Adapted and revised from Dionissios and Elefteria, 2004.)

For instance, it has been reported that organic macromolecules such as soluble polymers may simply be too large to permeate microbial cell walls, which limits their effective oxidation via biological processes (Mantzavinos and Psillakis, 2004). AOPs might break these molecules into smaller intermediate molecules (e.g., short-chain organic acids). These molecules then can enter cells and are likely to be more readily biodegradable than the original molecules, as the rate of biological oxidation is generally thought to increase with decreasing molecular size. Conversely, complete AOP-based oxidation of these intermediates to carbon dioxide and water may be difficult and requires severe oxidative conditions, since the rate of chemical C–C bond scission seems to decrease with decreasing molecular size (Mantzavinos and Psillakis, 2004). Therefore, putting these two separate observations together in a qualitative sense, an AOP-based pretreatment of complex effluent for biodegradability enhancement and its integration with biological process is a promising concept by exploiting synergies of chemical–biological processes.

11.2.2 Mechanism

AOPs rely on *in situ* production of highly reactive hydroxyl radicals (OH·) which are formed by using one or more oxidants (e.g., ozone, hydrogen peroxide, oxygen).

OH· reacts nonselectively once formed and bio-recalcitrant organic components will be quickly and efficiently fragmented and converted into small inorganic molecules.

The main mechanism of AOPs can be essentially divided into three parts:

a. Formation of OH· radicals—Initiation

b. Initial attack on organics by OH· and their subsequent breakdown—Propagation

c. Subsequent attacks by OH· until ultimate mineralization—Termination

The proposed/hypothesized reaction mechanisms can be represented as follows

a. Initiation

$$RH + O_2 \xrightarrow{K_1} R^\cdot + HO_2^\cdot \tag{11.1}$$

$$H_2O + O_2 \xrightarrow{K_2} HO_2^\cdot + OH^\cdot \tag{11.2}$$

$$O_2 \xrightarrow{K_3} O^\cdot + O^\cdot \tag{11.3}$$

b. Propagation

$$R^\cdot + O_2 \xrightarrow{K_4} ROO^\cdot \tag{11.4}$$

$$ROO^\cdot + RH \xrightarrow{K_5} ROOH + R^\cdot \tag{11.5}$$

$$RH + HO_2^\cdot \xrightarrow{K_6} R^\cdot + H_2O_2 \tag{11.6}$$

$$RH + HO^\cdot \xrightarrow{K_7} R^\cdot + H_2O \tag{11.7}$$

c. Termination

$$2ROO^{\bullet} \xrightarrow{\text{K}_8} ROOR + O_2 \tag{11.8}$$

$$HO_2^{\bullet} + OH^{\bullet} \xrightarrow{\text{K}_9} H_2O + O_2 \tag{11.9}$$

where RH is the organic substrate resulting in COD, R' is the substrate radical/reactant, and ROOR is the reoriented organic substrate.

The radical reaction efficiency evaluates the effectiveness of the reactive species in the destruction of chemical contaminants. The hydroxyl radical has high oxidation potential of 2.8 V. The oxidation reaction pathway involves the attack of hydroxyl radical onto organic compounds. This electrophillic addition of the hydroxyl radical to organic compounds containing a π bond leads to the formation of radical and electron transfer with reduction of hydroxyl radical to a hydroxyl anion by an organic substrate. Also, the peroxide can react with hydroxyl radicals and other intermediates formed (Tembhekar et al., 2014).

11.2.3 Methodology

While dealing with advanced oxidative pretreatment, especially in the context of complex effluents the most important objective is to limit the oxidation rate so that maximum COD is retained while still enhancing biodegradability index (BI; BOD:COD) of the effluent. Depending on the pretreatment option adopted, it is highly essential to assess the efficiency of pretreatment by selecting appropriate parameters that predict the system performance in a holistic manner. The most common parameters that are adopted are mostly lumped parameters such as COD, BOD, and TOC. In certain cases, additional parameters such as color, toxicity, etc., are also of immense importance in mapping the pretreatment efficiency. Parameters like AOSC are also used sometimes to correlate COD and TOC. However, this parameter often reflects the oxidation at a particular time only and therefore, may not be very appropriate in assessing the progress of the partial oxidation reactions occurring during pretreatment.

Assessment of enhancement in biodegradability is usually done by measuring BOD and change in biodegradability is expressed by measuring BI, and comparing this BI with that of the raw effluent. However, it is also advisable to assess the biodegradability parameters such as BOD critically as substantial COD reduction during the oxidative pretreatment may be the cause of higher BI than the actual increase in the BI values. Similarly, when an anaerobic option is preferred after oxidative pretreatment the efficiency can be mapped in terms of biochemical methane potential/biogas production potential.

Assessment of toxicity is also critical in certain cases, since pretreatment may lead to the formation of more toxic intermediates. This data then becomes useful in assessing the most appropriate pretreatment conditions since optimum pretreatment would not result in the formation of such intermediates.

Review of the recent literature on oxidative pretreatment clearly shows its beneficial effect on subsequent biodegradability vis-à-vis toxicity reduction. However in certain cases, it may lead to insignificant or even a negative effect on the quality of the original effluent which may by virtue of the formation of stable intermediates that cannot be oxidized further.

Such limitations warrant the need of establishing a detailed step-by-step methodology as depicted in Figure 11.2 for integrating AOP pretreatment with subsequent downstream biological process rationally. It is very essential to understand the mechanism by which

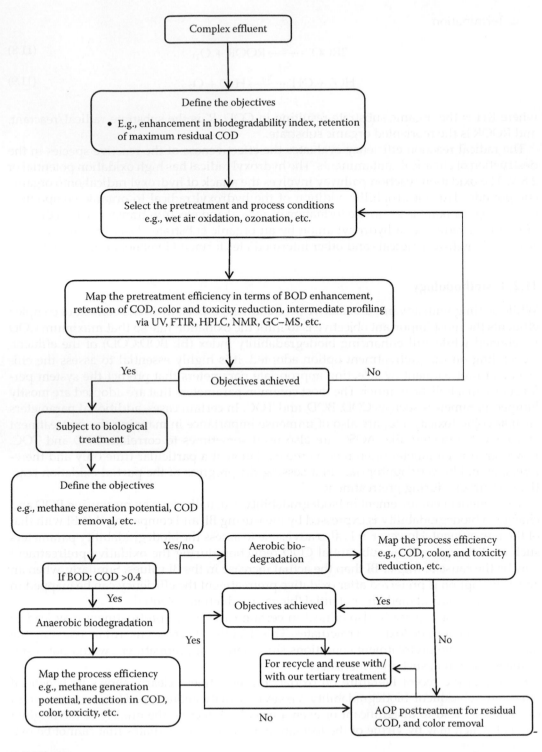

FIGURE 11.2
Methodology.

oxidative pretreatment enhances the biodegradability of a given effluent since the pretreatment step is the key to the overall process efficiency. Accurate studies with reference to process conditions, the properties of the pretreated stream (as evaluated from the posttreated spectrums using UV, FTIR, HPLC, NMR, GC–MS, etc.) is needed to understand the posttreatment scenario. These investigations would also enable in hypothesizing the reaction mechanism, pathway, and kinetics based on lumped parameters.

Depending on the extent of pretreatment, complexity of the effluent in question, and also the availability of the best analytical resources, characterization can be as simple as COD, TOC measurement to the highly complex such as monitoring of individual intermediate concentration profiles. In the real-world scenario of waste management such information gathered can be of immense benefit in understanding the mechanism and predicting close fit models, thereby bridging the gap between model and real effluents, validating the applicability of the model for similar effluents.

11.2.4 Recent Studies

In recent years, focus has been on various AOP-based treatment options. The literature on this aspect for the last two decades has already been reviewed elsewhere, most of which is targeted toward complete mineralization (Oller et al., 2011). Also, in the case of AOP for biodegradability enhancement most of the studies focus on enhancement of BOD: COD ratio and only limited studies evaluate their actual effect on downstream biological process (Scott and Ollis, 1995; Mantzavinos and Psillakis, 2004; Oller et al., 2011). There appears to be more attention towards using newer AOPs such as WAO, which is more feasible to be operated at mild conditions for multi-component streams and appears to be more feasible for practical applications. Recent reports on the evaluation of AOP-based treatment of complex effluent for biodegradability index enhancement and color removal and its effect on downstream aerobic and anaerobic processes and toxicity issues are reviewed and presented in Table 11.1. Also, one of the case studies of WAO and ozonation as AOP for pretreatment of complex distillery effluent (post biomethanation/biomethanated) has been discussed in detail in the subsequent section.

11.3 Case Study of Wet Oxidation Pretreatment of Complex Distillery Effluent (Biomethanated, BDWW) for Biodegradability Enhancement and Color Reduction

The cane molasses-based distilleries generate complex, troublesome and highly recalcitrant industrial organic effluents with high COD (90000–110000 mg/L), dark brown color and generated in huge volumes (8–15 L/L of alcohol produced) (Kumar, 2003; Sangve and Pandit, 2004). After conventional anaerobic digestion, the distillery wastewater (BDWW) still retains up to 40,000 mg/L COD and substantial color and becomes recalcitrant (biodegradability index <0.2) to further treatment by biological methods (Table 11.2). This problem provides an opportunity for the development of effective pretreatment methods to enhance biodegradability index and facilitate subsequent biogas generation potential from such complex effluent or enhanced aerobic biodegradation.

Wet Air Oxidation (WAO) was investigated as a pretreatment option for complex distillery effluent (BDWW) with emphasis on selectively enhancing the biodegradability and

TABLE 11. 1

Recent Reports on AOP-Based Pretreatment of Complex Effluents for Biodegradability Enhancement

References	Wastewater and Characteristics	AOP Applied as Pretreatment Option	Analysis Performed on Effluent	AOP Efficiency: BI Enhancement and COD Reduction	AOP Efficiency: Color Reduction	AOP Efficiency: Toxicity Reduction	Biological Treatment	Biological Treatment Efficiency	AOP Posttreatment for Residual COD and Color Removal	AOP Posttreatment Efficiency	Overall Integrated Treatment Process Efficiency
Suvidha et al. (2015)	Biomethanated distillery wastewater (COD: 40,000 mg/L, BOD: 6744 mg/L, pH: 7.6, color: dark brown)	Ozonation (2.8 g/h ozone dose, ozone concentration 28/Nm³, 20–240 min)	COD, BOD, and color	BI enhancement from 0.17 to 0.58, up to 33% COD reduction	up to 25.99%	up to 40%	Anaerobic and aerobic biodegradation	Anaerobic degradation: biogas with up to 62.8% methane content, up to 57.14% COD reduction. Aerobic degradation: maximum of 73.63% COD reduction	X	X	Overall 91.42% COD reduction after integrated ozonation, anaerobic–aerobic biodegradation
Sarat Chandra et al. (2014)	Biomethanated distillery wastewater (COD: 40,000mg/L, BOD: 6744 mg/L, pH: 7.61, color: dark brown)	Wet air oxidation temperature: 150–200°C, pressure: 6–12 bar, time: 0.5–2 h	COD, BOD, HPLC for mapping colouring compounds	BI enhancement from 0.2 to 0.88, 20%–70% COD reduction	up to 98%	80%–100% toxicity reduction	Anaerobic biodegradation	Up to 54.75% COD reduction. Biogas generation up to 64% methane content	X	X	Overall 62% COD reduction after integrated WAO (150°C, 6 bar and 30 min) and anaerobic biodegradation
Malik et al. (2014)	Biomethanated distillery wastewater (COD: 40,000 mg/L, BOD: 6744 mg/L, pH: 7.6, color: dark brown)	Wet air oxidation temperature: 150–200°C, pressure: 6–12 bars, Time: 0.5–2 h	COD, BOD	BI enhancement from 0.2 to 0.88, 20%–70% COD reduction	X	X	Aerobic and anaerobic bio degradation	Aerobic degradation: up to 67.7% COD reduction Anaerobic followed by aerobic degradation: up to 89.7% COD reduction	X	X	Overall 95% COD reduction after integrated WAO along with anaerobic–aerobic biodegradation
Espejo et al. (2014)	Urban wastewater spiked with 9 pharmaceutical compounds at 200 µg/L each	X	COD, BOD, TOC, HPLC	X	X	✓	Aerobic degradation	Only acetaminophen and caffeine completeley eliminated. 50% enhancement in biodegradability	Ozonation as a post treatment option	Biodegradability enhanced by 150%. All contaminants were eliminated. Non-toxic to daphnia magna	Integrated aerobic oxidation and ozonation increased biodegradability by 200%

(Continued)

TABLE 11.1 (*Continued*)

Recent Reports on AOP-Based Pretreatment of Complex Effluents for Biodegradability Enhancement

References	Wastewater and Characteristics	AOP Applied as Pretreatment Option	Analysis Performed on Effluent	AOP Efficiency			Biological Treatment	Biological Treatment Efficiency	AOP Posttreatment for Residual COD and Color Removal	AOP Posttreatment Efficiency	Overall Integrated Treatment Process Efficiency
				BI Enhancement and COD Reduction	Color Reduction	Toxicity Reduction					
Padoley et al. (2012)	Biomethanated distillery wastewater (COD: 40,000 mg/L, BOD: 6744 mg/L, pH: 7.61, color: dark brown)	Wet air oxidation temperature: 150–200°C, pressure: 6–12 bar, time: 0.5–2 h	COD, BOD, FTIR, NMR	BI enhancement from 0.2 to 0.88, 20%–70% COD reduction	X	X	Anaerobic biodegradation	34% COD reduction, biogas with up to 49.56% methane content	X	X	Overall 50% COD reduction after integrated WAO (175°C, 6 bar and 30 min) and anaerobic biodegradation X
Padoley et al. (2012)	Biomethanated distillery wastewater (COD: 35,000, BOD: 5000, pH: 7.61, color: brown)	Hydrodynamic cavitation at 5 bar and 13 bar	COD, BOD, TOC, color	BI enhancement up to 0.32. Up to 33% reduction in COD/TOC	Up to 48%	X	Anaerobic biodegradation (diluted effluent)	Higher biogas generation compared to untreated effluent with 70% COD reduction	X	X	X
Padoley et al. (2011)	Pyridine and 3-cyano pyridine (3-CP) plant wastewater (pyridine plant—COD: 65,000 mg/L, BI: 0.037, pyridine: 1700 mg/L, 3-CP plant—COD: 25,000 mg/L, BI: 0.125. 3-CP: 725 mg/L)	Fenton oxidation	COD, BOD	Pyridine plant ww: BI enhancement up to 0.79, 66% COD reduction, 62.4% pyridine removal 3-CP plant ww: BI enhancement up to 0.94, 84% COD reduction, 80% 3-CP removal	X	X	Aerobic degradation	—	X	X	Overall up to 82% COD and 84% pyridine reduction for pyridine plant ww Up to 94 % COD and 99% in 3-CP removal for 3-CP ww

Note: For reports on combination of AOP processes and biological treatments prior to 2011 refer Oller et al. (2011) and Mantzavinos and Psillakis (2004).

TABLE 11.2

Characteristics of Complex Wastewater (Biomethenated Distillery Wastewater)

Parameters	Value
pH	7.61
Color	Brown
COD (mg/L)	40,000
BOD (mg/L)	6744
TOC (mg/L)	18,700
Total solids (mg/L)	31,000
Total suspended solids (mg/L)	1600
Biomass (%)	1
VFA (mg/L)	180
BOD: COD ratio	0.17

Note: Values are average of 3 sets of observations with SD < 5%

retaining sufficient residual COD for biogas production potential or enhanced aerobic biodegradation. WAO becomes self-sustaining at COD >20,000 mg/L (Mishra et al. 1995). Biomethanated distillery wastewater (BDWW) was chosen as model complex effluent with characteristics detailed in Table 11.2. The biodegradability enhancement was evaluated in terms of biodegradability index which is defined as the ratio of BOD5: COD.

11.3.1 Wet Air Oxidation Principle and Mechanism

Wet air oxidation refers to the aqueous phase oxidation of organics and oxidizable inorganic compounds at elevated temperature and pressure using pure oxygen gas or air. Elevated temperatures are used to increase the oxidation rate by enhancing the solubility of oxygen in the aqueous solution, whereas elevated pressures are required to keep the water in liquid form. WAO can potentially break the long polymeric chain compounds and convert them to highly biodegradable low-molecular weight compounds. WAO process converts organic contaminant(s) present in industrial effluent(s) to carbon dioxide, water, and biodegradable short chain compounds such as acetic acid, and formaldehyde. This process occurs in the aqueous phase comprising suspended or soluble compounds in water that come in contact with either air or oxygen. In WAO process, water acts as a medium for dissolved oxygen to take part in the reaction as well as helps in hydrolysis.

Wet air oxidation can be achieved at moderate temperatures (125–320°C) and pressures in the range from 5 to 200 bar to prevent water from evaporating. These conditions are below the critical temperature of 374°C and critical pressure of 221 bar for water. The degree of oxidation achieved is a function of temperature, oxygen partial pressure, reaction time, and amenability to oxidation of the various organic components present in the effluent stream. It is often difficult to measure the concentration of individual components in a complex effluent mixture. Hence, the concept of lumped parameters to evaluate the WAO process (partial oxidation in the present case) is of great significance. The oxidation is said to be a free radical mechanism and can be also catalyzed using homogeneous or heterogeneous catalysts. The overall WAO mechanism incorporates two steps. One is the physical step, that is, the transfer of oxygen from the gas to the liquid phase and the transfer of carbon dioxide to the gas phase from the liquid phase. The second one being a chemical reaction between organic matter and oxygen dissolved in the liquid phase, producing carbon dioxide. WAO can have high destruction efficiency for breaking down higher molecular weight organic compounds, but the efficiency for destruction of total organic carbon (TOC) is not as high, because the organic carbon in the form of shorter chain compounds still exists. The chemical elements such as sulfur, chlorine, fluorine, phosphorus, and nitrogen in organic compounds also react in the WAO process and form respective ions such as sulfate, chloride, fluoride, phosphate, and nitrate or ammonium.

WAO being an exothermic process, the temperature of the entire reaction increases as the reaction proceeds. Hence, cooling needs to be provided during the reaction process.

Some of the limitations associate with WAO process are:

- Corrosion is a major problem occurring in case of incomplete neutralization of the chemical ions (highly acidic).
- Similarly, salts formed during neutralization may also create plugging problems.
- If the effluent added to the reactor for treatment contains a high concentration of oxidizable compounds then it needs dilution in order to control the reaction temperature.

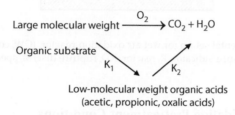

K_2: Oxidation of volatile fatty acids (VFAs) to carbon dioxide is the rate-limiting step in WAO (Mishra et al., 1995; Sarat Chandra et al., 2014).

11.3.2 Wet Air Oxidation Reactor and Operation for Distillery Effluent

The experimental studies were carried out in a WAO reactor, having 1.8 L capacity (Model–4578, Floor Stand HP/HT Reactor, Parr Instruments, 1 L, USA) made of SS-316. The reactor (ID 95 mm) was equipped with a four-bladed turbine type impeller (ID 50 mm). The agitation speed was varied between 150 and 500 rpm using a variable speed motor. The gas inlet gas release valve, cooling water feed line, and pressure valve was mounted on top of the reactor. The reactor was also equipped with rupture disc and a non-return valve at the gas inlet. 500 mL of the wastewater was used for pretreatment (equivalent to working volume of 0.5 L). The reactor was properly sealed, ensuring an absence of any leakage. The reaction mixture was then heated to the desired temperature and air was sparged using a gas sparger situated beneath the impeller, to a predetermined level (6–12 bar). The total pressure is the sum of air (applied) pressure and liquid vapor pressure (Figure 11.3).

The effluent stored at 4°C was allowed to attain room temperature. The wastewater was then used as such for pretreatment using WAO. The biomethanated distillery wastewater (BDWW) was subjected to WAO as a pretreatment in batch experiments. As no pH adjustment was made, the original wastewater pH was the initial pH for the pretreatment step. WAO was carried out at different conditions in the temperature range of 150–200°C, pressure 6–12 bar and time 15–120 min and agitation speed 150–500 rpm. After every run, the reactor was put in a cooling mode. The pretreated samples were analyzed for pH and COD (Shende and Mahajani, 1997; Gunale and Mahajani, 2007; Collado et al., 2010; Padoley et al., 2012; Tembhekar et al., 2014).

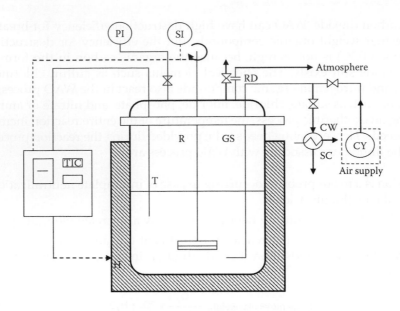

FIGURE 11.3
Schematic diagram of experimental set up for wet air oxidation reactor. (CW, cooling water; CY, cylinder; GS, gas sparger; H, Heater; PI, pressure indicator; R, reactor; RD, rupture disc; SI, speed indicator; TIC, temperature indicator and control.)

11.3.3 Effect of Wet Oxidation Pretreatment Conditions on Complex Distillery Effluent

11.3.3.1 Diffusion Mass Transfer Considerations

The overall WAO mechanism incorporates two steps. One is the physical step that is, the transfer of oxygen from the gas to the liquid phase and the transfer of carbon dioxide to gas phase from liquid phase. The second one being a chemical reaction between the organic matter and oxygen dissolved in the liquid phase, producing carbon dioxide as discussed earlier. While operating a WAO reactor, it is usually considered that gasses diffuse rapidly in to the gas phase. The only significant transfer resistance is located at the gas-liquid interface and the actual conditions within the reactor will depend on its hydrodynamics. For high mixing efficiencies, the oxygen concentration in the bulk liquid is close to the interface (or equilibrium) concentration and the overall rate is close to the chemical rate.

Owing to the high diffusivity of O_2 in the gas phase and its low solubility in water, the gas phase mass transfer resistance was estimated to be negligible in the range of operating temperatures used for present studies. To get a representative kinetic data, the reaction must take place in the slow-reaction regime with substantial kinetic control (in which case the overall rate is determined by the reaction rate in the bulk liquid) or in the fast-reaction regime (in which the reaction is complete in a very miniscule region close to the interface). In order to ascertain the true kinetics of the reaction it is necessary to eliminate diffusion resistance for which the effect of agitation was also accounted in this investigation, studies were conducted in the range of 150–500 rpm. The results are presented in Figure 11.4. In the present investigation, an impeller speed above 200 rpm does not yield any significant enhancement in COD removal (%), and therefore, for the present investigation, the liquid side mass transfer was found to be eliminated at ≥200 rpm (Tembhekar et al., 2014).

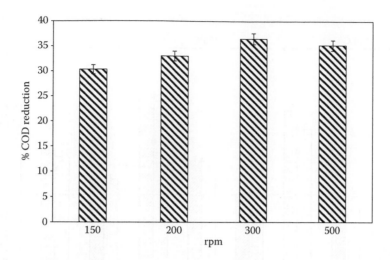

FIGURE 11.4
Effect of speed of agitation on COD removal during WAO of BDWW (initial COD: 40,000 mg/L, 175°C, 6 bar, 30 min).

11.3.3.2 Effect of Pressure

The effect of pressure on COD reduction vis-à-vis BI enhancement was studied in a range of 6–12 bar at three different temperatures (150–200°C) and various time intervals (15–120 min) for an initial COD of 40,000 mg/L (Figure 11.5). The COD concentration reduced in the range of 16%–36% at 6 bar pressure (150–200°C, 15 min), yielding BI values of 0.2–0.32.

Similarly, 50%–70% COD reduction and BI enhancement in the range of 0.55–0.88 was observed at 120 min. When the pressure was increased to 12 bar, the COD reduction from 17% to 36% (150–200°C, 15 min), with BI enhancing in the range of 0.3–0.4 was observed.

Similarly, a reduction in the COD was observed in the range of 55%–70% (150–200°C, 120 min), with enhanced BI in the range of 0.63–0.86. In the range of pressure considered in this study, the system behavior indicates that the pretreatment can be operated at lower range (6 bar) to get the minimum desirable BI (0.4 indicated by dotted horizontal line in the Figure 11.5).

This trend indicates that as pressure increases (in the range studied) there is a slight increase in the COD reduction (5%–7%) for all the experimental runs. From the results, it can be concluded that the experimental condition yielding around 40% COD reduction and a BI around 0.4 (175°C, 30 min and 6 bar) would be suitable for present setup where the primary objective was pretreatment (and not COD destruction) along with reorientation of complex waste molecules to become more amenable to enhance biodegradation (BI) (Padoley et al., 2012; Tembhekar et al., 2014).

The WAO reaction in the present study is most likely to proceed via the free radical mechanism and number of studies have been reported earlier by various authors (Li et al., 1991; Tufano, 1993; Collado et al., 2010).

11.3.3.3 Effect of Temperature

The effect of temperature (150–200 °C) was evaluated on the pretreatment of BDWW. The results were interpreted in terms of COD reduction and BI enhancement and are presented in Figure 11.5. As it is observed in WAO systems, a higher COD reduction was achieved at

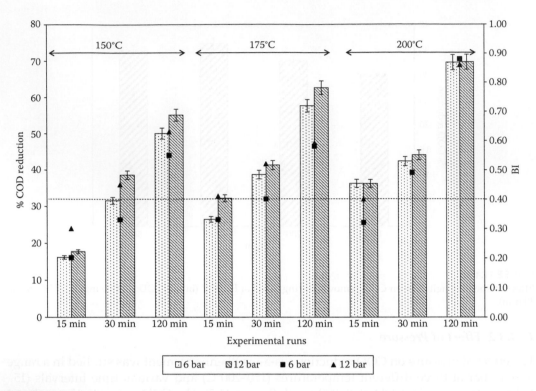

FIGURE 11.5
Effect of pressure and temperature of wet oxidation process on % COD reduction and BI profile in pretreated biomethanated wastewater.

higher temperatures. The COD reduction was observed in the range of 16%–69% for various experimental conditions and at minimum and maximum reaction time (15–120 min). The enhancement in BI was observed in the range of 0.2–0.88 (0.17 initial). In the present study, the effect of temperature is significant only up to 175°C indicating that higher temperature does not yield a linear effect on WAO efficiency after 175°C (Figure 11.5). For all practical purposes, a BI of 0.4 is good enough to make a waste amenable to biodegradation (Metcalf and Eddy, 1979), which in this study was observed at an experimental condition of 175°C, 6 bar, and 30 min of reaction time (Padoley et al. 2012; Tembhekar et al., 2014). This experimental condition of 175°C, 6 bar, and 30 min was observed to be optimum yielding 40% (Avg.) COD reduction and an enhanced BI of 0.4. This condition yields an effluent with sufficient C (COD) and enhanced BI, which can be further subjected to conventional biological treatment systems namely anaerobic digester to get additional energy (biogas) (Padoley et al., 2012; Sarat Chandra et al., 2014).

11.3.3.4 Effect of Initial COD

WAO studies at varying initial COD was also investigated at the observed optimum WAO condition (175 °C, 6 bar, and 30 min), in order to evaluate its effect on overall process efficiency. The results obtained indicate that at higher initial COD, the process efficiency (%) in terms of COD removal and BOD enhancement improves. Thus, maximum COD removal (33%) is observed at 40,000 mg/L initial COD (10,828 mg/L BOD). Similarly, minimum removal (17%) was observed at 15,000 mg/L initial COD (4672 mg/L BOD).

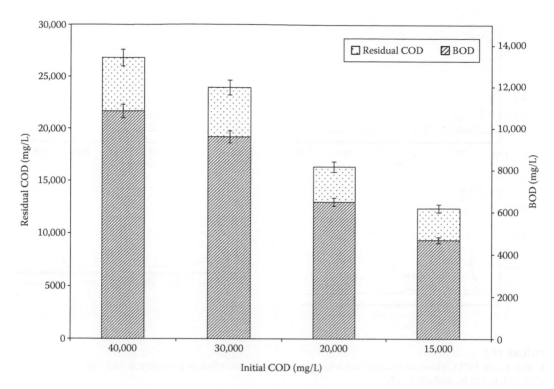

FIGURE 11.6
Effect of initial COD on wet air oxidation pretreatment of BDWW.

Several other literature reports (Pintar and Levec, 1992; Collado et al., 2010) available regarding single/pure compound also indicate inhibition effect of the initial higher substrate concentration on the rate of WAO (Figure 11.6) (Padoley et al., 2012; Tembhekar et al., 2014).

11.3.4 Effect of Wet Air Oxidation Pretreatment on Color Reduction of Distillery Effluent

In the case of BDWW effluent, the dark brownish color is mainly due to melanoidins produced by thermal degradation and condensation reactions of sugar (Plavsic et al., 2006). Melanoidins are color-containing compounds generated by the Maillard reaction and are major biopolymers that are nonbiodegradable by microorganism (Miyata et al., 2000). The quantification of melanoidins was done according to the protocol reported by Bharagava and Chandra (2010) through HPLC analysis. HPLC analysis was carried out for both WAO pretreated and untreated effluents. WAO-treated samples showed a reduction in peak areas compared with the control sample (Figure 11.7) indicating that decrease in color intensity might be mainly attributed to the WAO resulting in the degradation of color-containing compounds. In addition, the WAO-treated samples also showed shifting of peaks compared with control samples indicating the degradation as well as transformation of color-containing compounds by WAO.

The percentage decolorization of WAO pretreated effluent is represented in Figure 11.8. The WAO pretreated sample of BI 0.8 showed maximum decolorization of 97.8%, while pretreated effluent with BI 0.4, 0.5 showed 28.20% and 62.19% color reduction. The percentage color reduction of BDWW obtained in this study was comparable and was

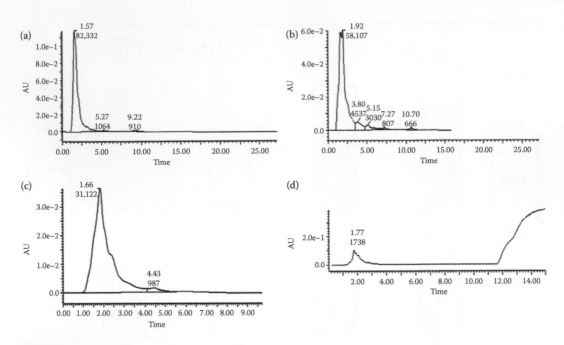

FIGURE 11.7
Comparative HPLC chromatogram of color containing compounds before ((a) control) and after pretreatment (b) 0.4 BI, (c) 0.5 BI, and (d) 0.8 BI.

found to be higher than that of other biological or physicochemical treatments (Sarat Chandra et al., 2014).

11.3.5 FTIR and NMR Analysis

FTIR analysis of WAO untreated and treated sample was performed using Fourier Transform Infrared Spectrometer (FTIR, Model: Vertex 70 - BRUKER). Similarly, the sample preparation for protonic nuclear magnetic resonances (H^1 NMR) spectrum was done

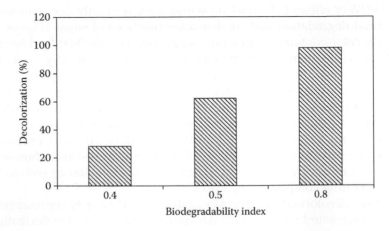

FIGURE 11.8
Decolorization of color containing compounds in WAO pretreated BDWW at different biodegradability index.

TABLE 11.3

FTIR Analysis of WAO-Treated B-DWW

Sample	Stretching Frequencies (λ_{max})	Assignments
Original B-DWW	3586 2853 1610 1420 1358 1044 865 772 731 692 651 615 544 431	—OH, —CH$_3$—O—, NH, CN, C—F, C—Cl, C—Br, C—I, S—S
Pretreated 0.2 BI (150°C, 6 bar, 15 min)	3614 3246 2850 2761 1612 1355 1043 690 654 613 508 432	OH, NH$_4^+$, RCOO$^-$, C—F, C—Br, S—S
Pretreated 0.4 BI (175°C, 6 bar, 30 min)	3579 3314 3245 2850 2762 1612 1424 1359 1301 1213 1045 693 654 544 508 427	OH, C$_n$H$_{2n-2}$, Polymeric OH, —CH$_3$—O—, NH$_4^+$, NO$_3^-$, RCOO$^-$, SO$_4^{2-}$, CN, C—Br, C—I, S—S
Pretreated 0.8 BI (200°C, 6 bar, 120 min)	3617 2851 2760 1648 1611 1417 1363 1044 871 732 692 652	OH, CH$_2$, R—NH$_2$, C=C, SO$_4^{2-}$, R—SO$_2$O—, CN, C—O—O—, C—Cl, C—Br

by extracting with D$_2$O under a nitrogen atmosphere and further transferred to a 5-mm NMR tube. The samples were analyzed at 25°C with a BRUKER AV III 500 MHz FT NMR Spectrometer and chemical shifts were recorded in ppm (δ) with TMS at 0.0 as an internal standard (Padoley et al., 2012).

FTIR analysis indicated the dissociation and reorientation of complex organic compounds in the untreated BDWW to new and simpler organic compounds. In comparison to the untreated sample, in the WAO-treated sample, main absorptions were found approximately near 2935 and 2860/cm and also at 1470 and 720/cm (Table 11.3), which is indicative of the presence of longer linear aliphatic chain compounds in the WAO-treated samples (Coates, 2000). Additionally in WAO treated sample, intense bands were found in the ranges 1600–1300, 1200–1000, and 800–600/cm, which showed the presence of simple hydrogen-bonded OH absorption of a hydroxyl (alcohol) functional group. Some sharp peaks occurred between 3670 and 3550/cm, which predicted non-hydrogen-bonded hydroxyl group. The absorption peaks between 1700 and 1650/cm in WAO-treated sample indicated the formation of carboxylate (carboxylic acid salt). From the IR spectra it can also be observed that strong OH bonds are produced in treated sample over control. An important structural feature of melanoidin (major coloring compound in the original sample) is the peak of C=N and C=C conjugated bonds. The WAO pretreatment aids in cleavage of the C=N (azomethane), and C=C (ethylinic) linkage producing more hydroxyl groups in (3400/cm) pretreated sample over the control. This finding substantiates that pretreatment converts complex molecules into simpler polymeric forms. The preliminary investigation on ^{1}H NMR spectrum of untreated and treated sample has shown signals at δ7.4 ppm δ8.0 ppm in only the untreated sample. This is indicative of the tentative presence of aromatic groups in the untreated sample, whereas these signals were not observed in WAO-treated samples, probably because of the breakdown of aromatic compounds by wet oxidation into simpler aliphatic compounds (Jacobsen, 2007).

11.3.6 Kinetics of WAO Pretreatment of Distillery Effluent

The kinetics of WAO pretreatment of BDWW as a model complex wastewater was studied in the temperature range 150–200°C and pressure range 6–12 bar at near neutral pH (7.6). The destruction of organics/COD via the wet oxidation technique is hypothesized to be a combination of various free radical reactions (Shende and Mahajani, 1997). In many cases, the oxidation goes through a very complicated pathway and leads to the formation of many different intermediates, such as (but not limited to) lower molecular weight carboxylic acids. In the case of industrial wastewater, many compounds are present in the waste stream. The oxidation of such a mixture is much more complex than the oxidation of single compound solutions. The reactivity of each component is different, and so are the intermediates formed from these compounds during the oxidation process. To model such a complicated process, the rate equation may be based on the lumped reactions and the lumped concentrations of all organic pollutants in industrial wastewater streams, that can be, COD in wastewater streams (Zhang and Chuang, 1999; Gunale and Mahajani, 2007; Garg and Mishra, 2010). The representation of such reactions is indicated in Equations 11.1 through 11.9 (Collado et al., 2010).

The reaction rates for various radicals can be represented (see Section 11.2.2)

$$\frac{dC_{OH^\cdot}}{dt} = K_2 C_{O_2} - K_7 C_{RH} C_{HO^\cdot} - K_9 C_{HO_2^\cdot} C_{OH^\cdot} \tag{11.10}$$

$$\frac{dC_{ROO^\cdot}}{dt} = K_4 C_{R^\cdot} C_{O_2} - K_9 C_{ROO^\cdot} C_{RH} - K_8 [C_{ROO^\cdot}]^2 \tag{11.11}$$

$$\frac{dC_{R^\cdot}}{dt} = K_1 C_{RH} C_{O_2} - K_4 C_{R^\cdot} C_{O_2} + K_5 C_{ROO^\cdot} C_{RH} + K_6 C_{RH} C_{HO_2^\cdot} + K_7 C_{RH} C_{HO^\cdot} \tag{11.12}$$

$$\frac{dC_{O^\cdot}}{dt} = K_3 C_{O_2} \tag{11.13}$$

$$\frac{dC_{HO_2^\cdot}}{dt} = K_2 C_{O_2} - K_6 C_{RH} C_{HO_2^\cdot} - K_9 C_{HO_2^\cdot} C_{OH^\cdot} + K_1 C_{RH} C_{O_2} \tag{11.14}$$

At steady state as there is no excess free radicals or accumulation and the organic substrate degradation can be represented as

$$\frac{dC_{RH}}{dt} = -K_1 C_{RH} C_{O_2} - K_5 C_{RH} C_{ROO^\cdot} - K_6 C_{RH} C_{HO_2^\cdot} - K_7 C_{RH} C_{HO^\cdot} \tag{11.15}$$

In the case of Equation 11.1 organic substrate resulting in COD (RH) represents one single component or a complex molecule. For complex wastewater such as DWW, there will be a number of such complex components (RH_1, RH_2, etc.).

As WAO of BDWW is complex and involves the formation of numerous reaction intermediates via many pathways, the modeling of each reaction and estimating K_1 to K_9 is impractical (Wu et al., 2003). Hence, for the development of rate equation for COD a lumped parameter model was used. The order with respect to the partial pressure of oxygen could vary between 0 and 1 depending upon the rate-determining reactions in the case of the free radical mechanism. In the present case, the results are presented in terms of COD since it is significant parameter from an environmental engineering point of view.

In the present investigation, the WAO of BDWW would have led to several reactions including COD reduction, VFA formation (acetic acid) as a byproduct (Equation 11.5), and conversion of COD into BOD (Equation 11.8); thereby enhancing the BI of the pretreated wastewater. All these observations are hypothesized in the reactions represented by Equations 11.1 through 11.9.

A power law model was used for representing the rate of reaction (Gunale and Mahajani, 2007) as

$$r = -\frac{d[COD]}{dt} = k[COD]^a[A]^b \qquad (11.16)$$

The oxygen concentration in the liquid phase [A] can be correlated to the partial pressure P_{O_2} and Equation 11.16 can be written as

$$r = -\frac{d[COD]}{dt} = k[COD]^a[P_{O_2}]^b \qquad (11.17)$$

Further, first-order behavior with respect to COD was presumed. For a given pressure, term $k[P_{O_2}]^b$ becomes constant and rate expression can be modified as

$$r = -\frac{d[COD]}{dt} = k'[COD]^a \qquad (11.18)$$

where $k' = k[P_{O_2}]^b$.
From Equation 11.18, for $a = 1$:

$$-\ln\frac{[C]}{[C_O]} = k't \qquad (11.19)$$

From Figure 11.9 and Equation 11.19, it can be concluded that the order with respect to [COD] is one. Figure 11.9 also gives the first-order kinetics plot at different temperatures.

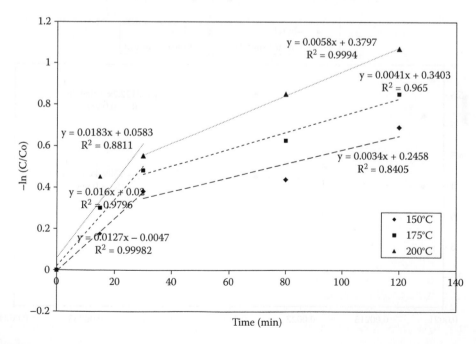

FIGURE 11.9
Determination of rate constant in WAO of B-DWW at different temperatures (6 bar, initial COD 40,000 mg/L, 200 rpm).

TABLE 11.4

Rate Constants for Fast and Slow Step at Different Temperatures

Temperature (K)	K_1 (1/min)	R^2	K_2 (1/min)	R^2
423	0.0127	0.9982	0.0034	0.8891
448	0.016	0.9796	0.0041	0.8788
473	0.0183	0.88	0.0058	0.9142

The reaction follows a two-step mechanism; the fast oxidation of organic substrate followed by slower oxidation of the low-molecular weight compounds formed, such as acetic acid and oxalic acid. The variation in k with respect to temperature and oxygen concentration can yield the energy of activation and order with respect to oxygen partial pressure. The slope of the curve from Figure 11.9 gives the values of k in 1/min. The rate constants and regression coefficients are presented in Table 11.4. For the fast step the reaction rate constant increased from 0.0127/min (at 150°C) to 0.0183/min (at 200°C), whereas the increase in k was from 0.0034 (at 150°C) to 0.0058/min (at 200°C) for slower step. It can be seen from the data presented in Table 11.4 that the k value at 200°C is 1.4 times that at 150°C in both steps.

Further, Arrhenius plot was developed by plotting log of rate constant values versus inverse of temperature (Figure 11.10). The activation energy, specific rate constant as a function of temperature was then calculated from the graph. Following equation was used to calculate the activation energy and specific rate constant:

$$k = k_o \exp\left[-\frac{E}{RT}\right] \tag{11.20}$$

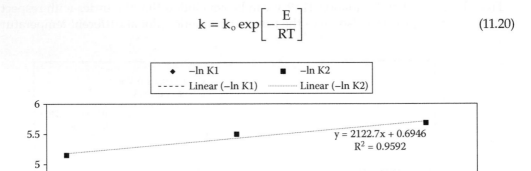

FIGURE 11.10
Arrhenius plot for WAO of B-DWW.

The activation energies were found to be 12.549 and 17.648 kJ/mol, respectively, for both the steps. Equation 11.17 for first and second step can be written as

$$-r_{COD_1} = 0.455 \exp\left[-\frac{1509.4}{T}\right][COD]^1[P_{O_2}]^b \tag{11.21}$$

$$-r_{COD_2} = 0.499 \exp\left[-\frac{2122.7}{T}\right][COD]^1[P_{O_2}]^b \tag{11.22}$$

In order to determine the order for partial pressure (b) of oxygen log of rate of reaction was plotted against log of the partial pressure. The order for partial pressure was calculated as $b = 0.42$ and 0.22. The combination of reactions (11.2), (11.3), (11.5), and (11.7) indirectly represents evidence for free radical mechanism, moreover the order with respect to oxygen partial pressure is observed to be 0.42 and 0.21, which reconfirms free radical mechanism. The final global rate expressions can be given as

$$-r_{COD_1} = 0.455 \exp\left[-\frac{1509.4}{T}\right][COD]^1[P_{O_2}]^{0.42} \quad \text{(Fast)} \tag{11.23}$$

$$-r_{COD_2} = 0.499 \exp\left[-\frac{2122.7}{T}\right][COD]^1[P_{O_2}]^{0.21} \quad \text{(Slow)} \tag{11.24}$$

As observed in Figure 11.9, the overall oxidation process can be divided into two steps—the fast oxidation of organic substrate followed by slower oxidation of the low-molecular weight compounds. It can be assumed that it is an irreversible reaction in the series and a consecutive unimolecular type first-order reaction can be given as (Levenspiel, 1998)

$$\text{Reactant (R)} \xrightarrow{k_1} \text{Intermediate (I = B + C)} \xrightarrow{k_2} \text{Mineralized product (P = CO}_2 + \text{H}_2\text{O)} \tag{11.25}$$

where Reactant (A) is the substrate/wastewater, which can be lumped to initial COD, Intermediate (I) consists of biodegradable entity (B: comprising low-molecular weight biodegradable products such as (but not limited to) VFAs which are known to be refractory to further WAO) and non-biodegradable entity (C) (Gunale and Mahajani, 2007), and Product (P) is the mineralized product (mainly CO_2 and H_2O which are not envisaged in the present study due to objective limited to pretreatment for biodegradability enhancement).

In this case, the rate of the reaction can be given as

$$r_R = \frac{dC_R}{dt} = -k_1 C_R \tag{11.26}$$

$$r_I = \frac{dC_I}{dt} = k_1 C_R - k_2 C_I \tag{11.27}$$

$$r_P = \frac{dC_P}{dt} = k_2 C_P \tag{11.28}$$

The changing concentration of intermediate (I) can be represented in the form of a differential equation as follows:

$$\frac{dC_I}{dt} + k_2 C_I = k_1 C_{R_O} e^{-k_1 t} \tag{11.29}$$

Integrating Equation 11.29 for initial conditions as $C_{I_O} = 0$ at $t = 0$, the final equation for changing intermediate concentration can be given as

$$C_I = C_{R_O} k_1 \left(\frac{e^{-k_1 t}}{k_2 - k_1} + \frac{e^{-k_2 t}}{k_1 - k_2} \right) \tag{11.30}$$

Further, in order to determine the maximum time required for completion of fast step or the time required attaining maximum intermediate concentration, Equation 11.30 needs to be differentiated followed by setting $dC_I/dt = 0$. This results in the following equation for t_{max}:

$$t_{max} = \frac{\ln(k_2/k_1)}{k_2 - k_1} \tag{11.31}$$

Also, maximum intermediate concentration $\left(C_{I_{max}} \right)$ can be found out by combining Equation 11.30 and 11.31

$$\frac{C_{I_{max}}}{C_{R_O}} = \left(\frac{k_1}{k_2} \right)^{k_2/(k_2 - k_1)} \tag{11.32}$$

Using Equations 11.31 and 11.32, the t_{max} and $C_{I_{max}}$ values were calculated as $t_{max} = 2.098$ min

$$\frac{C_{I_{max}}}{C_{R_O}} = 0.351$$

The proposed irreversible series reaction (Equation 11.25) indicates that intermediate (I) represents a biodegradable entity (B, which is residual BOD) and part of recalcitrant COD (C, which is retained till the end of reaction). The biodegradable fraction (B) is estimated to be obtained during fast step of the WAO process (i.e., during $t_{max} = 2.098$ min) and accordingly lumped COD at maximum intermediate concentration was predicted to be 14,040 mg/L for the inlet reactant lumped initial COD concentration that is, $C_{R_O} = C_O = 40,000$ mg/L. Further value of B also corresponds to 77% of the maximum biodegradable intermediate, which was experimentally observed to be 10,800 mg/L at optimized condition of WAO operation (175°C, 6 bar, 30 min and 200 rpm). Further, the ratio of BOD enhancement to the initial COD is experimentally observed to be 0.27, which on kinetic validation of rates is observed to be 0.22, indicating the broad agreement between two. This trend is very desirable from the point of view of partial destruction to induce biodegradability enhancement of the wastewater. The "C" may be corresponding to the VFA part (Figure 11.11) as observed in the present study (stable intermediates).

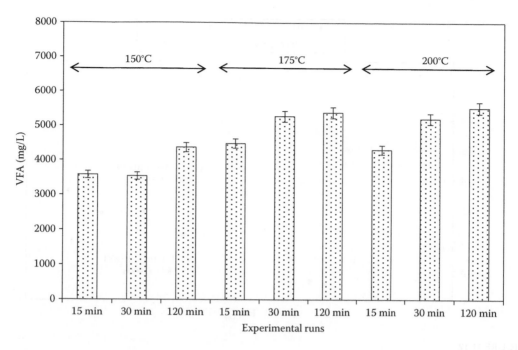

FIGURE 11.11
Profile of VFA generation at various WAO conditions for B-DWW (6 bar, original VFA 180 mg/L, initial COD 40,000 mg/L).

Further, mineralization of this highly biodegradable acid to CO_2 forms the slow step/rate limiting step for WAO process. The literature reports on WAO also indicate similar facts that the VFAs generated during the WAO are intermittent dead end-products, refractory to further WAO (Mishra et al., 1995; Shende and Mahajani, 1997; Gunale and Mahajani, 2007). This kinetic model is valid for the following operating condition ranges: 150–200°C temperature and 6–12 bar pressure.

Similarly, the rate of BI enhancement can be given as

$$r = \frac{d[BI]}{dt} \tag{11.33}$$

An attempt was made to correlate the rate of BI enhancement and rate of COD reduction, for which the following correlation was assumed:

$$\ln r_{BI} = \alpha \ln(-r_{COD}) \tag{11.34}$$

The proportionality constant α_1 and α_2 was graphically estimated as 0.95 ($R^2 = 0.84$) and 1.0016 ($R^2 = 0.9$), respectively, for fast and slow step (Figure 11.12). The BI and COD correlation can be represented as

$$\ln r_{BI_1} = [0.95 * \ln(-r_{COD_1})] - 10.649 \quad \text{(Fast step)} \tag{11.35}$$

$$\ln r_{BI_2} = [1.0016 * \ln(-r_{COD_2})] - 10.883 \quad \text{(Slow step)} \tag{11.36}$$

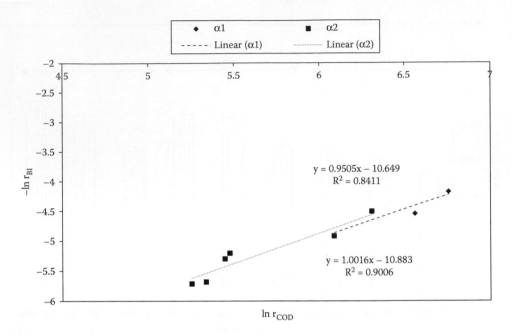

FIGURE 11.12
Determination of correlation between rate of BI enhancement and rate of COD destruction.

The concepts presented in Equations 11.1 through 11.9 were validated by analyzing/profiling VFA. Similar attempt was made to correlate the rate of BI enhancement and VFA increase during wet oxidation reaction (Figure 11.13).

$$\ln r_{BI} = \beta \ln r_{VFA} \tag{11.37}$$

Graphical analysis of Equation 11.29 gave proportionality constants β_1 and β_2 as 0.7454 ($R^2 = 0.806$) and 0.7153 ($R^2 = 0.944$) for fast and slow step, respectively (Figure 11.13). Following equation correlates BI and VFA:

$$\ln r_{BI_1} = \left[0.7454 * \ln r_{VFA_1} \right] - 8.607 \quad \text{(Fast step)} \tag{11.38}$$

$$\ln r_{BI_2} = \left[0.7153 * \ln r_{VFA_2} \right] - 8.3721 \quad \text{(Slow step)} \tag{11.39}$$

The VFA profile presented in Figure 11.11, indicate the concentration of VFA at each WAO pretreatment condition, the pretreatment lead to the generation of sufficient quantities of VFA (3540–5511 mg/L) suggesting the role of the pretreatment in transforming the parent complex pollutant molecules to simpler biodegradable form like acids (e.g., acetic acid) (Mishra et al., 1995; Tembhekar et al., 2014).

11.3.7 Effect of Wet Air Oxidation Pretreatment of Distillery Effluent on Toxicity Reduction via Seed Germination

The WAO pretreated effluent samples with enhanced biodegradability index and reduced color were subjected to seed germination assay of mustard seeds to assess the reduction in its

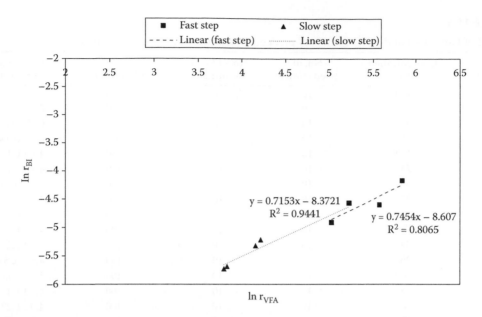

FIGURE 11.13
Determination of correlation between rate of BI enhancement and rate of VFA generation.

toxicity as per the protocol of Bharagava et al. (2008). The germination of seed was recorded with the appearance of radical in the petriplate and results are presented in Table 11.5. The untreated biomethanated distillery effluent of higher concentrations (75% and 100%) had shown 100% inhibitory effect on seed germination. In contrast, 40% seed germination was observed for an untreated effluent concentration of 25%. The biodegradability index of the effluent considered in this study is 0.172. The characteristics of distillery effluent depend upon the raw material used and various aspects of ethanol production process (Pant and Adholeya, 2007; Tewari et al., 2007). The pretreated effluent with an enhanced BI of 0.4 had shown 80% seed germination even at high concentrations (100%), while the pretreated effluents with enhanced BI of 0.5, 0.6, and 0.8 had shown 100% seed germination. The pretreated effluents at lower effluent concentrations (25%, 50%, and 75%) indicated 100% seed germination (Table 11.5). The favorable seed germination results with pretreated effluents are indicative of the potential of WAO pretreatment for facilitating toxicity reduction.

11.3.8 Anaerobic and Aerobic Biodegradation of WAO Pretreated Distillery Effluent

11.3.8.1 Anaerobic Biodegradation of WAO-Treated BDWW Effluent (BDWW)

The untreated and WAO-pretreated effluent with enhanced BIs (0.4–0.8) was subjected to anaerobic biodegradation to assess biogas generation as per the given protocol. Experiment was carried out for 12 days and biogas generation was analyzed for percent methane content and yield. The pretreated effluent samples demonstrated enhanced biogas generation with higher methane content when compared with the untreated effluent.

The cumulative biogas production from pretreated effluents at different enhanced BIs is shown in Figure 11.14. The results indicate that the WAO pretreatment transforms the complex compounds in the effluent to the simpler compounds, which are easily biodegradable, leading to biodegradability enhancement of the effluent.

TABLE 11.5

Effect of Untreated and WAO BMDWW on Seed Germination in Mustard Plants

Sample	Concentration (%)	No. of Seeds Sown	No. of Seeds Germinated	%Germination	Germination Index (GI)
Control	100	10	0	0	0
	75	10	0	0	0
	50	10	2	20	0.45 ± 0.0208
	25	10	4	40	0.9 ± 0.056
0.4 BI	100	10	8	80	4.1 ± 0.152
	75	10	10	100	4.9 ± 0.115
	50	10	10	100	4.8 ± 0.057
	25	10	10	100	4.8 ± 0.173
0.5 BI	100	10	10	100	4.9 ± 0.251
	75	10	10	100	4.9 ± 0.264
	50	10	10	100	4.8 ± 0.230
	25	10	10	100	4.8 ± 0288
0.6 BI	100	10	10	100	5.2 ± 0.215
	75	10	10	100	4.9 ± 0.1
	50	10	10	100	4.9 ± 0.225
	25	10	10	100	4.7 ± 0.55
0.8 BI	100	10	10	100	4.7 ± 0.4
	75	10	10	100	4.7 ± 0.36
	50	10	10	100	5.1 ± 0.1
	25	10	10	100	4.4 ± 0.22

Pretreated effluent with BI of 0.5 showed maximum methane content of 64.13%, followed by sample with BI of 0.6, 0.8, and 0.4 with a methane content of 56.99%, 52%, and 49.56%, respectively, when compared with untreated effluent with a methane content of only 15.3% (Figure 11.15).

Methane yield for the studied samples of different BI ranged from 28 to 79.31 mL/g of COD reduced (Figure 11.15). Hence, the results indicate that WAO pretreatment enhances

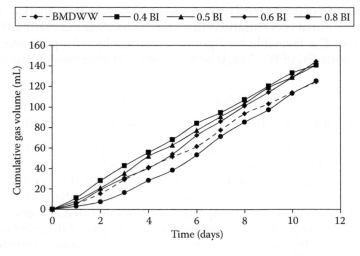

FIGURE 11.14

Cumulative gas generation of untreated and WAO pretreated effluent of different BI.

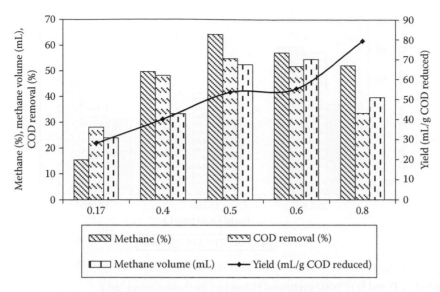

FIGURE 11.15
Effect of biodegradability index on methane content, yield, and COD removal via anaerobic digestion.

the biodegradability and retains sufficient residual COD for biogas generation (Sarat Chandra et al., 2014). During the biogas experiments, pH varied from 6.93 to 8.2. In all the experiments conducted, methane generation has substantially decreased after a period of 9–10 days. This observation was probably due to an increase in the pH to 8.2, where alkaline conditions are reported to affect methane generation due to the presence of ammonia (Shanmugam and Horan, 2009). The WAO-pretreated effluent with BI of 0.5 showed maximum reduction of COD (54.75%), followed by BI of 0.6, 0.8, and 0.4 a reduction of 51.69%, 48.06%, and 33.46%, respectively, when compared with the untreated effluent with only 28% reduction (Figure 11.15). The effect of COD concentration of WAO pretreated effluent on methane generation in a COD range 13,000 mg/L–20,000 mg/L indicated increase in methane generation with the increase in initial COD concentration (Figure 11.16).

11.3.8.2 Aerobic Biodegradation of WAO-Treated BDWW Effluent

The WAO pretreatment of the complex effluent under various pretreatment scenarios resulted in substantial biodegradability enhancement (BI up to 0.88) in comparison to the untreated effluent (BI = 0.17). The pretreated effluent of different BI (0.4, 0.6, 0.8) was further subjected to aerobic biodegradation for evaluation of biodegradability enhancement. The aerobic biodegradation profile of the WAO-pretreated and untreated effluent (control) is presented in Figure 11.17a. The results indicated enhanced biodegradability of WAO-pretreated effluent when compared with the untreated effluent. The percent COD reduction observed after aerobic biodegradation for ten days was 51.2%, 67.7%, and 64.9% for 0.4, 0.6, and 0.8 BI of pretreated effluent, respectively; when compared with only 22.5% COD removal from the untreated effluent. The aerobic biodegradation rate constant was enhanced for the WAO-pretreated effluent, and 0.6 BI sample yielded maximum percent COD reduction and higher value of biodegradation rate constant (Table 11.6), which was about three times higher than that of the untreated effluent. Although it is very difficult to characterize the complex effluent completely, the results are indicative of the fact that the chemical composition of the effluent was affected during the WAO treatment and possibly

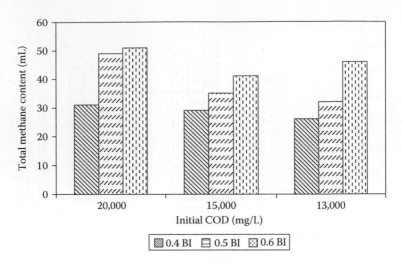

FIGURE 11.16
Effect of initial COD and BI of WAO pretreated effluent on methane content (mL).

resulted in a partial breakdown/reorientation of complex molecules to simpler biodegradable molecules (Malik et al., 2014; Sarat Chandra et al., 2014).

11.3.8.3 Combined Anaerobic–Aerobic Biodegradation of WAO-Treated BDWW

The model complex distillery effluent pretreated with WAO was subjected to anaerobic followed by aerobic biodegradation (AnD + AD). The rate of oxidation of the pretreated effluent was compared with that of the untreated, and also with the oxidation rate obtained when these techniques were employed individually. Figure 11.17c has shown the percent COD reduction of WAO pretreated (0.4, 0.6, 0.8 BI) and untreated effluent during anaerobic–aerobic biodegradation study conducted at batch scale. Results obtained indicated that the combined treatment technique enables significant enhancement in the COD removal. The combined biodegradation technique (AnD + AD) resulted in a COD reduction of up to 78.9%, 87.9%, 83.1%, and 43.1% in the 0.4, 0.6, 0.8 BI, and untreated effluent, respectively. Thus, when compared with 67.7% and 73.3% COD removal of 0.6 BI pretreated effluent for AD and AD after dilution, the combined biodegradation study (AnD + AD) for the same has yielded a higher COD reduction of 87.9%. Similarly, for other WAO-pretreated effluents (0.4, 0.8 BI), the anaerobic–aerobic batch study yielded maximum COD removal. And also, during the anaerobic degradation considerable amount of energy in the form of biogas was generated. WAO pretreated BDWW with BI of 0.6, showed biogas generation with a highest methane percentage of 57%, followed by effluent with BI of 0.8 and 0.4, having methane content of 53.5%, and 49.5%, respectively; while the control indicated a methane content of 4.8% only (Figure 11.18).

The difference in the rates of oxidation at various pretreatment conditions can be attributed to the formation of different kinds of intermediate products during the WAO pretreatment, which have further affected the biological treatment. We hypothesize that the difference in the rates of oxidation after WAO pretreatment is due to the formation of different intermediate low-molecular weight compounds, which are easily amenable to biodegradation. It has been reported in literature that WAO pretreatment enhances the substrate metabolic value by transforming higher molecular weight compounds into

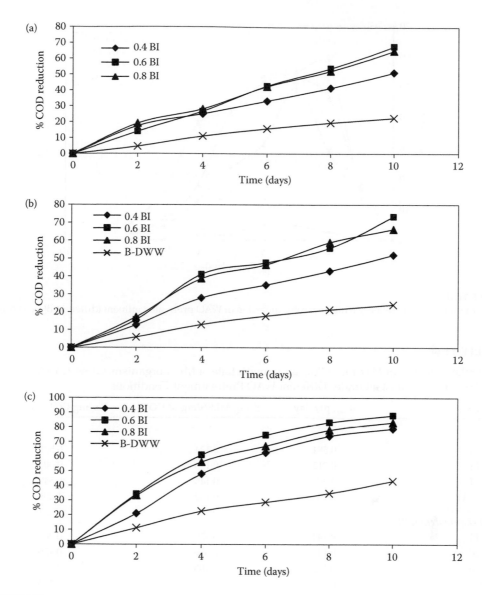

FIGURE 11.17
Effect of wet air oxidation pretreatment on biodegradability of the effluent during (a) aerobic oxidation, (b) aerobic oxidation at equivalent initial COD 13,000 mg/L, (c) anaerobic followed by aerobic oxidation.

low-molecular weight compounds with properties suitable to achieve better degradation rate (Gogate and Pandit, 2004; Pant and Adholeya, 2006). The values of first-order biodegradation rate constant for COD removal for WAO-pretreated effluent and untreated effluent under different treatment conditions are shown in Figure 11.17a–c. The values of biodegradation rate constants were higher for WAO-pretreated effluents when compared with the untreated effluent. Maximum biodegradation rate constant observed for 0.6 BI effluent indicated that the biodegradation rate was higher for 0.6 BI when compared with the effluent having 0.4 and 0.8 BI (Figure 11.17c and Table 11.6). The biodegradation rate constant at 0.6 BI for combined treatment (AnD + AD) was four times higher than the untreated

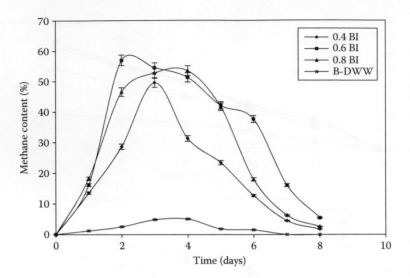

FIGURE 11.18

Profile of methane generation during anaerobic digestion of WAO pretreated effluent (different BI) and control.

TABLE 11.6

Calculated Values of Maximum Specific Growth Rate of Microorganism, Growth Yield and Saturation Constant under Different WAO Pretreatment Conditions

Batch	q_{max}, per day	Y, mg MLSS/mg of COD	Ks, mg COD/mg MLSS
AO			
0.4 BI	0.584	0.124	0.532
0.6 BI	0.212	0.196	0.0031
0.8 BI	0.285	0.145	0.549
B-DWW	0.156	0.102	1.039
AO (Equivalent COD)			
0.4 BI	0.144	0.475	0.268
0.6 BI	0.147	0.076	0.275
0.8 BI	0.182	0.403	0.269
B-DWW	0.089	1.045	0.333
AnD + AO			
0.4 BI	0.204	0.28	0.139
0.6 BI	0.237	0.32	0.241
0.8 BI	0.226	0.31	0.337
B-DWW	0.132	0.22	0.619

effluent during the same batch study. Also, the biodegradation rate constant at 0.6 BI for combined treatment (AnD + AD) is almost double than the values obtained for 0.6 BI during aerobic degradation (AD) and AD where initial COD was adjusted by dilution studies, respectively (Figure 11.17a,b). The difference in the values of rate of bio-oxidation for 0.4 BI, 0.6 BI, and 0.8 BI is attributable to the difference in the treatment conditions employed, possibly resulting in the formation of different kinds of intermediates during the WAO pretreatment. This difference in the rates also confirms the formation of various transformation intermediates or reoriented molecules under different WAO treatment conditions of temperature, pressure, and time.

11.4 Case Study of Ozonation-Based Pretreatment of Complex Distillery Effluent (Biomethanated, BDWW) for Biodegradability Enhancement and Color Reduction

11.4.1 Ozone Principle and Mechanism

Ozone is produced when dissociated oxygen molecules collide with another oxygen molecule resulting in the formation of an unstable gas molecule, ozone (O_3).

$$O_2 + hv \rightarrow O^{\cdot} + O^{\cdot} \quad O_2 + O^{\cdot} \rightarrow O_3$$

Ozone is a powerful oxidant having redox potential of 2.07. Ozone is generated by imposing a high voltage alternating current (6–20 kvolt) across a dielectric discharge gap, containing an oxygen-bearing gas. Ozone is highly reactive with a very short half-life. It can neither be stored as gas nor be transported. Hence, due to its unstable nature ozone is generated onsite. Ozone molecule is composed of a single and a double bond. The weak single bond results in the formation of free radicals. Ozone is a powerful and more reactive oxidant when compared with oxygen.

Free radicals of hydrogen peroxide (HO_2) and hydroxyl (OH) is generated on decomposition of ozone. These radicals are great oxidizing agents. Ozone molecule reacts instantly with contaminants present in wastewater, which comprise compounds containing amino groups, double bond compounds, or aromatic groups. In ozonation reaction, ozone is completely consumed releasing only oxygen. Many other oxidizing agents are often used in combination with ozone to achieve increased efficacy. Additional presence of peroxides, UV, and high pH enhance the efficiency of ozone.

After decomposition, the solution consists of a combination of ozone molecule and free radicals in highly reactive states. The free radicals in this solution are highly capable of withdrawing atoms from the substrate that further results into complete oxidation (Figure 11.19).

11.4.2 Ozonation Experimental Setup and Operation for Distillery Effluent

Ozone gas was generated using an ozone generator (PCI Ozone Corporation-Model INDO2-7, Ozone Research & Applications India Pvt. Ltd., Nagpur) by the Corona Discharge Method with water cooling. Oxygen used as feed gas was supplied from an oxygen cylinder at a uniform flow rate of 2 L/min. The experimental setup used for the ozonation experiments is shown in Figure 11.20. Ozone flow was regulated by monitoring gas flow rate. Ozonation was carried out in a semi-continuous mode by sparging the ozone into the effluent. The ozone concentration inlet and outlet gas was measured every 20 min during the oxidation process (using a UV analyzer BMT 964, Berlin). All the ozonation experiments conducted in well-ventilated laboratory and at room temperature, to avoid the ozone toxicity.

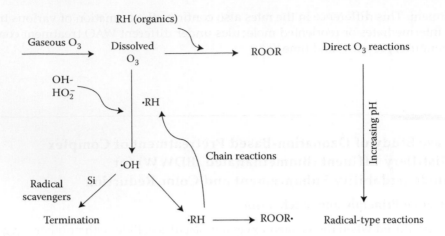

FIGURE 11.19
Ozone reactions with organics present in wastewater, where RH is organic substrate resulting in COD, R· is: substrate radical/reactant, ROOR· is: reoriented organic substrate, Si is radical scavenger species. (Adapted and revised from Alvares ABC, Daiper C, and Pasons SA. 2001. *Environmental Technology* 22, 409–427.)

The effluent (BDWW) stored at 4°C was allowed to attain room temperature, before pretreatment. For each batch, 2 L of raw effluent with known initial COD concentration (37000 mg/L) was treated in a glass reactor and kept above a magnetic stirrer. pH adjustments were not made before or during pretreatment. The ozone and oxygen mixture was bubbled at a flow rate of 2 L/min at an ozone dose of 2.8 g/h with ozone concentration of 4% w/w for 4 h. Pretreated samples (200 mL) were withdrawn after every 20 min and

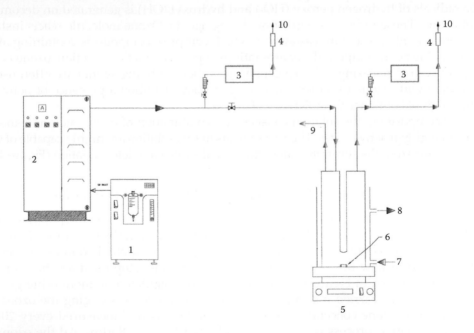

FIGURE 11.20
Experimental setup for ozonation: (1) oxygen cylinder, (2) ozone generator, (3) ozone gas analyser, (4) ozone destructor, (5) magnetic stirrer, (6) magnetic bead, (7) water in, (8) water out, (9) sample outlet, (10) vent to ambient.

analyzed for COD, BOD, color, and pH. Seed germination test was conducted to analyze the toxicity of the raw and pretreated effluent.

11.4.3 Effect of Ozonation Pretreatment on Distillery Effluent (BDWW)

The effect of ozonation was studied as a pretreatment option to enhance biodegradability. Ozonation was carried out for 4 h. The effect of ozonation on the biodegradability enhancement and COD reduction (%) of effluent is presented in Figure 11.21.

The results indicated that ozone pretreatment led to a reduction in the COD and enhancement in the biodegradability of the effluent, indicating partial mineralization during the pretreatment step. The ability of ozone to destroy recalcitrant compounds has been demonstrated previously in the literature (Gottschalk et al., 2000). Under the pretreatment conditions, ozonation led to a maximum BI enhancement of up to 0.58 and COD reduction of 33%. The ozonation of organic compounds in water usually produces oxygenated organic products and low-molecular weight organic acids that are relatively easily biodegradable (Gilbert, 1987; Contreras et al., 2003). Thus, apart from COD reduction, ozonation also improved amenability to biodegradation.

Ozonation led to a substantial drop in the pH value of the effluent (from pH 7.53–5.8) over the treatment period of 240 min, indicating the neutralization of carbonates due to the formation of acidic degradation products. Carboxylic acids have been indicated as products of melanoidin ozonation (Kim et al., 1985). Under alkaline conditions, ozone decompose to form highly oxidizing and nonselective hydroxyl radicals while in an acidic environment the direct reaction mechanism of ozone predominates (Staehelin and Hoigne, 1982; Gottschalk et al., 2000). Distillery spent wash effluent contains a high concentration of bicarbonate, which may act as a hydroxyl radical scavenger thus inhibiting the reaction of hydroxyl radicals with the organic matter (Glazeand Kang, 1987).

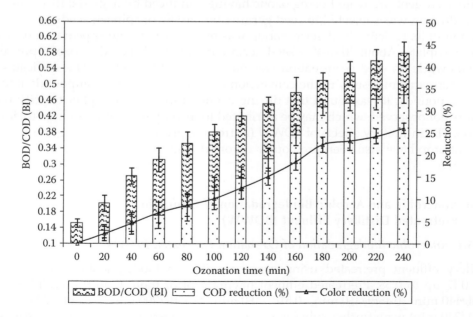

FIGURE 11.21
Effect of ozone pretreatment on biodegradability index (BI), COD and color removal of biomethanated distillery effluent.

Although ozonation pretreatment led to a reduction in COD, higher ozone treatment time did not enhance the COD removal efficiency and BI enhancement and ozone left the reactor without being consumed. Rate of COD removal consistently increased up to 180 min after which no significant COD removal was observed further till 240 min of reaction time.

11.4.4 Effect of Ozonation Pretreatment on Color Reduction of Distillery Effluent

During the pretreatment, ozone had a primary effect on the decolorization of the effluent coupled with some decrease in the COD. The color reduction of the BDWW has been attributed to the ability of ozone to directly attack the C=C bonds in aromatic and chomophoric molecules, leading to the formation of bleached products, such as aliphatic acids, ketones, and aldehydes (Sangave and Pandit, 2007). Absorbance measured at 254 nm is mostly the measure of the aromaticity present (maximum absorbance wavelength of organic compounds) (Beltran et al., 2000). Hence, this was chosen as the wavelength for monitoring absorption pre- and postozonation using UV–vis spectrophotometer that was incidentally also the absorption maxima for the wastewater sample under investigation. Ozone pretreatment led to a reduction in absorbance value from 0.377 to 0.279 (1000D) indicating the reduction in the colored pigments (melanoidins) from the effluent.

Under the pretreatment condition ozonation led to a maximum of 25.99% color reduction (Figure 11.21) as described earlier in the preceding section. The color removal effect was significant till 180 min similar to COD reduction after which no further effect of ozone on the effluent was observed.

11.4.5 Effect of Ozone Pretreatment of Distillery Effluent on Toxicity Reduction via Seed Germination

Distillery effluent pretreated using ozone having enhanced biodegradability index and color reduction was subjected to seed germination assay of spinach seeds to assess the reduction in its toxicity. Seed germination was recorded with the appearance of radical in the petriplate. About 30%–40% seed germination was observed in ozone pretreated effluents while no seed germination was observed for the control. This indicates that ozone pretreatment results in the conversion of toxic/recalcitrant compounds into less toxic compounds. It was found that during ozone pretreatment, higher toxicity reduction was not observed, which may be attributed to incomplete degradation of toxic compounds. The inhibitory effects of BDWW (untreated, pretreated) on seed germination of different crop have also been previously reported (Sangeeta et al., 2010; Sarat Chandra et al., 2014).

11.4.6 Anaerobic and Aerobic Biodegradation of Ozonation Pretreated Distillery Effluent (BDWW)

11.4.6.1 Anaerobic Biodegradation of Ozonation Pretreated BDWW

Distillery effluent pretreated using ozone resulted in biodegradability enhancement from 0.17 up to 0.58. Pretreated effluent having a BI of more than 0.4 (0.42BI/120 min, 0.45BI/140 min, 0.48BI/160 min, 0.51BI/180 min, 0.53BI/200 min, 0.56BI/220 min, and 0.58BI/240 min) were further subjected to anaerobic digestion. The ozone pretreated distillery effluent demonstrated biogas generation with high methane content compared with the untreated distillery effluent (Figure 11.22).

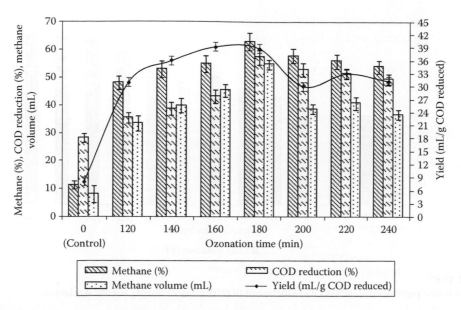

FIGURE 11.22
Effect of ozone pretreatment on methane content, yield, and COD removal via anaerobic digestion of biomethanated distillery effluent.

It can be seen that 180 min ozone pretreated effluent showed a maximum methane generation of 54.6 mL composing of a methane percentage of 62.8%, and methane yield of 39 mL/g COD reduced when compared with methane volume of 7.9 mL, methane percentage of 11.2% and yield of 7.97 mL/g COD reduced for the control. Extending the ozonation time beyond 180 min did not show significant enhancement in wastewater characteristics. Results indicated that ozone pretreatment substantially enhanced the BI while still retaining sufficient residual COD for methane/biogas production. During the course of the pretreatment, pH varied from 6.93 to 8.2, vis-à-vis, methane generation decreased after period of 9–10 days. This was probably due to the rise of pH up to 8.2. The optimum pH range for the generation of methane is reported to be 6.5–8 and at alkaline pH methane generation is affected due to ammonia generation. The ozone pretreated effluent of 180 min showed maximum COD reduction of 57.14% when compared with only 27.92% in the control indicating efficiency of ozonation in rendering the wastewater more amenable to biodegradation by the virtue of biodegradable intermediates formed during the course of the reaction (Figure 11.22).

11.4.6.2 Aerobic Biodegradation of Ozonation Pretreated BDWW

The effluent pretreated at 180 min using ozone was also subjected to aerobic oxidation. The biodegradability during the aerobic oxidation was evaluated in terms of COD reduction. The rates of degradation were compared with that of the control without any pretreatment (Figure 11.23).

Figure 11.23 shows the effect of ozonation as a pretreatment on the aerobic biodegradation of the BDWW effluent when compared with that of control. The results indicated enhanced biodegradability in case of ozone pretreated BDWW effluent when compared with the untreated effluent. The percent COD reduction observed after 15 days of aerobic

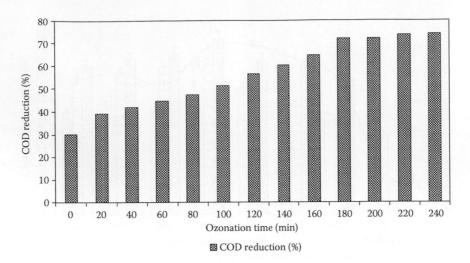

FIGURE 11.23
Effect of ozone pretreatment on % COD reduction of BDWW effluent during aerobic oxidation.

oxidation was up to 73.63% when compared with only 29.72% COD removal in the control. From Figure 11.23, it can be seen that COD removal was high up to initial 180 min of reaction after which minimum COD removal was observed when the reaction was continued further upto 240 min.

The experimental results of the study followed first-order kinetics and the results are presented in Figure 11.24.

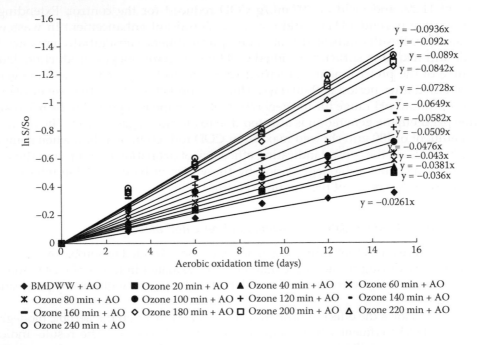

FIGURE 11.24
First order degradation plot for BDWW effluent and ozone pretreated BDWW effluent during aerobic oxidation.

The ozone pretreated BDWW effluent (180 min) yielded higher percent COD reduction and higher value of biodegradation rate constant during aerobic oxidation (Figure 11.24), which was about 3.23 times higher than that of the untreated effluent. Thus, these results confirmed that preozonation enhanced the fraction of readily biodegradable compounds leading to higher initial biodegradation rates.

Various factors such as concentration, chemical structure, and substituents of the target compound affects the biological degradation of the compound. Other factors such as pH and presence of any inhibitory compounds can also affect the biological degradation. Ozonation led to an improved biodegradability of the distillery effluent by the reaction of the ozone with the pollutant molecules and converting them into biologically degradable, more amenable products.

11.4.6.3 Combined Anaerobic–Aerobic Biodegradation of Ozonation-Treated BDWW

The ozone pretreated effluent was subjected to anaerobic followed by aerobic biodegradation (Ana + AO). The rate of oxidation of the ozone-pretreated BDWW effluent was compared with that of the untreated BDWW effluent, and also with the oxidation rate obtained when these techniques were employed individually.

Results obtained indicated that the combined treatment technique enables significant enhancement in the COD removal (Figure 11.25). The combined biodegradation technique (Ana + AO) resulted in a maximum COD reduction of 91.42% for 180 min ozone-pretreated BDWW effluent when compared with only 43.24% for untreated BDWW effluent. Thus, when compared with 71.42% COD removal of 180 min pretreated effluent for AO, the combined biodegradation study (Ana + AO) for the same has yielded a higher COD reduction of 91.42%. Similarly, for other ozone-pretreated effluents (120, 140, 160, 200, 220 and 240 min), the anaerobic–aerobic batch study yielded high COD removal and also, during the anaerobic degradation considerable amount of energy in the form of biogas was generated.

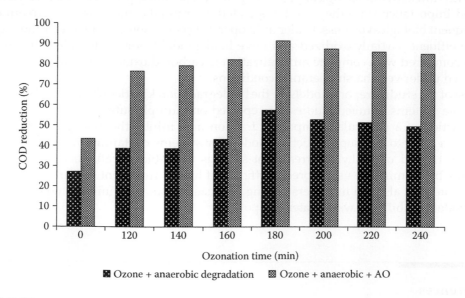

FIGURE 11.25
Effect of ozone pretreatment on biodegradability of the effluent during anaerobic, and anaerobic followed by aerobic oxidation in terms of COD reduction.

From Figure 11.25 it is clear that COD removal was high up to 180 min of ozone-pretreated effluent and after which enhancement in the COD removal decreased (Gupta et al., 2015).

11.5 Conclusion

A large number of studies conducted in the last two decades on the application of AOP for the treatment of industrial wastewater indicates that advanced oxidation-based processes have been used advantageously; mostly for complete mineralization and scantily as pretreatment option. Most of the literature shows that photo-catalytic oxidation and ozonation have by and large been popular pretreatment methods when compared with other AOPs. Efficiency of the oxidative treatment expressed using conventional parameters such as COD, BOD, and TOC have so far been reported for most of the industrial wastewaters. However, with ever-increasing complexity and stringent environmental norms, it has become mandatory to monitor parameters such as biodegradability, toxicity, and other important aspects such as color/pigments in case of highly colored wastewater.

In case of effluents that are not only complex with respect to organic content but also have color, application of AOPs as pretreatment option usually makes the complex wastewater amenable to further biological treatment. Thus, pretreated wastewater can be posttreated biologically with effective/efficient contaminant destruction. To achieve this process synergy the research focus over the past few years has been increasingly on evaluating hybrid/combined treatment systems to cover larger part of the biorecalcitrant components present in the wastewater, thereby reducing the complexity and impart desirable effluent quality.

The most common mechanism through which the AOPs function is a free radical mechanism, that has high oxidation potential and this property can be utilized for combining with downstream biological processing. In recent times, processes such as WAO has gained importance since they can be operated at optimal conditions as pretreatment to subsequent biological process. Under mild operating conditions, WAO render the organics in the effluent partially oxidized and more biodegradable for further biological options when compared with complete mineralization to carbon dioxide and water, which can be achieved under very harsh operating conditions.

Most of the studies report models on the biodegradation kinetics usually deal with single substrate systems. Limited studies are reported on multiple substrate systems, in which any fraction of a particular component fraction may inhibit the biodegradability. Thus, there is a need to develop predictive models for systems containing multiple substrates that are treated using hybrid treatment systems (AOP-biological). Such models would not only be of immense use to predict efficacy of hybrid treatment systems for industrial wastewater but also aid in understanding the mechanism pertaining to no or extremely low biodegradability of wastewater components.

References

Alvares ABC, Daiper C, and Pasons SA. 2001. Partial Oxidation by Ozone to Remove Recalcitrance from Wastewaters—A Review. *Environmental Technology* 22, 409–427.

Beltran FJ, Garcia-Araya JF, and Alvarez PM. 2000. Continuous flow integrated chemical (ozone)-activated sludge system treating combined agroindustrial-domestic wastewater. *Environmental Progress* 19: 28–35.

Bharagava RN and Chandra R. 2010. Biodegradation of the major color containing compounds in distillery wastewater by an aerobic bacterial culture and characterization of their metabolites. *Biodegradation* 21: 703–711.

Bharagava RN, Chandra R, and Rai V. 2008. Phytoextraction of trace elements and physiological changes in Indian mustard plants (*Brassica nigra* L.) grown in post methanated distillery effluent (PMDE) irrigated soil. *Bioresource Technology* 99: 8316–8324.

Coates J. 2000. Interpretation of infrared spectra: A practical approach. In: Meyers RA (ed.), *Encyclopedia of Analytical Chemistry*, John Wiley & Sons Ltd., Chichester, pp. 10,815–10,837.

Collado S, Garrido L, Laca A, Mario D. 2010. Wet oxidation of salicylic acid solutions. *Environmental Science and Technology* 44: 8629–8635.

Contreras S, Rodriguez M, Momani FA, Sans C, and Esplugas S. 2003. Contribution of the ozonation pre-treatment to the biodegradation of aqueous solution of 2,4-dichlorophenol. *Water Research* 37: 3164–3171.

Espejo A, Aguinaco A, Amat AM, and Beltran FJ. 2014. Some ozone advanced oxidation processes to improve the biological removal of selected pharmaceutical contaminants from urban wastewater. *Journal of Environmental Science and Health. Part A* 49(4): 410–421.

Garg A and Mishra A. 2010. Wet oxidation—An option for enhancing biodegradability of leachate derived from municipal solid waste (MSW) landfill. *Industrial and Engineering Chemistry Research* 49: 5575–5582.

Gilbert E. 1983. Investigations on the changes of biological degradability of single substances induced by ozonation. *Ozone Science and Engineering* 5: 137–49.

Gogate PR and Pandit AB. 2004. A review of imperative technologies for wastewater treatment II: Hybrid processes. *Advances in Environmental Research* 8: 553–597.

Gottschalk C, Libra JA, and Saupe A. 2000. *Ozonation of Water and Wastewater—A Practical Guide Understanding Ozone and its Application*. Wiley-VCH, Germany.

Gunale TL and Mahajani VV. 2007. Studies in mineralization of aqueous aniline using Fenton and wet oxidation (FENTWO) as a hybrid process. *Journal of Chemical Technology and Biotechnology* 82: 108–115.

Jacobsen N. 2007. *NMR Spectroscopy Explained*. John Wiley & Sons Inc., New Jersey.

Karaca G and Tasdemir Y. 2014. Application of advanced oxidation processes for polycyclic aromatic hydrocarbons removal from municipal treatment sludge. *Clean—Soil, Air, Water* 43: 191–196.

Kim SB, Hayase F, and Kato H. 1985. Decolorization and degradation products of melanoidins on ozonlysis. *Agricultural and Biological Chemistry* 49: 785.

Kumar A. 2003. *Handbook of Waste Management in Sugar Mills and Distilleries*. Somaiya Publication, India.

Levenspiel O. 1998. *Chemical Reaction Engineering*, 3rd ed. John Wiley & Sons, New York.

Li L, Chen P, and Gloyna EF. 1991. Generalized kinetic model for wet oxidation of organic compounds. *AIChE Journal* 37: 1687–1697.

Malik SN, Sarat Chandra T, and Tembhekar PD, Padoley KV, Mudliar SL, and Mudliar SN. 2014. Wet air oxidation induced enhanced biodegradability of distillery effluent. *Journal of Environmental Management* 1: 136–132.

Mantzavinos D and Psillakis E. 2004. Enhancement of biodegradability of industrial wastewaters by chemical oxidation pre-treatment. *Journal of Chemical Technology and Biotechnology* 79(4): 431–454.

Merayo N, Hermosilla D, Blanco L, Cortijo L, and Blanco Á. 2014. Assessing the application of advanced oxidation processes, and their combination with biological treatment, to effluents from pulp and paper industry. *Journal of Hazardous Materials* 262: 420–427.

Metcalf IE. 1979. *Wastewater Engineering: Treatment Disposal Reuse*, 2nd ed. Tata McGraw Hill, New Delhi.

Mishra VS, Mahajani VV, and Joshi JB. 1995. Wet air oxidation. *Industrial and Engineering Chemistry Research* 34: 2–48.

Miyata N, Mori T, Iwahori K, Iwahori K, Fujita M. 2000. Microbial decolorization of melanoidin containing wastewaters: Combined use of activated sludge and the fungus *Coriolus hirustus*. *Journal of Bioscience and Bioengineering* 89: 145–150.

Oller I, Malato S, and Sánchez-Pérez JA. 2011. Combination of Advanced Oxidation Processes and biological treatments for wastewater decontamination—A review. *Science of the Total Environment* 409, 4141–4166.

Padoley KV, Mudliar SN, Banerjee SK, Deshmukh SC, and Pandey RA. 2011. Fenton oxidation: A pretreatment option for improved biological treatment of pyridine and 3-cyanopyridine plant wastewater. *Chemical Engineering Journal* 166: 1–9.

Padoley KV, Saharan VK, Mudliar SN, Pandey RA, and Pandit AB. 2012. Cavitationally induced biodegradability enhancement of distillery effluent. *Journal of Hazardous Materials* 219: 69–74.

Padoley KV, Tembhekar PD, Sarat Chandra T, Pandit AB, Pandey RA, and Mudliar SN. 2012. Wet air oxidation as a pretreatment option for selective biodegradability enhancement and biogas generation potential from complex effluent. *Bioresource Technology* 120: 157–164.

Pant D and Adholeya A. 2006. Biological approaches for treatment of distillery wastewater: A review. *Bioresource Technology* 98: 2321–2334.

Pant D and Adholeya A. 2007. Enhanced production of ligninolytic enzymes and decolorization of molasses distillery wastewater by fungi under solid state fermentation. *Biodegradation* 18: 647–659.

Pintar A and Levec J. 1992. Catalytic liquid-phase oxidation of refractory organic in wastewater. *Chemical Engineering Science* 47: 2395–2400.

Plavsic M, Cosovic B, and Lee C. 2006. Copper complexing properties of melanoidins and marine humic material. *Science of the Total Environment* 366: 310–319.

Rizzo L, Selcuk H, Nikolaou AD, Meriç Pagano S, and Belgiorno V. 2014. A comparative evaluation of ozonation and heterogeneous photocatalytic oxidation processes for reuse of secondary treated urban wastewater. *Desalination and Water Treatment* 52: 1414–1421.

Sangave PC and Pandit AB. 2007. Combination of ozonation with conventional aerobic oxidation for distillery wastewater treatment. *Chemosphere* 68: 32–41.

Sangeeta Y, Chandra R, and Vibhuti R. 2010. Effect of biologically treated post methanated distillery effluent on seed germination and growth parameters of *Vicia faba*. *Indian Journal of Environmental Protection* 30: 253–365.

Sangve PC and Pandit AB. 2004. Ultrasound pretreatment for enhanced biodegradability of the distillery wastewater. *Ultrasonics Sonochemistry* 11 197–203.

Sarat Chandra T, Malik SN, and Suvidha G, Padmere ML, Shanmugam P, and Mudliar SN. 2014. Wet air oxidation pretreatment of biomethanated distillery effluent: Mapping pretreatment efficiency in terms color, toxicity reduction and biogas generation. *Bioresource Technology* 158: 135–410.

Scott JP and Ollis DF. 1995. Integration of chemical and biological oxidation processes for water treatment: Review and recommendations. *Environmental Progress* 14, 88–103.

Shanmugam P and Horan NJ. 2009. Optimizing the biogas production from leather fleshing waste by co-digestion with MSW. *Bioresource Technology* 100: 4117–4120.

Shende RV and Mahajani VV. 1997. Kinetics of wet oxidation of formic acid and acetic acid. *Industrial and Engineering Chemistry Research* 36: 4809–4814.

Staehelin J and Hoigne J. 1982. Decomposition of ozone in water. Rate of inhibition by the hydroxide ions and hydrogen peroxide. *Environmental Science and Technology* 16: 676–681.

Suvidha G, Sharma A, Sarat Chandra T, Malik SN, and Mudliar SN. 2015. Effect of ozone pretreatment on biodegradability enhancement and biogas production from biomethanated distillery effluent. Ozone Science and Engineering (In Press).

Tembhekar PD, Padoley KV, Mudliar SL, and Mudliar SN. 2014. Kinetics of wet air oxidation pretreatment and biodegradability enhancement of a complex industrial wastewater. *Journal of Environmental Chemical Engineering* 3: 339–348.

Tewari PK, Batra VS, and Balakrishnan M. 2007. Water management initiatives in sugarcane molasses based distilleries in India. *Res. Cons. Recycl* 52: 351–367.

Tufano VA. 1993. Multi-step kinetic model for phenol oxidation in high-pressure water. *Chem. Eng. Technol* 16: 186–190.

Wu Q, Hu X, and Yue P. 2003. Kinetics study on catalytic wet air oxidation of phenol. *Chemical Engineering Science* 58: 923–928.

Zhang Q and Chuang T. 1999. Lumped kinetic model for catalytic wet oxidation of organic compounds in industrial wastewater. *AIChE J* 45: 145–150.

12

The Role of Microbes in Plastic Degradation

Rajendran Sangeetha Devi, Velu Rajesh Kannan, Krishnan Natarajan, Duraisamy
Nivas, Kanthaiah Kannan, Sekar Chandru, and Arokiaswamy Robert Antony

CONTENTS

12.1 Introduction

Human society benefited from the use of polymers since approximately 1600 BC when
the ancient Mesoamericans first processed natural rubber into balls, figurines, and bands
(Hosler et al., 1999). In 1862, Alexander Parkes unveiled the first man-made plastic in the
Great International Exhibition at London. He dubbed the material as "Parkesine," now
called as celluloid. In 1907, chemist Leo Hendrik Baekland, while striving to produce a
synthetic varnish, stumbled upon the formula for a new synthetic polymer originating
from coal tar. He subsequently named the new substance "Bakelite." By 1909, Baekland
had coined "plastics" as the term to describe this completely new category of materials.
The first patent for polyvinyl chloride (PVC), a substance now used widely in vinyl

siding and water pipes, was registered in 1914. Plastics served as substitutes for wood, glass, and metal during World War I and–II. After the Second World War, newer plastics, such as polyurethane (PU), polyester, silicones, polypropylene (PP), and polycarbonate-joined polymethyl methacrylate and polystyrene (PS) and PVC in widespread applications. Plastics have come to be considered "common"—a symbol of the consumer society (American Chemistry Council, Inc., 2011). The history of manufactured plastics goes back more than 100 years; however, plastics are relatively modern when compared with other materials. Development of modern plastics really expanded in the first 50 years of the twentieth century, with the synthesis of 15 new classes of polymers. Plastics have a range of unique properties: they can be used at a very wide range of temperatures; they are chemical- and light resistant.

Plastics are synthetic or man-made organic polymers, similar in many ways to natural resins found in many trees and plants (Gorman, 1993). According to the American Chemistry Council Plastics Industry Producers' Statistics Group, US resin production increased 1.5% to 107.5 billion pounds in 2013, up from 105.9 billion pounds in 2012. Total sales for the year increased 1.5% to 108.7 billion pounds in 2013, up from 107.0 billion pounds in 2012. The versatility of these materials has led to a great increase in their use over the past three decades, and they have rapidly moved into all aspects of everyday life (Laist, 1987; Hansen, 1990). Plastics are lightweight, strong, durable, and cheap, characteristics that make them suitable for the manufacture of a very wide range of products (Laist, 1987; Pruter, 1987). Since they are also buoyant, an increasing load of plastic debris is being dispersed over long distances, and when they finally settle in sediments they may persist for centuries (Ryan, 1987; Hansen, 1990; Goldberg, 1995, 1997). With continuous growth for more than 50 years, global production in 2012 rose to 288 million tons—2.8% increase compared with 2011. However in Europe, in line with the general economic situation, plastics production decreased by 3% from 2011 to 2012 (Figure 12.1).

Although literally hundreds of plastic materials are commercially available, only a handful of these qualify as commodity thermoplastics in terms of their high volume and relatively low price. Low-density polyethylene (LDPE), high-density PE (HDPE), PP, PVC, PS, and PE terephthalate (PET) account for approximately 90% of the total demand (Figure 12.2). From 2009 to 2010 the global production of plastics increased by 15 million tons (6%) to 265 million tons, confirming the long-term trend of plastics production growth of almost 5% per year over the past 20 years. China remains the leading plastics producer with 23.9%, and the rest of Asia (incl. Japan) accounts for an additional 20.7%. Europe ranks second place in the global plastic production (PlasticsEurope, 2010). In 2010, Europe accounted for 57 million tons (21.5%) of the global production and China overtook Europe as the biggest production region at 23.5%.

12.2 Plastics and Its Types

The use of plastics has become a part in all sectors of the economy. The widespread applications of plastics are not only due to their favorable mechanical and thermal properties but also mainly due to their stability and durability (Rivard et al., 1995). Synthetic polymers are typically prepared by polymerization of monomers derived from oil or gas, and plastics are usually made from these by addition of various chemical additives. There are currently some 20 different groups of plastics, each with numerous grades and

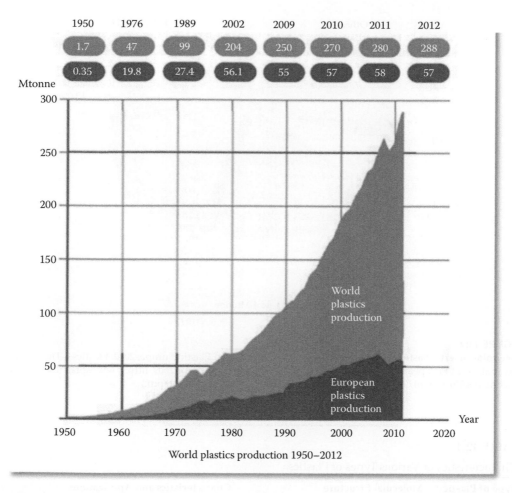

1950	1976	1989	2002	2009	2010	2011	2012
1.7	47	99	204	250	270	280	288
0.35	19.8	27.4	56.1	55	57	58	57

World plastics production 1950–2012

FIGURE 12.1
(**See color insert.**) World plastics production. (Adapted from PlasticsEurope, 2012. Plastics—The Facts 2012. An analysis of European plastics production, demand, and waste data for 2011. http://www.plasticseurope.org/documents/document/20121120170458-final_plasticsthefacts_ nov2012_ en_web_resolution.pdf.)

varieties (APME, 2006). Plastics are made from hydrocarbon monomers. They are produced by chemically modifying natural substances or are synthesized from inorganic and organic raw materials. On the basis of their physical characteristics, plastics are usually divided into thermosets, elastomers, and thermoplastics. These groups differ primarily with regard to molecular structure. The characteristics of the various types of plastics are listed in Table 12.1.

The majority of the plastics are thermoplastic; that is, once the plastic is formed it can be heated and reformed repeatedly. The other group is thermosets, which cannot be remelted. Synthetic plastics are widely used in packing of products such as pharmaceuticals, food, cosmetics, chemicals, and detergents. The most widely used plastics in packing are PE (LDPE, HDPE, MDPE, and LLDPE), PP, PS, PVC, PU, PET, polybutylene terephthalate, nylons, and so on (Table 12.2). In the 1980s, scientists started to explore whether plastics could be designed to become susceptible to microbial attack, making them degradable in a microbial active environment. Biodegradable plastics opened the way for new

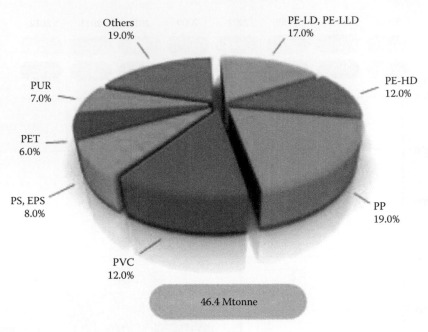

FIGURE 12.2
(See color insert.) Plastics demand by resin type. (Adapted from PlasticsEurope, 2010. Plastics—The Facts 2010. An analysis of European plastics production, demand, and recovery for 2009. http://www.plasticseurope.org/documents/document/20101006091310- final_ plasticsthefacts_28092010_lr.pdf.)

TABLE 12.1

Characteristics of Various Types of Plastics

Type of Plastic	Molecular Structure	Characteristics and Applications
Thermosets		Thermosets are hard and have a very tight-meshed, branched molecular structure. Curing proceeds during shaping, after which it is no longer possible to shape the material by heating. Further shaping may then only be performed by machining. Thermosets are used, for example, to make light switches.
Elastomers		While elastomers also have a cross-linked structure, they have a looser mesh than thermosets, giving rise to a degree of elasticity. Once shaped, elastomers also cannot be reshaped by heating. Elastomers are used, for example, to produce automobile tires.
Thermoplastics		Thermoplastics have a linear or branched molecular structure which determines their strength and thermal behavior; they are flexible at ordinary temperatures. At approx. 120–180°C, thermoplastics become a pasty/liquid mass. The service temperature range for thermoplastics is considerably lower than that for thermosets. The thermoplastics polyethylene, polyvinyl chloride, and polystyrene are used, for example, in packaging applications.

Source: From www.containerhandbook.de.

TABLE 12.2

Commonly Used Synthetic Plastics and Its Applications

Plastic	Symbol	Structure	Applications
Polypropylene terephthalate	01 PET		Soft drink, water and dressing bottles, peanut butter, and jam bars
High-density polyethylene	2 HDPE		Milk, juice and water bottles, trash and retail bags
Polyvinyl chloride	3 PVC		Juice bottles, cling films, raincoats, visors, shoe soles, garden hoses, and electricity pipes
Low-density polyethylene	4 LDPE		Frozen food bags, squeezable bottles, flexible container lids
Polypropylene	5 PP		Bottle caps, drinking straws, medicine bottles, car batteries, disposable syringes
Polystyrene	6 PS		Packing materials, laboratory ware, disposable cups, plates, trays, and cutlery
Others (often polycarbonate)	07 O		Beverage bottles, baby milk bottles, electronic casing

considerations of waste management strategies, since these materials are designed to degrade under environmental conditions or in municipal and industrial biological waste treatment facilities (Augusta et al., 1992; Witt et al., 1997).

Due to similar properties to conventional plastics, biodegradable plastics such as polyhydroxyalkanoates (PHAs), polylactides, polycaprolactone, aliphatic polyesters, and polysaccharides have been developed successfully over the last few years. The most important are poly (3-hydroxybutyrate) and poly(3-hydroxybutyrate-co-3-hydroxyvalerate). Bioplastics obtained from the growth of microorganisms or from plants, which are genetically engineered to produce such polymers, are likely to replace currently used plastics at least in some of the fields (Lee, 1996) (Table 12.3). The key properties of PHA were its biodegradability, apparent biocompatibility, and its manufacture from renewable resources. The global interest in PHAs is high as it is used in different packaging materials, medical devices, disposable personal hygiene, and also agricultural applications as a substitute for

TABLE 12.3

Biodegradable Plastics and Its Applications

Biodegradable Plastics	Structure	Applications
Polyglycolic acid (PGA)		Used for subcutaneous sutures, intracutaneous closures, abdominal and thoracic surgeries
Polyhydroxybutyrate (PHB)		Manufacturing disposable utensils and razors. Also used in medical applications, including sutures, wound dressings, and bone plates and screws. It can also be used for sustained drug delivery.
Polylactic acid (PLA)		Packaging and paper coatings; other possible markets include sustained release systems for pesticides and fertilizers, mulch films, and compost bag
Polycaprolactone (PCL)		Used in housing applications, drugs encapsulation, act as a scaffold for tissue repair via tissue engineering, in root canal filling, etc.
Polyhydroxyalkanoates (PHA)		Used for sutures, surgical mesh, repair patches, slings, cardiovascular patches, orthopedic pins, spinal fusion cages, implant materials, skin substitutes, wound dressings, etc.
Polyhydroxyvalerate (PHV)		Used in paper and film coatings; therapeutic drug delivery of worm medicine for cattle, and sustained release systems for pharmaceutical drugs and insecticides
Polyvinyl alcohol (PVOH)		Packaging and bagging applications which dissolve in water to release. Products such as laundry detergent, pesticides, and hospital washable
Polyvinyl acetate (PVAc)		Adhesives, the packaging applications include boxboard manufacture, paper bags, paper lamination, tube winding, and remoistenable labels

synthetic polymers such as PP, PE, and so on (Ojumu et al., 2004). Plastic shopping bags could be made from polylactic acid (PLA), a biodegradable polymer derived from lactic acid. This is one form of vegetable-based bioplastic. This material biodegrades quickly under composting conditions and does not leave toxic residue. However, bioplastic can have its own environmental impacts, depending on the way it is produced.

12.3 Hazards of Plastics

Each year approximately 140 million tons of synthetic polymers are produced (Shimao, 2001). The synthetic polymers are estimated to be approximately 20% of the municipal solid waste in the United States of America. In Australia, most household waste ends up in municipal landfill sites, estimated to be 25% of total waste by weight (Yayasekara et al., 2005). As per the United States Environmental Protection Agency, in 2011 plastics constituted over 12% of municipal solid waste. In the 1960s, plastics constituted less than 1% of municipal solid waste. The extensive use of plastic usage poses severe environmental threats to terrestrial and marine ecosystem, as they are hardly degradable and voluminously dumped after usage. Due to high production and wide usage of plastics, their disposal is a major problem. Accumulation of plastic products in the environment adversely affects wildlife, wildlife habitat, lands, waterways, and oceans.

Chlorinated plastic can release harmful soil, which can affect groundwater ecosystem. Methane gas, a highly powerful greenhouse gas produced during degradation process significantly causes global warming (Hester and Harrison, 2011). In the ocean, plastic pollution can kill marine mammals directly through entanglement in objects, such as fishing gear, but it can also kill through ingestion, by being mistaken for food. Studies have found that all kinds of species, including small zooplankton, large cetaceans, most seabirds, and all marine turtles, readily ingest plastic bits and trash items such as cigarette lighters, plastic bags, and bottle caps. Sunlight and seawater embrittle plastic, and the eventual breakdown of larger objects makes it available to zooplankton and other small marine animals (Plastic pollution, 2013).

Polyethylene, a form of plastic including shopping bags, disposable bottles and glasses, chewing gums, and toys, is believed to be carcinogenic. Phthalates, present in emulsions, inks, footwear, and toys among other products, is associated with hormonal disturbances, developmental issues, cancer, reduced sperm count, and infertility and weakened immunity (Plastic pollution, 2014). PVC, a form of plastic used in packaging, containers, utility items, and cosmetics has been linked to onset of cancer and birth and genetic conditions. It can also cause bronchitis, skin disease, deafness, and vision problems, and digestion and liver-related problems. Other hazardous poisonous chemicals released during the PVC life cycle, such as mercury, dioxins, and phthalates, may pose irreversible life-long health threats. When produced or burned, PVC plastic releases dioxins, a group of the most potent synthetic chemicals ever tested, which can cause cancer and harm the immune and reproductive systems. Some plastic water bottles contain Bisphenol A, a compound that is, believed to cause cancer, impair the immune system, lead to early puberty and trigger development of obesity and diabetes (Hester and Harrison, 2011).

12.4 Degradation of Plastics

Synthetic polymers are recognized as major solid waste environmental pollutants. Another problem is disposal of plastic wastes. Since 1990, the plastic industry has invested US$1 billion to support increased recycling, and to educate communities. Land filling is the most common method for disposing of municipal solid waste. Increasing amounts

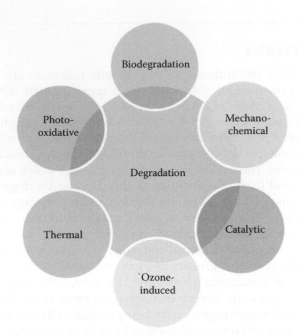

FIGURE 12.3
(See color insert.) Types of plastic degradation.

of synthetic polymers produced results in increasing interest in polymer biodegradation. Environmental factors such as light, heat, moisture, chemical conditions, and biological activity result in bond scission. The formation of structural homogeneities and formation of new functional groups also occurs during polymer degradation (Pospisil and Nespurek, 1997).

Depending upon the nature of the causing agents, polymer degradation has been classified as photo-oxidative degradation, thermal degradation, ozone-induced degradation, mechanochemical degradation, catalytic degradation, and biodegradation (Figure 12.3). The major degradation will either be photodegradation, thermal degradation, or biological degradation. Thermo degradation means the degradation of polymer by heat energy. It generally receives support from oxygen of the atmosphere and is known as thermo-oxidative degradation. The initial stage of degradation is the process of rupture in the bonds of macromolecules resulting in radical sites. Thermal degradation generally involves changes to molecular weight of polymer.

The degradation that is carried out in the presence of light is termed as "photodegradation." It is initiated first by the absorption of light energy by the appropriate group present in the polymer molecule. The light absorption results in the scission of the polymer molecule at an appropriate position of the chain leading to the conversion to smaller fragments. So, photodegradable polymers require either an in-built photo responsive group in the chain or an additive one (Kyrikou and Briassoulis, 2007). Photodegradation of polymer includes UV degradation and oxidation. During UV degradation, the UV light is used to degrade the end product. During the oxidation process, heat is used to break down the plastic. Many synthetic polymers are resistant to chemical and physical degradation. Both the thermal and physical methods of degradation reduce the molecular weight of the plastic and allow it to biodegrade.

12.5 Biodegradation of Plastics

Biodegradation is defined as the process in which organic substances are broken down by living organisms. It is expected to be the major mechanism of loss for most chemicals released into the environment. Biodegradation is the ability of microorganism to influence abiotic degradation through physical, chemical, or enzymatic action (Albertsson et al., 1987). Microorganisms are involved in the degradation and deterioration of both synthetic and natural polymers (Gu et al., 2000). The polymers are not directly utilized by microorganisms where most of the biochemical processes takes place. The degradation of plastics is a very slow process. Initially it is initiated by environmental factors, which include temperature, pH, and UV. In order to use such materials as a carbon and energy source, microorganisms have developed a special strategy.

Microorganisms such as bacteria and fungi are involved in the degradation of plastics. The microorganisms that are responsible for biodegradation differ from each other and have their own optimal growth conditions. Biodegradation is generally considered as consisting of both enzyme-catalyzed hydrolysis and nonenzymatic hydrolysis (Wackett and Hershberger, 2001). During degradation, exoenzymes from microorganisms break down complex polymers yielding short chains or smaller molecules, for example, oligomers, dimers, and monomers, that are smaller enough (water soluble) to pass the semi-permeable outer bacterial membranes and then to be utilized as carbon and energy sources (Gu, 2003). This initial process of polymer breaking down is called depolymerization. The degradative pathways associated with polymer degradation are often determined by the environmental conditions (Figure 12.4). The complete decomposition of a polymer produces organic acids, CO_2, CH_4, and H_2O (Alexander, 1977; Narayan, 1993).

The biological degradation of polymeric substances is a complex process involving several subsequent steps induced by the action of enzymes. The most important type of enzymatic polymer cleavage reaction is hydrolysis (Schink et al., 1992). The bonds such as glycosidic, ester, and peptide linkages are subjected to hydrolysis through nucleophilic attack on the carbonyl carbon atom. Generally, the biological degradation of polymeric

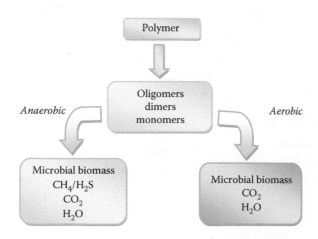

FIGURE 12.4

Polymer degradation under aerobic and anaerobic condition. (Adapted from Gu JD. 2003. *International Biodeterioration and Biodegradation* 52: 69–91.)

substances is influenced by the presence of enzymes and microorganisms, biotic availability of the polymeric structure and appropriate abiotic factors.

12.6 Factors Affecting Plastic Degradation

Biodegradation is governed by different factors that include polymer characteristics, type of organism, and nature of pretreatment. The polymer characteristics such as its mobility, crystallinity, molecular weight, type of functional groups, and substituent present in its structure, and plasticizers or additives added to the polymer all play an important role in its degradation (Gu et al., 2000; Artham and Doble, 2008). Biodegradability of polymers is affected by two main factors such as exposure conditions and characteristic features of polymers (Figure 12.5). Exposure conditions can be further categorized as abiotic and biotic factors. The main chain scission from photodegradation reduces the number of average molecular weight, which provides greater accessibility to the polymer chain by moisture and microorganisms (Albertsson, 1992; Gopferich, 1997; Stevens, 2003; Kijchavengkul and Auras, 2008). These smaller plastic molecules can be more easily hydrolyzed or utilized by microbes. In case of aliphatic aromatic polyesters, photodegradation can result in both main chain scission and cross-linking (Osawa, 1992; Schnabel, 1992). Photodegradation can affect the plastic films in two ways. First, it can cause random main chain scission, either via Norrish I or Norrish II mechanisms (Figure 12.6).

Abiotic factors such as temperature, pH, and moisture can affect the hydrolysis reaction rates during degradation. The increased temperature and moisture leads to increase

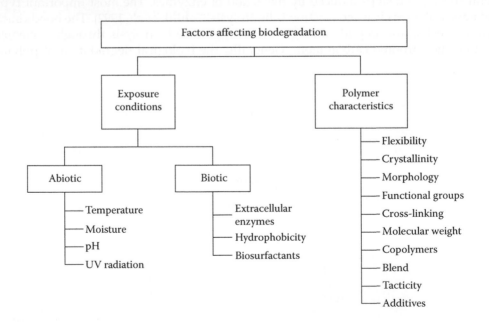

FIGURE 12.5

Factors affecting biodegradation. (Adapted from Kijchavengkul T and Auras R. 2008. *Polymer International* 57(6): 793–804.)

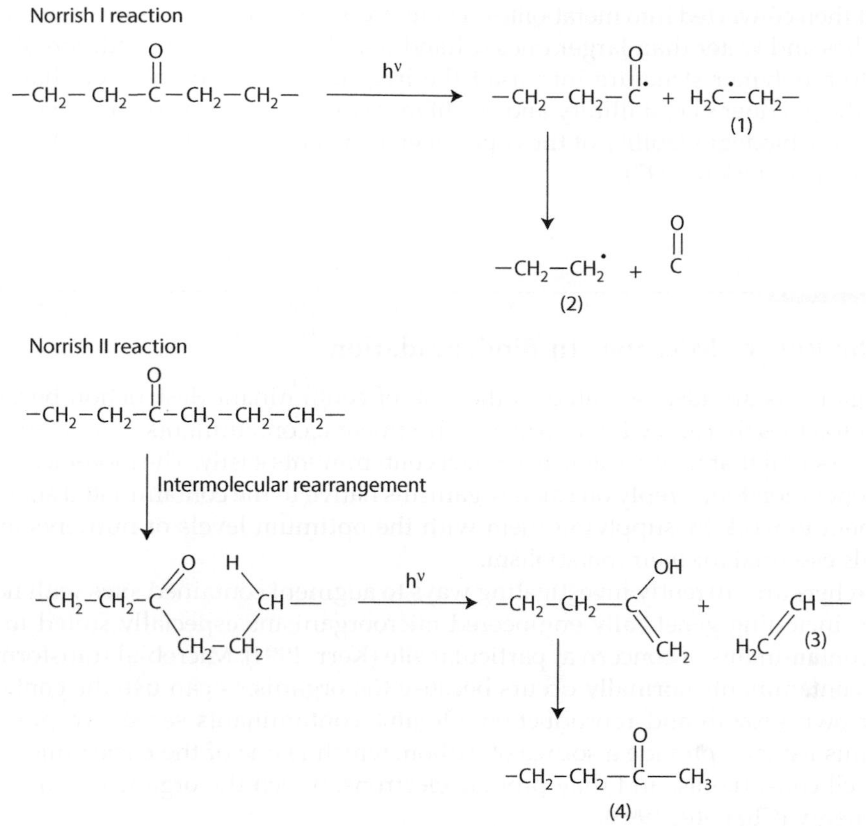

Norrish I reaction

(1)

(2)

Norrish II reaction

Intermolecular rearrangement

(3)

(4)

FIGURE 12.6
Norrish I and II reactions: (1) and (2) free radicals generated from Norrish I, (3) terminal double bond compound, and (4) methyl ketone compound. (Adapted from Kijchavengkul T. 2010. *Design of Biodegradable Aliphatic Aromatic Polyester Films for Agricultural Applications Using Response Surface Methodology.* School of Packaging, Doctoral Dissertation. Michigan State University, East Lansing, MI.)

in hydrolysis reaction rates and microbial activity (Henton et al., 2005). In high-moisture environmental conditions, there is an increase in hydrolysis reaction, which produces more chain scission and leads to an increase in the available sites for microorganisms to attack the polymer chains, causing faster degradation (Ho et al., 1999). Among biotic factors, extracellular enzymes produced by different microorganisms may have active sites with different shapes and hence more able to biodegrade certain polymers. For example, *Aspergillus niger* and *Aspergillus flavus* fungi produce enzymes that more easily digest aliphatic polyesters derived from 6–12-carbon di-acid monomers than those produced from other monomers (Chandra and Rustgi, 1997).

The kinetics of polymer degradation depend on whether the environment is dry air, humid air, soil, a landfill, a composting environment, sewage, freshwater, or a marine environment. Each environment has its own characteristic concentration profile of important factors: oxygen, water, and other chemicals, daylight and degrading microorganisms (Albertsson et al., 1987). Conformation flexibility plays an important role in polymer biodegradation. The more flexibility a polymer has, the more accessible it is for microbes. Microorganisms can digest only low-molecular weight portions, which are taken into the cells and then converted into metabolites. Therefore, smaller molecules are more accessible to microbes and water than larger ones (Chandra and Rustgi, 1997). Addition of comonomers into a polymer structure increased the irregularity of the polymer chains, which reduces the polymer's crystallinity and might increase accessibility to microbes and water. However, the biodegradability of the copolymer also depends on the types of comonomer introduced (Gopferich, 1997).

12.7 The Role of Microbes in Biodegradation

Microorganisms are ideally suited to the task of contaminant destruction because they possess enzymes that allow them to use environmental contaminants as food and because they are so small that they are able to contact contaminants easily. The bioremediation systems in operation today reply on microorganisms native to the contaminated sites, encouraging them to work by supplying them with the optimum levels of nutrients and other chemicals essential for their metabolism.

Researchers are currently investigating ways to augment contained sites with nonnative microbes including genetically engineered microorganisms especially suited to degrading the contaminants of concern at particular site (Kerr, 1994). Microbial transformation of organic contaminants normally occurs because the organisms can use the contaminants for their own growth and reproduction. Organic contaminants serve two purposes for the organisms: they provide a source of carbon, which is one of the basic building blocks of new cell constituents, and they provide electrons, which the organisms can extract to obtain energy (Chapelle, 1993).

Microorganisms gain energy by catalyzing energy-producing chemical reactions that involve breaking chemical bonds and transferring electrons away from the contaminant. This type of chemical reaction is called an oxidation–reduction reaction: the organic contaminant is oxidized, the technical term for losing electrons; correspondingly, the chemical that gains the electrons is reduced. The contaminant is called the electron donor, while the electron recipient is called the electron acceptor (National Research Council, 1993). The energy gained from these electron transfers is then "invested," along with some electrons and carbon from the contaminant, to produce more cells. The process of destroying organic compounds with the aid of O_2 is called aerobic respiration. In aerobic respiration, microbes use O_2 to oxidize part of the carbon in the contaminant to carbon dioxide (CO_2), with the rest of the carbon used to produce new cell mass. In the process the O_2 gets reduced, producing water. Thus, the major by-products of aerobic respiration are carbon dioxide, water, and an increased population of microorganisms (USEPA, 1987).

Many microorganisms can exist without oxygen, using a process called anaerobic respiration. In anaerobic respiration, nitrate (NO^{3-}), sulfate (SO_4^{2-}), metals such as iron (Fe^{3+}) and manganese (Mn^{4+}), or even CO_2 can play the role of oxygen, accepting electrons from the degraded contaminant (National Research Council, 1993). Thus, anaerobic respiration uses inorganic chemicals as electron acceptors. In addition to new cell matter, the by-products of anaerobic respiration may include nitrogen gas (N_2), hydrogen sulfide (H_2S), reduced forms of metals, and methane (CH_4), depending on the electron acceptor. Oxidative degradation is the main mechanism for the degradation of plastics. These mechanisms reduce the molecular weight of the material. The extracellular and intracellular enzymes that are produced by the microbes convert the polymer into monomer, dimer, and oligomer. The by-products produced during conversion enters into the microbial cell can be utilized as the energy source (Shimao, 2001). The different microorganisms that are responsible for the degradation of different groups of plastics are listed (Tables 12.4 through 12.6). A bacterium could constantly synthesis all of the enzymes required for degradation or else could activate enzyme synthesis as necessary to metabolize when needed or is thermodynamically favorable (Albertsson et al., 1987). Balasubramanian et al. (2014) reported that the environmental factors (physical and chemical) play a major role to initiate the HDPE degradation and also support the microorganisms to degrade PE (HDPE).

12.8 The Role of Enzymes in Biodegradation

The use of microbes for bioremediation is beset with many rate-limiting factors. Costly and time-consuming methods may be necessary to produce microbial cultures. Several factors include physical and chemical treatments, toxins, action of predators and high concentration of pollutants may damage or metabolically inactive the microbial cells. Enzymes exist in every living cell and hence in all microbes. Relative amounts of the various enzymes produced by the microorganisms vary with species and even between strains of the same species. Enzymes are very specific in their action on substrates, so the different enzymes help in the degradation of various types of enzymes (Underkofler, 1958). Laccase can help in the oxidation of the hydrocarbon backbone of PE. Gel permeation chromatography (GPC) determine whether cell-free laccase incubated with PE helps in the reduction of average molecular weight and average molecular number of PE by 20% and 15%, respectively (Sivan, 2011). Laccases are mostly present in lignin biodegrading fungi, where they catalyze the oxidation of aromatic compounds. Laccase activity is known to act on nonaromatic substrates (Mayer and Staples, 2002). Lignin and manganese-dependent peroxidases (LiP and MnP, respectively) and laccases are the three main enzymes of ligninolytic system (Hofrichter et al., 2001).

Some strains that are capable of degrading the PE are *Brevibacillus* spp., *Bacillus* spp., where proteases are responsible for degradation (Sivan, 2011). Papain and urease are the two proteolytic enzymes found to degrade medical polyester PU. Polymer degraded by papain was due to the hydrolysis of urethane and urea linkages producing free amine and hydroxyl groups (Phua et al., 1987). Lignin-degrading fungi and manganese peroxidase, partially purified from the strain of *Phanerochaete chrysosporium* also helps in the degradation of high-molecular weight PE under nitrogen and carbon limited conditions (Shimao, 2001). The enzymes responsible for the degradation of various types of plastics depict the substrates that utilize the plastics as carbon and energy sources and helps in biodegradation (Table 12.7). Microbial enzymes induce the rate of biodegradation of plastics very effectively without causing any harm to the environment.

12.9 Characterization of Plastic Degradation

The environmental conditions decide the group of microorganisms and the degradative pathway involved. The high-molecular weight polymers are degraded first into oligomers,

TABLE 12.4

List of Bacteria Used for Plastic Degradation

S.No.	Type of Plastic Used	Microorganisms	References
1.	Polyurethane	*Corynebacterium* and *Pseudomonas* sp.	Kay et al. (1991)
2.	Isotactic polypropylene	*Pseudomonas chlororaphis, Pseudomonas stutzeri,* and *Vibrio* sp.	Cacciari et al. (1993)
3.	Polyurethane	*Pseudomonas cepacia, Pseudomonas* sp., *and Arthrobacter globiformis*	El-Sayed et al. (1996)
4.	LDPE	*Rhodococcus ruber* C208	Chandra and Rustgi (1997)
5.	Polyurethane	*Bacillus* sp.	Blake and Howard (1998)
6.	PVC powder	*Pseudomonas aeruginosa*	Peciulyte (2002)
7.	Degradable polyethylene	*Rhodococcus rhodocorrous* ATCC 29672 and *Nocardia steroids* GK 911	Bonhomme et al. (2003)
8.	Polyethylene bags and plastic cups	*Streptococcus* sp., *Staphylococcus* sp., *Micrococcus* sp., *Moraxella* sp., and *Pseudomonas* sp.	Kathiresan (2003)
9.	LDPE	*Pseudomonas stutzeri*	Sharma and Sharma (2004)
10.	LDPE	*Brevibacillus borstelensis707*	Hadad et al. (2005)
11.	LDPE	*Rhodococcus ruber* C208	Sivan et al. (2006)
12.	Degradable polyethylene	*Bacillus mycoides*	Seneviratne et al. (2006)
13.	HDPE and LDPE	*Bacillus* sp., *Micrococcus* sp., *Listeria* sp., and *Vibrio* sp.	Kumar et al. (2007)
14.	Polyethylene carry bags	*Bacillius cereus, Pseudomonas* sp.	Aswale and Ade (2008)
15.	Polyethylene carry bags and cups	*Bacillus* sp., *Staphylococcus* sp., *Streptococcus* sp., *Diplococcus* sp., *Micrococcus* sp., *Pseudomonas* sp., and *Moraxella* sp.	Reddy (2008)
16.	Polyethylene carry bags	*Serratia marcescens*	Aswale and Ade (2009)
17.	HDPE, LDPE, and LLDPE	*Rhodococcus rhodochorus* ATCC 29672	Fontanella et al. (2009)
18.	Natural and synthetic polyethylene	*Pseudomonas* sp. (P1, P2, and P3)	Nanda et al. (2010)
19.	Polyethylene carry bags	*Serratia marcescens* 724, *Bacillus cereus, Pseudomonas aeruginosa, Streptococcus aureus* (B-324), and *Micrococcus lylae* (B-429)	Aswale (2010)
20.	HDPE	*Arthrobacter* sp. and *Pseudomonas* sp.	Balasubramanian et al. (2010)
21.	LDPE	*Staphylococcus epidermis*	Chatterjee et al. (2010)
22.	Polyethylene bags	*Pseudomonas aeruginosa, Pseudomonas putida,* and *Bacillus subtilis*	Nwachukwu et al. (2010)
23.	LDPE	*Bacillus cereus* C1	Suresh et al. (2011)
24.	LDPE powder	*Pseudomonas* sp., *Staphylococcus* sp., and *Bacillus* sp.	Usha et al. (2011)
25.	Degradable plastic	*Pseudomonas* sp., *Bacillus subtilis, staphylococcus aureus, Streptococcus lactis, Proteus vulgaris,* and *Micrococcus luteus*	Priyanka and Archana (2011)
26.	LDPE and LLDPE	*Baciilus cereus, Bacillus megaterium, Bacillus subtilis,* and *Brevibacillus borstelensis*	Abrusci et al. (2011)
27.	LDPE	*Pseudomonas aeruginosa* PAO1 (ATCC 15729), *Pseudomonas aeruginosa* (ATCC 15692), *Pseudomonas putida* (KT2440 ATCC 47054), and *Pseudomonas syringae* (DC3000 ATCC 10862)	Kyaw et al. (2012)

TABLE 12.5

List of Fungi Used for Plastic Degradation

S.No.	Type of Plastic Used	Microorganisms	References
1.	Degradable plastic	*Phanerochaete chrysosporium*	Lee et al. (1991)
2.	Polyurethane	*Chaetomium globosum* and *Aspergillus terreus*	Boubendir (1993)
3.	Polyurethane	*Curvularia senegalensis, Fusarium solani, Aureobasidium pullulans,* and *Cladosporium* sp.	Crabbe et al. (1994)
4.	Disposable plastic films	*Aspergillus flavus* and *Mucor rouxii* NRRL 1835	El-Shafei et al. (1998)
5.	HDPE	*Phanerochaete chrysosporium* ME-446 and *Trametes versicolor* (IFO 7043, IFO 15413)	Iiyoshi et al. (1998)
6.	PVC	*Poliporusversicolor, Pleurotus sajor caju, Phanerochaete chrysosporium,* ME 446, *Pleurotus ostreatus, Pleurotus sapidus, Pleurotus eryngii, Pleurotus florida*	Kirbas et al. (1999)
7.	Plasticized PVC	*Aureobasidium pullulans, Rhodotorula aurantiaca,* and *Kluyveromyces* spp.	Webb et al. (2000)
8.	Polyethylene	*Penicillium simplicissmum*	Yamada et al. (2001)
9.	PVC powder	*Aureobasidium pullulans, Geotrichum candidum, Alternaria alternaria, Cladosporium* sp., *Aspergillus* sp., *Penicillium* sp., and *Ulocladium atrum*	Peciulyte (2002)
10.	Low density polyethylene (LDPE) powder	*Penicillium pinophilium* and *Aspergillus niger*	Volke-Sepulveda et al. (2002)
11.	Degradable polyethylene	*Cladosporium cladosporides*	Bonhomme et al. (2003)
12.	Polyethylene bags and plastic cups	*Aspergillus niger* and *Aspergillus glaucus*	Kathiresan (2003)
13.	LDPE	*Aspergillus niger, Penicillium funiculosum, Chaetomium globosum, Gliocladium virens,* and *Pullularia pullulans*	Gilan et al. (2004)
14.	Degradable polyethylene	*Penicillium frequentans*	Seneviratne et al. (2006)
15.	Polyethylene carry bags and cups	*Aspergillus niger, Aspergillus ornatus, Aspergillus nidulans, Aspergillus cremeus, Aspergillus flavus, Aspergillus candidus,* and *Aspergillus glaucus*	Reddy (2008)
16.	LDPE	*Fusarium* sp. AF4	Shah et al. (2009)
17.	Polyethylene bags	*Aspergilllus niger*	Nwachukwu et al. (2010)
18.	High density polyethylene (HDPE)	*Aspergillus niger, Aspergillus oryzae,* and *Aspergillus flavus*	Konduri et al. (2010)
19.	Polyethylene carry bags	*Phanerochaete chrysosporium, Pleurotus ostretus, Aspergillus niger,* and *Aspergillus glaucus*	Aswale (2010)
20.	LDPE powder	*Aspergillus nidulans* and *Aspergillus flavus*	Usha et al. (2011)
21.	Polyethylene carry bags	*Aspergillus niger*	Aswale and Ade (2011)
22.	LDPE	*Aspergillus oryzae*	Konduri et al. (2011)
23.	Degradable plastic	*Aspergillus niger, Aspergillus nidulans, Aspergillus flavus, Aspergillus glaucus,* and *Penicillium* sp.	Priyanka and Archana (2011)
24.	LDPE powder	*Aspergillus versicolor* and *Aspergillus* sp.	Pramila and Ramesh (2011)
25.	LDPE	*Aspergillus* sp.	Raaman et al. (2012)

(Continued)

TABLE 12.5 (*Continued*)

List of Fungi Used for Plastic Degradation

S.No.	Type of Plastic Used	Microorganisms	References
26.	Plasticized PVC	*Aspergillus versicolor, Aspergillus niger, Aspergillus flavus, Chrysonilia setophila, and Penicillium* sp.	Sachin and Mishra (2013)
27.	PVC films	*Phanerochaete chrysosporium* PV1, *Lentinus tigrinus* PV2, *Aspergillus niger* PV3, and *Aspergillus sydowii* PV4	Ali et al. (2013)
28.	HDPE	*Aspergillus terreus* MF12	Balasubramaniyan et al. (2014)

TABLE 12.6

List of Actinomycetes Used for Plastic Degradation

S.No.	Type of Plastic Used	Microorganisms	References
1.	Starch polyethylene–prooxidant degradable polyethylene	*Streptomyces badius* 252	Lee et al. (1991)
2.	Starch polyethylene–prooxidant degradable polyethylene	*Streptomyces setonii* 75Vi2 and *Streptomyces viridosporus* T7A	Pometto et al. (1992)
3.	Polyurethane	*Actinetobacter calcoaceticus*	Halim et al. (1996)
4.	Disposable plastic films	*Streptomyces* spp. (8 strains)	El-Shafei et al. (1998)
5.	Polyethylene	*Streptomyces* sp.	Méndez et al. (2007)
6.	LDPE powder	*Streptomyces* KU8, *Streptomyces* KU5, *Streptomyces* KU1, and *Streptomyces* KU6	Usha et al. (2011)
7.	Polyethylene	*Streptomyces* sp., *Pseudonocardia, Actinoplanes, Sporichthya*	Sathya et al. (2012)
8.	Polyurethane	*Actinetobacter gerneri* P7	Howard et al. (2012)

some of which may be water soluble and further broken down into organic intermediates. The intermediate products may be acids, alcohols, ketones, and so on. The biodegradability of the plastic can be characterized using several strategies such as accumulation of biomass, oxygen uptake rate, carbon-dioxide evolution rate, surface changes, and changes in physical and chemical properties of plastics (Albertsson and Huang, 1995).

Several analytical techniques have been used to monitor the extent and nature of degradation. The study of changes in the physical and chemical properties of the plastics before and after degradation will help in understanding the mechanism of degradation. The study of mechanical properties comprises measuring the tensile strength of the plastic materials, elongation at fail and modulus of the polymer by using Instron. The physical properties of the plastic materials monitored are morphology, density, contact angle, viscosity, and so on. The morphology of the films can be monitored using scanning electron microscopy (SEM) and transmission electron microscopy. The molecular weight distribution of the plastic material can be monitored using GPC (Albertsson and Huang, 1995). The melting temperature (T_m) and glass transition temperature (T_g) can be measured using thermo gravimetric analysis and differential scanning calorimetry (DSC). The changes in crystalline and amorphous region were also monitored using x-ray diffraction, small angle x-ray scattering, and wide angle x-ray scattering.

The changes in the chemical properties such as the disappearance or formation of new functional groups could be measured by Fourier transform infrared spectroscopy (FTIR)

TABLE 12.7

Bacterial and Fungal Strains Depicting the Biodegradation of Plastics

Source	Enzymes	Microorganisms	Plastics Act as Substrate	References
Fungi	Manganese peroxidase	*Phanerochaete chrysosporium*	Polyethylene	Shimao (2001)
	Cutinase	*Fusarium*	PCL	Shimao (2001)
	Unknown	*Amycolaptosis* sp.	Polylactic acid (PLA)	Shimao (2001)
	Cutinase	*Aspergillus oryzae*	Polybutylene succinate (PBS)	Maeda and Yamagata (2005)
	Catalase, protease	*Aspergillus niger*	PCL	Tokiwa and Calabia (2009)
	Lipase	*Rhizopus delemar*	PCL	Tokiwa and Calabia (2009)
	Glucosidases	*Aspergillus flavus*	Polycaprolactone (PCL)	Tokiwa and Calabia (2009)
	Lipase	*Penicillium, Rhizopus arrizus*	Polyethylene adipate (PEA), PBS, PCL	Tokiwa and Calabia (2009)
	Urease	*Trichoderma* sp.	Polyurethane	Trevino and Garcia (2011)
	Serine hydrolase	*Pestalotiopsis microspora*	Polyurethane	Russell (2011)
Bacteria	Serine hydrolase	*Pseudomonas stutzeri*	Polyhydroxyalkanoate (PHA)	Shimao (2001)
	Unknown	*Protobacteria*	PHB, PCL, and PBS	Tokiwa and Calabia (2009)
	Unknown	*Streptomyces*	PHB, PCL	Tokiwa and Calabia (2009)
	Unknown	*Firmicutes*	PHB, PCL, and PBS	Tokiwa and Calabia (2009)

(Cacciari et al., 1993). The molecular weight and the molecular weight distribution of the degraded products or intermediates are characterized by techniques such as thin-layer chromatography (TLC), gas chromatography (GC), GC-mass spectrometry, chemilluminescence, matrix-assisted laser desorption/ionization-time of flight (MALDI-TOF), and nuclear magnetic resonance spectroscopy (Cacciari et al., 1993). The metabolic activity of the cells in the culture as well as in the biofilm can be determined by adenosine triphosphate assays, protein analysis, and fluorescein diacetate analysis (Gilan et al., 2004; Koutny et al., 2006).

12.10 Biodegradation of Natural Plastics

Polyhydroxyalkanoates are linear polyesters produced in nature by bacterial fermentation of sugar and lipids. Microorganisms that produce and store PHA under nutrient limited conditions may degrade and metabolize it when the limitation is removed (Williams and Peoples, 1996). Individual polymers are much too large to be transported directly across the bacterial cell wall. Therefore, bacteria must have evolved extracellular hydrolases capable of converting the polymers into corresponding hydroxyl acidmonomers (Gilmore et al., 1990). In general, no harmful intermediates or by-products are generated during PHA degradation. In fact, 3-hydroxybutyrate is found in all higher animals as blood plasma (Lee, 1996). For this reason, PHAs have been considered for medical applications, including long-term controlled drug release, surgical pins, sutures, and bone and blood vessel replacement.

A number of aerobic and anaerobic microorganisms that degrade PHA, particularly bacteria and fungi, have been isolated from various environments (Lee, 1996). *Acidovorax faecilis, Aspergillus fumigatus, Comamonas* sp., *Pseudomonas lemoignei,* and *Variovorax paradoxusare* are among those found in soil, while in activated sludge *Alcaligenes faecalis* and *Pseudomonas* have been isolated. *Comamonas testosterone* has been found in seawater, and *Ilyobacter delafieldii* is present in the anaerobic sludge. Because a microbial environment is required for degradation, PHA is not affected by moisture alone and is indefinitely stable in air (Luzier, 1992). PHAs have attracted industrial attention for use in the production of biodegradable and biocompatible thermoplastics (Takaku et al., 2006). Previous findings have reported a Streptomyces strain, *Streptoverticillium kashmeriense* AF1, capable of degrading PHB and PHBV, and a bacterial strain *Bacillus megaterium* AF3, capable of degrading PHBV, from the soil mixed with active sewage sludge on the basis of producing clear zones of hydrolysis on PHB and PHBV containing mineral salt agar plates (Shah et al., 2007).

Polyhydroxybutyrate polymer has attracted research and commercial interest worldwide, because it can be synthesized from renewable low-cost feedstocks and the polymerizations are operated under mild process conditions with minimal environmental impact. Furthermore, it can be biodegraded in both aerobic and anaerobic environments, without forming any toxic products. Chowdhury (1963) reported for the first time the PHB-degrading microorganisms from *Bacillus, Pseudomonas,* and *Streptomyces* species. Several aerobic and anaerobic PHB-degrading microorganisms have been isolated from soil (Lee, 1996). Sang et al. (2002) also reported various traces, cavities, and grooves as observed on the dented surface of PHBV films demonstrating that the degradation was a concerted effect of a microbial consortium colonizing the film surface, including fungi, bacteria, and actinomycetes. Numerous irregular erosion pits have also been observed by Molitoris et al. (1996) on the surface of PHA by *Comamonas* sp.

12.11 Biodegradation of Synthetic Plastics

The degradation of most synthetic plastics in nature is a very slow process that involves environmental factors, followed by the actions of microorganisms (Albertsson, 1980; Albertsson et al., 1994; Cruz-Pinto et al., 1994). The primary mechanism for the biodegradation of high-molecular weight polymer is the oxidation or hydrolysis by enzyme to create functional groups that improves the hydrophobicity. Physical properties such as crystallinity, orientation, and morphological properties such as surface area, affect the rate of degradation (Huang et al., 1992).

12.11.1 Polyethylene

Polyethylene is one of the synthetic polymers of high hydrophobic level and high-molecular weight. In natural form, it is not biodegradable. Thus, their use in the production of disposal or packing materials causes dangerous environmental problems (Kwpp and Jewell, 1992). To make this biodegradable, it requires the modification of its crystallinity molecular weight and mechanical properties that are responsible for PE resistance toward degradation (Albertsson et al., 1994). Biodegradation of PE is known to occur by two mechanisms: hydro-biodegradation and oxo-biodegradation (Bonhomme et al., 2003). These two mechanisms agree with the modifications due to the two additives, starch and pro-oxidant, used in the synthesis of biodegradable PE. Biodegradation of LDPE film was also reported as 0.2% weight loss in 10 years (Albertsson, 1980). Yamada-Onodera et al. (2001) isolated a strain of fungus *Penicillium simplicissimum* YK to biodegrade PE, without any additives.

UV light acts as an activator, used at the beginning of the degradation process for the activation of an inert material, PE. In a similar study, PE was treated by exposing it to UV light and also treated with nitric acid (Hasan et al., 2007). That pretreated polymer was then applied to microbial treatment using *Fusarium* sp. AF4 in a mineral salt medium containing treated plastic as a sole source of carbon and energy. An increase in the growth of fungus and some structural changes as observed by FTIR were observed in case of treated PE, which indicated the breakdown of polymer chain and the presence of oxidation products of PE (Koutny et al., 2006).

El-Shafei et al. (1998) investigated the ability of fungi and *Streptomyces* strains to attack degradable PE consisting of disposed PE bags containing 6% starch. They isolated eight different strains of *Streptomyces* and two fungi *Mucor rouxii* NRRL 1835 and *As. flavus*. In the process of biodegradation, the PE or paraffin molecules containing carbonyl group first get converted into an alcohol (containing -OH group) by a mono-oxygenase enzyme. The alcohol is then oxidized to an aldehyde (containing -CHO group) by alcohol dehydrogenase enzyme. An aldehyde dehydrogenase converts aldehyde to a fatty acid (containing -COOH group) (Albertsson, 1980). This fatty acid then undergoes β-oxidation pathway inside cells. Absorbance at 1710–1715/cm (corresponding to carbonyl compound), 1640/cm, and 830–880/cm (corresponding to $-C=C-$), which appeared after UV and nitric acid treatment, decreased during cultivation with microbial consortia (Hasan et al., 2007). Typical degradation of PE and formation of bands at 1620–1640/cm and 840–880/cm was also reported by Yamada-Onodera et al. (2001), attributed to oxidation of PE.

12.11.2 Polypropylene

Polypropylene is a thermoplastic polymer used in a wide variety of applications including packaging and labeling, textiles (e.g., ropes, thermal underwear, and carpets), stationery, plastic parts, and reusable containers of various types, laboratory equipment, loudspeakers, automotive components, and polymer banknotes. Most commercial PP is isotactic and has an intermediate level of crystallinity between that of LDPE and HDPE. PP is normally tough and flexible, especially when copolymerized with ethylene (Maier and Calafut, 1998).

Polypropylene is liable to chain degradation from exposure to heat and UV radiation such as that present in sunlight. Oxidation usually occurs in the tertiary carbon atom present in every repeat unit. A free radical is formed and then reacts further with oxygen, followed by chain scission to yield aldehydes and carboxylic acids (Morris, 2005). In external applications, it shows up as a network of fine cracks and crazes that become deeper and more severe with time of exposure. For external applications, UV-absorbing additives must be used. Carbon black also provides some protection from UV attack. The polymer can also be oxidized at high temperatures, a common problem during molding operations. Antioxidants are normally added to prevent polymer degradation. Microbial communities isolated from soil samples mixed with starch have been shown to be capable of degrading PP (Cacciari et al., 1993). Biodegradation of isotactic PP without any treatment is reported with one of the community designated as 3S among the four microbial communities (designated as 1S, 2S, 3S, and 6S) adapted to grow on starch containing PE obtained from enrichment culture. *Pseudomonas chlororaphis*, *Psuedomonas stuzeri*, and *Vibrio* species were identified in the community 3S. It is reported that UV-treated PP sample is more susceptible to degradation (Sameh et al., 2006). Outdoor soil burial tests were done on the samples of PP blend with differential biodegradable additives. DSC analysis of these polymers with different additives after a year showed no change in melting temperature and fraction of crystalline region. It concludes that the biodegradation begins at the amorphous region rather than at the crystalline region (Jakubowicz, 2003).

12.11.3 Polyvinyl Chloride

Polyvinyl chloride is a strong plastic that resists abrasion and chemicals and has low moisture absorption. PVC is used in construction because it is more effective than traditional materials such as copper, iron, or wood in pipe and profile applications. It can be made softer and more flexible by the addition of plasticizers, the most widely used being phthalates. Mostly, PVC is used in buildings for pipes and fittings, electrical wire insulation, floor coverings, and synthetic leather products. It is also used to make shoe soles, rigid pipes, textiles, and garden hoses. There are many studies about thermal and photodegradation of PVC (Owen, 1984; Braun and Bazdadea, 1986) but there are only few reports available on biodegradation of PVC. According to Kirbas et al. (1999), PVC having low-molecular weight can be exposed to biodegradation by the use of white-rot fungi. The fungus *As. fumigatus* effectively degrades plasticized PVC. *Phanerpchaete chrysosporium* was grown on PVC in a mineral salt agar. *Phanerochaete chrysosporium, Lentinus tigrinus, As. niger,* and *Aspergillus sydowii* can effectively degrade PVC (Ali, 2011).

12.11.4 Polystyrene

Polystyrene is a synthetic polymer that contains a repeating group, also considered to be highly stable and less susceptible for biodegradation. It is used in the production of disposable cups, packaging materials, in laboratory ware, and in certain electronic uses. PS is used for its lightweight, stiffness, and excellent thermal insulation. When it is degraded by thermal or chemical means it releases products such as styrene, benzene, toluene, and acrolein. There are very few reports on the biodegradation of PS, but the microbial decomposition of its monomer, styrene, have been reported by few researchers (Tsuchii et al., 1977).

Biodegradation experiments of PS, styrene oligomers, and PS copolymers have been tried using bacteria (Higashimura et al., 1983), fungi (Milstein et al., 1992), mixed culture and enzyme (Nakamiya et al., 1997) under different conditions. Enzymatic biodegradation of PS polymer was tried with hydro-quinone peroxidase enzyme with success (Nakamiya et al., 1997). The enzyme was extracted from lignin decolorizing bacterium *Azotobacter beijerinckii* HM121 and used in a two-phase (aqueous and solvent) system. Biodegradation results revealed that PS was degraded to small water-soluble molecules that could be detected by TLC. It was proposed that enzyme could have assisted in forming radical species that further underwent biodegradation to simpler compounds.

12.11.5 Polyurethane

Polyurethane is commonly utilized as a constituent material in many products including furniture, coating, construction materials, fibers, and paints. Structurally, PUR is the condensation product of polyisocyanate and polyol having intramolecular urethane bonds (carbonate ester bond, –NHCOO–) (Sauders and Frisch, 1964). Numerous published accounts report that polyester PUR is susceptible to biodegradation by bacteria, fungi, and enzymes, whereas polyether PUR has been found to be relatively more resistant to biodegradation. The breakdown products of the biodegradation are derived from polyester segment in PUR when ester bonds are hydrolyzed and cleaved (Nakajima-Kambe et al., 1999). Three types of PUR degradations have been identified in literature: fungal biodegradation, bacterial biodegradation, and degradation by polyurethanase enzymes (Howard et al., 2012). For example, four species of fungi *Curvularia senegalensis, Fusarium solani, Aureobasidium pullulans,* and *Cladosporium* sp. were obtained from soil and found to degrade ester-based PU.

Shah et al. (2007) has reported five bacterial strains that were isolated after soil burial of PUR film for 6 months. Those bacteria were identified as *Bacillus* sp. AF8, *Pseudomonas* sp. AF9, *Micrococcus* sp. AF10, *Arthrobacter* sp. AF11, and *Corynebacterium* sp. AF12. FTIR spectroscopy was used to confirm that the mechanism of PU biodegradation was the hydrolysis of the ester bond in PU. Results obtained by Nakajima-Kambe et al. (1995) and Howard and Blake (1999) have indicated that the PUR biodegradation was due to the hydrolysis of ester bonds. The decrease of the ratio of ester bond over ether bond was also approximately 50%, which agreed with the measured amount of PUR degraded. Pettit and Abbott (1975) reported that the decomposition of urea units by release of ammonia contributes to the degradation of PUR. Pathirana and Seal (1985) reported that some polyester-PUR degrading fungi produce extracellular esterases, proteases, or ureases in the presence of PUR. Esterase activity has been determined in the culture supernatant of *Corynebacterium* sp. (Kay et al., 1993), *Comamonas acidovorans* (Nakajima-Kambe et al., 1997), and in fungus such as *C. senegalensis* (Crabbe et al., 1994).

12.12 Composting

Most biodegradable plastics lend themselves well to composting systems. Composting and soil burial is the preferred disposal environment for most biodegradable plastics. Many cities around the world now compost garden organics, food wastes, cardboard, paper products, mixed municipal solid waste, and sewage sludge (European Bioplastics, 2011). The degradation mechanism of biodegradable plastics in a composting environment is primarily hydrolysis combined with aerobic and anaerobic microbial activity. A recent study by Japanese researchers found that when conventional LDPE film was under bioactive soil for almost 40 years, the surface of the film showed signs of biodegradation with the molecular weight dropping by half the original. The inner part of the sample was almost unchanged with the molecular weight being retained (Ohtaki et al., 1998). Anaerobic digestion, using thermophilic microbes to produce methane and compost, is also gaining support as an alternative to landfills. Methane production may be faster, more efficient, and more predictable in this system and a useful end-product.

Bioplastic products correctly certified as compostable can be diverted to the composting waste stream at the end of their life. Microbes, such as bacteria or fungi, with their enzymes, are able to "digest" the chain structure of compostable polymers as a source of nutrition (Bastioli, 2005). The resulting end-products are water, carbon dioxide, and a little biomass. For most operations, including plastic biodegradation, a combination of sludge and solid waste provides the best operation. Sewage is an excellent environment for biodegradation of biodegradable plastics as there is a preponderance of microbes and high levels of nitrogen and phosphorous. Activated sewage sludge will convert approximately 60% of a biodegradable polymer to carbon dioxide, while the remaining 40% will enter the sludge stream where, under anaerobic digestion, it will be converted to methane. Although the majority of biodegradable plastics are more easily degraded in composting and soil disposal environments than conventional plastics, environmentally degradable polymers could increase the capacity of landfill sites by breaking down in a relatively short time and freeing other materials for degradation, such as food scraps in plastic bags. Biodegradable plastics will not degrade appreciably in a dry landfill, however, unless they contain sensitizers and pro-oxidants (Garcia et al., 1998), which are transition metal catalysts such as manganese stearate or cobalt stearate.

Biodegradable plastics make a contribution to landfill gas production, and in landfills where gas is collected for use as an alternative energy source this can be a positive outcome. However, where gas utilization systems are not in place, the presence of biodegradable plastics will increase greenhouse gas emissions. Anaerobic microbes in the presence of water in the landfill will consume natural products and produce methane, CO_2, and humus. Typical landfill gas contains 50% methane and 45% CO_2, with the balance composed of water and trace compounds (Sullivan, 2011). To compliment landfill gas production, degradable polymers need to be consumed by anaerobic microbes to produce methane at rates comparable to those generated by degradation of natural products (i.e., lignin and cellulose in paper and garden waste). The plastic material must disintegrate during composting such that the residual plastic is not readily distinguishable from other organic materials in the finished compost. A plastic material or product is considered to have demonstrated satisfactory disintegration if after lab-scale composting, no more than 10% of its original dry weight remains after sieving. That is, 90% of the test sample must pass through a 2.0 mm sieve (Degli-Innocenti et al., 1998; Korner et al., 2005).

12.13 Molecular Aspects on Degradation

Molecular methods have application in biodegradation largely because they provide a direct means to detect, discriminate, and quantify species in a sample. Once the biodegradation pathway is well established, the organism could be genetically modified so that the desired enzyme is made to be secreted more. By applying genetic engineering, the time period required for biodegradation can be reduced or making the plastic more susceptible to microorganism to degrade the plastics (Wackett and Hershberger, 2001). Genetic engineering makes it possible to alter the properties of the existing degradative enzymes, to modify regulatory mechanisms, and to assemble within single organism degradative enzymes from phylogenetically distant organisms.

The latter possibility can be attempted by utilizing genetic engineering tools including DNA sequencing, protein sequencing, production of recombinant organism, and so on. Biosynthetic genes *phbA* (for 3-ketothiolase), *phbB* (NADPH-dependent acetoacetyl-CoA reductase), and *phbC* (PHB synthase) from acetyl-CoA have been cloned to produce PHA acid and poly(3-hidroxybutyric acid) (PHB) (Slater et al., 1988). These genes are clustered and are presumably organized in one operon. The genes have been expressed in *Escherichia coli* and in different species of *Pseudomonas*. As DNA sequences of genes that code for metabolic pathways become increasingly available, molecular procedures will continue to gain predominance in biodegradation protocols. mRNA-based methods will allow us to compare environmental expression of individual members of gene families and they may be useful in determining relationships between environmental conditions prevailing in microhabitats and particularly *in situ* activities of native microorganisms. The molecular techniques can provide a means for assessing overall community diversity and a species of particular interest (DNA) and *in situ* microbial activity (RNA) or any particular activity (mRNA). In order to assess the biodegradation potential of an environment, it is necessary to monitor the genetic potential of the environment.

Uncovering new genes with unknown functions provides new clues about the richness of microbial genetic diversity and gives us the pool from which novel biocatalysts (enzymes) can be isolated. New types of biochemical reactions (by enzymes) will not be discovered by DNA sequence analysis alone. Both reaction and gene screening need to occur in parallel. The enzymes involved in plastic degradation should be characterized and the genes responsible for those enzymes should be worked out. Once the genes responsible for the degradation would be known, then the genes would be used to enhance the polythene-degrading capacity (Aswale and Ade, 2008). The degradation study should be carried out and the efficient degrading microbes should be multiplied at large scale to decompose the polythene at commercial level.

12.14 Conclusions

Based on the literature survey, it can be concluded that plastics are very useful in our day-to-day life to meet our desired needs. Due to its good quality its use is increasing day by day and its degradation is becoming a great threat. In the natural environment, different kinds of microorganisms play an important role in various steps involved in the degradation of plastics. Studying the synergism between those microorganisms will give an insight for future efforts toward the biodegradation of these materials. The plastic materials have high-molecular weight and have hydrophobic surfaces making them difficult for the microorganisms to form stable biofilms and degrade them into small molecular oligomers. The screening of soil microorganisms, isolating microorganisms from marine environment and plastic waste dump site could lead to new explored strains, with superior performance. Various plastic-degrading methods are available but the cheapest, eco-friendly acceptable method is degradation using microbes. The microbe releases the extracellular enzymes to degrade the plastic but the detailed characterization of these enzymes is still needed to be carried out. Utilization of molecular techniques to detect specific groups of microorganisms involved in the degradation process will allow a better understanding of the organization of the microbial community involved in the attack of materials. The characterization of efficient plastic-degrading microbes at molecular level is still not available, so research should be focused in the field of genomics and proteomics, which could speed up the degradation.

References

Abrusci C, Pablos JL, Corrales T, Lopez-Marín J, Marín I, and Catalina F. 2011. Biodegradation of photo-degraded mulching films based on polyethylenes and stearates of calcium and iron as pro-oxidant additives. *International Biodeterioration and Biodegradation* 65: 451–459.

Albertsson AC. 1980. The shape of the biodegradation curve for low and high density polyethylenes in prolonged series of experiments. *European Polymer Journal* 16: 623–630.

Albertsson AC. 1992. Biodegradation of polymers. In: Hamid SH, Amin MB and Maadhah AG. (Eds.), *Handbook of Polymer Degradation*. Marcel Dekker, New York, pp. 345–363.

Albertsson AC and Huang SJ. 1995. Degradable Polymers, Recycling and Plastics Waste Management. Marcel Dekker, New York, pp. 205–208.

Albertsson AC, Andersson SO, and Karlsson S. 1987. The mechanism of biodegradation of polyethylene. *Polymer Degradation and Stability* 18: 73–87.

Albertsson AC, Barenstedt C, and Karlsson S. 1994. Abiotic degradation products from enhanced environmentally degradable polyethylene. *Acta Polymerica* 45: 97–103.

Alexander M. 1977. *Introduction to Soil Microbiology*, 2nd edn. Wiley, New York.

Ali MI. 2011. Microbial degradation of polyvinyl chloride plastics, PhD thesis. Quaid-i-Azam University, p. 122. Ali MI, Ahmed S, Robson G, Javed I, Ali N, Atiq N and Hameed A. 2013. Isolation and molecular characterization of polyvinyl chloride (PVC) plastic degrading fungal isolates. *Journal of Basic Microbiology* 54(1): 18–27.

American Chemistry Council. 2011. Life Cycle of a Plastic Product. http://www.americanchemistry.com/s_plastics/doc.asp?CID=1571&DID=5972 (accessed on: March 30, 2011).

APME. 2006. *An Analysis of Plastics Production, Demand and Recovery in Europe*. Association of Plastics Manufacturers, Brussels.

Artham T and Doble M. 2008. Biodegradation of aliphatic and aromatic polycarbonates. *Macromolecular Bioscience* 8(1): 14–24.

Aswale P. 2010. Studies on bio-degradation of polythene. PhD thesis, Dr. Babasaheb Ambedkar Marathwada University, Aurangabad, India.

Aswale P and Ade A. 2008. Assessment of the biodegradation of polythene. *Bioinfolet* 5: 239.

Aswale P and Ade A. 2011. Polythene degradation potential of *Aspergillus niger*. In: Sayed IU. (Ed.), Scholary Articles in Botany, Pune.

Aswale PN and Ade AB. 2009. Effect of pH on biodegradation of polythene by *Serretia marscence*. *The Ecotech* 1: 152–153.

Augusta J, Müller RJ, and Widdecke H. 1992. Biologisch abbaubare Kunststoffe: Testverfahren und Beurteilungskriterien. *Chemie Ingenieur Technik* 64: 410–415.

Balasubramanian V, Natarajan K, Hemambika B, Ramesh, N, Sumathi CS, Kottaimuthu R, and Rajesh Kannan V. 2010. High-density polyethylene (HDPE) degrading potential bacteria from marine ecosystem of Gulf of Mannar, India. *Letters in Applied Microbiology* 51: 205–211.

Balasubramanian V, Natarajan K, Rajesh Kannan V, and Perumal P. 2014. Enhancement of *in vitro* high-density polyethylene (HDPE) degradation by physical, chemical and biological treatments. *Environmental Science and Pollution Research* 21: 12,549–12,562.

Bastioli C. 2005. *Handbook of Biodegradable Polymers*. Smithers Rapra Press, Billingham.

Blake RC and Howard GT. 1998. Adhesion and growth of a *Bacillus* sp. on a polyesterurethane. *International Biodeterioration and Biodegradation* 42: 63–73.

Bonhomme S, Cuer A, Delort AM, Lemaire J, Sancelme M, and Scott C. 2003. Environmental biodegradation of polythene. *Polymer Degradation and Stability* 81: 441–452.

Boubendir A. 1993. Purification and biochemical evaluation of polyurethane degrading enzymes of fungal origin. *Dissertation Abstract International* 53: 4632.

Braun D and Bazdadea E. 1986. In: Nass LI and Heiberger CA. (Eds.), *Encyclopaedia of PVC*. 2nd edn, Vol. 2, Marcel Dekker, New York.

Cacciari I, Quatrini P, Zirletta G, Mincione E, Vinciguerra V, Lupatelli P, and Sermanni GG. 1993. Isotactic polypropylene biodegradation by a microbial community: Physicochemical characterization of metabolites produced. *Applied and Environmental Microbiology* 59(11): 3695–3700.

Chandra R and Rustgi R. 1997. Biodegradation of maleated linear low-density polyethylene and starch blends. *Polymer Degradation and Stability* 56: 185–202.

Chapelle F. 1993. *Ground-Water Microbiology & Geochemistry*. John Wiley & Sons, Inc., New York.

Chatterjee S, Roy B, Roy D, and Banerjee R. 2010. Enzyme-mediated biodegradation of heat treated commercial polyethylene by *Staphylococcal* species. *Polymer Degradation and Stability* 95: 195–200.

Chowdhury AA. 1963. Poly-β-hydroxybuttersaure abbauende Bakterien und exo-enzyme. *Archives in Microbiology* 47: 167–200.

Crabbe JR, Campbell JR, Thompson L, Walz SL, and Schultz WW. 1994. Biodegradation of colloidal ester-based polyurethane by soil fungi. *International Biodeterioration and Biodegradation* 33: 103–113.

Cruz-Pinto JJC, Carvalho MES, and Ferreira JFA. 1994. The kinetics and mechanism of polyethylene photo-oxidation. *Angewandte Makromolekulare Chemie* 216: 113–133.

Degli-Innocenti F, Tosin M, and Bastioli C. 1998. Evaluation of the biodegradation of starch and cellulose under controlled composting conditions. *Journal of Environmental Polymer Degradation* 6(4): 197–202.

El-Sayed AHMM, Mahmoud WM, Davis EM, and Coughlin RW. 1996. Biodegradation of polyurethane coatings by hydrocarbon- degrading bacteria. *International Biodeterioration and Biodegradation* 37: 69–79.

El-Shafei HA, El-Nasser NHA, Kansoh AL, and Ali AM. 1998. Biodegradation of disposable polyethylene by fungi and *Streptomyces* species. *Polymer Degradation and Stability* 62: 361–365.

European Bioplastics. 2011. Bioplastics at a Glance. Avaliable online: http://www.europeanbioplastics.org/index.php?id=182 (Accessed: March 30, 2011)

Fontanella S, Bonhomme S, Koutny M, Husarova L, Brusson JM, Courdavault JP, Pitteri S et al. 2009. Comparison of the biodegradability of various polyethylene films containing pro-oxidant Additives. *Polymer Degradation and Stability* 95: 1011–1021.

Garcia RA and Gho JG. 1998. Degradable/compostable concentrates, process for making degradable/compostable packaging materials and the products thereof. US Patent 5 854 304.

Gilan I, Hadar Y, and Sivan A. 2004. Colonization, biofilm formation and biodegradation of polyethylene by a strain of *Rhodococcus ruber*. *Applied Microbiology and Biotechnology* 65: 97–104.

Gilmore DF, Fuller RC, and Lenz R. 1990. Biodegradation of poly (beta-hydroxyalkanoates). In: Barenberg SA, Brash JL, Narayan R and Redpath AE. (Eds.), *Degradable Materials: Perspectives, Issues and Opportunities*. CRC Press, Bocan Raton, FL.

Goldberg ED. 1995. The health of the oceans—A 1994 update. *Chemical Ecology* 10: 3–8.

Goldberg ED. 1997. Plasticizing the seafloor: An overview. *Environmental Technology* 18: 195–202.

Gopferich A. 1997. Mechanisms of polymer degradation and elimination. In: Domb AJ, Kost J and Wiseman DM. (Eds.), *Handbook of Biodegradable Polymers*. Harwood Academic, Amsterdam, pp. 451–471.

Gorman M. 1993. *Environmental Hazards—Marine Pollution*. ABC-CLIO Inc., Santa Barbara.

Gu JD. 2003. Microbiological deterioration and degradation of synthetic polymeric materials: Recent research advances. *International Biodeterioration and Biodegradation* 52: 69–91.

Gu JD, Ford TE, Mitton DB, and Mitchell R. 2000. Microbial degradation and deterioration of polymeric materials. In: Revie W. (Ed.), *The Uhlig Corrosion Handbook*, 2nd edn. Wiley, New York, pp. 439–460.

Hadad D, Geresh S, and Sivan A. 2005. Biodegradation of polyethylene by the thermophilic bacterium *Brevibacillus borstelensis*. *Journal of Applied Microbiology* 98: 1093–1100.

Hansen J. 1990. Draft position statement on plastic debris in marine environments. *Fisheries* 15: 16–17.

Hasan F, Shah AA, Hameed A, and Ahmed S. 2007. Synergistic effect of photo and chemical treatment on the rate of biodegradation of low density polyethylene by *Fusarium* sp. AF4. *Journal of Applied Polymer Science* 105: 1466–1470.

Henton DE, Gruber P, Lunt J, and Randall J. 2005. Polylactic acid technology. In: Mohanty AK, Misra M and Drzal LT. (Eds.), *Natural Fibers, Biopolymers, and Biocomposites*. CRC Press, Boca Raton, FL, pp. 527.

Hester, RE and Harrison RM. (eds.) 2011. *Marine Pollution and Human Health*. Issues in Environmental Science and Technology. Royal Society of Chemistry, Cambridge, Vol. 33, pp. 84–85.

Higashimura T, Sawamoto M, Hiza T, Karaiwa M, Tsuchii A, and Suzuki T. 1983. Effect of methyl substitution on microbial degradation of linear styrene dimers by two soil bacteria. *Applied and Environmental Microbiology* 46: 386–391.

Ho KLG, Pometto AL, Gadea-Rivas A, Briceno JA, and Rojas A. 1999. Degradation of polylactic acid (PLA) plastic in costa rican soil and iowa university compos rows. *Journal of Environmental Polymer Degradation* 7(4): 173–177.

Hofrichter M, Lundell T, and Hatakka A. 2001. Conversion of milled pine wood by manganese peroxidase from *Phlebia radiata*. *Applied and Environmental Microbiology* 67: 4588–4593.

Hosler D, Burkett SL, and Tarkanian MJ. 1999. Prehistoric polymers: Rubber processing in ancient. *Mesoamerica Science* 284: 1988–1991.

Howard GT and Blake RC. 1999. Growth of *Pseudomonas fluorescens* on a polyester-polyurethane and the purification and characterization of a polyurethanase-protease enzyme. *International Biodeterioration and Biodegradation* 42: 213–220.

Howard GT, Norton WN, and Burks T. 2012. Growth of *Acinetobacter gerneri* P7 on polyurethane and the purification and characterization of a polyurethane enzyme. *Biodegradation* 23: 561–573.

Huang SJ, Roby MS, Macri CA, and Cameron JA. 1992. The effects of structure and morphology on the degradation of polymers with multiple groups. In: Vert M, Feijen J, Albertsson A, Scott G and Chiellini E. (Eds.), *Biodegradable Polymers and Plastic*. The Royal Society of Chemistry, Cambridge, pp. 149–157.

Iiyoshi Y, Tsutsumi Y, and Nishida T. 1998. Polyethylene degradation by lignin-degrading fungi and manganese peroxidase. *Journal of Wood Science* 44: 222–229.

Jakubowicz I. 2003. Evaluation of degradability of biodegradable polyethylene (PE). *Polymer Degradation and Stability* 80: 39–43.

Kathiresan K. 2003. Polythene and plastic degrading microbes from mangrove soil. *Revista de Biología Tropical* 51: 629–633.

Kay MJ, McCabe RW, and Morton LHG. 1993. Chemical and physical changes occurring in polyester polyurethane during biodegradation. *International Biodeterioration and Biodegradation* 31: 209–225.

Kay MJ, Morton LHG, and Prince EL. 1991. Bacterial degradation of polyester polyurethane. *International Biodeterioration Bulletin* 27: 205–222.

Kerr RS. 1994. *Handbook of Bioremediation*. CRC Press, Inc. Lewis Publishers, Boca Raton, FL, pp. 85–96.

Kijchavengkul T. 2010. *Design of Biodegradable Aliphatic Aromatic Polyester Films for Agricultural Applications Using Response Surface Methodology*. School of Packaging, Doctoral Dissertation. Michigan State University, East Lansing, MI.

Kijchavengkul T and Auras R. 2008. Perspective: Compostability of polymers. *Polymer International* 57(6): 793–804.

Kirbas Z, Keskin N, and Guner A. 1999. Biodegradation of polyvinylchloride (PVC) by white rot fungi. *Bulletin of Environmental Contamination and Toxicology* 63: 335–342.

Konduri MKR, Anupam KS, Vivek JS, Kumar RDB, and Narasu ML. 2010. Synergistic effect of chemical and photo treatment on the rate of biodegradation of high density polyethylene by indigenous fungal isolates. *International Journal of Biotechnology and Biochemistry* 6: 157–174.

Konduri MKR, Koteswarareddy G, Kumar DBR, Reddy BV, and Narasu ML. 2011. Effect of prooxidants on biodegradation of polyethylene (LDPE) by indigenous fungal isolate, *Aspergillus oryzae*. *Journal of Applied Polymer Science* 120: 3536–3545.

Korner I, Redemann K, and Stegmann R. 2005. Behavior of biodegradable plastics incomposting facilities. *Waste Management* 25: 409–415.

Koutny M, Sancelme M, Dabin C, Pichon N, Delort A, and Lemaire J. 2006. Acquired biodegradability of polyethylenes containing prooxidant additives. *Polymer Degradation and Stability* 91: 1495–1503.

Kumar S, Hatha AAM, and Christi KS. 2007. Diversity and effectiveness of tropical mangrove soil microflora on the degradation of polythene carry bags. *Revista de Biologia Tropical* 55: 777–786.

Kwpp LR and Jewell WJ. 1992. Biodegradability of modified plastic films in controlled biological environments. *Environmental Technology* 26: 193–198.

Kyaw BM, Champakalakshmi R, Sakharkar MK, Lim CS, and Sakharkar KR. 2012. Biodegradation of low density polythene (LDPE) by *Pseudomonas* species. *Indian Journal of Microbiology* 52(3): 411–419.

Kyrikou J and Briassoulis D. 2007. Biodegradation of agricultural plastic films: A critical review. *Journal of Polymers and the Environment* 15: 125.

Laist DW. 1987. Overview of the biological effects of lost and discarded plastic debris in the marine environment. *Marine Pollution Bulletin* 18: 319–326.

Lee B, Pometto AL, Fratzke A, and Bailey TB. 1991. Biodegradation of degradable plastic polyethylene by *Phanerochaete* and *Streptomyces* species. *Applied and Environmental Microbiology* 57: 678–685.

Lee SY. 1996. Bacterial polyhydroxyalkanoates. *Biotechnology and Bioengineering* 49: 1–14.

Luzier WD. 1992. Materials derived from biomass/biodegradable materials. *Proceedings of the National Academy of Sciences of the United States of America* 89: 839–842.

Maeda H and Yamagata Y. 2005. Purification and characterization of a biodegradable plastic-degrading enzyme from *Aspergillus oryzae*. *Applied Microbiology and Biotechnology* 67: 778–788.

Maier C and Calafut T. 1998. *Polypropylene: The Definitive User's Guide and Databook*. William Andrew Inc., Norwich, New York, p. 14.

Mayer AM and Staples RC. 2002. Laccase new functions for an old enzyme. *Phytochemistry* 60: 561–565.

Méndez CR, Vergaray G, Vilma R, Karina B, and Cárdenas J. 2007. Isolation and characterization of polyethylene-biodegrading mycromycetes. *Revista Peruana de Biología* 13(3): 203–205.

Milstein O, Gersonde R, Huttermann A, Chen MJ, and Meister JJ. 1992. Fungal biodegradation of lignopolystyrene graft copolymers. *Applied and Environmental Microbiology* 58: 3225–3232.

Molitoris HP, Moss ST, de Koning GJM, and Jendrossek D. 1996. Scanning electron microscopy of polyhydroxyalkanoate degradation by bacteria. *Applied Microbiology and Biotechnology* 46: 570–579.

Morris PJT. 2005. *Polymer Pioneers: A Popular History of the Science and Technology of Large Molecules*. Chemical Heritage Foundation, Vol. 5, p. 88.

Nakajima-Kambe T, Onuma F, Akutsu Y, and Nakahara T. 1997. Determination of the polyester polyurethane breakdown products and distribution of the polyurethane degrading enzyme of *Comamonas acidovorans* strain TB-35. *Journal of Fermentation Bioengineering* 83: 454–458.

Nakajima-Kambe, T, Onuma F, Kimpara N, and Nakahara T. 1995. Isolation and characterization of a bacterium which utilizes polyester polyurethane as a sole carbon and energy source. *FEMS Microbiology Letters* 129: 39–42.

Nakajima-Kambe T, Shigeno-Akutsu Y, Nomura N, Onuma F, and Nakahara T. 1999. Microbial degradation of polyurethane, polyester polyurethanes and polyether polyurethanes. *Applied Microbiology and Biotechnology* 51: 134–140.

Nakamiya K, Ooi S, and Kinoshita T. 1997. Non-heme hydroquinone peroxidase from *Azotobacter beijerinckii* Hm121. *Journal of Fermentation and Bioengineering* 84: 14–21.

Narayan R. 1993. Biodegradation of polymeric materials (anthropogenic macromolecules) during composting. In: Hoitink HAJ, Keener HM. (Eds.), *Science and Engineering of Composting: Design, Environmental, Microbiological and Utilization Aspects*. Renaissance Publishers, Washington, OH, pp. 339–362.

National Research Council. 1993. *In Situ Bioremediation When Does it Work?* National Academy Press, Washington, DC, pp. 224.

Nwachukwu S, Obidi O, and Odocha C. 2010. Occurrence and recalcitrance of polyethylene bag waste in Nigerian soils. *African Journal of Biotechnology* 9: 6096–6104.

Ohtaki A, Sato N, and Nakasaki K. 1998. Biodegradation of poly-ε-caprolactone under controlled composting conditions. *Polymer Degradation and Stability* 61: 499–505.

Ojumu TV, Yu J, and Solomon BO. 2004. Production of polyhydroxyalkanoates, a bacterial biodegradable polymer. *African Journal of Biotechnology* 3: 18–24.

Osawa Z. 1992. Photoinduced degradation of polymers. In: Hamid SH, Amin MB, Maadhah AG (Eds.), *Handbook of Polymer Degradation*. Marcel Dekker, New York, pp. 169–217.

Owen ED. 1984. *Degradation and Stabilisation of PVC*. Elsevier Applied Science Publishers, Barking.

Pathirana RA and Seal KJ. 1985. Studies on polyurethane deteriorating fungi. Part 3. Physico-mechanical and weight changes during fungal deterioration. *International Biodeterioration* 21: 41–49.

Peciulyte D. 2002. Microbial colonization and biodeterioration of plasticized polyvinyl chloride plastics. *Ekologija* 4: 7–15.

Pettit D and Abbott SG. 1975. Biodeterioration of footwear. In: Gilbert JR and Lovelock DW. (Eds.), *Microbial Aspects of the Deterioration of Material*. Academic, London, pp. 237–253.

Phua SK, Castillo E, Anderson JM, and Hiltner A. 1987. Biodegradation of polyurethane *in vitro*. *Journal of Biomedical Materials Research* 21: 231–246.

Plastic Pollution. 2013. *Encyclopædia Britannica*. *Online* Retrieved 1 August, 2013.

Plastic Pollution. 2014. *Encyclopaedia Britannica*. *Online* Retrieved 21 May, 2014.

PlasticsEurope, 2010. Plastics—The Facts 2010. An analysis of European plastics production, demand and recovery for 2009. http://www.plasticseurope.org/documents/document/20101006091310- final_ plasticsthefacts_28092010_lr.pdf.

PlasticsEurope, 2012. Plastics—The Facts 2012. An analysis of European plastics production, demand and waste data for 2011. http://www.plasticseurope.org/documents/document/20121120170458-final_plasticsthefacts_nov2012_ en_web_resolution.pdf.)

Pometto AL III, Lee BT, and Johnson KE. 1992. Production of an extracellular polyethylene-degrading enzyme(s) by *Streptomyces* species. *Applied and Environmental Microbiology* 58: 731–733.

Pospisil J and Nespurek S. 1997. Highlights in chemistry and physics of polymer stabilization. *Macromolecular Symposia* 115: 143–163.

Pramila R and Ramesh KV. 2011. Biodegradation of low density polyethylene (LDPE) by fungi isolated from marine water—A SEM analysis. *African Journal of Microbiological Research* 5: 5013–5018.

Priyanka N and Archana T. 2011. Biodegradability of Polythene and Plastic by the Help of Microorganism: A aay for brighter future. *Journal of Environmental and Analytical Toxicology* 1: 111.

Pruter AT. 1987. Sources, quantities and distribution of persistent plastics in the marine environment. *Marine Pollution Bulletin* 18: 305–310.

Raaman N, Rajitha N, Jayshree A, and Jegadeesh R. 2012. Biodegradation of plastic by *Aspergillus* spp. isolated from polythene polluted sites around Chennai. *Journal of Academia and Industrial Research* 1(6): 313–316.

Reddy RM. 2008. Impact of soil composting using municipal solid waste on biodegradation of plastics. *Indian Journal of Biotechnology* 7: 235–239.

Rivard C, Moens L, Roberts K, Brigham J, and Kelley S. 1995. Starch esters as biodegradable plastics: Effects of ester group chain length and degree of substitution on anaerobic biodegradation. *Enzyme and Microbial Technology* 17: 848–852.

Russell JR. 2011. Biodegradation of polyester polyurethane by endophytic fungi. *Applied and Environmental Microbiology* 77: 6076–6084.

Ryan PG. 1987. The incidence and characteristics of plastic particles ingested by seabirds. *Marine Environment Research* 23: 175–206.

Sachin SS and Mishra RL. 2013. Screening and identification of soil fungi with potential of plastic degrading ability. *Indian Journal of Applied Research* 3(11): 34–36.

Sameh AS, Alariqi Kumar AP, Rao BSM, and Singh RP. 2006. Biodegradation of γ-sterilized biomedical polyolefins under composting and fungal culture environments. *Journal of Polmer Degradation and Stability* 91: 1105–1116.

Sang BI, Hori K, Tanji Y, and Unno H. 2002. Fungal contribution to *in situ* biodegradation of poly(3-hydroxybutyrate-co-3-hydroxyvalerate) film in soil. *Applied Microbiology and Biotechnology* 58: 241–247.

Sathya R, Ushadevi T, and Panneerselvam A. 2012. Plastic degrading actinomycetes isolated from mangrove sediments. *International Journal of Current Research* 4(10): 001–003.

Sauders JH and Frisch KC. 1964. *Polyurethanes: Chemistry and Technology, Part II Technology*. Inter-Science Publishers, New York.

Schink B, Brune A, and Schnell S. 1992. Anaerobic degradation of aromatic compounds. In: Winkelmann, G. (Ed.), *Microbial Degradation of Natural Products*. Verlag Chemie, Weinheim, pp. 219–242.

Schnabel W. 1992. *Polymer Degradation: Principles and Practical Applications*. Hanser, New York.

Seneviratne G, Tennakoon NS, Weerasekara MLMAW, and Nandasena KA. 2006. Polyethylene bio-degradation by a developed *Penicillium–Bacillus* biofilm. *Current Science* 90: 20–22.

Shah AA, Hasan F, Akhter J, Hameed A, and Ahmed S. 2007. Degradation of polyurethane by novel bacterial consortium isolated from soil. *Annals of Microbiology* 58: 381–386.

Shah AA, Hasan F, Akhter J, Hameed A, and Ahmed S. 2009. Isolation of *Fusarium* sp. AF4 from sewage sludge, with the ability to adhere the surface of polyethylene. *African Journal of Microbiology Research* 3: 658–663.

Shah AA, Hasan F, Hameed A, and Ahmed S. 2007. Isolation and characterization of poly (3-hydroxy-butyrate-co-3-hydroxyvalerate) degrading bacteria and purification of PHBV depolymer-ase from newly isolated *Bacillus* sp. AF3. *International Biodeterioration and Biodegradation* 60: 109–115.

Sharma A and Sharma A. 2004. Degradation assessment of low density polythene (LDPE) and polythene (PP) by an indigenous isolate of *Pseudomonas stutzeri*. *Journal of Science and Industrial Research* 63: 293–296.

Shimao M. 2001. Biodegradation of plastics. *Current Opinion in Biotechnology* 12: 242.

Sivan A. 2011. New perspectives in plastic biodegradation. *Current Opinion in Biotechnology* 22: 422–426.

Sivan A, Szanto M, and Pavlov V. 2006. Biofilm development of the polyethylene-degrading bacterium *Rhodococcus ruber*. *Applied Microbiology and Biotechnology* 72: 346–352.

Slater SC, Voige WH, and Dennis DE. 1988. Cloning and expression in *Escherichia coli* of the *Alcaligenes eutrophus* H16 poly-β-hydroxybutyrate biosynthesis pathway. *Journal of Bacteriology* 170: 4431–4436.

Stevens ES. 2003. What makes green plastics green? *Biocycle* 24: 24–27.

Sullivan D. 2011. Compostable plastics and organic farming. *BioCycle* 52(3): 25.

Suresh B, Maruthamuthu S, Palanisamy N, Ragunathan R, Pandiyaraj KN, and Muralidharan VS. 2011. Investigation on biodegradability of polyethylene by *Bacillus cereus* strain Ma-Su isolated from compost soil. *International Research Journal of Microbiology* 2: 292–302.

Takaku H, Kimoto A, Kodaira S, Nashimoto M, and Takagi M. 2006. Isolation of a gram-positive poly (3-hydroxybutyrate) (PHB) degrading bacterium from compost, and cloning and characterization of a gene encoding PHB depolymerise of *Bacillus megaterium* N-18-25-9. *FEMS Microbiology Letters* 264: 152–159.

Tokiwa Y and Calabia BP. 2009. Biodegradability of plastics. *International Journal of Molecular Science* 10: 3722–3742.

Trevino AL and Garcia G. 2011. Polyurethane foam as substrate for fungal strains. *Advances in Bioscience and Biotechnology* 2: 52–58.

Tsuchii A, Suzuki T, and Takahara Y. 1977. Microbial degradation of styrene oligomer. *Agricultural and Biological Chemistry* 41: 2417–2421.

Underkofler LA. 1958. Production of microbial enzymes and their applications. *Applied Microbiology* 6: 212–221.

USEPA. 1987. Ground water: Office of Research and Development, Center for Environmental Research Information, Robert S. Kerr Environmental Research Laboratory, EPA/625/6-87/016.

Usha R, Sangeetha T, and Palaniswamy M. 2011. Screening of Polyethylene degrading microorganisms from garbage soil. *Libyan Agricultural Research Center Journal International* 2: 200–204.

Volke-Sepulveda T, Saucedo-Castaneda G, Gutierrez-Rojas M, Manzur A, and Favela-Torres E. 2002. Thermally treated low density polyethylene biodegradation by *Penicillium pinophilum* and *Aspergillus niger*. *Journal of Applied Polymer Science* 83: 305–314.

Wackett LP and Hershberger CD. 2001. *Biocatalysis and Biodegradation; Microbial Transformation of Organic Compounds*. ASM Press, Washington.

Webb JS, Nixon M, Eastwood IM, Greenhalgh M, Robson GD, and Handley PS. 2000. Fungal colonization and biodeterioration of plasticized polyvinyl chloride. *Applied and Environmental Microbiology* 66(8): 3194–3200.

Williams SF and Peoples OP. 1996. Biodegradable plastics from plants. *Chemtech* 38: 38–44.

Witt U, Muller RJ, and Deckwer WD. 1997. Biodegradation behavior and material properties of aliphatic/aromatic polyesters of commercial importance. *Journal of Environmental Polymer Degradation* l5: 81–89.

Yamada-Onodera K, Mukumoto H, Katsuyaya Y, Saiganji A, and Tani Y. 2001. Degradation of polyethylene by a fungus *Penicillium simplicissimum* YK. *Polymer Degradation and Stability* 72: 323–327.

Yayasekara R, Harding I, Bowater I, and Lonergan G. 2005. Biodegradability of selected range of polymers and polymer blends and standard methods for assessment of biodegradation. *Journal of Polymer Environment* 13: 231.

13

Biodegradation of Chemical Pollutants of Tannery Wastewater

Arumugam Gnanamani and Varadharajan Kavitha

CONTENTS

13.1 Introduction

Industrialization is the basic need for sustained economic growth for developing countries. Among the industries, leather industry occupies a place of prominence in the Indian economy in view of its massive potential for employment, growth, and exports. The recent developments in science and technology place this industry at the top-most level with a tremendous increase in production. Approximately 23 billion square feet of leather is produced worldwide per year and the total value is estimated as > US$70 billion (Ramasamy et al., 1999; Anonymous, 2011). The major leather production centers in the world are located in Mexico, Brazil, Japan, South Korea, China, India, and Pakistan. Korea, Japan, and Italy import hides from countries that have a large meat production industry, that is, the United States, Australia, and the European countries, whereas the South American countries, for example, Argentina and Brazil, process their own hides. Leather industry is prevalent in Brazil, especially in the state Rio Grande do Sul, where, more than 50% of Brazilian leather is produced (Basegio et al., 2002).

Leather processing has been considered as an important economic activity and remains as traditional, often not optimized for the chemical and water usage and thus facing serious challenges (Roš and Ganter, 1998). Since environment and development are inseparable such as two sides of a coin, the stringent environmental regulations mildly shook the leather industry during 1980s. But, intensive research initiatives for the past two decades on process modifications and wastewater treatment systems have repositioned the leather industry. However, the industry is being strongly challenged by the peoples' forum for its ability to comply with environmental standards.

Tanning processes, that is, leather-manufacturing processes involve series of steps with large input of chemicals and water. Scheme 13.1 details the descriptions on processes, percentage of input chemicals, wastes generated, and the pollution load for better understanding.

During processing, only 30%–40% of chemicals are taken up and the remaining unspent chemicals are released into the wastewater along with other wastes generated from each sequential step. According to Ahmed Basha et al. (2009), approximately 30–35 m^3 wastewater is disposed into the environment during the processing of every 1 ton of rawhide by world leather industry.

As described, a variety of chemicals, which includes, lime, sodium chloride, sodium carbonate, ammonium chloride, sulfuric acid, tannins, and dyes, in addition to a large volume of water are used for processing leather. Hence, the groundwater sources are exploited to their fullest potential and get polluted to a greater extent (Anbalagan et al., 1997). Followed by processing, the wastewater (effluent) is stored in large lagoons and the percolation of dissolved solids, mainly the NaCl, affects the surrounding soil exhaustively (Mandal et al., 2005), and its presence in ground water (Mondal et al., 2005), makes the water unsuitable for drinking, irrigation, and consumption by animals. The continuous discharge of effluent by the tanneries present on the banks of the Palar River, contaminate the river to the extent of it being unsuitabile for drinking or agricultural purposes (Sundar et al., 2010).

Because of high contamination of tannery effluent in land, river, and groundwater, the health of the lives in these areas are highly questionable. Both direct and indirect exposures cause severe illness and disability. Even relatively minor exposures occurring frequently, can eventually build up to toxic levels. Vapors from degreasing and finishing solvents are an obvious source of exposure. An important health risk factor for the tannery workers is occupational exposure to chromium, which is used as a basic tanning pigment. The workers on exposure to leather dust, which contains chromium in the protein-bound form, exhibited a higher mean concentration of urinary and blood chromium than the reference values as reported by Rastogi et al. (2008). Thus, a severe challenge has been faced by the manufacturers.

13.2 Characteristics of Tannery Wastewater

In general, tannery wastewater is a dark brown in color, basic with high content of organic substances that vary according to the chemicals used (Kongjao et al., 2008). Wastewater from tanneries usually contains high concentrations of chlorides, aliphatic sulfonates, sulfates, aromatic and aliphatic ethoxylates, sulfonated polyphenols, acrylic acid condensates, fatty acids, dyes, proteins, soluble carbohydrates, and Na$_2$S. These substances are either derived from hides and skins or obtained upon addition of reagents during the processes (Ahmed Basha et al., 2009). The pH of the combined stream of wastewater varies from 7.0 to 11.0 depending on the stages of operations and the product range as designed by the

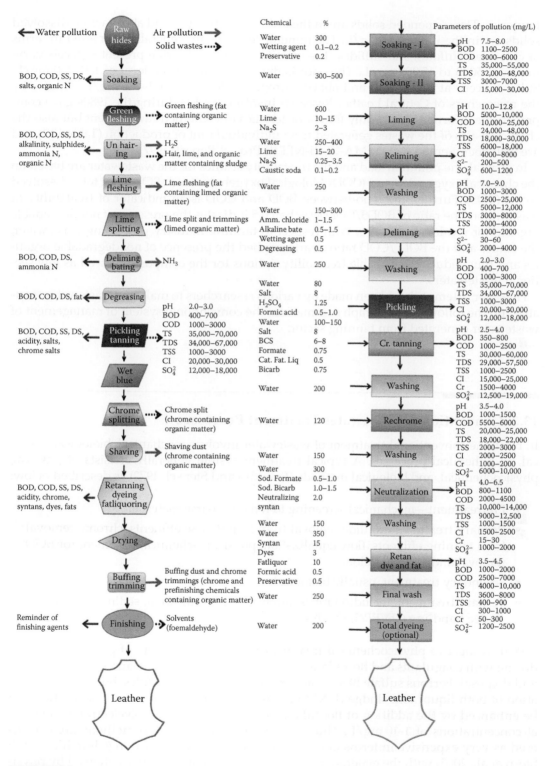

SCHEME 13.1

Schematic view on the processes, chemicals, and pollutants of leather processing.

tanneries. The suspended solids are in the order of 2000 mg/L and above while dissolved solids are in the range 10–20 g/L. The mineral contents are generally expressed in terms of chlorides, sulfates, and sulfides. The salinity is due to chloride present at levels varying from 5000 to 15,000 mg/L. The odor is due to sulfides at the level of 100–250 mg/L. Sulfates occur at 2000 mg/L and above, nitrogen at levels of 200–400 mg/L. Studies by the researchers of Central Leather Research Institute (CLRI) during the 1980s gave comprehensive information not only for characteristics of combined raw effluent but also the characteristics of the wastes generated from individual unit of production (Table 13.1) and are well documented (AISHTMA-CLRI-NEERI Project Report, 1997).

In general, the pollution load and the treatability studies for the wastewater are based on the chemical oxygen demand (COD), biological oxygen demand (BOD), and total dissolved solids (TDS). Further, the ratio between BOD and COD is an indicator of treatability of wastewater. The ratio of BOD5/COD less than 0.5 indicates the presence of nondegradable organics or recalcitrant agents. With respect to tanning industry processing wastewater, the values and the BOD_5/COD ratio clearly implied the presence of nondegradable organics and thus reduce the possible treatability options for the composite as well as few sectional wastewaters.

However, attempts have been made by various researchers to make the wastewater treatable and the following paragraph summarizes the conventional system of management of wastewater generated from tanning sector.

13.3 Conventional Wastewater Treatment Processes

In brief, the conventional treatment of wastewater involves mechanical, biological, physical, and chemical processes. A typical treatment system generally consists of physical, physicochemical, and biological methods (Franklin and Stensel, 1972) as described below:

- Pretreatment—mechanical screening to remove coarse material.
- Primary treatment—sulfide removal from beamhouse effluents, chrome removal from tanning effluents, flow equalization, physical—chemical treatment for BOD removal and neutralization.
- Secondary treatment, usually biological.
- Tertiary treatment, including nitrification and denitrification, sedimentation, and sludge handling (UNEP/IEO, 1994).

With respect to physicochemical treatment methods, it consists of pretreatment and dosing with coagulants and flocculants, followed by sedimentation and sludge handling and disposal. Ferrous sulfate has the advantage of removing sulfides, but imparts coloration of both liquor and sludge (UNEPIE/PAC, 1994). The efficiency of coagulation can be enhanced by the addition of flocculants, long-chain anionic polyelectrolyte, usually at concentrations of 1–10 mg/L. Thus, the physical and chemical methods are considered as very expensive in terms of energy and reagents consumption (Churchley, 1994; Stern et al., 2003) with the generation of excessive sludge (Chu, 2001). Followed by physicochemcial methods (Song et al., 2004), units of biological treatment (Vijayaraghavan and Murthy, 1997), oxidation processes (Sekaran et al., 1996; Lofrano et al., 2007), and

TABLE 13.1

Characteristics of Combined and Sectional Waste from Tanneries in India

Units of Production	Volume of Wastewater (cum)	Parameters of Pollution mg/L									BOD5/ COD Ratio
		pH	SS	TDS	COD	BOD	Cl⁻	SO₄⁻	S⁻	Cr	
Soaking	6–9	7.5–8.0	3000–7000	32,000–48,000	3000–6000	1100–2500	15,000–30,000	800–1500	—	—	0.366–0.416
Liming	3–5	10.0–12.8	6000–18,000	18,000–30,000	10,000–30,000	5000–10,000	4000–8000	600–1200	200–500	—	0.5–0.33
Deliming	1.5–2	7.0–9.0	2000–4000	3000–8000	2500–7000	1000–3000	1000–2000	2000–4000	30–60	—	0.4–0.42
Pickling	0.5–1	2.0–3.0	1000–3000	34,000–67,000	1000–3000	400–700	20,000–30,000	12,000–18,000	—	—	0.42–0.33
Tanning with chrome	1–2	2.5–4.0	1000–2500	29,000–57,500	1000–2500	350–800	15,000–25,000	12,500–19,000	—	1500–4000	0.32–0.35
Fatliquoring and dyeing	3–6	3.5–4.5	400–900	3600–8000	2500–7000	1000–2000	300–1000	1200–2500	—	50–300	0.4–0.28
Combined wastewater (flow composited)	30–40	7.0–9.0	2000–5000	9000–18,000	2500–8000	1200–3000	6000–9500	1600–2500	30–150	120–200	0.48–0.375

Source: Adapted from AISHTMA-CLRI-NEERI. 1997. Report on implementation of cleaner technologies in a cluster of tanneries at Tamilnadu in processing one ton of rawhides.

activated sludge or aerated lagoon systems (Rajamani et al., 1995; Jawahar et al., 1998; Eckenfelder, 2002; Tare et al., 2003; Dantas Neto et al., 2004) exist. The findings of these studies finally suggested that various parameters, viz., mean cell residence time, mixed liquor volatile suspended solids concentration, hydraulic detention time, aeration time, food to microorganisms (F:M) ratio, and the dissolved oxygen of the reactor decides the overall function of the treatment process. Further, the combination of UASB with an aerobic posttreatment enhanced the performance of the overall wastewater treatment process and COD removal efficiency. A major disadvantage, however, is the highly reduced performance during winter, high salinity of tannery wastewater, and presence of recalcitrant molecules (Ramasami and Rao, 1991). Generally, these conventional treatments are unable to reduce all of the polluting parameters such as COD, chlorides, sulfates, and ammonia (Cassano et al., 1997, 2001). Vijayaraghavan et al. (1999) suggested that treatment of high-strength organic waste can normally be achieved only by anaerobic processes, which are time-consuming and subject to environmental stress. Aerobic processes although faster are not economical for high-strength wastes because the aeration cost is prohibitive. According to Balakrishnan et al. (2002), anaerobic lagoons significantly reduces the pollutants level of wastewater of tanning sector, however, the significant air pollution, occupation of land and installation of mechanical surface aerators, corrosion of metallic bodies due to the generation of sulfides reduces the treatment feasibility at considerable level.

Recently, Farabegoli et al. (2004) suggested the effectiveness of sequencing batch reactor (SRB) on treatment of composite tannery effluent. The presence of high-tolerance bacteria was attributed for the performance of SRB. Ganesh et al. (2006) studied the treatability of tannery wastewater using SRB and reported that SRB is an effective tool to reduce COD of tannery effluent. But, still, the treated water could not reach the discharge limit.

13.4 Treatability Studies on Individual Streams: Biodegradation of Individual Components

From the above-said descriptions it has been clearly understood that treatment of tannery effluent is specific toward the composition of chemicals present in wastewater. Since most of the chemicals are not degradable and recalcitrant in nature, the treatability of composite effluent is challenging. Reports suggested that the degraded product of biodegradable chemicals may inhibit the degradation of other chemicals and advise treatability of individual sectional wastewater. Even in sectional streams, numbers of chemical inputs are found as nondegradable. For example, (i) in soaking stream, presence of wetting agent, preservatives and sodium chloride, (ii) in liming wastewater, the pH, lime, and sulfide, (iii) in tanning wastewater, chromium, tannins, and syntans, and (iv) in finishing wastewater, dyes, fat liquors, and preservatives, restrict the treatability at considerable level. Hence, researchers initiated the degradation studies on individual chemicals and explored the potential feasibilities to achieve the discharge norms of pollution control board authorities. The following paragraph summarizes the biodegradation studies explored on individual leather chemicals. The information provided emphasizes the possibilities on biodegradation of tannery effluent, however, it depends with organisms and environmental conditions.

13.4.1 Biodegradation Studies: Soaking Wastewater

13.4.1.1 Characteristics of Soak Wastewater

The wastewater generated from soaking operations is called as soak liquor; contains, in addition to salt, soluble proteins, dirt, dung, blood particles, hydrolysable fats, and preservatives. The pH of the soak liquor is in the range of 7.66 ± 0.4. The major pollution parameters measured as $BOD_5 2080 \pm 480$ mg/L, COD 3660 ± 400 mg/L, total solids 33000 ± 6000 mg/L, TDSs 29460 ± 8460 mg/L, and suspended solids 3540 ± 560 mg/L, chlorides as $Cl^- 15540 \pm 3280$ mg/L, and sulfate as $SO_4^{2-} 760 \pm 210$ mg/L. The BOD to COD ratio indicates (0.36–0.56) the possibilities for biodegradation of organics. But the presence of high salt and TDS are the major problem with the terrestrial microbes and thus infers that it is impractical for the treatability of soak liquor using microbes. However, some marine microorganisms with high salt tolerance can destruct the organics in the presence of salt. Ingram (1940) and Kincannon and Gaudy (1966) observed that the high salt content caused more severe effects on microorganisms, reduce respiration of BOD oxidation rate, and also reduce the sedimentation efficiencies due to high buoyancy forces. Lawton and Eggert (1957) observed lower BOD removal efficiencies in biological trickling filters at salt concentrations exceeding 2%. Stewart et al. (1962) observed temporary reduction in BOD of saline wastewater using bacterial suspension or biofilm cultures. Wilson et al. (1988) studied the role of RBC unit in the treatment of saline water with high TDS. Kapoor et al. (1999) and Kumar et al. (2005) suggested the necessity of adapted microbial cultures for the reduction of TDS in tannery wastewaters. Kargi and Dincer (1996a,b) reported an improved COD removal rate by the use of extreme halophile halobacter-activated sludge. Further, Lefebvre et al. (2005, 2006) observed more than 78% COD removal in soak liquor using Upflow Anaerobic Sludge Blanket (UASB). Recently, Sivaprakasam et al. (2008) studied the degradation of organics in saline water using mixed microbial consortia, which consists of *Pseudomonas aeruginosa*, *Bacillus flexus*, *Exiguobacterium homiense*, and *Staphylococcus aureus* and reported that about 80% COD reduction was achieved with 8% salinity. According to the literatures, it has been understood that (i) the adaptability of terrestrial microbes for high salt concentration requires extensive time period, (ii) cell disruption at high salinity reduces the degradability, (iii) rapid exposure to high salt concentration disturbs the native microbes to that extend that they could not rejuvenate and thus, the virginity of the soil and the quality of ground water have been affected. Only very meager reports were available on the role of marine high salt tolerance microbes, on treatability of saline water and thus suggest the use of potential salt tolerant microbes may solve most of the problems associated with the treatment of soak liquor. Further, it has been recommended that removal of salt before biological treatment could also be beneficial for the organics present in the soak liquor. Sanchez-Gonzalez et al. (2011) isolated a new facultative alkaliphilic *B. flexus* strain from maize processing wastewater, which were able to destruct the organics at high-saline conditions. Other than bacterial species some algal species showed appreciable reduction in BOD and COD in saline water. In heterotrophic bacteria, the release of nutrients in the form of ammonia, utilized by the algae, and also directly contribute to the reduction in the organic load of the saline wastewater by the uptake of organic compounds (Abeliovich and Weismann, 1978). Some microalgae have the ability to utilize high-saline waters in arid or semiarid zones, which are unsuitable for conventional agriculture. Halotolerance algae are very attractive candidates due to their ability to grow in saline water (Ben-Amotz et al., 1982). Laubscher (1991) reported the growth of *Dunaliella salina* in high-saline wastewater of tanning sector and suggested the suitability of this organism for the treatment of saline water. Oren (2010) reviewed the role of halophilic microorganisms for the treatment of hypersaline wastewater and reported that microorganisms belong to the family Halobacteriaceae, *Salinibacter ruber* unicellular green flagellate green algae *D. salina,* and Cyanobacteria shown growth in saline water. Halobacterium requires at least 15% salt for growth and lyses at lower concentrations. Even *Bacillus* spp. and *Halomonas* spp. showed growth at high-salt concentration (Hinteregger and Streichsbier, 1997; Kubo et al., 2001). The response to increased salinity involves some distinct physiological changes in the algal species. Gilmour et al. (1985) stated that there is an increased flux of ions across the thylakoid membrane when algae are subjected to salt stress. All the described literatures suggested the use of high salt tolerant microbes as a mixed consortium able to destruct the organics within the stipulated period of time, which considerably reduces the cost involved in the management or treatment of soak liquor and recovery of salt for reuse.

13.4.2 Biodegradation of Liming and Deliming Wastewater

In liming and deliming wastewater, the major pollution parameters identified are pH, BOD, COD. TS, TDS, and SS and the respective values are 11.0 ± 1.8, 8000 ± 140, $19,400 \pm 1200$, $36,000 \pm 1400$, $28,000 \pm 1300$, 8500 ± 400 mg/L. The sulfide content is measured as 350 ± 40 mg/L. Further, the obnoxious odor of composite tannery effluent is due to the mixing of liming process wastewater. The value of BOD/COD implies the unfeasibility of biodegradation of organics present in the liming wastewater. Further, the high alkaline pH destroys the native microbes and the tolerance toward the high alkaline pH and the sulfide content could not be expected from the normal terrestrial microbes. Alkaliphiles can tolerate two unit of pH above neutrality. However, *Bacillus* species, the major workhorse of industrial microorganisms (Gundala et al. 2013) and one of the species *B. flexus* displayed survival under extreme pH conditions. The presence of highly branched unsaturated fatty acids may increase the membrane fluidity and this imparts resistant at extreme conditions (Rothschild and Mancinelli, 2001). Most of the facultative alkaliphiles showed survival at pH > 12.0. Studies on the biodegradation of lime wastewater has been initiated in the recent past, however, more attention has been given to the management of fleshing a solid waste generated after liming.

It was estimated that for every ton of hides or skins processed into leather, 200–300 kg of limed fleshing was generated and the chemical analyses revealed that it is unsafe to dispose the fleshing and its treatability along with the municipal solid wastes. The volatile solids concentration was in the range of 0.65 ± 0.04 g/g dry weight, COD was in the range of 0.82 ± 0.06 g/g dry weight, total Kjeldahl nitrogen 0.09 ± 0.04 g/g, total carbon content was 35.12 ± 6.12 and $9.07 \pm 0.85\%$. Ravindranath (2012) studied the liquefaction of limed fleshing using microbial population of sewage sludge, anaerobic sludge, and anaerobically treated tannery effluent followed by biomethanation using UASB Reactor and reported the feasibility of biological treatment of limed fleshing effectively.

With respect to sulfide, the total sulfide in wastewater varyed from 10 to 5000 mg/L. Removal of sulfide is a high-energy process. Partial oxidation of sulfide to sulfur by microbes is a suitable alternative and the insoluble sulfur is recovered easily (Midha and Dey, 2008). Aerobic microorganism or chemotrophes such as *Thiobacillus, Pseudomonas, Beggiatoa,* and *Thiothrix* have been reported for the conversion of H_2S to sulfur (Subletta, 1987; Lizama and Sankey, 1993; Oh et al., 1998; Krishnakumar et al., 2005). Sublette and Sylvester (1987) studied the use of *Thiobacillus denitrificans* in the oxidation of sulfide to sulfate. The role of mixed culture of *Thiobacilli* on the aerobic oxidation of sulfide to elemental sulfur was studied by Buisman et al. (1990). Immobilized *Thiobacillus* species showed 95%

removal efficiency of hydrogen sulfide. With regard to anaerobic microorganism or photo-trophs such as *Chlorobium* and *Chromatium* species were used for conversion. Kleinjan et al. (2003) suggested that only those microorganisms that can store sulfur extracellularly for easy separation of sulfur should be used for sulfide removal.

13.4.3 Biodegradation of Chromium

The spent chrome liquor from tannery is one of the potential sources of pollution. Chromium is known to be highly toxic to the aquatic organisms in the hexavalent state and somewhat less toxic in the trivalent form. Hexavalent chromium is carcinogenic, even at less concentration; 10 mg/L can cause nausea, vomiting, skin irritation, and problems related to respiratory tract. It can also cause lung carcinoma due to chromium toxicity.

With respect to tanning sector, according to Ross et al. (1981), of the total chromium used in the processing of leather, 40% is retained in the sludge, disposal of which onto land and into water bodies has led to increased chromium levels reaching as high as 30,000 mg/kg. Biological reduction of chromium occurs under aerobic and anaerobic conditions. Microbial species such as *Euglena gracilis, P. aeruginosa, Enterobacter cloacae, Pseudomonas fluroscens,* and so on are involved in chromium bioreduction under aerobic conditions and some species reduce under anaerobic conditions also. Shen et al. (1996) reported reduction of Cr(VI) during the anaerobic degradation of benzoate. Gnanamani et al. (2010) reported that marine *Bacillus* sp. MTCC 5514 demonstrates hexavalent chromium reduction and also tolerance to trivalent chromium, which is attributed to the biosurfactant of the *Bacillus* strain as shown in Scheme 13.2. In brief, the initial chromium (VI) reduction by extracellular chromium reductase and the entanglement of reduced Cr(III) by the micelles of biosurfactant prevent the direct exposure of cells to chromium and thus impart tolerance to the bacterial cells.

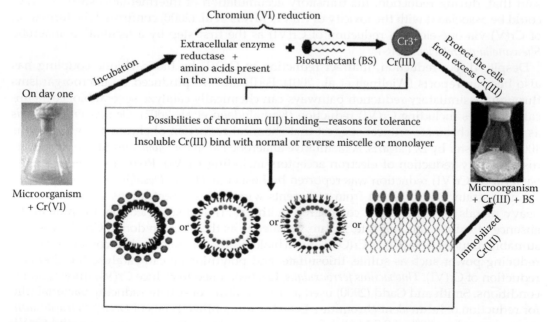

SCHEME 13.2
Chromium (VI) reduction and Cr(III) tolerance by MTCC 5514 mediated by biosurfactant.

The addition of organic sources to the soil or water may also have influence on the reduction in Cr(VI) (Losi et al., 1994; Cifuentes et al., 1996).

The role of algae on chromium reduction has also been in reports (Guha et al., 2001; Rehman and Shakoori, 2001; Sheng et al., 2005; El-Sikaily et al., 2007). With reference to anaerobic reduction of Cr(VI), organic contaminants and aromatic compounds act as suitable electron donors. Few genetically engineered microorganisms (Gonzalez, 2006) and a variety of heterotrophic microorganisms are involved in the reduction of chromium (Satarupa and Paul, 2013). Most of the soil microorganisms have the tendency to reduce Cr(VI) if the proper environmental conditions are available (Camargo et al., 2003). The major mechanistic pathway identified for chromium reduction by bacterial species is enzymatic pathway (Sharma, 2002; Hawley et al., 2004). It has been quite interesting that some species are resistant to Cr(VI), but they are not reducers and most of the metal-tolerant bacteria can reduce Cr(VI) only anaerobically. The plasmids of most of the *Pseudomonas* spp. are determinants for chromium reduction. Avudainayagam (2002) reported Gram-positive bacterial isolate of tannery effluent capable of reducing Cr(VI) as a terminal electron acceptor and with relatively high level of resistance. In nutrient-rich medium Gram-positive bacterium (ATCC 700729) reduces more than 87% of Cr(VI). Reduction of Cr(VI) by *Agrobacterium radiobacter* EPS-916, *Escherichia coli* ATCC 33456, *Aerococcus* sp., *Micrococcus* sp., *Aeromonas* sp. under both aerobic and anaerobic conditions are in reports. *Pseudomonas putida* PRS 2000, *Pseudomonas fluorescens* LB 303, and *Es. coli* AC80 aerobically reduce Cr(VI) to Cr(III). Even a wide range of both aliphatic and aromatic organic pollutants acts as electron donors for chromium reduction by the cocultures of *Es. coli* ATCC 33456 and *P. putida* DMP-1. The metabolites of one organism served as electron donors for Cr(VI) by *E. coli*. Other than chromate reductase enzyme, NADH, NADPH, or H2 act as electron donors with possible involvement of cytochrome *b*, *c*, and *d*. Membrane vesicles *of En. cloacae*, reduced with NADH and then exposed to Cr(VI), oxidized c and b cytochromes and reduced Cr(VI) by membrane vesicles (Chen and Hao, 1998). It has been said that, during reduction, the transitory accumulation of intermediates such as Cr(V) could be associated with the toxicity of Cr(VI). Myers et al. (2000) confirmed the formation of Cr(V) via one-electron reduction of Cr(VI) as the first step by a facultative anaerobe *Shewanella putrefaciens* MR-1.

Despite direct reduction, indirect reduction of Cr(VI) by biotic–abiotic coupling has also been in reports (Wielinga et al., 2001). Fe(II) and S^{2-} produced by microorganisms through dissimilatory reduction pathways can chemically catalyze several biogeochemical processes including Cr(VI) reduction (Kamaludeen et al., 2003). *Desulfovibrio vulgaris* as a model chromate reducer suggest that chemical reduction of chromate by Fe(II) was 100 times faster by *D. vulgaris*. A facultative anaerobe *Pantoea agglomerans* SP1 demonstrate dissimilatory reduction of electron acceptors including Cr(VI). Role of sulfate-reducing bacteria on Cr(VI) reduction was reported by Losi et al. (1994). *Desulfotomaculum reducens* sp. nov. strain MI-1, isolated from sediments with high concentrations of Cr and other heavy metals by enrichment, could grow with Cr(VI) as a sole electron acceptor in the absence of sulfate and butyrate, lactate, or valerate as the electron donor. Cr(VI) was presumably reduced to Cr(III) as $Cr(OH)_3$. Even biologically generated sulfur compounds with reducing power such as sulfite, thiosulfate, and polythionate can catalyze the chemical reduction of Cr(VI). *Thiobacillus ferrooxidans*, has been used to reduce Cr(VI) under aerobic conditions. Smith and Gadd (2000) used a mixed culture of sulfate-reducing bacterial film for reduction of hexavalent chromium. Even the immobilized cells of *Desulfovibrio desulfuricans* can reduce 80% of 0.5 M Cr(VI) with lactate or H_2 as the electron donor and Cr(VI) as the electron acceptor and the insoluble Cr(III) accumulated on the surface or interior of the gel. Rajwade and Paknikar (2003) developed an efficient chromate-reduction process using a strain of *Pseudomonas mendocina* MCM B-180 for treatment of chromate-containing wastewater. The bacterial strain used was resistant to 1600 mg Cr(VI)/L and reduced 2 mM chromate (100 mg Cr(VI)/L) in 24 h. Srivastava and Thakur (2006) identified five morphologically different fungi from leather tanning effluent in which *Aspergillus* sp. and *Hirsutella* sp. had higher potential (more than 70%) to remove chromium within 3 days.

13.4.4 Biodegradation of Fat Liquors

Fatliquoring is a process that imparts and influences the physical characteristics of leather, mainly the wetting properties, waterproofness, tensile strength, extensibility, and permeability to water/air through fat liquors in the form of oil or emulsions. Fat liquors of natural, synthetic, and semisynthetic nature are in use in leather industries. The composition of oil form is mainly the esters of fatty acids and the composition of emulsion is mainly composed of amphiphilic molecules. The amount of fat liquor used was about 10%–20% based on wet blue leather and the exhaustion was calculated as 85%–90% and 10%–15% of the unused fat liquor discharged along with the wastewater. With regard to biodegradation of fat liquors, Luo et al. (2010, 2011) reported that compared to mineral oil, natural oil degraded by native microorganisms and the difference in biodegradability is closely related to the content of double bonds and hydroxyl groups. Salam et al. (2012) found that the metabolic intermediates of biodegradation of canola oil did not affect the growth of the organisms. Lefebvre et al. (1998) observed that the formation of foam could be one of the reasons for the low biodegradability of lipids and reported that saponification of lipids enhances the microbial growth. Aggelis et al. (2001) reported that compared to anaerobic treatment, aerobic treatment was more efficient and in addition the authors suggested the combined anaerobic and aerobic treatment for better degradation.

In tanning sector, use of both natural and synthetic fat liquors is unavoidable. Kalyanaraman et al. (2012, 2013) studied the treatability of natural and synthetic fat liquors. The results of their study prove that the BOD/COD ratio of natural fat liquor calculated as 0.44 suggested some degradable nature of the fat liquors, and the ratio for synthetic liquor calculated as 0.077 indicates the nonbiodegradable nature of the synthetic fat liquor. The major component of synthetic fat liquor is chlorinated paraffin and accounts for about 10% for the fat liquor and after exhaustion more than 2% of chlorinated paraffin is released into the wastewater. It is well known that chlorinated paraffin is grouped under persistent organic pollutants and is toxic to both terrestrial and aquatic lives. Luo et al. (2011) suggested more double bonds or hydroxyl groups are essential for the biodegradation of lipids. And thus synthetic fat liquors require pretreatments such as advanced oxidation process before being subjected to biological treatment.

13.4.5 Biodegradation Studies on Tannins

According to Spencer et al. (1988), tannins are defined as naturally occurring water-soluble polyphenols of varying molecular weight, which differ from most other natural phenolic compounds in their ability to precipitate proteins from solutions. The presence of numbers of phenolic hydroxyl groups makes them easily complex with protein than cellulose and pectins (McLeod, 1974; Mueller Harvey and McAllan, 1992). Field and Lettinga (1992) reported the growth inhibitory effect of tannins and the recalcitrant behavior elaborately. Between the two types of tannins, the condensed ones are more resistant to the hydrolysable ones and thus the routine biodegradation through enzymes could not be expected

(William et al., 1986; Scalbert, 1991). Deschamps (1989) reported the resistance of few microorganisms toward tannins. Microbial degradation of tannins has been reviewed by a number of researchers (William et al., 1986; Deschamps, 1989; Field and Lettinga, 1992b; Saxena et al., 1995; Archambault et al., 1996; Hatamoto et al., 1996; Selinger et al., 1996; Lane et al., 1997; Lekha and Lonsane, 1997). Degradation of gallotannins by an aerobic bacterium *Achromobacter* sp. was reported first time by Lewis and Starkey (1969). During 1980, Deschamps et al. studied the degradation of tannic acid by 15 different bacterial strains belonging to the genera *Bacillus, Staphylococcus,* and *Klebsiella* by enrichment technique. Wattle tannin-degrading bacteria were identified as *Enterobacter aerogenes, Enterobacter agglomerans, Cellulomonas,* and *Staphylococcus,* whereas quebracho tannin-degrading bacteria were *Cellulomonas, Arthrobacter, Bacillus, Micrococcus, Corynebacterium,* and *Pseudomonas.* According to Field and Lettinga (1992) the eventual transformation of phenolic intermediates to cleaved products such as aliphatic acids, hydrogen, carbon dioxide, and methane requires a syntrophic association between the phenolic degraders and methanogenic bacteria.

The role of tannase enzyme mediated degradation of tannins was reported by Deschamps et al. (1983). Tannase production by the strains of *Bacillus pumilus, Bacillus polymyxia,* and *Klebsiella planticola* using chestnut bark as the sole source of carbon was also reported by Deschamps and Lebeault (1984). Tannase production by different *Aspergillus* strain and other fungal strain was reported by Sherief et al. (2011). These authors studied the role of nitrogen, carbon, pH, and minerals on tannase production by *Aspergillus* sp. Among the microorganism, *Aspergillus niger* (Knudson, 1913; Haslam and Stangroom, 1966; Barthomeuf et al., 1994; Lekha and Lonsane, 1994; Bajpai and Patil, 1996; Bradoo et al., 1996), *Aspergillus oryzae* (Iibuchi et al., 1967; Doi et al., 1973; Beverini and Metche, 1990; Bajpai and Patil, 1996; Bradoo et al., 1996), *Aspergillus japonicus* (Ganga et al., 1977; Bradoo et al., 1996), and *Aspergillus awamori* (Bradoo et al., 1996) were reported as the best producer of tannase.

Vijayaraghavan and Ramanujam (1999) studied anaerobic degradation of condensed tannin in the presence of chlorides and reported that the toxicity is due to tannin on anaerobic contact filter at 1180 mg/L. Gupta and Haslam (1989) reported that concentration of tannins of the order of 0.3–2 g/L strongly inhibited the formation of methanogens, and hydrolysable tannin concentration above 914 mg/L inhibited the performance of the treatment (Field et al., 1998). The possible degradative pathway for condensed tannins and hydrolysable tannins as explored by Bhat et al. (1998) is shown in Schemes 13.3 and 13.4.

With regard to degradation of tannin by fungus, tannic acid degradation by *A. niger* was first evidenced by Knudon (1913). Degradation of gallotannins (Nishira, 1961) by the species of *Aspergillus, Fusarium, Penicilum, Sporotrichum, Cylindrocarpon,* and *Trichoderma* was in reports. Growth of *Aspergillus, Penicillium, Fomes, Polyporus,* and *Trametes* were found better on gallotannins than condensed tannins (Lewis and Starkey, 1969). Mahadevan and Muthukumar (1980) reported the tannin degradative efficacy of species of *Chaetomium, Fusarium, Rhizoctonia, Cylindrocarpon,* and *Trichoderma.* Rajkumar and Nandy (1983) observed the growth of molds such as *Aspergillus* or *Penicillium* on the surface of tan liquors in tanneries and during 1985 these authors (Suseela and Nandy, 1985) studied the effect of temperature, pH, and carbon source on the degradation of tannic acid and gallic acid by *Penicillium chrysogenum.* Detannification of oak leaves by the fungus *Sporotrichum pulverulentum* was reported by Makkar et al. (1994). Lorusso et al. (1996) reported reduction in canola meal by an enzyme preparation of a white-rot fungus *Trametes versicolor.* Presence of other metabolites influences tannin degradation (Ganga et al., 1977) by fungus.

With regard to yeast, Aoki et al. (1976a,b) reported the tannin degrading enzymatic system of *Candida.* Further, condensed tannin degradation by *Candida guilliermondii, Candida tropicalis,* and *Torulopsis candida* were reported by Otuk and Deschamps (1983) and Vennat et al. (1986). Reduction in tannin content of pine and gaboon wood bark extracts by yeasts was reported by Otuk and Deschamp (1983). Gamble et al. (1996) reported the condensed tannin degradation potential of *Ceriporiopsis subvermispora* and *Cyathus stercoreus*. Between the years 1969 and 2001, the following fungal species were reported as condensed tannin degraders, viz., species of *Aspergillus A. flavus, A. fumigatus, A. niger, A. teneus, Calvatia gigantean, Chaetomium cupreum, Susarium* sp., *Penicillum* sp., *Rhizoctania batadicola, Trichoderma* sp.

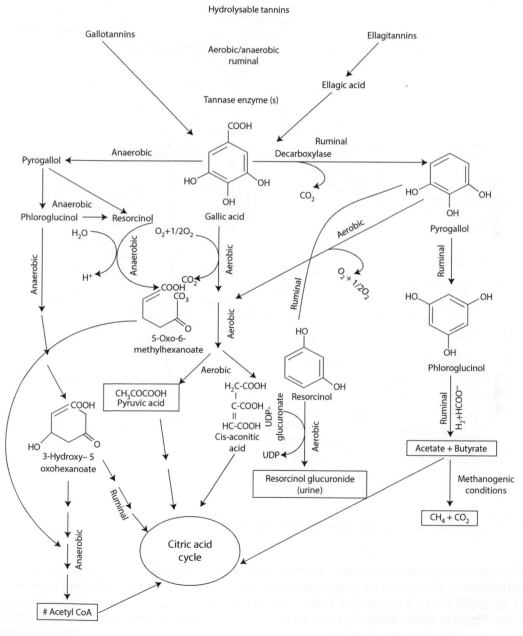

SCHEME 13.3

Possible microbial degradative pathway for hydrolysable tannins. (Adapted from Bhat TK, Singh B, and Sharma OP. 1998. *Biodegradation* 9(5): 343–357.)

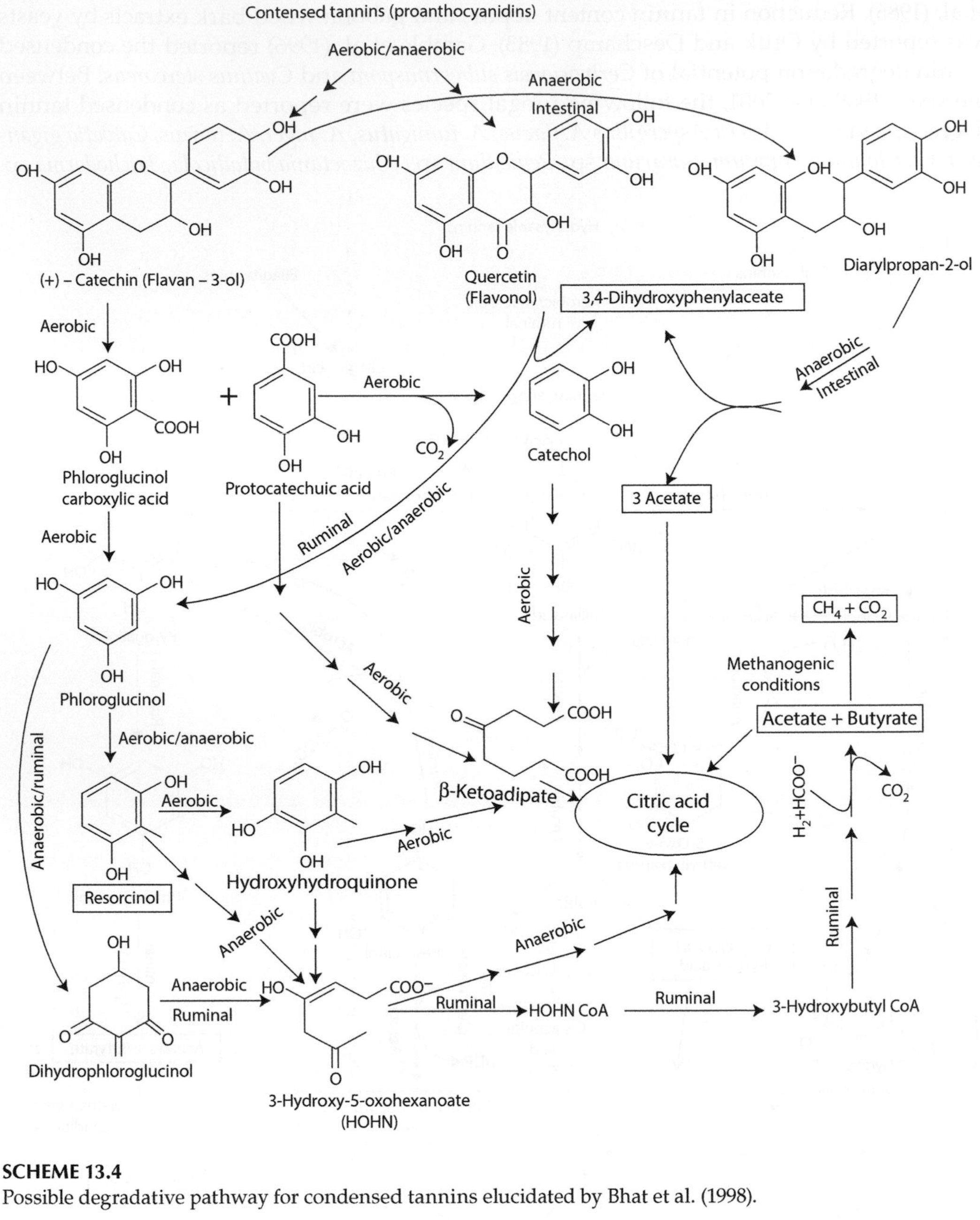

SCHEME 13.4
Possible degradative pathway for condensed tannins elucidated by Bhat et al. (1998).

(*T. viridae*), *Streptomyces* sp., and *Hymensocyphys ericae* of ericoid endophyte produce extra-cellular polyphenol oxidase to degrade gallotannins (Bending and Read, 1996). Gnanamani et al. (2001) studied removal of tannins by peroxidase system of *Phanerochaete chrysosporium*. These authors stated that peroxidases, if assumed to be a stronger oxidizer of tannin, complete the removal of tannin after cleavage of ether linkage, and therefore COD removal should also follow the same pattern of tannin removal. However, the reversed order of removal of COD, that is, high-percentage removal of tannin in condensed tannin over tannic acid, was observed by them and suggested that peroxidase activity on tannin degradation might be inhibited by certain experimental factors after the breakage of the ether bond. Condensed tannin has a molecular structure with delocalized Pi orbital system, which can generate radical cation intermediates and thus, the degradation of condensed tannin was affected by peroxidase system of Phanerochaete *chrysosporium.* Cruz-Hernández et al. (2005) studied the degradation efficacy of 11 fungal strains and found *A. niger* showed the highest tannin-degrading capacity with respect to hydrolysable tannins and condensed tannins and *Penicillium commune* degraded the condensed tannin catechin.

With regard to bacterial species, it is well known that tannins are toxic and bacteriostatic compounds making nonreversible reactions with proteins (Scalbert, 1991). Nevertheless, some bacteria may degrade many phenolic compounds including the catechin-degradation products such as catechol and protocatechuic acid (Deschamps, 1989; Field and Lettinga, 1992; Paller et al., 1995). Three species of *Pseudomonas* utilize catechin within 12 days (Lewis and Starkey, 1968, 1969). Deschamps et al. (1980) isolated catechin degrading *Bacillus, Staphylococcus,* and *Klebsiella* by enrichment culture technique. Bacteria, which degrade wattle tannin, were identified as *En. aerogenes, En. agglomerans, Cellulomonas,* and *Staphylococcus.* Species of *Rhizobium* such as *Rhizobium japonicum, R. leguminosarum, R. phaseoli, R. trifolii,* and so on utilized catechin as sole carbon source (Muthukumar et al., 1982, Gajendiran and Mahadevan, 1988). *Pseuodomonas solanacearum* utilized spectrum of phenolic compounds (Arunakumari and Mahadevan, 1984). Bacterial species such as *Acinetobacter, Alcaligenes, Aeromonas,* and *Pseudomonas,* isolated from tannery effluent utilized 25 mg/100 mL of tannin concentration (Lakshman Perumalsamy, 1993). Arunachalam et al. (2003) identified degradation of catechin by *Acinetobacter calcoaceticus.* Jones et al. (1994) identified *Prevotella ruminicola,* an anaerobic organism having special mechanism to degrade tannins in anaerobic conditions. Few bacterial species generates EPS to protect them from tannins and induction of EPS appears to be a bacterial defense mechanism that permits the bacterium to utilize tannins (Brooker et al., 1994). Gram-positive cocci, isolated from the feces of koalas and identified as *Streptococcus bovis* biotype I, formed a distinct clear zone on tannin-treated brain heart infusion agar, which suggested the anaerobic degradation of tannin by this bacterium (Osawa, 1990; Osawa et al., 1993). Substantial degradation and disappearance of labeled condensed tannins from *Desmodium intortum* in the gastrointestinal tract of sheep and goats was reported by Perez-Maldonado and Norton (1996). *Citrobactor freundii* isolated from tannery effluent show growth at concentrations as high as 50 g/L of gallotannins. *Bacillus pumilus, B. polymyxa, Corynebacterium* sp., and *Klebsiella pneumonia* are reported to degrade 10 g/L gallotannins. The extracellular enzyme tannase is involved in the process of degradation by these species (Saxena et al., 1995).

13.4.6 Degradation of Dyes

Leather industry consumes an extensive range of colored substances that can be applied as dyes and pigments. The major constituents of the dyes are azo, anthraquinone, phthalacyanines derivatives, in which, the azo dyes usage is immense because of the stability and wide variety of colors. Pollution control board authorities restricted the direct discharge of dyeing wastewater to the sewage stream due to aesthetic issues, toxicity, and safety. Decolorization or removal of dyes is necessary before release into the natural streams. Physical and chemical treatment methods such as precipitation, coagulation, adsorption, flocculation, flotation, electrochemical destruction, mineralization, and decolorization processes (Gogate and Pandit, 2004) are in general followed for the removal of color, which have disadvantages such as cost, time, and release of residues. All these techniques minimize the toxicity level but do not neutralize the toxicity (Cooper, 1993; Maier et al., 2004). To alternate these techniques, microorganism can be used to completely degrade the azo dyes (Verma and Madamwar, 2003; Moosvi et al., 2005; Pandey et al., 2007; Khalid et al., 2008), because microorganisms reduce the azo dyes by secreting enzymes such as laccase, azo reductase, peroxidase, and hydrogenase. The reduced forms of azo dyes are further mineralized into simpler compounds and are utilized as their energy source (Stolz, 2001). Based on the available literature, the microbial decolorization of azo dyes is more effective under combined aerobic and anaerobic conditions (Chang and Lin, 2000; Isik and Sponza, 2004; Van der Zee and Villaverde, 2005; Lodato et al., 2007). A wide range of microorganisms is capable of degrading a variety of azo dyes including bacteria, actinomycetes, fungi, and yeast (Gingell and Walker, 1971; Paszczynski et al., 1991; Martins et al., 1999; Kirby et al., 2000; Wesenberg et al., 2003; Olukanni et al., 2006). They have developed enzyme systems for the decolorization and mineralization of azo dyes under certain environmental conditions (Anjali et al., 2006).

The bacterial reduction of the azo dye is usually nonspecific and bacterial decolorization is normally faster. A wide range of aerobic and anaerobic bacteria such as *Bacillus subtilis*, *Pseudomonas* sp., *Es. coli*, *Rhabdobacter* sp., *Enterococcus* sp., *Staphylococcus* sp., *Xenophilus* sp., *Corneybaterium* sp., *Clostridium* sp., *Micrococcus dermacoccus*, *Acinetobacter* sp., *Geobacillus*, *Lactobacillus*, *Rhizobium*, *Proteus* sp., *Morganella* sp., *Aeromonas* sp., *Alcaligenes* sp., and *Klebsiellla* sp. have been extensively reported as degraders of azo dyes (Stolz, 2001; Pearce et al., 2003; Olukanni et al., 2006; Vijaykumar et al., 2007; Hsueh and Chen, 2008; Lin and Leu, 2008). Some strains of aerobic bacteria use azo dyes as sole source of carbon and nitrogen (Coughlin et al., 2002); others only reduce the azo group by special oxygen-tolerant azo reductases.

Azo dyes, in general, are not readily metabolized under aerobic condition, and as a result of metabolic pathways it degrades into intermediate compounds but is not mineralized. It can be completely degraded under coupled aerobic-anaerobic degradation (McMullan et al., 2001). In anaerobic condition, the azo bond undergoes cleavage to generate aromatic amines and is mineralized by nonspecific enzymes through ring cleavage under aerobic condition. Therefore, coupled anaerobic treatment followed by aerobic treatment can be an efficient degradation method of azo dyes (Feigel and Knackmuss, 1993). Chen et al. (2003) have described bacterial strains, which display a good growth in aerobic or agitation culture, but color removal was obtained with a high efficiency in anoxic or anaerobic culture (Cripps et al., 1990; Chung and Steven, 1993; Banat et al., 1996; Zissi et al., 1997; Delee et al., 1998; Kwang-Soo and Kim, 1998; Wong and Yuen, 1998; Swamy and Ramsay, 1999). Mixed bacterial culture can give a better degradation rate than the individual strain.

With regard to the role of fungal species on degradation of azo dyes, the most widely explored fungi are the ligninolytic fungi (Arora, 2003). Apart from this, *P. chrysosporium*, *Coriolus versicolar*, *Trametes versicolar*, *Funalia trogii*, *Penicillium geastrivous*, *Rhizopus oryzar*, *Pleurotus ostreatus*, *Rigidoporus lignosus*, *Pycnoporus sanguineus*, *Aspergillus flavus*, and *A. niger* have been reported, which are capable of degrading azo dyes (Fu and Viraraghavan, 2001; Wesenberg et al., 2003). However, long growth cycles and moderate decolorization rates limit the performance of fungal systems (Banat et al., 1996). In contrast, bacteria could reduce the color intensity more satisfactorily, but individual bacterial strains cannot mineralize azo dyes completely (Haug et al., 1991; Coughlin et al., 1997), and the intermediate products are carcinogenic aromatic amines, which need to be further decomposed. Gnanamani et al. (2004, 2005) studied the enzymatic degradation of azodyes using *Streptomyces* sp. SS07 and the intermediates released upon degradation were identified as amine products. The authors show the elucidated cleavage mechanism based on

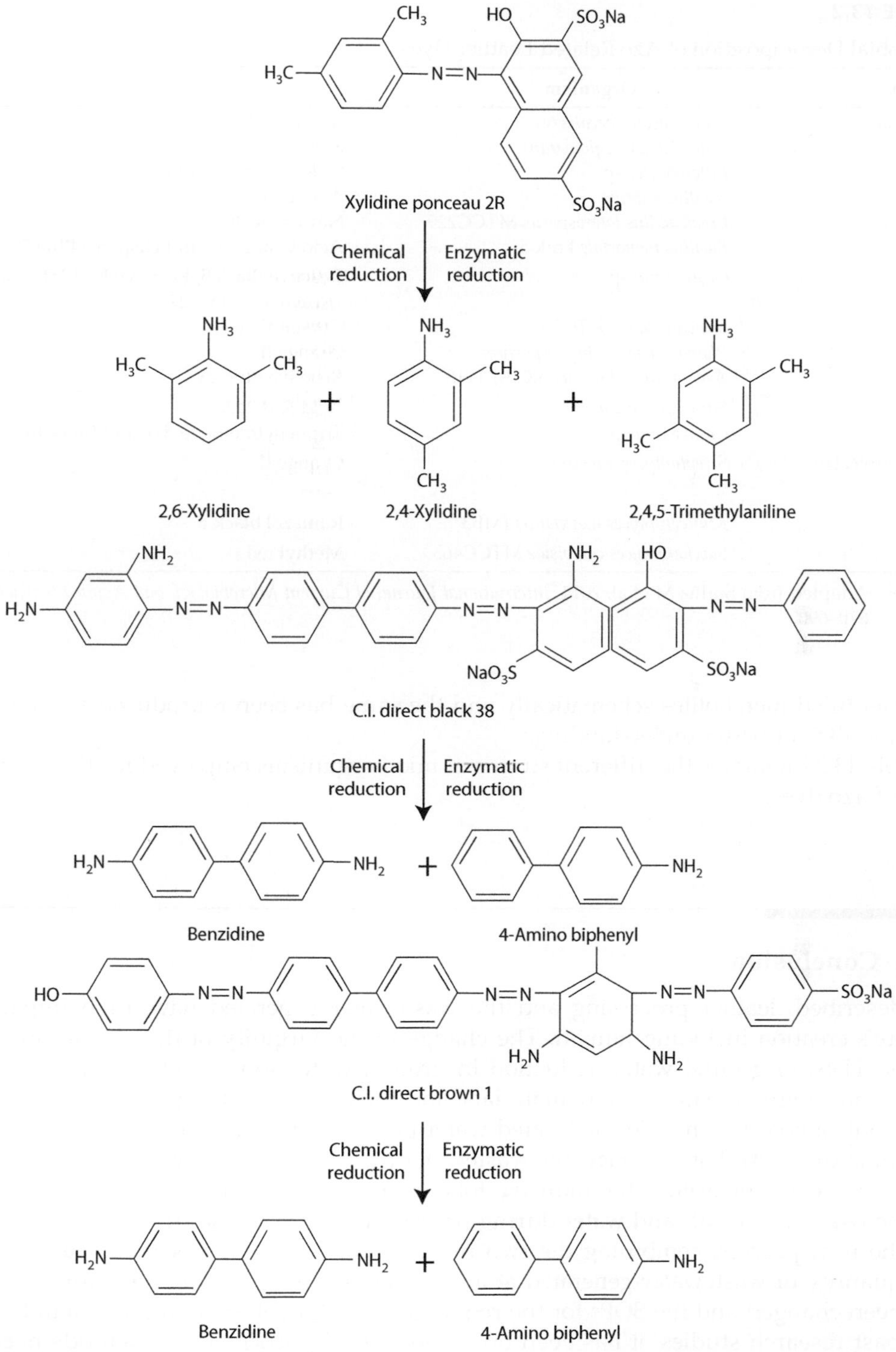

SCHEME 13.5
Enzymatic and chemical degradation of azo dyes to their respective amines. (Adapted from Gnanamani A et al. 2005. *Process Biochemistry* 40(11): 3497–3504.)

TABLE 13.2

Microbial Decomposition of Azo Related Leather Dyes

Strain	Organism	Dye
Bacteria	*Enterococcus faecalis* 66	Reactive Orange II
	Enterobacter agglomerans	Methyl Red
	Enterobacter sp.	CI Reactive Red 195
	Bacillus subtilis	Acid Blue 113
	Brevibacillus laterosporus MTCC229	Navy Blue 3G
	Bacillus fusiformis kmk 5	Acid Orange 10 and Disperse Blue 79
Fungi	*Geotrichum* sp.	Reactive Black 5, Reactive Red 158, and Reactive Yellow 27
	Shewanella sp. *NTOVI*	Crystal Violet
	Phanaerochaete chrysosporium	Orange II
	Aspergillus ochraceus NCIM-1146	Reactive Blue 25
Algae	*Spirogyra rhizopus*	Acid Red 247
	Cosmarium sp.	Triphenylmethane dye and Malachite green
Actinomycetes	*Streptomyces ipomoea*	Orange II
—	—	—
Yeast	*Kluyveromyces marxianus* IMB3	Ramazol black B
	Saccharomyces cerevisiae MTCC463	Methyl red

Source: Adapted from Sudha M et al. 2014. *International Journal of Current Microbiology and Applied Science* 3(2): 670–690.

the identified metabolites schematically and the same has been reproduced as shown in Scheme 13.5 for better understanding.

Table 13.2 illustrates the different strains of microorganisms employed for the degradation of azo dyes.

13.5 Conclusion

As described, leather processing and the wastewater generated have high impact on nature's creation including human. The change in the virginity of the soil, an increase in the TDSs of ground water, reduction in ground water sources, change in monsoon conditions, high demand for raw materials, and finally the discharge limit set by pollution control board authorities on treated water poses exorbitant challenges to the leather manufacturers. And at one stage, the wastewater treatment cost is higher than the manufacturing cost. However, if the industry follows the standard operating protocol (SOPs) on the use of chemicals and water during processing, reuse of one stream of wastewater for the next process, combining the two processes and reverse processing may reduce the quantity of wastewater generated at a considerable level. But the pollutant level has not been changed and the SOPs for the reduction in pollutant are in high demand. From the past research studies, it has been observed that physicochemical methods need an additional treatment option. Even primary and secondary treatment methods also need an additional management on sludge produced. Tertiary and advanced processes are cost effective and not affordable. Instead, natural degradation processes using microorganisms have been acceptable, despite the time consumed.

The review emphasized that the problems encountered with the composite effluent treatment can very well be managed with the attention given to individual streams. Since it is difficult for the natural microbial habitat to destruct or degrade the organic and inorganic–organic complexes at high pH and high salinity within the stipulated period of time, special enrichment cultures, or cultures designed to use the pollutants as their source of food and energy are required. The studies explored by number of researchers on various pollutants clearly indicated the possibility of microorganisms on complete destruction of pollutants, but limit with the species.

Most of the studies revealed that microbial consortia, that is, mixed microbial population showed tremendous pollutant reduction than individual species. Considering the view, microbial consortia with effective and efficient microbes if it is available to the treatment facility centers, solve the problems on destruction of organics in the composite effluent or individual effluent. What we need in the current scenario is a microbial consortia consisting of halophiles with high salt resistance, high pH tolerance, high tolerance toward different oxidation states of metals involved in the processing, able to generate enzymes such

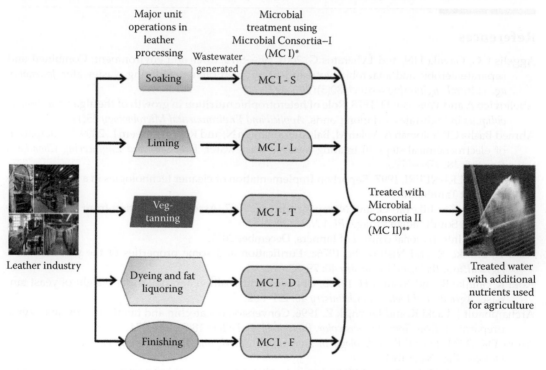

* MC I - Consortia of selected effective microbial species display growth at high pH, high salinity, phenolics, dyes and destruct the organics and refractory chemicals through its enzymatic system
MC I - S: responsible for destruction of organics in soak liquor
MC I - L: responsible for destruction of organics at high pH and able to remove sulphides
MC I - T: responsible for the destruction of phenolics
MC I - D: responsible for the destruction of dyes and fat liquors
MC I - F: destruction of other contaminants
** MC II - Consortia of selected effective microbial species display growth on degraded metabolites and destruct to simple molecules and nutrients

SCHEME 13.6
Schematic representation of possible treatability of wastewater of tanning sector using microbial of two different mixed groups.

as reductases, acyl hydrolyses, oxidases, and biosurfactant, able to act as absorbance and ready to destruct themselves (which avoid the sludge formation), and also act as nutrients for the other microbes. The degraded products can also be taken as a source of food and energy by the consortium. The following schematic representation on biological treatment of tannery wastewater to meet the discharge limit is shown in Scheme 13.6.

The microbial consortium I (MC I) and MCII may consist of both aerobic and facultative microbes of selected bacterial and fungal species able to survive and destruct the pollutants effectively. The residence or the retention time of these microbes with respect to pollutant may vary; however, the resultant treated water along with metabolites of pollutant and microbial metabolites acts as nutrients to the native microbes of soil and maintain soil virginity at the expected rate and at the same time save the ground water at considerable level.

References

Aggelis GG, Gavala HN, and Lyberatos G. 2001. SE—Structures and environment: Combined and separate aerobic and anaerobic biotreatment of green olive debittering wastewater. *Journal of agricultural engineering research* 80(3): 283–292.

Abeliovich A and Weisman D. 1978. Role of heterotrophic nutrition in growth of the alga *Scenedesmus obliquus* in high-rate oxidation ponds. *Applied and Environmental Microbiology* 35(1): 32–37.

Ahmed Basha CP, Soloman A, Velan M, Balasubramanian N, and Roohil Kareem L. 2009. Participation of electrochemical steps in treating tannery wastewater. *Industrial and Engineering Chemistry Research* 48: 9786–9796.

AISHTMA-CLRI-NEERI. 1997. Report on Implementation of cleaner technologies in a cluster of tanneries at Tamilnadu.

Anbalagan K, Karthikeyan G, and Narayanasamy N. 1997. Assessing pollution from tannery effluents in a South Indian village. *PLA Notes* 30: 3–6.

Anonymous, International council of tanners, December 2011.

Aoki K, Shinke R, and Nishira, H. 1976a. Purification and some properties of the yeast tannase. *Agricultural Biological Chemistry* 40: 79–85.

Aoki K, Shinke R, and Nishira H. 1976b. Chemical composition and molecular weight of yeast tannase. *Agricultural Biological Chemistry* 40: 297–302.

Archambault J, Lacki K, and Duvnjak Z. 1996. Conversion of catechin and tannic acid by an enzyme preparation from *Trametes versicolor*. *Biotechnology Letters* 18(7): 771–774.

Arora DK. 2003. *Fungal Biotechnology in Agricultural, Food, and Environmental Applications*. Marcel Dekker, Inc., New York.

Arunachalam M, Mohan N, and Mahadevan A. 2003. Cloning of *Acinetobacter calcoaceticus* chromosomal region involved in catechin degradation. *Microbiological Research* 158(1): 37–46.

Arunakumari A and Mahadevan A. 1984. Utilization of aromatic substances by *Pseudomonas solanacearum*. *Indian Journal of Experimental Biology* 2: 32–36.

Avudainayagam S. 2002. Long term tannery waste contamination: Effect on chromium chemistry. PhD dissertation. University of Adelaide, South Australia.

Bajpai B and Patil S. 1996. Tannin acyl hydrolase (EC 3.1.1.20) activity of *Aspergillus*, *Penicillium*, *Fusarium* and *Trichoderma*. *World Journal of Microbiology and Biotechnology* 12(3): 217–220.

Balakrishnan PA, Arunagiri A, and Rao PG. 2002. Ozone generation by silent electric discharge and its application in tertiary treatment of tannery effluent. *Journal of Electrostatics* 56(1): 77–86.

Banat IM, Nigam P, Singh D, and Marchant R. 1996. Microbial decolorization of textile-dye containing effluents: A review. *Bioresource Technology* 58(3): 217–227.

Barthomeuf C, Regerat F, and Pourrat H. 1994. Production, purification and characterization of a tannase from *Aspergillus niger* LCF 8. *Journal of Fermentation and Bioengineering* 77(3): 320–323.

Basegio T, Berutti F, Bernardes A, and Bergmann CP. 2002. Environmental and technical aspects of the utilisation of tannery sludge as a raw material for clay products. *Journal of the European Ceramic Society* 22(13): 2251–2259.

Ben-Amotz A, Sussman I, and Avron M. 1982. Glycerol production in the alga *Dunaliella*. In San Pietro A (ed.), *Biochemical and Photosynthetic Aspects of Energy Production*. Academic Press, New York, 191–208.

Bending GD and Read DJ. 1996. Effects of the soluble polyphenol tannic acid on the activities of ericoid and ecto mycorrhizal fungi. *Soil Biology and Biochemistry* 28(12): 1595–1602.

Beverini M and Metche M. 1990. Identification, purification and physicochemical properties of tannase of *Aspergillus oryzae*. *Science des Aliments* 10: 807–816.

Bhat TK, Singh B, and Sharma OP. 1998. Microbial degradation of tannins—a current perspective. *Biodegradation* 9(5): 343–357.

Bradoo S, Gupta R, and Saxena RK. 1996. Screening of extracellular tannase producing fungi: Development of a rapid and simple plate assay. *Journal of General Applied Microbiology* 42: 325–330.

Brooker JD, O'Donovan LA, Skene I, Clarke K, Blackall L, and Muslera P. 1994. *Streptococcus caprinus* sp. nov., a tannin-resistant ruminal bacterium from feral goats. *Letters in Applied Microbiology* 18(6): 313–318.

Buisman CJ, Wit B, and Lettinga G. 1990. Biotechnological sulphide removal in three polyurethane carrier reactors: Stirred reactor, biorotor reactor and upflow reactor. *Water Research* 24(2): 245–251.

Camargo FAO, Bento FM, Okeke BC, and Frankenberger WT. 2003. Chromate reduction by chromium-resistant bacteria isolated from soils contaminated with dichromate. *Journal of Environmental Quality* 32(4): 1228–1233.

Cassano A, Drioli E, and Molinari R. 1997. Recovery and reuse of chemical in unhearing, degreasing and chromium tanning processes by membranes. *Desalination* 113: 251–261.

Cassano A, Molinari R, Romano M, and Drioli E. 2001. Treatment of aqueous effluents of the leather industry by membrane processes: A review. *Journal of Membrane Science* 181(1): 111–126.

Chang, J-S and Lin Y-C. 2000. Fed-batch bioreactor strategies for microbial decolorization of azo dye using a *Pseudomonas luteola* Strain. *Biotechnology Progress* 16(6): 979–985.

Chen, JM and Hao OJ. 1998. Microbial chromium (VI) reduction. *Critical Reviews in Environmental Science and Technology* 28(3): 219–251.

Chen, K-C, Wu J-Y, Liou D-J, and John Hwang S-C. 2003. Decolorization of the textile dyes by newly isolated bacterial strains. *Journal of Biotechnology* 101(1): 57–68.

Chu W. 2001. Dye removal from textile dye wastewater using recycled alum sludge. *Water Research* 35(13): 3147–3152.

Chung, K-T and Stevens SE. 1993. Degradation azo dyes by environmental microorganisms and helminths. *Environmental Toxicology and Chemistry* 12(11): 2121–2132.

Churchley JH. 1994. Removal of sewage effluent- the use of a full scale ozone plant. *Water Science and Technology* 30(3): 275–284.

Cifuentes FR, Lindemann WC, and Barton LL. 1996. Chromium sorption and reduction in soil with implications to bioremediation. *Soil Science* 161(4): 233–241.

Cooper P. 1993. Removing colour from dyehouse waste waters—A critical review of technology available. *Journal of the Society of Dyers and Colourists* 109(3): 97–100.

Coughlin MF, Kinkle BK, and Bishop PL. 2002. Degradation of acid orange 7 in an aerobic biofilm. *Chemosphere* 46(1): 11–19.

Coughlin MF, Kinkle BK, Tepper A, and Bishop PL. 1997. Characterization of aerobic azo dye degrading bacteria and their activity in biofilms. *Water Science and Technology* 36(1): 215–220.

Cripps C, Bumpus JA, and Aust SD. 1990. Biodegradation of azo and heterocyclic dyes by *Phanerochaete chrysosporium*. *Applied and Environmental Microbiology* 56(4): 1114–1118.

Cruz-Hernández M, Contreras-Esquivel JC, Lara F, Rodriguez R, and Aguilar CN. 2005. Isolation and evaluation of tannin-degrading fungal strains from the Mexican desert. *Zeitschrift fur Naturforschung C-Journal of Biosciences* 60(11–12): 844–848.

Dantas Neto AA, de Castro Dantas TN, and Alencar Moura MCP. 2004. Evaluation and optimization of chromium removal from tannery effluent by microemulsion in the Morris extractor. *Journal of Hazardous Materials* 114(1): 115–122.

Delee W, O'Neill C, Hawkes FR, and Pinheiro HM. 1998. Anaerobic treatment of textile effluents: A review. *Journal of Chemical Technology and Biotechnology* 73(4): 323–335.

Deschamps AM. 1989. Microbial degradation of tannins and related compounds. In: Lewis NG and Paice MG (Eds.), *Plant Cell Wall Polymers Biogenesis and Biodegradation*. American Chemical Society, Washington, DC, pp. 559–566.

Deschamps AM and Lebeault J-M. 1984. Production of gallic acid from tara tannin by bacterial strains. *Biotechnology Letters* 6(4): 237–242.

Deschamps AM, Mahoudeau G, Conti M, and Lebeault J-M. 1980. Bacteria degrading tannic acid and related compounds. *Journal of Fermentation Technology* 58(2): 93–97.

Deschamps AM, Otuk G, and Lebeault J-M. 1983. Production of tannase and degradation of chestnut tannins by bacteria. *Journal of Fermentation Technology* 61: 55–59.

Doi S, Shinmyo A, Enatsu T, and Terui G. 1973. Growth associated production of tannase by a strain of *Aspergillus oryzae*. *Journal of Fermentation Technology* 61: 768–774.

Eckenfelder WW. 2002. *Industrial Water Pollution Control*. McGraw-Hill International Edition, Singapore.

El-Sikaily A, El Nemr A, Khaled A, and Abdelwahab O. 2007. Removal of toxic chromium from wastewater using green alga *Ulva lactuca* and its activated carbon. *Journal of Hazardous Materials* 148(1): 216–228.

Farabegoli G, Carucci A, Majone M, and Rolle E. 2004. Biological treatment of tannery wastewater in the presence of chromium. *Journal of Environmental Management* 71(4): 345–349.

Feigel BJ and Knackmuss H-J. 1993. Syntrophic interactions during degradation of 4-aminobenzene-sulfonic acid by a two species bacterial culture. *Archives of Microbiology* 159(2): 124–130.

Field JA and Lettinga G. 1992. Biodegradation of tannins. In: Helmut S and Astrid S (Eds.), *Metal Ions in Biological Systems. Degradation of Environmental in Pollutants and their Metalloenzymes*. Marcel Dekker, New York, pp. 61–98.

Field JA, Leyendeckers MJH, Sierra Alvarez R, Lettinga G, and Habets LHA. The methanogenic toxicity of bark tannins and the anaerobic biodegradability of water soluble bark matter. *Water Science and Technology* 20(1): 1988: 219–240.

Franklin LB and Stensel HD. 1972. Chapter 14, In: Tchobanoglous G (Ed.), *Wastewater Engineering: Treatment, Disposal and Reuse*. McGraw-Hill, New York.

Fu Y and Viraraghavan T. 2001. Fungal decolorization of dye wastewaters: A review. *Bioresource Technology* 79(3): 251–262.

Gajendiran N and Mahadevan A. 1988. Utilization of catechin by *Rhizobium* sp. *Plant and Soil* 108(2): 263–266.

Gamble GR, Akin DE, Makkar HP, and Becker K. 1996. Biological degradation of tannins in *Sericea lespedeza* (*Lespedeza cuneata*) by the white rot fungi *Ceriporiopsis subvermispora* and *Cyathus stercoreus* analyzed by solid-state ^{13}C nuclear magnetic resonance spectroscopy. *Applied and Environmental Microbiology* 62(10): 3600–3604.

Ganesh R, Balaji G, and Ramanujam RA. 2006. Biodegradation of tannery wastewater using sequencing batch reactor—Respirometric assessment. *Bioresource Technology* 97(15): 1815–1821.

Ganga PS, Nandy SC, and Santappa M. 1977. Effect of environmental factors on the production of fungal tannase. *Leather Science* 24: 8–16.

Gilmour DJ, Hipkins MF, Webber AN, Baker NR, and Boney AD. 1985. The effect of ionic stress on photosynthesis in *Dunaliella tertiolecta*. *Planta* 163(2): 250–256.

Gingell R and Walker R. 1971. Mechanism of azo reduction by *Streptococcus faecalis* II. The role of soluble flavines. *Xenobiotics* 13: 231–239.

Gnanamani A, Bhaskar M, Ganeshjeevan R, Chandrasekar R, Sekaran G, Sadulla S, and Radhakrishnan G. 2005. Enzymatic and chemical catalysis of xylidine ponceau 2R and evaluation of products released. *Process Biochemistry* 40(11): 3497–3504.

Gnanamani A, Bhaskar M, Ganga R, Sekaran G, and Sadulla S. 2004. Chemical and enzymatic inter-actions of Direct Black 38 and Direct Brown 1 on release of carcinogenic amines. *Chemosphere* 56(9): 833–841.

Gnanamani A, Kavitha V, Radhakrishnan N, Suseela Rajakumar G, Sekaran G, and Mandal AB. 2010. Microbial products (biosurfactant and extracellular chromate reductase) of marine microorgan-ism are the potential agents reduce the oxidative stress induced by toxic heavy metals. *Colloids and Surfaces B: Biointerfaces* 79(2): 334–339.

Gnanamani A, Sekaran G, and Babu M. 2001. Removal of tannin from cross-linked and open chain polymeric tannin substrates using heme peroxidases of *Phanerochaete chrysosporium. Bioprocess and Biosystems Engineering* 24(4): 211–217.

Gogate PR and Pandit AB. 2004. A review of imperative technologies for wastewater treatment II: Hybrid methods. *Advances in Environmental Research* 8(3): 553–597.

Gonzalez CG. 2006. Genetically modified organisms and justice: The international environmental justice implications of biotechnology. *Georgetown International Environmental Law Review* 19: 583.

Guha H, Jayachandran K, and Maurrasse F. 2001. Kinetics of chromium (VI) reduction by a type strain *Shewanella alga* under different growth conditions. *Environmental Pollution* 115(2): 209–218.

Gundala PB, Chinthala P, and Sreenivasulu B. 2013. A new facultative alkaliphilic, potassium solu-bilizing, *Bacillus* sp. SVUNM9 isolated from mica cores of Nellore District, Andhra Pradesh, India. *Research and Reviews: Journal of Microbiology and Biotechnology* 2(1): 1–7.

Gupta RK and Haslam E. 1989. Vegetable tannin structure and biosynthesis in polyphenol, in cereals and legumes. International Development Research Centre, Ottawa, Canada.

Haslam E and Stangroom JE. 1966. The esterase and depsidase activities of tannase. *Biochemical Journal* 99(1): 28–31.

Hatamoto O, Watarai T, Kikuchi M, Mizusawa K, and Sekine H. 1996. Cloning and sequencing of the gene encoding tannase and a structural study of the tannase subunit from *Aspergillus oryzae. Gene* 175(1): 215–221.

Haug W, Schmidt A, Nörtemann B, Hempel DC, Stolz A, and Knackmuss HJ. 1991. Mineralization of the sulfonated azo dye Mordant Yellow 3 by a 6-aminonaphthalene-2-sulfonate-degrading bacterial consortium. *Applied and Environmental Microbiology* 57(11): 3144–3149.

Hawley LE, Deeb AR, Kavanaugh CM, and Jacobs RGJ. 2004. *Treatment Technologies for Chromium (VI). Chromium (VI) Handbook.* CRC Press, Boca Raton, 275–310.

Hinteregger C and Streichsbier F. 1997. *Halomonas* sp., a moderately halophilic strain, for biotreat-ment of saline phenolic waste-water. *Biotechnology Letters* 19(11): 1099–1102.

Hsueh C-C and Chen B-Y. 2008. Exploring effects of chemical structure on azo dye decolorization characteristics by*Pseudomonas luteola. Journal of Hazardous Materials* 154(1): 703–710.

Iibuchi S, Minoda Y, and Yamada K. 1967, Studies on tannin acyl hydrolase of microorganisms. Part II. A new method determining the enzyme activity using the change of ultra violet absorption. *Agricultural Biological Chemistry* 31: 513–518.

Ingram M. 1940. The influence of sodium chloride and temperature on the endogenous respiration of *B. cereus. The Journal of General Physiology* 23(6): 773–780.

Isik M and Sponza DT. 2004. Decolorisation of azo dyes under batch anaerobic/aerobic conditions. *Journal of Environmental Science and Health* 39:1107–1127.

Jawahar AJ, Chinnadurai M, Ponselvan JK, and Annadurai G. 1998. Pollution from tanneries and options for treatment of effluent. *Indian Journal of Environmental Protection* 18(9): 672–678.

Jones GA, McAllister TA, Muir AD, and Cheng K-J. 1994. Effects of sainfoin (*Onobrychis viciifolia* Scop.) condensed tannins on growth and proteolysis by four strains of ruminal bacteria. *Applied and Environmental Microbiology* 60(4): 1374–1378.

Kalyanaraman C, Sri Bala Kameswari K, Sudharsan Varma V, Tagra S, and Raghava Rao J. 2013. Studies on biodegradation of vegetable-based fat liquor-containing wastewater from tanneries. *Clean Technologies and Environmental Policy* 15(4): 633–642.

Kalyanaraman C, Sri Bala Kameswari K, Vidya Devi L, Porselvam S, and Raghava Rao J. 2012. Combined advanced oxidation processes and aerobic biological treatment for synthetic fatliquor used in tanneries. *Industrial and Engineering Chemistry Research* 51(50): 16171–16181.

Kamaludeen SPB, Arunkumar KR, Avudainayagam S, and Ramasamy K. 2003. Bioremediation of chromium contaminated environments. *Indian Journal of Experimental Biology* 41(9): 972–985.

Kapoor A, Thiruvenkatachari V, and Roy Cullimore D. 1999. Removal of heavy metals using the fungus *Aspergillus niger*. *Bioresource Technology* 70(1): 95–104.

Kargi F and Dincer AR. 1996a. Effect of salt concentration on biological treatment of saline wastewater by fed-batch operation. *Enzyme and Microbial Technology* 19(7): 529–537.

Kargi F and Dincer AR. 1996b. Enhancement of biological treatment performance of saline wastewater by halophilic bacteria. *Bioprocess Engineering* 15(1): 51–58.

Khalid A, Arshad M, and Crowley DE. 2008. Decolorization of azo dyes by *Shewanella* sp. under saline conditions. *Applied Microbiology and Biotechnology* 79(6): 1053–1059.

Kincannon DF and Gaudy Jr. AF. 1966. Some effects of high salt concentrations on activated sludge. *Journal (Water Pollution Control Federation)* 38(7): 1148–1159.

Kirby N, Marchant R, and McMullan G. 2000. Decolourisation of synthetic textile dyes by *Phlebia tremellosa*. *FEMS Microbiology Letters* 188(1): 93–96.

Kleinjan WE, Arie de Keizer, and Albert JH Janssen. 2003. Biologically produced sulfur. In: Steudel R (Ed.), *Elemental Sulfur and Sulfur-Rich Compounds I*. Springer, Berlin, Heidelberg, pp. 167–188.

Knudson L. 1913. Tannic acid fermentation. *Journal of Biological Chemistry* 14: 159–202.

Kongjao S, Damronglerd S, and Hunsom M. 2008. Simultaneous removal of organic and inorganic pollutants in tannery wastewater using electrocoagulation technique. *Korean Journal of Chemical Engineering* 25(4): 703–709.

Krishnakumar B, Majumdar S, Manilal VB, and Haridas A. 2005. Treatment of sulphide containing wastewater with sulphur recovery in a novel reverse fluidized loop reactor (RFLR). *Water Research* 39(4): 639–647.

Kumar R, Deepa Kachroo T, and Poonam S. 2005. Aerobic method of removing total dissolved solids from tannery wastewaters. US Patent No. 6,905,863 B2.

Kubo M, Hiroe J, Murakami M, Fukami H, and Tachiki T. 2001. Treament of hypersaline containing wastewater with salt-tolerant microorganisms. *Journal of Bioscience and Bioengineering* 91(2): 222–224.

Kwang-Soo S and Kim C-J. 1998. Decolorisation of artificial dyes by peroxidase from the white-rot fungus, *Pleurotus ostreatus*. *Biotechnology Letters* 20(6): 569–572.

Lakshman Perumalsamy P. 1993. Studies on the utilization of tannin by bacteria. *Pollution Research* 12: 1–9.

Lane RW, Yamakoshi J, Kikuchi M, Mizusawa K, Henderson L, and Smith M. 1997. Safety evaluation of tannase enzyme preparation derived from *Aspergillus oryzae*. *Food and Chemical Toxicology* 35(2): 207–212.

Laubscher RK. 1991. The culture of *Dunaliella salina* and the production of p-carotene in tannery effluents. MSc Thesis, Rhodes University, Grahamstown.

Lawton GW and Eggert CV. 1957. Effect of high sodium chloride concentration on trickling filter slimes. *Sewage and Industrial Wastes* 29(11): 1228–1236.

Lefebvre X, Paul E, Mauret M, Baptiste P, and Capdeville B. 1998. Kinetic characterization of saponified domestic lipid residues aerobic biodegradation. *Water Research* 32(10): 3031–3038.

Lefebvre O, Vasudevan N, Torrijos M, Thanasekaran K, and Moletta R. 2005. Halophilic biological treatment of tannery soak liquor in a sequencing batch reactor. *Water Research* 39(8): 1471–1480.

Lefebvre O, Vasudevan N, Torrijos M, Thanasekaran K, and Moletta R. 2006. Anaerobic digestion of tannery soak liquor with an aerobic post treatment. *Water Research* 40(7): 1492–1500.

Lekha PK and Lonsane BK. 1994. Comparative titres, location and properties of tannin acyl hydrolase produced by *Aspergillus niger* PKL 104 in solid-state, liquid surface an submerged fermentations. *Process Biochemistry* 29(6): 497–503.

Lekha PK and Lonsane BK. 1997. Production and application of tannin acyl hydrolase: State of the art. *Advances in Applied Microbiology* 44: 216–260.

Lewis JA and Starkey RL. 1969. Decomposition of plant tannins by some soil microorganisms. *Soil Science* 107(4): 235–241.

Lewis JA and Starkey RL. 1968. Vegetable tannins, their decomposition and effects on decomposition of some organic compounds. *Soil Science* 106(4): 241–247.

Lin Y-H and Leu J-Y. 2008. Kinetics of reactive azo-dye decolorization by *Pseudomonas luteola* in a biological activated carbon process. *Biochemical Engineering Journal* 39(3): 457–467.

Lizama HM and Sankey BM. 1993. Conversion of hydrogen sulphide by acidophilic bacteria. *Applied Microbiology and Biotechnology* 40(2–3): 438–441.

Lodato A, Alfieri F, Olivieri G, Di Donato A, Marzocchella A, and Salatino P. 2007. Azo-dye convention by means of *Pseudomonas* sp. OXI. *Enzyme Microbiologal Technology* 41: 646–652.

Lofrano G, Meriç S, Belgiorno V, and Napoli R. 2007. Fenton's oxidation of various-based tanning materials. *Desalination* 211(1): 10–21.

Lorusso L, Lacki K, and Duvnjak Z. 1996. Decrease of tannin content in canola meal by an enzyme preparation from *Trametes versicolor*. *Biotechnology Letters* 18(3): 309–314.

Losi ME, Amrhein C, and Frankenberger WT. 1994. Factors affecting chemical and biological reduction of hexavalent chromium in soil. *Environmental Toxicology and Chemistry* 13(11): 1727–1735.

Luo Z, Xia C, Fan H, Chen X, and Peng B. 2011. The biodegradabilities of different oil-based fatliquors. *Journal of the American Oil Chemists' Society* 88(7): 1029–1036.

Luo Z, Yao J, Fan H, Xia C, and Wang S. 2010. The biodegradabilities of rape oil based fatliquors prepared from different methods. *Journal of American Leather Chemist Association* 105: 121–127.

Mahadevan A and Muthukumar G. 1980. Aquatic microbiology with reference to tannin degradation. *Hydrobiologia* 72(1–2): 73–79.

Maier J, Kandelbauer A, Erlacher A, Cavaco-Paulo A, and Gübitz M. 2004. A new alkali-thermostable azoreductase from *Bacillus* sp. strain SF. *Applied and Environmental Microbiology* 70(2): 837–844.

Makkar HPS, Singh B, and Kamra DN. 1994. Biodegradation of tannins in oak (*Quercus incana*) leaves by *Sporotrichum pulverulentum*. *Letters in Applied Microbiology* 18(1): 39–41.

Mandal OP, Sinha AK, and Sinha KMP. 2005. Studies on primary productivity of a wetland. In: Kumar A (Ed.), *Fundamental of Limnology*, API I publishing corporation, India, pp. 230–237.

Martins MAM, Cardoso MH, Queiroz MJ, Ramalho MT, and Campus AMO. 1999. Biodegradation of azo dyes by the yeast *Candida zeylanoides* in batch aerated cultures. *Chemosphere* 38(11): 2455–2460.

McLeod MN. 1974. Plant tannins- their role in forage quality. *Nutrition Abstracts and Reviews* 44: 803–815.

McMullan G, Meehan C, Conneely A, Kirby N, Robinson T, Nigam P, Banat I et al. 2001. Microbial decolourisation and degradation of textile dyes. *Applied Microbiology and Biotechnology* 56(1–2): 81–87.

Midha V and Dey A. 2008. Biological treatment of tannery wastewater for sulfide removal. *International Journal of Chemical Science* 6(2): 472–486.

Mondal NC, Saxena VK, and Singh VS. 2005. Impact of pollution due to tanneries on groundwater regime. *Current Science* 88(12): 1988–1994.

Moosvi S, Keharia H, and Madamwar D. 2005. Decolourization of textile dye Reactive Violet 5 by a newly isolated bacterial consortium RVM 11.1. *World Journal of Microbiology and Biotechnology* 21(5): 667–672.

Mueller Harvey I and McAllan AB. 1992. Tannins: Their biochemistry and nutritional properties. In: Morrison IM (ed.), *Advances in Plant Cell Biochemistry and Biotechnology*. JAI Press Ltd., London, UK, pp. 151–217.

Muthukumar G, Arunakumari A, and Mahadevan A. 1982. Degradation of aromatic compounds by *Rhizobium* spp. *Plant and Soil* 69(2): 163–169.

Myers CR, Carstens BP, Antholine WE, and Myers JM. 2000. Chromium (VI) reductase activity is associated with the cytoplasmic membrane of anaerobically grown *Shewanella putrefaciens* MR-1. *Journal of Applied Microbiology* 88(1): 98–106.

Nishira H. 1961. Studies on tannin decomposing enzyme of molds. X. Tannase fermentation by molds in liquid culture with phenolic substances. *Journal of Fermentation Technology* 39: 137–146.

Oh KJ, Kim D, and Lee I-H. 1998. Development of effective hydrogen sulphide removing equipment using *Thiobacillus* sp. IW. *Environmental Pollution* 99(1): 87–92.

Olukanni OD, Osuntoki AA, and Gbenle GO. 2006. Textile effluent biodegradation potentials of textile effluent-adapted and non-adapted bacteria. *African Journal of Biotechnology* 5(20): 1980–1984.

Oren A. 2010. Industrial and environmental applications of halophilic microorganisms. *Environmental Technology* 31(8–9): 825–834.

Osawa R. 1990. Formation of a clear zone on tannin-treated brain heart infusion agar by a *Streptococcus* sp. isolated from feces of koalas. *Applied and Environmental Microbiology* 56(3): 829–831.

Osawa RO, Walsh TP, and Cork SJ. 1993. Metabolism of tannin-protein complex by facultatively anaerobic bacteria isolated from koala feces. *Biodegradation* 4(2): 91–99.

Otuk G and Deschamps AM. 1983. Degradation of condensed tannin by several types of yeasts. *Mycopathologia* 83: 107–111.

Paller G, Hommel RK, and Kleber H-P. 1995. Phenol degradation by *Acinetobacter calcoaceticus* NCIB 8250. *Journal of Basic Microbiology* 35(5): 325–335.

Pandey A, Singh P, and Iyengar L. 2007. Bacterial decolorization and degradation of azo dyes. *International Biodeterioration and Biodegradation* 59(2): 73–84.

Paszczynski A, Pasti MB, Goszczynski S, Crawford DL, and Crawford RL. 1991. New approach to improve degradation of recalcitrant azo dyes by *Streptomyces* spp. and *Phanerochaete chrysosporium*. *Enzyme and Microbial Technology* 13(5): 378–384.

Pearce CI, Lloyd JR, and Guthrie JT. 2003. The removal of colour from textile wastewater using whole bacterial cells: A review. *Dyes and Pigments* 58(3): 179–196.

Perez-Maldonado RA and Norton BW. 1996. Digestion of C-labelled condensed tannins from *Desmodium intortum* in sheep and goats. *British Journal of Nutrition* 76(4): 501–513.

Suseela RG and Nandy SC. 1983. Isolation, purification, and some properties of *Penicillium chrysogenum* tannase. *Applied and Environmental Microbiology* 46(2): 525–527.

Rajamani S, Ramasami T, Langerwerf JSA, and Schappman JE. 1995. Environmental management in tanneries—feasible chromium recovery and reuse system. In: *Proceedings of the 3rd International Conferences on Appropriate Waste Management Technologies for Developing Countries* (AWMTDC, 95), Nagpur, India, pp. 965–969.

Rajwade JM and Paknikar KM. 2003. Bioreduction of tellurite to elemental tellurium by *Pseudomonas mendocina* MCM B-180 and its practical application. *Hydrometallurgy* 71(1): 243–248.

Ramasami T and Rao PG. 1991. International Consultation Meeting on Technology and Environmental Up gradation in Leather Sector, New Delhi, pp. T1-1–T1-30.

Ramasamy T, Raghava Rao J, Chandrababu NK, Parthasarathi K, Rao PG, Saravanan P, Gayatri R, and Sreeram KJ. 1999. Beamhouse and tanning operations: A revisit to process chemistry. *Journal of the Society of Leather Technologies and Chemists* 83: 39–45.

Rastogi SK, Amit P, and Sachin T. 2008. Occupational health risks among the workers employed in leather tanneries at Kanpur. *Indian Journal of Occupational and Environmental Medicine* 12(3): 132–135.

Ravindranath E. 2012. Studies on liquefaction of limed fleshings and enhancement of biomethanization from tannery waste. PhD Thesis submitted to Anna University under Faculty of Civil Engineering. URL: http://hdl.handle.net/10603/9626.

Rehman A and Shakoori AR. 2001. Heavy metal resistance *Chlorella* spp., isolated from tannery effluents, and their role in remediation of hexavalent chromium in industrial waste water. *Bulletin of Environmental Contamination and Toxicology* 66(4): 542–547.

Roš M and Gantar A. 1998. Possibilities of reduction of recipient loading of tannery wastewater in Slovenia. *Water Science and Technology* 37(8): 145–152.

Ross DS, Sjogren RE, and Bartlett RJ. 1981. Behavior of chromium in soils: IV. Toxicity to microorganisms. *Journal of Environmental Quality* 10(2): 145–148.

Rothschild LJ and Mancinelli RL. 2001. Life in extreme environments. *Nature* 409(6823): 1092–1101.

Salam DA, Naik N, Suidan MT, and Venosa AD. 2012. Assessment of aquatic toxicity and oxygen depletion during aerobic biodegradation of vegetable oil: Effect of oil loading and mixing regime. *Environmental Science & Technology* 46(4): 2352–2359.

Sanchez-Gonzalez M, Blanco-Gamez A, Escalante A, Valladares AG, Olvera C, and Parra R. 2011. Isolation and characterization of new facultative alkaliphilic *Bacillus flexus* strains from maize processing waste water (Nejayote). *Letters in Applied Microbiology* 52(4): 413–419.

Satarupa D and Paul AK. 2013. Hexavalent chromium reduction by aerobic heterotrophic bacteria indigenous to chromite mine overburden. *Brazilian Journal of Microbiology* 44(1): 307–315.

Saxena RK, Sharmila P, and Singh VP. 1995. Microbial degradation of tannins. *Progress in Industrial Microbiology* 32: 259–270.

Scalbert A. 1991. Antimicrobial properties of tannins. *Phytochemistry* 30(12): 3875–3883.

Sekaran G, Chitra K, Mariappan M, and Raghavan KV. 1996. Removal of sulphide in anaerobically treated tannery wastewater by wet air oxidation. *Journal of Environmental Science and Health Part A* 31(3): 579–598.

Selinger LB, Forsberg CW, and Cheng K-J. 1996. The rumen: A unique source of enzymes for enhancing livestock production. *Anaerobe* 2(5): 263–284.

Sharma K. 2002. Microbial Cr (VI) reduction: Role of electron donors, acceptors, and mechanisms, with special emphasis on *Clostridium* spp. PhD dissertation, University of Florida.

Shen H, Hap Pritchard P, and Sewell GW. 1996. Microbial reduction of Cr (VI) during anaerobic degradation of benzoate. *Environmental Science and Technology* 30(5): 1667–1674.

Sheng PX, Tan LH, Chen JP, and Ting Y-P. 2005. Biosorption performance of two brown marine algae for removal of chromium and cadmium. *Journal of Dispersion Science and Technology* 25(5): 679–686.

Sherief AA, EL-Tanash AB, and Nour A. 2011. Optimization of tannase biosynthesis from two local Aspergilli using commercial green tea as solid substrate. *Biotechnology* 10(1): 78–85.

Sivaprakasam S, Mahadevan S, Sekar S, and Rajakumar S. 2008. Biological treatment of tannery wastewater by using salt-tolerant bacterial strains. *Microbial Cell Factories* 7(1): 15.

Smith WL and Gadd GM. 2000. Reduction and precipitation of chromate by mixed culture sulphate-reducing bacterial biofilms. *Journal of Applied Microbiology* 88(6): 983–991.

Song Z, Williams CJ, and Edyvean RGJ. 2004. Treatment of tannery wastewater by chemical coagulation. *Desalination* 164(3): 249–259.

Spencer CM, Cai Y, Martin R, Gaffney SH, Goulding PN, Magnolato D, Lilley TH et al. 1988. Polyphenol complexation some thoughts and observations. *Phytochemistry* 27(8): 2397–2409.

Srivastava S and Thakur IS. 2006. Isolation and process parameter optimization of *Aspergillus* sp. for removal of chromium from tannery effluent. *Bioresource Technology* 97(10): 1167–1173.

Stern SR, Szpyrkowicz L, and Rodighiero I. 2003. Aerobic treatment of textile dyeing wastewater. *Water Science and Technology* 47(10): 55–59.

Stewart MJ, Ludwig HF, and Kearns WH. 1962. Effects of varying salinity on the extended aeration process. *Journal (Water Pollution Control Federation)* 34(11): 1161–1177.

Stolz A. 2001. Basic and applied aspects in the microbial degradation of azo dyes. *Applied Microbiology and Biotechnology* 56(1–2): 69–80.

Subletta KL. 1987. Aerobic oxidation of hydrogen sulfide by *Thiobacillus denitrificans*. *Biotechnology and Bioengineering* 29(6): 690–695.

Sublette KL and Sylvester ND. 1987. Oxidation of hydrogen sulfide by continuous cultures of *Thiobacillus denitrificans*. *Biotechnology and Bioengineering* 29(6): 753–758.

Sudha M, Saranya A, Selvakumar G, and Sivakumar N. 2014. Microbial degradation of Azo Dyes: A review. *International Journal of Current Microbiology and Applied Science* 3(2): 670–690.

Sundar K, Vidya R, Mukherjee A, and Chandrasekaran N. 2010. High chromium tolerant bacterial strains from Palar River Basin: Impact of tannery pollution. *Research Journal of Environmental and Earth Science* 2(2): 112–117.

Suseela RG and Nandy SC. 1985. Decomposition of tannic acid and gallic acid by *Penicillium chrysogenum*. *Leather Science* 32: 278–280.

Swamy J and Ramsay JA. 1999. The evaluation of white rot fungi in the decoloration of textile dyes. *Enzyme and Microbial Technology* 24(3): 130–137.

Tare V, Gupta S, and Bose P. 2003. Case studies on biological treatment of tannery effluents in India. *Journal of the Air and Waste Management Association* 53(8): 976–982.

UNEP/IEO. 1994. Tanneries and the environment: A technical guide to reducing the environmental impact of tannery operations. Technical Reports, Series 4, 120p.

UNEP IE/PAC. 1994. Tanneries and Environment, A Technical Guide.

Van der Zee FP and Villaverde S. 2005. Combined anaerobic–aerobic treatment of azo dyes—A short review of bioreactor studies. *Water Research* 39(8): 1425–1440.

Vennat B, Pourrat A, and Pourrat H. 1986. Production of a depolymerized tannin extract using a strain of *Saccharomyces rouxii*. *Journal of Fermentation Technology* 64: 227–232.

Verma P and Madamwar D. 2003. Decolourization of synthetic dyes by a newly isolated strain of *Serratia marcescens*. *World Journal of Microbiology and Biotechnology* 19(6): 615–618.

Vijayaraghavan K, Ramanujam TK, and Balasubramanian N. 1999. *In situ* hypochlorous acid generation for the treatment of distillery spent wash. *Industrial Engineering Chemistry Research* 38: 2264–2267.

Vijayaraghavan K and Murthy DVS. 1997. Effect of toxic substances in anaerobic treatment of tannery wastewaters. *Bioprocess Engineering* 16(3): 151–155.

Vijaykumar MH, Vaishampayan PA, Shouche YS, and Karegoudar TB. 2007. Decolourization of naphthalene-containing sulfonated azo dyes by *Kerstersia* sp. strain VKY1. *Enzyme and Microbial Technology* 40(2): 204–211.

Wesenberg D, Kyriakides I, and Agathos SN. 2003. White-rot fungi and their enzymes for the treatment of industrial dye effluents. *Biotechnology Advances* 22(1): 161–187.

Wielinga B, Mizuba MM, Hansel CM, and Fendorf S. 2001. Iron promoted reduction of chromate by dissimilatory iron-reducing bacteria. *Environmental Science and Technology* 35(3): 522–527.

William F, Boominathan K, Vasudevan N, Gurujeyalakshmi G, and Mahadevan A 1986. Microbial degradation of lignin and tannin. *Journal of Science Industrial Research* 45: 232–243.

Wilson JL, Conrad SH, Hagan E, Mason WR, and Peplinski W. 1988. The pore level spatial distribution and saturation of organic liquids in porous media. In: *Proceedings of the Petroleum Hydrocarbons and Organic Chemicals in Ground Water: Prevention, Detection and Restoration*, Houston, TX, pp. 107–133.

Wong PK and Yuen PY. 1998. Decolourization and biodegradation of N,N′-dimethyl-p-phenylenediamine by *Klebsiella pneumoniae* RS-13 and *Acetobacter liquefaciens* S-1. *Journal of Applied Microbiology* 85(1): 79–87.

Zissi U, Lyberatos G, and Pavlou S. 1997. Biodegradation of p-aminoazobenzene by *Bacillus subtilis* under aerobic conditions. *Journal of Industrial Microbiology and Biotechnology* 19(1): 49–55.

14

Microbial Degradation Mechanism of Textile Dye and Its Metabolic Pathway for Environmental Safety

Rahul V. Khandare and Sanjay P. Govindwar

CONTENTS

14.1 Introduction

Dyes are defined as colored substances which are applied to fibers giving them a permanent color which is able to resist fading upon exposure to sweat, light, water, and many chemicals, including oxidizing agents and microbial attack (Rai et al., 2005). The growth of the textile industry worldwide has seen a tremendous increase in the use of synthetic dyes, and this has been accompanied by a rise in pollution due to wastewater contaminated with dyestuffs (Pandey et al., 2007). The treatment of these high volumes of wastewater coming mostly from dyeing processes becomes essential as the presence of color renders these waters aesthetically unacceptable and unusable. The textile dye processors are one of the greatest generators of liquid effluent pollutants, due to the high quantities of water used in the dyeing processes. Moreover, the processing stages and types of synthetic dyes applied during this conversion determine the variable wastewater characteristics in terms of pH, dissolved oxygen, organic and inorganic chemical content (Saratale et al., 2009a).

Indiscriminate and uncontrolled release of textile dye effluent containing dyes and their products in aqueous ecosystems is aesthetically unacceptable and leads to serious hazards such as reduction in sunlight penetration, decreasing photosynthetic activity, dissolved oxygen concentration, and overall water quality, leading to acute toxic effects on aquatic flora and fauna, causing severe environmental problems worldwide. Dyes also have an adverse impact on the water bodies' TOC, BOD, COD, and many other important environmental parameters. Many synthetic dyes and their metabolites are toxic, carcinogenic, and mutagenic (Govindwar and Kagalkar, 2010). The recalcitrant nature of textile wastewater is due to the high content of dyestuffs, surfactants, dispersants, acids, bases, salts, detergents, humectants, and oxidants. Textile dyes are toxic, mutagenic and carcinogenic, posing threats to the environment, human health, and agricultural productivity (Kalme et al., 2007a,b), therefore, the treatment of these high volumes of wastewater coming mostly from the dyeing processes becomes essential as the presence of color renders these waters aesthetically unacceptable and unusable.

There are a number of physicochemical methods such as coagulation/flocculation, adsorption, membrane filtration, chemical precipitation, and an advanced oxidation process for the removal of dyes from effluents (Lopez-Grimau and Gutierrez, 2006; Gupta et al., 2009). These extensively used techniques produce large amounts of sludge which again requires safe disposal. The cost of the chemicals required for these processes is also a major drawback when considered by the small scale dye processor and by household industries. This financial burden probably plays a role in slowing down global efforts to eradicate pollution, particularly in developing countries where these techniques are clearly not affordable (Khandare et al., 2014). Workers throughout the world are trying to find a solution to this problem of textile dyes with an eye to cost effectiveness and environmentally friendly outcomes. The available methods for dye removal from textile effluents have been put together in Table 14.1. The biological methods have been thought to have the desired properties. The biological agents such as algae, fungi, actinomycetes, yeasts, and even plants have also been used for the degradation of textile dyes (Jadhav and Govindwar, 2006; Parshetti et al., 2007; Ghodake et al., 2009a). This approach of exploring living organisms for the treatment of pollutants such as textile dyes is called bioremediation.

TABLE 14.1

Methods for Removal of Dyes from Textile Wastewater

Physical Methods	Chemical Methods	Biological Methods
Precipitation/sedimentation	Catalytic degradation	Fungal degradation
Coagulation	Chemical precipitation	Algal treatment
Filtration	Reduction	Anaerobic digestion
Adsorption	Oxidation	Stabilization
Flotation	Neutralization	Trickling filters
Flocculation	Ion exchange	Activated sludge
Reverse osmosis	Electrolysis	Trickling filters
Membrane treatment	Advanced oxidation processes	Enzymatic processes
Distillation	Ozonation	Combinatorial systems
Solvent extraction	Electrochemical oxidation	Aerated lagoons
Photocatalytic degradation	Fenton reactions	Surface immobilization

Under bioremediation processes, the use of bacteria is the favored method because of their versatility, dynamic metabolisms, and potential enzymatic machineries. The use of bacteria of both aerobic and anaerobic types provides exciting avenues for treatment of dye-containing effluents and wastewater. Bacterial remediation is advocated as a non-hazardous, efficient, and environmentally friendly alternative method for the treatment of extremely toxic xenobiotic compounds such as textile dyes. This chapter focuses on the bacterial degradation of textile dyes in different conditions, their metabolism, laboratory scale and industrial scale reactors, and various product and toxicity analyses.

14.2 Water Consumption in Textile Dying Processes is a Key Problem

The textile industry requires huge volumes of water for dyeing and printing of textiles and other goods using various chemicals all of which, in solid or liquid form, contribute to the COD, BOD, and TOC and moreover, increase the TSS and TDS (Khandare et al., 2014). Specific water consumption for dyeing varies from 30 to 50 L/kg of cloth depending on the type of dye used. The overall water consumption of yarn dyeing is about 60 L/kg of yarn. The daily water consumption of an average sized textile mill with a production of about 8000 kg of fabric per day is about 1.6 million liters. Sixteen percent of this is consumed in dyeing and 8% in printing. The dyeing section contributes to 15%–20% of the total wastewater flow. Water is also required for washing the dyed and printed fabric and yarn to achieve fastness and bright color backgrounds (Kant, 2012). During all these processes, a number of other chemical such as mordants, fastners, fixing agents, surfactants, defoamers, acids, bases, and alkalis are utilized which contribute to the COD, BOD, TSS, TDS, and TOC of the wastewater making it even more polluted (Kurade et al., 2012a). The World Bank estimates that 17%–20% of industrial water pollution comes from textile dyeing and finishing treatments given to fabric. Some 72 toxic chemicals have been identified in water solely from textile dyeing, 30 of which cannot be removed (Chen and Burns, 2006). The textile industry has been condemned as the world's worst polluters of the environment. Water is needed to convey the chemicals into the fabric and to wash it at the beginning and end of every step. It becomes full of chemical additives and is then expelled as wastewater; which in turn pollutes the environment. Water, a finite resource is thus becoming scarce (Kant, 2012). It is a well known fact that the dyes even in very small quantity show brilliance making the colored water aesthetically unacceptable. Therefore the treatment of colored water is the need of the hour (Govindwar and Kagalkar, 2010).

14.3 Biological Methods for Dye Removal from Effluents

In the modern world, after the industrial revolution in Europe, most of the dye processing and manufacturing industries were established in the developing countries of the Indian subcontinent and other Asian nations. The textile wastewater as earlier mentioned is one of the most toxic in nature; therefore a number of technologies have been developed for its treatment including physical, chemical, biological, and combinatorial systems. The use of living organisms of various classes for the treatment of pollutants such as textile dyes is called "bioremediation." With the advent of modern environmentally friendly methods the use of micro-organisms has been globally advocated by environmentalists. Microbes, because of their diverse and dynamic nature, can get acclimatized to the toxic waste environment and new resistant strains capable of surviving under such stressful conditions develop naturally. This adaptation equips them with new weapons to utilize the pollutant as their food material and transform various toxic chemicals into less harmful products (Saratale et al., 2007). The use of microbes and their enzymatic treatment method for the complete decolorization and degradation of such dyes from textile effluent is advantageous in many ways, for example, being environmentally friendly, cost-effective, less sludge forming, yielding harmless end products that are nontoxic and/or could bring about complete mineralization and most importantly require less water consumption compared with physicochemical methods (Rai et al., 2005; Jadhav and Govindwar, 2006; Kalyani et al., 2008a).

The effectiveness of microbial decolorization depends on the adaptability and activity of the selected microorganisms. A large number of species have been tested for decolorization and mineralization of various textile dyes in recent years (Pandey et al., 2007). A wide range of microorganisms are capable of decolorizing a wide variety of dyes, for instances, fungi (Saratale et al., 2006; Humnabadkar et al., 2008), yeasts (Lucas et al., 2006; Jadhav et al., 2007, 2008a,b,c; Waghmode et al., 2011a,b), actinomycetes (Machado et al., 2006), algae (Parikh and Madamwar, 2005; Daeshwar et al., 2007; Aravindhan et al., 2007; Khataee et al., 2013) and plants (phytoremediation) (Ghodake et al., 2009a; Kagalkar et al., 2009, 2010, 2011; Kabra et al., 2011a,b, 2012, 2013; Khandare et al., 2011a,b, 2012, 2013a,b, 2014; Watharkar et al., 2013a,b). Moreover, plants are even capable of completely mineralizing many azo dyes under certain environmental conditions.

Yeast species with a potential for dye degradation have also been reported. *Galactomyces geotrichum* was able to decolorize reactive, azo, and triphenylmethane textile dyes (Jadhav et al., 2008a, 2009a). *Saccharomyces cerevisiae* MTCC-463 has been reported to have a role in the degradation of Methyl Red and Malachite Green (Jadhav and Govindwar, 2006; Jadhav et al., 2007). Moreover, *Saccharomyces cerevisiae* was also found to show bioaccumulation of reactive textile dyes such as Remazol Red RB, Remazol Black B, and Remazol Blue during growth in molasses-containing media (Aksu and Donmez, 2003). Decolorization of a highly sulfonated Navy Blue HER dye by *Trichosporon beigelii* NCIM-3326 has been proposed with the enzymatic mechanism, and toxicity of the degradation products was also reported (Saratale et al., 2009a).

Phytoremediation of textile dyes is comparatively a new approach, and to date, only a limited number of plant species for instance, *Phragmites australis, Typha angustifolia, Blumea malcommi, Tagetes patula, Typhonium flagelliforme, Eucalyptus* sp., *Glandullaria pulchella, Sesuvium portulacastrum, Gaillardia grandiflora, Petunia grandiflora, Zinnia angustifolia,* and *Aster amellus* have been explored for the degradation of textile effluents and dyes namely Brilliant Blue R, Direct Red 5B, Basic Red 46, Scarlett RR, Acid Blue 92, Remazol Red, Remazol Black B, Remazol Orange 3R, Green HE4B, Navy Blue HE2R, Navy Blue RX, and many simulated dye mixtures (Kagalkar et al., 2009, 2010; Patil et al., 2009, 2012; Govindwar and Kagalkar, 2010; Kabra et al., 2011a,b, 2012; Khandare et al., 2011a,b, 2012; Khataee et al., 2012; Watharkar et al., 2013a,b; Watharkar and Jadhav, 2014; Torbati et al., 2014).

Bacteria are also known to possess tremendous potential for dye degradation (Jadhav et al., 2007; Dawkar et al., 2008; Kalyani et al., 2008a,b, 2009; Patil et al., 2008; Telke et al., 2008; Saratale et al., 2009b; Tamboli et al., 2010a,b; Kurade et al., 2011; Phugare et al., 2011a,b).

14.3.1 Bacteria as Remediators of Textile Dyes

Bacteria as mentioned earlier have a diverse and dynamic metabolism. The mechanism of bacterial degradation of dyes involves various oxidoreductive enzymes which utilize these complex xenobiotic compounds as substrates and convert them to less complex metabolites. Most of the bacteria having the dye degradation mechanism are generally obtained from actual sites of dye disposal or from real textile effluents. The literature provides thorough information on the utilization of pure as well as mixed bacterial cultures.

14.3.1.1 Pure and Mixed Cultures of Bacteria for Dye Decolorization

As earlier mentioned, efforts to isolate pure bacterial cultures capable of degrading textile dyes started in the 1970s and *Bacillus cereus, Bacillus subtilis,* and *Aeromonas hydrophila* were shown to be potent remediators (Wuhrmann et al., 1980). Pure culture systems (Table 14.2) ensure reproducible data and the interpretation of experimental observations becomes practical. It also becomes easier to determine the detailed mechanisms. The understanding of the metabolism of dyes using pure culture based on the involvement of oxidoreductive enzymes' and various analytical methods has been well studied and documented (Kalme et al., 2007a,b; Dhanve et al., 2008; Jadhav et al., 2009b,c, 2011; Telke et al., 2009a,b, 2011; Dawkar et al., 2010; Tamboli et al., 2011). Mixed cultures of bacteria were proposed to be useful for industrial dye-containing effluents as the synergistic systems have proved extremely effective to collectively carry out biodegradation tasks that no individual pure strain can undertake successfully (Jadhav et al. 2010). However, mixed cultures do not provide the exact view of the dye metabolism and therefore the results are not easily reproduced making the interpretation of the results quite difficult.

Bacterial decolorization is efficacious and fast but individual pure cultures usually cannot degrade dyes completely and render some of the products as untreated which have been found to be carcinogenic aromatic amines which need to be further treated (Joshi et al., 2008). However, the mixed cultures are able to achieve higher degrees of degradation and mineralization due to synergistic activities during the metabolism (Chen and Chang, 2007; Saratale et al., 2010). In bacterial synergistic systems, individual strains may attack the dye molecule at different positions and/or could utilize metabolites produced by the other strains of the system (Chang et al., 2004; Jadhav et al., 2008b; Saratale et al., 2009b). Moreover, it is difficult to isolate pure bacterial strains from dye effluents and some of the bacteria are rendered uncultured as laboratory conditions are unfavorable for such bacterial strains. A number of reports on biodegradation studies of dyes using mixed and cocultures of bacteria have been listed in Table 14.3.

For the convenience of understanding, the decolorization of dyes by bacteria can be studied in three divisions viz. anaerobic, anoxic, and aerobic.

14.3.1.2 Decolorization of Dyes under Anaerobic Conditions

The research on bacterial dye reduction has been focused on the activity of (facultative) anaerobic bacteria (Telke et al., 2008). Bacterial strains from a number of origins have been used to investigate anaerobic dye degradation including pure and mixed cultures. Gingell and Walker (1971) explored *Pseudomonas faecalis* to degrade Red 2 G dye. *Bacillus subtilis* and *A. hydrophila* (Horitsu et al., 1977; Idaka and Ogawa, 1978) followed by *B. cereus* (Wuhrmann et al., 1980) were also found to degrade textile dyes. In case of azo dyes, the cleavage of –N=N– bond is the initial step. Decomposition of dyes under anoxic conditions is influenced by a variety of substrates. Substrates such as glucose, ethanol, starch, acetate, tapioca, and whey have been explored for dye degradation under methanogenic conditions (Isik and Sponza, 2005).

The anaerobic decolorization of dyes was thought to be because of the activities of methanogens (Razo Flores et al., 1997). Further studies showed that acidogenic as well as methanogenic bacteria play a key role in dye decolorization (Bras et al., 2001). Anaerobic-baffled reactors treating textile dye waste revealed that the g-*proteobacteria* and sulfate reducing bacteria were prominent members of mixed bacterial populations along with dominant bacteria such as the *Methanosaeta* species and *Methanomethylovorans hollandica* which acted as the main contributor (Plumb et al., 2001). Bacterial strains such as *B. subtilis, Pseudomonas luteola, Proteus mirabilis,* and *A. hydrophila,* sp. decolorized textile dyes under anoxic conditions (Chen et al., 1999, 2003; Chang et al., 2001; Yu et al., 2001). Although some of these cultures could grow aerobically, decolorization was achieved only under anaerobic conditions. Various large scale applications have also been proposed which focus on anaerobic sediments, digester sludge, anaerobic granular sludge, and activated sludge (Brown and Laboureur, 1983; Weber and Wolfe, 1987; Razo Flores et al., 1997; Beydilli et al., 1998; Bromly-Challenor et al., 2000). Anaerobic bacteria have also been used in the constructed bioreactors for the treatment of textile dyes and industrial effluent (Isik and Sponza, 2008; Mezohegyi et al., 2008).

Anaerobic treatments are known to cause the formation of aromatic amines. Therefore, the formation of aromatic amines during the anaerobic treatment of textile wastewater containing azo dyes still remains a matter of concern. The azo reductase plays the main part in breaking the azo bonds in the dye structure reducing equivalents such as FADH and NADH. Many azo dyes have sulfonate substituent groups and thus a high molecular weight which are unlikely to pass through the membranes of the bacterial cells. Therefore, the intracellular uptake of the dye for reducing activity becomes important (Robinson et al., 2001). The bacterial membranes are also known to be almost impermeable to flavin-containing cofactors and, thus restrict the transfer of reducing equivalents by flavins from the inside of the cell to the sulfonated azo dyes. Therefore, it becomes obvious that a mechanism other than reduction by reduced flavins produced by flavin-dependent azoreductases from the cells' internal must be responsible for sulfonated azo dye reduction (Russ et al., 2000). Myers and Myers (1992) have already suggested a mechanism which involves the electron transport linked reduction of the azo dyes extracellularly. For this, the bacterial cells must establish a connection between their intracellular electron transport systems and the dye molecules. To achieve this, the components of the

TABLE 14.2

Pure Cultures of Bacteria Degradation for Some of the Textile Dyes

Name of the Strain	Dye and Concentration	% Decolorization	Time (h)	Enzymes	References
Aeromonas hydrophila	Reactive Red 141, 3.8 g/L	70%–80%	24	NA	Chen et al. (2009a,b)
Sphingomonas sp. BN6	Acid azo dyes, Direct azo dyes, and Amaranth (0.1 µM)	NA	NA	Flavin Reductase	Russ et al. (2000)
Kocuria rosea MTCC 1532	Malachite Green, Methyl Orange (50 mg/L each)	80–100	5	DCIP Reductase, MG reductase, LiP, laccase, tyrosinase	Parshetti et al. (2006, 2010)
Pseudomonas desmolyticum	Direct Blue 6 and Red HE7B, Green HE4B (100 mg/L each)	70 and 90	72	LiP, laccase, tyrosinase	Kalm et al. (2007a,b, 2008, 2009)
Bacillus sp. VUS	Brown 3REL, Red HE7B, Reactive Orange 16, Navu Blue 2GLL, Orange T4LL (50 mg/L each)	70–100	5–48	DCIP Reductase, Riboflavin reductase, LiP, laccase, tyrosinase	Dawkar et al. (2008, 2009a,b, 2010), Jadhav et al. (2008b)
Pseudomonas sp. SUK1	Red BLI, Reactive Red 2, Reactive Blue 13 (50 mg/L each)	100	1	LiP, laccase tyrosinase, aminopyrine N-demethylase	Kalyani et al. (2008, 2009, 2010)
Rhizobium radiobacter	Reactive Red 141, Methyl Violet (10–50 mg/L)	90	8–49	LiP, MG reductase, DCIP reductase	Telke et al. (2008); Parshetti et al. (2009)
Comamonas sp. UVS	Direct Blue GL, Direct Red 5B, Red HE7B (50–1100 mg/L)	100	15–125	LiP, laccase tyrosinase, Veratryl alcohol oxidase	Jadhav et al. (2008c, 2009b,c)
Bacillus odysseyi SUK3, *Morganella morganii* SUK5 and *Proteus* sp. SUK7	Red HE3B, Reactive Blue 59 (50 mg/Leach)	97–99	12	LiP, laccase, aminopyrine N-demethylase and MG reductase, veratryl alcohol oxidase	Patil et al. (2008, 2010)
Acinetobacter calcoaceticus	Methyl Red, Methyl Orange (50 mg/L each)	90–98	24–48	LiP	Ghodake et al. (2009b,c)
Brevibacillus laterosporus	Direct Brown MR, Methyl orange, Blue-2B, Golden Yellow HER, Methyl Red, Remazol Red, Rubine GFL, Scarlett RR, Brilliant Blue, Brown 3R, Disperse Brown 118 (50–100 mg/L each)	80–95	12–72	LiP, aminopyrine N-demethylase and MG reductase, veratryl alcohol oxidase	Gomare et al. (2009), Gomare and Govindwar (2009), Kurade et al. (2011, 2013)

(Continued)

TABLE 14.2 (*Continued*)

Pure Cultures of Bacteria Degradation for Some of the Textile Dyes

Name of the Strain	Dye and Concentration	% Decolorization	Time (h)	Enzymes	References
Pseudomonas aeruginosa	Direct Orange 39, Remazol Black, Remazol Orange, Acid Violet 19, Remazol Red	60–90	5–48	LiP, DCIP reductase, tyrosinase, veratryl alcohol oxidase	Jadhav et al. (2011, 2012a,b, 2013a,b)
Proteus mirabilis	Reactive Blue 13 (100 mg/L each)	70–90	5	azo reductase, LiP, laccase tyrosinase,	Olukanni et al. (2010)
Pseudomonas sp. SU-EBT	Congo Red (100 mg/L each)	90	12	Laccase, azo reductase, DCIP reductase	Telke et al. (2010)
Sphingobacterium sp.	Direct Red 5B, Orange 3R, Direct Blue GLL (300 mg/L each)	45–65	8	Aryl Alcohol Oxidase	Tamboli et al. (2010a,b,c, 2011)
Bacillus sp. ADR	Methyl red, Reactive orange 16, Methyl orange (50 mg/L each)	90, 85	1	Laccase	Telke et al. (2011)
Providencia sp. SDS	Methyl Orange Remazol Black, Red HE8B, Red HE7B, Congo Red (50 mg/L each)	90–100	6–12	Veratryl alcohol oxidase, laccase, azo reductase, DCIP reductase	Phugare et al. (2011a,b)

electron transport system must be localized in the outer membrane of the bacterial cells, so that, they can make direct contact with the substrate or a redox mediator at surface of the cells. Gingell and Walker (1971) have proved that low redox mediator compounds can behave as electron shuttles between the dye and an NADH-dependent azo reductase which is situated in the outer membrane (Figure 14.1) (Keck et al., 1997; Chacko and Subramaniam, 2011).

The rate of anaerobic dye decolorization has been found to be nonspecific and mainly depends on organic carbon sources and the dye structure (Bromly-Challenor et al., 2000; Stolz, 2001). Anaerobic azo dye decolorization is a fortuitous process, where the dye might act as an electron acceptor of the electron transport chain. Van der Zee et al. (2001) proposed that the decolorization could be because of extracellular reactions occurring between reduced compounds generated by the anaerobic biomass. Generally, dye decolorization reactions follow first-order kinetics with respect to dye concentration and sometimes zero-order kinetics has also been observed (Isik and Sponza, 2005).

14.3.1.3 Decolorization of Dyes under Anoxic Conditions

Anoxic environments have less than 0.5 mg/L dissolved oxygen but are never completely oxygen-free as in anaerobic processes. Anoxic treatment can be carried out under operating conditions which are similar to aerobic treatments. Although, these treatment processes have a low efficacy of dye treatment they have been found to be useful for the removal of azo dyes from wastewater (Gottlieb et al. 2003). Anoxic decolorization of various dyes using mixed aerobic and facultative anaerobic microbial consortia has been

TABLE 14.3

Mixed and Co-Cultures Bacterial Degradation of Some of the Textile Dyes

Name of the Strains	Dye and Concentration	% Decolorization	Time (h)	Enzymes	References
Gloeocapsa pleurocapsoides and *Chroococcus minutus*	Direct Blue-15 (50 mg/L)	78–90	23 days	Oxidative	Parikh and Madamwar (2005)
Alcaligenes faecalis, Sphingomonas sp. EBD, *Bacillus subtilis, Bacillus thuringiensis,* and *Enterobacter cancerogenus*	Acid Red 97, FF Sky Blue, Amido Black 10B (100 mg/L)	92	24	NA	Kumar et al. (2007)
Pseudomonas aeruginosa and *Bacillus circulans*	Reactive Black 5 (100 mg/L)	70–90	48	Unknown enzyme secretion	Dafale et al. (2008)
Bacillus odysseyi SUK3, *Morganella morganii* SUK5, and *Proteus* sp. SUK7	Reactive Blue 59, Red HE3B (50 mg/L)	99–100	3–12	Oxidative, reductive and aminopyrine *N*-demethylase	Patil et al. (2008, 2010)
Unknown coculture	Direct Red 81 (200 mg/L),	90	35	NA	Madamwar et al. (2004)
Bacillus cereus, Pseudomonas putida, Pseudomonas fluorescence, and *Stenotrophomonas acidaminiphila*	Acid Red 88, Acid Red 119, Acid Red 97, Acid Blue 113, Reactive Red 120 (60 mg/L each)	78–90	24	Reductive	Khehra et al. (2005)
Bacillus cereus, Pseudomonas putida, Pseudomonas fluorescence, and *Stenotrophomonas acidaminiphila*	Acid Red 88 (100 mg/L)	98	3 months	Reductive	Khehra et al. (2006)
Pseudomonas vulgaris and *Micrococcus glutamicus*	Scarlet R and mixture of 8 dyes (50 mg/L each)	100	3	Reductive	Saratale et al. (2009b)
Proteus vulgaris and *Miccrococcus glutamicus*	Green HE4BD, mixture of six reactive dyes (50 mg/L each)	100	24	Oxidative and reductive	Saratale et al. (2010a)
Pseudomonas sp.	Reactive Orange 16 (100 mg/L)	100	48	Laccase and reductase	Jadhav et al. (2010)

Note: NA, Information not available.

reported (Nigam et al., 1996a; Kapdan et al., 2000; Padmavathy et al., 2003; Moosvi et al., 2005). Dye decolorization using pure and/or mixed cultures under anoxic conditions required complex organic sources such as yeast extract, peptone, or a combination of complex organic source and carbohydrate (Khehra et al., 2005). The economy of the process is the key factor when proposed for industries but the use of costly carbon and nitrogen sources increase the costs (Nigam et al., 1996b; Moosvi et al., 2005). Although glucose has

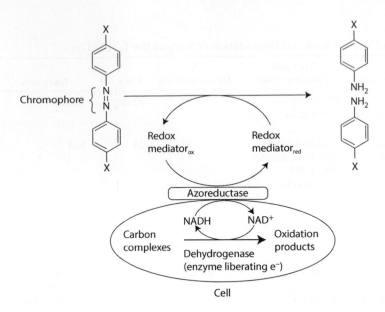

FIGURE 14.1
Mechanism of anaerobic reduction proposed by Keck et al. (1997). (Adapted from Chacko J and Subramaniam K. 2011. *International Journal of Environmental Science* 1: 1250–1260.)

been the choice of carbon source for anaerobic dye degradation it has been found to affect anoxic decolorization positively as well as negatively, showing fluctuating performances (Pandey et al., 2007).

14.3.1.4 Decolorization of Dyes under Aerobic Conditions

A number of bacterial strains that can aerobically decolorize various textile dyes have been isolated and explored under different conditions. Many of these strains require organic carbon sources for growth as they cannot utilize dye as the substrate (Stolz, 2001). *Pseudomonas aeruginosa* was able to decolorize Navitan Fast blue S5R, at a commercial tannery in the presence of glucose (Nachiyar and Rajkumar, 2005). There are only a very few bacteria that are able to grow on azo compounds as the sole carbon source (Pandey et al., 2007). These bacteria are able to reductively cleave –N=N– bonds and utilize amines as the carbon source for their growth and in fact utilize them as specific substrate. For instance, *Pigmentiphaga kullae* K24 and *Xenophilus azovorans* KF 46 which were found to grow aerobically on carboxy-orange II and carboxy-orange I dyes respectively (Zimmermann et al., 1982; Kulla et al., 1983). However, these organisms could not grow on structurally analogous dyes such as Acid Orange 20 and Acid Orange 7. *Sphingomonas* strain 1CX an obligate aerobic organism can grow on an azo dye Acid Orange 7 as sole carbon and nitrogen source. This strain could degrade only one of the formed amines namely 1-amino 2-naphthol during Acid Orange 7 decolorization. *Sphingomonas* ICX could also decolorize several other dyes with 1-amino-2-naphthol or 2-amino-1-naphthol bonded via azo bonds to a phenyl or naphthyl chemical species (Coughlin et al., 1999). *P. nitroreducens* and *Vibrio logei* were found to be able to use Methyl Red as the sole carbon source and no amine was detected in the degradation products (Adedayo et al., 2004). *Bacillus* sp. OY1-2, *Xanthomonas* sp. NR25-2, and *Pseudomonas* sp. PR41-1could utilize Acid Red 88 as the sole carbon source (Sugiura et al., 1999).

Anaerobic cleavage of azo dyes by involvement of redox mediators produced during the aerobic degradation of a xenobiotic compound was reported by Keck et al. (1997). *Sphingomonas* sp. strain BN6 Cell suspensions were grown aerobically in the presence of 2-naphthyl sulfonate and were found to show a 10–20 fold increase in the decolorization rate of Amaranth under anaerobic conditions. The mere addition of culture filtrates from these cells was found to enhance the anaerobic decolorization by cells 2-naphthyl sulfonate. A mechanism of the degradation of azo dyes was proposed based on these observations. Thus, the generated redox intermediates during the aerobic degradation of an aromatic compound could have a positive effect and enhance the decolorization process in anaerobic conditions (Keck et al., 1997). Aerobic bacteria are known to possess oxido-reductive enzymatic machineries and therefore can break the dye molecules symmetrically or asymmetrically and could also bring about deamination, desulfonation, hydroxylation, etc. Therefore, aerobic degradation can also achieve the breakdown of different dye structures. For instance, Kalyani et al. (2009) have proposed that *Pseudomonas* sp. SUK1 showed peroxidase activity which catalyzed the initially asymmetric cleavage of Reactive Red 2 dye. The resultant intermediate product was identified as phenylazo-2-naphthol 5,8-disulfonic acid and 1,3,5-triazine 2,4 diol revealing the oxidative asymmetric cleavage of the dye. The phenylazo-2-naphthol 5,8-disulfonic acid was then shown to be reduced by azo reductase giving rise to the reactive intermediate product in this reaction as 1-amino-8-naphthol 2,5-disulfonic acid and aniline. 1-Amino-8-naphthol 2,5-disulfonic acid was then shown to undergo desulfonation leading to formation of 1-amino-8-naphthol which after deamination gave 2-naphthol (Figure 14.2).

The involvement of oxidative and reductive enzymes is the real advantage in aerobic processes and unlike anaerobic degradation the aromatic amines are eliminated but the products formed still remain recalcitrant to degradation.

14.4 Combinatorial Systems of Bacteria with Fungi and/or Plants Put Up a Better Fight

As mentioned earlier, the mixed and cocultures of bacteria prove to be effective for bioremediation of textile dyes because of their synergistic metabolism. Similarly, bacteria along with other organisms such as fungi (Lade et al., 2012), yeast and plants have been proposed to be more efficient for textile dye removal.

A fungal-bacterial consortium can give enhanced degradation and detoxification of dyes and provides an alternate way for efficient removal of contaminants such as textile dyes as well as effluents (Khelifi et al., 2009; Su et al., 2009; Qu et al., 2010). Their eco-friendly nature, efficiency, and short degradation times are some of the important advantages of fungal–bacterial synergism over individual cultures. Such consortia are more effective due to the co-metabolic activities which allow the involved microbes to directly act on different positions of the dye and also on the intermediates (Keck et al., 2002; Chen and Chang, 2007). Some reports have demonstrated the potential of a fungal-bacterial consortium for enhanced degradation of textile dyestuff, for instance, free as well as the co-immobilized fungal strain *Penicillium* sp. QQ and bacterial strain *Exiguobacterium* sp TL were found to synergistically degrade Reactive Dark Blue K-R dye (Su et al., 2009; Qu et al., 2010), a developed consortium of *Pseudomonas* sp. SUK1 and *Aspergillus ochraceus* NCIM-1146 was

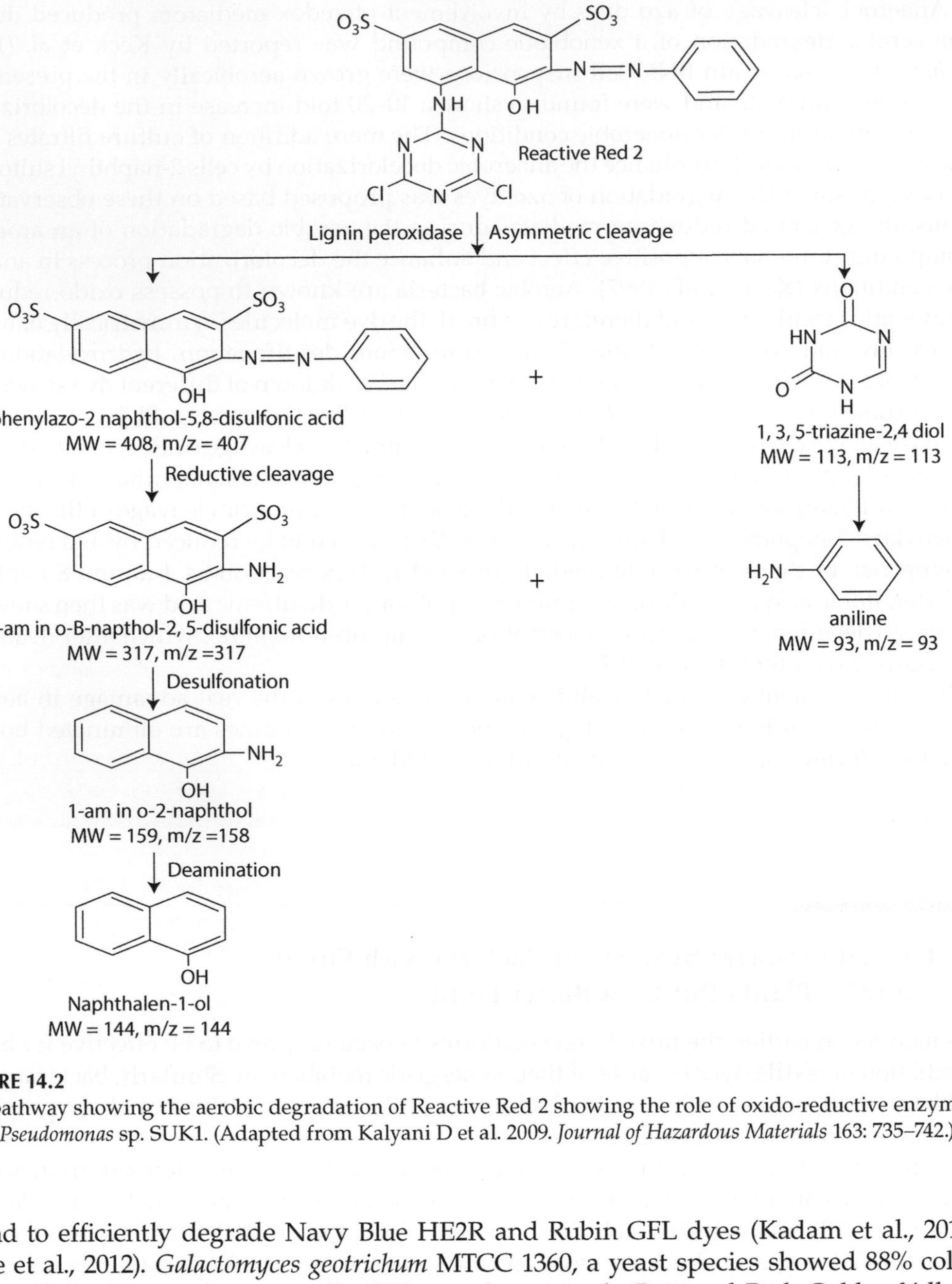

FIGURE 14.2
The pathway showing the aerobic degradation of Reactive Red 2 showing the role of oxido-reductive enzymes from *Pseudomonas* sp. SUK1. (Adapted from Kalyani D et al. 2009. *Journal of Hazardous Materials* 163: 735–742.)

found to efficiently degrade Navy Blue HE2R and Rubin GFL dyes (Kadam et al., 2011; Lade et al., 2012). *Galactomyces geotrichum* MTCC 1360, a yeast species showed 88% color removal of mixture of structurally different dyes namely Remazol Red, Golden Yellow HER, Rubine GFL, Scarlet RR, Methyl Red, Brown 3 REL, and Brilliant Blue in a consortium with *Brevibacillus laterosporus* (Waghmode et al., 2011a,b, 2012). *Bacillus cereus* was shown to decolorize adsorbed textile dyestuff on distillery industrial waste-yeast biomass under solid state fermentation (Kadam et al., 2013a). *Providensia* staurti strain EbtSPG was also able to degrade Solvent Red 5B dye adsorbed on sugar cane bagasse (Kadam et al., 2013b).

The bacterial-assisted plant-based treatments provide an ecofriendly, aesthetically pleasant, low cost and possible *in situ* alternative for textile dye removal. Tissue cultures of

Z. angustifolia plants and its root associated bacterium *Exiguobacterium aestuarii* showed enhanced degradation of Remazol Black B compared with the individual cultures in a developed consortium (Khandare et al., 2012). In another study, the *Portulaca grandiflora* and *Pseudomonas putida* consortium was found to show efficient degradation of Direct Red 5B dye (Khandare et al., 2013b). *Bacillus pumilus* was found to enhance the degradation of Navy Blue RX dye by *Pe. grandiflora* plants (Watharkar et al., 2013b). Recently, plant growth promoting rhizobacteria such as *Rhodobacter erythropholi, Azotobacter vinelandii, Rhizobium meliloti,* and *Bacillus megaterium* were found to show a great potential to treat textile dyes (Kadam et al., 2014).

14.5 Development of Bioreactors to Explore the Bacterial Dye Degradation Potential

Reactor development becomes crucial when laboratory observations have to be tried on actual industrial wastewater. Some workers have designed and developed such bioreactors and have shown the actual degradation of a number of dyes and effluents using them. *Pseudomonas mendocina* could effectively decolorize methyl violet in textile wastewater, with the use of a fixed-film reactor (Kanekar and Sarnaik, 1995; Kanekar et al., 1996). The synthetic dye Tartrazine was found to be readily decolorized in an anaerobic baffled reactor (Plumb et al., 2001). Disperse Blue 79 was also shown to be reduced in an anoxic sediment–water system (Weber and Adams, 1995). Azo dye decolorization in a continuous mode using fixed-bed bioreactors with gel-entrapped cells of *Ps. luteola* has been successfully demonstrated (Chen et al., 2005). A continuously running bacteria-fungi co-existence biofilm reactor with high efficiencies of dye degradation and textile wastewater treatment was successfully established with mixed cultures (Yang et al., 2009). It was found that, with constant feeding, dye concentration of 50 mg/L the beds with calcium alginate-immobilized cells had an optimal volumetric decolorization rate of 30.6 mg/h/L and a specific decolorization rate of 2.61 mg/g/cell/h when HRT and dye loading rate was 1.12 h and 2.25 mg/h, respectively. Sharma et al. (2004a) set up an upflow immobilized cell bioreactor having a consortium comprised of *Bacillus* sp., *Alcaligenes* sp., and *Aeromonas* sp. which formed a multispecies biofilm on refractory brick pieces used as support material and was further used to treat triphenylmethane textile dye Acid Blue-15. This dye was degraded to simple metabolic intermediates in the bioreactor with 94% decolorization at a flow rate of 4 mL/h. Khehra et al. (2006) evaluated the potential of a sequential anoxic aerobic bioreactor to decolorize azo dye Acid Red 88. In this approach, an upflow fixed-film column reactor having polyurethane foam as immobilization support was built using a consortium based on four acclimatized bacterial strains belonging to *Stenotrophomonas* sp., *Pseudomonas* sp., and *Bacillus* sp. that biotransformed azo dye AR-88 to nonaromatic metabolic intermediates. Likewise, aerobic mixed bacterial culture comprised of five isolates namely *Bacillus pumilus, B. cereus, B. vallismortis, B. megaterium,* and *B. subtilis* could efficiently decolorize the Direct Red 28 dye in an upflow immobilized packed bed bioreactor using marble chips as support matrix with an efficiency of 91% decolorization with 60 mL/h (Tony et al., 2009a,b). In an experiment, a combined anaerobic–aerobic treatment process based on a mixed culture of bacteria was used to degrade azo dyes. The experiment was integrated by exposing anaerobic granular sludge and aerobic aromatic amine-degrading bacterial enrichment cultures. The combined anaerobic–aerobic bioreactor was able to completely remove the AY-36 at a maximum

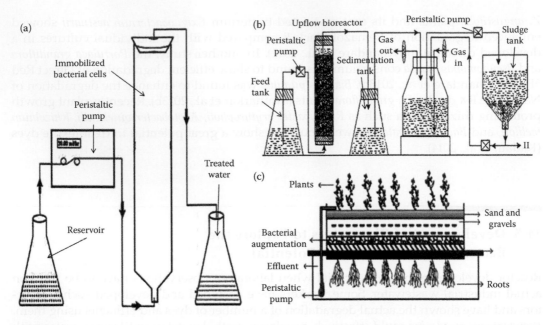

FIGURE 14.3
Different types of bacterial bioreactors (a) an upflow bioreactor with immobilized bacterial cells (Adapted from Jadhav S et al. 2013a. *Environmental Science and Pollution Research* 20: 2854–2866.), (b) an upflow and coupled bioreactor with sludge recycling facility (Adapted from Khehra M et al. 2006. *Dyes and Pigments* 70: 1–7.), (c) a phytoreactor with bacterial augmentation enhancing the efficacy of the system (Adapted from Kabra A, Khandare R, and Govindwar S. 2013. *Water Research* 47: 1035–1048.; Khandare R et al. 2013b. *International Biodeterioration and Biodegradation* 78: 89–97.).

loading rate of 100 mg dye/L/day (Ahmad et al., 2010). Recently, a sequential microaerophilic process has been shown to degrade textile dyes of various classes (Waghmode et al., 2011b, 2012). A batch and continuous upflow constructed upflow bioreactor was shown to degrade Amaranth in plain distilled water by *Ps. aeruginosa* BCH (Jadhav et al., 2013a,b). The immobilized bacterial cells of *Proteus vulgaris* in a *Luffa cylindrica* dried fruit net making it a constructed bioreactor with enhanced exposure to dyes with the cells treated Reactive Blue 172 dye to nontoxic levels (Saratale et al., 2011a). Microbial immobilization in a two-stage fixed bed reactor pilot plant for on-site anaerobic decolorization of textile wastewater was established and run successfully (Georgiou et al., 2005). The developed phytoreactors of *Po. grandiflora* and *G. pulchella* assisted with *Ps. putida* and *Pseudomonas montelii* respectively were found to decolorize different samples of dye mixtures and real textile effluents (Kabra et al., 2013; Khandare et al., 2013b). Thus mixed systems in reactors have proven to be effective for dye removal from wastewater. Different reactors with bacterial involvement have been shown in Figure 14.3.

14.6 Understanding the Mechanism of Dye Metabolism by Bacteria

Anaerobic biodegradation of textile dyes have been investigated since the 1970s. In this field, many anaerobic bacteria have been employed for the decolorization of the artificial solution of one or more dyes. In 1971, Walker and Ryan studied azo dye degradation by using intestinal anaerobes (Walker and Ryan, 1971). Rafii et al. (1990) reported the presence of azoreductases in the anaerobic bacteria *Clostridium* and *Eubacterium*. Further investigations showed that most azoreductases are known to be sensitive to oxygen, and later Rafii and Cerniglia (1995) proved that that azoreductase are capable of degrading nitro aromatic compounds. Many researchers have conducted interesting experiments with anaerobic degradation. Many of the investigations focused on one or several species isolated from an environment considered to be interesting, for example, textile effluent or contaminated soil. Their degradation performances were tested under different conditions (Kalyani et al., 2008a). These studies were usually performed during sterile laboratory conditions on artificial dye waters. Different strains of *Pseudomonas* and *Bacillus* were found to degrade several dyes (Dafale et al., 2008). However, color reduction is not enough and metabolites and chemicals can still linger in the water. The dyes are often cleaved into aromatic amines, which absorb light in other wavelengths than most dyes (Pinheiro et al., 2004). The presence of sulfate in dye molecules and in the water will affect the degradation process. The presence of sulfate reducing bacteria can contribute to release sulfide, which can chemically reduce some of the dyes (Dafale et al., 2008; Saratale et al., 2011b).

14.6.1 Bacterial Enzymes to Breakdown the Complex Dye Structure

Both aerobic and anaerobic bacteria have different mechanisms of degradation. Enzymes play the key role in these biotransformation mechanisms. Oxidizing enzymes such as LiP, veratryl alcohol oxidase, laccase, and tyrosinase are well known to degrade textile dyes. However, reducing enzymes such as azo reductase, riboflavin reductases, DCIP reductase, and Green HE4B reductase also break the complex dye structures (Saratale et al., 2011b).

14.6.1.1 Lignin Peroxidase (EC 1.11.1.14)

The IUPAC name of this enzyme is 1, 2-bis(3,4-dimethoxyphenyl)propane-1,3-diol: hydrogen-peroxide oxidoreductase. This heme protein is responsible for the oxidative breakdown of lignin. An enzyme oxidizes dye substrates at the heme iron as well as (3,4-dimethoxyphenyl)methanol (veratryl alcohol) to the radical cation. Molecular weight of the enzyme lies between 38 and 47 kDa. Peroxidases in particular, catalyze phenolic substrates resulting in radical formation by using hydrogen peroxide as the electron donor. Another group of peroxidases, versatile peroxidases (VP), have been invented in species of *Pleurotus* and *Bjerkandera* which is considered a hybrid of MnP and LiP. These VPs can oxidize not only Mn^{2+} but also phenolic and nonphenolic aromatic compounds including dyes (Heinfling et al., 1998a,b).

Purified LiP from *Brevibacillus laterosporous* MTCC 2298, *Bacillus* sp. VUS and *Acinetobacter calcoaceticus* NCIM 2890 have been found to efficiently decolorize several sulfonated textile dyes (Gomare et al., 2008; Dawkar et al., 2009a; Ghodake et al., 2009b). The catalytic action of peroxidase has been shown in Figure 14.4.

14.6.1.2 Aryl Alcohol Oxidase (EC 1.1.3.7)

Aryl-alcohol:oxygen oxidoreductase is the systematic name of this enzyme. Other common names are veratryl alcohol oxidase, and aromatic alcohol oxidase. This enzyme is member of the oxidoreductase family, specifically those acting on the CH–OH group of

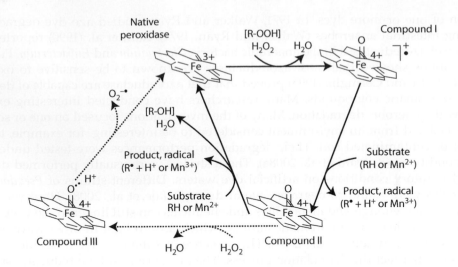

FIGURE 14.4
Catalytic cycle of peroxidase. (Adapted from Wesenberg D, Kyriakides I, and Agathos S. 2003. *Biotechnology Advances* 22: 161–187.)

donors with oxygen as acceptor. An aryl-alcohol oxidase catalyzes two substrates such as aromatic primary alcohol and O_2 and forms products such as aromaticaldehyde and H_2O_2. The bound radical of veratryl alcohol oxidase is known to bring about the oxidative cleavage of C–C and ether (C–O–C) bonds in lignin model compounds of the diarylpropane and arylpropane-aryl ether type.

$$\text{Veratryl alcohol} + O_2 \rightleftharpoons \text{an aromatic aldehyde} + H_2O_2$$

Enzyme veratryl alcohol oxidase is recognized to carry out oxidative cleavage along with desulfonation reactions for substrate breakdown. The purified form of this enzyme was able to decolorize textile dyes viz. Red HE7B and Direct Blue GLL, Remazol black, Methyl orange, Red HE3B (Jadhav et al., 2009c; Phugare et al., 2011b). An aryl alcohol oxidase action has been shown in Figure 14.5.

14.6.1.3 Laccase (EC 1.10.3.2)

Laccase is a copper-containing oxidase enzyme which performs oxidation of a wide range of organic pollutant substrates such as phenols, polyphenols, and anilines as well as highly

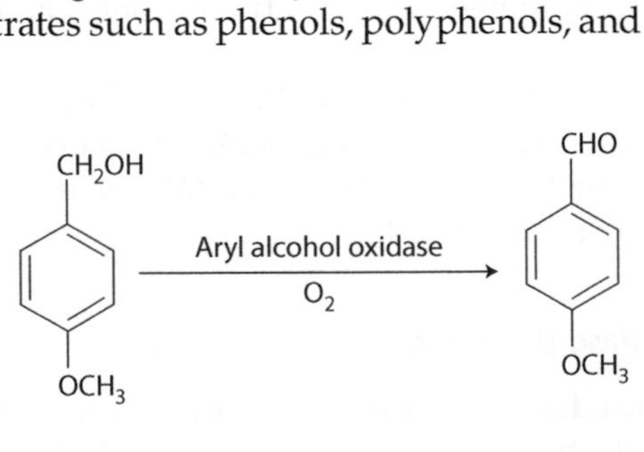

FIGURE 14.5
An aryl alcohol oxidase catalyzed reaction.

recalcitrant environmental pollutants on the basis of a one-electron transfer mechanism (Sakurai, 1992; Xu, 1996; Xu et al., 1996; Piontek et al., 2002). Laccase is therefore known to be a very useful biocatalyst in many biotechnological applications. The enzymes have been employed in the detoxification of different industrial effluents such as textile, paper, pulp, polymer synthesis, petrochemicals, bioremediation of contaminated soils, and wine and beverage stabilization. Laccase couples the oxidation of substituted phenolic and nonphenolic chemical moieties with oxygen as an electron acceptor to form free radicals. These free radicals further undergo demethylation, depolymerization, repolymerization, or quinone formation (Sharma et al., 2007). The presence of a redox mediator such as 2,2′-azinobis-[3-ethylthiazoline-6-sulfonate] (ABTS) accelerates the oxidation reaction by laccase of nonphenolic benzylalcohols (Bourbonnais and Paice, 1990). The catalytic action of laccase has been shown in Figure 14.6.

The purified laccase from *Pseudomonas desmolyticum* NCIM 2112 have been reported for 100% decolorization of various dyes such as Green HE4B, Direct Blue-6, and Red HE7B (Kalme et al., 2009). Telke et al. (2009b, 2011) have observed purified extracellular laccase involved in biodegradation processes from *Pseudomonas* sp. LBC1 and *Bacillus* sp. ADR. Laccase detoxifies azo dyes through a highly nonspecific free radical mechanism, and avoiding the formation of toxic aromatic amines (Chivukula and Renganathan, 1995).

14.6.1.4 Azo Reductase (EC 1.7.1.6)

It is located at the intracellular or extracellular site of the bacterial cell membrane. Azo reductase reduces an azo bond by NADH or NADPH or FADH which are electron donors (Russ et al., 2000). A ping-pong mechanism is a well-known mechanism in the azo bond breakdown reaction which requires 2 mol of NADH to reduce 1 mol of methyl Red (4-dimethylaminoazobenzene-2-carboxylic acid), a typical azo dye, into 2-aminobenzoic acid and *N,N*-dimethyl-*p* phenylenediamine. The mechanism of action of azo reductase is shown in Figure 14.7.

Rafii et al. (1990) first reported the presence of oxygen-sensitive azoreductase in anaerobic *Clostridium* and *Eubacterium* that decolorized sulfonated azo dyes during growth on solid

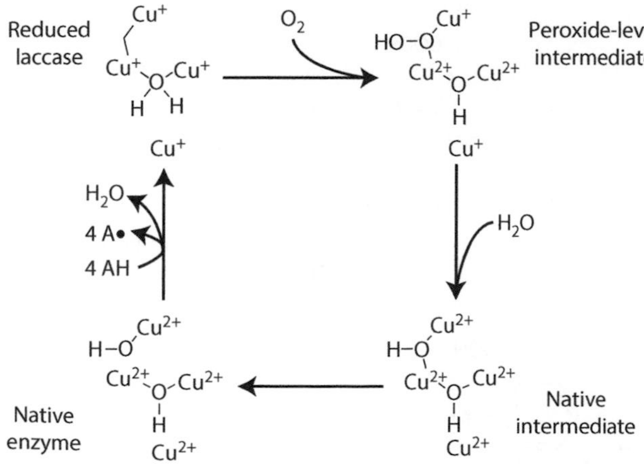

FIGURE 14.6
Catalytic activity of laccase. (Adapted from Wesenberg D, Kyriakides I, and Agathos S. 2003. *Biotechnology Advances* 22: 161–187.)

or complex media. The significant induction of azoreductase during decolorization of azo dyes under static condition was published earlier (Dawkar et al., 2008; Dhanve et al., 2008). The oxygen-sensitive Orange II azoreductase from *Pseudomonas* sp. KF46 exhibited the highest specificity toward the carboxyl group substituted sulfophenyl azo dyes. Azoreductase may generate toxic amines after reduction of an azo bond.

14.6.1.5 NADH-DCIP Reductase (EC 1.6.99.3)

This is a class of oxidase system which takes part in the detoxification of xenobiotic compounds including dyes (Salokhe and Govindwar, 1999). The NADH-DCIP reductase causes reduction of the DCIP using NADH as an electron donor. A blue colored oxidized form of DCIP becomes colorless after reduction. The catalytic activity of DCIP reductase is shown in Figure 14.8. In most of the bacterial decolorization of textile dye studies, a remarkable amount of activity of NADH-DCIP reductase can be easily observed. The reaction can be visually monitored as the blue color of DCIP clearly disappears in the given set of time conditions provided for the reaction to occur.

14.6.1.6 Tyrosinase (E.C. 1.14.18.1)

Every group of organisms including bacteria, fungi, insects, amphibians, aves, plants, and mammals possesses enzymes from this class of oxidoreductases as they have central importance in vital processes such as vertebrate's pigmentation and the browning of fruits and vegetables (Lorena et al., 2004).

Tyrosinase viz. diphenolase and catecholase catalyzes mainly two types of reactions, the oxidation of *o*-diphenols and *o*-quinones, respectively, using molecular oxygen and the *o*-hydroxylation of some monophenols (Chen and Flurkey, 2002). Tyrosine and catechol are the most prominent substrates of the enzyme tyrosinase that produces dopaquinone and *o*-benzoquinone, respectively. Figure 14.9 shows the reactions catalyzed by tyrosinase and catechol oxidase. Tyrosinase is potent in removing naturally occurring and xenobiotic aromatic compounds from aqueous suspension with efficiency (Duran and Esposito, 2000).

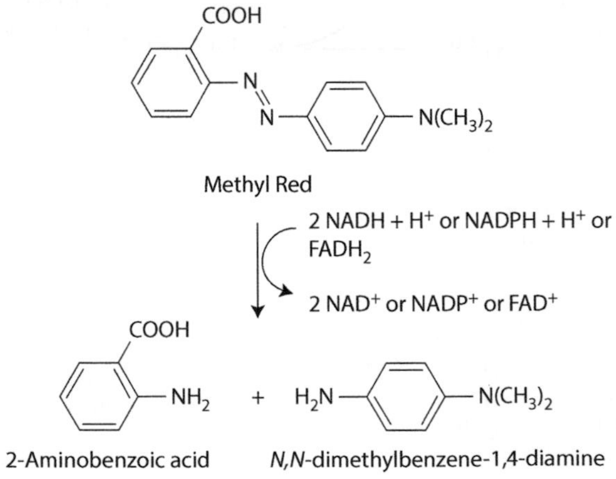

FIGURE 14.7
Mechanism of azo dye reduction by azo reductase.

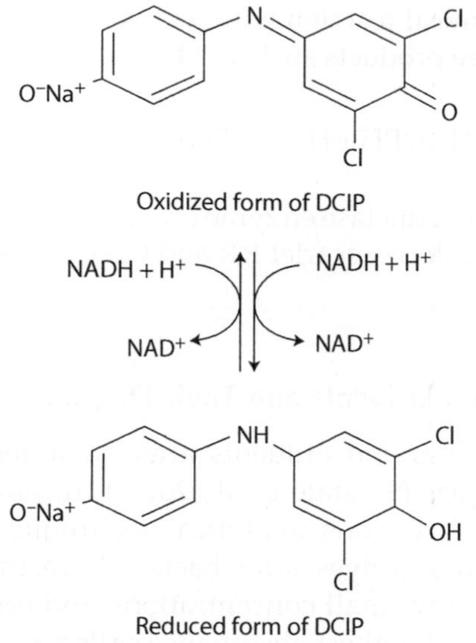

FIGURE 14.8
The catalytic activity of DCIP reductase.

14.6.1.7 Flavin Reductase (EC 1.5.1.30)

Reduced-riboflavin:NADP⁺ oxidoreductase is the systematic name of the class of this enzyme. This enzyme belongs to the family of oxidoreductases, specifically which acts on the CH–NH group of donors with NADP⁺ or NAD⁺ as acceptor. Other common names are NADPH:flavin oxidoreductase, flavin mononucleotide (FMN) reductase, riboflavin mononucleotide reductase, FMN reductase, riboflavin mononucleotide reductase, NADPH-specific FMN reductase, riboflavin mononucleotide reductase, FMN reductase (NADPH), NADPH2 dehydrogenase (flavin), NADPH-dependent FMN reductase, NADPH-FMN reductase, NADPH-flavin reductase, and NADPH:riboflavin oxidoreductase. A flavin

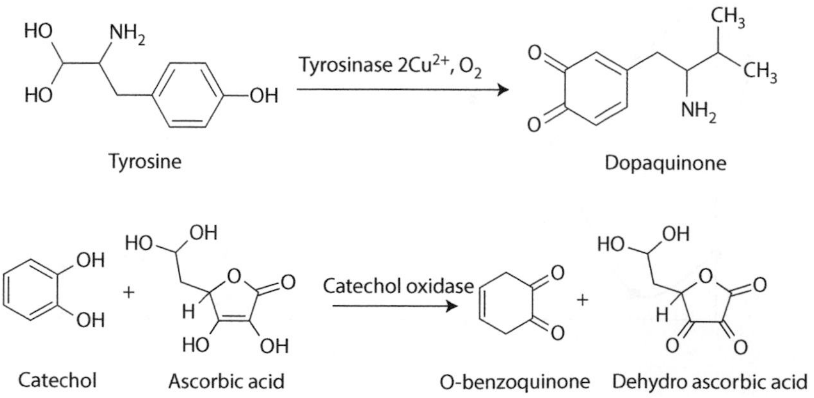

FIGURE 14.9
Reactions catalyzed by tyrosinase and catechol oxidase.

reductase catalyzes the chemical reaction with its two substrates such as reduced riboflavin and NADP⁺ to form three products such as riboflavin, NADPH, and H⁺.

$$\text{Riboflavin} + \text{NADPH} + \text{H}^+ \rightleftharpoons \text{Reduced riboflavin} + \text{NADP}^+$$

The ability of the riboflavin reductase enzyme present in consortium-GR and *Micrococcus glutamicus* NCIM-2168 to decolorize scarlet RR and Green HE4BD has been demonstrated (Saratale et al., 2009a,b).

14.6.2 Analyses of Dyes and Effluents and Their Degradation Metabolites

Bacterially treated textile dyes and effluents after treatment have been shown to be analyzed by many techniques (Saratale et al., 2011b). In order to find out the possible mechanism of dye removal, various analytical techniques are used to identify the metabolites generated from azo dyes after bacterial treatment. The dyes are known for their brilliance even at very small concentrations and each dye in pure form shows true color and a wavelength of highest absorbance called λ_{max} which could be measured by a wavelength scan (Govindwar and Kagalkar, 2010). UV–Vis spectroscopy is one of the preliminary analytical techniques to determine dye removal (Kalme et al., 2009). The disappearance of the single peak at λ_{max} is observed after decolorization of the dye and further evidence of the removal of the dye can be observed with an increase in absorbance toward the UV region (Saratale et al., 2009a). Figure 14.10 reveals the peak of Remazol Black B being lost after degradation by *E. aestuarii*. The decolorization percentage is calculated as follows

$$\text{Decolorization \%} = \frac{\text{Initial absorbance} - \text{Final absorbance}}{\text{Initial absorbance}} \times 100 \qquad (14.1)$$

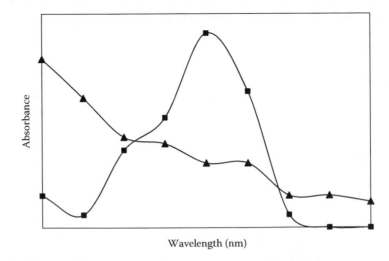

FIGURE 14.10
UV–Visible spectroscopic scan of Remazol Black B (■) and its metabolite (▲) after degradation by *Exiguobacterium aestuarii*.

In case of a mixture of dyes and industrial effluents the true colors are difficult to measure and therefore cannot be measured at a single wavelength. Therefore the color is measured using the American Dye Manufacturers Institute tristimulus filter method (ADMI 3WL) and a more specific ADMI 31WL using 31 different wavelengths which measures color independent of the hue (Chen et al., 2003; Tamboli et al., 2010a), thereby opening the way to a more accurate definition of water and wastewater color (dos Santos et al., 2007). The ADMI removal percentage is calculated as follows:

$$\text{ADMI removal \%} = \frac{\text{Initial ADMI} - \text{Final ADMI}}{\text{Initial ADMI}} \times 100 \qquad (14.2)$$

The thin-layer chromatography (TLC) has been proved to be the ideal method having fair separation power and the spot capacity necessary to resolve closely related dyes and their intermediates, and this is made possible by choosing different stationary phases from the almost unlimited variations in mobile phase mixtures (Dhanve et al., 2008). In recent times conventional TLC has been replaced by high-performance TLC (HPTLC) (Mohana et al., 2008; Kadam et al., 2011, 2013a,b; Kurade et al., 2011; Waghmode et al., 2011a, 2012; Kabra et al., 2012; Watharkar et al., 2013b) to analyze the various dye mixtures and textile effluents after their treatment. Although this technique relies on the same principles as TLC it has some advantages such as better sensitivity, automatic application devices, smaller plates, and greater precision. Recently, a mixture of five dyes namely, Brown 3REL, Scarlett RR, Direct Red 2B, Remazol Red and Malachite Green was shown to be metabolized by *Pseudomonas montelli* (Kabra et al., 2013) into different products by using HPTLC (Figure 14.11). Current liquid chromatography utilizes very small packing columns and a relatively high pressure and this is referred to as high-performance liquid chromatography (HPLC). In dye degradation studies (Lopez-Grimau and Gutierrez, 2006; Kalme et al., 2007a,b; Jadhav et al., 2008a,b, 2009a,b), the degradation was confirmed by the appearance of new HPLC peaks with different retention times (Rt) compared

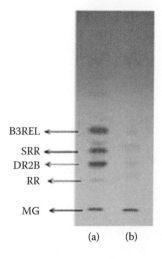

(a) (b)

FIGURE 14.11
HPTLC profile of the (a) mixture of Brown 3REL (B3REL), Scarlett RR (SRR), Direct Red 2B (DR2B), Remazol Red (RR), and Malachite Green (MG); and (b) the metabolites formed after the degradation by *Pseudomonas montelli*. (Adapted from Kabra A, Khandare R, and Govindwar S. 2013. *Water Research* 47: 1035–1048.)

with the original dye whose disappearance indicated the formation of new structural analogues. The Rt can be defined as the time required for a chemical species to travel the length of column and it is generally given in seconds or minutes. Fourier transform infrared spectroscopy (FTIR) is widely used in dye degradation studies. The FTIR spectrum enables the determination of both the type and bond strength of interactions that occur within textile dyes containing different functional groups after degradation by bacteria, and thus it has been known as a valuable analytical tool. A FTIR profile of Remazol Black B degraded by *E. aestuarii* with different peaks in the product sample reveals the degradation as shown in Figure 14.12. The generated unknown metabolites cannot be identified exactly until their mass spectra are obtained. In mass spectrometry spectra of the mixture components provide very powerful qualitative analytical tools. An HPLC profile of Direct Red 5B degraded by *Ps. putida* with different peaks is shown in Figure 14.13. Liquid chromatography–mass spectrometry (LC–MS) and gas chromatography–mass spectrometry (GC–MS) which utilizes liquid and gas respectively are well-known techniques for analyses of metabolites. Both these techniques are useful for the determination of metabolites of dyes formed after bacterial treatment. These techniques are also useful for the determination of molecular weights and other structural information of dye products and thus could help propose the microbial metabolic pathways of dyes

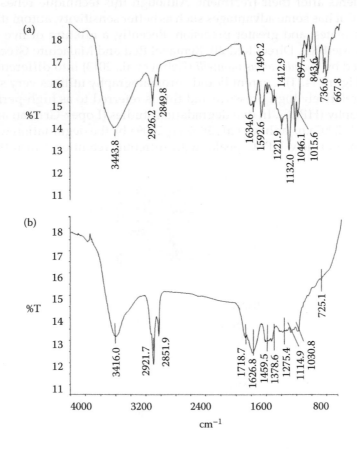

FIGURE 14.12
The FTIR profile of (a) Remazol Black B and (b) its metabolites after degradation by *Exiguobacterium aestuarii*. (Adapted from Khandare R et al. 2012. *Environmental Science and Pollution Research* 19: 1709–1718.)

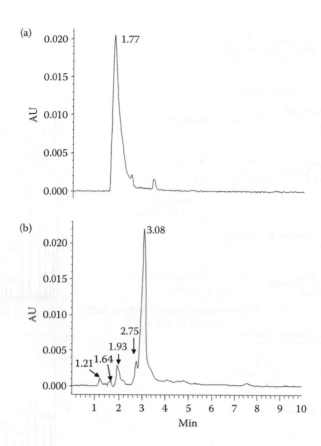

FIGURE 14.13
The HPLC profile of (a) Direct Red 5 B and (b) its metabolites after degradation by *Pseudomonas putida*. (Adapted from Khandare R et al. 2013a. *International Journal of Environmental Science and Technology* 10: 1039–1050.)

(Telke et al., 2009a; Saratale et al., 2009c). Nuclear magnetic resonance (NMR) is very efficient in providing structural information concerning molecular compounds. Moreover, the environmental parameters such as BOD, COD, TOC, TSS, TDS, conductivity, pH, etc., have also tested as per APHA (1998) after textile dye decolorization and biodegradation (Kabra et al., 2013; Khandare et al., 2013b, 2014).

14.6.3 Metabolic Pathway of Dyes Involving Bacterial Enzymes

Enzyme activities induced and the mass spectroscopic analyses after degradation usually help to predict the fate of their metabolism. A pathway of degradation of Direct Red 5B breakdown by veratryl alcohol oxidase and laccase has been shown in Figure 14.14. Veratryl alcohol oxidase from *Ps. putida* brings about the cleavage of Direct Red 5B into benzamide and intermediate (a). Intermediate (a) is then cleaved into intermediate (b) and 4-hydroxynaphthalene-2-sulfonic acid. Intermediate (b) undergoes asymmetric cleavage to form intermediate (c) and intermediate (d). Intermediate (c) after oxidative cleavage by laccase gives 4-hydroxybenzenesulfonic acid. The pathway also depicts the formation of intermediate and further metabolism. The GC–MS spectra are also shown in the Figure 14.14.

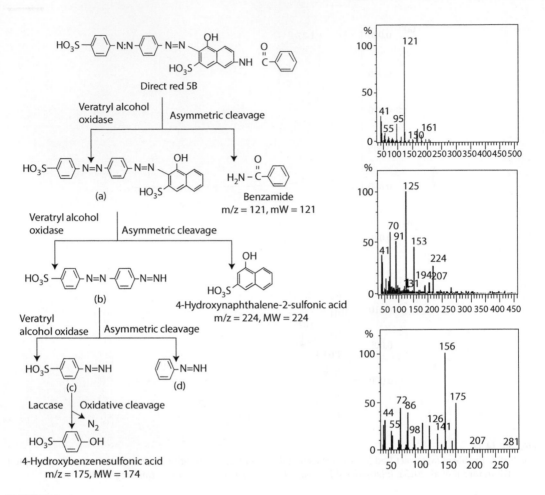

FIGURE 14.14
Pathway of degradation of Direct Red 5B by *Pseudomonas putida* showing the involvement of veratryl alcohol oxidase and laccase. The mass peaks of the metabolites are also shown on the right. (Adapted from Khandare R et al. 2013a. *International Journal of Environmental Science and Technology* 10: 1039–1050.)

14.7 Factors Affecting Bacterial Remediation of Textile Dyes

The complex chemical nature of textile wastewater is usually subject to daily and seasonal variations which include organics, sulfur compounds, nutrients, salts, metal ions, and different toxic substances (Pearce et al., 2003). The presence of such compounds generally affect the dye decolorization process. The efficacy of bacterial remediation is greatly influenced by the various physicochemical parameters such as pollutant bioavailability, level of aeration, temperature, pH, the dye structure and concentration, and nutrient availability. The exposed amount of biomass along with redox mediators are other important factors in the treatment of wastewater process. The dyes are intentionally designed to resist the degradation process; hence bacterial inoculum volume may have different degrees of removal efficiency for various dyes. Therefore, the effects of each of the factors on the dye decolorization must be understood by biological means prior to the treatment of the industrial wastewater.

14.7.1 Oxygen

Oxygen is the most important factors during the cell growth stage. Oxygen has a significant effect on the characteristics and physiology of the cells. Chang et al. (2004) have earlier reported the inhibition of the dye reduction mechanism if the extra cellular environment is aerobic due to the presence of the high-redox-potential electron acceptor, oxygen. The electrons liberated by the oxidation of the electron donor species are preferentially used to reduce the oxygen rather than the azo dye (Yoo et al., 2001). Similar results were seen in the studies on pure bacterial strains such as *Ps. luteola*, *Proteus mirabilis*, *Pseudomonas* sp. SUK1, *Pr. vulgaris* NCIM-2027, *M. glutamicus* NCIM-2168 (Chen et al., 1999; Chang et al., 2001; Kalyani et al., 2008a; Saratale et al., 2009a,b). From these investigations it has been recommended earlier that aeration and agitation processes should be avoided for efficient color removal (Chang and Lin, 2000). It is also noteworthy that if the air is replaced with oxygen-free nitrogen, the reduction of the azo dyes process is restored and occurs at a similar rate to that which was observed under continuous anaerobic conditions (Bragger et al., 1997).

On the contrary, aerobic conditions are required for the complete breakdown and mineralization of the reactive dye molecule. Oxygen plays a vital role in hydroxylation and ring opening of the simple aromatic compounds produced by the initial reduction of dyes (Chang et al., 2001; Pandey et al., 2007). Thus, for effective textile wastewater treatment, the balance between the aerobic and anaerobic steps is important in the treatment system so that the possibility of the re-aeration of a reduced dye solution (becoming dark again) is minimized. However, when the large scale processes are established, many other strains of bacteria have been found to become potent to achieve high levels of decolorization when used in a sequential anaerobic/aerobic treatment process (Khehra et al., 2006).

14.7.2 Pollutant Availability

Pollutant availability is another important aspect in the bacterial degradation processes. Generally, physicochemical properties of dye pollutants such as charge, hydrophobicity, and volatility affects its bioavailability. Hydrophobicity of any pollutant such as dye is measured as Log K_{ow}, that is, octanol:water distribution (Trapp and McFarlane, 1995). A high Log K_{ow} (Log $K_{ow} > 3$) represents the recalcitrant nature of pollutant while compared with pollutants with low Log K_{ow} (Log $K_{ow} < 3$). Reactive dyes are usually more water soluble than disperse and solvent dyes. Surfactants are very useful for the solubilization of organic dyes. Hence, the addition of a surfactant during the dye degradation process can scale up the treatment by increasing the bioavailability of dyes (Tehrani-Bagha and Holmberg, 2013).

14.7.3 Dye Structure and Concentration

The chemical structure and concentration of textile dyes also have significant impact on the bacterial degradation process. It has been observed that dyes with low molecular weights and simple structures at low concentrations are easily removed, showing higher removal rates. On the other hand, color removal gets lowered in the case of dyes with high-molecular weight having electron withdrawing groups like $-SO_3H$, $-SO_2NH_2$) (Sani and Banerjee, 1999). Azo compounds, hence, with a hydroxyl group or with an amino group are more likely to be degraded than those with a methyl, methoxy, sulfo, or nitro group (Nigam et al., 1996a,b). Moreover, Hu (2001) has proven that the color removal rate is also related to the number of azo bonds in the dye molecule. Monoazo dyes are removed at a faster rate than diazo or triazo dyes and this depicts that the decolorization rate is dependent on dye class such that (a) acid dyes exhibit low color removal due to the number of sulfonate groups in the dye, (b) direct dyes exhibit high levels of color removal that is, independent of the number of sulfonate groups in the dye, and (c) reactive dyes exhibit low levels of color removal (Hitz et al., 1978).

The concentration of dye substrate can influence the dye removal ability of bacteria due to its toxic nature (and co-contaminants) at higher concentrations. Also the decrease in decolorization rates may be caused due to inadequate biomass concentration (or improper cell to dye ratio) for the uptake of higher concentrations of dye in treatment processes (Saratale et al., 2006; Jadhav et al., 2008a). It was also known that reactive dyes having sulfonic acid (SO_3H) groups on aromatic rings greatly inhibited the growth of microorganisms at higher dye concentrations (Chen et al., 2003; Kalyani et al., 2008b). The dyes with concentrations of 1–10 μM were easily decolorized, but when the dye concentration was increased to 30 μM, color removal was reduced (Sani and Banerjee, 1999). Surprisingly, the absence of any effect of dye concentration on the reduction rate was published by Dubin and Wright (1975). This observation is compatible with a nonenzymatic reduction method that is, controlled by processes that are independent of the dye concentration.

14.7.4 Temperature

The temperature required to attain the maximum rate of color removal tends to correspond with the optimum cell culture growth temperature of 35–45°C. The rate of color removal was found to increase with an increasing temperature up to a certain limit with marginal reduction in the decolorization activity afterwards. The decline in color removal activity at higher temperatures can be linked to cell death or the denaturation of an azo reductase enzyme (Chang et al., 2001; Telke et al., 2008). However, it has been demonstrated that the azo reductase enzyme is relatively thermo stable and can remain active up to temperatures of 60°C over short periods of time (Pearce et al., 2003). Moreover, Chang et al. (2001) and Pearce et al. (2003) have revealed the phenomena of a shift in the optimum color removal temperature toward high values during dye degradation by immobilized cell culture as a support medium protects these cells.

14.7.5 pH

The pH of the medium as well as the wastewater is also an important factor to achieve better decolorization. Many scale-up studies have shown that the optimum pH for color removal is often neutral or slightly on the alkaline side. It is essential to maintain constant pH during decolorization as the rate of color removal increases only at optimum pH but tends to decrease rapidly at strongly acid or strongly alkaline pH. Interestingly, pH-independent complete decolorization was seen with the consortium-GB consisting of *G. geotrichum* MTCC1360 and *Bacillus* sp. VUS in the pH range from 5 to 9 (Jadhav et al., 2008b). A variety of azo dyes reduction, usually, results in an increase in pH due to the formation of basic aromatic amine metabolites. Chang et al. (2001) marked that the dye reduction rate increased nearly 2.5-fold with increase in pH from 5.0 to 7.0, while the rate became insensitive to pH in the range of 7.0–9.5.

14.7.6 Electron Donor

The oxidation and/or a reduction of organic compounds such as dyes correspond to the color removal process. The thermodynamics study shows that the reaction rate is likely to be influenced by the type of electron donor. It is also significant to verify the physiological electron donor for each biological color removal process, because it induces the reduction mechanism along with stimulation to the enzymatic system responsible for the reduction process (Van der Zee et al., 2001; Pearce et al., 2003). The addition of electron donors such as glucose or acetate ions apparently induces the reductive cleavage of azo bonds while formate acts as a most effective electron donor for the anaerobically induced electron transfer pathway to the dye molecule. It may be because the pathway shows involvement of formate dehydrogenase enzyme (Bras et al., 2001). Cell lysate as well as certain chemicals, such as thiomersal and *p*-chloromercuribenzoate inhibit the alcohol dehydrogenase of NADH-generating systems essential for the generation of reducing equivalents for dye reduction. Therefore, the rate of formation of NADH would affect the process leading to inhibition of azo dye reduction (Gingell and Walker, 1971; Yoo et al., 2001).

14.7.7 Redox Mediator and Its Potential

Redox mediators and their potential are the rate-limiting factors during dye degradation process. The redox potential can be considered as the ease with which a molecule will accept electrons and can be reduced. The more positive the redox potential, therefore, the more readily the molecule gets reduced (Bragger et al., 1997). It was found that redox mediator compounds, such as flavins, increase reduction of azo dyes by shuttling reduction equivalents from the cells to facilitate the nonenzymatic reduction of the extracellular azo dye (Plumb et al., 2001). The sulfonated group of azo dyes reduction reaction takes place by using extracellular reducing activity and the reduced dye will pass through the cell membrane (Keck et al., 1997). The addition of synthetic electron carriers enhances the rate of reduction of azo dyes by bacterial cells viz. quinone–hydroquinone are the most widely used redox mediators (Keck et al., 1997). A very small concentration of the redox mediator is adequate for such type of reactions. Intermediate compounds formed during the dye degradation process can act as redox mediators and help to enhance the overall degradation process (Kurade et al., 2011).

14.8 Toxicity Studies of Dyes and Their Degradation Products

A major crux confronts us as dyes and textile effluents cause a great threat to our valuable water resources. The presence of these dye stuffs in water bodies decreases the penetration of sunlight which reduces photosynthetic activity, dissolved oxygen concentration, water quality, and brings toxic effects to aquatic flora and fauna (Kalyani et al., 2008b). Other pollutants of textile effluents are toxic inorganic compounds, supplementary chemicals, dissolved solids, and residual chlorine. When residual chlorine combines with other pollutant compounds to form toxic substances, the dissolved oxygen from the receiving water body ultimately decreases. There are many bio-monitoring studies have been employed to investigate the ecotoxicology and risk assessment of untreated textile dyes as well as their degraded metabolites.

14.8.1 Microbial Toxicity Studies

Many natural and synthetic textile dyes are well known for their antibacterial ability. Diazo dyes containing 4,4′-diaminostilbene-2,2′-disulfonic acid and 4,4′-diaminobenzanilide as middle components were synthesized and evaluated for their antibacterial capacity against *Staphylococcus aureus, Streptococcus pyogenes, Escherichia coli, Ps. aeruginosa,* and *Pr. vulgaris* by the disk diffusion method. Saratale et al. (2011a,b) performed a microbial toxicity study against isolated *E. coli* DH5α, *Cellulomonas biazotea, Rhizobium radiobacter, Bacillus megatarium, Acinetobacter* sp., *Ps. desmolyticum, M. glutamicus,* and *Pr. vulgaris* to reveal the toxic nature of dye Reactive dye 172 and the nontoxic nature of its metabolite after degradation. Some investigators have measured the zone of inhibition in the presence of untreated dyes and their metabolites in the same concentration. This has particularly been tested on agriculturally important bacteria mainly the nitrogen and phosphate solubilizers (Kalme et al., 2007a,b; Gomare and Govindwar, 2009; Saratale et al., 2009c, 2010). The bacterial bioremediation was found to render toxic dyes into less toxic metabolites.

One more test called the plasmid nicking assay is a simple DNA damage test in which single strand breaks in the DNA can be brought about by electrophoresis. The differential mobility of the super coiled, open circle and linear forms of the plasmid causes a variety of band patterns. Covalently closed circular plasmid pBR322 DNA has been employed for the assay to investigate DNA damage by industrial wastewater (Siddiqui et al., 2011). In the Ames test study, the various strains of *Salmonella typhimurium* are employed for dye degradation under partially reductive conditions as well as oxidative conditions to understand the influence of the nitroreductase, azoreductase, cytochrome P450 enzymes, and O-acetyltransferase in the metabolic activation of this disperse dye (Prival and Mitchell, 1982). The *Salmonella* strains from group of *Enterobacteria* that are able to produce nitroreductase and azoreductase are used in the Ames test. However, for the significant induction of azoreductase in *Salmonella* a sufficient amount of mutagenic products is needed. The test must be done under reductive conditions, with the addition of FMN and other cofactors (Prival and Mitchell, 1982). The Ames assay has been reported earlier to investigate mutagenicity caused by textile dyes Disperse Blue 291 (Umbuzeiro et al., 2005).

14.8.2 Phytotoxicity Studies

This is an important toxicity parameter as textile wastewater is usually used for irrigation purposes in most developing countries (Kabra et al., 2013). Many reports have mentioned phytotoxicity studies that have been carried out using crop plants *Phaseolus mungo, Sorghum vulgare,* and *Triticum aestivum* seeds at room temperature. Germination (%) and the length of shoot and root of selected plants are recorded after few days. Microbial degradation of synthetic dye mixture by *Pseudomonas* sp. SUK 1 has been performed and detoxification has been confirmed by phytotoxicity studies (Chougule et al., 2014).

Cytogenotoxicity is the study used to understand the chromosomal aberrations caused by pollutants like dyes in model plant apical meristematic root cells. Uniformly sized and shaped *Allium cepa* bulbs were initially put in water for the development of roots. The bulbs were arranged in classified sets at $27 \pm 2°C$. Jadhav et al. (2011) have performed this study for textile dye Remazol Red by exposing *A. cepa* root cells to untreated dye solution and their respective treated that is, metabolites samples separately, while the root exposed to plain water was kept as a control. Toxicity analysis with *A. cepa* signifies that dye Red HE3B was found to exert oxidative stress and subsequently toxic effect on the root cells whereas the biodegradation metabolites of the dye by *Providencia* sp. SDS (PS) and *Pseudomonas aeuroginosa* consortium were found to be relatively less toxic in nature.

Comet assay is a single gel electrophoresis study used to quantify the extent of DNA damage in model plant apical root meristematic cells after the treatment of pollutants. Many literature surveys have endorsed the use of *Allium cepa* bulbs as the model plant for this study. A computerized image analysis system (Comet version 1.5) was employed to measure the percentage of DNA damage (% T) and tail length (TL). The method used for comet assay was carried out as explained by Achary et al. (2008).

Oxidative stress studies of textile dyes consisting of antioxidant enzymes and lipid peroxidation of model plant cells are examined in the bulbs of *A. cepa*. The bulbs in each case are exposed to the respective treatment of dyes and their metabolites for 72 h. Antioxidant enzymes namely catalase (CAT, E.C. 1.11.1.6), superoxide dismutase (SOD, E.C. 1.15.1.1), and guaiacol peroxidase (GPX, E.C. 1.11.1.7), and lipid peroxidation are known to give clear information on stress to the organisms.

14.8.3 Animal Toxicity Study

A histological study of fish gills (Ortiz-Ordoñez et al. 2011) after exposure of model fish to untreated and treated wastewater samples has shown the reduced toxicity of the dyes after degradation. Textile dyes have been appeared to be highly toxic to fishes like *Daphnia magna* (Bae and Freeman, 2005).

The pure inbred line of Swiss albino rats (*Rattus norvegicus* Berkenhout) has been utilized to examine the toxic effects of treated and untreated textile dye wastewater (Sharma et al., 2007). This study reports the acute toxicity of textile dye wastewater on the hematology and serum biochemistry of albino rats to predict stress on the human beings and cattle.

Frog embryo teratogenesis assay *Xenopus* (FETAX) bioassay is used for initial screening for developmental toxicants (ASTM, 2003). Birhanli and Ozmen (2005) have performed this assay for the toxicity and teratogenity evaluation of six commercial textile dyes using frog embryo of *Xenopus laevis*.

14.8.4 Cytotoxicity Studies

The MIT assay reveals the cytotoxic nature of untreated and treated dye solution by 3-(4, 5-dimethylthiazol-2-yl)-2, 5-diphenyltetrazolium bromide (MTT). The mouse fibroblast cell line from subcutaneous connective tissue (L929 cell line) grown as adherent culture in Dulbecco's Modified Eagle's Medium (DMEM) is used to expose to desired samples. In this assay, the cells are seeded in a 96 well microplate for cytotoxicity screening at a density of 1×10^5 cells per well with 24 h incubation. Untreated and treated dye samples are added in quadruplicates and incubated for 24 h. The intensity of the purple color formazan is measured on a microplate reader at 570 nm. Oturkar et al. (2013) have performed this assay to investigate the toxicity of dyes Reactive Red 141 and Reactive Red 2 and their metabolites.

A xanthene class of dye, erythrosine, widely used dye in textiles, drugs, cosmetics, and foods, is a highly toxic allergent, DNA damaging, carcinogenic, neurotoxic, and xenoestrogenic for humans and animals (Mittal et al., 2006). In Spain and Algeria, the spraying of a textile paint chemical has been found to cause a pulmonary disease known as Ardystil syndrome in many sprayers, leading to their deaths (Hoet et al., 1999). The International Agency for Research on Cancer (IARC) has explained that dyes like benzidine are powerfully carcinogenic to a variety of mammalian species, including humans (IARC, 1982). Two distinct human cell lines, viz. HEp-2 cells as well as Caco-2 cells, have been reported for toxicity of textile dye Malachite Green and its degraded products testings purpose (Stammati et al., 2005). The HEp-2 cell line has been established from a human carcinoma of the larynx and is sensitive to the toxic effects of xenobiotic of different origins (Moore et al., 1955). Caco-2 cells are derived from a human colon carcinoma and carry small intestinal enterocyte morphological, functional and enzymatic features, including phase I and II biotransformations (De Angelis et al., 1999). These cell lines are cultured and maintained in Dulbecco's modified Eagle's medium (DMEM). Both cell lines are kept in a CO_2 incubator at 37°C and used to estimate either the total protein content or the neutral red uptake (Borenfreund and Puerner, 1985). The proliferation capability can be measured by the colony-forming ability test. The cytotoxicity can be investigated by MTT assay on Caco-2 cells by measuring the conversion of yellow tetrazolium salt to the colored formazan (Trivedi et al., 1990).

14.9 Future Prospects

The degradation of textile dyes and effluents by bacteria has been proposed and advocated by the huge literature available. Bacteria for textile dye treatment have been explored in many ways. Most of the reactors involving bacterial treatment are still at the laboratory scale. the pilot-scale implementation of bacterial treatments still remain scarce. Therefore, studies at the actual site of disposal and at full pilot scale should be a great avenue for research. In order to find a solution, the combinatorial system of bacteria should be tried on large scales. The combined systems comprising aerobic and anaerobic bacteria can provide more efficient solutions for the sludge forming systems. The bacteria-assisted phytoremediation processes need to be motivated as this approach adds to the aesthetics of the system (Khandare et al., 2013b, 2014).

The solid-state fermentation approach reported by Kadam et al. (2013) can also be a promising way for the treatment of textile effluents at the actual site of disposal. This remediation strategy was recently coupled with bioenergy production showing the degradation of textile effluents absorbed on lignocellulosic agricultural waste which ultimately helped in the breakdown of lignin such as moieties favoring bioethanol production (Waghmare et al., 2014). This method if tried at the wastewater disposal site could ultimately provide huge amounts of pretreated lignocellulosic waste which would be readily available for bioethanol production.

The development of transgenic bacteria and plant species having the genes with coding for effective dye-degrading enzymes and stress-confronting abilities is also an important and exciting area of research. These bacteria could prove to be efficient tools for the degradation and detoxification of textile dyes and effluents. The area of bioremediation of textile dyes and wastewater using bacteria thus makes available a number of exciting opportunities for researchers around the globe.

14.10 Conclusion

Bacteria are an extremely potential bioremediation tool for textile dyes and allied effluents. The use of bacteria provide an ecofriendly, cost-effective, and efficacious alternative for dye degradation when compared with the available physicochemical methods. Different oxidoreductase enzymes from bacteria such as LiP, veratryl alcohol oxidase, laccase, tyrosinase, azo reductase, riboflavin reductase, DCIP reductase, and Malachite Green reductase are the key players which attack the complex structures of dyes. The bacterial strains from many habitats have been explored and utilized for dye removal from dye containing wastewater. Pure and mixed cultures of bacteria under different conditions are well documented and need to be utilized for further studies. The aerobic, anaerobic, anoxic, and combinatorial systems individually have their efficiencies but also possess certain shortcomings. The combined systems of bacteria fungi and/or plants are also able to provide better alternatives but need to be tried at the pilot and industrial scales. Various analytical techniques such as UV–Visibles spectroscopy, HPLC, HPTLC, HPLC, GC–MS, LC–MS, and NMR have earlier been used which have explored the degradation of many dyes and effluents to different metabolites. The pathways of dye degradation and the fate of metabolism have been proposed which are in agreement with the results obtained by other analytical techniques and also with the active enzymes from bacterial sources. There are many upcoming possibilities with bacterial degradation of textile dyes and effluents, covering pilot-scale reactor developments, building effective consortia with other classes of organisms, biodegradation coupled bioenergy production processes, and the development of transgenic bacteria and plants that have extremely potential dye degrading machineries.

References

Achary V, Jena S, Panda K, and Panda B. 2008. Aluminium induced oxidative stress and DNA damage in root cells of *Allium cepa* L. *Ecotoxicology and Environmental Safety* 70: 300–310.
Adedayo O, Javadpour S, Taylor C, Anderson W, and Moo-Young M. 2004. Decolorization and detoxification of methyl red by aerobic bacteria from a wastewater treatment plant. *World Journal of Microbiology and Biotechnology* 20: 545–550.
Ahmad R, Mondal P, and Usmani S. 2010. Hybrid UASFB-aerobic bioreactor for biodegradation of acid yellow-36 in wastewater. *Bioresource Technology* 101: 3787–3790.
Aksu Z and Donmez G. 2003. A comparative study on the biosorption characteristics of some yeasts for Remazol Blue reactive dye. *Chemosphere* 50: 1075–1083.
APHA. 1998. *Standard Methods for the Examination of Water and Wastewater*. 20th edn. American Public Health Association APHA–AWWA–WEF, Washington, DC, USA.
Aravindhan R, Rao JR, and Nair B. 2007. Removal of basic yellow dye from aqueous solution by sorption on green alga *Caulerpa scalpelliformis*. *Journal of Hazardus Material* 142: 68–76.
ASTM, American Society for Testing and Materials. 2003. Standard guide for conducting the Frog Embryo Teragonesis Assay-*Xenopus* (FETAX), E1439-98. In: *ASTM Standards on Biological Effects and Environmental Fate*, Vol. 11.05. Philadelphia, PA, pp. 447–457.
Bae J and Freeman H. 2005. Aquatic toxicity evaluation of copper complexed direct dyes to the *Daphnia magna*. *Dyes Pigments* 73: 126–132.
Beydilli M, Pavlostathis S, and Tincher W. 1998. Decolorization and toxicity screening of selected reactive azo dyes under methanogenic conditions. *Water Science and Technology* 38: 225–232.

Birhanli A and Ozmen M. 2005. Evaluation of the toxicity and teratogenity of Six commercial textile dyes using the frog embryo teratogenesis assay–*Xenopus*. *Drug Chemistry and Toxicology* 1: 51–65.

Borenfreund E and Puerner J. 1985. Toxicity determined *in vitro* by morphological alterations and neutral red absorption. *Toxicology Letters* 24: 119–124.

Bourbonnais R and Paice M. 1990. Oxidation of non-phenolic substrates: An expanded role for laccase in lignin biodegradation. *FEBS Letters* 267: 99–102.

Bragger J, Lloyd A, Soozandehfar S, Bloomfield S, Marriott C, and Martin G. 1997. Investigations into the azo reducing activity of a common colonic microorganism. *International Journal of Pharmaceutics* 157: 61–71.

Bras R, Ferra I, Pinheiro H, and Goncalves I. 2001. Batch tests for assessing decolourisation of azo dyes by methanogenic and mixed cultures. *Journal of Biotechnology* 89: 155–162.

Bromly-Challenor K, Knapp J, Zhang Z, Gray N, Hetheridge M, and Evans M. 2000. Decolorization of an azo dye by unacclimated activated sludge under anaerobic conditions. *Water Research* 34: 4410–4418.

Brown D and Laboureur P. 1983. The aerobic biodegradability of primary aromatic amines. *Chemosphere* 12: 405–414.

Chacko J and Subramaniam K. 2011. Enzymatic degradation of azo dyes—A review. *International Journal of Environmental Science* 1: 1250–1260.

Chang J, Chen B, and Lin Y. 2004. Stimulation of bacterial decolorization of an azo dye by extracellular metabolites from *Escherichia coli* strain NO3. *Bioresource Technology* 91: 243–248.

Chang J, Chou C, Lin Y, Ho J, and Hu T. 2001. Kinetic characteristics of bacterial azo-dye decolorization by *Pseudomonas luteola*. *Water Research* 35: 2041–2050.

Chang J and Lin Y. 2000. Fed-batch bioreactor strategies for microbial decolorization of azo dye using a *Pseudomonas luteola* strain. *Biotechnology Programs* 16: 979–985.

Chen B and Chang J. 2007. Assessment upon species evolution of mixed consortia for azo dye decolorization. *Journal of Chinese Institute of Chemical Engineering* 38: 259–266.

Chen B, Chen S, and Chang J. 2005. Immobilized cell fixed-bed bioreactor for wastewater decolorization. *Process in Biochemistry* 40: 3434–3440.

Chen B, Lin K, Wang Y, and Yen C. 2009a. Revealing interactive toxicity of aromatic amines to azo dye decolorizer *Aeromonas hydrophila*. *Journal of Hazardus Materials* 166: 187–194.

Chen H and Burns L. 2006. Environmental analysis of textile products. clothing textile. *Research Journal* 24: 248–261.

Chen H, Xu H, Heinze T, and Cerniglia C. 2009b. Decolorization of water and oilsoluble azo dyes by *Lactobacillus acidophilus* and *Lactobacillus fermentum*. *Journal of Indian Microbiology and Biotechnology* 36: 1459–1466.

Chen K, Huang W, Wu J, and Houng J. 1999. Microbial decolorization of azo dyes by *Proteus mirabilis*. *Journal of Indian Microbiology and Biotechnology* 23: 686–690.

Chen K, Wu J, Liou D, and Hwang S. 2003. Decolorization of textile dyes by newly isolated bacterial strains. *Journal of Biotechnology* 101: 57–68.

Chen L. and Flurkey W. 2002. Effect of protease inhibitors on the extraction of Crimini mushroom tyrosinase. *Curr. Top. Phytochem* 5: 109–120.

Chivukula M and Renganathan V. 1995. Phenolic azo dye oxidation by laccase from *Pyricularia oryzae*. *Applied and Environmental Microbiology* 61: 4374–4377.

Chougule A, Jadhav S, and Jadhav J. 2014. Microbial degradation and detoxification of synthetic dye mixture by *Pseudomonas* sp. SUK 1. *Proceedings of the National Academy of Science, India, Section B: Biological Science* DOI: 10.1007/s40011-014-0313-z.

Coughlin M, Kinkle B, and Bishop P. 1999. Degradation of azo dyes containing amino naphthol by Sphingomonas sp. strain ICX. *Journal of Indian Microbiology and Biotechnology* 23: 341–346.

Daeshwar N, Ayazloo M, Khataee A, and Pourhassan M. 2007. Biological decolorization of dye solution containing malachite Green by microalgae *Cosmarium* sp. *Bioresource Technology* 98: 1176–1182.

Dafale N, Rao N, Meshram S, and Wate S. 2008. Decolorization of azo dyes and simulated dye bath wastewater using acclimatized microbial consortium—Biostimulation and halo tolerance. *Bioresource Technology* 99: 2552–2558.

Dawkar V, Jadhav U, Jadhav S, and Govindwar S. 2008. Biodegradation of disperse textile dye Brown 3REL by newly isolated *Bacillus* sp. VUS. *Journal of Applied Microbiology* 105: 14–24.

Dawkar V, Jadhav U, Jadhav M, Kagalkar A, and Govindwar S. 2010. Decolorization and detoxification of sulphonated azo dye Red HE7B by *Bacillus* sp. VUS. *World Journal of Microbiology and Biotechnology* 26: 909–916.

Dawkar V, Jadhav U, Ghodake G, and Govindwar S. 2009b. Effect of inducers on the decolorization and biodegradation of textile azo dye Navy blue 2GL by *Bacillus* sp. VUS. *Biodegradation* 20: 777–787.

Dawkar V, Jadhav U, Telke A, and Govindwar S. 2009a. Peroxidase from *Bacillus* sp. VUS and its role in the decolorization of textile dyes. *Biotechnology and Bioprocess Engineering* 14: 361–368.

De Angelis I, Rossi L, Pedersen J, Vignoli A, Vincentini O, Hoogenboom L, Polman T, Stammati A, and Zucco F. 1999. Metabolism of furazolidone: Alternative pathways and modes of toxicity in different cell lines. *Xenobiotica* 29: 1157–1169.

Dhanve R, Shedbalkar U, and Jadhav J. 2008. Biodegradation of diazo reactive dye Navy blue HE2R (Reactive blue 172) by an isolated *Exiguobacterium* sp. RD3. *Journal of Bioprocess and Biotechnology Engineering* 13: 1–8.

dos Santos A, Cervantes F, and van Lier J. 2007. Review paper on current technologies for decolourisation of textile wastewaters: Perspectives for anaerobic biotechnology. *Bioresource Technology* 98: 2369–2385.

Dubin P and Wright K. 1975. Reduction of azo food dyes in cultures of *Proteus vulgaris. Xenobiotica* 59: 563–571.

Duran N and Esposito E. 2000. Potential applications of oxidative enzymes and phenoloxidase-like compounds in wastewater and soil treatment: A review. *Applied Catalysis B: Environmental* 28: 83–99.

Georgiou D, Hatires J, and Aivasidis A. 2005. Microbial immobilization in a two-stage fixed bed reactor pilot plant for on-site anaerobic decolorfization of textile wastewater. *Enzyme and Microbial Technology* 37: 597–605.

Gingell R and Walker R. 1971. Mechanisms of azo reduction by *Streptococcus faecalis*. II The role of soluble flavins. *Xenobiotica* 13: 231–239.

Gomare S and Govindwar S. 2009. *Brevibacillus laterosporus* MTCC 2298. A potential azo dye degrader. *Journal of Applied Microbiology* 106: 993–1004.

Gomare S, Jadhav J, and Govindwar S. 2008. Degradation of sulfonated azo dyes by the purified lignin peroxidase from *Brevibacillus laterosporus* MTCC 2298. *Biotechnology and Bioprocess Engineering* 13: 136–143.

Gomare S, Kalme S, and Govindwar S. 2009. Biodegradation of Navy Blue-3G by *Brevibacillus laterosporus* MTCC 2298. *Acta Chimica Slovenica* 56: 789–796.

Gottlieb A, Shaw C, Smith A, Wheatley A, and Forsythe S. 2003. The toxicity of textile reactive azo dyes after hydrolysis and decolourisation. *Journal of Biotechnology* 101: 49–56.

Govindwar SP and Kagalkar AN. 2010. *Phytoremediation Technologies for the Removal of Textile Dyes: An Overview and Future Prospectus*. Nova Science Publishers Inc., New York, USA.

Ghodake G, Jadhav S, Dawkar V, and Govindwar S. 2009c. Biodegradation of diazo dye Direct brown MR by *Acinetobacter calcoaceticus* NCIM 2890. *International Biodeterioration and Biodegradation* 63: 433–439.

Ghodake G, Kalme S, Jadhav J, and Govindwar S. 2009b. Purification and partial characterization of lignin peroxidase from *Acinetobacter calcoaceticus* NCIM 2890 and its application in decolorization of textile dyes. *Applied Biochemistry and Biotechnology* 152: 6–14.

Ghodake GS, Telke AA, Jadhav JP, and Govindwar SP. 2009a. Potential of *Brassica juncea* in order to treat textile effluent contaminated sites. *Int. J. Phytorem* 11: 297–312.

Gupta V, Carrott P, and Ribeiro Carrott M. 2009. Low-cost adsorbents: Growing approach to wastewater treatment—A review. *Critical Reviews in Environmental Science and Technology* 9: 783–842.

Heinfling A, Martinez M, Martinez A, Bergbauer M, and Szewzyk U. 1998a. Transformation of industrial dyes by manganese peroxidases from *Bjerkandera adusta* and *Pleurotus eryngii* in a manganese-independent reaction. *Applied and Environmental Microbiology* 64: 2788–2793.

Heinfling A, Ruiz-Dueñas F, Martínez M, Bergbauer H, Szewzyk U, and Martínez A. 1998b. A study on reducing substrates of manganese-oxidizing peroxidases from _Pleurotus eryngii_ and _Bjerkandera adusta_. _FEBS Letters_ 428: 141–146.

Hitz H, Huber W, and Reed R. 1978. The absorption of dyes on activated sludge. _Journal of the Society of Dyers and Colorists_ 94: 71–76.

Hoet P, Gilissen L, Leyva M, and Nemery B. 1999. _In vitro_ cytotoxicity of textile paints components linked to the "Ardystil syndrome." _Toxicology Science_ 52: 209–216.

Horitsu H, Takada M, ldaka E, Tomoyeda M, and Ogawa T. 1977. Degradation of p-aminoazobenzene by _Bacillus subtilis_. _European Journal of Applied Microbiology_ 4: 217–224.

Hu T. 2001. Kinetics of azoreductase and assessment of toxicity of metabolic products from azo dyes by _Pseudomonas luteola_. _Water Science Technology_ 43: 261–269.

Humnabadkar R, Saratale G, and Govindwar S. 2008. Decolorization of Purple 2R by _Aspergillus ochraceus_ (NCIM-1146). _Asian Journal of Microbiology, Biotechnology and Environmental Science_ 10: 693–697.

IARC Monographs. 1982. IARC monographs on the evaluation of the carcinogenic risk of chemicals to humans, suppl. 4, chemicals, industrial processes and industries associated with cancer in humans. _Lyon_ 1–29: 1–292.

Idaka E and Ogawa Y. 1978. Degradation of azo compounds by _Aeromonas hydrophila_ var. 2413. _Journal of the Society of Dyers and Colorists_ 94: 91–94.

Isik M and Sponza D. 2008. Anaerobic/aerobic treatment of a simulated textile wastewater. _Separation and Purification Technology_ 60: 64–72.

Isik M and Sponza DT. 2005. Substrate removal kinetics in an upflow anaerobic sludge blanket reactor decolorizing simulated textile wastewater. _Process Biochemistry_ 40: 1189–1193.

Jadhav J and Govindwar S. 2006. Biotransformation of malachite Green by _Saccharomyces cerevisiae_ MTCC 463. _Yeast_ 23: 315–323.

Jadhav J, Parshetti G, Kalme S, and Govindwar S. 2007. Decolourization of azo dye methyl red by _Saccharomyces cerevisiae_ MTCC 463. _Chemosphere_ 68: 394–400.

Jadhav JP, Kalyani DC, Telke AA, Phugare SS, and Govindwar SP. 2010. Evaluation of the efficacy of a bacterial consortium for the removal of color, reduction of heavy metals, and toxicity from textile dye effluent. _Bioresource Technology_ 101: 165–173.

Jadhav S, Ghodake G, Telke A, Tamboli D, and Govindwar S. 2009a. Degradation and detoxification of disperse dye Scarlet RR by _Galactomyces geotrichum_ MTCC 1360. _Journal of Microbiology and Biotechnology_ 19: 409–415.

Jadhav S, Jadhav M, Kagalkar A, and Govindwar S. 2008a. Decolorization of brilliant Blue G dye mediated by degradation of the microbial consortium of _Galactomyces geotrichum_ and _Bacillus_ sp. _Journal of the Chinese Institute of Chemical Engineering_ 39: 563–570.

Jadhav S, Jadhav U, Dawkar V, and Govindwar S. 2008b. Biodegradation of disperse dye Brown 3REL by microbial consortium of _Galactomyces geotrichum_ MTCC 1360 and _Bacillus_ sp. VUS. _Biotechnology and Bioprocess Engineering_ 13: 232–239.

Jadhav S, Patil N, Watharkar A, Apine O, and Jadhav J. 2013a. Batch and continuous biodegradation of Amaranth in plain distilled water by _P. aeruginosa_ BCH and toxicological scrutiny using oxidative stress studies. _Environmental Science and Pollution Research_ 20: 2854–2866.

Jadhav S, Phugare S, Patil P, and Jadhav J. 2011. Biochemical degradation pathway of textile dye Remazol red and subsequent toxicological evaluation by cytotoxicity, genotoxicity and oxidative stress studies. _International Biodeterioration and Biodegradation_ 65: 733–743.

Jadhav S, Surwase S, Kalyani D, Gurav R, and Jadhav J. 2012a. Biodecolorization of azo dye remazol Orange by _Pseudomonas aeruginosa_ BCH and toxicity (oxidative stress) reductionin _Allium cepa_ root cells. _Applied Biochemistry and Biotechnology_ 168: 1319–1334.

Jadhav S, Surwase S, Phugare S, and Jadhav J. 2013b. Response surface methodology mediated optimization of Remazol Orange decolorization in plain distilled water by _Pseudomonas aeruginosa_ BCH. _International Journal of Environmental Science and Technology_ 10: 181–190.

Jadhav S, Yedurkar S, Phugare S, and Jadhav J. 2012b. Biodegradation studies on acid violet 19, a triphenylmethane dye, by _Pseudomonas aeruginosa_ BCH. _Clean—Soil, Air, Water_ 2012: 551–558.

Jadhav U, Dawkar V, Ghodake G, and Govindwar S. 2008c. Biodegradation of direct Red 5B, a textile dye by newly isolated *Comamonas* sp. UVS. *Journal of Hazardus Materials* 158: 507–516.

Jadhav U, Dawkar V, Tamboli D, and Govindwar S. 2009c. Purification and characterization of veratryl alcohol oxidase from *Comamonas* sp. UVS and its role in decolorization of textile dye. *Biotechnology and Bioprocess Engineering* 14: 369–376.

Jadhav U, Dawkar V, Telke A, and Govindwar S. 2009b. Decolorization of direct Blue GLL with enhanced lignin peroxidase enzyme production in *Comamonas* sp. UVS. *Journal of Chemical Technology and Biotechnology* 84: 126–132.

Joshi T, Iyengar L, Singh K, and Garg S. 2008. Isolation, identification and application of novel bacterial consortium TJ-1 for the decolourization of structurally different azo Dyes. *Bioresource Technology* 99: 7115–7121.

Kabra A, Khandare R, and Govindwar S. 2013. Development of a bioreactor for remediation of textile effluent and dye mixture: A plant–bacterial synergistic strategy. *Water Res* 47: 1035–1048.

Kabra A, Khandare R, Kurade M, and Govindwar S. 2011a. Phytoremediation of a sulphonated azo dye Green HE4B by *Glandularia pulchella* (Sweet) Tronc. (Moss Verbena). *Environmental Science and Pollution Research* 18: 1360–1373.

Kabra A, Khandare R, Waghmode T, and Govindwar S. 2011b. Differential fate of metabolism of a sulfonated azo dye Remazol Orange 3R by plants *Aster amellus* Linn., *Glandularia pulchella* (Sweet) Tronc. and their consortium. *Journal of Hazardus Materials* 190: 424–431.

Kabra A, Khandare R, Waghmode T, and Govindwar S. 2012. Phytoremediation of textile effluent and mixture of structurally different dyes by *Glandularia pulchella* (Sweet) Tronc. *Chemosphere* 87: 265–272.

Kadam A, Kamatkar J, Khandare R, Jadhav J, and Govindwar S. 2013a. Solid-state fermentation: Tool for bioremediation of adsorbed textile dyestuff on distillery industry waste-yeast biomass using isolated *Bacillus cereus* strain EBT1. *Environmental Science and Pollution Research* 20: 1009–1020.

Kadam A, Kulkarni A, Lade H, and Govindwar S. 2014. Exploiting the potential of plant growth promoting bacteria in decolorization of dye disperse Red 73 adsorbed on milled sugarcane bagasse under solid state fermentation. *International Biodeterioration and Biodegradation* 86: 364–371.

Kadam A, Lade H, Patil S, and Govindwar S. 2013b. Low cost CaCl$_2$ pretreatment of sugarcane bagasse for enhancement of textile dyes adsorption and subsequent biodegradation of adsorbed dyes under solid state fermentation. *Bioresource Technology* 132: 276–284.

Kadam A, Telke A, Jagtap S, and Govindwar S. 2011. Decolorization of adsorbed textile dyes by developed consortium of *Pseudomonas* sp. SUK1 and *Aspergillus ochraceus* NCIM-1146 under solid state fermentation. *Journal of Hazardus Materials* 189: 486–494.

Kagalkar A, Jadhav M, Bapat V, and Govindwar S. 2011. Phytodegradation of the triphenylmethane dye malachite Green mediated by cell suspension cultures of *Blumea malcolmii* Hook. *Bioresource Technology* 102: 10312–10318.

Kagalkar A, Jagtap U, Jadhav J, Bapat V, and Govindwar S. 2009. Biotechnological strategies for phytoremediation of the sulphonated azo dye direct Red 5B using *Blumea malcolmii* Hook. *Bioresource Technology* 100: 4104–4110.

Kagalkar A, Jagtap U, Jadhav J, Govindwar S, and Bapat V. 2010. Studies on phytoremediation potentiality of *Typhonium flagelliforme* for the degradation of Brilliant Blue R. *Planta* 232: 271–285.

Kalme S, Ghodake G, and Govindwar S. 2007a. Red HE7B degradation using desulfonation by *Pseudomonas desmolyticum* NCIM 2112. *International Biodeterioration and Biodegradation* 60: 327–333.

Kalme S, Parshetti G, Jadhav S, and Govindwar S. 2007b. Biodegradation of benzidine based dye Direct blue 6 by *Pseudomonas desmolyticum* NCIM 2112. *Bioresource Technology* 98: 1405–1410.

Kalme S, Jadhav S, Jadhav M, and Govindwar S. 2009. Textile dye degrading laccase from *Pseudomonas desmolyticum* NCIM 2112. *Enzyme and Microbial Technology* 44: 65–71.

Kalme S, Parshetti G, Jadhav S, and Govindwar S. 2007b. Biodegradation of benzidine based dye Direct blue 6 by *Pseudomonas desmolyticum* NCIM 2112. *Bioresource Technology* 98: 1405–1410.

Kalyani D, Patil P, Jadhav J, and Govindwar S. 2008b. Biodegradation of reactive textile dye Red BLI by an isolated bacterium *Pseudomonas* sp. SUK1. *Bioresource Technology* 99: 4635–4641.

Kalyani D, Telke A, Dhanve R, and Jadhav J. 2008a. Ecofriendly biodegradation and detoxification of reactive Red 2 textile dye by newly isolated *Pseudomonas* sp. SUK1. *Journal of Hazardus Materials* 163: 735–742.

Kalyani D, Telke A, Dhanve R, and Jadhav J. 2009. Ecofriendly biodegradation and detoxification of reactive Red 2 textile dye by newly isolated *Pseudomonas* sp. SUK1. *Journal of Hazardus Materials* 163: 735–742.

Kanekar P and Sarnaik S. 1995. Microbial process for treatment. *Scientific Engineering and Toxic/Hazardus Substance Control A* 31: 1035–1041.

Kanekar P, Sarnaik S, and Kelkar A. 1996. Microbial technology for management of phenol bearing dyestuff wastewater. *Water Science and Technology* 33: 47–51.

Kant R. 2012. Textile dyeing industry an environmental hazard. *Natural Science* 4: 22–26.

Kapdan I, Kargi F, McMullan G, and Marchant R. 2000. Decolorization of textile dyestuffs by a mixed bacterial consortium. *Biotechnology Letters* 22: 1179–1181.

Keck A, Klein J, Kudlich M, Stolz A, Knackmuss H, and Mattes R. 1997. Reduction of azo dyes by redox mediators originating in the naphthalenesulfonic acid degradation pathway of *Sphingomonas* sp. strain BN6. *Applied and Environmental Microbiology* 63: 3684–3690.

Keck A, Rau J, Reemtsma T, Mattes R, Stolz A, and Klein J. 2002. Identification of quinoide redox mediators that are formed during the degradation of naphthalene-2-sulfonate by *Sphingomonas xenophaga* BN6. *Applied and Environmental Microbiology* 68: 4341–4349.

Khandare R, Kabra A, Awate A, and Govindwar S. 2013a. Synergistic degradation of diazo dye direct Red 5B by *Portulaca grandiflora* and *Pseudomonas putida*. *International Journal of Environmental Science and Technology* 10: 1039–1050.

Khandare R, Kabra A, Kadam A, and Govindwar S. 2013b. Treatment of dye containing wastewaters by a developed lab scale phytoreactor and enhancement of its efficacy by bacterial augmentation. *International Biodeterioration and Biodegradation* 78: 89–97.

Khandare R, Kabra A, Kurade M, and Govindwar S. 2011b. Phytoremediation potential of *Portulaca grandiflora* Hook. (Moss-Rose) in degrading a sulfonated diazo reactive dye Navy Blue HE2R (Reactive Blue 172). *Bioresource and Technology* 102: 6774–6777.

Khandare R, Kabra A, Tamboli D, and Govindwar S. 2011a. The role of *Aster amellus* Linn. in the degradation of a sulfonated azo dye Remazol Red: A phytoremediation strategy. *Chemosphere* 82: 1147–1154.

Khandare R, Rane N, Waghmode T, and Govindwar S. 2012. Bacterial assisted phytoremediation for enhanced degradation of highly sulfonated diazo reactive dye. *Environmental Science and Pollution Research* 19: 1709–1718.

Khandare R, Watharkar A, Kabra A, Kachole M, and Govindwar S. 2014. Development of a low cost phyto-tunnel system and its application for the treatment of dye containing wastewaters. *Biotechnology Letters* 36: 47–55.

Khataee A, Movafeghi A, Torbati S, SalehiLisar S, and Zarei M. 2012. Phytoremediation potential of duckweed (*Lemna minor* L.) in degradation of C.I. Acid Blue 92: Artificial neural network modeling. *Ecotoxicology and Environmenal Safety* 80: 291–298.

Khataee A, Vafaei F, and Jannatkhah M. 2013. Biosorption of three textile dyes from contaminated water by filamentous green algal *Spirogyra* sp.: Kinetic, isotherm and thermodynamic studies. *International Biodeterioration and Biodegradation* 83: 119–128.

Khehra M, Saini H, Sharma D, Chadha B, and Chimni S. 2005. Decolorization of various azo dyes by bacterial consortia. *Dyes and Pigments* 67: 55–61.

Khehra M, Saini H, Sharma D, Chadha B, and Chimni S. 2006. Biodegradation of a zodye C.I. Acid Red 88 by ananoxic–aerobics equential bioreactor. *Dyes and Pigments* 70: 1–7.

Khelifi E, Bouallagui H, Touhami Y, Godon J, and Hamdi M. 2009. Enhancement of textile wastewater decolourization and biodegradation by isolated bacterial and fungal strains. *Desalination and Water Treatment* 2: 310–316.

Kulla H, Klausener F, Meyer U, Ludeke B, and Leisinger T. 1983. Interference of aromatic sulfo groups in the microbial degradation of the azo dyes Orange I and Orange II. *Archives of Microbiology* 135: 1–7.

Kumar K, Devi S, Krishnamurthi K, Dutta D, and Chakrabarti T. 2007. Decolorization and detoxification of direct Blue-15 by a bacterial consortium. *Bioresource Technology* 98: 3168–3171.

Kurade M, Waghmode T, and Govindwar S. 2011 Preferential biodegradation of structurally dissimilar dyes from a mixture by *Brevibacillus laterosporus*. *Journal of Hazardus Materials* 192: 1746–1755.

Kurade M, Waghmode T, Kabra A, and Govindwar S. 2013. Degradation of a xenobiotic textile dye, disperse Brown 118, by *Brevibacillus laterosporus*. *Biotechnology Letters* 35: 1593–1598.

Kurade M, Waghmode T, Kagalkar A, and Govindwar S. 2012a. Decolorization of textile industry effluent containing disperses dye Scarlet RR by a newly developed bacterial-yeast consortium BL-GG. *Chemical Engineering Journal* 184: 33–41.

Lade H, Waghmode T, Kadam A, and Govindwar S. 2012. Enhanced biodegradation and detoxification of disperse azo dye Rubine GFL and textile industry effluent by defined fungal-bacterial consortium. *International Biodeterioration and Biodegradation* 72: 94–107.

Lopez-Grimau V and Gutierrez M. 2006. Decolorization of simulated reactive dyebath effluents by electrochemical oxidation assisted by UV light. *Chemosphere* 62: 106–112.

Lorena G, Maria J, Jose N, Garcia-Ruiz P, Francisco G, and Jose T. 2004. Deuterium isotope effect on the oxidation of monophenols and o-diphenols by tyrosinase. *Biochemistry Journal* 380: 643–650.

Lucas M, Amaral C, Sampaio A, Peres J, and Dias A. 2006. Biodegradation of the diazo dye reactive Black 5 by a wild isolate of *Candida oleophila*. *Enzyme and Microbial Technology* 39: 51–55.

Machado K, Compart L, Morais R, Rosa L, and Santos M. 2006. Biodegradation of reactive textile dyes by basidiomycetous fungi from Brazilian ecosystems. *Brazilian Journal of Microbiology* 37: 481–487.

Mezohegyi G, Bengoa C, Stubera F, Fonta J, Fabregat A, and Fortuny A. 2008. Novel bioreactor design for decolorisation of azo dye effluents. *Chemical Engineering Journal* 143: 293–298.

Mittal A, Mittal J, Kurup L, and Singh A. 2006. Process development for the removal and recovery of hazardous dye Erythrosine from wastewater by waste materials-bottom ash and de-oiled soya as adsorbents. *Journal of Hazardus Materials* 138: 95–105.

Mohana S, Shrivastava S, Divecha J, and Madamwar D. 2008. Response surface methodology for optimization of medium for decolorization of textile dye Direct Black 22 by a novel bacterial consortium. *Bioresource Technology* 99: 562–569.

Moore A, Sabachewsky L, and Toolan H. 1955. Culture characteristics of four permanent human cancer cell lines. *Cancer Research* 15: 598–602.

Moosvi S, Keharia H, and Madamwar D. 2005. Decolorization of textile dye 588 Reactive Violet 5 by a newly isolated consortium RVM 11. 1. *World Journal of Microbiology and Biotechnology* 21: 667–672.

Myers C and Myers J. 1992. Localization of cytochromes to the outer membrane of anaerobically grown Shewanella putrefaciens MR-1. *Journal of Bacteriology* 174: 3429–3438.

Nachiyar C and Rajkumar G. 2005. Purification and characterization of an oxygen insensitive azoreductase from *Pseudomonas aeruginosa*. *Enzyme and Microbial Technology* 36: 503–509.

Nigam P, Banat I, Singh D, and Marchant R. 1996a. Microbial process for the decolorization of textile effluent containing azo, diazo and reactive dyes. *Process Biochemistry* 31: 435–442.

Nigam P, McMullan G, Banat I, and Marchant R. 1996b. Decolorization of effluent from the textile industry by a microbial consortium. *Biotechnology Letters* 18: 117–120.

Olukanni O, Osuntokia A, Kalyani D, Gbenlea G, and Govindwar S. 2010. Decolorization and biodegradation of reactive Blue 13 by proteus mirabilis LAG. *Journal of Hazardus Materials* 184: 290–298.

Ortiz-Ordoñez E, Uría-Galicia E, Ruiz-Picos R, Sánchez-Duran A, Hernández-Trejo Y, Elías Sedeño-Díaz J, and López-López E. 2011. Effect of Yerbimat herbicide on lipid peroxidation, catalase activity, and histological damage in gills and liver of the freshwater fish *Goodea atripinnis*. *Archievs of Environmental Contamination and Toxicology* 61: 443–452.

Oturkar C, Patole M, Gawai K, and Madamwar D. 2013. Enzyme based cleavage strategy of *Bacillus lentus* BI377 in response to metabolism of azoic recalcitrant. *Bioresource Technology* 130: 360–365.

Padmavathy S, Sandhya S, Swaminathan K, Subrahmanyam Y, and Kaul S. 2003. Comparison of decolorization of reactive azo dyes by microorganisms isolated from various sources. *Journal of Environmental Science* 15: 628–633.

Pandey A, Singh P, and Iyengar L. 2007. Bacterial decolorization and degradation of azo dyes. *International Biodeterioration and Biodegradation* 59: 73–84.

Parikh A and Madamwar D. 2005. Textile dye decolorization using cyanobacteria. *Biotechnology Letters* 27: 323–326.

Parshetti G, Kalme S, Gomare S, and Govindwar S. 2007. Biodegradation of Reactive blue-25 by Aspergillus ochraceus NCIM-1146. *Bioresource Technology* 98: 3638–3642.

Patil A, Lokhande V, Suprasanna P, Bapat V, and Jadhav J. 2012. *Sesuvium portulacastrum* (L.) L.: A potential halophyte for the degradation of toxic textile dye, Green HE4B. *Planta* 235: 1051–1063.

Patil P, Desai N, Govindwar S, Jadhav J, and Bapat V. 2009. Degradation analysis of reactive Red 198 by hairy roots of *Tagetes patula* L. (Marigold). *Planta* 230: 725–735.

Patil P, Phugare S, Jadhav S, and Jadhav J. 2010. Communal action of microbial cultures for Red HE3B degradation. *Journal of Hazardus Materials* 181: 263–270.

Patil P, Shedbalkar U, Kalyani D, and Jadhav J. 2008. Biodegradation of reactive Blue 59 by isolated bacterial consortium PMB11. *Journal of Indian Microbiology and Biotechnology* 35: 1181.

Pearce C, Lloyd J, and Guthriea J. 2003. The removal of color from textile wastewater using whole bacterial cells: A review. *Dyes and Pigment* 58: 179–196.

Phugare S, Kalyani D, Patil A, and Jadhav J. 2011a. Textile dye degradation by bacterial consortium and subsequent toxicological analysis of dye and dye metabolites using cytotoxicity, genotoxicity and oxidative stress studies. *Journal of Hazardus Materials* 186: 713–723.

Phugare S, Waghmare S, and Jadhav J. 2011b. Purification and characterization of dye degrading of veratryl alcohol oxidase from *Pseudomonas aeruginosa* strain BCH. *World Journal of Microbiology and Biotechnology* 27: 2415–2423.

Pinheiro H, Touraud E, and Thomas O. 2004. Aromatic amines from azo dye reduction: Status review with emphasis on direct UV spectrophotometric detection in textile industry wastewaters. *Dyes and Pigments* 61: 121–139.

Piontek K, Antorini M, and Choinowski T. 2002. Crystal structure of a laccase from the fungus *Trametes versicolor* at 1.90-Å resolution containing a full complement of coppers. *Journal of Biological Chemistry* 277: 37663–37669.

Plumb J, Bell J, and Stuckey D. 2001. Microbial populations associated with treatment of an industrial dye effluent in an anaerobic baffled reactor. *Applied and Environmental Microbiology* 67: 3226–3235.

Prival M and Mitchell V. 1982. Analysis of a method for testing azo dyes for mutagenic activity in *Salmonella typhimurium* in the presence of flavin mononucleotide and hamster liver S9. *Mutation Research* 97: 103–116.

Qu Y, Shi S, Ma F, and Yan B. 2010. Decolorization of reactive dark blue K-R by the synergism of fungus and bacterium using response surface methodology. *Bioresource Technology* 101: 8016–8023.

Rafii F and Cerniglia C. 1995. Reduction of a zodyesandnitro aromatic compounds by bacteria lenzymes from the human in testinaltract. *Environmental Health Perspectives* 5: 17–19.

Rafii F, Franklin W, and Cerniglia C. 1990. Azoreductase activity of anaerobic bacteria isolated from human intestinal microflora. *Applied and Environmental Microbiology* 56: 2146–2151.

Rai H, Bhattacharya M, Singh J, Bansal T, Vats P, and Banerjee U. 2005. Removal of dyes from the effluent of textile and dyestuff manufacturing industry: A review of emerging techniques with reference to biological treatment. *Critical Reviews in Environmental Science and Technology* 35: 219–238.

Razo Flores E, Luijten M, Donlon B, Lettinga G, and Field J. 1997. Complete biodegradation of the azo dye azodisalicylate under anaerobic conditions. *Environmental Science and Technology* 31: 2098–2103.

Robinson T, McMullan G, Marchant R, and Nigam P. 2001. Remediation of dyes in textiles effluent: A critical review on current treatment technologies with a proposed alternative. *Bioresource Technology* 77: 247–255.

Russ R, Rau J, and Stolz A. 2000. The function of cytoplasmic flavin reductases in the reduction of azo dyes by bacteria. *Applied and Environmental Microbiology* 66: 1429–1434.

Sakurai T. 1992. Anaerobic reactions of *Rhus vernicifera* laccase and its type-2 copper-depleted derivatives with hexacyanoferrate (II). *Biochemistry Journal* 284: 681–685.

Salokhe M and Govindwar S. 1999. Effect of carbon source on the biotransformation enzymes in *Serratia marcescens*. *World Journal of Microbiology and Biotechnology* 15: 229–232.

Sani R and Banerjee U. 1999. Decolorization of triphenylmethane dyes and textile and dyestuff effluent by *Kurthia* sp. *Enzyme and Microbial Technology* 24: 433–437.

Saratale G, Bhosale S, Kalme S, and Govindwar S. 2007. Biodegradation of Kerosene in *Aspergillus ochraceus* (NCIM 1146). *Journal of Basic Microbiology* 47: 400–405.

Saratale GD, Kalme S, and Govindwar S. 2006. Decolorisation of textile dyes by *Aspergillus ochraceus* (NCIM-1146). *Indian Journal of Biotechnology* 5: 407–410.

Saratale R, Saratale G, Chang J, and Govindwar S. 2009a. Decolorization and biodegradation of textile dye navy Blue HER by *Trichosporon beigelii* NCIM-3326. *Journal of Hazardus Materials* 166: 1421–1428.

Saratale R, Saratale G, Chang J, and Govindwar S. 2009c. Ecofriendly decolorization and degradation of reactive Green 19 using *Micrococcus glutamicus* NCIM-2168. *Bioresource Technology* 110: 3897–3905.

Saratale R, Saratale G, Chang J, and Govindwar S. 2010. Decolorization and biodegradation of reactive dyes and dye wastewater by a developed bacterial consortium. *Biodegradation* 21: 999–1015.

Saratale G, Saratale R, Chang J, and Govindwar S. 2011a. Fixed-bed decolorization of Reactive Blue 172 by *Proteus vulgaris* NCIM-2027 immobilized on *Luffa cylindrica* sponge. *International Biodeterioration and Biodegradation* 65: 494–503.

Saratale R, Saratale G, Chang J, and Govindwar S. 2011b. Bacterial decolorization and degradation of azo dyes: A review. *Journal of the Taiwan Institute of Chemical Engineering* 42: 138–157.

Saratale R, Saratale G, Kalyani D, Chang J, and Govindwar S. 2009b. Enhanced decolorization and biodegradation of textile azo dye Scarlet R by using developed microbial consortium-GR. *Bioresource Technology* 100: 2493–2500.

Sharma D, Saini H, Singh M, Chimni S, and Chadha B. 2004a. Biodegradation of acid blue-15, a textile dye, by an up-flow immobilized cell bioreactor. *Journal of Indian Microbiology and Biotechnology* 31: 109–114.

Sharma S, Kalpana AS, Suryavathi V, Singh PRS, and Sharma K. 2007. Toxicity assessment of textile dye wastewater using swiss albino rats. *Australian Journal of Ecology* 13: 81–85.

Siddiqui A, Tabrez S, and Ahmad M. 2011. Short-term *in vitro* and *in vivo* genotoxicity testing systems for some water bodies of Northern India. *Environmental Monitoring and Assessment* 180: 87–95.

Stammati A, Nebbia C, Angelis I, Albo A, Carletti M, Rebecchi C, Zampaglioni F, and Dacasto M. 2005. Effects of malachite green (MG) and its major metabolite, leucomalachite green (LMG), in two human cell lines. *Toxicology In Vitro* 19: 853–858.

Stolz A. 2001. Basic and applied aspects in the microbial degradation of azo dyes. *Applied Microbiology and Biotechnology* 56: 69–80.

Su Y, Zhang Y, Wang J, Zhou J, Lu X, and Lu H. 2009. Enhanced biodecolorization of azo dyes by co-immobilized quinone-reducing consortium and anthraquinone. *Bioresource Technology* 100: 2982–2987.

Sugiura W, Miyashita T, Yokoyama T, and Arai M. 1999. Isolation of azo-dye degrading microorganisms and their application to whitedischarge printing of fabric. *Journal of Bioscience Bioengineering* 88: 577–581.

Tamboli D, Kagalkar A, Jadhav M, Jadhav J, and Govindwar S. 2010b. Production of polyhydroxyhexadecanoic acid by using waste biomass of *Sphingobacterium* sp. ATM generated after degradation of textile dye direct Red 5B. *Bioresource Technology* 101: 2421–2427.

Tamboli D, Gomare S, Kalme S, Jadhav U, and Govindwar S. 2010c. Degradation of Orange 3R, mixture of dyes and textile effluent and production of polyhydroxyalkanoates from biomass obtained after degradation. *International Biodeterioration and Biodegradation* 64: 755–763.

Tamboli D, Kurade M, Waghmode T, Joshi S, and Govindwar S. 2010a. Exploring the ability of *Sphingobacterium* sp. ATM to degrade textile dye Direct Blue GLL, mixture of dyes and textile effluent and production of polyhydroxyhexadecanoic acid using waste biomass generated after dye degradation. *Journal of Hazardus Materials* 182: 169–176.

Tamboli D, Telke A, Dawkar V, Jadhav S, and Govindwar S. 2011. Purification and characterization of bacterial aryl alcohol oxidase from *Sphingobacterium* sp. ATM and its uses in textile dye decolorization. *Biotechnology and Bioprocess Engineering* 16: 661–668.

Telke A, Ghodake G, Kalyani D, Dhanve R, and Govindwar S. 2011. Biochemical characteristics of a textile dye degrading extracellular laccase from a *Bacillus* sp. ADR. *Bioresource Technology* 102: 1752–1756.

Telke A, Joshi S, Jadhav S, Tamboli D, and Govindwar S. 2010. Decolorization and detoxification of Congo red and textile industry effluent by an isolated bacterium *Pseudomonas* sp. SU-EBT. *Biodegradation* 21: 283–296.

Telke A, Kalyani D, Dawkar V, and Govindwar S. 2009a. Influence of organic and inorganic compounds on oxidoreductive decolorization of sulfonated azo dye C.I. Reactive orange 16. *Journal of Hazardus Materials* 172: 298–309.

Telke A, Kalyani D, Jadhav J, and Govindwar S. 2008. Kinetics and mechanism of reactive red 141 degradation by a bacterial isolate *Rhizobium radiobacter* MTCC 8161. *Acta Chimica Slovenica* 55: 320–329.

Telke A, Kalyani D, Jadhav U, Parshetti G, and Govindwar S. 2009b. Purification and characterization of an extracellular laccase from a *Pseudomonas* sp. LBC1 and its application for the removal of bisphenol A. *Journal of Molecular Catalysis B: Enzymatic* 61: 252–260.

Tehrani-Bagha A and Holmberg K. 2013. Solubilization of hydrophobic dyes in surfactant solutions. *Materials* 6: 580–608.

Tony B, Goyal D, and Khanna S. 2009a. Decolorization of direct Red 28 by mixed bacterial culture in an up-flow immobilized bioreactor. *Journal of Indian Microbiology and Biotechnology* 36: 955–960.

Tony B, Goyal D, and Khanna S. 2009b. Decolorization of textile azo dyes by aerobic bacterial consortium. *International Biodeterioration and Biodegradation* 63: 462–469.

Torbati S, Khataee A, and Movafeghi A. 2014. Application of watercress (*Nasturtium officinale* R. Br.) for biotreatment of a textile dye: Investigation of some physiological responses and effects of operational parameters. *Chemical Engineering Research and Design* 92: 1934–1941.

Trapp S and McFarlane C. 1995. *Plant Contamination: Modeling and Simulation of Organic Processes.* Lewis, Boca Raton, FL, p. 254.

Trivedi A, Kitabatake N, and Doi E. 1990. Toxicity of dimethyl sulfoxide as a solvent in bioassay system with HeLa cells evaluated colorimetrically with 3-(4,5-dimethyl thiazol-2-yl)-2,5-diphenyltetrazolium bromide. *Agricultural and Biological Chemistry* 54: 2961–2966.

Umbuzeiro G, Freeman H, Warren S, Kummrow F, and Claxton L. 2005. Mutagenicity evaluation of the commercial product CI Disperse Blue 291 using different protocols of the *Salmonella* assay. *Food and Chemical Toxicology* 43: 49–56.

Van der Zee F, Bouwman R, Strik D, Lettinga G, and Field J. 2001. Application of redox mediators to accelerate the transformation of reactive azo dyes in anaerobic bioreactors. *Biotechnology and Bioengineering* 756: 691–701.

Waghmare P, Kadam A, Saratale G, and Govindwar S. 2014. Enzymatic hydrolysis and characterization of waste lignocellulosic biomass produced after dye bioremediation under solid state fermentation. *Bioresource Technology* 168: 136–141.

Waghmode T, Kurade M, and Govindwar S. 2011a. Time dependent degradation of mixture of structurally different azo and non azo dyes by using *Galactomyces geotrichum* MTCC 1360. *International Biodeterioration and Biodegradation* 65: 479–486.

Waghmode T, Kurade M, Khandare R, and Govindwar S. 2011b. A sequential aerobic/microaerophilic decolorization of sulfonated mono azo dye Golden yellow HER by microbial consortium GG-BL. *International Biodeterioration and Biodegradation* 65: 1024–1034.

Waghmode T, Kurade M, Lade H, and Govindwar S. 2012. Decolorization and biodegradation of rubine GFL by microbial consortium GG-BL in sequential aerobic/microaerophilic process. *Applied Biochemistry and Biotechnology* 167: 1578–1594.

Walker R and Ryan A. 1971. Some molecular parameters influencing rate of reduction of azo compounds by intestinal microflora. *Xenobiotica* 1: 483–486.

Watharkar A, Khandare R, Kamble A, Mulla A, Govindwar S, and Jadhav J. 2013a. Phytoremediation potential of *Petunia grandiflora* Juss. an ornamental plant to degrade a disperse, disulfonated triphenylmethane textile dye Brilliant Blue G. *Environmental Science and Pollution Research* 20: 939–949.

Watharkar A, Rane N, Patil S, Khandare R, and Jadhav J. 2013b. Enhanced phytotransformation of Navy Blue RX dye by *Petunia grandiflora* Juss. with augmentation of rhizospheric *Bacillus pumilus* strain PgJ and subsequent toxicity analysis. *Bioresource Technology* 142: 246–254.

Weber E and Adams R. 1995. Chemical and sediment-mediated reduction of the azo dye Disperse Blue 79. *Environmental Science and Technology* 29: 1163–1170.

Weber E and Wolfe L. 1987. Kinetic studies of the reduction of aromatic azo compounds in anaerobic sediment/water systems. *Environmental Toxicology and Chemistry* 6: 911–919.

Wesenberg D, Kyriakides I, and Agathos S. 2003. White-rot fungi and their enzymes for the treatment of industrial dye effluents. *Biotechnology Advances* 22: 161–187.

Wuhrmann K, Mechsner K, and Kappeler T. 1980. Investigations on rate determining factors in the microbial reduction of Azo Dyes. *European Journal of Applied Microbiology and Biotechnology* 9: 325–338.

Xu F. 1996. Oxidation of phenols, anilines, and benzenethiols by fungal laccases: Correlation between activity and redox potentials as well as halide inhibition. *Biochemistry* 35: 7608–7614.

Xu F, Shin W, Brown S, Wahleithner J, Sundaram U, and Solomon E. 1996. A study of a series of recombinant fungal laccases and bilirubin oxidase that exhibit significant differences in redox potential, substrate specificity, and stability. *Biochimica et Biophysica Acta* 1292: 303–311.

Yang Q, Li C, Li H, Li Y, and Yu N. 2009. Degradation of synthetic reactive azo dyes and treatment of textile wastewater by a fungi consortium reactor. *Biochemical Engineering Journal* 43: 225–230.

Yoo E, Libra J, and Adrian L. 2001. Mechanism of decolorization of azo dyes in anaerobic mixed culture. *Journal of Environmental Engineering* 127: 844–849.

Yu J, Wang X, and Yue P. 2001. Optimal decolorization and kinetic modeling of synthetic dyes by *Pseudomonas* strains. *Water Research* 35: 3579–3586.

Zimmermann T, Kulla H, and Leisinger T. 1982. Properties of purified orange II azoreductase, the enzyme initiating azo dye degradation by pseudomonas KF46. *European Journal of Biochemistry* 129: 197–203.

15

Isolation of Pure DNA for Metagenomic Study from Industrial Polluted Sites: A New Approach for Monitoring the Microbial Community and Pollutants

Ram Chandra and Sheelu Yadav

CONTENTS

15.1 Introduction

It is widely recognized that 93%–98% of the microbes present in the environment are not easily culturable. Oceans, glaciers, deserts, and almost every other environment on earth are being sampled for the identification of microorganisms. In all natural environments, soil and sludge are probably the most challenging environments for microbiologists, with respect to the microbial community size and the diversity of species present. About 4×10^7 prokaryotic cells are present in one gram of forest soil whereas the same amount of cultivated soil and grasslands contains nearly 2×10^9 prokaryotic cells. Estimating the total microbial diversity/dynamics in any environment is a persisting challenge especially during bioremediation to clean up contaminants. In actual fact, most of the species in different environments have never been described, and this condition will not change until fresh culture techniques are developed. At present the numerous approaches that are used to walk around the diversity of microbial communities are not satisfactory because of the limitations of culturing methods. "Metagenomics" tries to conquer this blockage by developing and using culture-independent approaches. Metagenomics is a cultural independent genomics analysis of a majority of microorganisms growing in the natural environment which has the potential to respond to fundamental queries in the field of microbiology. Metagenomics, also called ecological genomics, environmental genomics, or community genomics, is the whole genome shotgun approach to sequence genome from entire communities of microorganisms in different environmental samples.

During the last two decades, the development of methods to extract nucleic acids from environmental sources has opened up a new area in an earlier unknown range of microorganisms and provided prokaryotic genetic information available in environmental samples, independent of culturing the bacterial cells. Study of nucleic acids directly extracted from environmental samples allows researchers to study natural microbial communities without the need of cultivation. Culture independent methods or direct extraction methods can give more information about the bacterial genome.

The pulp and paper industries are the main source of environmental pollution such as water and soil pollution. There is a lot of difference in the metagenomic methods of normal soil and pulp paper mill waste contaminated sludge. The major known pollutants present in this contaminated site are chlorolignin, chlorophenol and some heavy metals, which is hazardous to the environment, animals, and human life due to various persistent toxic organic pollutants. The fact that some plants grow on the contaminated site means that they have a strategy to detoxify such pollutants and it has been reported that some bacteria are responsible for the degradation of lignin and chlorolignin by lignolytic enzymes (Chandra et al., 2009).

Additionally, some organic compounds such as polycyclic aromatic hydrocarbons (PAHs) and polychlorinated biphenyls are known as persistent organic pollutants that remain in sludge after secondary treatment, but they can be removed during the pretreatment process. To identify such bacterial strains we need to isolate them from the pulp and paper waste contaminated site and provide them natural conditions for growth.

All the procedures have been described to extract and purify DNA from soil and sediments, but not from pulp paper mill waste or contaminated sludge. Optimization of the extraction procedure is essential to release nucleic acids capably from a complex microbial contaminated site. However, the extraction of pure DNA from contaminated sludge is not simple. The toughest step for a sludge sample is isolation of DNA from humic substances, heavy metals, and other organic pollutants because they interfere with polymerase chain reaction (PCR) amplification (Tebbe and Vahjen, 1993). The differences in cell wall structures and in the adhesion nature of microbes with material, enzymatic, and biological characteristics of sludge affect the isolation of DNA during sample treatment.

Coextraction of other humic and fulvic acids along with nucleic acid act as inhibitor for obtaining pure DNA. Because, due to polycondensation of sludge organic matters from plants, microorganisms and animals which have three dimensional structure and they have strong binding ability with nucleic acid active site. This inhibits the reaction of Taq polymerase with DNA. It is reported that some inhibitors like heavy metals, organic substances, and humic substances (humic acid and fulvic acid) interfere with restriction endonuclease also (Holben et al., 1988; Porteous and Armstrong, 1991; Tsai and Olson, 1991; Jacobsen and Rasmussen, 1992). Size exclusion chromatography has also been used to remove humic substances and heavy metals (Matheson et al., 2009).

DNA purification can be optimized by the removal of inhibitors and protein also that gives clear PCR amplification. DNA as well as RNA are generally removed from sludge samples using two different methods modified from the simple soil extraction method. It has been also reported by Yu and Mohn (1999) that DNA and RNA can be extracted simultaneously from activated sludge samples and their method combines the bead-beating system (to break bacterial cells) and the use of diethyl pyrocarbonate (DEPC) to protect RNA and the removal of contaminants by ammonium acetate treatment. However, activated sludge is dissimilar from ordinary soil and sediments in at least the following characteristics: elevated biomass, low humic acid inside, and the existence of bacterial flocs. To consider the above aspects, we modified the simple extraction method of DNA extraction (Figure 15.1).

In this chapter, we discuss the recent strategies and developments related to identify and extract novel nucleic acid from sludge of pulp and paper mill contaminated sites for different advantages. Further, we mentioned the identification of bacterial species from sludge

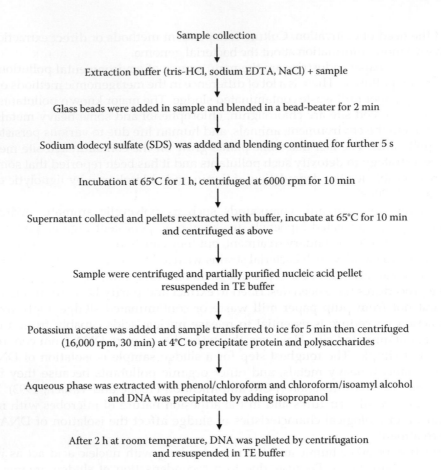

Sample collection

↓

Extraction buffer (tris-HCl, sodium EDTA, NaCl) + sample

↓

Glass beads were added in sample and blended in a bead-beater for 2 min

↓

Sodium dodecyl sulfate (SDS) was added and blending continued for further 5 s

↓

Incubation at 65°C for 1 h, centrifuged at 6000 rpm for 10 min

↓

Supernatant collected and pellets reextracted with buffer, incubate at 65°C for 10 min
and centrifuged as above

↓

Sample were centrifuged and partially purified nucleic acid pellet
resuspended in TE buffer

↓

Potassium acetate was added and sample transferred to ice for 5 min then centrifuged
(16,000 rpm, 30 min) at 4°C to precipitate protein and polysaccharides

↓

Aqueous phase was extracted with phenol/chloroform and chloroform/isoamyl alcohol
and DNA was precipitated by adding isopropanol

↓

After 2 h at room temperature, DNA was pelleted by centrifugation
and resuspended in TE buffer

FIGURE 15.1
Common protocol for DNA isolation from soil sample. (Adapted from Yeates C et al. 1998. *Biological Procedures Online* 1(1): 40–47.)

through cell lysis, an inhibitor removal is necessary for PCR amplification to polymerize the genomic DNA. Further restriction fragment length polymorphism (RFLP) is done by the ligation of PCR product with a cloning vector that transformed into a host and forms clones. Additionally we have discussed the protocol of DNA extraction with different methods and steps.

The common methods used in metagenomics to sequence genomes from all microbes in a specified environment involve isolating DNA directly from an environmental sample without requiring culture of the bacteria or virus. Such advancement is necessary because it is often difficult to maintain the multiple collection of growth conditions the microbes require surviving in culture. By the gene sequencing such as capillary electrophoresis (CE)-based Sanger sequencing, scientists gained the ability to elucidate genetic information from any given biological system. This technology has become widely adopted in laboratories around the world, yet has always been hampered by inherent limitations in throughput, scalability, speed, and resolution that often preclude scientists from obtaining the essential information they need for their course of study. To overcome these barriers, an entirely new technology is required—next-generation sequencing (NGS), a fundamentally different approach to sequencing that triggered numerous ground-breaking discoveries and ignited a revolution in genomic science. The concept behind NGS technology is similar to CE—the bases of a small fragment of DNA are sequentially identified from signals emitted as each fragment is resynthesized from a DNA template strand.

15.2 Characteristics of Some Industrial Waste in Environments

The majority of industrial waste contains several organic and inorganic pollutants which contribute to the BOD, COD, and TDS of both effluent and sludge. Besides, some industrial waste contains characteristic pollutants, for example, melanoidin in distillery waste, lignin and chlorolignin in pulp paper mill waste, and chromium and tannin in tannery waste. Heavy metals are also released from several industrial wastes (Chandra et al., 2009).

The pollutant parameter of some industrial waste has been shown in Table 15.1. Furthermore, several industrial wastes contain very complex organic pollutants which are still unknown in the literature and have recently been reported in studies by GC/MS-MS analysis (Chandra and Singh, 2012). The complex organic pollutants from industrial waste contaminated sites have been reported in Table 15.2.

TABLE 15.1

Different Pollutants Parameters and Characteristics of Various Industrial Waste

S. No.	Parameters	Pulp Paper Mill	Distillery	Tannery
1	pH	9.2 ± 1.72	8.5 ± 0.17	8.2 ± 0.2
2	BOD	$72{,}143 \pm 141.89$	$18{,}600 \pm 160$	3650 ± 132
3	COD	$213{,}136 \pm 164.81$	$32{,}860 \pm 517$	$14{,}216 \pm 406$
4	TSS	17.06 ± 20.5	$10{,}112 \pm 294.3$	90.92 ± 4.2
5	TDS	$13{,}402 \pm 96.32$	$10{,}212 \pm 42.67$	113.12 ± 11.01
6	Sulfate	1762 ± 41.11	3786 ± 78.09	1519.98 ± 26.5
7	Phenol	38.54 ± 2.61	510.58 ± 47.20	38.32 ± 3.24
8	Chloride	406	48.60 ± 2.45	500
9	Heavy metals			
i.	Cr	0.255 ± 0.04	3.03 ± 0.21	9.38 ± 0.9
ii.	Cd	0.06 ± 0.03	2.37 ± 0.06	1.22 ± 0.06
iii.	Cu	0.511 ± 0.10	3.12 ± 0.24	1.86 ± 0.04
iv.	Fe	3.99 ± 0.91	184.01 ± 1.63	3.02 ± 0.18
v.	Mn	0.8750 ± 0.03	5.03 ± 0.92	0.69 ± 0.02
vi.	Ni	2.84 ± 0.06	2.24 ± 0.3	0.44 ± 0.03
vii.	Zn	1.5 ± 0.3	14.11 ± 1.16	0.62 ± 0.04
10	Lignin	663 ± 4.23 mg/L	—	—
11	Melanoidin	—	2000 mg/L	—
12	Tannic acid	—	—	10 g/L

TABLE 15.2

Other Organic Pollutants Detected from Common Industrial Waste

S. No.	Pulp Paper Mill Waste Sludge	Distillery Waste Sludge	Tannery Waste Sludge
1.	4-Isopropoxy-butyric acid	3-Amino-2-oxazolidinone	Pentanoic acid
2.	Butane-1-ol	Cyclopropylmethanol acetate	2-Methylbutanoic acid
3.	Propane	4-Pyridinecarboxlic acid	Lactic acid
4.	Cyclohexanecarboxylic acid	2-Ethylpyridine	Hexanoic acid
5.	1-Phenyl-1-nonyne	n-Methyl-2-nitro-1-oxide	4-Methylbutanoic acid
6.	Tetradecanoic acid	3-Ethylpyridine	Propanoic acid
7.	1-Chlorooctadecane	Isonicotinyl formaldoxime	Acetic acid
8.	4,8-Dimethyl undecane	1,2-Benzenedicarboxylic acid	Butanidoic acid
9.	Butanoic acid	5,5-Dimethyl hexane	Benzoic acid
10.	Propanoic acid	Benzyl butyl phthalate	Tartaric acid

Note: Above compounds are detected in the GC/MS-MS analysis from different industries.

15.3 PCR Inhibitors

In the environment or soil, there are several pollutants which have a binding tendency with nucleic acid and inhibit PCR amplification. It is tough to determine all the causes of inhibition in DNA isolation from natural soil for PCR amplification. The PCR process can be affected by compounds that interfere with the interaction between DNA and Taq polymerase and thus inhibit the reaction, known as inhibitors. Common inhibitors are humic compounds, heavy metals, and other organic substances. The details are shown in Figure 15.2. The coextraction of contaminants such as humic substances is a major problem in the extraction process related to metagenomic study. The traditional DNA extraction protocols such as detergent, phenol-chloroform extraction, and protease treatment, are not able to complete the removal of such contaminants. The main inhibitors are the compounds or substances that inhibit PCR amplification.

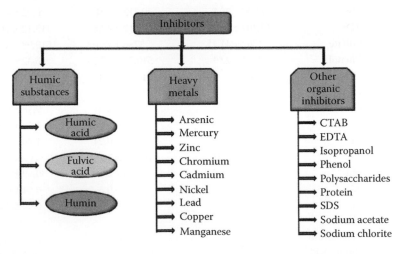

FIGURE 15.2
Inhibitor's classification present in the environment.

15.3.1 Humic Compounds

Humic compounds are formed by the breakdown of plants, animals, and microbial biomass range from 0.1 to 300 kDa. They are structurally complex, polyelectrolytic, and yellow to dark brown in color and are divided into following three classes based on solubility in acids and bases:

I. Humic acid: They are soluble in alkali and insoluble in acid.

II. Fulvic acid: They are soluble in alkali and acid.

III. Humin: They are insoluble in both alkali and acid.

Coextracted humic substances are the main contaminant when DNA is extracted from soil. These compounds absorb at 230 nm whereas DNA absorbs at 260 nm and protein at 280 nm. To evaluate the purity of the extracted DNA, absorbance ratios at 260/230 nm (DNA/humic acids) and 260/280 nm (DNA/protein) are determined.

Mostly all natural organic compounds can become bound or absorbed to humic substances (Fortin et al., 2004). In addition, humic acids have physicochemical properties similar to DNA. Hence, humic substances along with the adsorbed organic molecules are usually copurified with nucleic acid. Nearly all molecular methods such as hybridization, restriction digestions of DNA, PCR, and bacterial transformation are affected by humic acid (Tebbe and Vahjen, 1993). Dilution of samples provides a rapid and straightforward way of permitting PCR amplification. Diluting the sample sometime enhances the sensitivity of PCR by reducing the concentration of inhibitors relative to target DNA. Although, in some cases, the dilution is not sufficient to reduce the humic acids inhibition (Erb and Wagnerdobler, 1993).

15.3.2 Heavy Metals

Heavy metals also have a binding tendency with nucleic acid molecules. Heavy metals are classified into two classes based on the charge present in them. Cationic metals (metallic elements whose forms in soil are positively charged cations, e.g., Pb^{2+}) are mercury, cadmium, lead, nickel, copper, zinc, chromium, and manganese. The most common anionic compounds (elements whose forms in soil are combined with oxygen and are negatively charged, e.g., $MoO4^{2-}$) are arsenic, molybdenum, selenium, and boron. Humic acids have a metal binding affinity and have been reported to form complexes with Pb and Cu (Logan et al., 1997). Powdered activated charcoal is used in the removal of different contaminants because of its large surface area and pore volume that provide it special adsorption capacity. Amberlite ion exchangers have also the absorbing capacity for the metal cations that are Zn^{2+} and Cd^{2+}.

Nickel, nitrate hexahydrate (98.5 w/w%), copper, nitrate pentahydrate (99.0 w/w%), zinc, nitrate hexahydrate (99.0 w/w%), cadmium, nitrate tetrahydrate (99.0%), and lead, nitrate (99.0 w/w%), have been used to spike soil sub-samples to stimulate artificial contamination with Ni, Cu, Zn, Cd, and Pb, respectively. Acetic acid (99.5 w/v%), hydroxylamine hydrochloride (99.0 w/w%), ammonium acetate (98.0 w/w%), hydrogen peroxide (30% w/v%), hydrochloric acid (37% w/v%), and nitric acid (99.5% w/v%) have been used to prepare extracting solutions for sequential chemical fractionation. Citric acid (99.5% w/w%), tartaric acid (99.5 w/w%), and ethylenediaminetetraacetic acid (EDTA) (99.0 w/w%), have been used to prepare solutions for soil washing (Wuana et al., 2010).

Cationic metals are more soluble at lower pH levels. Increasing the soil pH to 6.5 or higher can remove the cations. Raising pH has the opposite effect on anionic elements. Heavy phosphate also reduces the presence of cations but increases the arsenic. Amberlite cations exchanger effectively removes cations. The drainage process improves soil aeration and this will allow metals to oxidize, making them less soluble. Therefore, when aerated, these metals are less available. Active organic matter is effective in reducing the availability of chromium in chromium containing soil or sludge.

15.4 Other Organic Inhibitors

Reagents generally used in the purification of nucleic acid inactivate the DNA polymerases (Table 15.3). Phenol or C helex resin left with the extracted DNA can inhibit the PCR process such as phenolic compounds from the organic extraction chelex resin, salts, guanidine, proteases organic solvents, phosphate buffers, and detergents (such as sodium dodecyl sulfate, SDS).

15.4.1 Mechanism of PCR Inhibition

A large number of mechanisms can cause a PCR to be unsuccessful. Binding of an inhibitor to the active site of a DNA polymerase, blocks the activity of DNA polymerase. Researchers

TABLE 15.3

Some other Reported PCR Inhibitors that Inhibit PCR Amplification

S. No.	Inhibitors	Concentration that Inhibit PCR	Method to Minimize the Effect of Inhibitors
1.	CTAB	≥0.005% (w/v) [49, 51]; 0.01%	70% ethanol wash
2.	EDTA	≥0.5 mM; 1 mM	Decrease the concentration of EDTA to 0.1 mM in the TE buffer or simply use Tris-HCl (10 mM) to bring DNA in solution. DNA can also be brought in pure water (but the DNA cannot be stored for long-term use)
3.	Ethanol	>1% (v/v)	Dry pellet and resuspend
4.	Fat	—	Lipase or hexane treatment and chloroform extraction
5.	Isopropanol	>1% (v/v)	Dry pellet and resuspend
6.	Phenol	>2% (v/v, ; >0.2%). Chlorogenic acid—plant phenol (0.24–0.36 µg/µL)—inhibited PCR in potato	Use PVP, PVP/ammonium acetate. Incorporation of 1.2% citric acid at the DNA extraction step neutralized the inhibitory effect of chlorogenic acid
7.	Polysaccharides	Acidic polysaccharides such as dextran sulfate are inhibitory Dextran sulfate: >0.1% ≥ 0.001% Pectin: >0.5% Xylan: >0.0025%	Use CTAB buffer and chloroform extraction. Treatment with enzymes such as pectinase, cellulase, hemicellulase, and α-amylase can be used to remove polysaccharides in high salt precipitation
8.	Protein	1% casein hydrolysate in PCR mixture caused inhibition	Use SDS, CTAB/guanidinium buffers, proteinase K
9.	SDS	≥0.005% (w/v)	Wash with 70% ethanol
10.	Sodium acetate	≥5 mM	Wash with 70% ethanol
11.	Sodium chloride	≥25 mM	Wash with 70% ethanol or use silica-based purification

Source: Adapted from Demeke T and Jenkins GR. 2010. *Analytical and Bioanalytical Chemistry* 396(6): 1977–1990.

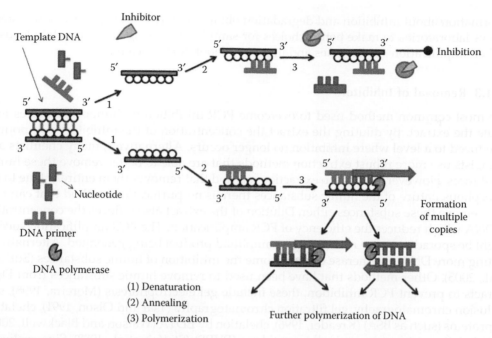

FIGURE 15.3
(See color insert.) Mechanism of inhibition in PCR.

mostly focused on the activity of the polymerase enzyme that can be inhibited by blocking an active site by various inhibitors, denatured by phenol and other detergents and degraded by proteinases. Other reasons include segregation of essential cofactor such as Mg^{2+}, for example, by chelators such as EDTA used in TE buffer. It was found that the inhibitory effect of divalent ions (Ca^{2+} and Mg^{2+}) are more than monovalent ions (K^+ and Na^+). Ca^{2+} ions in milk compete with Mg^{2+} for the binding sites on the polymerase, thereby inhibiting PCR (Bickley et al., 1996). Many PCR also do not succeed due to complicated template sequences, for example, GC rich, which causes the polymerization process to be inefficient. In some cases, adding more templates can reduce the amount of the PCR product obtained because inhibitors comes into the PCR reaction via DNA template preparation. Thus by decreasing the amount of templates PCR amplification can be enhanced. Phenols or polyphenols cross-link nucleic acids slow down resuspension of pellets and chelate metal ions. Polysaccharides sequester nucleic acids and inhibit Taq or reverse transcriptase. Humic substances (fulvic and humic acid) from dead organic matter, crosslink or adsorb nucleic acids, bind or adsorb to enzymes. Interactions between proteins and DNA polymerases are a major cause of inhibition. Humic, fulvic, and tannic acids, together with other polyphenolic compounds, are potent PCR inhibitors present in soil. Humics probably affect PCR both by binding to the polymerase via hydrogen bonds (as shown in Figure 15.3) and by changing the melting temperature of dsDNA, preventing primer annealing (Sutlovic et al., 2005).

15.4.2 Detection of Inhibitors

Real-time PCR data can also be used to detect inhibitors by analyzing target amplification efficiency (Kontanis and Reed, 2006). The IPC (Internal PCR Control) strategy has been used in combination with two autosomal targets of differing sizes to simultaneously assess both inhibitors and template degradation (Swango et al., 2006). The additional information about inhibition and degradation obtained by real-time quantitation systems allows laboratories to make better choices for sample processing and ultimately leads to higher amplification success rates and improved laboratory efficiency.

15.4.3 Removal of Inhibitors

The most common method used to overcome PCR inhibition by humic substances is to dilute the extract. By diluting the extract the concentration of the inhibitory compounds is reduced to a level where inhibition no longer occurs. Alternatively, many chemists and biologists use more robust extraction methods that are able to better remove these humic substances. However, there is no extraction method that removes them entirely. Due to the amorphous nature of the humic substances there is no purification method that can reliably remove these substances either. Dilution of the extract also reduces the concentration of DNA which reduces the efficiency of PCR amplification. The PCR amplification success might be sporadic or may result in a less amplified product being generated. Alternatively adding more DNA polymerase can overcome the inhibition of humic substances (Sutlovic et al., 2005). Other methods that have been used to remove humic substances from DNA extracts to prevent PCR inhibition, these include gel electrophoresis (Moreira, 1998), size exclusion chromatography, gel filtration chromatography (Tsai and Olson, 1991), chelation by proteins (such as BSA) (Kreader, 1996), chelation by EDTA (Watson and Blackwell, 2000), and chelation with polyvinylpolypyrrolidone (PVPP) (Steffan et al., 1988). Size exclusion chromatography has also been used to remove the humic substance to demonstrate the removal of inhibition by removal of the humic substance. Size exclusion chromatography has also been used to remove metal ion inhibition in hidden biological remains (Matheson et al., 2009). Adding a larger amount of templates and the ratio of template and primer is also an important parameter to optimize PCR amplification.

15.4.4 Sample Pretreatment

For the isolation of DNA, the segregation of different pollutants is a must and for that pretreatment of the sample is a necessary step before cell lysis. Recently, some simple and rapid purification methods have been reported for the successful removal of inhibitors from contaminated waste sites for pure metagenomic DNA. The use of PVPP and hexaadecyltrimethylammonium bromide (CTAB) (Zhou et al., 1996) before cell lysis reduces the coprecipitation of humic substances and enhances the purity of metagenomic DNA. This study has shown different category of soil samples used to evaluate the effect of CTAB and PVPP in the extraction buffer and it was compared in three different conditions: (i) no CTAB and no PVPP, (ii) 1% CTAB and no PVPP, and (iii) no CTAB and 2 g of PVPP. Further, the result showed that both CTAB and PVPP effectively removed the humic substances but in comparison with PVPP, CTAB resulted in low DNA loss as shown in Table 15.4. Accordingly the absorbance A_{260}/A_{280} ratio of isolated DNA in different conditions showed a positive correlation. In another study by using microwave-based method, tubes containing pellet with lysis solution were heated in a microwave oven at 600 ± 700 W for 45 s. Addition of PVPP at high concentration in the extraction buffer improved the purity of the extracted DNA by minimizing coextraction of humic substances. DNA extracted by the microwave method never inhibited Taq polymerase activity whereas the enzymatic method required a further purification step. The above processes have been effectively used on different environmental samples, for example, soil, sediments, and activated sludge (Orsini and Romano-spica, 2001).

TABLE 15.4

Comparison of DNA Yield and Purity in Different Conditions

S. No.	Treatment	DNA Yield	A_{260}/A_{280} Ratio	A_{260}/A_{230} Ratio
1.	No PVPP, no CTAB	17.1 ± 0.9	1.17 ± 0.02	0.72 ± 0.03
2.	CTAB, no PVPP	17.5 ± 1.2	1.35 ± 0.04	0.91 ± 0.03
3.	PVPP, no CTAB	10.9 ± 1.5	1.23 ± 0.05	0.88 ± 0.03
4.	Pure culture		1.89	1.57

Likewise, the use of skim milk in the extraction buffer (at 40 mg/g soil) shows a high yield of DNA. Skim milk is known to minimize the adsorption and degradation of nucleic acid in soil (Takada-Hoshino and Matsumoto, 2004). In another study, 0.2 M sodium ascorbate added in the homogenization solution is also reported to decrease the coextraction of humic acids (Holben et al., 1988). Chemical flocculation is one of the potential methods for copurification of inhibitors. The addition of multivalent cations such as magnesium chloride ($MgCl_2$), ferric chloride ($FeCl_3$), calcium chloride ($CaCl_2$), and aluminum ammonium sulfate ($AlNH_4(SO_4)_2$) during cell lysis and the protein precipitation step has been done, as shown in Figure 15.4. Further, the result showed that $AlNH_4(SO_4)_2$ and $FeCl_3$ gives the most effective results by improved PCR amplification. $MgCl_2$ and $CaCl_2$ gives the lowest results when added during cell lysis in comparison with $AlNH_4(SO_4)_2$ and $FeCl_3$ but the DNA yield was greater than control which was not treated with any chemical during either cell lysis or the precipitation step. Another set of tubes in which multivalent cation was added during protein precipitation gave no DNA recovery. It has been proposed

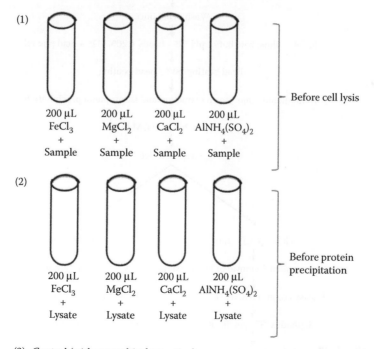

(1) — Before cell lysis

200 µL FeCl₃ + Sample   200 µL MgCl₂ + Sample   200 µL CaCl₂ + Sample   200 µL AlNH₄(SO₄)₂ + Sample

(2) — Before protein precipitation

200 µL FeCl₃ + Lysate   200 µL MgCl₂ + Lysate   200 µL CaCl₂ + Lysate   200 µL AlNH₄(SO₄)₂ + Lysate

(3) Control (without multivalent cation)

FIGURE 15.4
Addition of multivalent cation in two different steps of DNA extraction procedure.

that the aluminum cations may preferentially interact with the open and random structure of humic substances (Braid et al., 2003), but the mechanism of organic inhibitors by $AlNH_4(SO_4)_2$ is still not clear.

Pretreatment of the sample with sodium pyrophosphate (NaPP) along with gravel is an essential step and it is then centrifuged, followed by washing which has been recommended to remove inhibitors (Duarte et al., 1998). In this study the simultaneous study of DNA and RNA has been done (Figure 15.5).

Pretreatment with an advanced oxidative process has been described by Martens and Frankenberger (1995). This study showed that Fenton's reagent could enhance the PAH degradation. Recently, it has been reported in a study that the pretreatment of a sludge sample with calcium carbonate ($CaCO_3$) yields PCR well-suited metagenomic DNA.

Sample washing improves the recovery of DNA from polluted and high organic content sediments. Fortin et al. (2004) reported washing the sample three times with a different amount of buffer in every wash, using 10 mL of buffer but reducing the amount of tris-HCL and Na_2EDTA after the second wash from 50 mL to 10 mL and 5–0.1 mM,

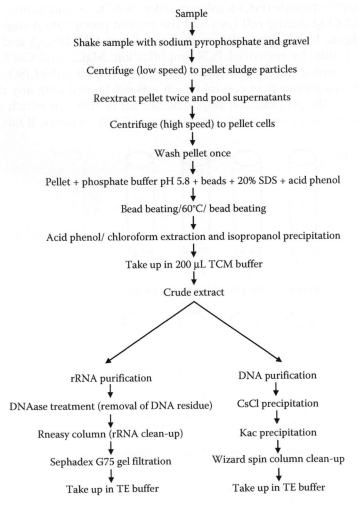

FIGURE 15.5
Synchronal indirect extraction of ribosomal RNA and genomic DNA from industrial waste contaminated site.

respectively. The integration of the soil washing step before the extraction of DNA from a polluted environment solubilize and eradicate the contaminants when a high quality of DNA is required.

15.5 Cell Lysis

The method of cell lysis is designed to release the DNA by breaking the cell wall and the membrane of the microbes. Many researchers have investigated different protocols for the lysis of cells because no single method is appropriate for all soil and sludge types, so different combinations of method need to be standardized. There are two types of methods of cell lysis—the direct method of cell lysis and the indirect method of cell lysis.

15.5.1 Direct Method of Cell Lysis

In this method, nucleic acid is directly extracted from the sludge sample without isolating the cell. In other words, cell lysis occurs within the sample and DNA is released within it. Direct cell lysis includes various physical and chemical factors which help in the lysis of the cell.

15.5.1.1 Mechanical Method

The mechanical lysis includes bead beating, microwave heating, bead mill homogenization, thermal shock, and ultrasonication. Glass beads are used in order to facilitate crushing of the cell walls of bacteria. In the bead-beating system, beads are added in the sludge sample and blended in a mini-bead beater for 2–3 min. Extensive bead-beater treatment results in increasingly extra shearing of DNA while the DNA yield increases proportionately (Van Elsas et al., 1997). Liquid homogenization has also been done to shear the sludge by forcing it into small tubes.

The ultrasonication method disrupts the bacterial cell efficiently (Ramsay, 1984). The method uses pulsed, high-frequency sound waves to disturb and lyse cells, bacteria, spores, and fine tissue. A key consideration when ultrasonicating is done is sample heating. The cavitations caused by the ultrasonicator can move the temperature up rapidly, and the solution may get extremely hot in the vicinity of the tip. For this reason, one can do many short pulses rather than one long continuous pulse. Also, it is important to make sure that the cell suspension is chilled and that it is packed with crushed ice during sonication. Thermal shock includes a number of repeated freezing–thawing processes, since it is less brutal than other mechanical methods and can effectively lyse the bacterial cell. The main disadvantage of mechanical methods such as ultrasonication and the bead-beating system is the shearing of nucleic acid. In grinding, sludge is frozen in liquid nitrogen and then crushed using a mortar and pestle. Because of the tensile strength of the cellulose and other polysaccharides comprising the cell wall, this method is the fastest and most proficient way to access plant proteins and DNA (Table 15.5).

15.5.1.2 Chemical/Enzymatic Method

Chemical or enzymatic lysis is comparatively moderate and often distinguishes against particular cell types and does not completely break soil or sediment samples. Chemicals

TABLE 15.5

Showing Different Mechanical Disruption Methods of Cell Lysis

Lysis Method	Apparatus	Description
Bead-beating	Bead beater	Beads are added in the sample and blended
Ultrasonication	Sonicator	High-frequency sound waves shear cells
Thermal shock	Freezer or dry ice with ethanol	Repeated cycles of freezing and thawing disrupt cells through ice crystal formation
Liquid homogenization	Homogenizer French Press	Cell or sludge suspensions are sheared by forcing them through a narrow space
Grinding	Mortar and pestle	Grinding sludge sample, frozen in liquid nitrogen

involved in lysis protocol, are CTAB, PVPP, Tween 80, Tween 20, Triton X 100, Tris-EDTA, NaCl, and SDS. In enzymes, lysozymes are also used to break the sample. The treatment of SDS and lysozyme together also has been done. With this method, lysozyme has added to the cell suspension and lysis allowed to continue for 30 min to 1 h. Lysozyme (200 µg/mL) can be used to digest the polysaccharide component of bacterial cell walls. After adding of each chemical or enzyme and also mechanical thawing, the sample should be centrifuged at $10,000g$ for 10 min at 4°C.

Since mechanical and chemical or enzymatic methods have some drawbacks, it has been reported that both chemical and mechanical methods can be combined to get a higher yield of nucleic acid (Gray and Herwig, 1996). In this study, the incubation of the sample at 70°C was for 1 h in SDS buffer and then homogenized in a bead beater.

15.5.2 Indirect Method of Cell Lysis

In the indirect method of cell lysis, bacterial cells can be released by the use of a cation exchange resin and subsequently separated from sludge particles by low speed centrifugation (Jacobsen and Rasmussen, 1992). The sample is then treated with lysozyme and ionic detergent and DNA is then isolated.

The indirect method specifically breaks the prokaryote DNA, minimizes the extraction of extracellular DNA, and provides larger fragment DNA with a high level of purity (Leff et al., 1995). However, it has been reported that the DNA obtained only corresponds to about 25%–35% of the total number of bacteria present in the soil. Different bacterial groups strongly hold to soil particles, which might be the picture of the composition of the microbial community in the sample (Steffan et al., 1988).

15.6 Quantification of Metagenomic DNA

A diphenylamine (DPA) indicator will confirm the presence of DNA. This procedure involves chemical hydrolysis of DNA: when heated (e.g., ≥95°C) in acid, the reaction requires a deoxyribose sugar and therefore is specific for DNA. Under these conditions, the 2-deoxyribose is converted to w-hydroxylevulinyl aldehyde, which reacts with the compound, DPA, to produce a blue-colored compound. DNA concentration can be determined measuring the intensity of absorbance of the solution at the 600 nm with a spectrophotometer and compared with a standard curve of known DNA concentrations.

Measuring the intensity of absorbance of the DNA solution at wavelengths 260 and 280 nm is used as a measure of DNA purity. DNA absorbs UV light at 260 and 280 nm, and aromatic proteins absorb UV light at 280 nm; a pure sample of DNA has a ratio of 1.8 at 260/280 and is relatively free from protein contamination. A DNA preparation that is contaminated with protein will have a 260/280 ratio lower than 1.8. DNA can be quantified by cutting the DNA with a restriction enzyme, running it on an agarose gel, staining with ethidium bromide or a different stain and comparing the intensity of the DNA with a DNA marker of known concentration. Using the southern blot technique, this quantified DNA can be isolated and examined further using PCR and RFLP analyses. These procedures allow differentiation of the repeated sequences within the genome. These techniques are used in forensic science for comparison, identification, and analysis of DNA samples.

15.7 Isolation of Nucleic Acid

DNA isolation is a process done for the purification of DNA after cell lysis using physical and chemical methods or a combined method.

15.7.1 Deproteinization

When cell lysis is completed protein should be removed for the purification of metagenomic DNA. It has been reported that protein can be salted out by saturated salt solution. Different studies have used different salts such as sodium chloride (Selenska and Klingmüller, 1991), ammonium acetate (Xia et al., 1995), potassium acetate (Smalla et al., 1993), and sodium acetate (Holben et al., 1988). Many researchers used the traditional in deproteinization method whereby organic solvents such as phenol, phenol-chloroform (P:C), chloroform-isoamyl alcohol (C:I), and phenol-chloroform-isoamyl alcohol (P:C:I) are used to separate the protein from the sample. Use of NaCl allows coprecipitation of protein with sludge particles and cell debris (Harry et al., 1999). The pH should be alkaline (pH 9) for the effective extraction of nucleic acid (Frostegard et al., 1999).

15.7.2 DNA Purification

DNA is purified by the precipitation method after the isolation of protein. Precipitation is carried out with ethanol or isopropanol. Cullen and Hirsch (1998) compared the precipitation of soil metagenomic DNA with ethanol, isopropanol, and polyethylene glycol (PEG) 8000. They have reported that precipitation with ethanol gave a lower recovery of DNA and more humic substances than PEG 8000 or Isopropanol. Arbeli and Fuentes (2007) reported precipitation using 5% PEG yielded significantly less humic acids without affecting PCR and hence 5% PEG has been recommended for the precipitation of soil metagenomic DNA. Porteous et al. (1997) and Yeates et al. (1997) stated that the use of alcohol in precipitation favors the coextraction of contaminants such as humic acids but PEG greatly reduces the coextraction of humic substances. Various chromatographic methods for the purification of soil metagenomic DNA are mentioned below.

15.7.2.1 Gel Filtration

Different approaches have also been developed for the purification of soil nucleic acid. Silica gel is most widely used for the purification step. The silica gel binds to humic materials and DNA initially, and some humic acids can be sequentially eluted from the matrix. This same technology has been found in the Elutip d syringe tip filters used by Picard et al. (1992) but they did not find the silica gel protocols alone to be effective in removing sufficient amounts of humic materials from DNA extracts of several soil types. Generally cesium chloride density gradient centrifugation is an applied and efficient approach for the isolation of DNA from lignocellulosic contaminants (Holben et al., 1988; Jacobsen and Rasmussen, 1992; Tien et al., 1999). Although, due to the extended processing time, this method is not appropriate for purification of various samples. Tsai and Olson (1991) use Polyacrylamide Bio-gel P-6 and P-30 for gel filtration. Sephadex 100 has also purified DNA successfully (Leff et al., 1995). Cullen and Hirsch (1998) reported that Sephadex G-75 purified metagenomic DNA from soil. Frostegard et al. (1999) used Sephacryl S200 and S400 together. A combination of Sephadex G200 and G150; Sepharose 6B, 4B, and 2B; Bio-Gel P100, P200; and Toyopearl HW 55, HW 65, and HW 75 is reported by Miller (2001). Jackson et al. (1997) compared the effectiveness of Sepharose 4B, Sephadex G-200, and Sephadex G-50 with a diverse set of soils and found Sepharose 4B to be superior.

15.7.2.2 Ion Exchange

After organic extraction and alcoholic precipitation, DNA is purified by DEAE cellulose column, reprecipitated, and then passed through a Wizard DNA clean-up column and a Bio101 Fast DNA spin column. Although not commonly used, we found ion exchange chromatography to be an excellent way to remove humic materials from soil DNA extracts. DEAE cellulose columns are commonly used for purifying tRNAs and have been used for DNA extracts. Hydroxyapatite (HA) column chromatography has been effectively used as well (Ogram et al., 1987; Steffan et al., 1988). These column steps are relatively laborious, but the great amount of humic materials removed makes them worthwhile. Syringe-tip ion-exchange columns are available commercially and have been applied to DNA purification from soils (Tebbe and Vahjen, 1993). In most cases, the DNA extracts obtained after DEAE cellulose chromatography are not sufficiently pure for PCR or restriction enzyme digestion. The eluted DNA is precipitated with alcohol to reduce its volume and to remove salts in preparation for the final silica gel spin column procedures. Amberlite ion exchangers are known to absorb metal cations such as Zn^{2+} and Cd^{2+} under alkaline conditions and metal oxyanions such as CrO_4^{2-} under acidic pH (Desai and Madamwar, 2007).

15.7.2.3 Adsorption

One of the recommended methods for HA removal, is adsorption on activated carbon (AC). The adsorption of HA is affected by many factors, including the porous texture and surface chemistry of AC (Bjelopavlic et al., 1991), the macromolecular size and chemical structure of HA, and solution parameters such as ionic strength and pH. Various forces, electrostatic and nonelectrostatic ones, are considered in the adsorption process of natural organic matter from water. Adsorption is the adhesion of atoms, ions, or molecules from a gas, liquid, or dissolved solid to a surface. This process creates a film of the ascorbate on the surface of the adsorbent. This process differs from absorption, in which a fluid (the absorbate) permeates or is dissolved by a liquid or solid (the absorbent). Adsorption is a surface-based process while absorption involves the whole volume of the material. The term sorption encompasses both processes, while desorption is its reverse. Adsorption is a surface phenomenon. A promising method of upgrading the adsorption capacity of AC is the modification of the chemical character of the surface (Cheng et al., 2005). The HA molecules are negatively charged at a typical water treatment condition, that is, neutral pH of solution. Hence, for enhanced uptake of HA the activated carbon used should be characterized by a positively charged surface.

Miller et al. (1999) have compared and statistically evaluated the effectiveness of nine DNA extraction procedures for the isolation and purification of soil metagenomic DNA. They have studied the effects of different chemical extractants and physical disruption methods in combination with four different DNA purification methods (silica-based DNA binding, agarose gel electrophoresis, ammonium acetate precipitation, and Sephadex G-200 gel filtration) for DNA recovery and the removal of PCR inhibitors. The results indicated that the optimum DNA recovery requires brief, low speed bead mill homogenization in the presence of a phosphate buffered SDS-chloroform mixture, followed by Sephadex G-200 column purification. The comparative evaluation of various methods by various investigators also indicated that only a combination of two or more methods render sufficiently pure DNA from soil (Harry et al., 1999). In addition, the method of lysis and extraction also determines the efficiency of purification strategies. Thus, the optimized purification protocol confirmed that only a combination of methods would be efficient in yielding pure environmental DNA from humic-rich soils.

15.7.2.4 Other DNA Extraction Kits

For the extraction and purification of soil metagenomic DNA, some kits that are easy to use are available and also claim an increased yield of inhibitor free metagenomic DNA from soil and sludge samples. These kits are UltraClean™, PowerSoil™, PowerMax™ (Mo Bio Laboratories Inc., Carlsbad, CA, USA), SoilMaster™ (Epicentre Biotechnologies, Madison, WI, USA), and FastDNA® Spin kit for soil (MP Biomedicals, Solon, OH, USA). Recently, Whitehouse and Hottel (2007) compared the efficiency of five different commercial DNA extraction kits for the recovery of DNA from soil samples spiked with Gram-negative bacterium, *Francisella tularensis*. Using real-time PCR analysis, it was reported that UltraClean™ and PowerMax™ kits were more efficient than SoilMaster™ and FastDNA® kits. Thus, the efficiencies of commercial kits reported by various researchers are inconsistent.

15.8 PCR Amplification

PCR allows the isolation of DNA fragments from genomic DNA by selective amplification of a specific region of DNA. This use of PCR augments many methods, such as generating hybridization probes for Southern or Northern hybridization and DNA cloning, which require larger amounts of DNA, representing a specific DNA region. PCR supplies these techniques with high amounts of pure DNA, enabling analysis of DNA samples even from very small amounts of starting material. Because PCR amplifies the regions of DNA that it targets, PCR can be used to analyze extremely small amounts of sample. Quantitative PCR methods allow the estimation of the amount of a given sequence present in a sample—a technique often applied to quantitatively determine levels of gene expression. Quantitative PCR is an established tool for DNA quantification that measures the accumulation of the

TABLE 15.6

Various Temperatures and Time in Different Steps of PCR Amplification

Steps	Temperature (°C)	Time (min)	Number of Cycles
Initial denaturation	95	1–3	1
Denaturation	95	0.5	25–40
Annealing	68	0.5	
Extension	72	1 min/kb	
Final extension	72	5–15	1

DNA product after each round of PCR amplification, that is, denaturation at 94–96°C, annealing at 68°C, and extension at 72°C (Table 15.6).

15.9 Genomic Separation (RFLP)

Many different methods and technologies are available for the separation of genomic DNA. In RFLP analysis, the DNA sample is broken into pieces (digested) by restriction enzymes and the resulting restriction fragments are separated according to their lengths by gel electrophoresis. Although now largely obsolete due to the rise of inexpensive DNA sequencing technologies, RFLP analysis was the first DNA profiling technique inexpensive enough to see widespread application.

Restriction enzymes are proteins isolated from bacteria that recognize specific short sequences of DNA and cut the DNA at those sites. The normal function of these enzymes in bacteria is to protect the organism by attacking foreign DNA, such as viruses.

The restriction enzyme is added to the DNA being analyzed and incubated for several hours, allowing the restriction enzyme to cut at its recognition sites. The DNA is then run through a gel, which separates the DNA fragments according to size. You can then visualize the size of the DNA fragments and assess whether or not the DNA was cut by the enzyme, and transferred to a membrane via the Southern blot procedure. Hybridization of the membrane to a labeled DNA probe then determines the length of the fragments which are complementary to the probe. An RFLP occurs when the length of a detected fragment varies between individuals. The different steps in RFLP protocol are shown in Figure 15.6.

15.9.1 Ligation

The majority of ligation reactions involve DNA fragments that have been generated by restriction enzyme digestion. Most restriction enzymes digest DNA asymmetrically across their recognition sequence, which results in a single stranded overhang on the digested end of the DNA fragment. The overhangs, called "sticky ends," are what that allow the vector and insert to bind to each other. Different kinds of DNA vectors are given in Table 15.7 with their target gene. When the sticky ends are compatible, meaning that the overhanging base pairs on the vector and insert are complementary, the two pieces of DNA connect and ultimately are fused by the ligation reaction. Usually, scientists select two different enzymes for adding an insert into a vector (one enzyme on the 5' end and a different enzyme on the 3' end). This ensures that the insert will be added in the correct orientation and prevents the vector from ligating to itself during the ligation process. If the sticky ends

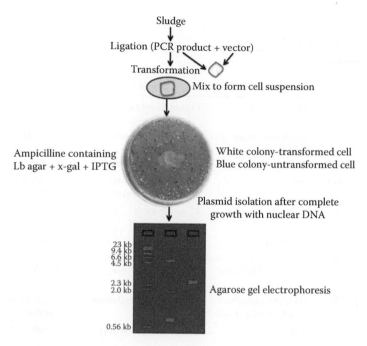

FIGURE 15.6
(See color insert.) Schematic diagram of RFLP.

on either side of the vector are compatible with each other, the vector is much more likely to ligate to itself rather than to the desired insert. If you are in this situation, it is important to treat the digested vector backbone with a phosphatase before performing the ligation reaction (phosphatase removes the 5′ phosphate and therefore prevents the ligase from being able to fuse the two ends of the vector together). After ligation of the PCR product with the vector, this reaction mixture is incubated at 4°C (overnight).

15.9.2 Transformation and Cloning

"Transformation" may also be used to describe the insertion of new genetic material into nonbacterial cells. The recombinant vector with the PCR product can be transformed into bacterial cells by incubation in ice for 15 min, then heat shock at 42°C for 1 min, then again keeping on ice for 5 min, as shown in Figure 15.6. Then a number of copies of the recombinant cell are formed, and this process is known as cloning. There are two types of gene cloning: *in vivo*, which involves the use of restriction enzymes and ligase using vectors and cloning the fragments into host cells (as in Figure 15.6). The other type is *in vitro* which is using the PCR method to create copies of fragments of DNA.

15.9.3 Blue/White Screening

Whether the bacteria are transformed or not can be identified by blue–white screening. In the simple cloning system, we used antibiotic resistance to select the desired clone. This was possible because our desired clones all carried antibiotic resistant genes which untransformed bacteria did not. The blue–white screen is a screening technique that allows for the rapid and convenient detection of recombinant bacteria in vector-based molecular cloning experiments. DNA of interest is ligated into a vector. The vector is then inserted

TABLE 15.7

Different Types of Cloning Vectors

S. No.	Vector Used for Library Construction	Source	Target Gene	Host Strain	References
1.	pBluescript SK	Meadow soil	Lipase, Esterase	*E. coli* DH5	Henne et al. (2000)
2.	pBeloBAC11	North American soil	Antibacterial, hemolytic activities, lipase, amylase, nuclease	*E. coli* DH10B	Rondon et al. (2000)
3.	pBluescript SK	Soil	4-Hydroxybutyrate dehydrogenase	*E. coli* DH5 alpha	Henne et al. (1999)
4.	pBluescript SK	Soil	H$^+$ antiporters	*E. coli* KNabc	Majernik et al. (2001)
5.	Fosmid	Forest topsoil	Unique lipolytic activity	*E. coli*	Lee et al. (2004)
6.	Cosmid	Unplanted field soil	Novel biocatalysts	*E. coli*	Voget et al. (2003)
7.	pBluescript SK	Meadow, sugarbeet field, cropland soil	Alcohol oxidoreductase	*E. coli* DH5 alpha	Knietsch et al. (2003)
8.	n.r	Soil	Carbonyls formation	*E. coli*	Knietsch et al. (2003)
9.	pBluescript SK	Soil	Coenzyme B(12)-dependent glycerol and diol dehydratases	*E. coli* DH5 alpha	Knietsch et al. (2003)
10.	pZero-2	Soil and enrichment culture	Amidase	*E. coli* TOP10	Gabor et al. (2004)
11.	pUC19	Soil	Amylase	*E. coli* DH5 alpha	Yun et al. (2004)
12.	pJN105, pCF430	Soil	Aminoglycoside and tetracycline antibiotic resistance	DH10B, DH5 alpha	Riesenfeld et al. (2004)
13.	Cosmid	Soil	Deoxyviolacein and broad spectrum antibiotic violacein	*E. coli*	Brady et al. (2004)

into a competent host cell (bacteria), viable for transformation, which are then grown in the presence of x-gal. Cells transformed with vectors containing recombinant DNA will produce white colonies; cells transformed with nonrecombinant plasmids (i.e., only the vector) grow into blue colonies. pBLU and other specially designed cloning vectors make use of the lac operon. pBLU carries the gene for β-galactosidase (also known as lacZ). This enzyme catalyzes the breakdown of lactose as a food source. It can also degrade an artificial substrate called x-gal, which turns blue when it is broken down by β-galactosidase. Colonies that produce β-galactosidase and are fed will turn blue. On the other hand colonies that do not produce β-galactosidase remain white in color, even in the presence of x-gal. Successful ligation of a foreign DNA into a multiple cloning site of vector interrupts lac Z and abolishes the production of functional β-galactosidase. Different white colonies are selected and incubated in 5 mL LB broth with ampicilline (100 µg/mL). Incubated up to 16 h/overnight at 37°C, cells having PCR product as plasmid are grown in this media along with nuclear DNA.

15.9.4 Plasmid Isolation

Plasmid is isolated by using MINI PREB kit from all tubes suppose 1–10. From this process, approximately 5 μg plasmid is obtained from 5 mL cell culture. About 1 μg is taken out of the 5-μg plasmid product to confirm clones either having 1.5 kb of PCR product or not. The plasmid is digested with EcoR1 restriction enzyme. For restriction digestion reaction mixture (plasmid DNA, restriction enzyme EcoR1, buffer, and water to make up the volume as desired) is prepared and digested.

15.9.5 Agarose Gel Electrophoresis

Agarose gel electrophoresis is a method of gel electrophoresis used in biochemistry, molecular biology, and clinical chemistry to separate a mixed population of DNA or proteins in a matrix of agarose. The proteins may be separated by charge and or size (IEF agarose, essentially size independent), and the DNA and RNA fragments by length Kryndushkin et al. (2003).

15.10 Analysis of Metagenomic Data by Construction of Genomic Library

Two approaches have been emerged to extract biological information from metagenomic libraries.

15.10.1 Sequence-Based Analysis

Sequence-driven analysis relies on the use of conserved DNA sequences to design hybridization probes or PCR primers to screen metagenomic libraries for clones that contain sequences of interest. Significant discoveries have also resulted from random sequencing of metagenomic clones. Sequencing of clones carrying phylogenetic anchors, such as the 16S rRNA gene and the archaeal DNA repair gene *rad A*. Beja et al. (2000) led to functional information about the organisms from which these clones were derived. The sequence conservation of regions of phylogenetic anchors facilitates their isolation without prior knowledge of the full gene sequence. By contrast, the sequences of most genes of practical importance are too divergent to make it possible to identify new homologues by PCR or hybridization. However, a few classes of genes contain sufficiently conserved regions to facilitate their identification by sequence instead of function. Two hotly pursued examples are the genes encoding polyketide synthases (PKSs) and peptide synthetases, which contribute to synthesis of complex antibiotics. The PKSs are modular enzymes with repeating domains containing divergent regions that provide the variation in chemical structures of the products. These regions are flanked by highly conserved regions, which have provided the basis for designing probes to identify PKS genes among metagenomic clones (Courtois et al., 2003).

15.10.2 Function-Based Analysis

Function-driven analysis is initiated by identification of clones that express a desired trait, followed by characterization of the active clones by sequence and biochemical analysis. This approach quickly identifies clones that have potential applications in medicine, agriculture, or industry by focusing on natural products or proteins that have useful activities. The limitations of the approach are that it requires expression of the function of interest in the host cell and clustering of all of the genes required for the function. It also depends on the availability of an assay for the function of interest that can be performed efficiently on vast libraries, because the frequency of active clones is quite low. Many approaches are being developed to mitigate these limitations. Improved systems for heterologous gene expression are being developed with shuttle vectors that facilitate the screening of the metagenomic DNA in diverse host species and with modifications of *Escherichia coli* to expand the range of gene expression.

15.11 Applications of Metagenomics

Metagenomics has the potential to advance knowledge in a wide variety of fields. It can also be applied to solve practical challenges in medicine, engineering, agriculture, sustainability, and ecology. Increased knowledge of how microbial communities cope with pollutants improved the assessments of the potential of contaminated sites to recover from pollution and increased the chance of bioaugmentation or biostimulation trials to succeed (George et al., 2010).The major goals of metagenomics may be the identification of functional genes, estimation of the microbial diversity, understanding of the population dynamics of a whole community or assembly of the complete genome of an uncultured organism.

15.11.1 Metagenomics Expression Libraries

Discovery of metabolites from DNA requires the construction and screening of metagenomics expression DNA libraries in suitable cloning vectors and host strains. The classical approach described by Henne et al. (1991) includes the construction of small insert libraries (<10 kb) in a standard sequencing vector and in *E. coli* as a host strain. In the detection of large gene clusters and operons, a small insert library is not allowed. To overcome this drawback, researchers have constructed large insert libraries, such as cosmid DNA libraries (mostly in the PWE15 vector of the strata gene) with insert sizes ranging from 25 to 35 kb (Entcheva et al., 2001) or bacterial artificial chromosome (BAC) libraries up to 200 kb. The construction of fosmid with insert size of 40 kb of foreign DNA has also been reported (Beja et al., 2000).

15.11.1.1 Microbial Diversity Analysis

The metagenomics based microbial diversity analysis and direct analysis of 16S rRNA gene sequences in the environment can be used to study the diversity of microorganisms without culturing (Lane et al., 1985). Development of PCR and cyclic sequencing technology resulted in large-scale sequencing of complete 16S rDNA genes, which accelerated the discovery of diverse taxa from various environmental sources (Handelsman, 2004). Steffan et al. (1988) reported higher DNA from larger fractions of indigenous microbial populations yielded by direct lysis methods. In some studies, indirect methods have shown more diversity than direct lysis methods (Gabor et al., 2003). The amount of coextracted eukaryotic DNA may also influence the yield and diversity correlation. Therefore, non-bacterial DNA pools should be considered to assess population dynamics and diversity. Burgmann et al. (2001) have taken standard quantity of soil DNA extracted with different bead-beating settings and subjected them to PCR targeting 16S rRNA genes.

15.11.1.2 Direct Extraction of rRNA

Moran et al. (1993) reported that the use of rRNA, rather than genes coding for rRNA (rDNA), as the target nucleic acid for 16S rRNA probes has been proposed to be advantageous, since multiple ribosomes in bacterial cells may provide an increased sensitivity of detection. During the DNA extraction method, it is very difficult to recover RNA from RNases in comparison with DNA. The study by Miller (2001) stated that gel filtration resins such as Sepharose 2B and others are not so suitable for RNA purification because of the low-molecular weight of RNA coextract with humic substances. After the cell lysis, ribosomes are separated by centrifugation steps and rRNA then extracted, which avoids humic acid contamination and RNA degradation (Felske et al., 1996).

15.11.1.3 Estimation of Soil Microorganisms

Some studies have been proposed to estimate the numbers of soil microorganisms. Taylor et al. (2002) reported recently 4′,6-diamino-2-phenylindole staining method for counting numbers of soil microorganisms. But this technique cannot to be used due to operational complexity and high cost. Aoshima et al. (2006) reported a slow-stirring method for the isolation of metagenomic DNA from various kinds of soil with minimal shearing. They have obtained a linear proportional relationship between soil bacterial biomass and the amount of DNA isolated by this method. Therefore, the bacterial biomass could be evaluated by quantifying levels of environmental DNA. However, coextraction of extracellular DNA should be considered, which may lead to the overestimation of the number of living bacteria.

15.11.1.4 Estimation of Active versus Total Biodiversity

The important indicators of the metabolic status of microbial communities are recovery of intact RNA from environmental samples and nucleic acid ratio. The amount of DNA and RNA is used to measure the number of cells present and the metabolic activity respectively as the amount of cellular DNA is stable when compared with RNA that changes with growth rate. A wide range of RNA/DNA ratio varies from 0.96 to 0.04 for various marine sediment samples (Dell'Anno et al., 1998). The activity of a particular species can be measured by species to specific probes based quantification of 16S rRNA, 16S rDNA, and the rRNA:rDNA ratio (Muttray and Mohn, 2000). To obtain a reliable RNA/DNA ratio, both RNA and DNA should be recovered from environmental samples without bias (Hurt et al., 2001). The cell extract can be divided into two aliquots and processed for the isolation of DNA and RNA simultaneously (Griffiths et al., 2000). Half of the sample was incubated at 37°C with RNase A to obtain pure DNA and the other half with treated with RNase-free DNase to obtain RNA (Figure 15.7).

15.11.2 Single-Cell Genomics of Nonculturable Microorganisms

Kvist et al. (2007) reported that eliminating the community complexity by initial selection and picking the single cell is used to construct the genomic library of specific microorganism. The intact cells were isolated from the environmental sample. The detection and collection of a single cell of interest can be done by fluorescence-activated cell sorting (Huber et al., 1995). Recently two methods have been reported: micromanipulation (Ishoy et al., 2006) and microfluidics (Marcy et al., 2007). However, isolating single cells alone has not

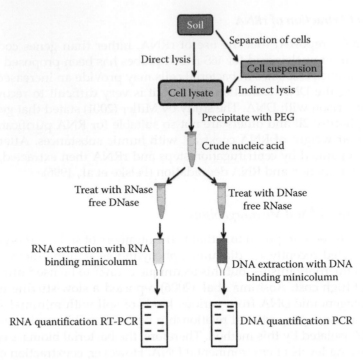

FIGURE 15.7
(**See color insert.**) Isolation and purification of nucleic acid (DNA and RNA) from soil bacteria.

solved the problem of the uncultured organism, because single cells do not provide sufficient DNA for genomic sequencing. The sequencing of a genome from a single cell has been possible by the development of whole genome amplification methods (Hosono et al., 2003). The few femtograms of DNA in a bacterial cell up to micrograms can be multiplied by the multiple displacement amplification (MDA) method. Raghunathan et al. (2005) successfully demonstrated that the genome of a single *E. coli* cell or a single human sperm could be amplified by MDA. The random hexamers bind to the denatured DNA and the phi29 DNA polymerase extends the strand until it reaches newly synthesized double-stranded DNA. The enzyme proceeds polymerization by displacing the double strand. The random primers bind to the newly synthesized single-stranded region and polymerization starts on the new strands, resulting in whole genome amplification of the DNA of single cells as shown in Figure 15.8.

15.11.3 Mobile Metagenomics

The mobile metagenome or "mobilome" is an exciting extension of metagenomics. According to Ou et al. (2007) the mobile metagenome is the genomics of mobile elements from uncultured organisms. Plasmids, transposons, insertion sequences, and integrons, which may move between bacterial cells in a population or mobilize into a new host species and introduce new genetic material, are mobile genetic elements (MGE) (Jones and Marchesi, 2007a). Gene flow, or horizontal gene transfer, between species are a driving force in bacterial adaptation and evolution and can be promoted by MGEs. But the mobile metagenome of many ecosystems have not been efficiently explored. Very few reports are available on the soil mobile metagenome (Stokes et al., 2006; Ono et al., 2007).

(a)

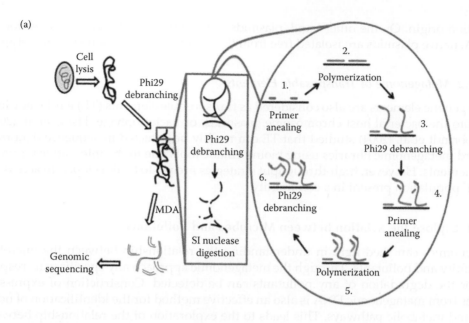

(b)   Random hexamer

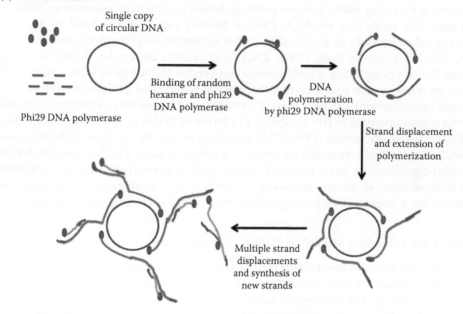

FIGURE 15.8
(See color insert.) (a) MDA of a linear DNA fragment, (b) MDA of a circular single-stranded DNA.

15.11.3.1 Plasmid Metagenome

In any environment autonomously replicating plasmids play a chief role in metabolic activities among uncultured organisms. For accessing the plasmids from the environment, a recent strategy has been developed that is a transposon-aided capture (TRACA) system (Jones and Marchesi, 2007b). In the TRACA system, plasmids are obtained from metagenomic DNA extracts independent of plasmid encoded traits such as selectable markers or replication origin. On the other hand, plasmids are captured by *in vitro* transposition. In TRACA, native plasmids are isolated free from their origin of replication and phenotypes.

15.11.3.2 *Metagenome of Transposable Elements*

Mobile genetic elements are also considered as transposable elements (TE) while they integrate into the bacterial host chromosome, plasmids, or bacteriophage. Hsiao et al. (2003) and Dobrindt et al. (2004) studied that TE can readily be detected in sequence data from standard metagenomic libraries using bioinformatic tools due to the integrative nature of these elements. However, high-throughput strategies are yet to be developed to access the total TE population present in a community.

15.11.4 Explore the Relation between Microbes and Pollutants

Metagenomics can shed light in understanding the relationship between the microbial community and pollutants. Through the metagenomic approach any specific gene responsible for the degradation of any pollutants can be detected. Construction of expression libraries from metagenomic DNA is also an effective method for the identification of novel genes and metabolic pathways. This leads to the exploration of the relationship between microbial communities and pollutants at specific sites. Another procedure for novel gene discovery is by using stable isotope probing, based on the incorporation of stable-labeled substrates into molecular markers. Once a labeled substrate is added in a community, those microbes capable of metabolizing this substrate or pollutants are likely to incorporate labeled atoms in their DNA, RNA, and protein molecules (Malik et al., 2008). It has also been reported that naphthalene dioxygenase (NDO) gene which have multicomponent enzyme system and initiates the metabolism of low molecular weight polycyclic aromatic hydrocarbons (PAHs). Thus, this enzyme is frequently described as a dominant enzyme group involved in the aerobic degradation of PAHs in the environment (Gomes et al., 2007). This study revealed that PAH pollution is capable of shaping NDO gene diversity within microbial communities present at polluted sites. Further, studies focusing on microbial communities from polluted aquatic and terrestrial environments to determine their composition, structure, metabolic capacities, and ecological relationships are still necessary for a deep understanding of how this complex environment functions.

15.11.5 Understanding Microbial Cell–Cell Interaction

Bacteria have evolved sophisticated mechanisms to coordinate gene expression at population and community levels in the face of different environmental constraints. Numerous bacteria, including plant pathogens can synthesize and sense diffusible molecules which act as signals to trigger gene expression when they reach a threshold concentration. In confined environments, accumulation of the signals reflects bacterial cell density (i.e., the quorum) hence this regulatory process was termed quorum sensing (QS) by Fuqua et al. (1994).

15.11.5.1 *Quorum Sensing*

Quorum sensing is a process of cell to cell communication in bacteria. QS bacteria produce and release chemical signal molecules termed as autoinducers whose external concentration increases as a function of increasing cell-population density. Beyond controlling gene expression on a global scale, QS allows bacteria to communicate within and between

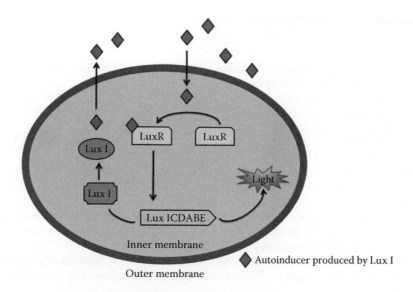

FIGURE 15.9
Quorum sensing in *Vibrio fischeri*; a Lux I and Lux R signaling circuit.

species. A variety of different molecules are pligopeptides in Gram-positive bacteria, *N*-acyl homoserine lactones (AHL) in Gram-negative bacteria, and a family of autoinducers known as autoinducer-2 (AI-2) in both Gram-negative and Gram-positive bacteria.

Specifically, luxS encoding the AI-2 synthase is present in roughly half of all sequenced bacterial genomes; AI-2 production has been verified in a large number of these species, and AI-2 controls gene expression in a variety of bacteria. Together, these findings have led to the hypothesis that bacteria use AI-2 to communicate between species (reviewed in Xavier and Bassler, 2003). Similar to this mechanism, Gram-negative bacterium such as *Vibrio fischeri* use two proteins, luxI and luxR, and control expression of the luciferase operon (luxICDABE) required for light production as shown in Figure 15.9. Several genes involved in the biosynthesis of QS-signals were identified such as *N*-acyl homoserine lactones (AHSLs) (Williamson et al., 2005). Some of them exhibit a weak homology with those previously identified in cultured bacteria and encode the synthesis of unsaturated and substituted long chain AHSLs that have not been described earlier. Furthermore, functional metagenomics allowed the identification of QS mimics.

15.11.5.2 Quorum Quenching

Quorum quenching is the process of preventing QS by disrupting the signaling. This may be achieved by degrading the signaling molecule. Using KG medium, quorum quenching bacteria can be isolated from various environments including those which have previously been considered unculturable. Recently, a well-studied quorum quenching bacterium has been isolated and its AHL degradation kinetic has been studied by using rapid resolution liquid chromatography (RRLC). Many *Bacillus* species secrete an enzyme, AiiA that cleaves the lactone rings from the acyl moieties of AHLs and renders the AHLs inactive in signal transduction. AiiA is extremely nonspecific with regard to the AHL acyl side chain, which suggests that this strategy interferes generally with AHL-mediated communication between Gram-negative bacteria (Dong et al., 2001).

References

Angly FE, Felts B, Breitbart M, Salamon P, Edwards RA, Carlson C et al. 2006. The marine viromes of four oceanic regions. *PLoS Biology* 4:e368.

Aoshima H, Kimura A, Shibutani A, Okada C, Matsumiya Y, and Kubo M. 2006. Evaluation of soil bacterial biomass using environmental DNA extracted by slow-stirring method. *Applied Microbiology and Biotechnology* 71: 875–880.

Arbeli Z and Fuentes CL. 2007. Improved purification and PCR amplification of DNA from environmental samples. *FEMS Microbiology Letters* 272: 269–275.

Beja O, Suzuki MT, Koonin EV, Aravind L, Hadd A, Nguyen LP et al. 2000. Construction and analysis of bacterial artificial chromosome libraries from a marine microbial assemblage. *Environmental Microbiology* 2: 516–529.

Bickley J, Short JK, McDowell DG, and Parkes HC. 1996. Polymerase chain reaction (PCR) detection of *Listeria* monocytogenes in diluted milk and reversal of PCR inhibition caused by calcium ions. *Letters in Applied Microbiology* 22: 153–158.

Bjelopavlic M, Newcombe G, and Hayes R. 1999. Adsorption of NOM onto activated carbon: Effect of surface charge, ionic strength, and pore volume distribution. *Journal of Colloid Interface Science* 210: 271–280.

Braid MD, Daniels LM, and Kitts CL. 2003. Removal of PCR inhibitors from soil DNA by chemical flocculation. *Journal of Microbiological Methods* 52: 389–393.

Brady SF, Chao CJ, and Clardy J. 2004. Longchain *N*-acyltyrosine synthases from environmental DNA. *Applied and Environmental Microbiology* 70: 6865–6870.

Burgmann H, Pesaro M, Widmer F, and Zeyer J. 2001. A strategy for optimizing quality and quantity of DNA extracted from soil. *Journal of Microbiological Methods* 45: 7–20.

Chandra R, Bharagava RN, Yadav S, and Mohan D. 2009. Accumulation and distribution of toxic metals in wheat (*Triticum aestivum L.*) and Indian mustard (*Brassica campestris L*) irrigated with distillery and tannery effluents. *Journal of Hazardous Materials* 162(2): 1514–1521.

Chandra R and Singh R. 2012. Decolourisation and detoxification of rayon grade pulp paper mill effluent by mixed bacterial culture isolated from pulp paper mill effluent polluted site. *Biochemical Engineering Journal* 61: 49–58.

Cheng W, Dastgheib SA, and Karanfil T. 2005. Adsorption of dissolved natural organic matter by modified activated carbons. *Water Research* 39: 2281–2290.

Courtois S, Cappellano CM, Ball M, Francou FX, Normand P, Helynck G et al. 2003. Recombinant environmental libraries provide access to microbial diversity for drug discovery from natural products. *Applied and Environmental Microbiology* 69: 49–55.

Cullen DW and Hirsch PR. 1998. Simple and rapid method for direct extraction of microbial DNA from soil for PCR. *Soil Biology and Biochemistry* 30: 983–993.

Dean FB, Hosono S, Fang L, Wu X, Faruqi AF, Bray-Ward P et al. 2002. Comprehensive human genome amplification using multiple displacement amplification. *Proceedings of the National Academy of Science of the United States of America* 99: 5261–5266.

Dell'Anno A, Fabiano M, Duineveld GCA, Kok A, and Danovaro R. 1998. Nucleic acid (DNA, RNA) quantification and RNA/DNA ratio determination inmarine sediments: Comparison of spectrophotometric, fluorometric, and high performance liquid chromatography methods and estimation of detrital DNA. *Applied and Environmental Microbiology* 64: 3238–3245.

Demeke T and Jenkins GR. 2010. Influence of DNA extraction methods, PCR inhibitors and quantification methods on real-time PCR assay of biotechnology-derived traits. *Analytical and Bioanalytical Chemistry* 396(6): 1977–1990.

Desai C and Madamwar D. 2007. Extraction of inhibitor-free metagenomic DNA from polluted sediments, compatible with molecular diversity analysis using adsorption and ionexchange treatments. *Bioresearch Technology* 98: 761–768.

Dobrindt U, Hochhut B, Hentschel U, and Hacker J. 2004. Genomic islands in pathogenic and environmental microorganisms. *Nature Review and Microbiology* 2: 414–424.

Dong YH, Wang LH, Xu JL, Zhang HB, Zhang XF, and Zhang LH. 2001. Quenching quorum sensing-dependent bacterial infection by an N-acyl homoserine lactonase. *Nature* 411: 813–817.

Duarte, GF, Rosado AS, Seldin L, Keijzer-Wolters AC, and van Elsas JD. 1998. Extraction of ribosomal RNA and genomic DNA from soil for studying the diversity of the indigenous bacterial community. *Journal of Microbiological Methods* 32(1): 21–29.

Entcheva P, Liebl, W, Johann, A, Hartsch T, and Streit WR. 2001. Direct cloning from enrichment cultures, a reliable strategy for isolation of complete operons and genes from microbial consortia. *Applied and Environmental Microbiology* 67: 89–99.

Erb RW and Wagnerdobler I. 1993. Detection of polychlorinated biphenyl degradation genes in polluted sediments by direct DNA extraction and polymerase chain reaction. *Applied and Environmental Microbiology* 59: 4065–4073.

Felske A, Engelen B, Nubel U, and Backhaus H. 1996. Direct ribosome isolation from soil to extract bacterial rRNA for community analysis. *Applied and Environmental Microbiology* 62: 4162–4167.

Fortin N, Beaumier D, Lee K, and Greer CW. 2004. Soil washing improves the recovery of total community DNA from polluted and high organic content sediments. *Journal of Microbiological Methods* 56: 181–191.

Frostegard A, Courtois S, Ramisse V, Clerc S, Bernillon D, Le Gall F et al. 1999. Quantification of bias related to the extraction of DNA directly from soils. *Applied and Environmental Microbiology* 65: 5409–5420.

Fuqua C, Winans SC, and Greenberg EP. 1994. Quorum sensing in bacteria: The LuxR-LuxI family of cell density-responsive transcriptional regulators. *Journal of Bacteriology* 176: 269–275.

Gabor EM, de Vries EJ, and Janssen DB. 2003. Efficient recovery of environmental DNA for expression cloning by indirect extraction methods. *FEMS Microbiology and Ecology* 44: 153–163.

George I et al. 2010. Application of metagenomics to bioremediation. *Metagenomics: Theory, Methods and Applications.* Caister Academic Press, Norfolk, UK.

Gomes NC, Borges LR, Paranhos R, Pinto FN, Krögerrecklenfort E, Mendonça-Hagler LC, and Smalla K. 2007. Diversity of ndo genes in mangrove sediments exposed to different sources of polycyclic aromatic hydrocarbon pollution. *Applied and Environmental Microbiology* 73: 7392–7399.

Gray JP and Herwig RP. 1996. Phylogenetic analysis of the bacterial communities in marine sediments. *Applied and Environmental Microbiology* 62: 4049–4059.

Griffiths RI, Whiteley AS, O'Donnell AG, and Bailey MJ. 2000. Rapid method for coextraction of DNA and RNA from natural environments for analysis of ribosomal DNA- and rRNA-based microbial community composition. *Applied and Environmental Microbiology* 66:5488–5491.

Handelsman J. 2004. Metagenomics: Application of genomics to uncultured microorganisms. *Microbiology and Molecular Biology Review* 68: 669–685.

Harry M, Gambier B, Bourezgui Y, and Garnier-Sillam E. 1999. Evaluation of purification procedures for DNA extracted from organic rich samples: Interference with humic. *Analysis* 27: 439–442.

Henne A, Daniel R, Schmitz RA, and Gottschalk G. 1999. Construction of environmental DNA libraries in *Escherichia coli* and screening for the presence of genes conferring utilization of 4-hydroxybutyrate. *Applied and Environmental Microbiology* 65:3901–3907.

Henne A, Schmitz RA, Bomeke M, Gottschalk G, and Daniel R. 2000. Screening of environmental DNA libraries for the presence of genes conferring lipolytic activity on *Escherichia coli. Applied and Environmental Microbiology* 66: 3113–3116.

Holben WE, Jansson JK, Chelm BK, and Tiedje JM. 1988. DNA probe method for the detection of specific microorganisms in the soil bacterial community. *Applied and Environmental Microbiology* 54: 703–711.

Hosono S, Faruqi AF, Dean FB, Du Y, Sun Z, Wu X et al. 2003. Unbiased whole-genome amplification directly from clinical samples. *Genome Research* 13: 954–964.

Hsiao W, Wan I, Jones SJ and Brinkman FSL. 2003. IslandPath: Aiding detection of genomic islands in prokaryotes. *Bioinformatics* 19: 418–420.

Huber R, Burggraf S, Mayer T, Barns SM, Rossnagel P, and Stetter KO. 1995. Isolation of a hyperthermophilic archaeum predicted by *in situ* RNA analysis. *Nature* 376: 57–58.

Hurt RA, Qiu X, Wu L, Roh Y, Palumbo AV, Tiedje AM, and Zhou J. 2001. Simultaneous recovery of RNA and DNA from soils and sediments. *Applied and Environmental Microbiology* 67: 4495–4503.

Ishoy T, Kvist T, Westermann P, and Ahring BK. 2006. An improved method for single cell isolation of prokaryotes from meso-, thermo- and hyperthermophilic environments using micromanipulation. *Applied Microbiology and Biotechnology* 69: 510–514.

Jackson CR, Harper JP, Willoughby D, Roden EE, and Churchill PF. 1997. A simple, efficient method for the separation of humic substances and DNA from environmental samples. *Applied and Environmental Microbiology* 63: 4993–4995.

Jacobsen CS and Rasmussen OF. 1992. Development and application of a new method to extract bacterial DNA from soil based on separation of bacteria from soil with cation exchange resin. *Applied and Environmental Microbiology* 58: 2458–2462.

Jones BV and Marchesi JR. 2007a. Accessing the mobile metagenome of the human gut microbiota. *Molecular Biosystems* 3: 749–758.

Jones BV and Marchesi JR. 2007b. Transposon-aided capture (TRACA) of plasmids resident in the human gut mobile metagenome. *Nature Methods* 4: 55–61.

Knietsch A, Waschkowitz T, Bowien S, Henne A, and Daniel R. 2003. Metagenomes of complex icrobial consortia derived from different soils as sources for novel genes conferring formation of carbonyls from short-chain polyols on *Escherichia coli*. *Journal of Molecular Microbiology and Biotechnology* 5: 46–56.

Kontanis EJ and Reed FA. 2006. Evaluation of real-time PCR amplification efficiencies to detect PCR inhibitors. *Journal of Forensic Sciences* 51(4): 795–804.

Kreader C. 1996. Relief of amplification inhibition in PCR with bovine serum albumin or T4 Gene 32 protein. *Applied and Environmental Microbiology* 62:1102–1106.

Kryndushkin DS, Alexandrov IM, Ter-Avanesyan MD, and Kushnirov VV 2003. Yeast [PSI +] prion aggregates are formed by small Sup35 polymers fragmented by Hsp104. *Journal of Biological Chemistry* 278(49): 49636–49643. doi: 10.1074/jbc.M307996200.PMID14507919.

Kvist T, Ahring BK, Lasken RS, and Westermann P. 2007. Specific single cell isolation and genomic amplification of uncultured microorganisms. *Applied Microbiology Biotechnology* 74:926–35.

Lane DJ, Pace B, Olsen GJ, Stahl DA, Sogin ML, and Pace NR. 1985. Rapid determination of 16S ribosomal RNA sequences for phylogenetic analyses. *Proceedings of the National Academy of Science of the United States of America* 82: 6955–6959.

Lee S-W, Won K, Lim HK, Kim J-C, Choi GJ, and Cho KY. 2004. Screening for novel lipolytic enzymes from uncultured soil microorganisms. *Applied Microbiology and Biotechnology* 65(6): 720–726.

Leff LG, Dana JR, McArthur JV and Shimkets LJ. 1995. Comparison of methods of DNA extraction from stream sediments. *Applied and Environmental Microbiology* 61: 1141–1143.

Lim HK, Chung EJ, Kim J-C, Choi GJ, Jang KS, Chung YR et al. 2005. Characterization of a forest soil metagenome clone that confers indirubin and indigo production on *Escherichia coli*. *Applied and Environmental Microbiology* 71(12): 7768–7777.

Logan EM, Pulford ID, Cook GT, and MacKenzie AB. 1997. Complexation of Cu^{2+} and Pb^{2+} by peat and humic acid. *European Journal of Soil Science* 48: 685–696.

Majernik A., Gottschalk E, and Daniel R. 2001. Screening of environmental DNA libraries for the presence of genes conferring Na^+ (Li^+)/H^+ antiporter activity on *Escherichia coli*: Haracterization of the recovered genes and the corresponding gene products. *Journal of Bacteriology* 183, 6645–6653.

Malik S, Beer M, Megharaj M, and Naidu R. 2008. The use of molecular techniques to characterize the microbial communities in contaminated soil and water. *Environment International* 34: 265–276.

Marcy Y, Ouverney C, Bik EM, Losekann T, Ivanova N, Martin HG et al. 2007. Dissecting biological "dark matter" with single-cell genetic analysis of rare and uncultivated TM7 microbes from the human mouth. *Proceedings of the National Academy of Science of the United States of America* 104: 11889–11894.

Martens DA and Frankenberger Jr. WT. 1995. Enhanced degradation of polycyclic aromatic hydrocarbons in soil treated with an advanced oxidative process—Fenton's reagent. *Soil and Sediment Contamination* 4(2): 175–190.

Matheson CD, Marion TE, Hayter S, Esau N, Fratpietro R, and Vernon KK. 2009. Technical note: Removal of metal ion inhibition encountered during DNA extraction and amplification of copper-preserved archaeological bone using size exclusion chromatography. *American Journal of Physics and Anthropology* 140:384–391.

Miller DN, Bryant JE, Madsen EL, and Ghiorse WC. 1999. Evaluation and optimization of DNA extraction and purification procedures for soil and sediment samples. *Applied and Environmental Microbiology* 65: 4715–4724.

Miller DN. 2001. Evaluation of gel filtration resins for the removal of PCR-inhibitory substances from soils and sediments. *Journal of Microbiological Methods* 44: 49–58.

Moran MA, Torsvik VL, Torsvik T, and Hodso RE. 1993. Direct extraction and purification of rRNA for ecological studies. *Applied and Environmental Microbiology* 59: 915–918.

Moreira D. 1998. Efficient removal of PCR inhibitors using agarose-embedded DNA preparations. *Nucleic Acids Research* 26: 3309–3310.

Muttray AF and Mohn WW. 2000. Quantitation of the population size and metabolic activity of a resin acid degrading bacterium in activated sludge using slot–blot hybridization to measure the rRNA:rDNA ratio. *Microbiology and Ecology* 38: 348–357.

Ogram A, Sayler GS, and Barkay T. 1987. The extraction and purification of microbial DNA from sediments. *Journal of Microbiological Methods* 7: 57–66.

Ono A, Miyazaki R, Sota M, Ohtsubo Y, Nagata Y, and Tsuda M. 2007. Isolation and characterization of naphthalene-catabolic genes and plasmids from oil-contaminated soil by using two cultivation-independent approaches. *Applied Microbiology and Biotechnology* 74: 501–510.

Orsini M and Romano-Spica V. 2001. A microwave-based method for nucleic acid isolation from environmental samples. *Letters in Applied Microbiology* 33: 17–20.

Ou H, He X, Harrison EM, Kulasekara BR, Thani AB, Kadioglu A et al. 2007. MobilomeFINDER: Web-based tools for in silico and experimental discovery of bacterial genomic islands. *Nucleic Acids Research* 35: W97–W104.

Picard C, Ponsonnet C, Paget E, Nesme X, and Simonet P. 1992. Detection and enumeration of bacteria in soil by direct DNA extraction and polymerase chain reaction. *Applied and Environmental Microbiology* 58: 2717–2722.

Porteous LA and Armstrong JL. 1991. Recovery of bulk DNA from soil by a rapid, small-scale extraction method. *Current Microbiology* 22: 345–348.

Porteous LA, Seidler RJ, and Watrud LS. 1997. An improved method for purifying DNA from soil for polymerase chain reaction amplification and molecular ecology applications. *Molecular Ecology* 6: 787–791.

Raghunathan A, Ferguson Jr. HR, Bornarth CJ, Song W, Driscoll M, and Lasken RS. 2005. Genomic DNA amplification from a single bacterium. *Applied and Environmental Microbiology* 71: 3342–3347.

Ramsay AJ. 1984. Extraction of bacteria from soil: Efficiency of shaking or ultrasonication as indicated by direct counts and autoradiography. *Soil Biology and Biochemistry* 16: 475–481.

Riesenfeld CS, Goodman RM, and Handelsman J. 2004. Uncultured soil bacteria are a reservoir of new antibiotic resistance genes. *Environmental Microbiology* 6: 981–989.

Rondon MR, August PR, Bettermann AD, Brady SF, Grossman TH, Liles MR et al. 2000. Cloning the soil metagenome: A strategy for accessing the genetic and functional diversity of uncultured microorganisms. *Applied and Environmental Microbiology* 66: 2541–2547.

Selenska S and Klingmüller W. 1991. Direct detection of nif-gene sequences of *Enterobacter agglomerans* in soil. *FEMS Microbiology Letters* 80: 243–246.

Smalla K, Cresswell N, Mendoca-Hagler LC, Wolters A, and Van Elsas JD 1993. Rapid DNA extraction protocol from soil for polymerase chain reaction-mediated amplification. *Journal of Applied Bacteriology* 74: 78–85.

Sosio M, Giusino F, Cappellano C, Bossi E, Puglia AM, and Donadio S 2000. Artificial chromosomes for antibiotic-producing actinomycetes. *Nature Biotechnology* 8: 343–345.

Steffan RJ, Goksoyr J, Bej AK, and Atlas R 1988. Recovery of DNA from soils and sediments. *Applied and Environmental Microbiology* 54: 2908–2915.

Stokes HW, Nesbo CL, Holley M, Bahl MI, Gillings MR, and Boucher Y. 2006. Class 1 integrons potentially predating the association with Tn402-like transposition genes are present in a sediment microbial community. *Journal of Bacteriology* 188: 5722–5730.

Sutlovic D, Definis GM, Andelinovic S, Gugic D, and Primorac D. 2005. Taq polymerase reverses inhibition of quantitative real time polymerase chain reaction by humic acid. *Croatian Medical Journal* 46:556–562.

Swango, KL, Timken, MD, Chong MD, and Buoncristiani MR. 2006. A quantitative PCR assay for the assessment of DNA degradation in forensic samples. *Forensic Science International* 158(1): 14–26.

Takada-Hoshino Y and Matsumoto N. 2004. An improved DNA extraction method using skim milk from soils that strongly adsorb DNA. *Microbes and Environment* 19: 13–19.

Taylor JP, Wilson B, Mills MS, and Burns RG. 2002. Comparison of microbial numbers and enzymatic activities in surface soils and subsoils using various techniques. *Soil Biology and Biochemistry* 34: 387–401.

Tebbe CC and Vahjen W. 1993. Interference of humic acids and DNA extracted directly from soil in detection and transformation of recombinant DNA from bacteria and a yeast. *Applied and Environmental Microbiology* 59: 2657–2665.

Tien CC, Chao CC, and Chao WL 1999. Methods for DNA extraction from various soils: A comparison. *Journal of Applied Microbiology* 86: 937–943.

Tsai Y and Olson BH. 1991. Rapid method for direct extraction of DNA from soil and sediments. *Applied and Environmental Microbiology* 57: 1070–1074.

Van Elsas JD, Mäntynen V, and Wolters AC. 1997. Soil DNA extraction and assessment of the fate of *Mycobacterium chlorophenolicum* strain PCP-1 in different soils by 16S ribosomal RNA gene sequence based most-probable-number PCR and immunofluorescence. *Biology and Fertility of Soils* 24: 188–195.

Voget S, Leggewie C, Uesbeck A, Raasch C, Jaeger KE, and Streit WR 2003. Prospecting for novel biocatalysts in a soil metagenome. *Applied and Environmental Microbiology* 69, 6235–6242.

Watson RJ and Blackwell B. 2000. Purification and characterization of a common soil component which inhibits the polymerase chain reaction. *Canadian Journal of Microbiology* 46:633–642.

Whitehouse CA and Hottel HE. 2007. Comparison of five commercial DNA extraction kits for the recovery of *Francisella tularensis* DNA from spiked soil samples. *Molecular and Cellular Probes* 21: 92–96.

Williamson NR, Simonsen HT, Ahmed RAA, Goldet G, Slater H, Woodley L et al. 2005. Biosynthesis of the red antibiotic, prodigiosin, in Serratia: Identification of a novel 2-methyl-3-n-amyl-pyrrole (MAP) assembly pathway, definition of the terminal condensing enzyme, and implications for undecylprodigiosin biosynthesis in Streptomyces. *Molecular Microbiology* 56(4): 971–989.

Wuana RA, Okieimen FE, and Imborvungu JA. 2010. Removal of heavy metals from a contaminated soil using organic chelating acids. *International Journal of Environmental Science and Technology* 7(3): 485–496.

Xavier KB and Bassler BL. 2003. LuxS quorum sensing: More than just a numbers game. *Current Opinion in Microbiology* 6: 191–117.

Xia X, Bollinger J, and Ogram A. 1995. Molecular genetic analysis of the response of three soil microbial communities to the application of 2,4-D. *Molecular Ecology* 4: 17–28.

Yeates C, Gillings MR, Davison AD, Altavilla N, and Veal DA. 1997. PCR amplification of crude microbial DNA extracted from soil. *Letters in Applied Microbiology* 25: 303–307.

Yeates C, Gillings MR, Davison AD, Altavilla N, and Veal DA. 1998. Methods for microbial DNA extraction from soil for PCR amplification. *Biological Procedures Online* 1(1): 40–47.

Yu, Z and Mohn WW. 1999. Killing two birds with one stone: Simultaneous extraction of DNA and RNA from activated sludge biomass. *Canadian Journal of Microbiology* 45(3): 269–272.

Yun J, Seowon K, Sulhee P, Hyunjin, Y, Myo- Jeong K, Sunggi H, and Sangyeol R. 2004. Characterization of a novel amylolytic enzyme encoded by a gene from a soilderived metagenomic library. *Applied and Environmental Microbiology* 70(12): 7229–7235.

Zhou J, Bruns MA, and Tiedje JM 1996. DNA recovery from soils of diverse composition. *Applied and Environmental Microbiology* 62: 316–322.

16

Biotransformation and Biodegradation of Organophosphates and Organohalides

Ram Chandra and Vineet Kumar

CONTENTS

16.1 Introduction

Organophosphorus compounds or organophosphates (OPs) form a large group of chemicals that are most widely used around the world for protecting agricultural crops from insects, pests and weeds, for livestock, human health and as warfare nerve agents. They are also used as plasticizers, stabilizers in lubricating and hydraulic oils, flame retardants, and gasoline additives. One of the main distribution routes for OPs into the environment are believed to be wastewater and discharges from wastewater treatment plants (WWTP). The excessive use of OPs have generated a number of environmental problems such as contamination of air, water and terrestrial ecosystems, harmful effects on different biota, and the disruption of biogeochemical cycling. It is believed that poisoning by between 750,000 and 3,000,000 OPs occur globally every year. OPs act as acetylcholinesterase inhibitors resulting in an accumulation of acetylcholine and the continued stimulation of acetylcholine receptors. Therefore, they are also called anticholinesterase agents.

Besides this, organohalides are compounds that contain one or more halogen atom. Organohalides of natural and anthropogenic origin are ubiquitous in the environment. Over 1500 organohalides are known to be produced naturally (Ballschmiter, 2003). Synthetic organohalides have found uses as solvents, refrigerants, insecticides, degreasing agents, pesticides, pharmaceuticals, plasticizer polymers, and medicines. Organohalides are also produced as by products in various industrial processes. The majority of these compounds are chlorinated. However, their uncontrolled release into the environment has caused environmental damage, as many (xenobiotic) organohalides are not only toxic, but also highly recalcitrant to biodegradation, and they readily accumulate in lipids leading to bioaccumulation, often posing the greatest health risk to humans.

Chemical and physical methods of decontamination of OPs and organohalides are not only expensive and time consuming, but also in most cases do not provide a complete solution. The elimination of a wide range of OPs and organohalides from the environment is an absolute requirement to promote the sustainable development of our society with low environmental impact. Interest in the microbial biodegradation of these pollutants has intensified in recent years as mankind strives to find sustainable ways to clean up contaminated environments. This biodegradation and biotransformation methods endeavor to harness the astonishing, naturally occurring, microbial catabolic diversity to degrade, transform or accumulate a huge range of organic compounds. This chapter focusses on biodegradation and biotransformation of OPs and organohalides through microorganisms in the environment, with an emphasis on how microbial enzymes play a major role in the degradation and detoxification of environmental pollutants. This book chapter explores global challenges on the issues of OPs and organohalides and their fate in the environment.

16.2 Biotransformation

Biotransformation is defined as a biological process whereby a substance is changed from one chemical to another (transformed). It involves simple, chemically defined reactions catalyzed by enzymes present in the cells (i.e., microorganisms, plants, and animals). Biotransformation processes are often preferred to chemical processes when high specificity is required, to attack specific sites on the substrate and for a single isomer of the product. However, the biotransformation process modifies not only the physicochemical properties of compounds, such as solubility or bioavailability, but also the toxicity level of the given xenobiotics. The terms biotransformation and metabolism (metabolic transformation) are often used synonymously; particularly when applied to drugs. The term metabolism is often used to describe the total fate of a xenobiotic which includes absorption, distribution, biotransformation, and excretion. Metabolism is commonly used to mean biotransformation from the standpoint that the products of xenobiotic biotransformation are called metabolites. There are two major types of biotransformations: (i) xenobiotic biotransformation and (ii) biosynthetically directed biotransformations. In xenobiotic biotransformations, the substrate is foreign to the biological system. Biosynthetically directed biotransformations can be also used to reveal features of the biosynthesis. Biotransformation of xenobiotics occurs in two forms: (i) mammalian biotransformation and (ii) microbial biotransformation.

16.2.1 Mammalian Biotransformation

In mammals, biotransformation of xenobiotics and other foreign chemicals catalyzed by enzymes are widely distributed throughout the body. However, the liver is the primary biotransforming organ due to its large size and high concentration of biotransforming enzymes. The kidneys and lungs are next with 10%–30% of the liver's capacity. A low capacity exists in the skin, intestines, testes, and placenta. Since the liver is the primary site for biotransformation, it is also potentially quite vulnerable to the toxic action of a xenobiotic that is activated to a more toxic compound in the liver and to a lesser extent also in the lung and intestine. An important consequence of mammalian biotransformation is that the physical properties of a xenobiotic are generally changed from those favoring absorption (lipophilicity) to those favoring excretion in urine or feces (hydrophilicity). Without biotransformation, lipophilic xenobiotics would be excreted from the body so slowly that they would eventually overwhelm and kill an organism. However, some chemicals stimulate the synthesis of enzymes involved in xenobiotic biotransformation. This process, known as enzyme induction, is an adaptive and reversible response to xenobiotic exposure. Enzyme induction enables some xenobiotics to accelerate their own biotransformation and elimination. The structure (i.e., amino acid sequence) of a given biotransforming enzyme may differ among individuals, which can give rise to differences in the rates of xenobiotic biotransformation. The reactions catalyzed by xenobiotic biotransforming enzymes is shown in Table 16.1 (Williams, 1971).

There are many different processes that can occur and the pathways of drug metabolism can be divided into two phases: phase I and phase II.

16.2.1.1 Phase I

Phase I reactions involve hydrolysis, reduction, and oxidation of xenobiotics, as shown in Table 16.1. These reactions expose or introduce a functional group (–OH, –NH_2, –SH, or –COOH) and usually result in only a small increase in the hydrophilicity of xenobiotics. In general, phase I biotransformation is often required for subsequent phase II biotransformation as shown in Figure 16.1.

Although several enzyme systems participate in phase I metabolism of xenobiotics, perhaps the most notable enzyme in the xenobiotic transformation pathway is catalyzed by cytochrome P450s (CYPs; P450s). The highest concentration of P450 enzymes involved in xenobiotic biotransformation is found in liver endoplasmic reticulum (microsomes), but

TABLE 16.1

General Pathways of Xenobiotic Biotransformation and Their Major
Subcellular Locations

Reaction	Enzyme	Localization
Phase I		
Hydrolysis	Esterase	Microsomes, cytosol, lysosomes, blood
	Peptidase	Blood, lysosomes
	Epoxide hydrolase	Microsomes, cytosol
Reduction	Azo- and nitro-reduction	Microflora, microsomes, cytosol
	Carbonyl reduction	Cytosol, blood, microsomes
	Disulfide reduction	Cytosol
	Sulfoxide reduction	Cytosol
	Quinone reduction	Cytosol, microsomes
	Reductive dehalogenation	Microsomes
Oxidation	Alcohol dehydrogenase	Cytosol
	Aldehyde dehydrogenase	Mitochondria, cytosol
	Aldehyde oxidase	Cytosol
	Xanthine oxidase	Cytosol
	Monoamine oxidase	Mitochondria
	Diamine oxidase	Cytosol
	Prostaglandin H synthase	Microsomes
	Flavin-monooxygenases	Microsomes
	Cytochrome P450	Microsomes
Phase II		
	Glucuronide conjugation	Microsomes
	Sulfate conjugation	Cytosol
	Glutathione conjugation	Cytosol, microsomes
	Amino acid conjugation	Mitochondria, microsomes
	Acylation	Mitochondria, cytosol
	Methylation	Cytosol, microsomes, blood

P450 enzymes are present in virtually all tissues. All P450 enzymes are heme-containing proteins. The heme iron in cytochrome P450 is usually in the ferric (Fe^{3+}) state. When reduced to the ferrous (Fe^{2+}) state, cytochrome P450 can bind ligands such as O_2 and carbon monoxide (CO). The CYPs detoxify and/or bioactivate a vast number of xenobiotic chemicals and conduct functionalization reactions that include *N*- and *O*-dealkylation,

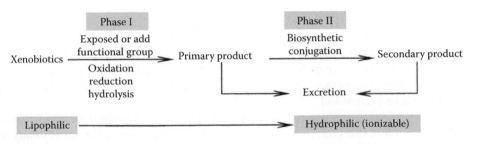

FIGURE 16.1
Biotransformation of xenobiotic in mammalian system.

aliphatic and aromatic hydroxylation, *N-* and *S*-oxidation, and deamination. The basic reaction catalyzed by cytochrome P450 is monooxygenation in which one atom of oxygen is incorporated into a substrate, designated RH, and the other is reduced to water with reducing equivalents derived from NADPH, as follows:

$$Substrate\ (RH) + O_2 + NADPH + H + Product\ (ROH) + H_2O + NADP+$$

Cytochrome P450 catalyzes several types of oxidation reactions, such as hydroxylation of an aliphatic or aromatic carbon, epoxidation of a double bond, heteroatom (S-, N-, and I-), oxygenation and *N*-hydroxylation, heteroatom (*O-, S-, N-,* and Si-) dealkylation, oxidative group transfer, cleavage of esters, and dehydrogenation.

16.2.1.2 *Phase II*

Phase II biotransformation is catalyzed often by the "transferase" enzymes that perform conjugating reactions. Included in the phase II reaction schemes are glucuronidation, sulfonation (more commonly called sulfation), methylation, acetylation, glutathione conjugation, and amino acid conjugation (such as glycine, taurine, and glutamic acid). The products of phase II conjugations are typically more hydrophilic than the parent compounds and therefore usually more readily excretable. Phase II biotransformation of xenobiotics may or may not be preceded by phase I biotransformation. For example, morphine, heroin, and codeine are all converted to morphine-3-glucuronide. In the case of morphine, this metabolite forms by direct conjugation with glucuronic acid. In the other two cases, however, conjugation with glucuronic acid is preceded by phase I biotransformation: hydrolysis (deacetylation) in the case of heroin and *O*-demethylation (involving oxidation by cytochrome P450) in the case of codeine. Similarly, acetaminophen can be glucuronidated and sulfated directly, whereas phenacetin must undergo phase I metabolism (involving *O*-deethylation to acetaminophen) prior to undergoing phase II biotransformation. Specific families of phase II xenobiotic-metabolizing enzymes include the UDP-glucuronosyltransferases (UGTs), sulfotransferases (STs), *N*-acetyltransferases (arylamine *N*-acetytransferase; NATs), glutathione *S*-transferases (GSTs), and various methyltransferases, such as thiopurine *S*-methyl transferase and catechol *O*-methyl transferase. The GSTs function as cytosolic dimeric isoenzymes of 45–55 kDa size that have been assigned to at least four classes: alpha, mu, pi, theta, and zeta; humans possess > 20 distinct GST family members. The conjugation of certain xenobiotics with glutathione is catalyzed by all class of GSTs. For example, the alpha, mu, and pi classes of human GSTs all catalyze the conjugation of 1-chloro-2,4-dinitrobenzene. In concert with the phase I enzymatic machinery, the phase II enzymes metabolize, detoxify, and at times bioactivate xenobiotic substrates in coordination.

16.2.2 Microbial Biotransformation

Microbial transformation may be defined as when the transformation of organic compounds occurs by microorganisms. The microorganisms have the ability to chemical modify a wide variety of organic compounds. These microorganisms during the biotransformation process synthesize a wide range of enzymes which act on organic compounds and convert them into other compounds or modify them. Biotransformation processes are mediated by two group of microorganisms especially bacteria and fungi.

16.2.2.1 Biotransformation of Lindane

Hexachlorocyclohexane (HCH), formerly referred to as gamma benzene hexachloride (BHC), exists in several isomeric forms, including alpha (α), beta (β), gamma (γ), delta (δ), Zeta (ζ), eta (η), and theta (θ). The gamma isomer is commonly known as lindane [gamma-hexachlorocyclohexane (γ-HCH)], as shown in Figure 16.2. The Dutch scientist Dr Teunis van der Linden discovered its insecticidal properties (Hardie, 1964). It can be used as an insecticide and has been used to kill soil-dwelling and plant-eating insects. Lindane and other HCH isomers are highly chlorinated hydrocarbons. The presence of a large number of electron withdrawing chlorine groups makes some of the HCH isomers rather recalcitrant in an oxic environment. The other isomers can be formed during the synthesis of lindane, and have been used either as fungicides or to synthesize other chemicals. In 2005, the production and agricultural use of lindane was banned under the Stockholm convention on persistent organic pollutants (POPs) (Hanson, 2005). The use of lindane had been banned in more than 50 countries and restricted in 33 countries (Humphreys et al., 2008). Further, in 2006, the US EPA cancelled agricultural uses of lindane (ATSDR, 2005). Lindane has a half-life of about two weeks in soil and water. Once in the soil, lindane adsorbs strongly to organic matter and is therefore relatively immobile in the soil. Lindane in soil with especially low organic matter content or subject to high rainfall can leach into surface and even ground soil microflora and aquatic microflora are adversely affected by lindane. Lindane significantly reduced the growth and activity of nitrifying and denitrifying bacteria in soil (Martinez-Toledo et al., 1993; Sáez et al., 2006).

In general, the HCH isomers can be biodegraded to a series of less chlorinated organic compounds under both aerobic and anaerobic conditions, and in some cases, HCHs can be used as the sole carbon source for bacterial growth (Rubinos et al., 2007). During the microbial degradation, the chloride atoms of HCHs, which are usually considered to be toxic and xenobiotic, may most commonly be replaced by hydrogen or hydroxyl groups. The efficiency of its microbial degradation in the environment depends on certain biotic and abiotic factors such as the availability of HCH degrading microbes, temperature, pH, moisture, texture and organic content of soil, etc.

16.2.2.1.1 Aerobic Degradation

The aerobic degradation pathway of lindane has been studied in some detail for *Sphingobium japonicum* UT26 (formerly *Sphingobium paucimobilis* SS86) that was able to use lindane as the sole source of carbon and energy and was isolated from an experimental field to which lindane had been applied (Nagata et al., 1999, 2007). Other HCH-degrading *Sphigobium* strains, such as *S. indicum* B90 (Kumari et al., 2002), *S. indicum* B90A (Dogra et al., 2004) from India and *S. francense* Sp+ (Ceremonie et al., 2006) from France were also characterized. Lindane is degraded under both aerobic and anaerobic conditions, but it is generally mineralized only under aerobic conditions (Phillips et al., 2005). The aerobic

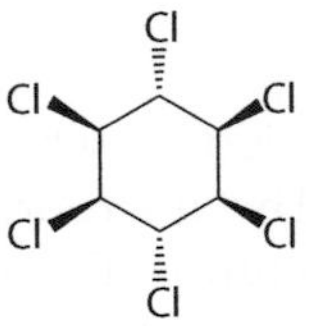

FIGURE 16.2
Chemical structure of lindane (γ-HCH).

degradation of lindane is divided into two pathways: (i) upstream pathway and (ii) downstream pathway.

16.2.2.1.1.1 Upstream pathway   In this pathway two initial dehydrochlorination reactions produce the putative product 1,3,4,6-tetrachloro-1,4-cyclohexadiene (1,3,4,6-TCDN) via the observed intermediate γ-pentachlorocyclohexene (γ-PCCH). Subsequently 2,5-dichloro-2,5-cyclohexadiene-1,4-diol (2,5-DDOL) is generated by two rounds of hydrolytic dechlorinations via a second putative metabolite, 2,4,5-trichloro- 2,5-cyclohexadiene-1-ol (2,4,5-DNOL). 2,5-DDOL is then converted by a dehydrogenation reaction to 2,5-dichlorohydroquinone (2,5-DCHQ). The major upstream pathway reactions described above are enzymatically catalyzed by LinA (dehydrochlorinase), LinB (halidohydrolase), and LinC (dehydrogenase) proteins, but two other, minor products, 1,2,4-trichlorobenzene (1,2,4-TCB) and 2,5-dichlorophenol (2,5-DCP), are produced, presumptively by spontaneous dehydrochlorinations of the two putative metabolites, 1,3,4,6-TCDN and 2,4,5-DNOL. Both 1,2,4-TCB and 2,5-DCP appear to be dead-end products that are not degraded by UT26.

16.2.2.1.1.2 Downstream pathway   Downstream degradation pathway is a reductive dechlorination of 2,5-DCHQ to chlorohydroquinone (CHQ) which is catalyzed by LinD (reductive dechlorinase) protein. The pathway then bifurcates, with the minor route being a further reductive dechlorination by LinD protein to produce hydroquinone (HQ), which is then ring cleaved to γ-hydroxymuconic semialdehyde (γ-HMSA). The conversion of HQ γ-HMSA is catalyzed by LinE (ring cleavage dioxygenase). The major route involves the direct ring cleavage of CHQ to an acylchloride by LinE, which is further transformed to maleylacetate (MA) through the action of LinF (reductase). MA is onverted to β-ketoadipate and then to succinyl coenzyme A (CoA) and acetyl-CoA, which are both metabolized in the tricarboxylic acid (TCA). The complete aerobic degradation pathway of lindane (γ-HCH) is shown in Figure 16.3.

16.2.2.1.1.3 Anaerobic Degradation   Several bacterial species such as *Clostridium sphenoides*, *Clostridium rectum* and several other representatives of Bacillaceae and Enterobacteriaceae actively metabolize lindane under anaerobic conditions (Heritage and Mac Rae, 1977; Haider, 1979; Ohisa et al., 1980). The degradation of lindane is initiated with a dechlorination to form pentachlorocyclohexane (PCCH), from which the 1,2-di-chlorobenzene (DCB) and 1,3,-di-chlorobenzene (DCB) isomers and finally mono-chlorobenzene (CB) are formed as shown in Figure 16.4 (Quintero et al., 2005). In the degradation of lindane, intermediate metabolites such as tetrachlorocyclohexene (TCCH) and tetrachlorocyclohexenol (TCCOL) have been detected. During the degradation of lindane by *Xanthomonas* sp. ICH12, formation of two intermediates, γ-2,3,4,5,6-pentachlorocyclohexene (γ-PCCH) and 2,5-dichlorobenzoquinone (2,5-DCBQ), were identified by gas chromatography-mass spectrometric (GC–MS) analysis (Quintero et al., 2005). γ-PCCH and and 2,5-dichlorohydroquinone were a novel metabolites from HCH degradation (Manickam et al., 2007).

16.2.2.2 Biotransformation of DDT

1,1,1-Trichloro-2,2-di(*p*-chlorophenyl)-ethane (DDT) is a very important persistent organochlorine pesticide and widely used to control agricultural pests and vectors of malaria, plague, dengue and other insect-borne disease since the 1940s (Wong et al., 2005). The contamination of DDT is of great concern due to its long half-life, recalcitrance to degradation, bioaccumulation and biomagnifications in food chains, and potential toxicity to humans

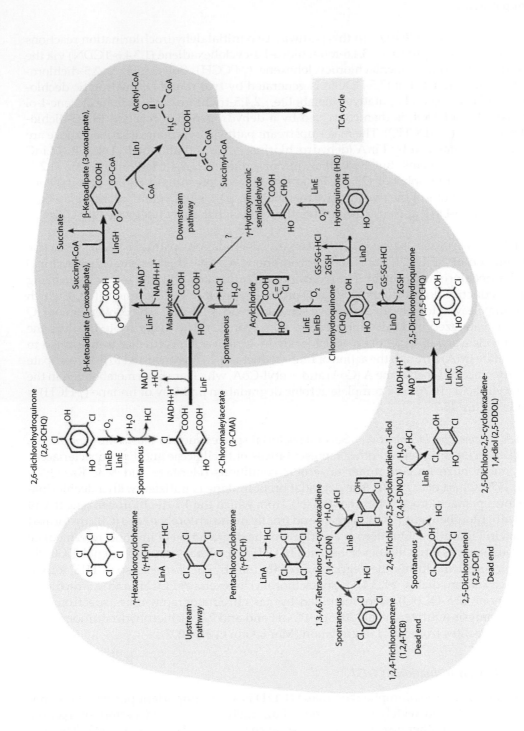

FIGURE 16.3

Aerobic degradation pathways of lindane (γ-HCH) in *Sphingobium japonicum* UT26. (Adapted from Nagata Y et al. 2007. *Applied Microbiology and Biotechnology* 76: 741–752.)

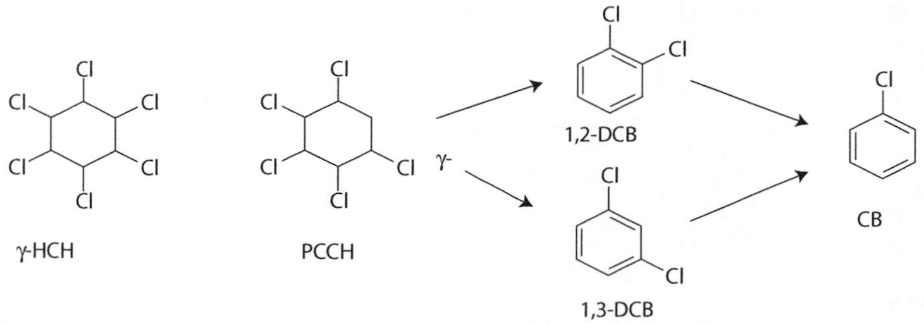

FIGURE 16.4
Degradation routes for lindane (γ-HCH) in *Xanthomonas* sp. ICH12 under anaerobic conditions.

and wildlife. Its production and use have been banned or restricted in many several countries in the 1970s. Since DDT residues are lipophilic, they tend to accumulate in the fatty
tissues of the ingesting organisms along the food chain. DDT is a colorless crystalline
substance which is nearly insoluble in water but highly soluble in fats and most organic
solvents. Banned for agricultural use worldwide by the 2001 Stockholm Convention on
Persistent Organic Pollutants, the use of DDT is still permitted in small quantities in countries that need it, with support mobilized for the transition to safer and more effective
alternatives. DDT residues in water and soil are of concern as their uptake can lead to the
accumulation of primary products. Their removal from water and soil is therefore a priority.
DDT residues have been shown in Figure 16.5 to persist in the environment predominantly
in the form of DDT; 1,1,1-trichloro-2-*o*-chlorophenyl-2-*p*-chloro-phenylethane (*o,p*-DDT);
1,1-dichloro-2,2-bis(4-chlorophenyl)ethane (*o,p*-DDD); 1,1-dichloro-2,2-bis(4-chlorophenyl)
ethylene (*p,p*-DDE). DDD and DDE are the transformation products of DDT formed due
to microbial action or due to chemical or photochemical reactions. The chemical structure
of DDT contains chlorinated aliphatic and aromatic structures that impart great chemical
stability. The toxicity of DDT is mainly due to its chlorine atoms, and dechlorination may
reduce its environmental risk and biotoxicity. The chemical structure of DDT is shown in
Figure 16.5.

Microorganisms play an important role in the fate of DDT in natural and controlled
environments. A range of bacteria and white rot fungi, such as *Eubacterium limosum* (Yim
et al., 2008), *Alcaligenes eutrophus* A5 (Nadeau et al., 1994), *Boletus edulis* (Huang et al., 2007),
Fusarium solani (Mitra et al., 2001), and *Phanerochaete chrysosporium* (Bumpus and Aust,
1987) have been demonstrated to degrade DDT in both pure culture and natural soil. DDT
can be biodegraded by two distinct microbial processes, aerobic oxidative degradation and
anaerobic reductive dechlorination.

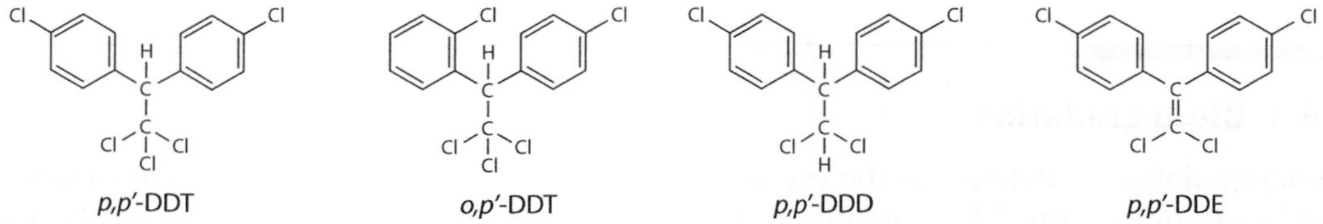

FIGURE 16.5
DDT and its residues.

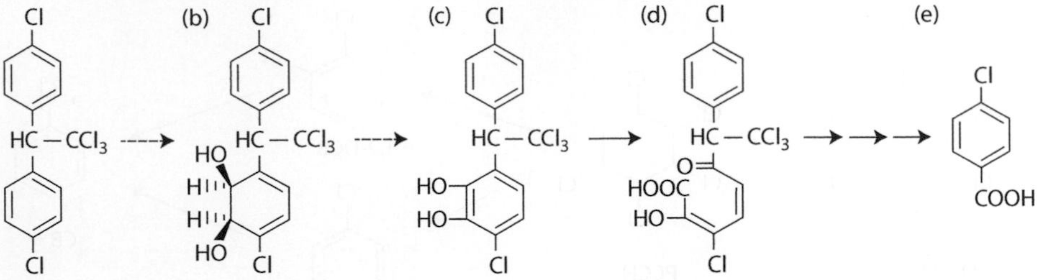

FIGURE 16.6
Biotransformation of DDT under aerobic condition by *Alcaligenes eutrophus* A5. (a) DDT, (b) 2,3-dihydrodiol-DDT, (c) 2,3-dihydroxy-DDT, (d) ring fission product, and (e) 4-chlorobenzoic acid.

16.2.2.2.1 Aerobic Transformation

The aerobic transformation pathway of DDT by *Alcaligenes eutrophus* A5 has been studied in detail as shown in Figure 16.6 (Nadeau et al., 1994). *A. eutrophus* A5 initially oxidizes DDT (compound a) at the ortho and meta positions to form a 2,3-dihydrodiol-DDT intermediate (compound b). It is proposed that this is a dioxygenase type of attack resulting in the transient production of a DDT dihydrodiol (compound b). The dihydrodiol compound is unstable and easily dehydrates into two hydroxylated compounds under weak acidic conditions (pH < 7.0). The dihydrodiol-DDT would be further degraded to 2,3-dihydroxy-DDT (compound c) by a dehydrogenase. 2,3-Dihydroxy-DDT would be further metabolized through meta cleavage to form the yellow ring-fission product (compound d) which would then be catabolized to 4-chlorobenzoic acid (4-CBA) (compound e). 4-CBA appears to be a terminal product formed from DDT by *A. eutrophus* A5 because it has lost the ability to further degrade 4-CBA.

16.2.2.2.2 Anaerobic Transformation

In anaerobic conditions, DDT transformation commonly follows a pathway of reductive dechlorination (You et al., 1996). Reductive dechlorination is the major mechanism for the microbial conversion of both the o,p'-DDT and p,p'-DDT isomers of DDT to DDD. *Proteus vulgaris* was one of the first pure cultures of bacteria observed to reduce DDT to DDD under reducing conditions. The anaerobic transformation pathway of DDT in bacteria is shown in Figure 16.7. In this pathway, DDD undergoes reductive dechlorination to 2,2-bis (4-chloro-phenyl) ethylene (DDNU), which is successively oxidized to 2,2-bis (4-chlorophenyl) ethanol (DDOH) and 2,2-bis (4-chlorophenyl) acetic acid (DDA). DDA is decarboxylated to 2,2-bis (4-chlorophenyl) methane (DDM), which is oxidized to DBP, which is not further metabolized under anaerobic conditions. Recent studies have shown that 4-chlorobiphenyl-degrading Gram-negative bacteria primarily transform DDT to DDD, which is then dehydrogenated to DDE, followed by dechlorination to DDMU under aerobic conditions.

16.3 Biodegradation

Biodegradation is defined as the process by which organic substances are broken down by living organisms. Microorganisms have the ability to interact, both chemically and physically, with substances, leading to structural changes or complete degradation of the target molecule. A huge number of bacteria, fungi, and actinomycetes genera possess the

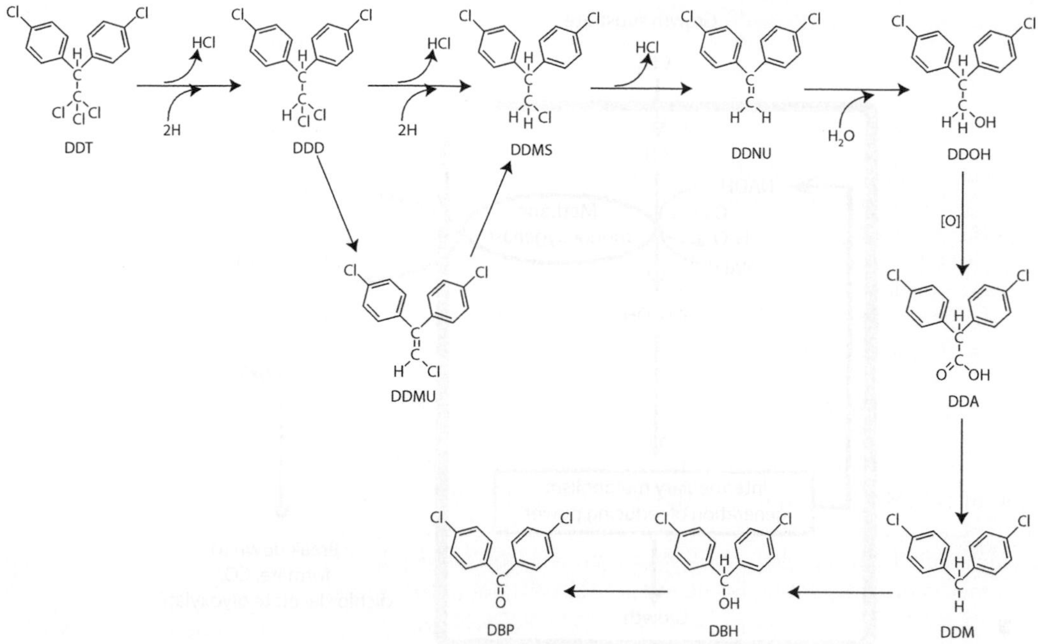

FIGURE 16.7
Bacterial transformation of DDT via reductive dechlorination.

capability to degrade organic pollutants in the environment. It is degradation based on two processes: (i) growth and (ii) metabolism. In the case of growth, organic pollutants are used by microorganisms as the sole source of carbon, nitrogen, phosphorus or sulfur and energy. This process results in complete degradation. It often happens in the natural environment that those degradation processes are accompanied by transformations of other compounds, other xenobiotics. This phenomenon is defined using various terms, such as cometabolism, gratuitous metabolism, cooxidation, accidental or free metabolism. Cometabolism is defined as the metabolism of an organic compound without nutritional benefit in the presence of a growth substrate which is used as the primary carbon and energy source. It is a common phenomenon of microbial activities. Microorganisms growing on a particular substrate gratuitously oxidize a second substrate (cosubstrate). The cosubstrate is not assimilated, but the product may be available as substrate for other organisms of a mixed culture. This metabolism is of no benefit to the organism. The prerequisites of cometabolic transformation are the enzymes of the growing cells and the synthesis of cofactors necessary for enzymatic reaction, for example, of hydrogen donors (NADH) for oxygenase. The basic principle is shown in Figure 16.8.

16.3.1 Organophosphates

Organophosphates (OPs) are usually esters, amides, or thiol derivatives of phosphoric, phosphonic, phosphinic acids or thiophosphoric acids. OPs have the general structure with a terminal oxygen atom (or sulfur atom) connected to phosphorus by a double bond, that is a phosphoryl group, and two lipophilic groups as well as a leaving group bonded to the phosphorous as is shown in Figure 16.9. Usually, R_1 and R_2 are aryl or alkyl groups that are bonded to the phosphorus atom either directly (forming phosphinates), or through

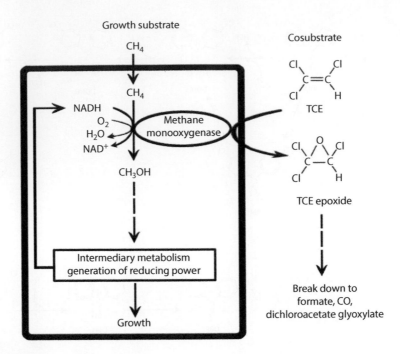

FIGURE 16.8
Commetabolic degradation of trichloroethylene.

an oxygen or sulfur atom (forming phosphates or phosphothioates). The X group can be diverse and may belong to a wide range of aliphatic, aromatic, or heterocyclic groups. The X group, also binding to the phosphorus atom through an oxygen or sulfur atom is called the "leaving group" because on hydrolysis of the ester bond it is released from phosphorus. On the basis of structural characteristics OPs are divided into at least 13 types, including phosphates, phosphonates, phosphinates, phosphorothioates (S=), phosphonothioates (S=), phosphorothioates (S substituted), phosphonothioates (S substituted), phosphorodithioates, phosphorotrithioates, phosphoramidothioates (Gupta, 2006).

The first OPs were synthesized in 1873 by von Hoffman. He synthesized methyl phosphorus chloride, which led to the synthesis of a number of insecticides. In 1903, Michaelis introduced a compound with PCN bond, which led to the synthesis of a number of insecticides and the nerve agent tabun. Further, Schrader developed sarin and tabun in 1937 and in 1944 the Germans developed soman. The British developed VX in 1952. After the

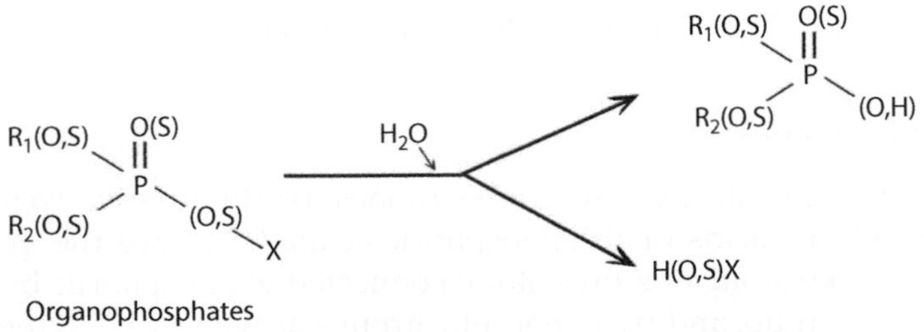

FIGURE 16.9
General structure of organophosphates and its major degradation pathway.

war, in the 40's and 50's, the study of OPs was again oriented towards the development of less toxic compounds (Gupta, 2006). The chemical structure of some OPs is shown in Figure 16.10. Further, the physical properties of various OPs such as tabun, sarin, soman, VX, glyphosate, chlorpyrifos, parathion, methyl parathion, diazinon, coumaphos, monocrotophos, fenamiphos, and phorate are listed in Table 16.2.

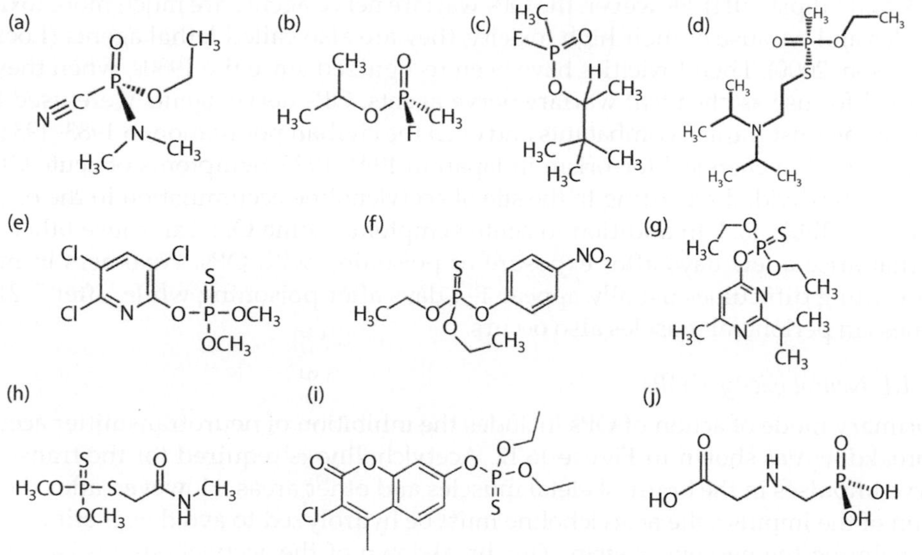

FIGURE 16.10
The structure of some organophospahtes: (a) tabun, (b) sarin, (c) soman, (d) VX, (e) chlorpyrifos, (f) parathion, (g) diazinon, (h) dimethoate, (i) coumaphos, (j) glyphosphate.

TABLE 16.2

Use, Toxicity, and Half-Life of Some Organophospahtes

Name	Type	Half-Life (Days)	Mammalian LD50 (mg/kg)
Tabun	CWA	1.5–2.5	150–400
Sarin	CWA	1.5–2.5	75–100
Soman	CWA	1.5–2.5	35–50
VX	CWA	4–42	10
Chlorpyrifos	Insecticide	16–120	135–163
Parathion	Insecticide	30–180	2–10
Methyl Parathion	Insecticide	25–130	3–30
Glyphosate	Herbicide	30–174	3530–5600
Coumaphos	Acaricide	24–1,400	16–41
Fenamiphos	Nematicide	12–28	6–10
Monocrotophos	Insecticide	40–60	18–20
Dicrotophos	Insecticide	45–60	15–22
Diazinon	Insecticide	11–21	80–300
Dimethoate	Insecticide	2–41	160–387
Fenitrothion	Insecticide	12–28	1700
Ethoprophos	Nematicide	3–30	146–170

16.3.1.1 Fate in Environment and Mode of Action of Toxicity

OPs are not ideal pesticides because of the lack of target vector selectivity, and severe toxicity and even death in humans and domestic animals. The World Health Organization (WHO) estimates that every year 3 million people experience acute poisoning by OPs (WHO, 1990). The main routes of OPs exposure are shown in Figure 16.11. Humans are exposed to OPs via ingested food and drink and by breathing polluted air (WHO, 2001; Eleršek and Filipič, 2011). However, the OPs warfare nerve agents are much more toxic than pesticides and because of their high toxicity, they are also called lethal agents (Lockridge and Masson, 2000). Their toxicities have been recognized since the 1930s, when they were developed for use as chemical warfare nerve agents. OPs nerve agents were used by the Iraqi army against Iranian combatants and even the civilian population in 1983–1988. They were also used in chemical terrorism in Japan in 1994–1995. Symptoms of acute OPs poisoning can be divided according to the site of acetylcholine accumulation in the organism as shown in Table 16.3. In addition to acute symptoms, some OPs can cause other symptoms that arise a few days after exposure or poisoning with OPs. Weakness in muscles and breathing difficulties usually appear 1–4 days after poisoning while, after 7–21 days, weakness in peripheral muscles also occurs.

16.3.1.1.1 Neurotoxicity of OPs

The primary mode of action of OPs includes the inhibition of neurotransmitter acetylcholine breakdown as shown in Figure 16.12. Acetylcholine is required for the transmission of nerve impulses in the brain, skeletal muscles and other areas. However, after the transmission of the impulse, the acetylcholine must be hydrolyzed to avoid over stimulating or overwhelming the nervous system. This breakdown of the acetylcholine is catalyzed by an enzyme called acetylcholine esterase (AChE). AChE converts acetylcholine into choline and acetyl CoA by binding the substrate at its active site at serine 203 to form an enzyme substrate complex. Further reactions involve release of choline from the complex and then rapid reaction of acylated enzymes with water to produce acetic acid and the regenerated AChE. It has been estimated that one enzyme can hydrolyze 300,000 molecules of acetylcholine every minute.

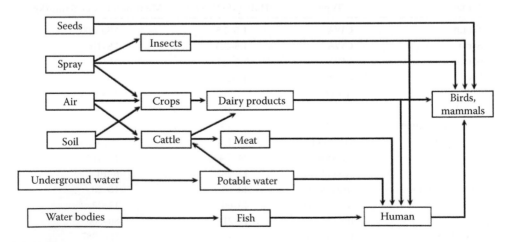

FIGURE 16.11
Routes of exposure to organophosphates. (Adapted from Eleršek T and Filipič M. 2011. *Pesticides—The Impacts of Pesticides Exposure.* InTech, China, pp. 243–260.)

TABLE 16.3

Sign and Symptoms of Organophosphates Poisoning

Organophosphates Accumulation Site	Symptoms
Muscarine manifestations	Ophtalmic: Conjunctival injection, lacrimation, miosis, blurred vision, diminished visual acuity, ocular pain Respiratory: Rhinorrhea, stridor, cough, chest tightness, dyspnea, apnea Gastrointestinal: Nausea, vomiting, salivation, diarrhea, abdominal cramping, fecal incontinence Genitourinary: Frequency, urgency, incontinence
Central nervous system manifestations	Anxiety, restlessness, depression, confusion, ataxia, tremors, convulsion, comma, areflexia
Nicotinic manifestations	Cardiovascular, trachydysrhythmias, hypertension

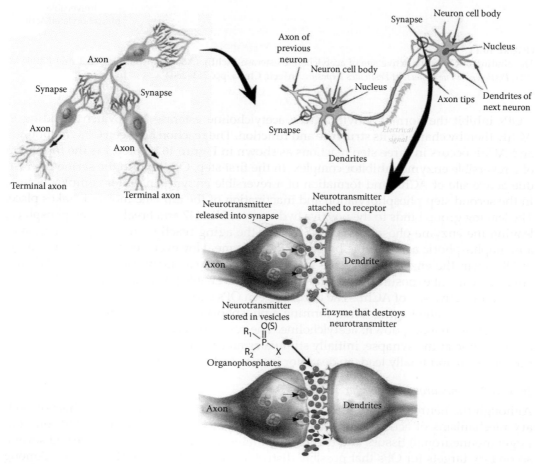

FIGURE 16.12

(See color insert.) Mode of action of organophosphates on the mammalian nervous system.

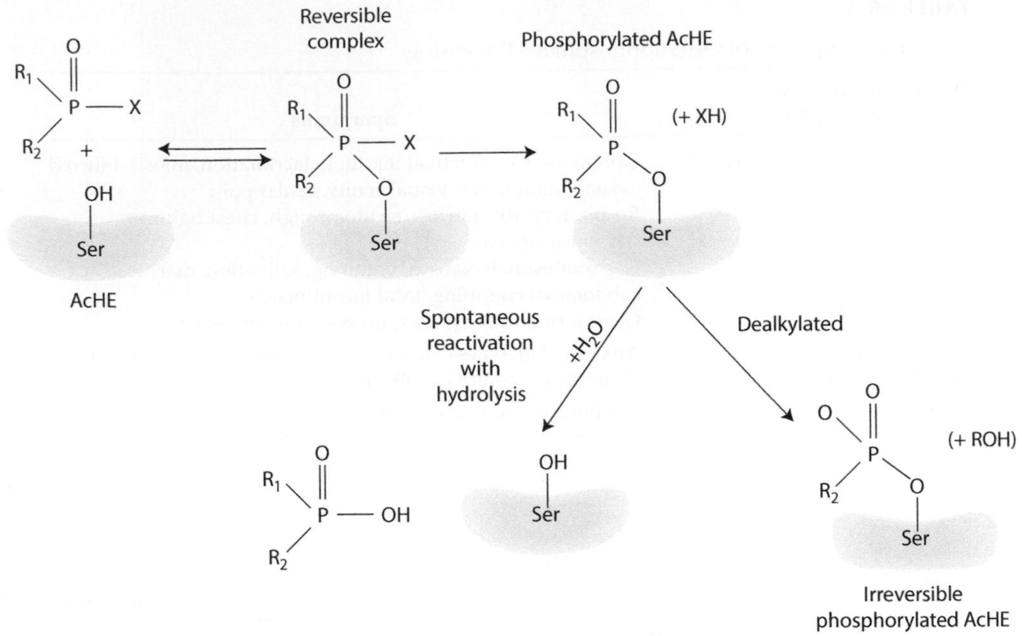

FIGURE 16.13
The binding of OP to the active site of acetylcholine esterase (AchE). (Adapted from Eleršek T and Filipič M. 2011. *Pesticides—The Impacts of Pesticides Exposure.* InTech, China, pp. 243–260.)

OPs inhibit the normal activity of the acetylcholine esterase by covalent bonding to AChE, thereby changing its structure and function. The reaction between OPs compounds and AChE occurs in three step reactions as shown in Figure 16.13. Step 1 is the formation of a reversible enzyme inhibitor complex. In the first step, OPs bind to the serine 203 residue active site of AChE and formation of a reversible enzyme inhibitor complex occurs. In the second step phosphorylation and inactivation of the enzyme molecule takes place. The leaving group binds to the positive hydrogen of His 447 and breaks off the phosphate, leaving the enzyme phosphorylated. Step 3 is the aging reaction involving formation of a monophosphoric acid residue bound to the enzyme. However, spontaneous hydrolysis of OPs from the enzyme active site is very slow. The toxic manifestations and lethality after nerve agent exposure appear to follow the irreversible phosphorylation of the serine containing active site of AChE. The time between OPs exposure and the irreversible phosphorylation is called aging. Thus, forming a phosphorylating protein that is inactive and incapable of breaking down: acetylcholine. The resulting accumulation of toxic levels of acetylcholine at the synapse, initially stimulates and then paralyzes cholinergic synaptic transmission and finally leads to death for insects and mammals.

16.3.1.1.2 Nonneuronal Molecular Targets of OPs

Although the neurotoxicity of OPs is well described, little is known about their secondary mechanisms of activity and the consequences of chronic exposure to OPs on nontarget (nonneuronal) tissues and organs in humans. Recent studies have revealed several secondary targets for OPs that possibly disturb a variety of biological processes. Among the enzymes that are inhibited by OPs are carboxylases, which take part in xenobiotic metabolism. Their inhibition with OPs can block metabolic transformation of various substances. OPs can also influence xenobiotic metabolism via the active sulfur atom that arises from desulfuration in phase I of metabolism and strongly inhibits CYP enzymes (Hodgson and Rose, 2006). OPs also inhibit lipases, which play an important role in cell signaling (Quistad et al., 2006).

16.3.1.1.3 Immunotoxicity of OPs

Toxic effects of OPs on the immune system can be direct or indirect and are reflected in different immune organ pathologies and lowered humoral and/or cell immunity.

Direct immunotoxic effects of OPs can be due to

- Inhibition of serine hydrolases (complement system) or esterases (lymphocyte and monocyte membranes) in the immune system
- Oxidative damage of immune system organs
- Changes in signal transduction pathways that control proliferation and immune cell differentiation
- Indirect immunotoxicity of OPs is expressed as changes in the nervous system or chronic effects of altered metabolism on the immune system

16.3.1.1.4 Genotoxicity and Carcinogenicity of OPs

Because of the chronic exposure of human populations to low OP concentrations, it is very important to study the influence of OPs on cancer development and progress, and to elucidate the underlying mechanisms. Experimental *in vitro* and *in vivo* studies have shown that several OPs exert genotoxic activity (Bolognesi, 2003), and there are also reports showing that OPs can induce neoplastic transformation of cells (Isoda et al., 2005). OPs have been reported to be *weakly mutagenic* in bacteria, but mutagenic in yeast (IARC, 1987).

- To induce *DNA damage* in peripheral lymphocytes *in vitro* (Ündeger and Basaran, 2005) and *in vivo* in occupationally exposed workers (Garaj-Vrhovac et al., 2000)
- To induce *micronuclei formation* in bone marrow (Mathew et al., 1990)
- To cause *sperm abnormalities in* OP exposed mice (Mathew et al., 1992)

Hreljac et al. (2008) have recently studied the mechanisms of genotoxicity and potential carcinogenicity of selected model OPs [parathion (PT), paraoxon (PO), and dimefox (DF)] in the *in vitro* experimental model with human hepatoma (HepG2) cells. They demonstrated that OPs act on several targets:

- Low concentrations of parathion and paraoxon were *genotoxic*, while dimefox acted as a *mitogen*
- The three model OPs induced numerous *variations of gene expression*, particularly of genes involved in stress response, inseparably connected to basic cell processes
- Important in cancer development (cell cycle, apoptosis, xenobiotic metabolism, DNA repair)
- In addition, *changes in phosphorylation of kinases*, which are connected to stress response, were observed

16.3.1.2 Biodegradation of OPs

Incineration and chemical hydrolysis have been widely used to destroy OPs. However, bio-degradation is considered to be a reliable cost-effective technique resulting in less second-ary pollutants for the removal of OPs. Various microorganisms capable of biodegrading OPs and able to use OPs as the sole source of carbon, nitrogen or energy have been isolated from polluted environments. Though these OPs are relatively alien for the microbes, they have evolved novel degradation enzymes and pathway(s) for their metabolism. In 1973, the first bacterium to degrade OPs was isolated from a soil sample from the Philippines and was identified as *Flavobacterium* sp. ATCC 27551 (Sethunathan and Yoshida, 1973). Since then, several bacteria, a few fungi and cyanobacteria that can use OPs as a source of carbon, nitrogen or phosphorus have been isolated. A list of microorganisms capable of degrading these compounds is presented in Table 16.3. Most microorganisms can degrade one OP or a narrow range of Ops. In general, OPs do not adversely affect bacteria, because bacteria do not possess AChE, and some microorganisms can even use OPs as an energy source (Singh and Walker, 2006). The biodegradation of OPs are also influenced by soil properties and the chemical structure of the OPs. Alkaline soils have been shown to be conducive to a higher degradation rate of two OP insecticides, the widely used chlorpyrifos (Singh et al., 2003b) and fenamiphos (Singh et al., 2003a). These observations led to the hypothesis that OPs-degrading genes might have evolved from genes that are required for survival in an alkaline pH (Singh et al., 2003b). OPs share similar chemical structures, and therefore soil that developed enhanced degradation for one OP also rapidly degraded other OPs, in a well-known phenomenon called cross-enhanced degradation (Singh et al., 2005). A list of micro-organisms capable of degrading OPs are presented in Table 16.4.

16.3.1.2.1 Parathion

Parathion (O,O-diethyl-O-p-nitrophenyl phosphorothioate) is one of the most toxic insecticides registered by the EPA and is classified as a Restricted Use Pesticide (RUP). Parathion has received extensive attention among the OPs because of its widespread use and the ready detection of its hydrolytic product (*p*-nitrophenol). It is rapidly degraded in biologically active conditions. Several studies have documented the involvement of soil microorganisms in the degradation of parathion in soil under aerobic and anaerobic conditions. Several bacterial species have been isolated either from parathion enrichment or other OP enriched environments, which can hydrolyze parathion as shown in Table 16.3. Sethunathan and Yoshida (1973) isolated the first organophosphorus degrading bacterium, *Flavobacterium* strain ATCC 27551, which was able to hydrolyze parathion leading to the accumulation of *p*-nitrophenol. In another study, Siddaramappa et al. (1973) isolated two bacteria *Bacillus* sp. and *Pseudomonas* sp. from parathion amended flooded alluvial soil. *Pseudomonas* sp. hydrolyzed parathion and then released nitrite from *p*-nitrophenol while *Bacillus* sp. was unable to hydrolyze parathion but was able to use *p*-nitrophenol as a sole carbon and energy source as soon as it was formed.

Munnecke and Hsieh (1974) reported that the isolation of a mixed culture of four bacteria consisting of *Pseudomonas* sp., *Xanthomonas* sp., *Azotomonas* sp., and *Brevibacterium* sp. was able to hydrolyze parathion. Further, metabolic studies revealed that only one of the bacteria was able to metabolize parathion to *p*-nitrophenol. Complementary studies by Munnecke and Hsieh (1976) concluded that parathion degradation by the mixed culture of nine bacterial isolates, five of which were fluorescent pseudomonads and four species of *Xanthomonas* sp., *Azotomonas*, *Brevibacterium* sp., and one unknown followed different degradation pathways under aerobic and anaerobic conditions. Under aerobic conditions, the

TABLE 16.4

Microorganisms Isolated for the Degradation of Organophosphates (OPs)

Organophosphates	Microorganisms	Mode of Degradation	Metabolites
Chlorpyrifos	Bacteria		
	Enterobacter sp.	Catabolic (C, P)	Diethylthiophospshate (DETP) and 3,5,6-trichloro-2-pyridinol (TCP)
	Flavobacterium sp. ATCC27551	Cometabolic	Diazinon and methylparathion
	Micrococcus sp.	Cometabolic	Diethylthiophospshate (DETP) and 3,5,6-trichloro-2-pyridinol (TCP)
	Fungi		
	Phanerochaete chrysosporium	Catabolic (C)	Diethylthiophospshate (DETP) and 3, 5, 6-trichloro-2-pyridinol (TCP)
Parathion	Bacteria		
	Flavobacterium sp. ATCC27551	Cometabolic	*p*-Nitrophenol
	Pseudomonas stutzeri	Cometabolic	Diethyl thiophosphate and *p*-nitrophenol
	Bacillus spp.	Cometabolic	*p*-Nitrophenol
	Pseudomonas sp.	Catabolic (C, N)	*p*-Nitrophenol
Methyl parathion	Bacteria		
	Pseudomonas sp.	Cometabolic	*p*-Nitrophenol
	Plesimonas spM6	Cometabolic	*p*-Nitropheno
	Pseudomonas putida	Catabolic (C)	*p*-Nitrophenol
	Pseudomonas sp. *A3*	Catabolic (C, N)	*p*-Nitrophenol
	Flavobacterium balustinum	Catabolic (C)	*p*-Nitrophenol
Glyphosate	Bacteria		
	Pseudomonas sp.	Catabolic (P)	Aminomethylphosphonic acid
	Alcaligene sp.	Catabolic (P)	Aminomethylphosphonic acid
	Geobacillus caldoxylosilyticus T20	Catabolic (P)	Aminomethylphosphonate
	Flavobacterium sp.	Catabolic (P)	Aminomethylphosphonate
	Fungi		
	Penicillium chrysogenum	Catabolic (N)	Aminomethylphosphonate
	Penicillium citrinum	Catabolic (P)	2-Aminoethylphosphonic and 2-xoalkylphosphonic acid
Monocrotophos	Bacteria		
	Pseudomonas spp.	Catabolic (C)	Carbon dioxide, ammonia, and phosphates
	Arthrobacter spp.	Catabolic (C)	Carbon dioxide, ammonia, and phosphates
	Bacillus megaterium	Catabolic (C)	Carbon dioxide, ammonia, and phosphates
	Arthrobacter atrocyaneus	Catabolic (C)	Carbon dioxide, ammonia, and phosphates
Fenitrothion	Bacteria		
	Flavobacterium sp.	Cometabolic	4-Nitrophenol
	Burkholderia sp. NF100	Catabolic (C)	Methylhydroquinone
Diazinon	Bacteria		
	Flavobacterium sp.	Catabolic (P)	2-Isopropyl-6-methyl-4-hydroxy-pyrimidine
	Pseudomonas sp.	Cometabolic	Diethyl phosphorothioate

(Continued)

TABLE 16.4 (*Continued*)

Microorganisms Isolated for the Degradation of Organophosphates (OPs)

Organophosphates	Microorganisms	Mode of Degradation	Metabolites
Chemical warfare agents			
G agent	Bacteria		
	Pseudomonas diminuta	Cometabolic	—
	Altermonas spp.	Cometabolic	—
V agent	Bacteria		
	Pseudomonas diminuta	Cometabolic	—

Note: Symbol in brackets after mode of degradation represents the type of nutrient that the pesticide provides to degrading microorganisms. C, carbon; N, nitrogen; P, phosphorus.

primary pathway involved an initial hydrolysis of parathion yielding *p*-nitrophenol and DETP, which was further metabolized to simple nonaromatic producers. Under anaerobic conditions, parathion was metabolized by a reductive pathway. The aromatic nitro group of parathion was reduced to form aminoparathion, which subsequently undergoes hydrolysis to yield *p*-aminophenol and DETP. These compounds were further metabolized by other bacterial strains as a sole carbon and energy source. The complete metabolic pathway for degradation of parathion is shown in Figure 16.14.

16.3.1.2.2 Chlorpyrifos

Chlorpyrifos (*O,O*-diethyl *O*-(3,5,6-trichloro-2-pyridyl) phosphorothioate) is one of the most widely used insecticides effective against a broad spectrum of insect pests of economically important crops. It is used throughout the world to control a variety of chewing and sucking insect pests and mites on a range of economically important crops, including citrus fruit, bananas, vegetables, potatoes, coffee, cocoa, tea, cotton, wheat, rice, and so on (Thengodkar and Sivakami, 2010). It is also registered for use on lawns, ornamental plants, animals, domestic dwellings as well as commercial establishments.

The EPA classifies chlorpyrifos as class II toxicity (moderately toxic). It does not mix well with water, so it is usually mixed with oily liquids before it is applied to crops or animals. The use of chlorpyrifos has been vastly restricted in the United States and some European countries, even for agricultural purposes. However, it is still widely used in developing countries such as India, where in the year 2000, it was the fourth highest consumed pesticide after monocrotophos, acephate and endosulfan (Ansaruddin and Vijayalakshmi, 2003). Extensive use of chlorpyrifos contaminates air, groundwater, rivers, lakes, rainwater, and fog water. The contamination has been found up to about 24 km from the site of application. Moreover, much evidence suggests that chlorpyrifos may affect the endocrine system, respiratory system, cardiovascular system, nervous system, immune system, as well as the reproductive system due to its high mammalian toxicity. It is therefore essential to eliminate these pollutants from the environment. As the major degradation product of chlorpyrifos is 3,5,6-trichloro-2-pyridinol (TCP). TCP has greater water solubility than its parent molecule and causes widespread contamination of soils and aquatic environments (Xu et al., 2008). TCP is classified as persistent and mobile by the US Environmental Protection Agency (EPA) with a half-life ranging from 65 to 360 days in soil, depending on the soil type, climate, and other conditions (Li et al., 2010). Li et al. (2010) suggested that the accumulated TCP in liquid medium or soil, which has antimicrobial property, prevents the proliferation

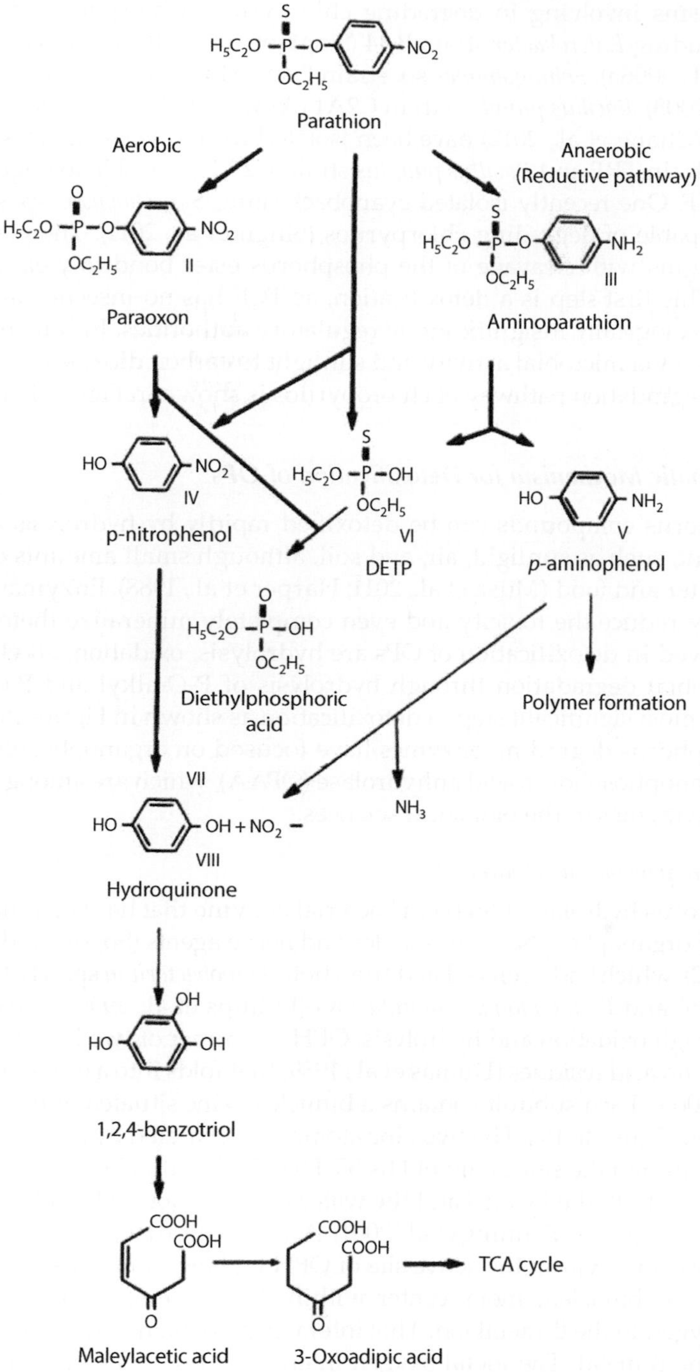

FIGURE 16.14
Degradation pathway of parathion by mixed bacterial culture. Heavy arrows denote definite pathways; light arrows denote postulated pathways.

of microorganisms involving in degrading chlorpyrifos. Chlorpyrifos-degrading bacterial strains including *Enterobacter* strain B-14 (Singh et al., 2004), *Stenotrophomonas* sp. strain YC-1 (Yang et al., 2006a), *Sphingomonas* sp. strain Dsp-2 (Li et al., 2007), *Paracoccus* sp. strain TRP (Xu et al., 2008), *Bacillus pumilus* strain C2A1 (Anwar et al., 2009), and *Bacillus laterosporus* strain DSP (Zhang et al., 2012) have been isolated from diverse sources; however, only *Paracoccus* sp. strain TRP and *Bacillus pumilus* strain C2A1 were able to degrade both chlorpyrifos and TCP. One recently isolated cyanobacterium, *Synechocystis* sp. strain PUPCCC 64, was also capable of degrading chlorpyrifos (Singh et al., 2011). The major pathway of degradation begins with cleavage of the phosphorus ester bond to yield the breakdown product TCP. This first step is a detoxification, as TCP has no insecticidal activity and is considered toxicologically insignificant by regulatory authorities. In soil and water, TCP is further degraded via microbial activity and sunlight to carbon dioxide and organic matter. The complete degradation pathway of chloropyrifos is shown in Figure 16.15.

16.3.1.3 Enzymatic Mechanism for Detoxification of OPs

Organophosphorus compounds can be detoxified rapidly by hydrolysis on exposure to the environment, such as sunlight, air, and soil, although small amounts can be detected in drinking water and food (Musa et al., 2011; Harper et al., 1988). Enzymatic hydrolysis of OPs can greatly reduce the toxicity and even completely mineralize them. The principal reactions involved in detoxification of OPs are hydrolysis, oxidation, alkylation and dealkylation. Microbial degradation through hydrolysis of P-O-alkyl and P-O-aryl bonds is considered the most significant step in detoxification as shown in Figure 16.9. Most studies of organophosphorus degrading enzymes have focused on organophosphorus hydrolase (OPH) and organophosphorus acid anhydrolase (OPAA), which are among the most extensively studied enzymes in the biological sciences.

16.3.1.3.1 Organophosphorus Hydrolase

Organophosphorus hydrolase (OPH) is a bacterial enzyme that has been shown to degrade a wide range of organophosphorus pesticides and nerve agents (Sogorb and Vilanova, 2002; Kang et al., 2012), which had been isolated from both *Flavobacterium* sp. ATCC 27551 (Mulbry and Karns, 1989) and *Pseudomonas diminuta* MG (Phillips et al., 1990). The OPs are mainly detoxified through oxidation and hydrolysis. OPH is a dimer of two identical subunits containing 336 amino acid residues (Dumas et al., 1989) that folds into a $(\alpha\beta)_8$-barrel motif (Gerlt and Raushel, 2003). Each subunit contains a binuclear zinc situated at the C-terminal portion as shown in Figure 16.16a. The two zinc atoms are separated by about 3.4 A° and linked to the protein through the side chain of His 55, His 57, His 201, His 230, Asp 301, and a carboxylated Lys 169. Both the Lys 169 and the water molecule (or hydroxide ion) act to bridge the two zinc ions together (Benning et al., 2001). A schematic diagram of the structure of the binuclear metal center within the active site of OPH is shown in Figure 16.16b.

OPs bind to the binuclear metal center within the active site via coordination of the phosphoryl oxygen to the β-metal ion. This interaction weakens the binding of the linking hydroxide to the β-metal. The metal-oxygen interaction polarizes the phosphoryl oxygen bond and creates a more electrophilic phosphorus center. Subsequently, nucleophilic attack by the bound hydroxide is assisted by proton abstraction from Asp 301. The hydroxide attacks the phosphorus center, resulting in weakening of the bond to the leaving group (Raushel, 2002). A working model for the OPH reaction mechanism is shown in Figure 16.17.

In summary, the role of one metal ion in the active site of OPH is to increase the electrophilicity of the phosphorus center through coordination with the nonester oxygen atom

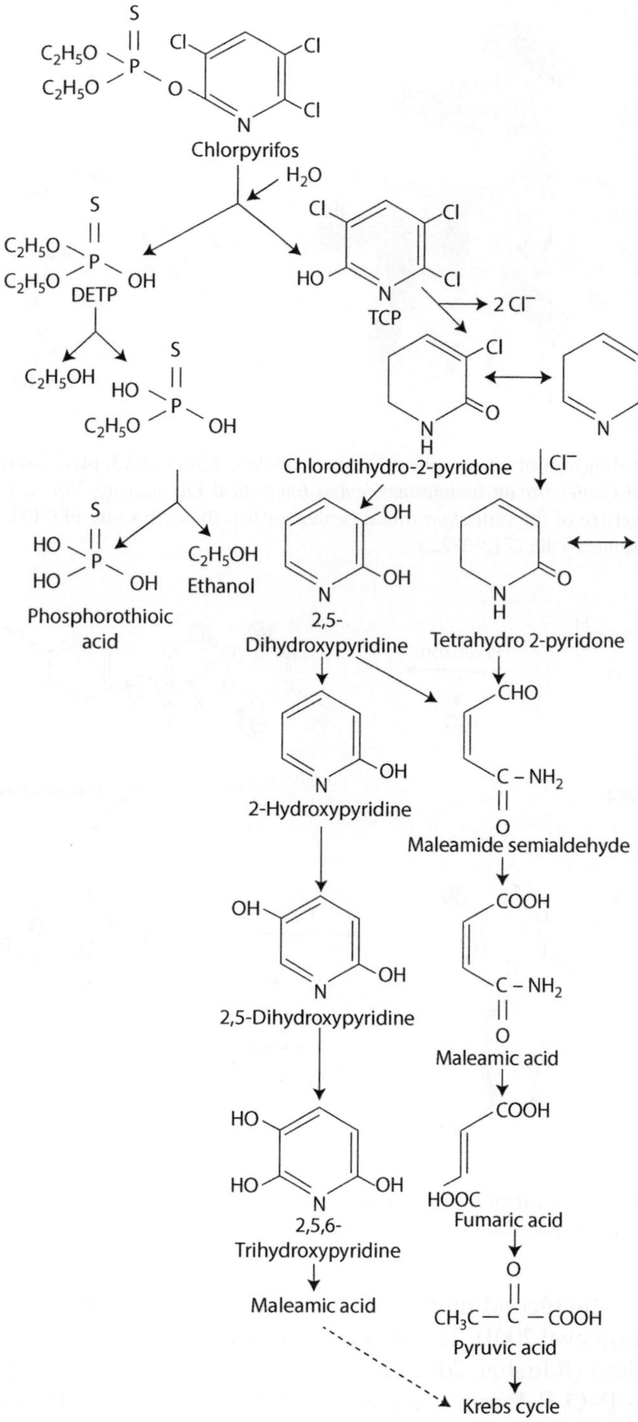

FIGURE 16.15
Proposed pathways for chlorpyrifos degradation by microorganisms. When the conversion of one compound to another is believed to occur through a series of intermediates, the steps are indicated by dotted arrows. DETP, diethylthiophosphate; TCP, trichloropyridinol. (Adapted from Singh BK and Walker A. 2006. *FEMS Microbiological Reviews* 30: 428–471.)

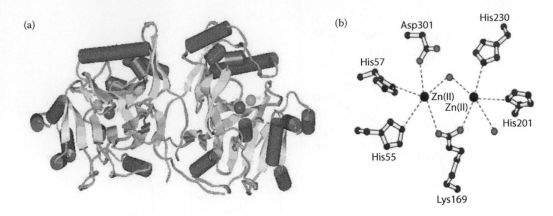

FIGURE 16.16
(See color insert.) (a) Ribbon diagram of organophosphorus hydrolase (OPH). (Adapted from Zheng Y et al. 2008. In: *The 2nd International Conference on Bioinformatics and Biomedical Engineering*, Vol. 1, IEEE, pp. 13–16.) (b) Representation of the structure of the binuclear metal center within the active site of OPH. (Adapted from Benning MM et al. 2001. *Biochemistry* 40: 2712–2722.)

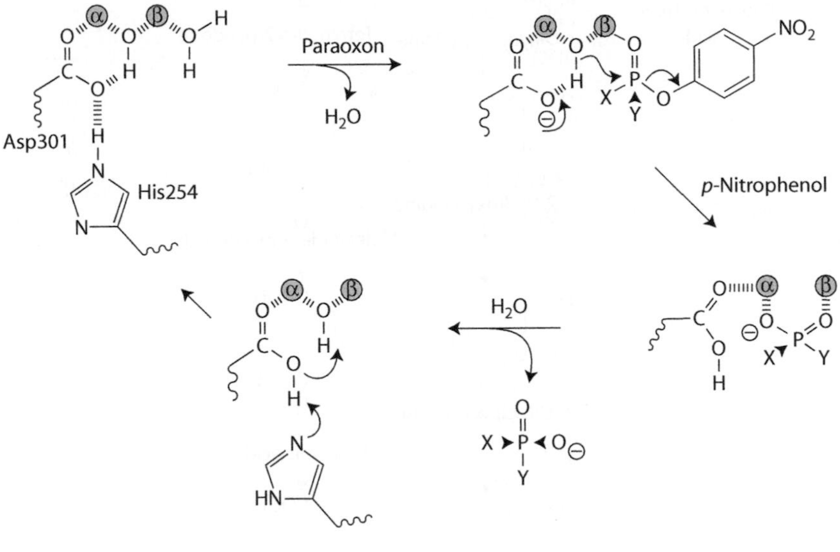

FIGURE 16.17
Working model for the catalytic mechanism for hydrolysis of organophosphorus nerve agents by organophosphorus hydrolase (OPH). (Adapted from Raushel FM. 2002. *Current Opinion in Microbiology* 5: 288–295.)

of the substrate, whereas the second metal ion acts as a promoter of the attacking nucleophile (Efrmenko and Sergeeva, 2001). The two Zn ions within the active site have distinct functions or act in tandem (Raushel, 2002). However, OPH has a wide range of substrate specificity. It hydrolyzes P–O, P–F, and P–S bonds to different extents. The lowest specificity is for the P–S bond. However, the enzyme does not catalyze the cleavage of carbonyl groups such as those found in pnitrophenyl acetate. The effects of metal substitution on the catalytic activity of OPH were studied by removing the native metal (Zn) from purified OPH and reconstitution with a series of divalent cations which include Co, Cd, Cu, Fe, Mn, and Ni (Benning et al., 2001). Further enzymatic assays showed that Co21 had the greatest activity against paraoxon (Omburo et al., 1992). It was suggested that divalent cations increased the activity of enzyme by assisting folding of expressed enzyme in the medium (Manavathi et al., 2005).

16.3.1.3.2 *Organophosphorus Acid Anhydrolase*

Organophosphorus acid anhydrolases (OPAA) are a class of bimetalloenzymes that hydrolyze a variety of toxic acetylcholinesterase-inhibiting organophosphorus (OP) compounds, including pesticides and fluorine-containing chemical nerve agents. A variety of microorganisms can detoxify organophosphorus compounds by hydrolyzing them using OPAA. The bacterial enzyme OPAA is able to catalyze the hydrolysis of both proline dipeptides and several types of OPs. Defrank and Cheng (1991) have isolated halophilic bacteria. They found that these isolated bacteria possess high levels of enzymatic activity against several highly toxic organophosphorus compounds. The predominant enzyme is designated organophosphorus acid anhydrase 2 (OPAA2). The enzyme is a single polypeptide with a molecular weight of 60 KDa. Further, a highly active OPAA from *Alteromonas undina* was isolated and purified and is composed of a single polypeptide with molecular weight 53 kDa (Cheng et al., 1993). Vyas et al. (2010) determined the crystal structure of native OPAA (58 kDa) from *Alteromonas* sp. strain JD6.5 as shown in Figure 16.18a. The OPAA structure is composed of two domains, a small N-terminal domain (from residue 1 to approximately 160) and a large C-terminal domain or C-domain (residues approximately 161–440). Although both domains of OPAA consist of mixed β-sheets with an almost equal number of β-strands (six and five in the N- and C-domains, respectively), they differ in overall topology and shape. The latter exhibits a "pita bread" architecture and harbors the active site with the binuclear Mn^{2+} ions. The native OPAA structure revealed unexpectedly the presence of a well-defined nonproteinaceous density in the active site whose identity could not be definitively established but is suggestive of a bound glycolate, which is isosteric with a glycine (Xaa) product. All three glycolate oxygens coordinate the two Mn^{2+}

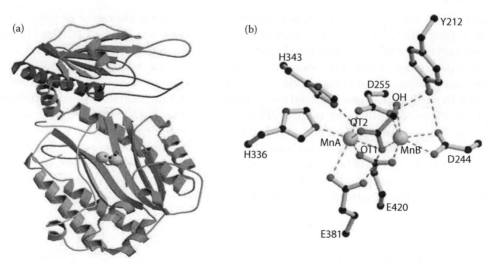

FIGURE 16.18

(See color insert.) Ribbon diagram of organophosphorus acid anhydrolase (OPAA) determined from *Alteromonas* sp. strain JD6.5. (a) Ball-and-stick model of OPAA active site residues (yellow bonds) that are involved in the coordination of the two metal ions (yellow spheres) and in the interaction with the glycolate molecule (green bonds). Metal coordinations are shown as green dashed lines, while hydrogen bonds and other contacts with the glycolate are shown as red dashed lines (b). (Adapted from Vyas NK et al. 2010. *Biochemistry* 49(3): 547–559.)

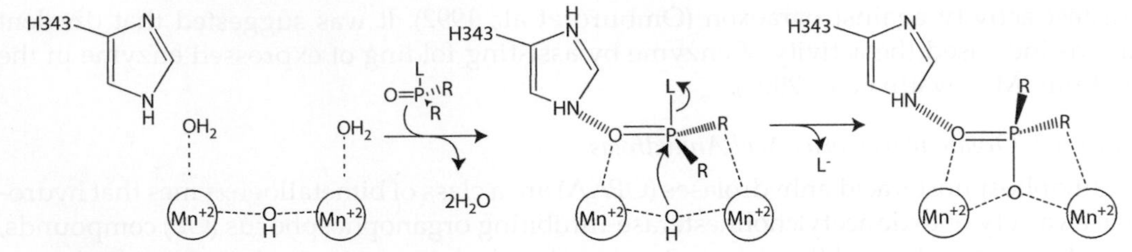

FIGURE 16.19
Proposed catalytic mechanism for OPAA cleavage of phosphotriesters. L is leaving group which can be fluoride, or *p*-nitrophenol. R is an ester or methyl group. (Adapted from Bigley NA and Raushel FM. 2013. *Biochimica et Biophysica* Acta 1834(1): 443–453.)

atoms. The two Mn^{2+} atoms identified as MnA and MnB as shown in Figure 16.18b are separated by 3.3(0.1A). This demonstration of the presence of the two metal ions not only provides conclusive direct evidence of the observation of metal cation requirement for OPAA activity but also establishes the location of the catalytic center.

The proposed chemical mechanism of OPAA shown in Figure 16.19 is similar to that of OPH but with some significant differences (Vyas et al., 2010; Bigley and Raushel, 2013). The most significant difference in the proposed mechanisms for OPH and OPAA is in the initial binding of the substrate. Structural evidence shows that the free phosphoryl oxygen binds to the more solvent exposed A-metal of OPAA, and the ester oxygen extending to the small pocket ligates to the B-metal. The bidentate ligation of the substrate is proposed to allow the hydroxide to attack the phosphorus center without dissociation from one of the metals. The leaving group is expelled, and the phosphoryl product is bound in a tridentate manner.

16.3.2 Organohalides

Oragnohalides are halogen-substituted hydrocarbons with a wide range of physical and chemical properties. They may be saturated (alkyl halides), unsaturated (alkenyl halides), or aromatic (aryl halides) as shown in Figure 16.20. Organohalides are widely used as herbicides, insecticides, antibiotics, medicines, plastics, solvents, refrigerants, synthetic precursors, and so on. These compounds are produced in large quantities by the chemical industry. In addition to these artificial compounds, more than 3800 kinds of organohalides are produced biologically or by natural abiogenic processes such as the eruption of volcanoes, forest fires, and other geothermal processes. Marine organisms such as seaweeds, sponges, corals, tunicates, and bacteria are the major biological producers of organohalides. Terrestrial plants, fungi, lichen, bacteria, insects, some higher animals, and even humans also produce various organohalogen compounds. Most organohalides are chlorides (chlorocarbons and chlorohydrocarbons). Organohalide compounds such as chloroethenes,

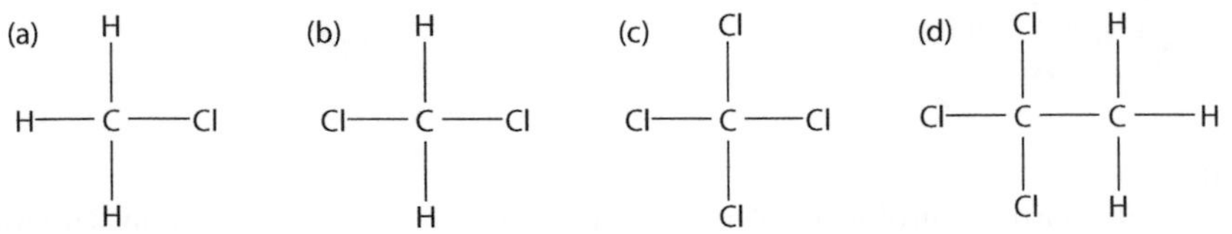

FIGURE 16.20
Some typical alkyl halides: (a) chloromethane, (b) dichloromethane, (c) carbon tetrachloride, and (d) 1,1,1-trichloroethane.

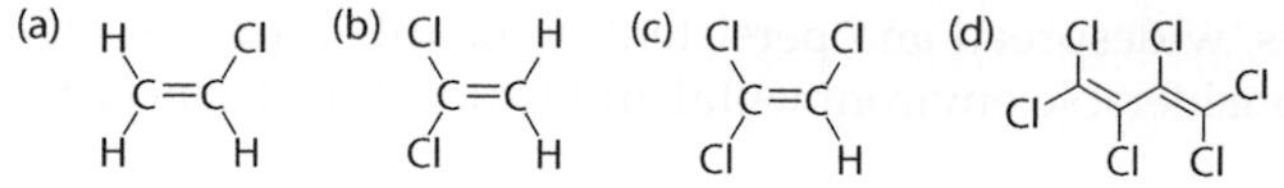

FIGURE 16.21
The more common low molecular weight alkenyl chlorides: (a) monochloroethylene (vinyl chloride), (b) 1,1 dichloroethylene (vinylidene chloride), (c) cis-1,2- dichloroethylene, and (d) hexachlorobutadiene.

chloroethanes, and polychlorinated benzenes are among the most significant pollutants in the world. These compounds are often found in contamination plumes with other pollutants such as solvents, pesticides, and petroleum derivatives. However, their uncontrolled release into the environment has caused environmental damage, as many (xenobiotic) organohalides are not only toxic, but highly recalcitrant to biodegradation, and they readily accumulate in lipids leading to bioaccumulation and affect the food chain.

16.3.2.1 Alkyl Halides

Alkyl halides are compounds in which halogen atoms are substituted for hydrogen on an alkyl group. The structural formula of some typical halides is shown in Figure 16.21.

16.3.2.2 Alkenyl Halides

The alkenyl or olefinic organohalides contain at least one halogen atom and at least one carbon–carbon double bond. The most significant of these are the lighter chlorinated compounds as shown in Figure 16.21.

16.3.2.3 Aryl Halides

Aryl halide (also known as haloarene or halogenoarene) is an aromatic compound in which one or more hydrogen atoms directly bonded to an aromatic ring are replaced by a halide. The structural formula of some important aryl halides as shown in Figure 16.22. These compounds are made by substitution chlorination of aromatic hydrocarbons as shown, for example, by the reaction below for the synthesis of a PCB.

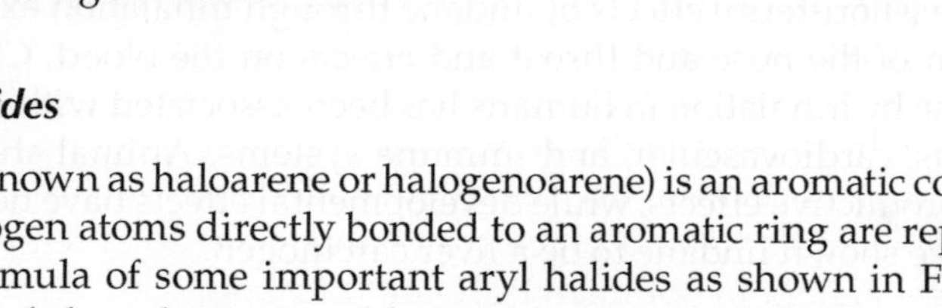

This is substitution reaction of an aromatic compound (biphenyl) to produce an organo-chlorine product (2,3,5,2',3'-PCB, a PCB compound). The product is one of 210 possible

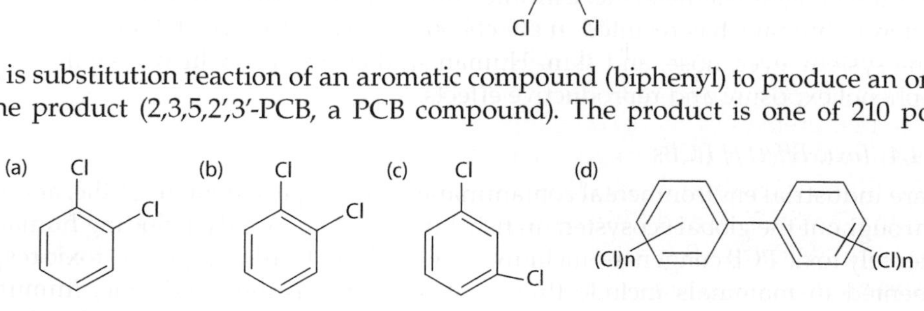

FIGURE 16.22
Some of the more important aryl halides (a) monochlorobenzene, (b) 1,2 dichlorobenzene, (c) 1,3 dichlorobenzene, (d) polychlorinated biphenyls.

congeners of PCBs, widespread and persistent pollutants found in the fate tissue of most humans and of considerable environmental and toxicological concern.

16.3.2.4 Fate in Environment and Mode of Action of Toxicity of Organohalides

Organohalide insecticides exhibit a wide range of toxic effects and varying degrees of toxicity. Many of these compounds are neuropoisons and their most prominent acute effects are upon the central nervous system, manifested by symptoms of CNS poisoning including tremors, irregular jerking of the eyes, change in personality, and loss of memory. Some of the toxic effect of specific organohalides insecticides and classes of these compounds are discussed below.

16.3.2.4.1 Toxic Effect of DDT

As described previously, DDT is an organochloride man-made chemical widely used to control insects on agricultural crops. It is a persistent environmental contaminant and its widespread use has resulted in worldwide contamination. The tendency of DDT to concentrate in the fat of humans, livestock, and wildlife contributes to its ability to adversely affect organisms. In particular, greater accumulation in species higher on the food chain, known as biomagnification, has resulted in severe adverse effects on many forms of wildlife, especially predatory species. DDT has caused chronic effects on the nervous system, liver, kidney, and immune systems in experimental animals (ATSDR, 1994).

16.3.2.4.2 Toxic Effect of Lindane (γ-HCH)

Lindane is used as an insecticide on fruit and vegetable crops. Lindane is quite toxic to humans. The acute (short-term) effects of lindane through inhalation exposure in humans consist of irritation of the nose and throat and effects on the blood. Chronic (long-term) exposure to lindane by inhalation in humans has been associated with effects on the liver, blood, and nervous, cardiovascular, and immune systems. Animal studies indicate that lindane causes reproductive effects, while developmental effects have not been noted. Oral animal studies have shown lindane to be a liver carcinogen.

16.3.2.4.3 Toxic Effect of PCP

PCP is strongly toxic to plants. It is rapidly absorbed through the gastrointestinal tract following ingestion. If deposition in the tissues occurs, the major sites are the liver, kidneys, plasma protein, brain, spleen and fat but accumulation is not common. Unless kidney and liver functions are impaired, PCP is rapidly eliminated from blood and tissues, and is excreted unchanged via the urine. Chronic (long-term) exposure to pentachlorophenol by inhalation in humans has resulted in effects on the respiratory tract, blood, kidney, liver, immune system, eyes, nose, and skin. Human studies are inconclusive regarding pentachlorophenol exposure and reproductive effects.

16.3.2.4.4 Toxic Effect of PCBs

PCBs are industrial environmental contaminants that are persistent, lipohilic, and ubiquitous throughout the global ecosystem in fish, birds, and mammals including humans. The most acutely toxic PCB congeners, including 3,3′,4,4′,5-pentachlorobiphenyl toxic responses documented in mammals include thymic atrophy, a wasting syndrome, immunotoxic effects, reproductive impairment, porphyria and related liver damage, and induction of specific isozymes of the cytochrome P450 system. The persistency and lipophilicity of PCBs permit them to biomagnify once entering the aquatic food chain.

16.3.2.5 Biodegradation of Organohalides

The recalcitrance of organic pollutants increase with the increase of halogination. The substitution of halogen as well as nitro and sulfo groups at the aromatic ring is accomplished by an increasing electrophilicity of the molecule. The biodegradation of anthropogenic organohalides has been the object of intense study in recent years (Lovley, 2003). Microbial bioremediation of contaminated sites has become commonplace whereby key processes involved in bioremediation include anaerobic degradation and transformation of these organohalides by organohalide respiring bacteria and also via hydrolytic, oxygenic, and reductive mechanisms by aerobic bacteria. Whereas organic molecules with few halogen substituents can often be mineralized under aerobic conditions, highly chlorinated compounds such as tetrachloroethene, hexachlorobenzene, polychlorinated dibenzo-*p*-dioxins and polychlorinated biphenyls (PCBs) are often persistent. Organohalides are resistant to electrophilic attack by oxygenase of aerobic bacteria. The only documented microbial process leading to a transformation of such highly halogenated compounds is reductive dehalogenation under anaerobic conditions, the reductive attack of anaerobic bacteria is of significance. Under these conditions, the reduction of organohalide molecules is a favorable process leading to a less halogenated product. Bacteria metabolize organohalides in several different ways. First, they may use organohalides as a source of carbon and energy. Second, bacteria use organohalides as an acceptor for electrons generated during metabolic oxidation reactions for example, *Desulfomonile tiedjei* oxidizes pyruvate coupled with a metabolic reduction of 3-chlorobenzoate to yield benzoate and chloride anion. These reactions have been linked to ATP formation, but benzoate is not assimilated to build up intracellular organic molecules. Last, bacteria cometabolize organohalides without accompanying ATP formation or assimilation of the carbon structure into biomolecules. Cometabolic degradation of chlorinated compounds, nevertheless, can be extremely rapid. Yet, no organisms have been found that oxidatively degrade important environmental chemicals such as chloroform, trichloroethylene, 1,1,1-trichloroethane, 1,2-dichloropropane, and 1,2,3-trichloropropan (Janssen et al., 2005) and use them as a carbon source. Attempts to obtain enrichments or pure cultures that aerobically grow on these chemicals have met with no success. However, some other halogenated chemicals are easily biodegradable, and cultures that utilize chloroacetate, 2-chloropropionate and 1-chlorobutane can be readily enriched from almost any soil sample. For still other compounds, degradative organisms can be isolated, but only after prolonged adaptation or if a suitable inoculum is used in which the catabolic activity that is sought is apparently already enriched due to preexposure to halogenated chemicals in the environment. Compounds of this class of intermediate degradability include dichloromethane, 1,2-dichloroethane and the nematocides 1,2-dibromoethane and 1,3-dichloropropene as presented in Table 16.5. A critical step in the degradation of organohalides is cleavage of the carbon–halogen bond. Bacteria halogenases play a major role for the cleavage of the carbon–hydrogen bond. These halogenases make use of a variety of distinctly different catalytic mechanisms to cleave carbon–halogen bonds. A list of micro-organisms capable of degrading OPs is presented in Table 16.6.

16.3.2.5.1 Pentacholorophenol

Chlorophenols (CPs) are aromatic ring structures containing at least one chlorine atom (Cl) and one hydroxyl (OH) group at the benzene rings. They may be used on large scale as wood preservatives, fungicides, herbicides, and general biocides. Five groups of CPs have been recognized on the basis of their chemical structures, monochlorophenols (MCPs),

TABLE 16.5

Degradability of Some (Halogenated) Aliphatic and Aromatic Compounds

Degradable	Difficult	Impossible
Methylchloride, chloroacetic acid, toluenes, chlorobenzene, naphthalene, 1,2-dichloroethane, vinyl chloride, epichlorohydrin, pentachlorophenol	Dichloroethylene, trichloroethylene, 1,2-dibromoethane, 1,3-dichloropropylene, hexachlorocyclohexane	Chloroform, 1,1,1-trichloroethane, 1,2-dichloropropane, 1,2,3-trichloropropane

TABLE 16.6

Microorganisms Isolated for the Degradation of Organophohalides

Organohalides	Microorganisms	References
PCP	Bacteria	
	Bacillus cereus (AY927692)	Singh et al. (2009)
	Seeratia marcescens (DQ002384)	Singh et al. (2009)
	Bacillus sp. (AY952465)	Chandra et al. (2007)
	Flavobacterium sp.	Chanama and Crawford (1997)
	Pseudomonas fluorescens	Shah and Thakur (2002)
	Sphingobium chlorophenolicum ATCC 39723	Cai and Xun (2002)
	Mycobacterium chlorophenolicus	Haggblom et al. (1988)
	Mycobacterium fortuitum	Haggblom et al. (1988)
	Rhodococcus sp.	Apajalahti and Salkinoje-Salonen (1986)
	Arthrobacter sp. ATCC 33790	Schenk et al. (1989)
PCB	Bacteria	
	Burkholderia sp. strain LB400	Pieper and Seeger (2008)
	Burkolderia xenovorans LB400	Seeger et al. (2001)
	Rhodococcus jostii RHA1	Warren et al. (2004)
	Dehalococcoides	Maymo-Gatell et al. (1999)
	Desulfitobacterium	Sanford et al. (1996)
	Dehalobacter	Holliger et al. (1998)
	Desulfuromonas	Krumholz (1997)
DDT	Bacteria	
	Eubacterium limosum	Yim et al. (2008)
	Alcaligenes eutrophus A5	Nadeau et al. (1994)
	Boletus edulis	Huang et al. (2007)
	Fungi	
	Fusarium solani	Mitra et al. (2001)
	Phanerochaete chrysosporium	Bumpus and Aust (1987)

polychlorophenols (poly-CPs), chloronitrophenols (CNPs), chloroaminophenols (CAPs), and chloromethylphenols (CMPs). However, poly-CPs such as dichlorophenols (DCPs), trichlorophenols (TCPs), tetrachlorophenols (TeCPs), and pentachlorophenols (PCP) are more recalcitrant to bacterial degradation than MCPs due to the presence of the two or more chlorine atoms at the phenolic rings.

Aerobic degradation of CPs and their derivatives have been extensively investigated in bacteria, and many bacteria with the ability to utilize CPs as their sole carbon and energy

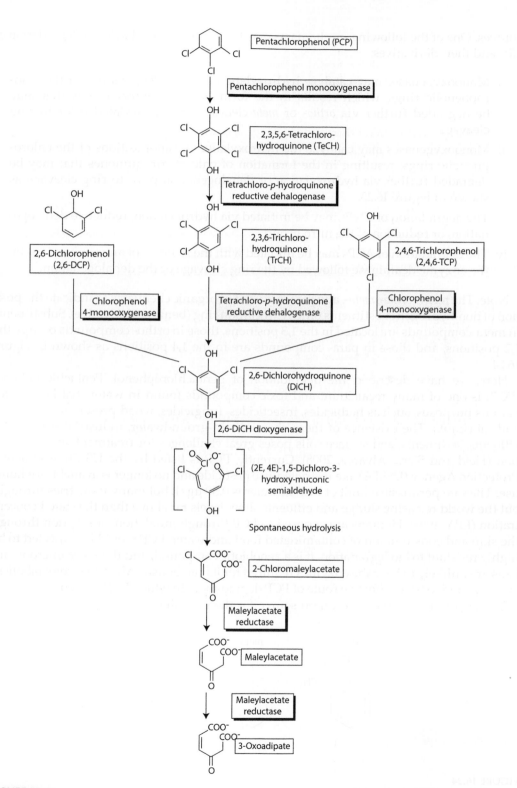

FIGURE 16.23
General degradation pathway of pentachlorophenol, 2,6-dichlorophenol, and 2,4,6-trichlorophenol.

sources. One of the following mechanisms may be involved in the bacterial degradation of CPs and their derivatives:

i. Monooxygenases may catalyze hydroxylation at the *ortho*-positions of the chlorophenolic rings, which results in the formation of chlorocatechols that may be degraded further via *ortho*- or *meta*-cleavage or hydroxylated prior to ring cleavage.

ii. Monooxygenases may catalyze the hydroxylation at *para*-positions of the chlorophenolic rings, resulting in the formation of chlorohydroquinones that may be degraded further via hydroxylation or dehalogenation prior to ring cleavage as shown in Figure 16.23.

iii. The degradation of CNPs may be initiated via hydroxylation, reductive dehalogenation, or reduction of the nitro group.

iv. The degradation of ACPs may be initiated with the removal of ammonium ions by the enzyme deaminase followed by the ring cleavage or the dehalogenation.

Note: The terms *ortho*-, *meta*-, and *para*- are used in organic chemistry to indicate the position of non-hydrogen substituents on a hydrocarbon ring (benzene derivative). Substituents in meta-compounds are located in the 1,3 positions, those in ortho- compounds occupy the 1,2 positions, and those in para- compounds are in the 1,4 positions as shown in Figure 16.24.

Here, we have described the degradation of pentachlorophenol. Pentachlorophenol (PCP) is one of many recalcitrant and toxic compounds found in water that is used for various purposes, such as herbicides, insecticides, fungicides, wood preservatives, resins, and lubricants. The existence of these chemicals in groundwater, industrial wastewater effluents, sediments, and surface soils poses great challenges for treatment and remediation (Field and Sierra-Alvarez, 2008). Currently, PCP is listed by the US Environmental Protection Agency (USEPA) as a restricted-use pesticide and no longer is available for home use. The safe permissible limit of PCP in water is 0.30 µg/L but many industries throughout the world releasing sludge and effluents show levels far above than the stated concentration (EPA, 1999). Humans are exposed to PCP through inhalation, absorption through the skin and consumption of contaminated food and water. PCP would be expected to be highly resistant to biodegradation. It is a xenobiotic compound, and therefore microorganisms are unlikely to have the enzymes required for its catabolism. Microbial cometabolism is considered to be an effective route of PCP degradation, in which PCP is used as a second substrate and may serve as a carbon source for the growth of microorganisms. PCP can

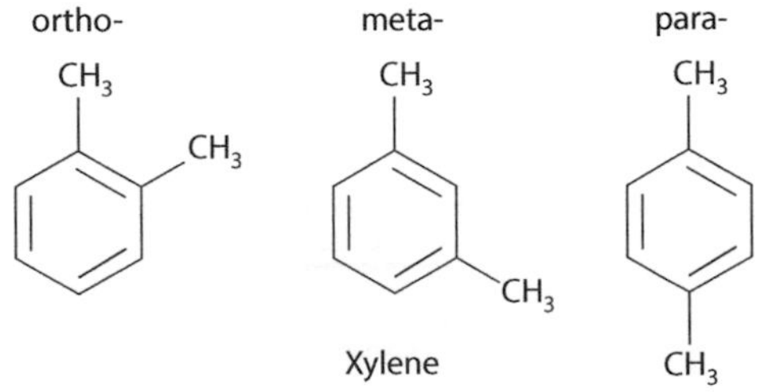

FIGURE 16.24
Chemical structure shows the difference between ortho-, meta-, and para-xylene.

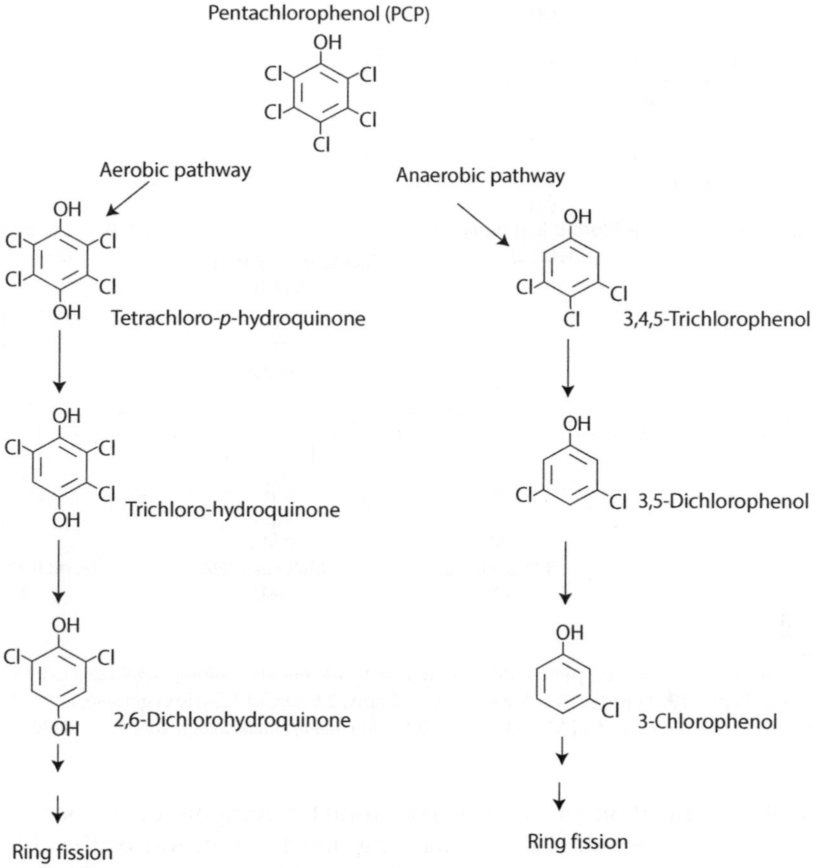

FIGURE 16.25
Microbial degradation pathway of PCP under aerobic and anaerobic conditions.

be degraded by soil bacteria such as *Arthrobacter* sp., *Sphingomonas* sp., *Mycobacterium* sp., *Pseudomonas cepacia,* and *Flavobacterium* sp. PCP has been degraded both aerobically and anaerobically in the environment through the action of microorganisms (Chaudhry and Chapalamadugu, 1991). A simplified aerobic and anaerobic microbial degradation pathway of PCP is shown in Figure 16.25.

16.3.2.5.1.1 Aerobic Degradation   Aerobic degradation of PCP has been extensively investigated in bacteria and several bacterial isolates such *Sphingobium chlorophenolicum L-I* (formerly *Sphingomonas chlorophenolica* ATCC 39723) (Takeuchi et al., 2001), *Kocuria* sp. CL2 (Karn et al., 2011), *Bacillus* sp. (Karn et al., 2010), *Novosphingobium lentum* (Dams et al., 2007), *Pseudomonas* sp. (Kao et al., 2005), *Sphingomonas chlorophenolica* (Yang et al., 2006a,b), and *Enterobacter* sp. SG1 (Karn and Geetanjali, 2014) were found to utilize PCP as a sole source of carbon and energy. PCP is initially biotransformed under aerobic conditions to tetrachloro-p-hydroquinone (TeCHQ) due to hydroxylation at para-position by either PCP-4-monooxygenase (PcpA) or cytochrome P-450 type enzyme and then to trichloro-*p*-hydroquinone (TrCHQ) and dichlorohydroquinone (DiCHQ).

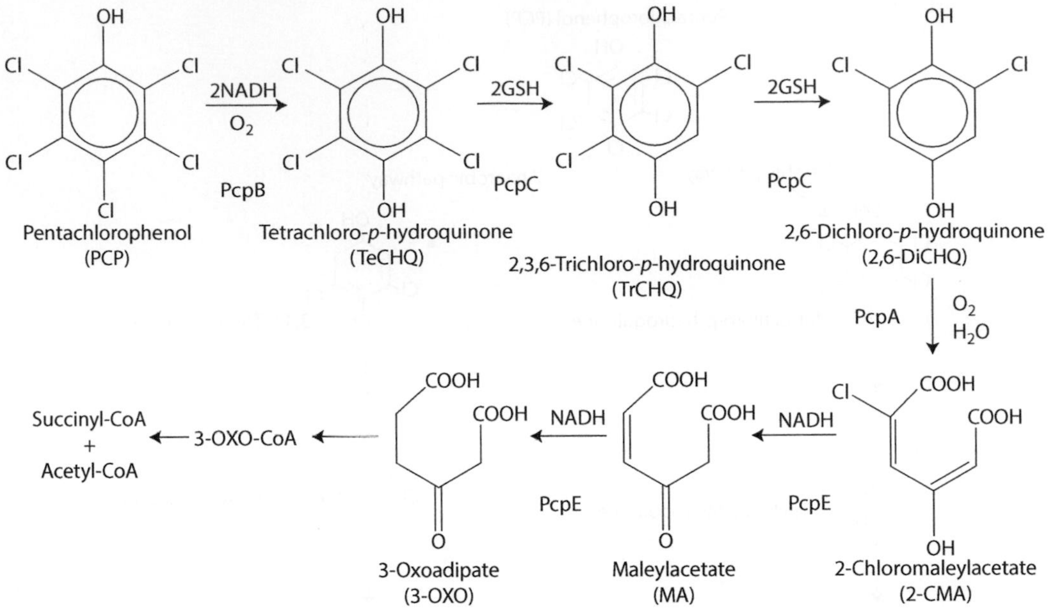

FIGURE 16.26
Bacterial degradation pathways for pentachlorophenol in *Sphingomonas chlorophenolicum* L-1 PCP. PcpB, PCP 4-monooxygenase; PcpC, TeCH reductive dehalogenase; PcpA, 2,6-DiCH 1,2-dioxygenase; PcpE, MA reductase; CoA, coenzyme A. (Adapted from Cai M and Xun L. 2002. *Journal of Bacteriology* 184: 4672–4680.)

For complete degradation of chlorinated aromatic compounds to occur, two steps are necessary, the cleavage of the aromatic ring and the removal of the chloride atom. In *S. chlorophenolicum* L-1 (previously known as *S. chlorophenolicum* ATCC 39723), PCP-4-monooxygenase (PcpB) catalyzes the conversion of PCP to TeCHQ via the removal of chloride ions. In the next step, TeCHQ is sequentially dehalogenated to 2,3,6-trichloro-*p*-hydroquinone (2,3,6 TrCHQ) by a TeCHQ-reductive dehalogenase (PcpC). Further degradation of 2,6-DiCHQ occurs via ring cleavage by the 2,6- DCHQ-1,2-dioxygenase (PcpA), leading to formation of 2-chloromaleylacetate (2-CMA) that is further degraded via the TCA cycle. The four gene products PcpB, PcpC, PcpA, and PcpE were responsible for the metabolism of PCP to 3-oxoadepate. The complete degradation pathway of PCP is shown in Figure 16.26.

The degradation of PCP has been reported in another bacteria *Mycobacterium chlorophenolicum* PCP-1 and *Mycobacterium fortuitum* CG-2 (formerly *Rhodoccocus* strains). In these bacteria PCP is hydroxylated to TeCHQ by a membrane bound cytochrome P-450 type enzyme (Uotila et al., 1991, 1992). Subsequently, TeCHQ undergoes hydrolytic dehalogenation followed by reductive dehalogenation to form dichloro-1,2,4-trihydroxybenzene, which produces BT after two successive reductive dehalogenation (Apajalahti and Salkinoja-Salonen, 1987). The complete degradation pathway of PCP is shown in Figure 16.27.

16.3.2.5.1.2 Anaerobic Degradation The reductive dechlorination of PCP has been demonstrated in anaerobic sediments (Mikesell and Boyd, 1986) and the complete degradation of PCP has been observed in anaerobic sludges. PCP may be reductively dehalogenated to 2,3,4,5-tetra-chlorophenol (2,3,4,5-TeCP), 3,4,5-trichlorophenol (3,4,5-TCP),

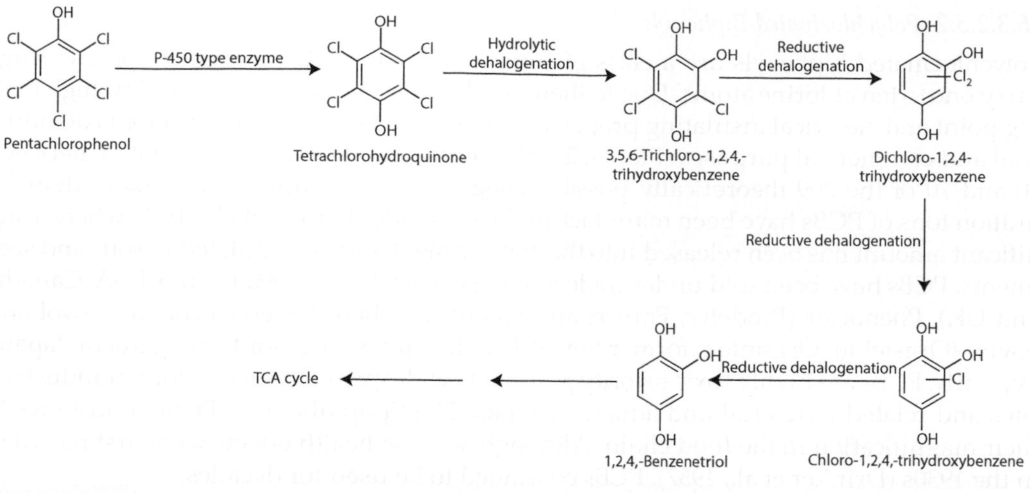

FIGURE 16.27
Bacterial degradation pathways for pentachlorophenol (PCP) in *Mycobacterium fortuitum* CG-2. (Adapted from Arora PK and Bae H. 2014. *Microbial Cell Factories* 13: 31.)

3,5-dichlorophenol, 3DCP and finally to phenol of which the further benzene ring is then broken to produce methane (CH_4) and carbon dioxide (CO_2) by anaerobic bacteria (Mikesell and Boyd, 1986). A pathway of anaerobic PCP degradation is shown in Figure 16.28. Ortho-dechlorinations are most common, sometimes leading to the accumulation of 3,5-dichlorophenol (3,5-DCP). The combination of phenol-dehalogenating and phenol-degrading cultures was used for complete mineralization of PCP under anaerobic conditions. In this process, a phenol-dehalogenating culture dehalogenates PCP to phenol under anaerobic conditions. The phenol is then further degraded by phenoldegrading culture under iron reducing or sulfate reducing conditions (Yang et al., 2009).

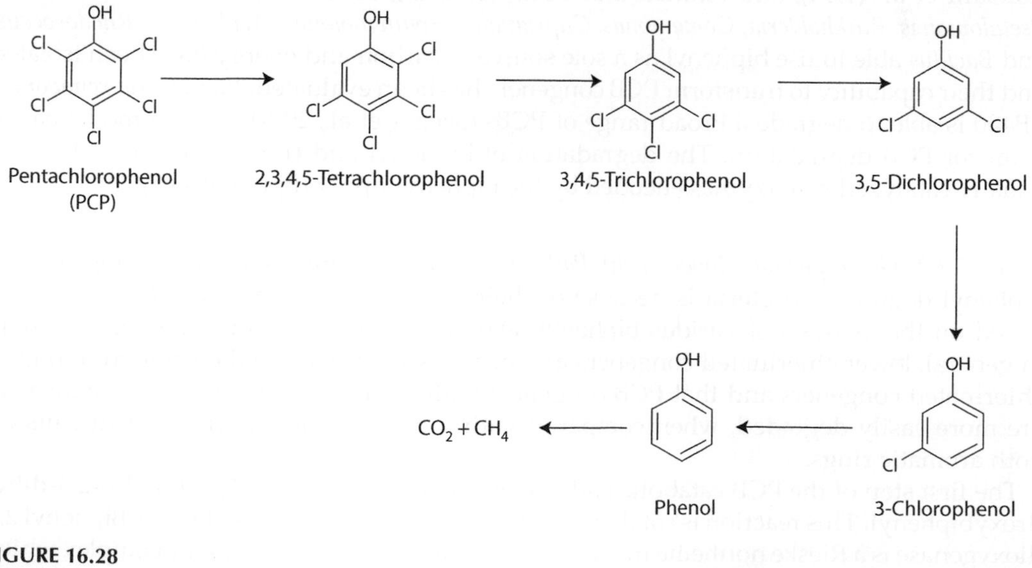

FIGURE 16.28
Anaerobic microbial dehalogenation and degradation of pentachlorophenol (PCP).

16.3.2.5.2 Polychlorinated Biphenyls

Polychlorinated biphenyls are a class of compounds where the aromatic biphenyl rings carry one to ten chlorine atoms. Due to their nonflammability, chemical stability, high boiling point and electrical insulating properties, PCBs have been used for hundreds of industrial and commercial purposes (Pieper, 2005). Commercial PCB mixtures contain between 20 and 70 of the 209 theoretically possible congeners. It is estimated that more than 1.5 million tons of PCBs have been manufactured worldwide (Faroon et al., 2003), where a significant amount has been released into the environment and accumulated in soils and sediments. PCBs have been sold under trade names such as Aroclor (Monsanto, USA, Canada, and UK), Phenoclor (Prodelec, France, and Spain), Clophen (Bayer, Germany), Sovol and Sovtol (Orgsteklo, Orgsintez, former Soviet Union), and Kanechlor (Kanegafuchi, Japan). Typically, PCBs are highly toxic priority pollutants widespread in several former industrial sites and related terrestrial and aquatic habitats. The lipophilicity of PCBs contributes to their magnification in the food chain. Although adverse health effects were first recorded in the 1930s (Drinker et al., 1937), PCBs continued to be used for decades.

Since then, PCBs have been shown to cause cancer and have a number of serious effects on the immune, reproductive, nervous and endocrine systems (Aoki, 2001). Thus, their removal from soils and sediments is a priority of great relevance in several industrialized countries. Metabolic breakdown of PCBs by microorganisms are considered to be one of the major routes of environmental degradation for these widespread pollutants. Environmental monitoring has demonstrated that the PCBs found in environmental samples, are those containing five or more chlorine atoms. This indicates that the less chlorinated biphenyls are more rapidly degraded than highly chlorinated biphenyl. The degree of chlorination seems to be the most significant factor affecting the biodegradation of PCBs, but other factor such as water solubility, pH, temperature, oxygen, nutrient supplies etc. by various materials all combine to varying extents to affect biodegradability. The ability to degrade PCB in found in several genera of Gram-positive and Gram-negative aerobic and anaerobic bacteria.

16.3.2.5.2.1 Aerobic Degradation Since the pioneering studies of Lunt and Evans (1970), Catelani et al. (1971), and Ahmed and Focht (1973a,b), several bacterial strains such as *Pseudomonas, Burkholderia, Comamonas, Cupriavidus, Sphingomonas, Acidovorax, Rhodococcus,* and *Bacillus* able to use biphenyl as a sole source of carbon and energy have been isolated and their capability to transform PCB congeners has been evaluated. *Burkolderia xenovorans* LB400 is able to degrade a broad range of PCBs (Seeger et al., 2001) and is a model bacterium for PCB degradation. The degradation of biphenyl and transformation of PCBs is usually catalyzed by enzymes encoded by the biphenyl upper (bph) pathway.

16.3.2.5.2.2 The Biphenyl Upper (bph) Pathway A major interest for analyzing aerobic biphenyl degrading bacteria is because of their capability to transform PCB congeners. Based on the analysis of various biphenyl degrading isolates, it could be deduced that, in general, lower chlorinated congeners are more easily transformed compared to higher chlorinated congeners and that PCB congeners with chlorines only on one aromatic ring are more easily degraded, when compared with those bearing chlorine substituents on both aromatic rings.

The first step of the PCB catabolic pathway is the formation of cis-2,3-dihydro-2,3-dihydroxybiphenyl. This reaction is catalyzed by biphenyl 2,3 dioxygenase (BphA). Biphenyl 2,3 dioxygenase is a Rieske nonheme iron oxygenase, multicomponent enzyme complex which catalyzes the incorporation of two oxygen atoms into the aromatic ring to form arene cis

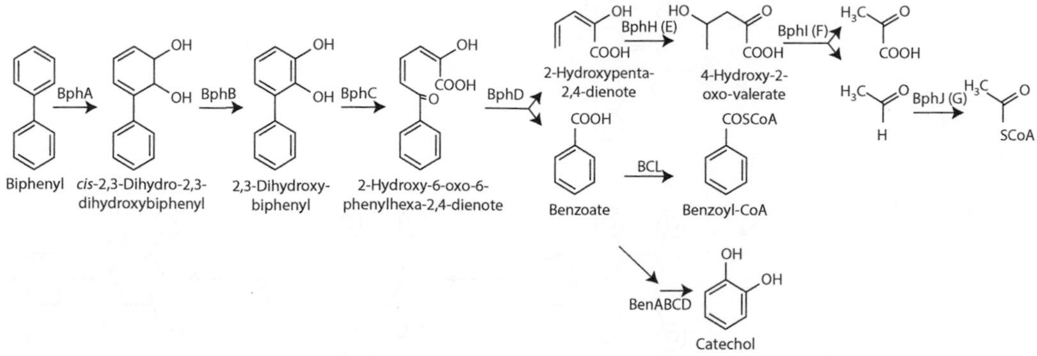

FIGURE 16.29
Pathway for biphenyl degradation BphA: Biphenyl 2,3-dioxygenase; BphB: cis-2,3-dihydro-2,3-dihydroxy-biphenyl dehydrogenase; BphC: 2,3-dihydroxybiphenyl 1,2-dioxygenase; BphD: 2-hydroxy-6-phenyl-6-oxo-hexa-2,4-dieneoate (HOPDA) hydrolase; BphH(E): 2-hydroxypenta-2,4-dienoate hydratase; BphI: acylating acetaldehyde dehydrogenase; BphJ: 4-hydroxy-2-oxovalerate aldolase. BCL: benzoate-CoA ligase; BenABCD: benzoate 1,2-dioxygenase and benzoate dihydrodiol dehydrogenase. (Adapted from Pieper DH and Seeger M. 2008. *Journal of Molecular Microbiology and Biotechnology* 15: 121–138.)

-diols (Gibson and Parales, 2000) where a ferredoxin and a ferredoxin reductase act as an electron transport system to transfer electrons from NADH to the terminal oxygenase. The biphenyl 2,3-dioxygenases are of crucial importance for the successful metabolism of PCBs (Figure 16.29). The next step of PCB catabolism is the formation of 2,3 dihydroxybiphenyl which is catalyzed by a cis-2,3 dihydro-2,3-dihydroxybiphenyl dehydrogenase (BphB) and regenerate NADH. *cis* -Dihydrodiol dehydrogenases are involved in various aromatic degradation pathways and exhibit common features such as an absolute requirement for NAD+. 2,3 Dihydroxybiphenyl is attacked by meta-clevaing 2,3-Dihydroxybiphenyl 1,2-dioxygenases (BphC), and 2-hydroxy-6-oxo-phenylhexa-2,2-dienoic acid is formed.

The fourth step of the bph pathway is catalyzed by 2-hydroxy-6-phenyl-6-oxohexa-2,4-dieneoate (HOPDA) hydrolase BphD, which hydrolyzes HOPDA to yield 2-hydroxy-penta-2,4-dienoate and benzoate. The last step in PCB catabolism, leading to the formation of chlorinated benzoic acid, is catalyzed by hydrolase. The complete degradation pathway of PCB is shown in Figure 16.29.

16.3.2.5.2.3 Anaerobic Degradation The first report on bacterial metabolism of PCBs under anaerobic conditions was in sediments of the Hudson River (Brown et al., 1987). Numerous studies have since reported on the microbial dechlorination of PCBs *in situ* and in laboratory experiments with sediment slurries. PCBs undergo reductive dechlorination under anaerobic conditions, leading to the formation of less-chlorinated congeners. Anaerobic reductive dechlorination is an important step toward the bioremediation of PCBs since it reduces the chlorine content of PCBs mixtures thereby reducing their bioaccumulation potential and, in some respects, toxicity. Dechlorination of PCBs in anaerobic environments is generally considered to be slow, with major change occurring over a period of months or years. Anaerobic reductive dechlorination of PCBs involves replacement of a chlorine atom with a hydrogen atom. The reduction potential of PCBs increases with increasing chlorine numbers. However, the positions of the chlorine substituents also influence the ease of reductive dechlorination. Generally, chlorines in the meta and para position are most readily removed. The reductive dehalogenation of highly and moderately chlorinated PCBs by anaerobic microorganisms generally involves selective dechlorination from

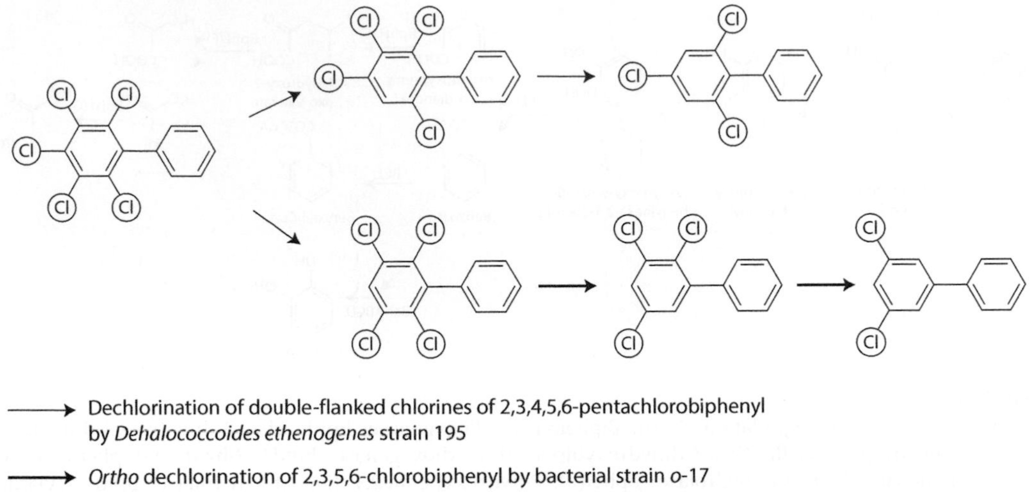

———▶ Dechlorination of double-flanked chlorines of 2,3,4,5,6-pentachlorobiphenyl
        by *Dehalococcoides ethenogenes* strain 195

———▶ *Ortho* dechlorination of 2,3,5,6-chlorobiphenyl by bacterial strain *o*-17

FIGURE 16.30
Anaerobic reductive dehalogenation of PCBs by *Dehalococcoides* and related bacteria and by an enrichment culture. (Adapted from Pieper DH and Seeger M. 2008. *Journal of Molecular Microbiology and Biotechnology* 15: 121–138.)

ortho and meta positions as shown in Figure 16.30. *Dehalococcoides ethenogenes* 195 was the first *Dehalococcoides* isolated based on its capability to dehalogenate tetrachloroethene to ethene (Maymo-Gatell et al., 1999). Several anaerobic bacteria, such as the *Dehalococcoides* (Maymo-Gatell et al., 1999), *Desulfitobacterium* (Sanford et al., 1996), *Dehalobacter* (Holliger et al., 1998), *Desulfomonile* (de Weerd and Suflita, 1990), *Desulfuromonas* (Krumholz, 1997), and *Sulfospririllum* (Boyle et al., 1999), have been identified as being able to reductively dehalogenate chlorinated phenols, benzoates, and trichloroethene and to couple this reaction to the synthesis of ATP via a chemiosmotic mechanism. Cutter et al. (2001) found that the bacterial strain *o*-17 catalyzes the reductive dechlorination of a 2,3,5,6-tetrachlorobiphenyl. The organism, bacterium *o*-17, has high sequence similarity with the green nonsulfur bacteria and with a group that includes *Dehalococcoides ethenogenes*.

16.3.2.6 Enzymatic Mechanism for Detoxification of Organohalides

Microbial growth on halogenated substrates requires the production of catabolic enzymes that cleave the carbon–halogen bond. Such enzymes are commonly called dehalogenases. Various metabolic pathways for organohalides involving dehalogenases, which catalyze cleavage of the carbon–halogen bond, have been discovered. The substrate range of dehalogenases determines to a large extent the range of synthetic halogenated organic compounds that can be used as a growth substrate by microbial cultures. This critical role of dehalogenases has made them an important target for microbiological research on the bacterial degradation of xenobiotic compounds. Here, we focus on bacterial dehalogenases. These enzymes make use of a variety of distinctly different catalytic mechanisms to cleave the carbon–halogen bonds. In general, different types of bacterial dehalogenases are capable of cleavage of halogen substituents from haloalkanes, haloalcoholes, haloaromatics and haloalkanoic acids as shown in Table 16.7.

Concerning the enzymatic cleavage of the carbon–halogen bond, seven mechanisms of dehalogenation are known so far: (i) Reductive dehalogenation (Figure 16.31a). In the course of a reductive dehalogenation, the halogen substituent is replaced by hydrogen. (ii)

TABLE 16.7

Bacterial Dehalogenases and Their Role in Detoxification and Degradation of Organohalides

Enzyme	Reaction
Haloalkane dehalogenases	The haloalkane dehalogenases catalyze the irreversible hydrolysis of a variety of haloalkanes to the corresponding alcohol, halide, and a hydrogen ion
Haloacid dehalogenases	Haloacid dehalogenases catalyze the hydrolysis of halogenated carboxylic acids, such as 2-chloroacetate, which is an intermediate in the degradation of 1,2- dichloroethane
Haloalcohol dehalogenases	These enzymes are able to displace a halogen from a vicinal haloalcohol substrate
4-Chlorobenzoyl-CoA dehalogenases	These enzymes catalyze the formation of 4-hydroxybenzoylCoA and chloride from 4-chlorobenzoyl-CoA
3-Chloroacrylic acid dehalogenases	The chloroacrylic acid dehalogenases displace a halogen from an sp2-hybridized carbon atom

Oxygenolytic dehalogenation (Figure 16.31b). Oxygenolytic dehalogenation reactions are catalyzed by monooxygenases (or dioxygenases), which incorporate one (or two) atoms of molecular oxygen into the substrate. (iii) Hydrolytic dehalogenation (Figure 16.31c). In the course of hydrolytic dehalogenation reactions, catalyzed by halidohydrolases, the halogen substituent is replaced in a nucleophilic substitution reaction by a hydroxy group which is derived from water. (iv) "Thiolytic" dehalogenation (Figure 16.31d). In dichloromethane-utilizing bacteria, a dehalogenating glutathione S-transferase catalyzes the formation of a S-chloromethyl glutathione conjugate, with a concomitant dechlorination taking place. (v) Intramolecular substitution (Figure 16.31e). Intramolecular nucleophilic

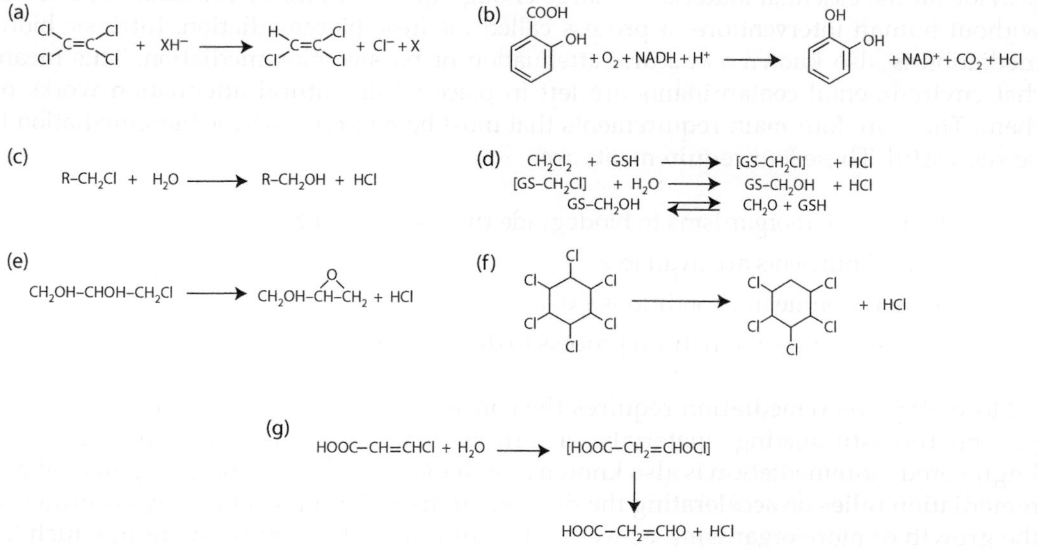

FIGURE 16.31

Dehalogenation mechanisms. (a) Reductive dehalogenation; (b) oxygenolytic dehalogenation; (c) hydrolytic dehalogenation; (d) "thiolytic" dehalogenation; (e) intramolecular substitution; (f) dehydrohalogenation; (g) hydration.

displacement yielding epoxides is a mechanism involved in the dehalogenation of vicinal haloalcohols. (vi) Dehydrohalogenation (Figure 16.31f). In dehydrohalogenation, HCl is eliminated from the molecule, leading to the formation of a double bond. (vii) Hydration (Figure 16.31g). A hydratase-catalyzed addition of a water molecule to an unsaturated bond can yield dehalogenation of vinylic compounds, such as 3-chloroacrylic acid, by chemical decomposition of an unstable intermediate. The carbon–halogen bond between a halogen and an arenic or vinylic carbon atom is more difficult to cleave than the one between a halogen and sp^3-hybridized carbon atom. Therefore, most haloaromatics are dehalogenated after ring cleavage.

Five dehalogenases that are members of different protein superfamilies have been studied by X-ray crystallography. This has provided detailed insight into their catalytic mechanisms and evolutionary relationships.

16.4 Other Processes of Transformation of Pollutants in Environment

16.4.1 Bioremediation

Bioremediation is a cost-effective, sustainable, natural approach to cleaning up contaminated soil and groundwater through the use of biological agents such as bacteria, fungi, and other organisms or their enzymes (Strong and Burgess, 2008). The enzymes and the microorganisms themselves have also been much used for the bioremediation processes. The bioremediation process involves biotransformation and biodegradation by transforming contaminants to nonhazardous or less hazardous chemicals through enzymatic action. Enzymatic bioremediation is potentially a rapid method of removing contaminants from soil and water. However, in a few cases the natural conditions at the contaminated site provide all the essential materials in large enough quantities for bioremediation to occur without human intervention—a process called intrinsic bioremediation. Intrinsic bioremediation is also known as natural attenuation or passive bioremediation. This means that environmental contaminants are left in place while natural attenuation works on them. There are four main requirements that must be met for intrinsic bioremediation to be successful. These four requirements are

1. Sufficient microorganisms to biodegrade the contaminant
2. Required nutrients are available
3. Good environmental conditions exist
4. The time to allow the natural process to degrade the contaminant

More often, bioremediation requires the construction of engineered systems to supply microbe-stimulating materials—a process called engineered bioremediation. Engineered bioremediation is also known as enhanced bioremediation. Engineered bioremediation relies on accelerating the desired biodegradation reactions by encouraging the growth of more organisms, as well as by optimizing the environment in which the organisms must carry out the detoxification reactions. Generally, it is used when any one of the four necessary conditions for intrinsic bioremediation is not available or when the process needs to be completed faster. On the basis of removal and transportation of

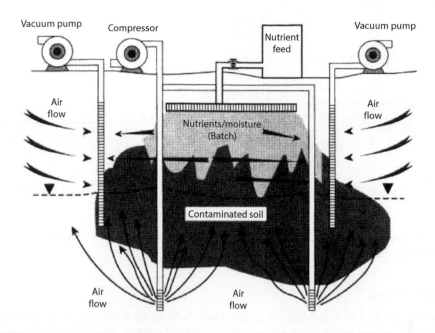

FIGURE 16.32
In situ bioremediation of contaminated soil. Bioremediation process enhancing in soil with injection of oxygen, moisture, and nutrients, for effective treatment.

wastes for treatment there are basically two methods: (i) *in situ* bioremediation and (ii) *ex situ* bioremediation.

i. *In situ* bioremediation:

In situ ("in place") bioremediation refers to the clean up of contaminated soils and groundwater without removing contaminated media from the subsurface, typically through the use of physical and/or chemical processes. *In situ* bioremediation of contaminated soil is shown in Figure 16.32.

ii. *Ex situ* bioremediation:

Ex situ bioremediation requires excavation of contaminated soil or pumping of groundwater to facilitate microbial degradation. This technique has more disadvantages than advantages. Several treatment strategies are involved in the bioremediation processes as follows.

16.4.2 Biosparging

Biosparging is an *in situ* remediation technology that involves the injection of air under pressure below the water table to increase groundwater oxygen concentrations and to enhance the rate of the biological degradation of contaminants by naturally occurring bacteria. It increases the mixing in the saturated zone and thereby increases the contact between soil and groundwater. Biosparging can be used to reduce concentrations of petroleum constituents that are dissolved in groundwater, adsorbed to soil below the water table, and within the capillary fringe. However, biosparging can also treat constituents

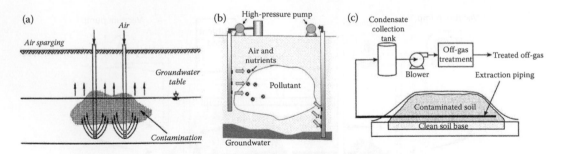

FIGURE 16.33
Treatment strategies involve in bioremediation processes: biosparging (a), bioventing (b), biopiling (c).

adsorbed to soils in the unsaturated zone. The schematic view of biosparging is shown in Figure 16.33a.

16.4.3 Bioventing

Bioventing is an *in situ* remediation technology deliver air from the atmosphere into the soil above the water table through injection wells placed in the ground where the contamination exists. It employs low air flow rates and provides only the amount of oxygen necessary for biodegradation while minimizing volatilization and release of contaminants to the atmosphere. Bioventing primarily assists in the degradation of adsorbed fuel residuals, but also assists in the degradation of volatile organic compounds (VOCs) as vapors move slowly through biologically active soil. The rate of natural degradation is generally limited by the lack of oxygen and other electron acceptors (i.e., a compound that gains electrons during biodegradation) rather than by the lack of nutrients (i.e., electron donors). In conventional bioventing systems, oxygen is delivered by an electric blower to subsurface wells. The schematic view of biventing is shown in Figure 16.33b.

16.4.4 Biopiling

Biopiling remediation is an *ex situ* full-scale remediation technology, where contaminated soil is constructed into engineered piles or cells with the aim of enhancing conditions required for biodegradation through greater control of oxygen, with nutrients such as phosphorus and nitrogen and water. The microbial activity of these microorganisms degrades hydrocarbon contaminants such as diesel that are adsorbed to the soil particles. The contaminants are reduced into carbon dioxide and water. A typical biopile sytem is shown in Figure 16.33c.

16.4.5 Biostimulation

Biostimulation is the addition of electron acceptors, electron donors, or nutrients to a polluted site in order to encourage the growth of naturally occurring chemical-degrading microorganisms. Comprehensively, biostimulation could be perceived as including the introduction of adequate amounts of water, nutrients, and oxygen into the soil, in order to enhance the activity of indigenous microbial degraders or to promote cometabolism.

16.4.6 Bioaugmentation

Bioaugmentation is the applications of indigenous or allochthonous wide type or genetically modified microorganisms to polluted hazardous waste sites in order to accelerate the removal of undesired compounds. It is thought that the bioaugmentation approach should be applied when the biostimulation and bioattenuation have failed.

16.5 Challenges

For more than 100 years, a large variety of OPs and organohalides have found widespread and massive application in industry, agriculture, and private households, for example, solvents, refrigerants, insecticides, degreasing agents, pesticides, pharmaceuticals, plasticizer polymers, and medicines. The degradation of OPs and organohalides has attracted considerable attention because of the accidental and deliberate release of large quantities of these chemicals into the environment, and their high mammalian toxicity.

A large number of microorganisms (i.e., bacteria and fungi) that can degrade these OPs and organohalides by cometabolism or mineralization have been isolated and characterized. These microorganisms utilized OPs and organohalides as the sole carbon, nitrogen and phosphorous source. Some microorganisms can degrade a large number of OPs and organohalides while some can degrade only one or few structurally similar OPs and organohalides. Because hydrolysis of OPs and organohalides reduces mammalian toxicity by several orders of magnitude, the environmental fate of degradation products has not received much attention from the scientific community. Complete degradation pathways for parathion, chlorophyrifos, lindane, DDT, PCP, and PCB are known, but the pathways for several other OPs and organohalides are not yet fully known. Our knowledge about the fate of OPs and organohalides in the environment is still incomplete. This area of research needs concerted efforts as degradation products of several compounds are pollutants and may have deleterious effects on the environment and nontarget organisms. In addition bioremediation technology uses microorganisms and their enzymes to reduce, eliminate, or transform such OPs and organohalides present in soils, sediments, water, or air to benign products. There is a wide diversity of highly toxic and persistent OPs and organohalides in the environment and the implementation of bioremediation techniques is needed to remove them or render them harmless. There has been considerable interest in the use of OPs and organohalides degrading enzymes prophylactically and therapeutically for poisoning by OPs and organohalides. Future areas of research include increasing enzyme activity against poor substrates and improving enzyme catalytic activities in mixtures of chemicals. The biodegradation and biotransformation of several persistent OPs and organohalides have recently been a major area of research because only a few isolated microorganisms have the capacity to degrade persistent OPs and organohalides. Another challenge for the scientific community is the scaling-up of laboratory access to the field because of differential behavior of isolated microorganisms in the environment. The application of genetic engineering and biochemical techniques to improve and evolve natural biodegradative capabilities will ultimately create strains capable of degrading complex mixtures of compounds. The use of modern technologies, such as metagenomics and nanotechnology, in association with more conventional biochemical and molecular analyses, can help to achieve this goal and will further our understanding of the biology that underpins the degradation of OPs and organohalides in relevant environmental contexts.

References

Aoki Y. 2001. Polychlorinated biphenyls, polychlorinated dibenzo-p-dioxins, and polychlorinated dibenzofurans as endocrine disrupters—what we have learned from Yusho disease. *Environmental Research* 86: 2–11.

Agency for Toxic Substances and Disease Registry (ATSDR). 2005. Toxicological profile for hexachlorocyclohexanes update [online].

Agency for Toxic Substances and Diseases Registry (ATSDR)/US Public health Service, Toxicological Profile for 4,4'-DDT, 4,4'-DDE, 4, 4'-DDD (Update). 1994. ATSDR. Atlanta, GA.

Ahmed M and Focht DD. 1973a. Degradation of polychlorinated biphenyls by two species of *Achromobacter*. *Canadian Journal of Microbiology* 19: 47–52.

Ahmed M and Focht DD. 1973b. Oxidation of polychlorinated biphenyls by *Achromobacter* pCB. *Bulletin of Environmental Contamination and Toxicology* 10: 70–72.

Apajalahti JHA and Salkinoja-Salonen MS. 1987. Complete dechlorination of tetrachlorohydroquinone by cell-extracts of pentachlorophenol-induced *Rhodococcus chlorophenolicus*. *Journal of Bacteriology* 169: 5125–5130.

Apajalahti JHA and Salkinoje-Salonen MS. 1986. Degradation of polychlorinated phenols by *Rhodococcus chlorophenolicus*. *Applied Microbiology and Biotechnology* 25: 62–67.

Ansaruddin PA and Vijayalakshmi K (Eds.). 2003. The womb is not safe anymore. *Indegenous Agriculture News* 2(1): 3.

Arora PK and Bae H. 2014. Bacterial degradation of chlorophenols and their derivatives. *Microbial Cell Factories* 13: 31.

Anwar S, Liaquat F, Khan QM, Khalid ZM, and Iqbal S. 2009. Biodegradation of chlorpyrifos and its hydrolysis product 3,5,6-trichloro-2-pyridinol by *Bacillus pumilus* strain C2A1. *Journal of Hazardous Materials* 168: 400–405.

Ballschmiter K. 2003. Pattern and sources of naturally produced organohalogens in the marine environment: Biogenic formation of organohalogens. *Chemosphere* 52: 313–324.

Benning MM, Sims H, Raushel FM, and Holden HM. 2001. High resolution X-ray structures of different metal-substituted forms of phosphotriesterase from *Pseudomonas diminuta*. *Biochemistry* 40: 2712–2722.

Bigley NA and Raushel FM. 2013. Catalytic mechanisms for phosphotriesterases. *Biochimica et Biophysica Acta* 1834(1): 443–453.

Bolognesi C. 2003. Genotoxicity of pesticides: A review of human biomonitoring studies. *Mutation Research* 543 (3): 251–272.

Boyle AW, Phelps CD, and Young LY. 1999. Isolation from estuarine sediments of a *Desulfovibrio* strain which can grow on lactate coupled to the reductive dehalogenation of 2,4,6-tribromophenol. *Applied and Environmental Microbiology* 65: 1133–1140.

Brown JF, Bedard DL, Brennan MJ, Carnahan JC, Feng H, and Wagner RE. 1987. Polychlorinated biphenyl dechlorination in aquatic sediments. *Science* 236: 709–712.

Bumpus JA and Aust SD. 1987. Biodegradation of DDT [1,1,1-trichloro-2,2-bis(4-chlorophenyl)ethane) by the white rot fungus *Phanerochaete chrysosporium*. *Applied and Environmental Microbiology* 53: 2001–2008.

Cai M and Xun L. 2002. Organization and regulation of pentachlorophenol-degrading genes in *Sphingobium chlorophenolicum* ATCC 39723. *Journal of Bacteriology* 184: 4672–4680.

Ceremonie H, Boubakri H, Mavingui P, Simonet P, and Vogel TM 2006. Plasmid-encoded gamma-hexachlorocyclohexane degradation genes and insertion sequences in *Sphingobium francense* (ex-*Sphingomonas paucimobilis* Sp+). *FEMS Microbiology Letters* 257: 243–252.

Catelani D, Sorlini C, and Treccani V. 1971. The metabolism of biphenyl by *Pseudomonas putida*. *Experientia* 27: 1173–1174.

Chaudhry GR and Chapalamadugu S. 1991. Biodegradation of halogenated organic compounds. *Microbiological Reviews* 55(1): 59–79.

Cheng TC, Harvey SP, and Stroup AN. 1993. Purification and properties of a highly active organo-phosphorus acid anhydrolase from *Alteromonas undina*. *Applied and Environmental Microbiology* 59: 3138–3140.

Chanama S and Crawford RL. 1997. Nutritional analysis of pcpA and its role in pentachlorophe-nol degradation by *Sphingomonas* (*Flavobacterium*) *chlorophenolica* ATCC 39723. *Applied and Environmental Microbiology* 63: 4833–4838.

Chandra R, Raj A, Purohit HJ, and Kapley A. 2007. Characterization and optimization of three poten-tial aerobic bacterial strains for kraft lignin degradation from pulp paper waste. *Chemosphere* 67: 839–846.

Cutter LA, Watts JEM, Sowers KR, and May HD. 2001. Identification of a microorganism that links its growth to the reductive dechlorination of 2,3,5,6-chlorobiphenyl. *Environmental Microbiology* 3: 699–709.

Dams RI, Paton GI, and Killham K. 2007. Rhizomediation of pentachlorophenol by *Sphingobium chlo-rophenolicum* ATCC 39723. *Chemosphere* 68: 864–870.

de Weerd KA and Suflita JM. 1990. Anaerobic aryl reductive dehalogenation of halobenzoates by cell extracts of *Desulfomonile tiedjei*. *Applied and Environmental Microbiology* 56: 2999–3005.

Defrank JJ and Cheng T. 1991. Purification and properties of an organophosphorus acid anhydrase from a halophilic bacterial isolate. *Journal of Bacteriology* 173(6): 1938–1943.

Dumas DP, Caldwell SR, Wild JR, and Raushel FM. 1989. Purification and properties of the phosphot-riesterase from *Pseudomonas diminuta*. *Journal of Biological Chemistry* 264: 19,659–19,665.

Drinker CK, Warren MF, and Bennet GA. 1937. The problem of possible systemic effects from certain chlorinated hydrocarbons. *Journal of Industrial Hygiene and Toxicology* 19: 283–311.

Dogra C, Raina V, Pal R, Suar M, Lal S, Gartemann KH, Holliger C et al. 2004. Organization of lin genes and IS6100 among different strains of hexachlorocyclohexane degrading *Sphingomonas paucimobilis*: Evidence for horizontal gene transfer. *Journal of Bacteriology* 186: 2225–2235.

Efrmenko EN and Sergeeva VS. 2001. Organophosphate hydrolase—An enzyme catalyzing degrada-tion of phosphorus-containing toxins and pesticides. *Russian Chemical Bulletin* (International Edition) 50: 1826–1832.

Eleršek T and Filipič M. 2011. Organophosphorus pesticides—Mechanisms of their toxicity. In: Stoytcheva M (Ed.), *Pesticides—The Impacts of Pesticides Exposure*. InTech, China, pp. 243–260.

EPA. 1999. Integrated risk information system (IRSI) on pentachlorophenol. In: *Environmental Protection Agency, National Centre for Environmental Assessment*. Office of Research and Development, Washington, DC, USA.

Field JA and Sierra-Alvarez R. 2008. Microbial degradation of chlorinated phenols. *Reviews in Environmental Science and Biotechnology* 7: 211–241.

Faroon O, Keith L, Smith-Simon C, and De Rosa C. 2003. *Polychlorinated Biphenyls. Human Health Asp ects. Concise International Chemical Assessment Document* 55. World Health Organization, Geneva, 2003.

Gerlt, JA and Raushel FM. 2003. Evolution of function in (b/a)8-barrel enzymes. *Current Opinion in Chemical Biology* 7: 252–264.

Gibson DT and Parales RE. 2000. Aromatic hydrocarbon dioxygenases in environmental biotechnol-ogy. *Current Opinion in Biotechnology* 11: 236–243.

Gupta RC. 2006. Classification and uses of organophosphates and carbamates. In: Gupta RC (Ed.), *Toxicology of Organophosphate and Carbamate Compounds*, Elsevier, Academic Press, USA, pp. 5–17.

Garaj-Vrhovac V and Zeljezic D. 2000. Evaluation of DNA damage in workers occupationally exposed to pesticides using single-cell gel electrophoresis (SCGE) assay: Pesticide genotoxicity revealed by comet assay. *Mutation Research* 469: 279–285.

Hanson DJ. 2005. Five chemicals pass hurdle for control under POP treaty. *Chemical and Engineering News* 83(48): 23.

Häggblom M. 1998. Reductive dechlorination of halogenated phenols by a sulphate-reducing consortium. *FEMS Microbiology and Ecology* 26: 35–41.

Haider K. 1979. Degradation and metabolization of lindane and other hexachlorocyclohexane isomers by anaerobic and aerobic soil microorganisms. *Zeitschrift für Naturforschung Teil* 34: 1066–1069.

Hardie DWF. 1964. Benzene hexachloride. In: Kirk RE and Othmer DF (Eds.), *Encyclopedia of Chemical Technology.* John Wiley and Sons, New York, pp. 267–281.

Harper LL, Mcdaniel CS, Miller CE, and Wild JR. 1988. Dissimilar plasmids isolated from *Pseudomonas diminuta* MG and a *Flavobacterium* sp. (ATCC27551) contains identical opd genes. *Applied and Environmental Microbiology* 54: 2586–2589.

Heritage AD and Mac Rae IC. 1977. Identification of intermediates formed during the degradation of hexachlorocyclohexanes by *Clostridium sphenoides*. *Applied and Environmental Microbiology* 33: 1295–1297.

Hodgson E and Rose RL. 2006. Organophosphorus chemicals: Potent inhibitors of the human metabolism of steroid hormones and xenobiotics. *Drug Metabolism Reviews* 38 (1–2): 149–162.

Holliger C, Hahn D, Harmsen H, Ludwig W, Schumacher W, Tindall B, Vazquez F et al. 1998. *Dehalobacter restrictus* gen. nov. and sp. nov., a strictly anaerobic bacterium that reductively dechlorinates tetra- and trichloroethene in an anaerobic respiration. *Archives of Microbiology* 169: 313–321.

Hreljac I, Zajc I, Lah T, and Filipic M. 2008. Effects of model organophosphorous pesticides on DNA damage and proliferation of HepG2 cells. *Environmental and Molecular Mutagenesis* 49(5): 360–367.

Huang Y, Zhao X, and Luan SJ. 2007. Uptake and biodegradation of DDT by 4 ectomycorrhizal fungi. *Science of the Total Environment* 385: 235–241.

Humphreys E, Janssen S, Hell A, Hiatt P, Solomon G, and Miller MD. 2008. Outcome of the California ban on pharmaceutical lindane: Clinical and ecological impacts. *Environmental Health Perspectives* 116: 297–302.

Isoda H, Talorete TP, Han J, Oka S, Abe Y, and Inamori Y. 2005. Effects of organophosphorous pesticides used in china on various mammalian cells. *Environmental Science* 12: 9–19.

IARC. 1987. Evaluation of the carcinogenic risk of chemicals to humans: Overall evaluations of carcinogenicity. IARC Monograph Supplment 7, International Agency for Research on Cancer, Lyon, p. 392.

Janssen DB, Dinkla Inez JT, Poelarends GJ, and Terpstra P. 2005. Bacterial degradation of xenobiotic compounds: Evolution and distribution of novel enzyme activities. *Environmental Microbiology* 7(12): 1868–1882.

Kang, DG, Kim CS, Seo JH, Kim IG, Choi SS, Ha JH, Nam SW et al. 2012. Coexpression of molecular chaperone enhances activity and export of organophosphorus hydrolase in *Escherichia coli*. *Biotechnology Progress* 28(4): 925–930.

Kao CM, Liu JK, Chen YL, Chai CT, and Chen SC. 2005. Factors affecting the biodegradation of PCP by *Pseudomonas Mendocina* NSYSU. *Journal of Hazardous Materials* 124: 68–73.

Karn SK and Geetanjali. 2014. Pentachlorophenol remediation by *Enterobacter* sp. SG1 isolated from industrial dump site. *Pakistan Journal of Biological Sciences* 17: 388–394.

Karn SKR, Chakrabarti SK, and Reddy MS. 2010. Characterization of pentachlorophenol degrading *Bacillus* strains from secondary pulp-and-paper-industry sludge. *International Biodeterioration and Biodegradation* 64:609–613.

Karn SKR, Chakrabarti SK, and Reddy MS. 2011. Degradation of pentachlorophenol by *Kocuria* sp. CL2 isolated from secondary sludge of pulp and paper mill. *Biodegradation* 22: 63–69.

Krumholz LR. 1997. *Desulfuromonas chloroethenica* sp. nov. uses tetrachloroethylene and trichloroeth-ylene as electron acceptors. *International Journal of Systemic Bacteriology* 47: 1262–1263.

Kumari R, Subudhi S, Suar M, Dhingra G, Raina V, Dogra C, Lal S et al. 2002. Cloning and charac-terization of lin genes responsible for the degradation of hexachlorocyclohexane isomers by *Sphingomonas paucimobilis* strain B90. *Applied and Environmental Microbiology* 68: 6021–6028.

Krumholz LR. 1997. *Desulfuromonas chloroethenica* sp. nov. uses tetrachloroethylene and trichloroeth-ylene as electron acceptors. *International Journal of Systemic Bacteriology* 47: 1262–1263.

Lunt D and Evans WC. 1970. The microbial metabolism of biphenyl. *Biochemistry Journal* 118: 54–55.

Lockridge O and Masson P. 2000. Pesticides and susceptible populations: People with butyrylcholin-estrase genetic variants may be at risk. *Neurotoxicology* 21: 113–126.

Li XH, He J, and Li SP. 2007. Isolation of a chlorpyrifos-degrading bacterium, Sphingomonas sp. strain Dsp-2, and cloning of the mpd gene. *Research in Microbiology* 158: 143–149. doi: 10.1016/j.resmic.2006.11.007.

Li JQ, Liu J, Shen WJ, Zhao XL, Hou Y, Cao H, and Cui Z. 2010. Isolation and characterization of 3,5,6-trichloro-2-pyridinol-degrading *Ralstonia* sp. strain T6. *Bioresource Technology* 101: 7479–7483.

Lovley DR. 2003. Cleaning up with genomics: Applying molecular biology to bioremediation. *Nature Reviews Microbiology* 1: 35–44.

Mathew G, Rahiman MA, and Vijayalaxmi KK. 1990. *In vivo* genotoxic effects in mice of Metacid 50, an organophosphorus insecticide. *Mutagenesis* 5: 147–149.

Mathew G, Vijayalaxmi KK, and Abdul-Rahiman M. 1992. Methyl parathion-induced sperm shape abnormalities in mouse. *Mutation Research* 280(3): 169–173.

Mitra J, Mukherjee PK, Kale SP, and Murthy NBK. 2001. Bioremediation of DDT in soil by genetically improved strains of soil fungus *Fusarium solani*. *Biodegradation* 12: 235–245.

Mulbry WW and Karns JS. 1989. Parathion hydrolase specified by the *Flavobacterium* opd gene: Relationship between the gene and protein. *Journal of Bacteriology* 171: 6740–6746.

Manickam N, Misra R, and Mayilraj S. 2007. A novel pathway for the biodegradation of gamma-hexachlorocyclohexane by a *Xanthomonas* sp. strain ICH12. *Journal of Applied Microbiology* 102: 1468–1478.

Manavathi B, Pakala SB, Gorla P, Merrick M, and Siddavattam D. 2005. Influence of zinc and cobalt on expression and activity of parathion hydrolase from *Flavobacterium* sp. ATCC27551. *Pesticide Biochemistry and Physiology* 83: 37–45.

Martinez-Toledo MV, Salmeron V, Rodelas B, Pozo C, and Gonzalez-Lopez J. 1993. Studies on the effects of a chlorinated hydrocarbon insecticide, lindane, on soil microorganisms. *Chemosphere* 27: 2261–2270.

Maymo-Gatell X, Anguish T, and Zinder SH. 1999. Reductive dechlorination of chlorinated eth-enes and 1,2-dichloroethane by "*Dehalococcoides ethenogenes*" 195. *Applied and Environmental Microbiology* 65: 3108–3113.

Mikesell MD and Boyd SA. 1986. Complete reductive dechlorination and mineralization of penta-chlorophenol by anaerobic microorganisms. *Applied and Environmental Microbiology* 52: 861–865.

Munnecke DM and Hsieh DPH. 1974. Microbial decontamination of parathion and p-nitrophenol in aqueous media. *Applied Microbiology* 28: 212–217.

Munnecke DM and Hsieh DPM. 1976. Pathways of microbial metabolism of parathion. *Applied and Environmental Microbiology* 31: 63–69.

Musa S, Gichuki JW, Raburu PO, and Aura CM. 2011. Risk assessment for organochlorines and organo-phosphates pesticide residues in water and sediments from lower Nyando/Sondu-Miriu river within Lake Victoria Basin, Kenya. *Lakes and Reservoirs: Research and Management* 16: 273–280.

Nadeau LJ, Menn FM, Breen A, and Sayler GS. 1994. Aerobic degradation of 1,1,1-trichloro-2,2-bis(4-chlorophenyl)ethane (DDT) by *Alcaligenes eutrophus* A5. *Applied and Enviromental Microbiology* 60: 51–55.

Nagata Y, Endo R, Ito M, Ohtsubo Y, and Tsuda M. 2007. Aerobic degradation of lindane (γ-hexachlorocyclohexane) in bacteria and its biochemical and molecular basis. *Applied Microbiology and Biotechnology* 76: 741–752.

Nagata Y, Miyauchi K, and Takagi M. 1999. Complete analysis of genes and enzymes for γ-hexachlorocyclohexane degradation in *Sphingomonas paucimobilis* UT26. *Journal of Industrial Microbiology and Biotechnology* 23: 380–390.

Ohisa N, Yamaguchi M, and Kurihara N. 1980. Lindane degradation by cell-free extracts of *Clostridium rectum*. *Archives of Microbiology* 25: 221–225.

Omburo GA, Kuo JM, Mullins LS, and Raushel FM. 1992. Characterization of zinc binding site of bacterial phosphotriestearase. *Journal of Biological Chemistry* 267: 13,278–13,283.

Phillips TM, Seech AG, Lee H, and Trevors JT. 2005. Biodegradation of hexachlorocyclohexane (HCH) by microorganisms. *Biodegradation* 16: 363–392.

Phillips JP, Xin JH, Kirby K, Milne JCP, Krell P, and Wild JR. 1990. Transfer and expression of an organophosphate insecticide-degrading gene from *Pseudomonas* in *Drosophila melanogaster*. *Proceedings of the National Academy of Sciences of the United States of America* 87: 8155–8159.

Pieper DH. 2005. Aerobic degradation of polychlorinated biphenyls. *Applied Microbiology and Biotechnology* 67: 170–191.

Pieper DH and Seeger M. 2008. Bacterial metabolism of polychlorinated biphenyls. *Journal of Molecular Microbiology and Biotechnology* 15: 121–138.

Quintero JC, Moreira MT, Feijoo G, and Lema JM. 2005. Anaerobic degradation of hexachlorocyclohexane isomers in liquid and soil slurry systems. *Chemosphere* 61: 528–536.

Quistad GB, Liang SN, Fisher KJ, Nomura DK, and Casida JE. 2006. Each lipase has aunique sensitivity profile for organophosphorus inhibitors. *Toxicological Sciences* 91(1): 166–172.

Raushel FM. 2002. Bacterial detoxification of organophosphate nerve agents. *Current Opinion in Microbiology* 5: 288–295.

Rubinos DA, Villasuso R, Muniategui S, Barral MT, and Diaz Fierros F. 2007. Using the landfarming technique to remediate soils contaminated with hexachlorocyclohexane isomers. *Water, Air, and Soil Pollution* 181: 385–399.

Sanford RA, Cole JR, Loffler FE, and Tiedje JN. 1996. Characterization of *Desulfitobacterium chlororespirans* sp nov, which grows by coupling the oxidation of lactate to the reductive dechlorination of 3-chloro-4-hydroxybenzoate. *Applied and Environmental Microbiology* 62: 3800–3808.

Sáez F, Pozo C, Gómez MA, Martínez-Toledo MV, Rodelas B, and Gónzalez-López J. 2006. Growth and denitrifying activity of *Xanthobacter autotrophicus* CECT 7064 in the presence of selected pesticides. *Applied Microbiology and Biotechnology* 71: 563–567.

Schenk T, Muller R, Morsberger F, Otto MK, and Lingens F. 1989. Enzymatic dehalogenation of pentachlorophenol by extracts from *Arthrobacter* sp. strain ATCC 33790. *Journal of Bacteriology* 171: 5487–5491.

Seeger M, Cámara B, and Hofer B. 2001. Dehalogenation, denitration, dehydroxylation, and angular attack on substituted biphenyls and related compounds by a biphenyl dioxygenase. *Journal of Bacteriology* 183: 3548–3555.

Sethunathan N and Yoshida T. 1973. *Flavobacterium* sp. that degrades diazinon and parathion. *Canadian Journal of Microbiology* 19: 873–875.

Shah S and Thakur IS. 2002. Enrichment and characterization of pentachlorophenol degrading microbial community for the treatment of tannery effluent. *Pollution Research* 20: 353–363.

Siddaramappa R, Rajaram KP, and Sethunathan NN. 1973. Degradation of parathion by bacteria isolated from flooded soil. *Applied Microbiology* 26: 846–849.

Singh DP, Khattar JIS, Nadda J, Singh Y, Garg A, Kaur N, and Gulati A. 2011. Chlorpyrifos degradation by the cyanobacterium *Synechocystis* sp. strain PUPCCC 64. *Environmental Science and Pollution Research* 18: 1351–1359.

Singh BK and Walker A. 2006. Microbial degradation of organophosphorus compounds. *FEMS Microbiological Reviews* 30: 428–471.

Singh BK, Walker A, Morgan JAW, and Wright DJ. 2003a. Role of soil pH in the development of enhanced biodegradation of fenamiphos. *Applied and Environmental Microbiology* 69: 7035–7043.

Singh BK, Walker A, Morgan JAW, and Wright DJ. 2003b. Effects of soil pH on the biodegradation of chlorpyrifos and isolation of a chlorpyrifos-degrading bacterium. *Applied and Environmental Microbiology* 69: 5198–5206.

Singh BK, Walker A, and Wright DJ. 2005. Cross enhancement of accelerated biodegradation of organophosphorus compounds in soils: Dependence on structural similarity of compounds. *Soil Biology and Biochemistry* 37: 1675–1682.

Singh S, Singh BB, Chandra R, Patel DK, and Rai V. 2009. Synergistic biodegradation of pentachlorophenol by *Bacillus cereus* (DQ002384), *Serratia marcescens* (AY927692) and *Serratia marcescens* (DQ002385). *World Journal of Microbiology and Biotechnology* 25(10): 1821–1828.

Singh BK, Walker A, Morgan JAW, and Wright DJ. 2004. Biodegradation of chlorpyrifos by Enterobacter strain B-14 and its use in the bioremediation of contaminated soils. *Applied and Environmental Microbiology* 70: 4855–4863.

Sogorb MA and Vilanova E. 2002. Enzymes involved in the detoxification of organophosphorus, carbamate and pyrethroid insecticides through hydrolysis. *Toxicology Letters* 128(1): 215–228.

Strong PJ and Burgess JE. 2008. Treatment methods for wine-related ad distillery wastewaters: A review. *Bioremediation Journal* 12: 70–87.

Takeuchi M, Hamana K, and Hiraishi A. 2001. Proposal of the genus *Sphingomonas sensu stricto* and three new genera, *Sphingobium*, *Novosphingobium* and *Sphingopyxis*, on the basis of phylogenetic and chemotaxonomic analyses. *International Journal of Systemic Evolution and Microbiology* 51: 1405–1417.

Thengodkar RRM and Sivakami S. 2010. Degradation of chlorpyrifos by an alkaline phosphatase from the cyanobacterium *Spirulina platensis*. *Biodegradation* 21: 637–644.

Uotila JS, Salkinoja-Salonen MS, and Apajalahti JHA. 1991. Dechlorination of pentachlorophenol by membrane boundenzymes of *Rhodococcus chlorophenolicus* PCP-I. *Biodegradation* 2: 25–31.

Uotila JS, Kitunen VH, Saastamoinen T, Coote T, Haggblom MM, and Salkinoja-Salonen MS. 1992. Characterization of aromatic dehalogenases of *Mycobacterium fortuitum* CG-2. *Journal of Bacteriology* 174: 5669–5675.

Ündeger U and Basaran N. 2005. Effects of pesticides on human peripheral lymphocytes in vitro: Induction of DNA damage. *Archives of Toxicology* 79: 169–176.

Vyas NK, Nickitenko A, Rastogi VK, Shah SS, and Quiocho FA. 2010. Structural insights into the dual activities of the nerve agent degrading organophosphate anhydrolase/prolidase. *Biochemistry* 49(3): 547–559.

Warren R, Hsiao WL, Kudo H, Myhre M, Dosanjh M et al. 2004. Functional characterization of a catabolic plasmid from polychlorinatedbiphenyl-degrading *Rhodococcus* sp. Strain RHA1. *Journal of Bacteriology* 186: 7783–7795.

WHO. 1990. *Public Health Impact of Pesticides Used in Agriculture*. WHO, Geneva.

WHO. 2001. *Organophosphorous Pesticides in the Environment-Integrated Risk Assessment*. WHO, Geneva.

Williams RT. 1971. *Detoxification Mechanisms*, 2nd edition. Wiley, New York.

Wong MH, Leung AOW, Chan JKY, and Choi MPK. 2005. A review on the usage of POP pesticides in China, with emphasis on DDT loadings in human milk. *Chemosphere* 60: 740–752.

Yang C, Liu N, Guo XM, and Qiao CL. 2006a. Cloning of *mpd* gene from a chlorpyrifos-degrading bacterium and use of this strain in bioremediation of contaminated soil. *FEMS Microbiology Letters* 265: 118–125.

Yang CF, Lee CM, and Wang CC. 2006b. Isolation and physiological characterization of the pentachlorophenol degrading bacterium *Sphingomonas chlorophenolica*. *Chemosphere* 62: 709–714.

Yang S, Shibata A, Yoshida N, and Katayama A. 2009. Anaerobic mineralization of pentachlorophenol (PCP) by combining PCP-dechlorinating and phenol-degrading cultures. *Biotechnology and Bioengineering* 102(1): 81–90.

Yim YJ, Seo J, Kang SI, Ahn JH, and Hur HG. 2008. Reductive dechlorination of methoxychlor and DDT by human intestinal bacterium *Eubacterium limosum* under anaerobic conditions. *Archives of Environmental Contamination and Toxicology* 54: 406–411.

You GR, Sayles GD, Kupferle MJ, Kim IS, and Bishop PL. 1996. Anaerobic DDT biotransformation: Enhancement by application of surfactants and low oxidation reduction potential. *Chemosphere* 32: 2269–2284.

Zhang Q, Wang BC, Cao ZY, and Yu Y.L. 2012. Plasmid-mediated bioaugmentation for the degradation of chlorpyrifos in soil. *Journal of Hazardous Materials* 221: 178–184.

Zheng Y, Liu D, Wang B, Zhang Q, Wan J, Li XL, and Li W. 2008. Bioinformatics analysis of methyl parathion hydrolase MPH and the structure prediction with homology modeling. In: *The 2nd International Conference on Bioinformatics and Biomedical Engineering,* Vol. 1, IEEE, pp. 13–16.

Xu GM, Zheng W, Li YY, Wang SH, Zhang JS, and Yan Y. 2008. Biodegradation of chlorpyrifos and 3,5,6-trichloro-2-pyridinol by a newly isolated *Paracoccus* sp. TRP. *International Biodeterioration and Biodegradation* 62: 51–56.

17

Petroleum Hydrocarbon Stress Management in Soil Using Microorganisms and Their Products

Rajesh Kumar, Amar Jyoti Das, and Shatrohan Lal

CONTENTS

17.1 Introduction

The majority of hydrocarbons found on earth occur naturally in crude oil, where decomposed organic matter provides an abundance of carbon and hydrogen which, when bonded, can catenate to form seemingly limitless chains. These hydrocarbons can be gases (e.g., methane and propane), liquids (e.g., hexane and benzene), waxes or low melting solids (e.g., paraffin wax and naphthalene), or polymers (e.g., polyethylene, polypropylene, and polystyrene). Hydrocarbons existing on the planet based on their nature and properties are classified as saturated, unsaturated, cycloalkanes, and aromatic hydrocarbons. Saturated hydrocarbons (alkanes) are the basis of petroleum fuels and are found as either linear or branched species. These saturated hydrocarbons or petroleum hydrocarbons have gained importance during the last century because of their use as a source of energy. The modern lifestyle and the world's economy today depend to a large extent on the availability of this "black gold" and its derivatives. Today the fuels derived from crude oil, supply more than half of the world's total supply of energy (OECD, 1998). Gasoline, kerosene, and diesel are used as fuel for cars, tractors, trucks, aircrafts, and ships. Hexane [6C] is a widely used nonpolar, nonaromatic solvent, as well as a significant fraction of common gasoline. The [6C] through [10C] alkanes, alkenes, and isomeric cycloalkanes are the top components of gasoline, naphtha, jet fuel, and specialized industrial solvent mixtures. With the progressive addition of carbon units, the simple nonring-structured hydrocarbons have higher viscosities, lubricating indices, boiling points, solidification temperatures, and deeper color. At the opposite extreme from [1C] methane lies the heavy tars that remain as the *lowest fraction* in a crude oil refining retort. They are collected and widely utilized as roofing compounds, pavement composition, wood preservatives (the creosote series), and as extremely high viscosity shear-resisting liquids. Heating oil and natural gas are used to heat homes and commercial buildings as well as to generate electricity. Besides this, crude oil products are the basic materials used in manufacturing of synthetic fibers for clothing and in plastics, paints, fertilizers, insecticides, soaps, and synthetic rubber (Speight, 1999). A number of typical chemical compounds present in petroleum products are listed in Table 17.1.

TABLE 17.1

Typical Chemical Compounds in Petroleum Products

Petroleum Products	Compound Classes
Gasoline	High concentration of BTEXs, monoaromatic and branched alkanes. Lower concentration of *n*-alkanes, alkenes, cycloalkanes, and naphthalenes.
Kerosene	High concentration of cycloalkanes and *n*-alkanes; lower concentration of monoaromatic and branched alkanes and very low concentration of BTEXs and PAHs.
Diesel	High concentration of *n*-alkanes; lower concentration of branched alkanes, cycloalkanes, monoaromatics, naphthalene, PAHs, and very low concentration of BTEXs.
Fuel oil	High concentration of *n*-alkanes, cycloalkanes, lower concentration of naphthalenes, PAHs, and very low concentration of BTEXs.
Lubricating and motor oil	High concentration of cycloalkanes, branched alkanes, and low concentration of barium, BTEXs, and PAHs.

Source: Adapted from Potter TL and Simmons KE. 1998. *Composition of Petroleum Mixtures*, vol. 2. Amherst Scientific Publishers, Amherst, MA.

This overdependence of the world on petroleum hydrocarbons to meet their energy and other requirement has resulted in more extraction and refinement of crude oils. However, there is another side to this usefulness, as crude oil prospecting is sometimes accompanied with spillage. These spillages are a source of environmental contamination.

Oil spills may occur for numerous reasons such as equipment failure, disasters, deliberate acts, or human error (Anderson and LaBelle, 2000). The places where crude oil is found are not always the places where it is refined and needed. Sabotage and accidents lead to release of these crude oils in the environment. In the refining process, the release of oil into refinery effluents as waste disposal, is although practically negligible and of a lower order of magnitude but it can affect the soil properties, microbial population and if the magnitude is large enough it can also seep into the groundwater. Discharge and wash waters from tankers and vessels are a kind of oil pollution that is fairly unnoticed, but is common. Besides this, oil spills can take place near coastlines or in the open sea. Air and ocean currents can also transport pollutants for thousands of kilometers; therefore, oil spills affect more than just isolated locations. In many spills involving tankers or offshore oil wells, some of the oil spilled initially catches fire. When crude oil burns, the combustion results in atmospheric emissions of gasses, which contribute to global warming (CO_2) and acid rain (SO_2, NO_x), as well as large quantities of toxic ash. The toxic ash is made up of microscopic particles, which can travel for hundreds of kilometers. Humans inhaling these particles may experience allergic reactions, which result in sore throats and breathing problems. The less dense (lighter) components of the spilled oil are more volatile and eventually evaporate into the atmosphere. The petroleum hydrocarbon then reacts with sunlight and oxygen to form greenhouse and acidic gases similar to those from the combustion of oil. The negative impacts of oil that burns or evaporates is more diffuse (spread out) than that of oil which ends up on shore but still causes appreciable damage to the natural environment. Kvenvolden et al. (2000) reported that the oil mixes with sediments on the ocean floor and turns into a thick tar-like mass, which can destroy the habitat of many bottom-dwelling organisms. These tar-like clumps can also drift with tides and currents eventually washing up on beaches far away from the spill. If a spill occurs near a coastline, beached oil can leak into fresh groundwater reservoirs that often extend under beaches, contaminating local wells.

Although the pollution caused by oil spills in the air and oceans cannot be easily controlled using biological means, the soil contaminated by these can be remediated using different techniques. Still, this is a new field which has attracted the attention of environmentalists and policy makers throughout the world as environmental petroleum contamination and its constituents is an inevitable problem that affects many geographical regions to a variable extent depending on the local environmental law (Graj et al., 2013).

Soil contamination with petroleum and its products is a serious worldwide concern (Banks et al., 2003; Chaineau et al., 2003; Hentati et al., 2013). Petroleum-contaminated soil pollutes local ground water, renders potable water unsafe, limits ground water use, and causes ecological toxicity (Wang et al., 2008). The toxicity of petroleum hydrocarbon on soil organisms has been widely studied, but research regarding its ecotoxicity is still assaying behind (Cermak et al., 2010; Tang et al., 2011). Petroleum is a natural product resulting from the anaerobic conversion of biomass under high temperature and pressure. Although most of its components are subject to biodegradation, this occurs at relatively slow rates. Moreover, petroleum hydrocarbons are poorly degraded and have thus become the most widespread environmental contaminant (Margesin et al., 2000). The four classes of petroleum hydrocarbons (saturates, aromatics, asphaltenes, and resins; Sanchez et al., 2006) differ in their susceptibility to microbial attack (bacteria and fungi). Bacteria and fungi generally work by breaking down petroleum hydrocarbons into less harmful substances (Medina-Bellver et al., 2005). Petroleum hydrocarbons bioremediation depends on the presence of microorganisms with the appropriate metabolic capabilities. Most of the bacteria frequently isolated from hydrocarbon-polluted sites belong to the genera *Pseudomonas, Sphingomonas, Acinetobacter, Alcaligenes, Micrococcus, Bacillus, Flavobacterium, Arthrobacter, Alcanivorax Mycobacterium, Rhodococcus,* and *Actinobacter* (Atlas, 1992; Okoh and Trejo-Hernandez, 2006; Obayori et al., 2009). Optimal rates of hydrocarbon biodegradation by microorganism can be sustained by the adequate concentration of nutrients, oxygen, and pH values (Perry, 1984; Atlas, 1992; Amund and Nwokoye, 1993; Ulrici, 2000). But, low solubility and high hydrophobic nature of petroleum hydrocarbon compounds make them highly unavailable to microorganisms. Release of biosurfactants is one of the strategies used by microorganisms to influence the uptake of petroleum hydrocarbon and hydrophobic compounds (Marin et al., 1996; Johnsen et al., 2005; Obayori et al., 2009). Biosurfactants are surface active materials, increase the surface area of hydrophobic water insoluble substrates, resulting in increase in their bioavailability, thereby enhancing the growth of bacteria and rate of bioremediation (Hou et al., 2001; Ron and Rosenberg, 2002). The present chapter incorporates an overview of different strategies being used for managing petroleum hydrocarbon stress through remediation by microorganisms and their products especially biosurfactants.

17.2 Various Conventional Methods for Management of Petroleum Oil Hydrocarbon-Contaminated Soil

17.2.1 Soil Washing

Soil washing is an *ex situ* remedial method to treat soil contaminated with organic or inorganic compounds. In this method, solvents or mechanical processes are employed to scrub contaminant from the soil (Melanie Fortune—CHEE 484). Selection of solvents for washing the soil depends on the basis of their ability to solubilize specific contaminants (Asante-Duah, 1996; Feng et al., 2001; Chu and Chan, 2003; Urum et al., 2003; Khan et al., 2004). The soil-washing method separates fine soil (clay and silt) from coarse soil (Khan et al., 2004). Since hydrocarbon contaminants tend to bind and sorb to smaller soil particles, hence the soil-washing methods separates them from the larger ones, and reduces the volume of the contaminated soil (Riser-Roberts, 1998; Khan et al., 2004). Further, the smaller soil particles which contain the contaminants can be treated by other incineration or bioremediation or disposed in accordance with federal regulations (Khan et al., 2004).

17.2.2 Landfarming

Landfarming is one of the successful methods for treating petroleum hydrocarbons (FRTR, 1999a; Khan et al., 2004). It has been practiced worldwide for over 100 years, and by the petroleum industry and refineries for more than 25 years (Riser-Roberts, 1998). Landfarming is a method, which reduces the concentration of petroleum constituents present in soil through processes associated with bioremediation (Khan et al., 2004). In this method, excavated contaminated soil is spread in a thin layer not more than 1.5 m over the ground surface of the treatment site and aerobic microbial activity is enhanced within the soil through aeration or by addition of nutrients, minerals, and water. Microbes, which have been selected for their success in breaking down hydrocarbons are often added to the soil to increase the degradation rate of adsorbed petroleum products (Riser-Roberts, 1998; USEPA, 1998a; Hejazi, 2002; Khan et al., 2004).

17.2.3 Thermal Desorption

Thermal desorption is a method where contaminated soil is excavated, screened, and heated to release petroleum from the soil (USEPA, 1995; Khan et al., 2004). Generally, heating of the soils is done at 100–600°C range temperatures, so that those contaminants with boiling points in this range separate from the soil through vaporization (Khan et al., 2004). Further, the vaporized contaminants are collected and treated by some other means (Dermatas and Meng, 2003; Khan et al., 2004). Thermal desorption is an effective method for the treatment of hydrocarbon-contaminated soil, but its effect varies on the full range of organics (FRTR, 1999b; Khan et al., 2004).

17.2.4 Phytoremediation

Phytoremediation is a biological method to remove the contaminant from the environment by using plants (Raskin, 1996; Ndimele et al., 2010). It is a broad term that has been in use since 1991 to describe the use of plants to reduce the volume, mobility or toxicity of contaminants in soil and groundwater (McCutcheon and Schnoor, 2003). The phytoremediation method adopts various mechanisms to remove contaminants. Such mechanisms are phytoextraction, phytovolatilization, phytodegradation, rhizodegradation, rhizofiltration, phytostabilization, and hydraulic control (Figure 17.1). Each of these mechanisms will have an effect on the volume, mobility, or toxicity of contaminants, as the application of

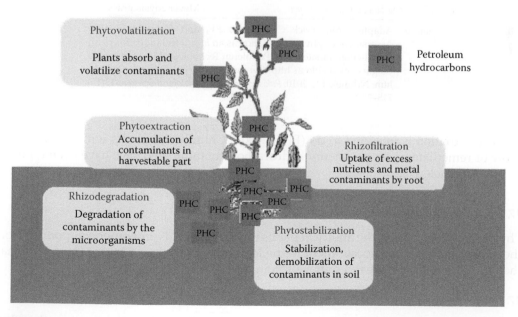

FIGURE 17.1
Mechanisms of petroleum hydrocarbons remediation by plants.

TABLE 17.2

Plants with Petroleum Hydrocarbons Phytoremediation Potential

	Scientific Name	Common Name
1.	*Agropyron smithii*	Western wheatgrass
2.	*Andropogon geradi*	Big bluestem
3.	*Bouteloua gracilis*	Blue grama
4.	*Boutelova curtipendula*	Side oats grama
5.	*Buchloe dactyloides*	Common buffalo grass
6.	*Chloris gayana*	Bell rhodegrass
7.	*Cynodon dactylon*	Bermuda grass
8.	*Daucas carota*	Carrot
9.	*Elymus canadensis*	Canada wild rye
10.	*Fetusca rubra* var. *arctared*	Arctared red fescue
11.	*Glycine max*	Soybean
12.	*Lemna gibba*	Duckweed
13.	*Lolium multiflorum*	Annual ryegrass
14.	*Lolium perenne* L.	Ryegrass
15.	*Medicago sativa* L.	Alfalfa
16.	*Panicum virgatum*	Switch grass
17.	*Panicum coloratum*	Verde kleingrass
18.	*Phaseolus vulgaris* L.	Bush bean
19.	*Populus deltoids* × *nigra*	Poplar tree
20.	*Secale cereal* L.	Winter rye
21.	*Schizachyrium scoparius*	Little bluestem
22.	*Sorghastrum nutans*	Indian grass
23.	*Sorghum bicolor*	Sorghum
24.	*Sorghum vulgare* L.	Sudangrass
25.	*Zoysia japonica* var. *meyer*	Meyer zoysia grass

Source: Adapted from Frick CM, Farrell RE, and Germida JJ. 1999. Assessment of phytoremediation as an *in situ* technique for cleaning oil-contaminated sites. Petroleum Technology Alliance Canada, Calgary. Available at http://www.rtdf.org/pub/phyto/phylinks. htm; Ndimele PE. 2010. *Pakistan Journal of Biological Sciences* 13(15): 715–722.

phytoremediation is intended to do (EPA, 2000). A number of plants with a potential to grow or remediate under petroleum hydrocarbon-contaminated sites have been reported as listed in Table 17.2.

17.2.4.1 Phytoextraction

Phytoextraction refers to the ability of plants to remove metals and other compounds from the subsurface and translocate them to the leaves or other plant tissues. The plants may then need to be harvested and removed from the site (EPA, 2000).

17.2.4.2 Phytovolatilization

Phytovolatilization may also entail the diffusion of contaminants from the stems or other plant parts that the contaminant travels through before reaching the leaves (McCutcheon, 2003).

17.2.4.3 Phytodegradation

When the phytodegradation mechanism is at work, contaminants are broken down after they have been taken up by the plant. As with phytoextraction and phytovolatilization, plant uptake generally occurs only when the contaminants' solubility and hydrophobicity fall into a certain acceptable range (EPA, 2000).

17.2.4.4 Rhizodegradation

Rhizodegradation refers to the breakdown of contaminants within the plant root zone or rhizosphere. Rhizodegradation is believed to be carried out by bacteria or other microorganisms whose numbers typically flourish in the rhizosphere. Studies have documented up to 100 times as many microorganisms in rhizosphere soil as in soil outside the rhizosphere (McCutcheon, 2003). Microorganisms may be so prevalent in the rhizosphere because the plant exudes sugars, amino acids, enzymes, and other compounds that can stimulate bacterial growth (Lynch, 1990; Marilley and Aragno, 1999; García et al., 2001). The roots also provide additional surface area for microbes to grow on and a pathway for oxygen transfer from the environment. The localized nature of rhizodegradation means that it is primarily useful in contaminated soil, and it has been investigated and found to have at least some success in treating a wide variety of most organic chemicals, including petroleum hydrocarbons, polycyclic aromatic hydrocarbons (PAHs), chlorinated solvents, pesticides, polychlorinated biphenyls (PCBs), and benzene, toluene, ethylbenzene, and xylenes (BTEX) (EPA, 2000).

17.2.4.5 Rhizofiltration

In the rhizofiltration process, contaminants are taken up by the plant and removed from the site when the plant is harvested. However, in this case, the contaminant is removed from the dissolved phase and concentrated in the root system. Rhizofiltration is typically exploited in groundwater (either *in situ* or extracted), surface water, or wastewater for removal of metals or other inorganic compounds (EPA, 2000).

17.2.4.6 Phytostabilization

Phytostabilization takes advantage of the changes that the presence of the plant induces in soil chemistry and the environment. These changes in soil chemistry may induce adsorption of contaminants onto the plant roots or soil or cause metals precipitation onto the plant root. The physical presence of the plants may also reduce contaminant mobility by reducing the potential for water and wind erosion. Phytostabilization has been successful in addressing metals and other inorganic contaminants in soil and sediments (EPA, 2000).

17.2.5 Biopiles

Biopiles is a soil treatment method generally employed for treating petroleum-contaminated soil (Khan et al., 2004). This treatment involves the piling of contaminated soils (petroleum) into piles and then enhancing aerobic microbial activity by aeration and the addition of minerals, nutrients, and moisture (Filler et al., 2001; Khan et al., 2004). Most of the biopiles have an underground system through which air passes and have a covering to prevent runoff, evaporation, and volatilization (Khan et al., 2004). This is a short-term treatment method that last from a few weeks to a few months (Khan et al., 2004). Biopiles treat most petroleum products (USEPA, 1998b). Lighter petroleum products such as gasoline tend to be removed during aeration by evaporation and midrange products such as diesel or kerosene have a greater tendency of biodegradation in this method as they contain lower amounts of volatile components, whereas heating and lubricating has a lower biodegradation tendency as they consist of heavier compounds which do not evaporate (Chaineau et al., 2003; Khan et al., 2004).

17.2.6 Bioventing

In the bioventing technique, air is injected into the contaminated media at a rate designed to increase *in situ* biodegradation and reduce the off-gassing of volatilized contaminants to the atmosphere (Khan et al., 2004). Baker and Moore (2000) have studied the optimized performance and effectiveness of *in situ* bioventing. Mihopoulos et al. (2002) and Diele et al. (2002) have discussed numerical models and their applications in bioventing system design and operation. This method is most successful on mid-weight petroleum products such as diesel, because lighter products tend to volatilize quickly and the heavier products generally take longer to biodegrade (Khan et al., 2004).

17.3 Microbial Degradation and Microbial Uptake of Petroleum Hydrocarbons

All the conventional methods mentioned above have some or the other limitation and therefore a new field for degradation of petroleum hydrocarbons is gaining importance which employs microbes or their products for remediation of petroleum-contaminated sites. Microbial degradation is one of the most promising mechanisms for removing the petroleum hydrocarbon pollutants from the environment (Atlas, 1985; Lal and Khanna, 1996). The microbial degradation and removal of contaminants in the soil occurs through two distinct but interrelated processes, biodegradation and microbial uptake. Microbial degradation is the "microbially mediated chemical transformation of organic compounds," whereas microbial uptake is the direct removal of the contaminant by adsorbing compounds to the membrane surface or absorbing compounds through the membrane (Lyman et al., 1992). Both are correlated processes in which the contaminant taken up may either be in original form or a biotransformed product. There are different types of microorganisms responsible for biodegradation of petroleum hydrocarbon pollutants such as bacteria, fungi, and yeast, but bacteria are the most promising and active agents in petroleum degradation (Jahangeer and Kumar, 2013).

17.3.1 Aerobic Degradation

The complete and rapid degradation of the petroleum hydrocarbon pollutant (organic pollutant) is brought about under aerobic conditions. Peroxidase and oxygenases are the main enzymes which are involved in the intracellular attack of organic pollutants and activation as well as incorporation of oxygen in the enzyme key reaction. Organic pollutants are converted step by step into intermediates of the central intermediary metabolism for example, the tricarboxylic acid cycle (Figure 17.2). The end products of hydrocarbon mineralization are CO_2, H_2O, and energy, although complete degradation does not always occur.

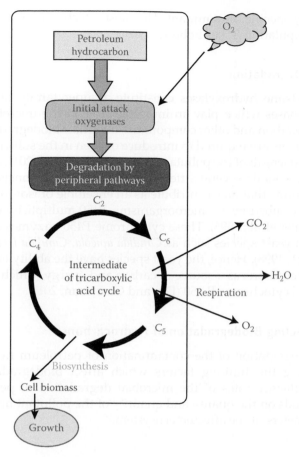

FIGURE 17.2
Main principle of aerobic degradation of hydrocarbons (aerobic) by microorganisms.

Subsequent end products may be directly taken up by microbes and not degraded further or may be degraded to smaller, simpler, more stable intermediaries, and incorporated into the soil as humus or soluble acids, ketones, and alcohols (Lyman et al., 1992). Biosynthesis of cell biomass occurs from the central precursor metabolites such as acetyl-CoA, succinate, pyruvate. Sugars required for various growth and biosyntheses are synthesized by a process termed gluco-neo-genesis (Das and Chandran, 2011).

The potential of a particular petroleum hydrocarbon to be degraded, independent of soil properties, depends on its chemical structure. The main considerations for biodegradation are the size of the contaminant and the type and geometry of its bonds. Some petroleum hydrocarbons have bonds that are difficult to assess or difficult to be broken down by microbes. Different molecular configurations also affect the degradability for example, linear alkanes are more readily degradable than branched alkanes (Riser-Roberts, 1998).

The number of different microbes able to degrade a specific contaminant decreases as the contaminant becomes more difficult to degrade. Also, not all microbes are able to directly take up all the hydrocarbon contaminants. This results in variations of petroleum hydrocarbons and microbial population composition over time with the most readily degradable hydrocarbons and associated microbes being replaced by less degradable hydrocarbons and associated microbes. It should be noted that one type of microorganism is rarely able to fully degrade any specific contaminant. The most effective remediation occurs with a diverse microbial population (Riser-Roberts, 1998).

17.3.2 Enzymatic Degradation

Cytochrome P450 alkane hydroxylases constitute a super family of ubiquitous heme-thiolate monooxygenases which play an important role in the microbial degradation of petroleum oil hydrocarbon and other compounds. To initiate biodegradation of any hydrocarbon, enzyme systems are required to introduce oxygen in the substrate which basically depends on the chain length of the pollutant (Das and Chandran, 2011; Figure 17.3). Higher eukaryotes generally contain several different P450 families that consist of a large number of individual P450 forms that may contribute as an ensemble of isoforms to the metabolic conversion of a given substrate. In microorganisms, P450 multiplicity can only be seen in a few species (Zimmer et al., 1996). These cytochrome P450 enzymes had been predominately isolated from yeast species such as *Candida apicola, Candida tropicalis,* and *Candida maltose* (Scheuer et al., 1998). Hence, the yeast species have the ability to use *n*-alkanes and other aliphatic hydrocarbons as a source of carbon and energy, which is governed by the multiple microsomal cytochrome P-450 (Das and Chandran, 2011).

17.3.3 Factors Affecting Biodegradation of Hydrocarbons

For successful biodegradation of the contamination of petroleum hydrocarbons proper knowledge regarding the limiting factors which affect biodegradation is necessary. Limiting factors influence rates of the microbial degradation of petroleum pollutants which in turn depends on the quality and quantity of the pollutant mixture and physical and chemical parameters of the affected ecosystem.

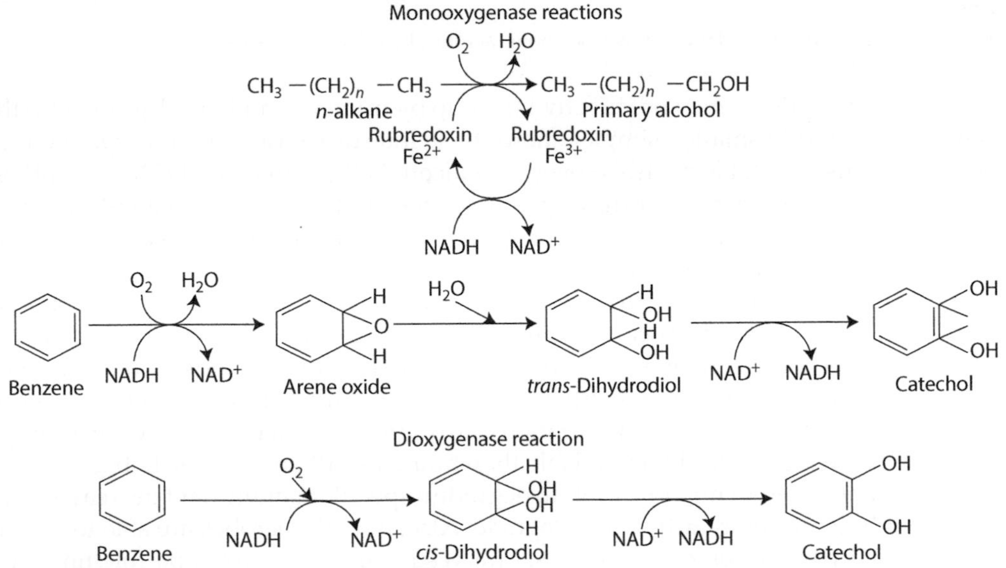

FIGURE 17.3
Enzymatic reactions involved in the processes of hydrocarbons degradation. (Adapted from Das N and Chandran P. 2011. *Biotechnology Research International*, Article ID 941810, 13p. DOI: 10.4061/2011/941810.)

17.3.3.1 Temperature

Temperature plays a very important role in biodegradation of petroleum hydrocarbons through its direct effect on the chemistry of the pollutants and on the diversity and physiology of the microorganisms (Atlas, 1975; Das and Chandran, 2011). At low temperatures, the viscosity of the oil is increased, while the volatility of toxic low-molecular weight hydrocarbons is reduced, delaying the onset of biodegradation (Atlas, 1975). Temperature affects the solubility of hydrocarbons to a great extent (Foght et al., 1996). Various research findings regarding the effect of temperature on biodegradation of hydrocarbons have been reported at wide ranges of temperature so far. However, highest biodegradation rates generally occur in the range of 30–40°C in soil environments, 20–30°C in some fresh water environments and 15–20°C in marine environments (Bartha and Bossert, 1984; Cooney et al., 1985; Das and Chandran, 2011). There are many reports which demonstrate the biodegradation efficiency of microbes in a mesophilic environment, but there are very few reports regarding biodegradations of hydrocarbons in a psychrophilic environment as petroleum contamination is recognized as a significant threat to polar environments. (Yumoto et al., 2002; Delille et al., 2004; Pelletier et al., 2004; Okoh, 2006).

17.3.3.2 pH

An important factor which affects the activity of introduced bacteria in biodegradation of petroleum hydrocarbon is the pH of the soil or the environment (Maren et al., 1998). The pH of the soil is highly variable, ranging from 2.5 in mine spoils to 11 in alkaline deserts. Most heterotrophic bacteria favor a pH of 7.0 but fungi prefer acidic conditions. Therefore, an extreme pH of soils would have a negative influence on the ability of microbial populations to degrade hydrocarbons. Dibble and Bartha (1976) observed an optimal pH of 7.8, in the range 5.0–7.8 for the mineralization of oily sludge in soil. Kastner et al. (1998) reported that the shift of pH from 5.2 to 7.0 enables poly aromatic hydrocarbons degradation by *Sphingomonas paucimobilis*. This led to the conclusion that the neutral pH of soil favors biodegradation of hydrocarbons and small pH shifts have dramatic negative effects on the degradation rate.

17.3.3.3 Nutrients

The nutrient status of the contaminated ecosystem has direct impacts on biodegradation of petroleum hydrocarbon and microbial activity. Nitrogen, phosphorus, and iron play a vital role as the sole nutrient for biodegradation of hydrocarbon pollutants (Das and Chandran, 2011). Nitrogen and phosphorus are necessary for cellular metabolism and can be found in low concentrations in many soils (Mohn and Stewart, 2000). Therefore, the addition of nutrients is necessary to enhance the biodegradation of oil pollutants (Choi et al., 2002; Kim et al., 2005; Das and Chandran, 2011). However, excessive nutrient concentrations can inhibit biodegradation activity (Chaillan et al., 2006; Das and Chandran, 2011).

17.4 Biosurfactant Assisted Bioremediation of Petroleum Hydrocarbons

Petroleum oil hydrocarbon degrading microorganisms play a vital role in the biological treatment of petroleum oil contamination as they adapt to grow and thrive in oil-containing environments. One of the limiting factors in this process is the bioavailability of many fractions of the oil. This limitation can be overcome by the biosurfactants of diverse chemical nature and molecular size produced by hydrocarbon degrading microorganisms. Biosurfactants increase the surface area of hydrophobic water insoluble substrates by increasing their bioavailability and enhancing the growth of bacteria and the rate of bioremediation (Ron and Rosenberg, 2002).

17.4.1 Biosurfactants

Biosurfactants are heterogeneous group of surface active agents produced by microorganisms, which possess both hydrophobic and hydrophilic moieties that reduce surface and interfacial tensions by accumulating at the interface between two immiscible fluids such as oil and water (Ilori et al., 2005; Mahmound et al., 2008; Muthusamy et al., 2008; Kiran et al., 2009; Obayori et al., 2009; Das and Chandran, 2011). Structurally, they contain a hydrophobic moiety of unsaturated or saturated hydrocarbon chains or fatty acids and a hydrophilic moiety, comprising an acid, peptide cations or anions, mono-, di-, or polysaccharides. Biosurfactants are generally grouped into two classes, one is low-molecular weight surface active agents called biosurfactants and other is high molecular weight surface active agents called bioemulsifiers (Karanth et al., 1999). They efficiently reduce surface and interfacial tensions (Karanth et al., 1999). These are the compounds with vast potential for use in the environment, and in petroleum, food, pharmaceutical and other industries as these are environmentally friendly, easily degradable, economical and stable at elevated temperatures, pH and salt concentrations when compared with their chemical counterparts (Banat et al., 2000; Borjana et al., 2001).

17.4.2 Types of Biosurfactants

The major classes of low-mass surfactants include glycolipids, lipopeptides, and phospholipids, whereas high-mass surfactants include polymeric and particulate surfactants (Kappeli and Finnerty, 1979; Nitschke and Coast, 2007) as listed in Table 17.3.

17.4.3 Properties of Biosurfactants

Low-molecular-mass biosurfactants are efficient in lowering surface and interfacial tensions, whereas high-molecular-mass biosurfactants are more effective in stabilizing oil-in-water emulsions (Pacwa-Płociniczak et al., 2011; Geys et al., 2014). The biosurfactants accumulate at the interface between two immiscible fluids or between a fluid and a solid by reducing surface tension (Figure 17.4). The most active biosurfactants can lower the surface tension of water from 72 dynes/cm to 25–30 dynes/cm and also the interfacial tension between water and *n*-hexadecane (Desai and Banat, 1997; Soberón-Chávez and Maier, 2011). Biosurfactant activity depends on the concentration of the surface-active compounds until the critical micelle concentration (CMC) is obtained. At concentrations above the CMC, biosurfactants molecules associate to form micelles, bilayers and vesicles. Micelle formation enables biosurfactants to reduce the surface and interfacial tension and increase the solubility and bioavailability of hydrophobic organic compounds (Whang et al., 2008; Figure 17.5). CMC is commonly used to measure the efficiency of the surfactant. Efficient surfactants have a low CMC, which means that less surfactant is required to decrease the surface tension (Desai and Banat, 1997). Micelle formation has a significant role in micro-emulsion formation. Micro-emulsions are clear and stable

TABLE 17.3

Types of Biosurfactants and their Uses

Groups	Biosurfactant Class	Microorganism	Applications	References
Glycolipids: Glycolipids are low molecular weight biosurfactants in which carbohydrates are attached to a long-chain aliphatic acid	Rhamnolipids: Rhamnolipids are glycolipids that are composed of one or two L-rhamnose molecules coupled to a mono or dimer of β-hydroxy fatty acids	*Pseudomonas aeruginosa, Pseudomonas* sp., *Burkholderia* sp.	Degradation and dispersion of different classes of hydrocarbons and their emulsification, removal of heavy metals from soil	Geys et al. 2014; Pacwa Plociniczak et al. 2011; Ron and Rosenberg 2002
	Trehalolipids: Trehalolipids are glycolipids that contain trehalose lipids as hydrophilic moiety	*Mycobacterium tuberculosis, Rhodococcus erythropolis, Arthrobacter* sp., *Nocardia* sp., *Corynebacterium* sp.	Enhancement in bioavailability of hydrocarbons	Shao, 2011; Plociniczak et al. 2011
	Sophorolipids: Sophorolipids are glycolipids in which dimeric carbohydrate sophorose linked to a long-chain hydroxylfatty acid by glycosidic linkage	*Torulopsis bombicola, Torulopsis petrophilum, Torulopsis apicola, Starmerella bombicola, Wickerhamiella domercqiae, Candida battistae*	Recovery of hydrocarbons from dregs and muds (microbial enhanced oil recovery), heavy metal removal from sediments	Sen, 2008; Kapadia Sanket and Yagnik 2013; Geys et al. 2014; Plociniczak et al. 2011
Lipopetides: Lipopetides are biosurfactants of low molecular weight in which consist of a lipid attached to a polypeptide chain	Surfactin: Surfactin are cyclic lipopetides which of consist of a seven amino-acid-ring structure coupled to a fatty-acid chain via lactone linkage	*Bacillus subtilis*	Enhancement of the biodegradation of hydrocarbons, removal of heavy metals. Antimycoplasmal activity, anti-adhesive application, antibacterial and anti-inflammatory application	Kapadia Sanket and Yagnik et al. 2013; Plociniczak et al. 2011; Wang et al. 2008; Shaligram and Singhal, 2010
	Lichenysin: Lichenysin anionic cyclic lipoheptapeptide biosurfactants produced by *Bacillus licheniformis*.	*Bacillus licheniformis*	Enhancement of oil recovery	Nerurkar, 2010; Plociniczak et al. 2011
Fatty acids, phospholipids, and neutral lipids	Corynomycolic acid	*Corynebacterium lepus*	Bitumen recovery improvement	Plociniczak et al. 2011

(Continued)

TABLE 17.3 (*Continued*)

Types of Biosurfactants and their Uses

	Biosurfactant	Microorganism	Applications	References
	Spiculisporic acid	*Penicillium spiculisporum*	Metal ion sequestration from aqueous solution; preparation of new emulsion-type organogels, superfine microcapsules (vesicles or liposomes)	Plociniczak et al. 2011
	Phosphatidylethanolamine	*Acinetobacter sp., Rhodococcus erythropolis*	Increasing the tolerance of bacteria to heavy metals	Plociniczak et al. 2011
Polymeric biosurfactants	Emulsan	*Acinetobacter calcoaceticus* RAG-1	Stabilization of the hydrocarbon-in-water emulsions	Plociniczak et al. 2011
	Alasan	*Acinetobacter radioresistens* KA-53	Stabilization of the hydrocarbon-in-water emulsions	Plociniczak et al. 2011
	Biodispersan	*Acinetobacter calcoaceticus* A2	Dispersion of limestone in water	Plociniczak et al. 2011
	Liposan	*Candida lipolytica*	Stabilization of hydrocarbon-in-water emulsions	Plociniczak et al. 2011
	Mannoprotein	*Saccharomyces cerevisiae*	Stabilization of hydrocarbon-in-water emulsions	Plociniczak et al. 2011
Particulate biosurfactants		*Acinetobacter sp.*	Helps in uptake of hydrocarbon such as alkane uptake	Kapadia Sanket and Yagnik et al. 2013

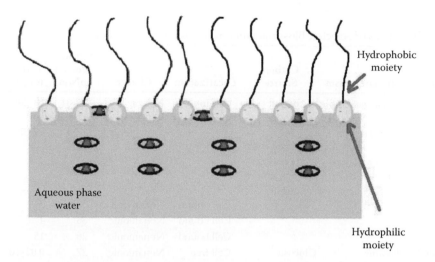

FIGURE 17.4
(See color insert.) Accumulation of biosurfactants at the interface between liquid and air. (Modified from Pacwa-Płociniczak M et al. 2011. *International Journal of Molecular Science* 12: 633–654.)

liquid mixtures of water and oil domains separated by monolayer or aggregates of bio-surfactants (Desai and Banat, 1997; Nguyen et al., 2008). Micro-emulsions are formed when one liquid phase is dispersed as droplets in another liquid phase, for example oil dispersed in water (direct micro-emulsion) or water dispersed in oil (reverse micro-emulsion) (Desai and Banat, 1997). The effectiveness of microbial biosurfactants is determined by the carbon source use for its production, its charge, and its ability to change surface and interfacial tensions. Some of the important properties of biosurfactants are listed in Table 17.4.

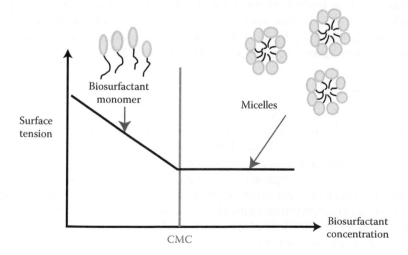

FIGURE 17.5
The relationship between biosurfactant concentration, surface tension, and formation of micelles. (Modified from Whang LM et al. 2008. *Journal of Hazardous Material.* 151: 155–163.)

TABLE 17.4

Important Properties of Selected Biosurfactants from a Range of Microorganisms

Biosurfactant	Microorganism	Carbon Source	Localization	Charge	Surface Tension (mN/m)	CMC (mg/I)	Interfacial Tension (mN/m)
Rhamnolipids	*Pseudomonas* sp.	*n*-Alkanes glycerol	Cell free	Anionic	25–31	10–200	4
Sophorolipids	*Torulopsis* sp	Glucose, vegetable oil		Nonanionic/ anionic	25–35	60–82	1–9
Trehalose coryno mycolates	*Rhodococcus erthropolis*	*n*-Alkanes, carbohydrates					
Mono			Cell bound	Nonanionic	32	3	16
di			Cell bound	Nonanionic	36	4	17
tri			Cell bound	Nonanionic	26	15	<1
Lipopeptides	*Bacillus licheniformis*	Glucose	Cell free	Nonanionic	27	0.02–10	<1
Surfactant	*Bacillus subtilis*	Glucose	Cell free	Nonanionic	27	5	1

Source: Adapted from Sijneriz F, Hommel RK, and Kleber HP. 2011. *Encyclopedia of Life Support Systems*. EOLSS Publications-E books. http:www.eolss.net/sample-chapters/c17/e6-58-05-06.pdf

17.4.4 Factors Affecting Biosurfactant Production

17.4.4.1 Carbon Source

Two basic types of carbon sources, carbohydrates and hydrocarbons are used for the production of biosurfactants. When microorganisms grow at the expense of water immiscible substrates, a spontaneous release of biosurfactants is observed (Hisatsuka et al., 1971; Ito and Inoue, 1982). Carbohydrate substrates are used due to the relatively lower power requirement for dispersion and easy down streaming operations when compared with hydrocarbons, whereas hydrocarbons generate more heat of reaction during cultivation which require extensive cooling surfaces within the bioreactor system (Guerra-Santos et al., 1986; Shodhganga, 2014). Various microorganisms produce different types of biosurfactant using different carbon sources as listed in Table 17.5.

17.4.4.2 Nitrogen Source

Yeast extract is considered as the best nitrogen source for biosurfactant production by *Bacillus strains* isolated from the marine sediments of the Tamil Nadu (India) coastal area (Gnanamani et al., 2010; Pacwa-Płociniczak et al., 2011). Apart from yeast and beef extract other nitrogen sources are often used for the production of biosurfactants. *Pseudomonas fluorescence* grown on olive oil as the carbon source proved to be a more efficient biosurfactant producer with ammonium nitrate as the nitrogen source when compared with ammonium chloride and sodium nitrate (Abouseoud et al., 2007; Pacwa-Płociniczak et al., 2011). Soniyamby et al. (2011) demonstrated that *Pseudomonas sp* grown on a medium as sodium nitrate as the nitrogen source gives better yields of biosurfactant compared with ammonia and urea. According to Khopade et al. (2012) phenylalanine is a more efficient nitrogen source for production of biosurfactants by *Nocardiopsis* sp. *B4* when used in combination with olive oil as the carbon source. Wu et al. (2008) demonstrated sodium nitrate as the best nitrogen source among ammonium

TABLE 17.5

Biosurfactant Producing Microorganism Using Different Carbon Sources

Microorganism	Sources of Isolation	Biosurfactant	Carbon Source	References
Pseudomonas aeruginosa J4	Oil-contaminated site	Rhamnolipid	Glucose, diesel, kerosene, glycerol, olive oil, sunflower oil, grape seed oil	Wei et al. (2005), Pacwa-Płociniczak et al. (2011)
Pseudomonas aeruginosa EM1	Waste water of petrochemical factory	Rhamnolipid	Glucose, glycerol, sucrose, hexane, olive oil, oleic acid, soybean oil	Wu et al. (2008), Pacwa-Płociniczak et al. (2011)
Bacillus subtilis	Crude oil	Surfactins with an heptapeptide moiety	odium acetate, sodium citrate, fructose, glucose, glycerol, lactose, meat extract, paraffin, sucrose, tryptone	Pereira et al. (2013)
Bacillus amyloliquefaciens XZ-173	Tomato rhizosphere soil	Lipopeptides	Soybean flour, rice straw	Zhu et al. (2012)
Pseudomonas sp.	Oil-spilled soil	Rhamnolipid	Glucose, molasses, cheese, whey	Anandaraj and Thivakaran (2010), Pacwa-Płociniczak et al. (2011)
Pseudomonas sp.	Used edible oil	Rhamnolipid	Used edible oil, rice water, diesel, petrol, whey	Pacwa-Płociniczak et al. (2011), Soniyamby et al. (2011)
Pseudomonas aeruginosa LBI	Petroleum-contaminated soil	Rhamnolipids	Soap stock	Pacwa-Płociniczak et al. (2011), Benincasa et al. (2002)
Serratia marcescens	Petroleum-contaminated soil	Lipopeptide	Glycerol	Anyanwu (2011), Pacwa-Płociniczak et al. (2011)
Pseudomonas aeruginosa SP4	Petroleum-contaminated soil	Rhamnolipid	Palm oil	Sarachat et al. (2010), Pacwa-Płociniczak et al. (2011)
Bacillus subtilis	Oil-contaminated soil	Surfactin	Vegetable oil/kerosene/petrol/diesel	Priya and Usharani (2009), Pacwa-Płociniczak et al. (2011)

nitrate, ammonium chloride, urea, and yeast extract for the production of biosurfactants by *Pseudomonas aeruginosa* EM1.

17.4.4.3 Environmental Factors

Environmental factors such as temperature, pH, oxygen, and agitation affect the biosurfactant production to a great extent. There are a few research studies which demonstrate the effect of temperature on biosurfactant production. Different organisms possess varied biosurfactant production ability for example, *Pseudomonas aeruginosa* grown in a salt medium shows increased rhamnolipid production at temperatures between 25 and 37°C and at about 42°C, production decreases (Yu-Hong et al., 2005) while *Serratia marcescens* produces a lipopeptide type of biosurfactant at a temperature range up to 100°C and 12% NaCl concentration with a wide range of pH (Anyanwu et al., 2011). The agitation rate affects the mass transfer efficiency of both oxygen and medium components and is considered crucial for cell growth and biosurfactant production. With an increase in agitation rate from 50 to 200 rpm, the growth rate of the aerobic strain of *Pseudomonas aeruginosa* increases from 0.22 to 0.72/h. A little more research needs to be undertaken on these aspects of environmental effects on biosurfactant production (Maqsood and Jamal, 2011).

17.4.5 Enhancement of Petroleum Hydrocarbons Degradation by Biosurfactants

The use of biosurfactants to enhance the bioremediation of petroleum hydrocarbon-contaminated environments is regarded as the effective method nowadays (Pacwa-Płociniczak et al., 2011). Biosurfactants can enhance hydrocarbon bioremediation by two mechanisms. The first includes the increase of substrate bioavailability for microorganisms, while the second involves interaction with the cell surface which increases the hydrophobicity of the surface allowing hydrophobic substrates to associate more easily with bacterial cells (Mulligan and Gibbs, 2004; Pacwa-Płociniczak et al., 2011). Hydrocarbons are hydrophobic organic chemicals which exhibit limited solubility in water and tend to partition to the soil matrix. This partitioning can account for as much as 90%–95% or more of the total contaminant mass (Pacwa-Płociniczak et al., 2011). As a result, the hydrocarbon contaminants exhibit moderate to poor recovery by physicochemical treatments; limited bioavailability to microorganisms and limited availability to oxidative and reductive chemicals when applied to *in situ* and/or *ex-situ* applications. Due to their amphipathic/amphiphilic nature, biosurfactants enhance the emulsification of hydrocarbons where they form micelles that accumulate at interphase between liquids of different polarities such as water and oil. Hence, biosurfactants reduce surface tension and facilitate hydrocarbon uptake. Biosurfactants are widely used in the remediation of petroleum hydrocarbon as they have the ability to stimulate growth on the hydrophobic surface and can increase the nutrient uptake of hydrophobic substrates. This leads to overcoming of the poor availability of hydrocarbon contaminants to microorganisms and thus increases the chances of biodegradation of hydrocarbons. Therefore, the addition of biosurfactants is expected to enhance biodegradation by mobilization, solubilization or emulsification. The mobilization mechanism occurs at concentrations below the biosurfactants' CMC. At such concentrations, biosurfactants reduce the surface and interfacial tension between air/water and soil/water systems. Due to reduction of the interfacial force, contact of biosurfactants with soil/oil system increases the contact angle and reduces the capillary force holding oil and soil together. In turn, above the biosurfactant CMC, the solubilization process takes place. At these concentrations, biosurfactants molecules associate to form micelles, which dramatically increase the solubility of oil (Déziel et al., 1996; Urum and Pekdemir, 2004; Nguyen et al., 2008; Nievas et al., 2008; Pacwa-Płociniczak et al., 2011). The hydrophobic ends connect together inside the micelle, creating an environment compatible for hydrophobic organic molecules. This is known as solubilization. Emulsification is a process that forms a liquid, known as an emulsion, containing very small droplets of fat or oil suspended in a fluid, usually water. High molecular weight biosurfactants are efficient emulsifying agents. Cameotra and Singh (2009) reported the role of rhamnolipid biosurfactant synthesized by *Pseudomonas aeruginosa* in the uptake of *n*-alkane. They reported an exciting and new mechanism for hydrocarbon uptake involving the internalization of hydrocarbons inside the cell for subsequent degradation. According to their mechanism, biosurfactants disperse hydrocarbons into microdroplets, increasing the availability of hydrocarbon to the bacterial cells. Thereafter, biosurfactant-coated hydrocarbon droplets are uptaken by the bacterial cells by adopting the similar mechanism to *pinocytosis* (a process through which liquid droplets are ingested by living cells) as explained in Figure 17.6. Volkering et al. (1998) reviewed the interactions between microorganisms, soil, pollutant, and surfactants which are explained in Figure 17.7.

There are reports where bacterial strains (PGPR) and the consortium producing rhamnolipid class of biosurfactant have been used for cultivation of medicinal plants such as *Withania sonmifera* in hydrocarbon-contaminated soil without any toxicity in the active ingredient and reduction in the medicinal properties (Kumar et al., 2014). In case of wheat and mustard rhizosphere these biosurfactant producing bacterial strains in consortia have been used for enhancing petroleum hydrocarbon degradation in the rhizosphere (Kumar et al., 2013).

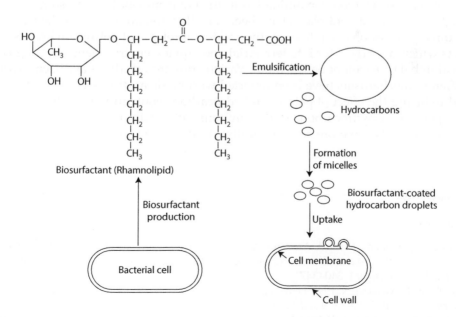

FIGURE 17.6
Involvement of biosurfactant (rhamnolipid) produced by *Pseudomonas* sp in the uptake of hydrocarbons. (Modified from Das N and Chandran P. 2011. *Biotechnology Research International*, Article ID 941810, 13 p. DOI: 10.4061/2011/941810.)

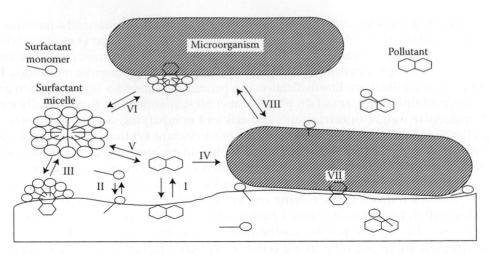

FIGURE 17.7
Schematic overview of the interactions between microorganisms, soil, pollutant, and surfactants: I, sorption of pollutant; II, sorption of surfactant molecules onto soil; III, solubilization of pollutant; IV, uptake of pollutant from the water phase by microorganisms; V, partitioning of pollutant between the water phase and the micelles; VI, sorption of micelles to microorganisms; VII, direct uptake of pollutant from the solid phase by microorganisms; VIII, sorption of microorganisms onto soil. (Adapted from Volkering F, Breure AM, and Rulkens WH. 1998. *Biodegradation* 8: 401–417.)

17.5 Conclusion

Petroleum oil hydrocarbon contamination in the environment is increasing day by day. Although various treatment plants have been installed and are in use for the treatment of petroleum hydrocarbons by refineries, the accidental release of petroleum and its various products which directly enter into terrestrial and aquatic environments creating eco-toxicological imbalance cannot be ruled out. Various research studies have been carried out and different mechanisms have been reported worldwide as discussed above for analyzing and reducing the effect of petroleum hydrocarbons contamination. The management of these petroleum hydrocarbons in the environment is still in its infancy, there is a lot more going on and even more needs to be done in this area to make them more specific.

References

Abouseoud M, Maachi R, and Amrane A. 2007. Biosurfactant production from olive oil by *Pseudomonas fluorescens*. *Communicating Current Research and Educational Topics and Trends in Applied Microbiology* 1: 340–347.

Amund OO and Nwokoye N. 1993. Hydrocarbon potentials of yeast isolates from a polluted Lagoon. *Journal of Scientific Research and Development* 1: 65–68.

Anandaraj B and Thivakaran P. 2010. Isolation and production of biosurfactant producing organism from oil spilled soil. *J Biosciene Technology* 1: 120–126.

Anderson CM and LaBelle RP. 2000. Update of comparative occurrence rates for offshore oil spills. *Spill Science Technology Bulletin* 6(5/6): 303–321.

Anyanwu CU, Obi SKC, and Okolo BN 2011. Lipopeptide biosurfactant production by *Serratia marcescens* NSK-1 strain isolated petroleum-contaminated soil. *Journal of Applied Sciences Research* 7: 79–87.

Asante-Duah DK. 1996. *Managing Contaminated Sites: Problem Diagnosis and Development of Site Restoration.* Wiley, New York, NY.

Atlas RM. 1975. Effects of temperature and crude oil composition on petroleum biodegradation. *Journal of Applied Microbiology* 30(3): 396–403.

Atlas RM. 1985. Effects of Hydrocarbons on Micro-Organisms and Biodegradation in Arctic Ecosystems. In: Petroleum Effects in the Arctic Environment, Engelhardt, F.R. (Ed.). Elsevier, London, UK, pp. 63–99.

Atlas RM. 1992. Petroleum Microbiology. In: *Encyclopedia of Microbiology.* Academic Press, Baltimore, MD, pp. 363–369.

Baker RS and Moore AT. 2000. Optimizing the effectiveness of *in situ* bioventing. *Pollution Engineering* 32(7): 44–47.

Banat IM, Makkar RS, and Cameotra SS. 2000. Potential commercial applications of microbial surfactants. *Applied Environmental Microbiology* 53: 495–508.

Banks MK, Mallede H, and Rathbone K. 2003. Rhizosphere microbial characterization in petroleum contaminated soil. *Soil and Sediment Contamination* 12: 371–385.

Bartha R and Bossert I. 1984. The treatment and disposal of petroleum wastes. In: Atlas RM. (Ed.), *Petroleum Microbiology.* Macmillan, New York, pp. 553–578.

Benincasa M. 2002. Rhamnolipid production by *P. aeruginosa* LBI growing on soap-stock as the sole carbon source. *Journal of Food Engineering* 54: 283–288.

Borjana KT, George RI, and Nelly EC. 2001. Biosurfactant production by a new *Pseudomonas putida* strain. *Zeitschrift für Naturforschung* 57: 356–360.

Cameotra SS and Singh P. 2009. Synthesis of rhamnolipid biosurfactant and mode of hexadecane uptake by *Pseudomonas* species. *Microbial Cell Factories* 8: 1–7.

Cermak JH, Stephenson GL, Birkholz D, Wang Z, and Dixon DG. 2010. Toxicity of petroleum hydrocarbon distillates to soil organisms. *Environmental Toxicology and Chemistry* 29(12): 2685–2694.

Chaillan F, Chaîneau CH, Point V, Saliot A, and Oudot J. 2006. Factors inhibiting bioremediation of soil contaminated with weathered oils and drill cuttings. *Environmental Pollution* 144(1): 255–265.

Chaineau CH, Yepremian C, Vidalie JF, Ducreux J, and Ballerini D. 2003. Bioremediation of a crude oil-polluted soil: Biodegradation leaching and toxicity assessments. *Water, Air, and Soil Pollution* 144(1–4): 419–440.

Choi SC, Kwon KK, Sohn JH, and Kim SJ. 2002. Evaluation of fertilizer additions to stimulate oil biodegradation in sand seashore mesocosms. *Journal of Microbiology and Biotechnology* 12(3): 431–436.

Chu W and Chan KH. 2003. The mechanism of the surfactant-aided soil washing system for hydrophobic and partial hydrophobic organics. *Science of the Total Environment* 307(1–3): 83–92.

Cooney JJ, Silver SA, and Beck EA. 1985. Factors influencing hydrocarbon degradation in three freshwater lakes. *Microbial Ecology* 11(2): 127–137.

Das N and Chandran P. 2011. Microbial degradation of petroleum hydrocarbon contaminants: An overview. *Biotechnology Research International*, Article ID 941810, 13p. DOI: 10.4061/2011/941810.

Delille D, Coulon F, and Pelletier E. 2004. Effects of temperature warming during a bioremediation study of natural and nutrient-amended hydrocarbon-contaminated sub-Antarctic soils. *Cold Regions Science and Technology* 40: 61–70.

Dermatas D and Meng X. 2003. Utilization of fly ash for stabilization/solidification of heavy metal contaminated soils. *Engineering Geology* 70(3–4): 377–394.

Desai JD and Banat IM. 1997. Microbial production of surfactants and their commercial potential. *Microbiology and Molecular Biology Reviews* 61: 47–64.

Déziel E, Paquette G, Villemur R, Lépine F, and Bisaillon JG. 1996. Biosurfactant production by a soil *Pseudomonas* strain growing on polycyclic aromatic hydrocarbons. *Applied and Environmental Microbiology* 62: 1908–1912.

Dibble JT and Bartha R. 1976. The effect of iron on the biodegradation of petroleum in seawater. *Applied and Environmental Microbiology* 31: 544–550.

Diele F, Notarnicola F, and Sgura I. 2002. Uniform air velocity field for a bioventing system design: Some numerical results. *International Journal of Engineering Science* 40(11): 1199–1210.

EPA. 2000. A Citizen's Guide to Phytoremediation. EPA 542-F-98-011. United States Environmental Protection Agency, p. 6. Available at: http//www.bugsatwork.com/XYCLONYX/EPA_GUIDES/PHYTO.PDF

Feng D, Lorenzen L, Aldrich C, and Mare PW. 2001. Ex-situ diesel contaminated soil washing with mechanical methods. *Minerals Engineering* 14(9): 1093–1100.

Filler DM, Lindstrom JE, Braddock JF, Johnson RA, and Nickalaski R. 2001. Integral biopile components for successful bioremediation in the Arctic. *Cold Regions Science and Technology* 32(2–3): 143–156.

Foght JM, Westlake DWS, Johnson WM, and Ridgway HF. 1996. Environmental gasoline-utilizing isolates and clinical isolates of *Pseudomonas aeruginosa* are taxonomically indistinguishable by chemotaxonomic and molecular techniques. *Microbiology* 142(9): 2333–2340.

Frick CM, Farrell RE, and Germida JJ. 1999. Assessment of Phytoremediation as an *in situ* technique for cleaning oil-contaminated sites. Petroleum Technology Alliance Canada, Calgary. Available at: http://www.rtdf.org/pub/phyto/phylinks.htm

FRTR. 1999a. Thermal desorption. Federal Remediation Technologies Roundtable. USEPA, 401 M Street, S.W., Washington, DC, http://www.frtr.gov/matrix2/section4/4_29.html

FRTR. 1999b. Landfarming. Federal Remediation Technologies Roundtable. USEPA, 401 M Street, S.W., Washington, DC, http://www.frtr.gov/matrix2/section4/4_15.html

García JL, Probanza A, Ramos B, and Mañero FJG. 2001. Ecology, genetic diversity and screening strategies of plant growth promoting rhizobacteria. *Journal of Plant Nutrition and Soil Science* 164: 1–7.

Geys R, Soetaert W, and Bogaert IV. 2014. Biotechnological opportunities in biosurfactant production. *Current Opinion in Biotechnology* 30: 66–72.

Gnanamani A, Kavitha V, Radhakrishnan N, and Mandal AB. 2010. Bioremediation of crude oil contamination using microbial surface active agents: Isolation, production and characterization. *Journal of Bioremediation and Biodegradation* 3: 1–8.

Graj W, Lisiecki P, and Szulc A. 2013. Bioaugmentation with petroleum-degrading consortia has a selective growth-promoting impact on crop plants germinated in diesel oil-contaminated soil. *Water, Air and Soil Pollution* 224: 1676.

Guerra-Santos LH, Kappeli O, and Fiechter A. 1986. Dependence of *Pseudomonas aeruginosa* continuous culture biosurfactant production on nutritional and environmental factors. *Applied Microbiology and Biotechnology* 24: 443–448.

Hejazi RF. 2002. Oily sludge degradation study under arid conditions using a combination of landfarm and bioreactor technologies. PhD thesis, Faculty of Engineering and Applied Science, Memorial University of Newfoundland, St John's, Canada.

Hentati O, Lachhab R, Ayadi M, and Ksibi M. 2013. Toxicity assessment for petroleum-contaminated soil using terrestrial invertebrates and plant bioassays. *Environmental Monitoring and Assessment* 185: 2989–2998.

Hou FS, Milke MW, Leung DW, and MacPherson DJ. 2001. Variation in phytoremediation performance with diesel contaminated soil. *Environmental Technology* 22: 215–222.

Ilori MO, Amobi CJ, and Odocha AC. 2005. Factors affecting biosurfactant production by oil degrading *Aeromonas* spp. isolated from a tropical environment. *Chemosphere* 61(7): 985–992.

Ito S and Inoue S. 1982. Sophorolipids from Torulopsis bombicola: Possible relation to alkane uptake. *Applied Environmental Microbiology* 43: 1278–1283.

Jahangeer KV. 2013. An overview on microbial degradation of petroleum hydrocarbon contaminants. *International Journal of Engineering and Technical Research* 1(8): 35–37.

Johnsen AR, Wick LY, and Harms H. 2005. Principles of microbial PAH degradation in soil. *Environmental Pollution* 133: 71–84.

Kapadia Sanket G and Yagnik BN. 2013. Current trend and potential for microbial biosurfactants. *Asian Journal of Experimental Biological Science* 4(1): 1–8.

Kappeli O and Finnerty WR. 1979. Partition of alkane by an extracellular vesicle derived from hexadecane grown Acinetobacter. *Journal of Bacteriology* 140: 707–712.

Karanth NGK, Deo PG, and Veenanadig NK. 1999. Microbial production of biosurfactants and their importance. *Current Science* 77: 116–123.

Khan FI, Husain T, and Hejazi R. 2004. An overview and analysis of site remediation technologies. *Journal of Environmental Management* 71: 95–122.

Khopade A, Biao R, Liu X, Mahadik K, Zhang L, and Kokare C. 2012. Production and stability studies of the biosurfactant isolated from marine *Nocardiopsis* sp. B4. *Desalination* 285: 198–204.

Kim SJ, Choi DH, Sim DS, and Oh YS. 2005. Evaluation of bioremediation effectiveness on crude oil-contaminated sand. *Chemosphere* 59(6): 845–852.

Kiran GS, Hema TA, Gandhimathi R, Selvin J, Thomas TA, Rajeetha Ravji T, and Natarajaseenivasan K. 2009. Optimization and production of a biosurfactant from the sponge-associated marine fungus *Aspergillus ustus* MSF3. *Colloids and Surfaces B* 73(2): 250–256.

Kumar R, Bharagava RN, Kumar M, Singh SK, and Kumar G. 2013. Enhanced biodegradation of mobil oil hydrocarbons by biosurfactant producing bacterial consortium in wheat and mustard rhizosphere. *Petroleum and Environmental Biotechnology* 4(5): 1–8.

Kumar R, Das AJ, and Juwarkar A. 2015. Reclamation of petrol oil contaminated soil by rhamnolipids producing PGPR strains for growing *Withania somnifera* a medicinal shrub. *World Journal of Microbiology and Biotechnology* 31: 307–313.

Kvenvolden KA, Rosenbauer RJ, Hostettler FD, and Lorenson TD. 2000. Application of organic geochemistry to coastal tar residues from central California. *International Geology Review* 42: 1–14.

Lal B and Khanna S. 1996. Degradation of crude oil by *Acinetobacter calcoaceticus* and *Alcaligenes odorans. Journal of Applied Bacteriology* 81(4): 355–362.

Lyman WJ, Reidy PJ, and Levy B. 1992. *Mobility and Degradation of Organic Contaminants in Subsurface Environments.* C.K. Smoley, Chelsea, p. 395.

Lynch JM (Ed.). 1990. *The Rhizosphere.* John Wiley and Sons Ltd, Chichester, p. 458.

Mahmound A, Aziza Y, Abdeltif A, and Rachida M. 2008. Biosurfactant production by *Bacillus* strain injected in the petroleum reservoirs. *Journal of Industrial Microbiology and Biotechnology* 35: 1303–1306.

Maqsood MI and Jamal A. 2011. Factors affecting the rhamnolipid biosurfactant production. *Pakistan Journal of Biotechnology* 8(1): 1–5.

Maren MK, Breuer J, and Mahro B. 1998. Salinity, and pH on the degradation of polycyclic aromatic hydrocarbons (PAHs) and survival of PAH-degrading bacteria introduced into soil. *Applied Environmental Microbiology* 64(1): 359–362.

Margesin R, Walder G, and Schinner F. 2000. The impact of hydrocarbon remediation (diesel oil and polycyclic aromatic hydrocarbons) on enzyme activities and microbial properties of soil. *Acta Biotechnologica* 20: 313–333.

Marilley L and Aragno M. 1999. Phytogenetic diversity of bacterial communities differing in degree of proximity of *Lolium perenne* and *Trifolium repens* roots. *Applied Soil Ecology* 13: 127–136.

Marin M, Pedregosa A, Rios S, and Laborda F 1996. Study of factors influencing the degradation of heating oil by *Acinetobacter calcoaceticus* MM5. *International Biodeterioration and Biodegradation* 38: 67–75.

McCutcheon SC and Schnoor JL. 2003. *Phytoremediation: Transformation and Control of Contaminants.* John Wiley & Sons, Inc., Hoboken, NJ. ISBN: 0-471-39435-1. 987 pp.

Medina-Bellver JI, Marín P, Delgado A, Rodríguez-Sánchez A, Reyes E, Ramos JL, and Marqués S. 2005. Evidence for *in situ* crude oil biodegradation after the Prestige oil spill. *Environmental Microbiology* 7(6): 773–779.

Melanie Fortune—CHEE 484, Remediation Technologies—Soil Washing. Assignment. Available at www.frtr.gov/matrix2/section4/4-19.html (last accessed 17-12-2014).

Mihopoulos PG, Suidan MT, Sayles GD, and Kaskassian S. 2002. Numerical modeling of oxygen exclusion experiments of anaerobic bioventing. *Journal of Contaminant Hydrology* 58(3–4): 209–220.

Mohn WW and Stewart GR. 2000. Limiting factor for hydrocarbon biodegradation at low temperature in arctic soils. *Soil Biology and Biochemistry* 32: 1161–1172.

Mulligan CN and Gibbs BF. 2004. Types, production and applications of biosurfactants. *Proceedings of the Indian National Science Academy* 1: 31–55.

Muthusamy K, Gopalakrishnan S, Ravi TK, and Sivachidambaram P. 2008. Biosurfactants: Properties, commercial production and application. *Current Science* 94(6): 736–747.

Ndimele PE. 2010. A review on the phytoremediation of petroleum hydrocarbon. *Pakistan Journal of Biological Sciences* 13(15): 715–722.

Nerurkar AS. 2010. Structural and molecular characteristics of lichenysin and its relationship with surface activity. *Advances in Experimental Medicine and Biology* 672: 30415.

Nguyen TT, Youssef NH, McInerney MJ, and Sabatini DA. 2008. Rhamnolipid biosurfactant mixtures for environmental remediation. *Water Research* 42: 1735–1743.

Nievas ML, Commendatore MG, Estevas JL, and Bucalá V. 2008. Biodegradation pattern of hydrocarbons from a fuel oil-type complex residue by an emulsifier-producing microbial consortium. *Journal of Hazardous Material* 154: 96–104.

Nitschke M and Coast SG. 2007. Biosurfactants in food industry. *Trends in Food Science and Technology* 18: 252–259.

Obayori OS, Ilori MO, Adebusoye SA, Oyetibo GO, Omotayo AE, and Amund OO. 2009. Degradation of hydrocarbons and biosurfactant production by *Pseudomonas* sp. strain LP1. *World Journal of Microbiology and Biotechnology* 25: 1615–1623.

OECD. 1998. *Global Energy: The Changing Outlook*. IEA/OECD, Paris.

Okoh AI. 2006. Biodegradation alternative in the cleanup of petroleum hydrocarbon pollutants. *Biotechnology and Molecular Biology Review* 1(2): 38–50.

Okoh AI and Trejo-Hernandez MR. 2006. Remediation of petroleum hydrocarbon polluted systems: Exploiting the bioremediation strategies. *African Journal of Biotechnology* 5(25): 2520–2525.

Pacwa-Płociniczak M, Płaza GA, Piotrowska-Seget Z, and Cameotra SS. 2011. Environmental applications of biosurfactants: Recent advances. *International Journal of Molecular Science* 12: 633–654.

Pelletier E., Delille D, and Delille B. 2004. Crude oil bioremediation in sub-Antarctic intertidal sediments: Chemistry and toxicity of oiled residues. *Marine Environmental Research* 57: 311–327.

Perry JJ. 1984. Microbial metabolism of cyclic alkanes. In: Atlas RM. (Ed.), *Petroleum Microbiology*. Macmillan, New York, NY, pp. 61–98.

Potter TL and Simmons KE. 1998. *Composition of Petroleum Mixtures*, Vol. 2. Amherst Scientific Publishers, Amherst, MA.

Priya T and Usharani G. 2009. Comparative study for biosurfactant production by using *Bacillus subtilis* and *Pseudomonas aeruginosa*. *Botany Research International* 2: 284–287.

Raskin I. 1996. Plants genetic engineering may help with environmental clean-up. *Proceedings of the National Academy of Science of the United States of America* 93: 3164–3166.

Riser-Roberts E. 1998. *Remediation of Petroleum Contaminated Soil: Biological, Physical, and Chemical Processes*. Lewis Publishers, Boca Raton, FL.

Ron EZ and Rosenberg E. 2002. Biosurfactants and oil bioremediation. *Current Opinion in Chemical Biology* 13: 249–252.

Sanchez O, Isabel F, Núria V, Tirso GdO, Joan OG, and Jordi M. 2006. Presence of opportunistic oil-degrading microorganisms operating at the initial steps of oil extraction and handling. *International Microbiology* 9: 119–124.

Sarachat T, Pornunthorntawee O, Chavadej S, and Rujiravanit R. 2010. Purification and concentration of rhamnolipid biosurfactant produced by *Pseudomonas aeruginosa* SP4 using foam fractionation. *Bioresource Technology* 101: 324–330.

Scheuer U, Zimmer T, Becher D, Schauer F, and Schunck WH. 1998. Oxygenation cascade in conver-
sion of n-alkanes to α, ω-dioic acids catalyzed by cytochrome P450 52A3. *Journal of Biological Chemistry* 273(49): 32528–32534.

Sen R. 2008. Biotechnology in petroleum recovery: The microbial EOR. *Progress in Energy and Combustion* 34: 714–724.

Shaligram NS and Singhal RS. 2010. Surfactin—A review on biosynthesis, fermentation, purification and applications. *Food Technology and Biotechnology* 48(2): 119–134.

Shao Z. 2011. Trehalolipids. *Microbiology Monographs* 20: 121–143.

Shodhganga. 2014. shodhganga.inflibnet.ac.in/bitstream/10603/9983/.../07_chapter%202.pdf (December 1, 2014).

Sijneriz F, Hommel RK and Kleber HP. 2011. Production of biosurfactants. In: *Encyclopedia of Life Support Systems*. EOLSS Publications-E books. http:www.eolss.net/sample-chapters/c17/e6-58-05-06.pdf.

Soberón-Chávez G and Maier RM. 2011. Biosurfactants: A general overview. In: Soberón-Chávez G, (Ed.), *Biosurfactants*. Springer-Verlag, Berlin, Germany, pp. 1–11.

Soniyamby AR, Praveesh BV, Vimalin Hena J, Kavithakumari P, Lalitha S, and Palaniswamy M. 2011. Enhanced production of biosurfactant from isolated Pseudomonas sp growing on used edible oil. *Journal of American Science* 7: 50–52.

Speight JG. 1999. *The Desulfurization of Heavy Oils and Residua*, 2nd ed. Marcel Dekker, New York.

Tang JC, Wang M, Wang F, Sun Q, and Zhou QX. 2011. Eco-toxicity of petroleum hydrocarbon contaminated soil. *Journal of Environmental Sciences* 23(5): 845–851.

Ulrici W. 2000. Contaminated soil areas, different countries and contaminants, monitoring of contaminants. In: Rehm H-J and Reed G. (Eds.), *Biotechnology: Environmental Processes II*, Vol. 11b, Second Edition, Wiley-VCH Verlag GmbH, Weinheim, Germany, pp. 1–5.

Urum K and Pekdemir T. 2004. Evaluation of biosurfactants for crude oil contaminated soil washing. *Chemosphere* 57: 1139–1150.

Urum K, Pekdemir T, and Gopur M. 2003. Optimum conditions for washing of crude oil-contaminated soil with biosurfactant solutions. *Process Safety and Environmental Protection: Transactions of the Institution of Chemical Engineers, Part B* 81(3): 203–209.

USEPA. 1995. How to evaluate alternative cleanup technologies for underground storage tank sites. *Office of Solid Waste and Emergency Response*. US Environmental Protection Agency, Publication # EPA 510-B-95-007, Washington, DC.

USEPA. 1998a. Landfarming. Office of the Underground Storage Tank, US Environmental Protection Agency, Publication # EPA 510-B-95-007. Available at http://www.epa.gov/swerust/cat/land-farm.htm

USEPA. 1998b. Biopiles. Office of the Underground Storage Tank, US Environmental Protection Agency, Publication # EPA 510-B-95-007. Available at http://www.epa.gov/swerust1/cat/bio-piles.htm

Volkering F, Breure AM, and Rulkens WH. 1998. Microbiological aspects of surfactants used for biological soil remediation. *Biodegradation* 8: 401–417.

Wang D, Liu Y, Lin Z, Yang Z, and Hao C. 2008. Isolation and identification of surfactin producing Bacillus subtilis strain and its effect of surfactin on crude oil. *Wei Sheng Wu Xue Bao* 48(3): 304–311.

Wei YH, Chou CL, and Chang JS. 2005. Rhamnolipid production by indigenous Pseudomonas aeruginosa J4 originating from petrochemical wastewater. *Biochemical Engineering Journal* 27: 146–154.

Whang LM, Liu PWG, Ma CC, and Cheng SS. 2008. Application of biosurfactant, rhamnolipid, and surfactin, for enhanced biodegradation of diesel-contaminated water and soil. *Journal of Hazardous Material.* 151: 155–163.

Wu JY, Yeh KL, Lu WB, Lin CL, and Chang JS. 2008. Rhamnolipid production with indigenous *Pseudomonas aeruginosa* EM1 isolated from oil-contaminated site. *Bioresource Technology* 99: 1157–1164.

Yu-Hong W, Chou CL, and Chang JS. 2005. Rhamnolipid production by indigeneous *P. aeruginosa* J4 orignating from petrochemicals wastes. *Biochemical Engineering Journal* 27: 146–154.

Yumoto I, Nakamura A, Iwata H, Kojima K, Kusumoto K, Nodasaka Y, and Matsuyama H. 2002. *Dietzia psychralcaliphila* sp. nov., a novel facultatively psychrophilic alkaliphile that grows on hydrocarbons. *International Journal of Systemic Evolution and Microbiology* 52: 85–90.

Zhu Z, Zhang G, Luo Y, and Ran W. 2012. Qirong Shen. Production of lipopeptides by Bacillus amyloliquefaciens XZ-173 in solid state fermentation using soybean flour and rice straw as the substrate. *Bioresource Technology* 112: 254–260.

Zimmer T, Ohkuma M, Ohta A, Takagi M, and Schunck WH. 1996. The CYP52 multigene family of *Candida maltosa* encodes functionally diverse n-alkane-inducible cytochromes p450. *Biochemical and Biophysical Research Communications* 224(3): 784–789.

18

Recent Advances in the Expression and Regulation of Plant Metallothioneins for Metal Homeostasis and Tolerance

Preeti Tripathi, Pradyumna Kumar Singh, Seema Mishra, Neelam Gautam, Sanjay Dwivedi, Debasis Chakrabarty, and Rudra Deo Tripathi

CONTENTS

18.1 Introduction

The status of nutrient elements in soil determines the growth and nutritional value of field crops. Heavy metals such as Cu and Zn act as key regulators for a range of plant physiological processes such as Cu and Zn-dependent enzymatic reactions, involvement of Zn in protein sequences as Zn^{2+}-binding structural domains, and Cu as a vital component for electron transfer reactions (Cobbet and Goldsbrough, 2002). Moreover, various heavy metal ions such as As, Cr, Cd, and Pb are highly toxic to living cells. Excessive accumulation of these metals in plants has resulted in the alteration of absorption and transport of essential elements, interfering with normal metabolism, and having an impact on growth and reproduction (Xu and Shi, 2000). Further excessive accumulation of essential metals in plants can also lead to the alteration of physiological processes, membrane damage, imbalance in reactive oxygen species (ROS), and crop yield reduction (Xu and Shi, 2000). Consequently, plants have evolved a range of underlying mechanisms for homeostasis/detoxification of excess essential metal and toxic metals. The homeostasis involves a complex mechanism network that controls the uptake, accumulation, trafficking, and detoxification of metals. These mechanisms mainly contain chelation of toxic metals by metal ligands, vacuolar sequestration, and activation of antioxidant systems (Verbruggen et al., 2009). The proper regulation of these mechanisms maintains the concentration of metal ions at cellular and organism level, achieving the level of metal tolerance in plants. Metal chelators such as organic acids, amino acids, phytochelatins, and metallothioneins (MTs) play important roles in metal detoxification (Hall, 2002). Various approaches have been used to pin down the physiological roles of these metabolites, yet studies on plant MTs are required to understand their specific function, regulation, and expression in plants during metal stress. Therefore, detailed emphasis has been given here on the MT-oriented research work with special reference to tissue- and metal-specific expression and regulation of MTs under various types of stress to improve our understanding of their multifunctional role for the advancement of MT related future research work.

18.2 General Facts about MTs

Metallothioneins are low-molecular-mass (>10,000 Da), cysteine-rich proteins which are widely distributed in animals, plants, fungi, as well as bacteria including cyanobacteria (Cobbet and Goldsbrough, 2002; Leszczyszyn et al., 2013). MT was discovered in 1957 by Vallee and Margoshe from purification of a Cd-binding protein from horse (equine) renal cortex (Cobbet and Goldsbrough, 2002). MTs ardently bind to metals such as Zn, Cu, and Cd to reduce their concentration to a physiological or nontoxic level at different developmental stages and in various tissues of plants. In addition to essential metal homeostasis and (toxic) metal detoxification, various evidences show that plant MTs also act as powerful antioxidants (Hassinen et al., 2011; Xia et al., 2012; Kim et al., 2013). There are more than 50 plant MT-like proteins included in various databases, about one third of all known MTs (Liu et al., 2000). Several *MT* genes have been isolated and characterized from plants. MTs are not only constitutively expressed, but the production of different types of plant MT is also stimulated by numerous endogenous and exogenous factors in a temporally and spatially regulated manner (Leszczyszyn et al., 2013). Transcriptional induction of the *MT* gene is mediated by metal responsive transcription factor 1 (*MTF-1*) an essential zinc finger protein that binds to specific DNA motifs known as metal responsive elements. It has been reported that while transcriptional activation by Zn can be achieved by elevated Zn concentration alone, induction by Cd, Cu, and H_2O_2 additionally needs Zn saturated MTs. This is elaborated by preferential binding of Cd or Cu to MT or its oxidation by H_2O_2, concomitantly releasing Zn which may lead to activation of transcriptional factor *MTF-1*. Contrastingly, the apo-MT (metal free form) inhibits the activation of *MTF-1* (Zhang et al., 2003; Waldron et al., 2009) established by animal models, yet to be validated in plants.

18.3 MTs, Classification, and Structure

The primary structures of MTs comprise several highly conserved CC, CXC, and CXXC motifs, where C = cysteine, X = another amino acid. Therefore, MTs have the capacity to bind heavy metals, both physiological (such as Zn, Cu, Se, and Ni) and xenobiotic (such as Cd, Hg, Ag, and As), through the thiol group of its cysteine residues, which represents nearly the 30% of its amino acidic residues (Cobbet and Goldsbrough, 2002; Du et al., 2012).

　　The first MT identified in plant was the wheat EcMT protein, isolated from mature embryos and shown to bind Zn^{2+} (Lane et al., 1987), and since then more than 140 MT sequences have been recorded from various species (Guo et al., 2003; Zhou et al., 2005). Plant MTs typically contain two Cys-rich domains (four to eight cysteine each) which are separated by a cysteine-poor linker/spacer (30–50 residues) that varies in length depending on the type and source of the MTs. Based on the arrangement of cysteine residues and the length of the spacer region, plant MTs are classified into four types, viz. type 1 to type 4 (Cobbett and Goldsbrough, 2002), illustrated in Figure 18.1. It has been hypothesized that the differences in Cys motifs could account for differences in metal specificities (Robinson et al., 1993; Cobbett and Goldsbrough, 2002). The number of *MT* genes vary in different plant species, for instance, the rice gene family consists of 11 genes. On the basis of structural and phylogenetic analysis, the rice (*Oryza sativa*) *MT* gene family was classified into two classes. Class I comprising 10 of 11 genes was further divided into four types viz., *OsMT-I-1a* to *OsMT-I-4c*. Class II rice MT contains one member of the family *OsMT-II-1a* represented (Zhou et al., 2006).

Unlike mammalian MTs containing highly conserved 20 Cys residues, the cysteine arrangement patterns in plants are more complicated (Cobbett and Goldsbrough, 2002). Further, the genomic arrangement of plant *MTs* also differ significantly from mammalian *MTs*. Such as, all the 11 putative OsMT proteins contain cysteine-rich signature motifs; however, analysis of the chromosomal distribution of these genes revealed that three genes in type 4 are distributed in the same chromosome but are not closely linked, while

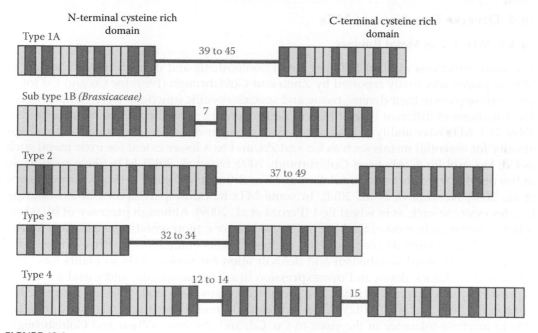

FIGURE 18.1
(See color insert.) Classification of plant MT (Type 1, Type 2, Type 3, and Type 4) according to the cysteine motifs in their N- and C-terminal regions and the length of the C corresponding spacer regions. Blue and yellow boxes represent cysteine (Cys) and other amino acids (X), respectively. Red line and number indicate the spacer region containing the amino acid number. Type 1 contains six Cys-X-Cys motifs that are equally distributed among two domains. Type 2 contains Cys–Cys motif in amino acid positions 3 and 4 of these types and Cys-X-X-Cys motif is present at the end of the N-terminal cysteine-rich domain. Overall, the N-terminal domain of type 2 MTs are highly conserved and the C-terminal domain contains three Cys-X-Cys motifs. Type 3 contains only four Cys residues in the N-terminal domain and six Cys residues in the C-terminal are arranged in the Cys-X-Cys motif. Type 4 has three cysteine rich domains, each containing 5 or 6 Cys residues. Most of the cysteines present as Cys-X-Cys motif.

other genes from the same type are located on different chromosomes as a single copy (Zhou et al., 2006). The similar dispersed pattern of *MT* genes over the whole genome was found in tomato (Giritch et al., 1998) and *Arabidopsis* (Guo et al., 2003).

In contrast, the functional *MT* genes of mammals, for example, mice and humans, are tightly linked to each other on chromosomes 8 and 16, respectively (Palmiter et al., 1992). Since the three genes of type 4 *OsMT* are located on the same chromosome and are located within a 140-kb fragment of genomic DNA and share high sequence homology, thus, they might have evolved from gene duplications of more recent unequal recombination events. Similar features have been found in other plant *MT* genes family such as in cotton and *Arabidopsis* (Hudspeth et al., 1996; Guo et al., 2003). Thus, it seems that the linked organization of *MT* genes is a typical characteristic associated with mammals. Analysis of *MT3* gene duplication in *Silene vulgaris*, a metal-tolerant plant species showed two function genes: *SvMT3a* and *SvMT3b* (Nevrtalova et al., 2014). While *SvMT3a* has specific expression pattern in Cu tolerant ecotypes, *SvMT3b* shared common expression pattern in all metallophyte ecotypes. This suggests that there is a selective pressure to maintain both copies of *MT3* genes functional in *S. vulgaris* for Cu tolerance.

18.4 Diverse Functions of MTs

18.4.1 MTs Act as Metal Binders

The main functions of plant MTs are metal homeostasis and detoxification. The role of MTs in plants was firstly reported by Zhou and Goldsbrough (1995) for Cu and Cd tolerance. Subsequently their diverse tissue and species-specific functions were characterized. The functions of different type of MTs in various plant species have been summarized in Table 18.1. MTs have ability to bind a variety of mono and divalent metal ions with a high affinity for essential metals such as Cu and Zn, and to a lesser extent for toxic metal such as Cd, Hg, and Pb (Cobbet and Goldsbrough, 2002; Du et al., 2012). MTs likely participate in the uptake, transport, and regulation of Zn and Cu in biological systems (Leszczyszyn et al., 2007; Nevrtalova et al., 2014). In some MTs histidine participates in Zn binding besides cysteine such as in wheat Ec-1 (Peroza et al., 2009). Although presence of histidine is not so common in most of the MT families, at least one representative of each sub-family of plant MTs contains one or more His residues (Tomas et al., 2014). Further evidence for the function of metal metabolism and detoxification for various MTs in plants has been confirmed by knock down and overexpression lines, tissue-specific and metal-regulated expression, and characterization of MT–metal complexes (Guo et al., 2003; Freisinger, 2007; Yuan et al., 2008; Du et al., 2012). When expressed in metal-sensitive yeast, plant MTs are able to increase tolerance of the yeast to Cu, Cd, and Zn stress (Zhou and Goldsbrough, 1995; Roosens et al., 2004). Guo et al. (2003) proposed distinct functions of plant MTs in *Arabidopsis* during heavy metal homeostasis, particularly for Cu, such as *MT1a* and *MT2b* were involved in the distribution of Cu through phloem, while *MT2a* and *MT3* chaperone excess metals in mesophyll cells and root tips. They concluded that MTs may be involved in remobilization of metal ions from senescing leaves and the sequestration of excess metal ions in trichomes.

In many plant species, such as *Arabidopsis*, rice, wild rice (*Porteresia coarctata*), as well as metal hyperaccumulator *Thlaspi caerulescens*, *MT* gene expression is strongly enhanced by heavy metals such as Cu, Cd, and Zn, although by Cu to a higher extent (Zhou and

TABLE 18.1

Different Functions of *MT* Genes in Various Plant/Fungal Species

Species	Gene	Function	References
Arabidopsis thanliana	*AtMT*	Tissue specific expression and induction during senescence and in response to Cu	Guo et al. (2003)
	AtMT1, AtMT2	Enhances tolerance to Cd in yeast	Lee et al. (2004)
	AtMT2a, AtMT3	Enhances resistance to Cd in *Vicia faba* guard cells	Lee et al. (2004)
	AtMT4a	Enhances Cu and Zn tolerance in *Arabidopsis thanliana*	Rodríguez-Llorente et al. (2010)
	AtMT2b	Enhances root to shoot as translocation in *Nictiana tobaccum*	Grispen et al. (2009)
Brassica juncea	*BjMT2*	Enhances Cu and Cd tolerance in *Escherichia coli* and *A. thaliana*	An et al. (2006)
Citrullus lanatus	*ClMT2*	Drought stress induced	Akashi et al. (2004)
Fagopyrum esculentum	*FeMT3*	Enhances drought and oxidative tolerance in buckwheat	Samardzic et al. (2010)
Oryza sativa	*OsMT1a*	Plays roles in Zn homeostasis and enhance drought tolerance in transgenic rice	Yang et al. (2009)
	OsMT2b	Plays roles in root development and zygotic embryo germination of rice	Yuan et al. (2008)
	OsMT1e-P	Multiple abiotic stress tolerance and ROS scavenging	Kumar et al. (2012)
	OsMT-1-IIa, OsMT-1-IIc, OsMT-1-IId	Higher expression in as tolerant rice	Gautam et al. (2012)
	OsMT-1-IIb, OsMT-1-IId	Higher expression during salt stress	Gautam et al. (2012)
Porteresia coarctata	*PcMT3*	Cd, Cu, and Zn induced	Usha et al. (2011)
Thlaspi caerulescens	*MT2*	Contributes to the metal adapted phenotype in *T. caerulescens*	Hassinen et al. (2009)
	MT3	Cu homeostasis functions in *T. caerulescens*	Roosens et al. (2004)
Hebeloma cylindrosporum	*HcMT1 and HcMT2*	Functional validation in yeast for Cu and Cd tolerance, respectively	Ramesh et al. (2009)
Avicennia marina	*AmMT2*	Enhances Zn, Cd, Cu, Pb tolerances in transgenic *A. marina*	Huang and Wang (2010)
Hordeum vulgare	*MT4*	Zn storage in developing and mature grain	Hegelund et al. (2012)
Populus alba L	*MT1*	Zn induced	Castiglione et al. (2007)
	MT2	Induced by Cd, Cu, Zn	Macovei et al. (2010)
	MT3	Water deficit induced and as ROS scavenger	Berta et al. (2009)
Tamarix hispida	*ThMT3*	Enhances Cd, Zn, Cu, and NaCl tolerance in transgenic yeast	Yang et al. (2011)
Silene vulgaris	*SvMT3*	Controlling Cu homeostasis in plants	Nevrtalova et al. (2014)
Sedum alfredii	*SaMT2*	Increased Cd tolerance and accumulation in yeast and tobacco	Zhang et al. (2014)

Goldsbrough, 1994; Hsieh et al., 1995; Roosens et al., 2004; Usha et al., 2011). Jacob et al. (2004) showed that complexation of Cd by MTs is a key mechanism for Cd tolerance in the ectomycorrhizal fungus *Paxillus involutus*.

18.4.2 MTs Ameliorate Oxidative Stress

In addition to metal homeostasis, MTs are shown to act as strong ROS scavengers (Xue et al., 2009). Recent researches show that MTs can protect plants against oxidative damage and other abiotic (e.g., drought, salinity, and low temperature) stresses which result in generation of ROS (Xue et al., 2009; Freisinger, 2011). Study by Xue et al. (2009) showed that type 3 plant MT in cotton (*GhMT3a*) was induced in seedlings by several abiotic stresses such as salinity, drought, low temperature, and heavy metal ions. These induced expression patterns of *GhMT3a* could be inhibited in the presence of antioxidants. Further, *GhMT3a* recombinant protein exerted *in vitro* ROS scavenging ability. *GhMT3a* overexpressing transgenic tobacco displayed increased tolerance to environmental stresses through MT-mediated ROS balance. Akashi et al. (2004) suggested that (*Citrullus lanatus* sp.) *MT2* (*ClMT2*) induction contributes against drought and high light stress in wild watermelon and they demonstrated for the first time the ROS scavenging properties of plant MT. Kumar et al. (2012) showed that ectopic expression of *OsMT1e-P* protects against oxidative stress primarily through efficient scavenging of ROS in rice. The involvement of plant MT3 in maintaining the redox balance either by sequestering heavy metals or by directly scavenging deleterious oxygen radicals has also been proposed in wild rice (*P. coarctata*) exposed to heavy metals (Usha et al., 2011). While the research on the antioxidant properties of plant MTs is increasing, the interaction between metal binding and ROS scavenging is still not clear. When ROS would bind to the Cys residues of the MTs during the ROS scavenging process, metals would be released (Wong et al., 2004). Some authors have proposed that the exagrate released metals might be involved in a signalling cascade.

Besides the above-mentioned metal detoxification and ROS ameliorating functions, MTs have been reported to be induced during senescence (Guo et al., 2003). Bhalerao et al. (2003) reported that *MTs* were abundant in a cDNA library derived from senescing (*Populus tremula × tremuloides*) leaves. The induction of MTs by senescence may also be related to ROS removal or signaling.

18.5 MTs Expression and Regulation

As discussed above, *MT* gene expression may be induced by a variety of abiotic and biotic stresses, thus, MTs can be considered as general stress proteins (Brkljačić et al., 2004; Jin et al., 2006; Dąbrowska, 2012). The level of the response to these inducers depends on the *MT* gene. The expression pattern analysis of *MT* genes can help to reveal their possible biological functions. However, so far it is not possible to unambiguously assign a specific function to a given MT as proposed functions overlap, and are complementary to each other, or also a given type of MT may perform different roles in different biological contexts. Organ specificity and developmental dependence have been reported for *MT* genes in many plant species. There are some general trends in the expression of these four types of MTs, with type-1 MTs expressed predominantly in roots, type-2 MTs in leaves, type-3 MTs in fruits, and type-4 MTs in seeds (Zhou and Goldsbrough, 1994, 1995; Guo et al., 2003); however, the expression characteristics of each type of plant *MT* gene differs from species to species. For instance, pea MT (*PsMT*) was expressed more in roots compared with leaves and seeds (Evans et al., 1990), while a soybean *MT* was expressed more strongly in leaves than in roots (Kawashima et al., 1991). Expression of the *Arabidopsis MT* genes showed that the expression of *MT1a* and *MT2b* genes was particularly high in phloem tissues of all organs and *MT2a* and *MT3* were expressed in mesophyll cells, while *MT4* expression was confined to seeds (Guo et al., 2003).

It is clear from various reports that MTs play a differential role during different developmental stages. Expression profiling of the rice *MT* gene family, consisting of 11 genes, revealed that each gene has different tissue expression patterns and thus they may perform different functions in specific tissues (Zhou et al., 2006). In another study (Gautam et al., 2012) in rice, four *MT* genes were specifically expressed in developing roots, with the highest expression of *OsMT-I-Ic* gene followed by *OsMT-I-If* and *OsMT-I-If* and *OsMT-I-Ic*. While seven *MTs* were expressed during seed development (*OsMT-1-IIa, OsMT-1-IIb, OsMT-1-IIc, OsMT-1-Ia, OsMT-1-IIIa, OsMT-1-Ib*, and *OsMT-1-If*), among them four were also expressed in leaves (*OsMT-1-IIa, OsMT-1-IIb, OsMT-1-IIIa*, and *OsMT-1-Ib*). However, the wheat *EcMT* and the rice *OsMT-II-1a* genes are preferentially expressed in developing seeds (Lane et al., 1987; Zhou et al., 2005). In *Arabidopsis*, *MT2a* and *MT3* were induced in young leaves and at root tips in response to Cu (Guo et al., 2003). In kiwifruit and pineapple, the expression of *MT* genes is confined to specific stages of fruit development (Ledger and Gardner, 1994; Moyle et al., 2005). Furthermore, the expression of *MT* genes has been found to be induced by ethylene in *Sambucus nigra* (Coupe et al., 1995) and to increase during senescence in *Arabidopsis* (García-Hernández et al., 1998; Guo et al., 2003; Navabpour et al., 2003). In all cases, the differential expression of *MT* genes strongly suggests that each *MT* isoform may have specialized functions in different tissues.

Expressions of several *OsMT* genes from class I have also been observed during reproductive development, suggesting their role in accumulation and storage of metals in reproductive development. These MT proteins may provide the mechanism to produce fortified grains enriched in metal concentrations in seeds (Lucca et al., 2001). In another study (Usha et al., 2011), the increased expression of *PcMT3* in wild rice was observed in root and shoot tissues upon exposure to heavy metals suggesting MT-mediated heavy metal tolerance. Investigation into different types of MTs will help understand their role in heavy metal tolerance of different plant species.

In a study, six *Arabidopsis MTs* (*MT1a, MT2a, MT2b, MT3, MT4a*, and *MT4b*; including all four types of plant MTs) were expressed in the Cu- and Zn-sensitive yeast mutants, *Dcup1* and *Dzrc1 Dcot1*, respectively, to determine the function of plant MTs as metal chelators (Guo et al., 2008). Study revealed that all four types of *Arabidopsis* MTs provided similar levels of Cu tolerance and accumulation to the *Dcup1* mutant, while type-4 MTs (*MT4a* and *MT4b*) conferred greater Zn tolerance and higher accumulation of Zn than other MTs to the *Dzrc1 Dcot1* mutant. Under elevated Cu, *MT1a* was found to contribute in Cu homeostasis in the roots. They showed that MTs and phytochelatins function cooperatively against Cu and Cd toxicity in plants. Among fungi, *Candida glabrata* contains a large family of *MT* genes similar to higher eukaryotes. However, in contrast to plant *MT* genes, the transcription of *C. glabrata MT* genes did not activate by Cd but only by Cu and Ag (Mehra et al., 1989). Ramesh et al. (2009) characterized two *MT* genes, *HcMT1* and *HcMT2*, from the ectomycorrhizal fungus *Hebeloma cylindrosporum*. Expression of *HcMT1* and *HcMT2* in *H. cylindrosporum* under metal stress conditions was studied by competitive reverse transcription-PCR analysis in response to various heavy metals. They showed that while Cu and Cd induced the transcription of *HcMT1* and *HcMT2*, Zn, Pb, and Ni did not.

MTs are induced by various chemical and physical agents acting directly or indirectly on multiple *cis*-acting elements in the regulatory regions of *MT* genes (Haq et al., 2003). Analysis of multiple *cis*-acting motifs in the regulatory regions of *MT* genes may give an overall view of various biotic and abiotic factors regulating basal *MT* gene transcription and induction/repression of *MT* gene activity. Several putative regulatory elements were identified in the promoter region of the *PcMT3* gene (Usha et al., 2011). Promotor-specific sequences as metal response elements (*MRE*), abscicic acid response element (*ABRE*), and Cu response element (*CuRE*) are responsible for the regulation of *MT* gene expression. *MREs*, well characterized in animals, are essential for binding transcription factors that regulate the expression of *MTs* and they have been shown to be essential for heavy metal response and to be involved in some tissue development and physiological reactions (Palmiter, 1998; Coyle et al., 2002; Haq et al., 2003; Waldron et al., 2009). *MRE*-like elements have been reported in the upstream sequences of a few plants such as pea, tomato, and oil palm (Evans et al., 1990; Whitelaw et al., 1997; Abdullah et al., 2002).

Analysis of the *PcMT3* promoter and screening of type-3 *MT* isoforms in wild rice revealed the lack of *MRE* core sequences in the promoter regions (Zhou et al., 2006; Usha et al., 2011). Other metal regulatory elements such as *CuRE* are also absent in the *PcMT3* promoter. This suggests that there may be unique and unidentified *MREs* present in the *PcMT3* promoter or other regulatory elements such as the *CGCG* box, *STRE* (stress responsive element) that may be involved in the indirect regulation of *PcMT3* gene under heavy metal stress. The presence of several *cis*-acting elements identified in the 5′ upstream regions of *PjMT* genes correlates with their expression patterns under heavy metal/oxidative stress conditions (Usha et al., 2009). In *Prosopis juliflora* as well no *MREs* was found in the upstream regions of *PjMTs* but *CuRE* was present in its putative promoter region and was induced in response to Cu.

Another example is from a rice *MT* gene (*ricMT*) promoter, which contains three regions that are primarily responsible for its tissue-specific expression. These include the 5′-distal region (−1382/−910), which is important for the aerial parts of transgenic seedlings, a 5′-proximal region (−194/−1) for the roots and an internal region (−909/−708) for the floral organs in transgenic *Arabidopsis* plants (Lü et al., 2007). One *cis*-element has been identified in the green algae *Chlamydomonas reinhardtii*: a Cu response element (*CuRE*) with the conserved core sequence 5′-GTAC-3′ (Quinn and Merchant, 1995; Quinn et al., 2000). A Cu response regulator, *CRR1*, binds to *CuRE* sites and mediates the expression of downstream genes under Cu-deficient conditions (Quinn et al., 2000). In addition, another report shows that both *CuRE* and *CRR1* are required for Ni response (Quinn et al., 2003). In yeast, it has been reported that Cu *MT* genes are induced by an *MRE*, with the consensus core sequence 5′-HTHNNGCTGD-3′ and its corresponding metal-responsive transcription factor, *ACE1*. *ACE1* can bind to *MRE* sequences and regulate the downstream *MT* genes (Fürst et al., 1988; Dixon et al., 1996). Dong et al. (2010) reported that the −1052/−583 region of the *OsMT-I-4b* promoter, containing four *CuREs*, can lead to efficient Cu^{2+} induction and the region −1052/−914, which contains two *CuREs* (localized at −1041/−1038 and −1014/−1011), is responsible for the induction of *OsMT-I-4b* promoter activity by Al^{3+} (Dong et al., 2010). Thus, it suggests that different combinations of metal responsive elements (*CuREs* and *MREs*) are involved in the response to different metals.

Promoter sequences analysis also indicates diverse functions of plant MTs. Ramesh et al. (2009) demonstrated that both *HcMT1* and *HcMT2* promoters contained the standard stress response elements implicated in metal response, such as *MRE* and general stress response (*GATA*), response to phosphate starvation (*PHO*), response to nitrogen utilization (*NIT*), and heat shock (*HSF*) elements. However, only the *HcMT1* promoter contained *STRE* and *GCN*, which are known to be responsible for the multiple stress response and amino acid- and nitrogen starvation-related stress, respectively. On the other hand, the *HcMT2* promoter contained glucocorticoid response elements (*GCR*), DNA replication-related element (*DRE*) and *MCM* which are involved in regulating glycolytic enzymes, drug or heavy metal efflux mechanisms and osmo-sensing in heavy metal tolerance-induced pathways, respectively. The differences in and locations of potential regulatory elements in *HcMT1* and *HcMT2* promoter regions might be the reasons for the different regulation patterns. The *ABRE*, *ERE*, *LTRE*, and *MRE*-like motifs found in *Arabidopsis* and *MT* promoters have also been reported in the promoters of Cd-regulated rice *miR* (micro RNA) genes with target genes that encode transcription factors and metabolic proteins controlling plant development and the stress response (Ding et al., 2011). The presence, in the promoters of many of the *MT* genes, of regulatory sequences associated with the response of plants to jasmonates, abscisic acid and fungal elicitors, and with activation of meristematic cells, indicates potential involvement of MTs in many processes enabling the proper growth and development of plants and environmental adaptation to changing conditions. *In silico* analyses of the promoter sequences of genes encoding MTs provide a platform for learning more about the functions of MTs in higher plants and represent a direction for future research.

18.6 Conclusion and Future Prospects

Plants appear to contain a diversity of MTs with the potential to perform distinct roles in the metabolism of different metal ions. Although many recent studies have revealed the roles of MTs in plants, validation of the multifunctional role needs to be validated in a suitable model system such as yeast or *Arabidopsis*. This review has given emphasis to the recent advancement for potential molecular mechanisms that may regulate the metal- and tissue-specific expression of different types of MTs with special reference to metal tolerance and the homeostasis role. One possible role of MTs in toxic metals chelation and sequestration needs to be investigated for its potential utilization in phytoremediation purposes. To throw light on these areas, more experiments should be conducted to investigate whether metal–MT complexes transported from root to shoot within the plant achieve higher accumulation or not.

In the view of the multifarious roles a given MT may perform in different biological contexts, it would be highly desirable to accelerate the efforts to study the speciation of plant MTs from the respective plant tissues, the nature of the bound metal ions, and metallation degree under relevant conditions. Another important area for future research is to investigate the relationship between the role of chaperones in the sequestration and intracellular trafficking of various nutritionally important (especially Cu and Zn) metals for the purpose of homeostasis and biofortification in crop plants, and toxic metals (such as Cd, Hg, As, and Pb) for tolerance and detoxification. Further, there is scarcity of information on the signal transduction pathways involved in response to heavy metal and MTs interaction in plants. Detailed studies on *MT* deficient mutants and over-expression of certain *MT* key genes will provide more clear insight about functional evidence in relation to tolerance and detoxification mechanisms.

Acknowledgments

Dr. Preeti Tripathi is thankful to the UGC, New Delhi for the award of the Dr. D.S. Kothari Postdoctoral Fellowship. Pradyumna Kumar Singh and Neelam Gautam are grateful to the Academy of Scientific and Innovative Research (AcSIR), New Delhi for their registration in the Ph.D. program. Dr. Seema Mishra is thankful to the Department of Science and Technology, New Delhi for the award of Young Scientist of the Fast Track Scheme.

References

Abdullah SNA, Cheah SC, and Murphy DJ. 2002. Isolation and characterisation of two divergent type 3 metallothioneins from oil palm, *Elaeis guineensis. Plant Physiology and Biochemistry* 40(3): 255–263.

Akashi K, Nishimura N, Ishida Y, and Yokota A. 2004. Potent hydroxyl radical-scavenging activity of drought-induced type-2 metallothionein in wild watermelon. *Biochemical and Biophysical Research Communications* 323(1): 72–78.

An ZG, Li CJ, Zu YG, Du YJ, Andreas W, Roland G, and Thomas R. 2006. Expression of *BjMT2*, a metallothionein from *Brassica juncea*, increases copper and cadmium tolerance in *Escherichia coli* and *Arabidopsis thaliana*, but inhibits root elongation in *Arabidopsis thaliana* seedlings. *Journal of Experimental Botany* 57: 3575–3582.

Berta M, Giovannelli A, Potenza E, Traversi ML, and Racchi ML. 2009. Type 3 metallothioneins respond to water deficit in leaf and in the cambial zone of white poplar (*Populus alba*). *Journal of Plant Physiology* 166: 521–530.

Bhalerao R, Keskitalo J, Sterky F, Erlandsson R, Björkbacka H, Birve SJ, and Jansson S. 2003. Gene expression in autumn leaves. *Plant Physiology* 131(2): 430–442.

Brkljačić JM, Samardžić JT, Timotijević GS, and Maksimović VR. 2004. Expression analysis of buckwheat (*Fagopyrum esculentum Moench*) metallothionein-like gene *MT3* under different stress and physiological conditions. *Journal of Plant Physiology* 161(6): 741–746.

Castiglione S, Franchin C, Fossati T, Lingua G, Torrigiani P, and Biondi S. 2007. High zinc concentrations reduce rooting capacity and alter metallothionein gene expression in white poplar (*Populus alba* L. cv. *villafranca*). *Chemoshpere* 67: 1117–1126.

Cobbett C and Goldsbrough P. 2002. Phytochelatins and metallothioneins: Roles in heavy metal detoxification and homeostasis. *Annual Reviews of Plant Biology* 53(1): 159–182.

Coupe SA, Taylor JE, and Roberts JA. 1995. Characterisation of an mRNA encoding a metallothionein-like protein that accumulates during ethylene-promoted abscission of *Sambucus nigra* L. leaflets. *Planta* 197(3): 442–447.

Coyle P, Philcox JC, Carey LC, and Rofe AM. 2002. Metallothionein: The multipurpose protein. *Cellular and Molecular Life Science* 59(4): 627–647.

Dąbrowska G. 2012. Plant metallothioneins: Putative functions identified by promoter analysis in silico. *Acta Biologica Cracoviensia Series Botanica* 54(2): 109–120.

Ding Y, Chen Z, and Zhu C. 2011. Microarray-based analysis of cadmium-responsive microRNAs in rice (*Oryza sativa*). *Journal of Experimental Botany* 62: 3563–3573.

Dixon WJ, Inouye C, Karin M, and Tullius TD. 1996. CUP$_2$ binds in a bipartite manner to upstream activation sequence c in the promoter of the yeast copper metallothionein gene. *Journal of Biological Inorganic Chemistry* 1(5): 451–459.

Dong CJ, Wang Y, Yu SS, and Liu JY. 2010. Characterization of a novel rice metallothionein gene promoter: Its tissue specificity and heavy metal responsiveness. *Journal of Integrative Plant Biology* 52(10): 914–24.

Du J, Yang JL, and Li CH. 2012. Advances in metallotionein studies in forest trees. *Plant Omics* 5(1): 46–51.

Evans IM, Gatehouse LN, Gatehouse JA, Robinson NJ, and Croy RR. 1990. A gene from pea (*Pisum sativum* L.) with homology to metallothionein genes. *FEBS Letters* 262(1): 29–32.

Freisinger E. 2007. Spectroscopic characterization of a fruit-specific metallothionein: *M. acuminate* MT3. *Inorganica Chimica Acta* 360(1): 369–380.

Freisinger E. 2011. Structural features specific to plant metallothioneins. *Journal of Biological Inorganic Chemistry* 16: 1035–1045.

Fürst P, Hu S, Hackett R, and Hamer D. 1988. Copper activates metallothionein gene transcription by altering the conformation of a specific DNA binding protein. *Cell* 55(4): 705–717.

García-Hernández M, Murphy A, and Taiz L. 1998. Metallothioneins 1 and 2 have distinct but over-lapping expression patterns in *Arabidopsis*. *Plant Physiology* 118(2): 387–397.

Gautam N, Verma PK, Verma S, Tripathi RD, Trivedi PK, Adhikari B, and Chakrabarty D. 2012. Genome-wide identification of rice class I metallothionein gene: Tissue expression patterns and induction in response to heavy metal stress. *Functional and Integrative Genomics* 12(4): 635–647.

Giritch A, Ganal M, Stephan UW, and Bäumlein H. 1998. Structure, expression and chromosomal localisation of the metallothionein-like gene family of tomato. *Plant Molecular Biology* 37(4): 701–714.

Grispen VMJ, Irtelli B, Hakvoort HWJ, Vooijs R, Bliek T, Bookum WM, Verkleij JAC, and Schat H. 2009. Expression of the *Arabidopsis* metallothionein 2b enhances arsenite sensitivity and root to shoot translocation in tobacco. *Environmental and Experimental Botany* 66: 69–73.

Guo WJ, Bundithya W, and Goldsbrough PB. 2003. Characterization of the *Arabidopsis* metallothio-nein gene family: Tissue specific expression and induction during senescence and in response to copper. *New Phytology* 159(2): 369–381.

Guo WJ, Meetam M, and Goldsbrough PB. 2008. Examining the specific contributions of individ-ual *Arabidopsis* metallothioneins to copper distribution and metal tolerance. *Plant Physiology* 146(4): 1697–706.

Hall JL. 2002. Cellular mechanisms for heavy metal detoxification and tolerance. *Journal of Experimental Botany* 53(366): 1–11.

Haq F, Mahoney M, and Koropatnick J. 2003. Signaling events for metallothionein induction. *Mutation Researche—Fundamental and Molecualar Mechanisms* 533(1): 211–226.

Hassinen VH, Tervahauta AI, Schat H, and Kärenlampi SO. 2011. Plant metallothioneins–metal che-lators with ROS scavenging activity? *Plant Biology* 13(2): 225–232.

Hassinen VH, Tuomainen M, Peraniemi S, Schat H, Kärenlampi SO, and Tervahauta AI. 2009. Metallothioneins 2 and 3 contribute to the metal-adapted phenotype but are not directly linked to Zn accumulation in the metal hyperaccumulator, *Thlaspi caerulescens*. *Journal of Experimental Botany* 60: 187–196.

Hegelund JN, Schiller M, Kichey T, Hansen TH, Pedas P, Husted S, and Schjoerring JK. 2012. Barley metallothioneins: MT3 and MT4 are localized in the grain aleurone layer and show differen-tial zinc binding. *Plant Physiology* 159: 1125–1137.

Hsieh HM, Liu WK, and Huang PC. 1995. A novel stress-inducible metallothionein-like gene from rice. *Plant Molecular Biology* 28(3): 381–389.

Huang GY and Wang YS. 2010. Expression and characterization analysis of type 2 metallothionein from grey mangrove species (*Avicennia marina*) in response to metal stress. *Aquatic Toxicology* 99: 86–92.

Hudspeth RL, Hobbs SL, Anderson DM, Rajasekaran K, and Grula JW. 1996. Characterization and expression of metallothionein-like genes in cotton. *Plant Molecular Biology* 31(3): 701–705.

Jacob C, Courbot M, Martin F, Brun A, and Chalot M. 2004. Transcriptomic responses to cadmium in the ectomycorrhizal fungus *Paxillus involutus*. *FEBS Letters* 576(3): 423–427.

Jin S, Cheng Y, Guan Q, Liu D, Takano T, and Liu S. 2006. A metallothionein-like protein of rice (rgMT) functions in *E. coli* and its gene expression is induced by abiotic stresses. *Biotechnology Letters* 28(21): 1749–1753.

Kawashima I, Inokuchi Y, Chino M, Kimura M, and Shimizu N. 1991. Isolation of a gene for a metallothionein-like protein from soybean. *Plant Cell Physiology* 32(6): 913–916.

Kim YO, Jung S, Kim K, and Bae HJ. 2013. Role of pCeMT, a putative metallothionein from *Colocasia esculenta*, in response to metal stress. *Plant Physiology and Biochemistry* 64: 25–32.

Kumar G, Kushwaha HR, Panjabi-Sabharwal V, Kumari S, Joshi R, Karan R, Mittal S et al. 2012. Clustered metallothionein genes are co-regulated in rice and ectopic expression of OsMT1e-P confers multiple abiotic stress tolerance in tobacco via ROS scavenging. *BMC Plant Biology* 12(1): 107.

Lane B, Kajioka R, and Kennedy T. 1987. The wheat-germ Ec protein is a zinc-containing metallothionein. *Biochemistry and Cell Biology* 65(11): 1001–1005.

Ledger SE and Gardner RC. 1994. Cloning and characterization of five cDNAs for genes differentially expressed during fruit development of kiwifruit (*Actinidia deliciosa* var. *deliciosa*). *Plant Molecular Biology* 25(5): 877–886.

Lee J, Shim D, Song WY, Hwang I, and Lee Y. 2004. *Arabidopsis* metallothioneins 2a and 3 enhance resistance to cadmium when expressed in *Vicia faba* guard cells. *Plant Molecular Biology* 54: 805–815.

Leszczyszyn OI, Imam HT, and Blindauer CA. 2013. Diversity and distribution of plant metallothioneins: A review of structure, properties and functions. *Metallomics* 5(9): 1146–1169.

Leszczyszyn OI, Schmid R, and Blindauer CA. 2007. Toward a property/function relationship for metallothioneins: Histidine coordination and unusual cluster composition in a zinc-metallothionein from plants. *Proteins* 68(4): 922–935.

Liu JY, Lu T, and Zhao NM. 2000. Classification and nomenclature of plant metallothionein-like proteins based on their cysteine arrangement patterns. *Acta Botanica Sinica* 42: 649–652.

Lucca P, Hurrell R, and Potrykus I. 2001. Genetic engineering approaches to improve the bioavailability and the level of iron in rice grains. *Theory of Applied Genetics* 102(2–3): 392–397.

Lü S, Gu H, Yuan X, Wang X, Wu AM, Qu L, and Liu JY. 2007. The GUS reporter-aided analysis of the promoter activities of a rice metallothionein gene reveals different regulatory regions responsible for tissue-specific and inducible expression in transgenic *Arabidopsis*. *Transgenic Research* 16(2): 177–191.

Macovei A, Ventura L, Dona M, Fae M, Balestrazz A, and Carbonera D. 2010. Effect of heavy metal treatment on metallothionein expression profiles white poplar (*Populus alba* L.) cell suspension cultures. *Analele Universitatii Din Oradea—Fascicula Biologie* 2: 274–279.

Mehra RK, Garey JR, Butt TR, Gray WR, and Winge DR. 1989. *Candida glabrata* metallothioneins. Cloning and sequence of the genes and characterization of proteins. *Journal of Biological Chemistry* 264(33): 19747–19753.

Moyle R, Fairbairn DJ, Ripi J, Crowe M, and Botella JR. 2005. Developing pineapple fruit has a small transcriptome dominated by metallothionein. *Journal of Experimental Botany* 56(409): 101–112.

Navabpour S, Morris K, Allen R, Harrison E, Soheila A, and Buchanan-Wollaston V. 2003. Expression of senescence-enhanced genes in response to oxidative stress. *Journal Experimental Botany* 54(391): 2285–2292.

Nevrtalova E, Baloun J, Hudzieczek V, Cegan R, Vyskot B, Dolezel J, and Hobza R. 2014. Expression response of duplicated metallothionein 3 gene to copper stress in *Silene vulgaris* ecotypes. *Protoplasma* 251(6): 1427–1439.

Palmiter RD. 1998. The elusive function of metallothioneins. *Proceedings of the National Academy of Science of the United States of America* 95(15): 8428–8430.

Palmiter RD, Findley SD, Whitmore TE, and Durnam DM. 1992. MT-III, a brain-specific member of the metallothionein gene family. *Proceedings of the National Academy of Science of the United States of America* 89(14): 6333–6337.

Peroza EA, Kaabi AA, Meyer-Klaucke W, Wellenreuther G, and Freisinger E. 2009. The two distinctive metal ion binding domains of the wheat metallothionein Ec-1. *Journal of Inorganic Biochemistry* 103: 342–353.

Quinn JM, Barraco P, Eriksson M, and Merchant S. 2000. Coordinate copper- and oxygen-responsive Cyc6 and Cpx1 expression in *Chlamydomonas* is mediated by the same element. *Journal of Biological Chemistry* 275(9): 6080–6089.

Quinn JM, Kropat J, and Merchant S. 2003. Copper response element and Crr1-dependent Ni²⁺ responsive promoter for induced, reversible gene expression in *Chlamydomonas reinhardtii*. *Eukaryotic Cell* 2(5): 995–1002.

Quinn JM and Merchant S. 1995. Two copper-responsive elements associated with the Chlamydomonas Cyc6 gene function as targets for transcriptional activators. *The Plant Cell Online* 7(5): 623–628.

Ramesh G, Podila GK, Gay G, Marmeisse R, and Reddy MS. 2009. Different patterns of regulation for the copper and cadmium metallothioneins of the ectomycorrhizal fungus *Hebeloma cylindrosporum*. *Applied Environmental Microbiology* 75(8): 2266–2274.

Robinson NJ, Tommey AM, Kusket C, and Jacksont PJ. 1993. Plant metallothioneins, *Biochemistry Journal* 295: 1–10.

Rodríguez-Llorente ID, Pérez-Palacios P, Doukkali B, Caviedes MA, and Pajuelo E. 2010. Expression of the seed-specific metallothionein mt4a in plant vegetative tissues increases Cu and Zn tolerance. *Plant Science* 178: 327–332.

Roosens NH, Bernard C, Leplae R, and Verbruggen N. 2004. Evidence for copper homeostasis function metallothionein of metallothionein (MT3) in the hyperaccumulator *Thlaspi caerulescens*. *FEBS Letters* 577: 9–16.

Samardzic JT, Nikloic DB, Timotijevic GS, Jovanovic ZS, Milisavljevic MD, and Maksimaic VR. 2010. Tissue expression analysis of *FeMT3*, a drought and oxidative stress related metallothionein gene from buckwheat (*Fagopyrum esculentum*). *Journal of Plant Physiology* 167: 1407–1411.

Tomas M, Carlos SA, and Roger B. 2014. His-containing plant metallothioneins: Comparative study of divalent metal–ion binding by plant *MT3* and *MT4* isoforms. *Journal of Biological Inorganic Chemistry* 19(7): 1149–1164.

Usha B, Keeran NS, Harikrishnan M, Kavitha K, and Parida A. 2011. Characterization of a type 3 metallothionein isolated from *Porteresia coarctata*. *Biologia Plantarum* 55(1): 119–124.

Usha B, Venkataraman G, and Parida A. 2009. Heavy metal and abiotic stress inducible metallothionein isoforms from *Prosopis juliflora* (SW) DC show differences in binding to heavy metals in vitro. *Molecular Genetics and Genomics* 281(1): 99–108.

Verbruggen N, Hermans C, and Schat H. 2009. Molecular mechanisms of metal hyperaccumulation in plants. *New Phytology* 181: 759–776.

Waldron KJ, Rutherford JC, Ford D, and Robinson NJ. 2009. Metalloproteins and metal sensing. *Nature* 460: 823–830.

Whitelaw CA, Le Huquet JA, Thurman DA, and Tomsett AB. 1997. The isolation and characterisation of type II metallothionein-like genes from tomato (*Lycopersicon esculentum* L.). *Plant Molecular Biology* 33(3): 503–511.

Wong HL, Sakamoto T, Kawasaki T, Umemura K, and Shimamoto K. 2004. Down-regulation of metallothionein, a reactive oxygen scavenger, by the small GTPase *OsRac1* in rice. *Plant Physiology* 135: 1447–1456.

Xia Y, Lv Y, Yuan Y, Wang G, Chen Y, Zhang H, and Shen Z. 2012. Cloning and characterization of a type 1 metallothionein gene from the copper-tolerant plant *Elsholtzia haichowensis*. *Acta Physiologie Plantarum* 34: 1819–1826.

Xu Q and Shi G. 2000. The toxic effects of single Cd and interaction of Cd with Zn on some physiological index of *Oenanthe javanica* (Blume) DC. *Journal of Nanjing Normal University (Natural Science)* 23(4): 97–100.

Xue T, Li X, Zhu W, Wu C, Yang G, and Zheng C. 2009. Cotton metallothionein *GhMT3a*, a reactive oxygen species scavenger, increased tolerance against abiotic stress in transgenic tobacco and yeast. *Journal of Experimental Botany* 60(1): 339–349.

Yang JL, Wang YC, Liu GF, Yang CP, and Li CH. 2011. *Tamarix hispida* metallothionein-like *ThMT3*, a reactive oxygen species scavenger, increases tolerance against Cd²⁺, Zn²⁺, Cu²⁺, and NaCl in transgenic yeast. *Molecular Biology Report* 38: 1567–1574.

Yang Z, Wu YR, Ling HQ, and Clu CC. 2009. *OsMT1a*, a type 1 metallothionein, plays the pivotal role in zinc homeostasis and drought tolerance in rice. *Plant Molecular Biology* 70: 219–229.

Yuan J, Chen D, Ren Y, Zhang X, and Zhao J. 2008. Characteristic and expression analysis of a metallothionein gene, *OsMT2b*, down-regulated by cytokinin suggests functions in root development and seed embryo germination of rice. *Plant Physiology* 146(4): 1637–1650.

Zhang B, Georgiev O, Hagmann M, Günes C, Carmer M, Faller P et al. 2003. Activity of metal-responsive transcription factor 1 by toxic heavy metals and H_2O_2 *in vitro* is modulated by metallothionein. *Molecular Cell Biology* 23: 8471–8485.

Zhang J, Zhang M, Tian S, Lu L, Shohag MJI, and Yang X. 2014. Metallothionein 2 (*SaMT2*) from *Sedum alfredii* Hance confers increased Cd tolerance and accumulation in yeast and tobacco. *PLoS One* 9(7): e102750.

Zhou G, Xu Y, Li J, Yang L, and Liu JY. 2006. Molecular analyses of the metallothionein gene family in rice (*Oryza sativa* L.). *Journal of Biochemistry and Molecular Biology* 39(5): 595–606.

Zhou GK, Xu YF, and Liu JY. 2005. Characterization of a rice class II metallothionein gene: Tissue expression patterns and induction in response to abiotic factors. *Journal of Plant Physiology* 162(6): 686–696.

Zhou J and Goldsbrough P. 1994. Functional homologs of fungal metallothionein genes from *Arabidopsis*. *The Plant Cell Online* 6: 875–884.

Zhou J and Goldsbrough PB. 1995. Structure, organization and expression of the metallothionein gene family in *Arabidopsis*. *Molecular General Genetics* 248(3): 318–328.

Index

A

AAD, *see* Aryl-alcohol dehydrogenases (AAD)
AAO, *see* Aryl alcohol oxidase (AAO)
ABR, *see* Anaerobic baffled reactor (ABR)
ABTS, *see* 2,2'-Azinobis-[3-ethylthiazoline-6-sulfonate] (ABTS)
AC, *see* Activated carbon (AC)
Acetogenesis, 160
Acetogens, 144
Acetylcholine esterase (AChE), 488
AChE, *see* Acetylcholine esterase (AChE)
Acidogenesis, 160
Actinomycetes, 22–23; *see also* Microorganism
    in compost
    for plastic degradation, 356
Activated carbon (AC), 456
Activated sludge (AS); *see also* Bioaugmentation;
    Bioremediation; Industrial
    wastewaters; Wastewater treatment
    plant (WWTP)
    bioaugmentation of, 192
    and wastewater-metagenomic projects,
      206–207
Activated sludge process (ASP), 44, 45–46,
      159, 191; *see also* Aerobic biological
      processes; Tannery wastewater
      treatment methods
    applications, 234
    CASP, 232–233
    contact stabilization, 233
    equation, 233
    extended aeration system, 234
    foaming, 235
    limitations, 234–235
    oxidation ditch, 234
    SBR, 234
    sludge bulking, 234–235
    for TWW treatment, 231
    types of, 233
    in WWTPs, 186
Activated sludge process-partial nitrification
      (AS/PN), 106; *see also* Biological
      nitrogen removal
    –Anammox, 101, 106–107
N-Acyl homoserine lactones (AHSLs), 467
AD, *see* Anaerobic digesters (AD)
ADH, *see* Alcohol dehydrogenase (ADH)

ADMI, *see* American Dye Manufacturers
    Institute (ADMI)
Adsorbable organic halide (AOX), 38
    removal mechanism for, 46
Adsorbents, 52; *see also* Physicochemical
    wastewater treatment
Adsorption, 456; *see also* Nucleic acid isolation
Advanced bio-processes, 160; *see also* Industrial
    wastewater treatment
    anammox process, 161
    biohydrogen production, 160
    biological nitrogen cycle, 161
    photo fermentation, 160–161
Advanced oxidation processes (AOPs), 239, 300;
    see also Complex effluent AOP; Tannery
    wastewater treatment methods
AerAOB, *see* Aerobic ammonium-oxidizing
    bacteria (AerAOB)
Aeration retention time (ART), 101–102
Aerobic ammonium-oxidizing bacteria
    (AerAOB), 104
Aerobic biological processes, 45; *see also* End of
    pipe treatment
    activated sludge process, 45–46
    low sludge bioprocess, 47
    membrane bioreactor, 47–48
    moving bed biofilm reactor system, 48–49
Aerobic composting, 12; *see also* Composting
    in-vessel systems, 16
    passive composting piles, 12
    windrow composting, 13–15
Aerobic processes, 159
    for TWW treatment, 230–231
Aerobic respiration, 352
AF, *see* Anaerobic filters (AF)
Agrochemicals, 59; *see also* Pesticides
AHL, *see N*-Acyl homoserine lactones (AHSLs)
AHSLs, *see N*-Acyl homoserine lactones (AHSLs)
AI-2, *see* Autoinducer-2 (AI-2)
Alcohol dehydrogenase (ADH), 142
Aldehydes, 132
Algae, 131; *see also* Cyanobacteria; Microalgae
    biofuel production from, 133
    biogas production, 135
Alkalinity of water, 188
Alkenyl halides, 501; *see also* Organohalide
Alkyl halides, 500, 501; *see also* Organohalide

Printed and bound by CPI Group (UK) Ltd, Croydon, CR0 4YY

24/10/2024

01778285-0010